W9-ANQ-907

FOOD PROTEINS AND THEIR APPLICATIONS

FOOD SCIENCE AND TECHNOLOGY

A Series of Monographs, Textbooks, and Reference Books

1. Flavor Research: Principles and Techniques, *R. Teranishi, I. Hornstein, P. Issenberg, and E. L. Wick*
2. Principles of Enzymology for the Food Sciences, *John R. Whitaker*
3. Low-Temperature Preservation of Foods and Living Matter, *Owen R. Fennema, William D. Powrie, and Elmer H. Marth*
4. Principles of Food Science
   Part I: Food Chemistry, *edited by Owen R. Fennema*
   Part II: Physical Methods of Food Preservation, *Marcus Karel, Owen R. Fennema, and Daryl B. Lund*
5. Food Emulsions, *edited by Stig E. Friberg*
6. Nutritional and Safety Aspects of Food Processing, *edited by Steven R. Tannenbaum*
7. Flavor Research: Recent Advances, *edited by R. Teranishi, Robert A. Flath, and Hiroshi Sugisawa*
8. Computer-Aided Techniques in Food Technology, *edited by Israel Saguy*
9. Handbook of Tropical Foods, *edited by Harvey T. Chan*
10. Antimicrobials in Foods, *edited by Alfred Larry Branen and P. Michael Davidson*
11. Food Constituents and Food Residues: Their Chromatographic Determination, *edited by James F. Lawrence*
12. Aspartame: Physiology and Biochemistry, *edited by Lewis D. Stegink and L. J. Filer, Jr.*
13. Handbook of Vitamins: Nutritional, Biochemical, and Clinical Aspects, *edited by Lawrence J. Machlin*
14. Starch Conversion Technology, *edited by G. M. A. van Beynum and J. A. Roels*
15. Food Chemistry: Second Edition, Revised and Expanded, *edited by Owen R. Fennema*
16. Sensory Evaluation of Food: Statistical Methods and Procedures, *Michael O'Mahony*
17. Alternative Sweetners, *edited by Lyn O'Brien Nabors and Robert C. Gelardi*
18. Citrus Fruits and Their Products: Analysis and Technology, *S. V. Ting and Russell L. Rouseff*
19. Engineering Properties of Foods, *edited by M. A. Rao and S. S. H. Rizvi*

20. Umami: A Basic Taste, *edited by Yojiro Kawamura and Morley R. Kare*
21. Food Biotechnology, *edited by Dietrich Knorr*
22. Food Texture: Instrumental and Sensory Measurement, *edited by Howard R. Moskowitz*
23. Seafoods and Fish Oils in Human Health and Disease, *John E. Kinsella*
24. Postharvest Physiology of Vegetables, *edited by J. Weichmann*
25. Handbook of Dietary Fiber: An Applied Approach, *Mark L. Dreher*
26. Food Toxicology, Parts A and B, *Jose M. Concon*
27. Modern Carbohydrate Chemistry, *Roger W. Binkley*
28. Trace Minerals in Foods, *edited by Kenneth T. Smith*
29. Protein Quality and the Effects of Processing, *edited by R. Dixon Phillips and John W. Finley*
30. Adulteration of Fruit Juice Beverages, *edited by Steven Nagy, John A. Attaway, and Martha E. Rhodes*
31. Foodborne Bacterial Pathogens, *edited by Michael P. Doyle*
32. Legumes: Chemistry, Technology, and Human Nutrition, *edited by Ruth H. Matthews*
33. Industrialization of Indigenous Fermented Foods, *edited by Keith H. Steinkraus*
34. International Food Regulation Handbook: Policy • Science • Law, *edited by Roger D. Middlekauff and Philippe Shubik*
35. Food Additives, *edited by A. Larry Branen, P. Michael Davidson, and Seppo Salminen*
36. Safety of Irradiated Foods, *J. F. Diehl*
37. Omega-3 Fatty Acids in Health and Disease, *edited by Robert S. Lees and Marcus Karel*
38. Food Emulsions: Second Edition, Revised and Expanded, *edited by Kåre Larsson and Stig E. Friberg*
39. Seafood: Effects of Technology on Nutrition, *George M. Pigott and Barbee W. Tucker*
40. Handbook of Vitamins: Second Edition, Revised and Expanded, *edited by Lawrence J. Machlin*
41. Handbook of Cereal Science and Technology, *Klaus J. Lorenz and Karel Kulp*
42. Food Processing Operations and Scale-Up, *Kenneth J. Valentas, Leon Levine, and J. Peter Clark*
43. Fish Quality Control by Computer Vision, *edited by L. F. Pau and R. Olafsson*
44. Volatile Compounds in Foods and Beverages, *edited by Henk Maarse*
45. Instrumental Methods for Quality Assurance in Foods, *edited by Daniel Y. C. Fung and Richard F. Matthews*
46. Listeria, Listeriosis, and Food Safety, *Elliot T. Ryser and Elmer H. Marth*
47. Acesulfame-K, *edited by D. G. Mayer and F. H. Kemper*
48. Alternative Sweeteners: Second Edition, Revised and Expanded, *edited by Lyn O'Brien Nabors and Robert C. Gelardi*
49. Food Extrusion Science and Technology, *edited by Jozef L. Kokini, Chi-Tang Ho, and Mukund V. Karwe*
50. Surimi Technology, *edited by Tyre C. Lanier and Chong M. Lee*
51. Handbook of Food Engineering, *edited by Dennis R. Heldman and Daryl B. Lund*
52. Food Analysis by HPLC, *edited by Leo M. L. Nollet*
53. Fatty Acids in Foods and Their Health Implications, *edited by Ching Kuang Chow*
54. *Clostridium botulinum*: Ecology and Control in Foods, *edited by Andreas H. W. Hauschild and Karen L. Dodds*
55. Cereals in Breadmaking: A Molecular Colloidal Approach, *Ann- Charlotte Eliasson and Kåre Larsson*
56. Low-Calorie Foods Handbook, *edited by Aaron M. Altschul*

57. Antimicrobials in Foods: Second Edition, Revised and Expanded, *edited by P. Michael Davidson and Alfred Larry Branen*
58. Lactic Acid Bacteria, *edited by Seppo Salminen and Atte von Wright*
59. Rice Science and Technology, *edited by Wayne E. Marshall and James I. Wadsworth*
60. Food Biosensor Analysis, *edited by Gabriele Wagner and George G. Guilbault*
61. Principles of Enzymology for the Food Sciences: Second Edition, *John R. Whitaker*
62. Carbohydrate Polyesters as Fat Substitutes, *edited by Casimir C. Akoh and Barry G. Swanson*
63. Engineering Properties of Foods: Second Edition, Revised and Expanded, *edited by M. A. Rao and S. S. H. Rizvi*
64. Handbook of Brewing, *edited by William A. Hardwick*
65. Analyzing Food for Nutrition Labeling and Hazardous Contaminants, *edited by Ike J. Jeon and William G. Ikins*
66. Ingredient Interactions: Effects on Food Quality, *edited by Anilkumar G. Gaonkar*
67. Food Polysaccharides and Their Applications, *edited by Alistair M. Stephen*
68. Safety of Irradiated Foods: Second Edition, Revised and Expanded, *J. F. Diehl*
69. Nutrition Labeling Handbook, *edited by Ralph Shapiro*
70. Handbook of Fruit Science and Technology: Production, Composition, Storage, and Processing, *edited by D. K. Salunkhe and S. S. Kadam*
71. Food Antioxidants: Technological, Toxicological, and Health Perspectives, *edited by D. L. Madhavi, S. S. Deshpande, and D. K. Salunkhe*
72. Freezing Effects on Food Quality, *edited by Lester E. Jeremiah*
73. Handbook of Indigenous Fermented Foods: Second Edition, Revised and Expanded, *edited by Keith H. Steinkraus*
74. Carbohydrates in Food, *edited by Ann-Charlotte Eliasson*
75. Baked Goods Freshness: Technology, Evaluation, and Inhibition of Staling, *edited by Ronald E. Hebeda and Henry F. Zobel*
76. Food Chemistry: Third Edition, *edited by Owen R. Fennema*
77. Handbook of Food Analysis: Volumes 1 and 2, *edited by Leo M. L. Nollet*
78. Computerized Control Systems in the Food Industry, *edited by Gauri S. Mittal*
79. Techniques for Analyzing Food Aroma, *edited by Ray Marsili*
80. Food Proteins and Their Applications, *edited by Srinivasan Damodaran and Alain Paraf*

Additional Volumes in Preparation

Food Emulsions: Third Edition, Revised and Expanded, *edited by Stig E. Friberg and Kåre Larsson*

Milk and Dairy Product Technology, *Edgar Spreer*

FOOD PROTEINS AND THEIR APPLICATIONS

edited by

Srinivasan Damodaran

University of Wisconsin—Madison
Madison, Wisconsin

Alain Paraf

Institut National de la Recherche Agronomique
Centre de Recherches de Tours
Nouzilly, France

MARCEL DEKKER, INC. NEW YORK • BASEL

Library of Congress Cataloging-in-Publication Data

Food proteins and their applications / edited by Srinivasan Damodaran,
Alain Paraf.
p. cm.
Includes index.
ISBN 0–8247–9820–1 (alk. paper)
1. Proteins. I. Damodaran, Srinivasan. II. Paraf, Alain.
TP453.P7F665 1997
664--dc21 97-4028
CIP

The publisher offers discounts on this book when ordered in bulk quantities. For more information, write to Special Sales/Professional Marketing at the address below.

This book is printed on acid-free paper.

Marcel Dekker, Inc.
270 Madison Avenue, New York, New York 10016

Current printing (last digit):
10 9 8 7 6 5 4 3 2

PRINTED IN THE UNITED STATES OF AMERICA

Preface

Proteins are highly complex biopolymers. In biological systems, proteins perform various functions, serving as biocatalysts, structural elements of cells (cytoskeletons), mechanical devices (muscles), messengers (hormones), taxis (transport proteins), metal scrapers (chelators), soldiers (antibodies), and protectors (toxins). This functional diversity, which requires the ability to assume various spatial structural forms, is attributable mainly to their intricate chemical makeup.

Proteins are one of the main constituents of foods. They perform several critical functions in food systems that contribute to the sensorial properties of foods. Typical functions of proteins in food systems include thickening, gelation, emulsification, foaming, texturization, water binding, adhesion and cohesion, and lipid and flavor binding and retention. These functional properties of proteins in food systems are also very much influenced by their structural states in foods. One of the oldest, most cherished wishes of food protein chemists is to develop a fundamental understanding of the structure-function relationships of food proteins, so that the functionality and uses of underutilized plant proteins can be enhanced through physical, chemical, enzymatic, and genetic modifications. Much progress has been made in the past decade. The goal of this book is to provide up-to-date information on the developments in this important area of food research.

This book is intended primarily as a textbook for graduate students and a reference book for food scientists in academia and industry. The chapters are grouped into three parts. Chapters 1–6 describe fundamental physicochemical aspects of protein structure, denaturation, protein-stabilized foams and emulsion, gels, protein–polysaccharide interactions, and protein–lipid interactions. Chapters 7–12 provide an in-depth analysis of structure-function relationships of major food proteins, such as caseins, whey proteins, soy proteins, wheat proteins, egg proteins, and muscle proteins. Chapters 13–17 discuss chemical, genetic, enzymatic, and physical modification methods to improve the functional properties of proteins. Chapter 18 describes the unique film-forming properties of proteins, and Chapters 19–22 analyze processing-induced changes in nutritional value of proteins and various extraction and analytical procedures, including immunological methods, used in protein isolation and characterization. Thus, we have attempted to cover a wide spectrum of topics, both fundamental and applied. Although care has been taken to minimize duplication, some inadvertent overlap between chapters is inevitable.

The contributors, internationally recognized experts in their area of research, have produced a remarkably comprehensive work on food proteins. We thank them for their hard work.

Srinivasan Damodaran
Alain Paraf

Contents

Contributors

Carl A. Batt Department of Food Science, Cornell University, Ithaca, New York

C. Y. Boquien Centre International de Recherches Daniel Carasso, Le Plessis-Robinson, France

J. I. Boye Food Research Institute, Agriculture Canada, Ottawa, Ontario, Canada

Ralph W. Burley CSIRO Division of Food Science Technology, North Ryde, New South Wales, Australia

Philippe Cayot Département de Biochimie, Physico-chimie et Properiétés Sensorielles de l'Aliment, ENSBANA, Université de Bourgogne, Dijon, France

Jean-Marc Chobert LEIMA, Institut National de la Recherche Agronomique, Nantes, France

Douglas G. Dalgleish Department of Food Science, University of Guelph, Guelph, Ontario, Canada

Srinivasan Damodaran Department of Food Science, University of Wisconsin—Madison, Madison, Wisconsin

Etsushiro Doi[†] Research Institute for Food Science, Kyoto University, Kyoto, Japan

P. A. Finot Nestec Ltd. Research Centre, Lausanne, Switzerland

Thomas Haertlé LEIMA, Institut National de la Recherche Agronomique, Nantes, France

V. R. Harwalkar Food Research Institute, Agriculture Canada, Ottawa, Ontario, Canada

K. Heremans Department of Chemistry, University of Leuven, Leuven, Belgium

A. Huyghebaert Department of Food Science and Technology, University of Ghent, Ghent, Belgium

---

[†]Deceased.

Naofumi Kitabatake Research Institute for Food Science, Kyoto University, Kyoto, Japan

John M. Krochta Department of Food Science and Technology, University of California, Davis, California

D. Lafiandra University of Tuscia, Viterbo, Italy

Denis Lorient Département de Biochimie, Physico-chimie et Propriétés Sensorielles de l'Aliment, ENSBANA, Université de Bourgogne, Dijon, France

C.-Y. Ma Food Research Institute, Agriculture Canada, Ottawa, Ontario, Canada

F. MacRitchie CSIRO Plant Industry, North Ryde, New South Wales, Australia

Yasuki Matsumura Research Institute for Food Science, Kyoto University, Kyoto, Japan

J. L. Maubois Laboratoire de Recherches de Technologie Laitière, Institut National de la Recherche Agronomique, Rennes, France

Tomohiko Mori Research Institute for Food Science, Kyoto University, Kyoto, Japan

P. K. Nandi Institut National de la Recherche Agronomique, Centre de Recherches de Tours, Nouzilly, France

Per Munk Nielsen Novo Nordisk A/S, Bagsvaerd, Denmark

David Oakenfull CSIRO Division of Food Science and Technology, North Ryde, New South Wales, Australia

G. Ollivier Laboratoire de Recherches de Technologie Laitière, Institut National de la Recherche Agronomique, Rennes, France

Paul Paquin Faculté des Sciences de l'Agriculture et de l'Alimentation, Université Laval, Sainte-Foy, Quebec, Canada

Alain Paraf Institut National de la Recherche Agronomique, Centre de Recherches de Tours, Nouzilly, France

John Pearce CSIRO Division of Food Science and Technology, North Ryde, New South Wales, Australia

Christian Sanchez Laboratoire de Physico-chimie et Genie Alimentiares, ENSRIA-INPL, Vandoeuvre-les-Nancy, France

Klaus D. Schwenke Plant Protein Chemistry Research Group, Universität Potsdam, Bergholz-Rehbrücke, Germany

Vladimir B. Tolstoguzov Department of Food Science, Nestec Ltd. Research Center, Lausanne, Switzerland

Shigeru Utsumi Research Institute for Food Science, Kyoto University, Kyoto, Japan

J. Van Camp Department of Food Technology and Nutrition, University of Ghent, Ghent, Belgium

Youling L. Xiong Food Science Section, Department of Animal Sciences, University of Kentucky, Lexington, Kentucky

1

Food Proteins: An Overview

Srinivasan Damodaran
University of Wisconsin—Madison
Madison, Wisconsin

I. INTRODUCTION

Proteins play several important roles in biological and food systems. Some of these include biocatalysts (enzymes), structural components of cells and organs (e.g., collagen, keratin, elastin, etc.), contractile proteins (actin, myosin, tubulin), hormones (insulin, growth factor, etc.), transport proteins (serum albumin, transferrin, hemoglobin), metal chelation (phosvitin, ferritin), antibodies (immunoglobulins), protective proteins (toxins and allergens), and storage proteins (seed proteins, casein micelles, egg albumen) as nitrogen and energy source for embryos.

Proteins are highly complex polymers, and their functional diversity mainly arises from their chemical make-up. For instance, while other biopolymers, such as polysaccharides and nucleic acids, are made up of one or a few monomers, proteins and polypeptides are made up of combinations of 20 different amino acids. In some proteins, some of the amino acid residues are enzymatically modified by the cytoplasmic enzymes. Examples of such modifications include glycosylation (ovalbumin, κ-casein, lectins, and vicilin-type proteins of legumes) and phosphorylation (α- and β-caseins, phosvitin, kinases, phosphorylases). In addition, unlike the ether and phosphodiester bonds that link the monomers in polysaccharides and nucleic acids, respectively, the substituted amide linkage in proteins is a partial double bond; this adds to the structural complexity of protein polymers. The various types of noncovalent interactions between the amino acid constituents and the specific properties of the amide linkage impart a multitude of spatial structural forms to proteins with diverse biological functions. Literally innumerable numbers of proteins with distinct structures and functions can be synthesized by varying the amino acid composition and sequence.

The functional properties of proteins in foods are related to their structural and other physicochemical characteristics. A fundamental understanding of the physical, chemical, and functional properties of proteins and the changes these properties undergo during processing is essential if the performance of proteins in foods is to be improved and if underutilized proteins, such as plant proteins and whey proteins, are to be increasingly used in traditional and processed food products.

Several of the chapters in this book deal with the physicochemical bases of protein functionality and the structure-function relationships of specific food proteins. In this chapter, a general overview of protein structure will be presented and its influence on the physical and chemical properties and functional properties of proteins will be discussed.

II. PROTEIN STRUCTURE

Proteins are polymers made up of 19 different α-amino acids and one imino acid linked via amide bonds, also known as peptide bonds. The constituent amino acids differ only in the chemical nature of the side-chain group at the α-carbon atom. The physicochemical properties, such as charge, solubility, and chemical reactivity, of the amino acids (hence proteins) are dependent on the chemical nature of the side-chain group. Amino acids with aliphatic (Ala, Ile, Leu, Met, Pro, and Val) and aromatic (Phe, Trp, and Tyr) side chains are nonpolar; they exhibit limited solubility in water. Amino acids with charged (Arg, Lys, His, Glu, and Asp) and uncharged (Ser, Thr, Asn, Gln, and Cys) side chains are quite soluble in water. Proline is the only imino acid present in proteins. The net charge of a protein at any pH is determined by the relative numbers of basic (Arg, Lys, and His) and acidic (Glu and Asp) amino acid residues in the protein.

One of the major factors influencing the properties, such as conformational stability, solubility, surface activity, fat binding, etc., of proteins is the overall hydrophobicity of the constituent amino acid residues. Hydrophobicity is generally defined as the excess free energy of a solute in water compared to that in an organic solvent. Thus, the hydrophobicities of amino acids can be determined by measuring their solubilities in water and in a reference organic solvent, such as ethanol [1], and using the relation:

$$\Delta G_{t(Et \to W)} = -RT \ln \left(\frac{S_{AA,Et}}{S_{AA,W}} \right) \qquad (1)$$

where $\Delta G_{t(Et \to W)}$ is the transfer free energy of the amino acid from ethanol to water, $S_{AA,Et}$ and $S_{AA,W}$ are the solubilities of the amino acid in ethanol and water, respectively, T is the temperature, and R is the gas constant.

Because an amino acid can be considered as a derivative of glycine, and since $\Delta G_{t(Et \to W)}$ is an additive function, the transfer free energy of an amino acid can be considered to be sum of the transfer free energies of glycine and the side chain, i.e.,

$$\Delta G_{t,\text{side chain}} = \Delta G_{t,AA} - \Delta G_{t,Gly} \qquad (2)$$

The hydrophobicities of amino acid side chains obtained in this manner are listed in Table 1. Amino acid side chains with large positive ΔG_t values are hydrophobic; they prefer to be in an organic phase or in the protein interior rather than in an aqueous environment. Amino acid side chains with negative ΔG_t values, especially the charged amino acid side chains, prefer to be on the protein exterior exposed to the aqueous environment.

The functional behaviors of biologically important proteins and food proteins are dependent on their structures. Four levels of structural hierarchy, namely, primary, secondary, tertiary, and quaternary structures, exist in proteins.

TABLE 1 Hydrophobicity of Amino Acid Side Chains at 25°C

Amino acid	ΔG_t (Ethanol → Water) (kJ/mol)
Alanine	2.09
Arginine	—
Asparagine	0
Aspartic acid	2.09
Cysteine	4.18
Glutamic acid	2.09
Glutamine	−0.42
Glycine	0
Histidine	2.09
Isoleucine	12.54
Leucine	9.61
Lysine	—
Methionine	5.43
Phenylalanine	10.45
Proline	10.87
Serine	−1.25
Threonine	1.67
Tryptophan	14.21
Tyrosine	9.61
Valine	6.25

Source: Ref. 1.

A. Primary Structure

The primary structure of a protein denotes the linear sequence in which the constituent amino acids are linked via peptide bonds. The chain length and the amino acid sequence of the polypeptide determines its ultimate three-dimensional structure in solution.

One of the structural features that distinguishes proteins from other biopolymers, such as polysaccharides and nucleic acids, is the partial double-bond character of the peptide bond, resulting from its resonance structure:

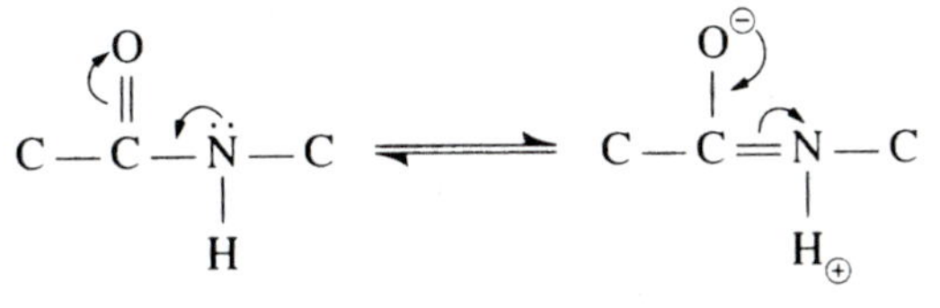

$$\mathrm{C-\overset{\overset{\displaystyle O}{\|}}{C}-\underset{\underset{\displaystyle H}{|}}{\ddot{N}}-C} \rightleftharpoons \mathrm{C-\overset{\overset{\displaystyle O^{\ominus}}{|}}{C}=\underset{\underset{\displaystyle H_{\oplus}}{|}}{N}-C}$$

Because of this resonance structure, the rotation of the peptide bond is restricted to a maximum of about 6°, the six atoms of the peptide unit becomes planar, protonation of the peptide N—H group becomes impossible, the oxygen and hydrogen atoms of C=O and N—H groups acquire partial negative and positive charges, respectively, and the peptide unit can exist in either *cis* or *trans* configuration. Almost all peptide bonds in proteins exist in the *trans* configuration, because it is more stable than the *cis* configuration. Because of the partial charges of the C═O and N—H groups, interchain and intrachain hydrogen bonding between these groups is possible under appropriate conditions.

Because the peptide bonds constitute one third of all covalent bonds of the polypeptide backbone, the restriction on their rotational freedom drastically reduces the flexibility of the polypeptide chain. Among the covalent bonds of the peptide backbone, only the N—C_α and the C_α—C bonds have rotational freedoms, and these are termed ϕ (phi) and ψ (psi) dihedral angles, respectively. However, steric hindrance arising from bulk side chains attached to the α-carbon atom restricts rotational freedoms of the ϕ and ψ angles as well. Because of these constraints, a majority of proteins do not exist in flexible random coil conformations in solution. On the contrary, most proteins assume a highly compact ordered structure because of steric factors and other noncovalent interactions among the amino acid residues. The selection of the 20 different amino acids and the choice of peptide (amide) bond as the primary linkage appear to be a deliberate design by nature to precisely control the flexibility and/or rigidity of polypeptides so that they may be used to perform several biological functions.

B. Secondary Structure

The secondary structure relates to periodic structures in polypeptides and proteins, in which the consecutive amino acid residues of the segment assume the same set of ϕ and ψ angles. The most commonly found secondary structures in proteins are the α-helix and the β-sheet. The α-helix is characterized by a pitch of 5.4 Å involving 3.6 amino acid residues. It is stabilized by intrachain hydrogen bonding between the ith N—H group and the C═O group of the i-4th residue.

Recent studies indicate that the instruction for α-helix formation is coded as a binary code (related to the arrangement of polar and nonpolar residues) in the amino acid sequence [2]. Peptide segments with repeating heptet sequences of [—P—N—P—P—N—N—P—], where P and N are polar and nonpolar residues, respectively, readily form α-helices. In this binary code, the precise identities of the polar and nonpolar residues are irrelevant. Slight variations in the binary code are tolerated, provided other interactions in the protein are favorable for α-helix formation. A good example of this is tropomyosin, which exists entirely in a coiled-coil α-helical rod form. In this protein, the repeating heptet sequence is [—N—P—P—N—P—P—P—]; despite this variation, tropomyosin exists in a α-helix form because of other favorable interactions in the coiled-coil rod [3].

The α-helical structures in protein are predominantly amphiphilic, that is, one half of the helical surface is hydrophilic and the other half is hydrophobic. In fact, if an α-helix is made out of the heptet sequence mentioned above, one would find that all the hydrophobic amino acid residues would lay on one side of the helix surface and all the hydrophilic residues would lay on the other side. Generally, in native proteins, the hydrophobic surface of the α-helix faces the interior of the protein and is engaged in hydrophobic interactions with other nonpolar groups in the interior. Such interactions generally contribute to the stability of the folded form of the protein.

The β-sheet structure is an extended structure in which the C═O and N—H groups are oriented perpendicular to the direction of the backbone. In this configuration, hydrogen bonding can occur only between sheets. Depending on the directions of the sheets, two types of β-pleated sheet structure, namely, parallel and antiparallel β-sheets, can form. The binary code in the amino acid sequence that specifies β-sheet formation is [—N—P—N—P—N—P—N—. . . .]. In other words, segments containing alternating polar and nonpolar amino acid residues exhibit a high probability of forming β-sheets. Generally, β-type proteins are more hydrophobic than the α-type proteins. They exhibit high denaturation temperatures. Examples are β-lactoglobulin (51% β-sheet) and soy 11S globulin (64% β-sheet), which have thermal denaturation temperatures of 75.6 and 84.5°C, respectively. In contrast, bovine serum albumin, which is an α-type protein, has a thermal denaturation temperature of only about 64°C [4,5]. When protein solutions are heated, more often than not, α-helix is converted to β-sheet [4], but conversion of β-sheet to α-helix has not been reported.

Polypeptide segments in which the consecutive amino acid residues assume random combinations of ϕ and ψ angles have disordered or aperiodic structures. Proteins containing high levels of proline residues usually assume a random structure. This is because of their pyrrolidine ring structure, in which the ϕ angle is fixed at 70°C, and their inability to form hydrogen bonds. A good example is casein: α- and β-caseins consist of about 8.5 and 17% proline residues, respectively. The uniform distribution of these residues in its sequence hinders with formation of α-helical and β-sheet structures, and these proteins exist predominantly in disordered states. About one third of the residues in collagen are either proline or hydroxyproline, which exists in a helical form in which three polypeptide chains are entwined to form a triple helix; the triple helix is stabilized by interchain hydrogen bonds. The geometry of this triple helix is different from that of the α-helix. Other food proteins that contain a large amount of proline residues include cereal proteins, such as gliadins and glutenins. Since the main biological function of caseins and cereal proteins is to be nitrogen and energy sources for infants and germinating seeds, a disordered structure is necessary to ensure high susceptibility of these proteins to proteolytic digestion.

C. Tertiary Structure

The tertiary structure refers to the spatial arrangement of the entire polypeptide chain with secondary structure segments into a compact three-dimensional folded form. Metamorphosis of a protein from a linear primary configuration into an intricate tertiary structure form in solution is a complex process; this is driven by the thermodynamic requirement to minimize the free energy of the molecule through optimization of various noncovalent interactions (hydrophobic, electrostatic, and van der Waals interactions and hydrogen bonding) between various groups in the protein. The most important geometric rearrangement that occurs during tertiary structure formation is the relocation of the nonopolar residues at the interior and disposal of the hydrophilic residues at the exterior of the protein molecule. Although a majority of hydrophobic groups are buried in the protein interior, not all of them can be buried. In fact, analysis of the surfaces of several globular proteins indicates that about 40–50% of the water-accessible protein surface is occupied by apolar residues [6]. Likewise, some polar residues are inevitably buried in the interior of proteins; however, these buried polar groups are usually hydrogen bonded to each other or engaged in electrostatic interaction with oppositely charged residues, such that their free energies are minimized in the apolar environment of the protein interior. Being very strong in low dielectric environments, these buried ion-pair interactions contribute significantly to the stability of proteins. Mounting evidence indicates that one of the major contributors to the thermostability of proteins from thermophilic organisms is such buried salt bridges.

In essence, the folding of a protein into a compact tertiary structure is accompanied by a reduction in the interfacial area between nonpolar groups of the protein and the surrounding solvent water. The distribution of hydrophilic and hydrophobic residues and their relative fraction in the amino acid sequence influences several physicochemical properties, such as shape, surface topography, and solubility (related to hydrophilicity/hydrophobicity of the protein surface) of the protein. Proteins containing a large number of hydrophilic residues distributed uniformly in its sequence tend to assume an elongated rodlike shape; this is because an elongated shape has a large surface area–to–volume ratio, so that more hydrophilic groups can be placed on the surface exposed to the solvent. In contrast, proteins containing a large number of hydrophobic residues tend to assume a spherical (globular) shape; this is because a spherical shape has the least surface area–to–volume ratio, so that more hydrophobic groups can be buried in the protein interior.

The tertiary structures of several single polypeptide proteins contain features known as domains. Domains (or "miniproteins") are the regions of the polypeptide chain that fold into a tertiary form independently. The domains then interact, giving rise to the unique tertiary structure of the protein. The number of domains in a protein is dependent on its molecular weight. The approximate size of a domain is about 100–150 amino acid residues. Thus, small proteins, such as lysozyme and β-lactoglobulin, usually contain only one domain, whereas large proteins, such as bovine serum albumin, contain more than one domain. The structural stability of each domain is usually independent of the others; differences in structural stabilities of domains may cause sequential unfolding of various parts of a protein molecule during thermal or interfacial denaturation.

D. Quaternary Structure

The quaternary structure refers to spatial arrangement of a protein containing several polypeptide chains. Each polypeptide chain is known as a subunit, and the quaternary complex is referred to oligomeric structure. Proteins that contain more than 28% hydrophobic amino acid residues (Val, Leu, Ile, Phe, Pro) generally form oligomeric structures (Fig. 1) [7]. The reason for this is that when the hydrophobic content is greater than 28%, it becomes physically impossible to bury all of the hydrophobic groups in the protein interior. Consequently, the tertiary structure of such proteins contains nonpolar patches on the surface, and hydrophobic interaction of these patches with adjacent protein molecules in aqueous solution leads to formation of oligomeric structures. Thus, formation of quaternary or oligomeric structure is primarily driven by the thermodynamic requirement to bury exposed hydrophobic patches. Other noncovalent interactions, such as electrostatic interactions and hydrogen bonding, at the interface of the subunits may also contribute to the stability of the quaternary structure.

Many food proteins, such as cereal, legume, and oilseed proteins, are oligomeric proteins with several subunits. Cereal proteins typically contain more than 35% hydrophobic amino acid residues and a high level (6–12%) of proline residues, and they exist in complex oligomeric states [8]. Soy globulins (7S and 11S) contain more than 40% hydrophobic amino acid residues. Because of this, the 7S protein exists as a trimeric protein with three different subunits and the 11S protein exists as a dodecamer with six

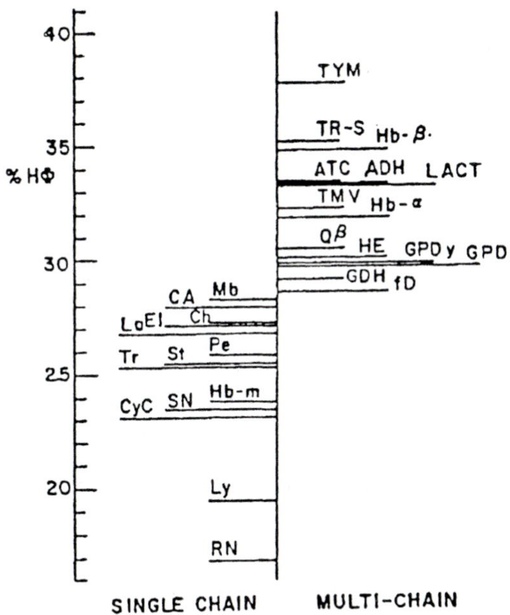

FIGURE 1 Mole percent of hydrophobic amino acids (Val, Pro, Leu, Ile, Phe) for a number of single-chain (left) and multichain (right) globular proteins of exactly known composition and known structure. TYM, Turnip yellow mosaic virus coat protein; TR, tryptophan synthetase (A chain, *E. coli*); ADH, alcohol dehydrogenase (horse); LACT, bovine β-lactoglobulin; TMV, tobacco mosaic virus coat protein; Hb, hemoglobin α-chain (human); Qβ, phage Qβ coat protein; He, hemerythrin (sipunculid); GPDy, glycerol-3-phosphate dehydrogenase (yeast); GDP, glycerol-3-phosphate dehydrogenase (pig); GDH, glutamate dehydrogenase (bovine); fD, phage fD coat protein; Mb, myoglobin (human); CA, carbonic anhydrase (human); Ch, chymotrypsinogen (bovine); El, elastase (pig); La, bovine lactalbumin; Pe, bovine pepsinogen; St, subtilisin; Tr, bovine trypsin; Hb-m, hemoglobin (monomeric, blowfly); SN, staphylococcal nuclease; CyC, cytochrome C (human); Ly, lysozyme (chicken); RN, ribonuclease (bovine).

acidic and six basic subunits. Both proteins exhibit complex dissociation and association behavior as a function of solution conditions, such as pH and ionic strength [9].

III. CONFORMATIONAL STABILITY

Folding of a protein from a linear primary structure to a complex tertiary or quaternary structure is driven by several noncovalent interactions, such as van der Waals, electrostatic, steric, and hydrophobic interactions and hydrogen bonding. The stability of the native folded structure is expressed as the net difference in free energy between the native and unfolded (denatured) states of the protein, and can be expressed as

$$\Delta G_{D \rightarrow N} = \Delta G_{H\text{-bond}} + \Delta G_{ele} + \Delta G_{H\phi} + \Delta G_{vdW} - T\Delta S_{conf} \quad (3)$$

where $\Delta G_{H\text{-bond}}$, ΔG_{ele}, $\Delta G_{H\phi}$, and ΔG_{vdW}, respectively, are free energy changes for hydrogen bonding, electrostatic, hydrophobic, and van der Waals interactions, ΔS_{conf} is the change in the conformational entropy of the polypeptide, and T is the temperature. The net stability, $\Delta G_{D \to N}$, of most proteins, irrespective of their size, is in the range of 10–20 kcal/mol; this indicates that, in spite of a multitude of interactions within protein molecules, they are only marginally stable. This metastable state can easily undergo conformational change when a few hydrogen bonds or electrostatic interactions or hydrophobic interactions are broken.

Disulfide bonds between cysteine residues in proteins can form as a result of protein folding when two cysteine residues are brought to close proximity with proper orientation. Proteins that require high structural stability in biological systems, especially in extracellular environments, usually contain disulfide bonds; the disulfide bonds stabilize the structure by reducing the tendency of the polypeptide chain to unfold.

Because proteins are inherently in a metastable state with a net stability of only about 10–20 kcal/mol, changes in a protein's environment, such as pH, ionic strength, temperature, and solvent composition, can easily induce conformational changes in the protein. Such environment-induced conformational changes in proteins affect biological functions in the case of enzymes and functional properties of food proteins. Enzymes lose their activity upon denaturation, and denaturation of food proteins usually causes insolubilization and loss of some functional properties. In some instances, partial denaturation of food proteins is desirable. For example, thermal denaturation of trypsin and amylase inhibitors of legumes improves digestibility and bioavailability of proteins and carbohydrates in legume-containing foods. Partially denatured proteins are generally more digestible and possess better functional properties, such as foaming and emulsifying properties, than do native proteins.

Protein denaturation is often considered to be a two-state transition phenomenon. That is, once a protein molecule begins to unfold, or once a few interactions in the protein are broken, the other interactions in the protein are weakened and the whole molecule completely unfolds when the denaturant concentration is increased. This cooperative nature of unfolding, observed with several small molecular weight proteins, suggests that globular proteins can exist only in the native and denatured states. However, recent studies with several proteins have shown that proteins can exist in a stable intermediate folded state known as a "molten globule state" [10–14]. The molten globule state is characterized by a relatively compact globule with nativelike secondary structure and a disrupted tertiary structure. One of the most extensively studied molten globule state is that of α-lactalbumin [14]. α-Lactalbumin assumes a molten globule state at extremes of pH, i.e., in the pH range 4.2–3.0 and above pH 9.5 [15], at moderate concentrations (~2 M) of guanidine hydrochloride at neutral pH [10], and when one of the four disulfide bonds of the protein is reduced [16]. The state of the molecule under all of these denaturing conditions is identical, implying that unfolding of α-lactalbumin under equilibrium conditions follows a general scheme:

$$N \rightleftharpoons A \rightleftharpoons D$$

where N, A, and D are the native, molten globule, and denatured states, respectively. The rate of transition between A and D is more rapid than that between N and A. It has been observed that the far-UV circular dichroic spectrum of the molten globule state of α-lactalbumin is similar to that of the native protein, indicating that it has the same backbone secondary structure as that of the native state, whereas NMR, difference absorption spectrum, fluorescence, and near-UV circular dichroic spectra of the molten globule state were quite different from those of the native state, indicating that the tertiary structure was different from that of the native state. The hydrodynamic size of the molten globule state was only slightly larger than that of the native state (i.e., 50 Å vs. 35 Å) (Fig. 2), suggesting that the molecule is in a slightly expanded but compact shape.

The molten globule state of α-lactalbumin formed at acid pH 4.2–3.0 is found to be susceptible to aggregation. This is presumably because of greater exposure of hydrophobic surface through alterations in the tertiary structure. This behavior has been successfully exploited in the fractionation of α-lactalbumin from cheese whey, which involves precipitation of the protein by heating at 55°C for 30 minutes at pH 3.8 [17]. The molten globule state exhibits strong affinity for hydrophobic ligands [18] and readily forms complexes with phospholipid vesicles [19,20]. Perhaps this latter property is the basis for formation of an insoluble lipoprotein complex between α-lactalbumin and whey phospholipids when the pH of the cheese whey is decreased to pH 3.8 at elevated temperature.

It is very likely that several food proteins may exhibit stable molten globule states under appropriate conditions. These partially denatured states may possess unique functional properties different from those of the native and fully denatured states. For example, in the native state, whey proteins (β-lactoglobulin and α-lactalbumin) are extremely soluble in the pH range 3–10; they do not precipitate at their isoelectric pH. However, when an average of one in six disulfide bonds in whey proteins is reduced by oxidative sulfitolysis, the reduced proteins show a U-shaped pH-solubility profile with minimum solubility at about pH 4.5 [21]. This dramatic change in the pH-solubility characteristics of whey proteins might be attributable to formation of a molten globule–like state upon partial reduction of disulfide bonds in these proteins [16].

Protein denaturation can be caused by several agents, such as heat, pressure, interfacial forces, mechanical shear, extremes of pH, chaotropic salts, detergents, and organic solvents. Heat and pressure (e.g., extrusion) are the most commonly used agents in food processing. The effects of these agents, especially thermal and pressure effects, on protein denaturation are discussed in detail in other chapters.

The structural stability of proteins is mainly maintained by noncovalent interactions, such as hydrogen bonding, electrostatic, hydrophobic, and van der Waals inter-

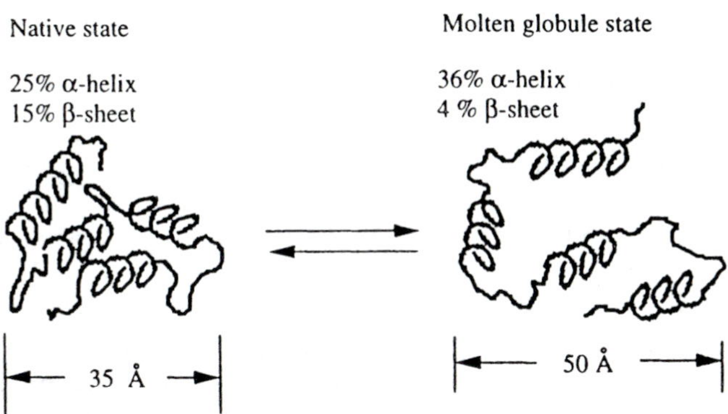

FIGURE 2 A schematic representation of transitions between the native, molten globule, and denatured states of α-lactalbumin. (Adapted from Ref. 10.)

actions, and also by interactions involving binding of prosthetic groups/ligands. The relative contributions of these various interactions to stability depend on the type of proteins and their functions in biological systems.

Hydrogen bonding in proteins predominantly exist in secondary structures, i.e., α-helix and β-sheets, of the polypeptide chain. Intersegment hydrogen bonding in proteins occurs predominantly among glutamine and asparagine side chains, for example, in gluten and legume proteins, which are rich in Gln and Asn. Since a majority of hydrogen bonding is associated with α-helix and β-sheets, correlations between thermostability and the secondary structure contents of proteins have been analyzed. Examination of several proteins from thermophilic and mesophilic organisms showed a positive correlation between secondary structure content and thermostability [22]. However, it has been observed that proteins from thermophilic organisms generally contain lower β-sheet content than their counterparts in mesophilic organisms [23], implying that β-sheets are not necessarily responsible for thermostability.

Electrostatic interactions between oppositely charged groups (salt bridges), partially buried in a nonpolar environment, have been shown to immensely contribute to thermostability of proteins [24]. For instance, the tetrameric glyceraldehyde-3-phosphate dehydrogenase from *Bacillus stearothermophilus* has greater thermostability than that from lobster; this difference arises mainly from two extra salt bridges found at the subunit interfaces of the enzyme from *B. stearothermophilus* [25]. No specific examples of food proteins containing such salt bridges contributing to their thermostability are known.

Prosthetic groups, such as metal ions and specific ligands, bound to protein molecules often contribute to their thermostability. For example, alkaline phosphatase is a zinc-binding protein. In the holo form, the conformational stability, ΔG_D, of this enzyme is about 129 kcal/mol, whereas in the apo form the stability is about 20 kcal/mol [31]. Binding of palmitic acid or retinol to β-lactoglobulin affects its thermostability as measured by differential scanning calorimetry [32]. The thermal denaturation temperature increases by about 8°C when palmitic acid binds, whereas it increases by about 4°C when retinol binds to β-lactoglobulin. On the other hand, whereas binding of palmitic acid significantly increases the apparent enthalpy of denaturation of β-lactoglobulin, binding of retinol significantly decreases it. It is not evident whether both palmitic acid and retinol bind to the same binding site. Nonetheless, it appears that binding of various small molecular weight hydrophobic ligands influence the structural stability of β-lactoglobulin.

Among the noncovalent interactions, hydrophobic interactions immensely contribute to conformational stability of proteins [26–28]. Unlike hydrogen bonding and electrostatic interactions, hydrophobic interactions are endothermic and entropy-driven, and their strength increases with temperature as high as 100°C [29]. This positive effect of temperature on hydrophobic interactions indicates its quintessential role in the thermostability of proteins. Several investigators have attempted to establish a quantitative correlation between protein hydrophobicity and thermal stability. Bull and Breese [30] found that thermal melting points of 14 proteins were positively correlated with Tanford's hydrophobicity index ($r^2 = +0.622$). Although this correlation was statistically significant, the low value of the correlation coefficient suggests that it is only an approximate one. On the other hand, the average residue volumes showed a much better correlation with the melting temperatures ($r^2 = 0.960$); since hydrophobic residues have large volumes, it was suggested that hydrophobic index may have a role in thermostability of proteins. However, when the least-square relation between the melting points of the proteins and hydrophobic index and the average residue volumes as the independent variables were analyzed, a positive coefficient for the average residue volume and a negative coefficient for the hydrophobicity index were found [30]. These analyses at best suggest that the mean residue hydrophobicity is not a meaningful index to relate stability to hydrophobicity. One of the factors contributing to this unreliability is the fact that, in a typical globular protein, about 40–50% of a protein's surface is occupied by nonpolar residues. Evidently, these exposed residues, instead of stabilizing the protein structure, may in fact tend to destabilize it. More often the exposed nonpolar residues are smaller residues, such as alanine, with small residue volume; the large nonpolar amino acid residues, such as leucine and isoleucine, have a greater tendency than the small apolar residues to be buried in the protein interior. This might be one of the reasons for a strong correlation between the average residue volume and the melting temperature, but not between the average hydrophobicity and the melting temperature.

Arginine is often found in abundance in proteins of thermophilic organisms [22], although its precise role in thermostability of proteins is not known. Combinations of certain groups of amino acid residues have been shown to influence thermostability of proteins. For example, statistical analysis of 15 different proteins has shown that thermal denaturation temperatures of these proteins were positively correlated ($r_2 = 0.96$) to the number percent of Asp, Cys, Glu, Lys, Leu, Arg, Trp, and Tyr residues [33]. On the other hand, thermal denaturation temperatures of the same set of proteins were negatively correlated ($r_2 = -0.95$) to the number percent of Ala, Asp, Gly, Gln, Ser, Thr, Val, and Tyr (Fig. 3). Other amino acid residues or other combinations of amino acid residues have been found to have poor correlation with thermal denaturation temperatures. It is notable that negatively and positively charged amino acid residues, in addition to large hydrophobic residues, appear in the stabilizing group of amino acids. This may mean that a combination of salt bridges and hydrophobicity is essential for flexibility/rigidity and thermostability of proteins.

IV. FUNCTIONAL PROPERTIES

In addition to their nutritional function, proteins play important roles in the expression of sensory attributes of foods. Food preferences by consumers are predominantly based upon organoleptic properties, such as color, flavor, and texture, of foods. Protein is one of the major ingredients that contribute to sensory properties of foods. For instance, the textural properties of bakery products are manifestations of the viscoelastic properties and dough-forming properties of wheat gluten; muscle proteins impart unique textural and succulence characteristics to meat products; the curd-forming properties of dairy products are due to the unique colloidal structure of casein micelles; the sensory properties of cakes and desserts are attributable to the properties of egg proteins. Some of the important functions performed by different food proteins in various food products and the mechanisms of expression of those functions are listed in Table 2.

Several definitions for functional properties of food proteins exist. Kinsella [34] defined functional properties as "those physical and chemical properties which affect the behavior of proteins in food systems during processing, storage, preparation and consumption." The physical and chemical properties that influence functional behavior of proteins in foods include their size, shape, amino acid composition and sequence, net charge, charge distribution, hydrophobicity, hydrophilicity, structures (secondary, tertiary, and quaternary), molecular flexibility and rigidity in response to external environment

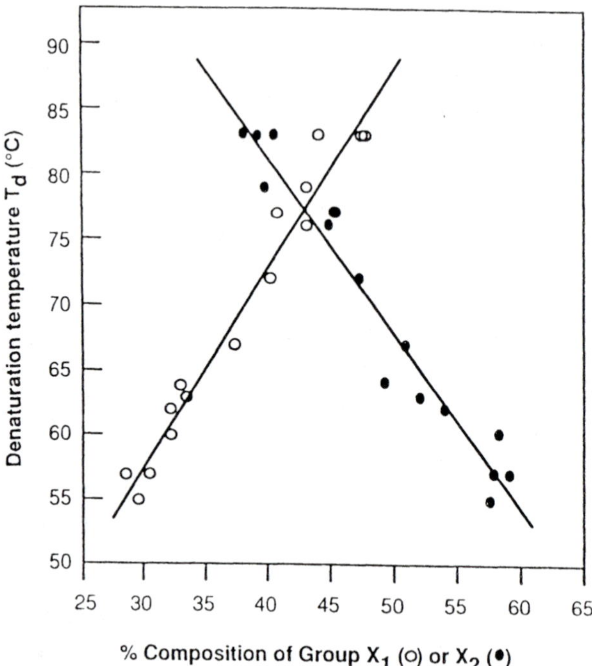

FIGURE 3 Group correlations of amino acid residues to thermal stability of globular proteins. Group X_1 represents Asp, Cys, Glu, Lys, Leu, Arg, Trp, and Tyr. Group X_2 represents Ala, Asp, Gly, Gln, Ser, Thr, Val, and Tyr. (Adapted from Ref. 33.)

(such as pH, temperature, and salt concentration), or interaction with other food constituents.

Although much is known about the physicochemical properties of several proteins and their functional properties in simple model systems, prediction of their behavior in real food systems has not been successful. One of the main reasons for this is denaturation of proteins during food processing and preparation. A combination of factors such as pH, temperature, ionic environment, and interaction of other food constituents (lipids, carbohydrates, salts, etc.) with proteins can cause unpredictable structural changes in proteins and thus affect their functional behaviors. In addition, the methods used for protein isolation can also cause variations in the initial conformation of proteins, which in turn may affect their behavior in a particular food product. Despite these difficulties, studies on functional properties of proteins in model systems have provided considerable insight into structure-function relationships of food proteins.

The structure-functionality relationships of various proteins with regards to many of the functions listed in Table 2 will be covered in detail in other chapters. Therefore, only those functional properties not discussed in other chapters will be discussed in the following sections.

A. Protein-Water Interactions

Water is an essential constituent of all living matter and of course of foods as well. Since the native structure of a protein to begin with is a consequence of its interaction with

Table 2 Functional Roles of Food Proteins in Food Systems

Function	Mechanism	Food	Protein type
Solubility	Hydrophilicity	Beverages	Whey proteins
Viscosity	Water binding, hydrodynamic size and shape	Soups, gravies, salad dressings, desserts	Gelatin
Water binding	Hydrogen bonding, ionic hydration	Meat sausages, cakes, breads	Muscle and egg proteins
Gelation	Water entrapment and immobilization, network formation	Meats, gels, cakes, bakeries, cheese	Muscle, egg, and milk proteins
Cohesion-Adhesion	Hydrophobic, ionic, hydrogen bonding	Meats, sausages, pasta, baked goods	Muscle, egg, and whey proteins
Elasticity	Hydrophobic bonding, disulfide cross-links	Meats, bakery	Muscle and cereal proteins
Emulsification	Adsorption and film formation at interfaces	Sausages, bologna, soup, cakes, dressings	Muscle, egg, and milk proteins
Foaming	Interfacial adsorption and film formation	Whipped toppings, ice cream, cakes, desserts	Egg and milk proteins
Fat and flavor binding	Hydrophobic bonding, entrapment	Low-fat bakery products, doughnuts	Milk, egg, and cereal proteins

Source: Ref. 52.

solvent water, its function in a food product is intimately related to its interaction with water. In fact, on a empirical level, the various functional properties of proteins can be viewed as manifestations of its interaction with water. For example, dispersibility, wettability, swelling, and solubility are related to thermodynamics of protein-water interactions. Properties such as thickening/viscosity, gelation, coagulation, etc., are hydrodynamic properties of protein polymers affected both by their size, shape, and molecular flexibility and their interaction with solvent water. Similarly, surface-active properties, such as foaming and emulsification, are simply the result of thermodynamically unfavorable interaction of exposed nonpolar patches of proteins with solvent water.

Water molecules bind to both polar and nonpolar (hydrophobic hydration) groups in proteins via dipole-dipole, charge-dipole, and dipole-induced dipole interactions. Charged amino acid residues, such as lysyl, arginyl, glutamic, and aspartic residues, bind about 6 moles of water per residue; uncharged polar residues, such as Ser, Thr, Gln, and Asn, bind about 2 moles of water per residue; the nonpolar amino acid residues bind about 1 mole of water per residue. The hydration capacity of a protein therefore is related, in part, to its amino acid composition [35]; the greater the number of charged residues, the greater the hydration capacity. Because a majority of nonpolar amino acid residues and a significant number of polar groups (e.g., peptide bond groups and some Gln and Asn residues) are buried in the protein interior, which are not hydrated, the hydration capacity of proteins must arise predominantly from binding of water to amino acid residues on the protein's surface.

Typically, proteins bind about 0.3–0.5 g water/g protein at a water activity of 0.9. This primarily represents an unfreezable, monomolecular layer of water on a protein's surface. This water is often referred to as "bound" water, with the connotation that it is immobile. Thermodynamic data, however, suggest otherwise. For instance, in the monolayer hydration range of 0.07–0.27 g water/g protein, the free energy change for desorption of water molecules from the protein surface is about 0.18 kcal/mol at 25°C [36], which is significantly lower than the thermal kinetic energy of water at 25°C (~0.6 kcal/mol). Therefore, the "bound" water must be reasonably mobile. Unlike globular proteins, casein micelles bind about 4 g water/g protein. This is attributable to the enormous amount of void spaces within the casein micelle structure, which imbibes water through capillary action and physical entrapment. The hydration capacity of a denatured globular protein is only about 10% greater than that of the native protein. This is because of the fact that even in the denatured state the protein retains much of its folded structure, although in a nonnative state, with the result that the surface area–to–volume (or mass) ratio of the protein is not drastically altered. Aggregation of the denatured protein also buries a significant amount of the protein's surface area.

Solution conditions, such as pH, ionic strength, and temperature, affect hydration of proteins. The hydration capacity is minimal at the isoelectric pH of a protein, where the net charge is zero and protein-protein interaction is maximum. At low concentration, i.e., <0.2 M, salts increase the hydration capacity of proteins. At this low ionic strength, although weak binding of salt counterions to proteins results in screening of the charges, it does not affect the hydration shells of charged groups, and the increase in hydration capacity essentially comes from the hydration shells of the bound ions.

In food applications, the water-holding capacity or water-uptake capacity of a protein is more important than its hydration capacity. Water-holding capacity refers to the ability of a protein matrix, such as protein particles, protein gels, or muscle, to absorb and retain water against gravity. This water includes bound water, hydrodynamic water, capillary water, and physically entrapped water. The physically entrapped water, however, is the largest fraction. It imparts juiciness and tenderness, particularly in comminuted meat products.

B. Solubility

Several functional properties, such as thickening, foaming, emulsification, and gelation, of proteins are affected by protein solubility. Solubility of a protein is fundamentally related to its hydrophilicity/hydrophobicity balance. Thus, the amino acid composition of a protein inherently affects its solubility characteristics. Because solubility of a protein is a manifestation of the differences in the energetics of protein-protein and protein-solvent interactions, Bigelow [37] suggested that the average hydrophobicity of amino acid residues and charge frequency are the two most important factors that determine solubility of proteins: the lower the average hydrophobicity and the higher the charge frequency, the higher the solubility. However, it has been pointed out that although the solubility characteristics of several proteins follow this empirical relationship, there are several exceptions [38]. On a fundamental level, the solubility characteristics of proteins should be related to composition of the protein's surface (and not necessarily to its overall amino acid composition) and the thermodynamics of its interaction with the solvent.

Conversely, surface hydrophobicity and surface hydrophilicity characteristics of a protein's surface are the most important factors that affect its solubility characteristics.

Protein solubility in aqueous solutions is dependent on pH. At pH values above and below the isoelectric pH, proteins carry a net charge; electrostatic repulsion and ionic hydration promote solubilization of the protein. For most proteins, minimum solubility occurs at the isoelectric pH, where the electrostatic repulsion and ionic hydration are minimum and hydrophobic interaction between surface nonpolar patches is maximum. However, some proteins, e.g., whey proteins (α-lactalbumin, β-lactoglobulin, and bovine serum albumin) (pI = 4.8–5.2) are highly soluble at their isoelectric pH. This is primarily because the exposed surfaces of these proteins contain a high ratio of hydrophilic to hydrophobic groups. Although these proteins are electrically neutral at their isoelectric pH, they contain a large number of charged and uncharged hydrophilic residues on the surface, and hydration of these polar residues creates hydration repulsion forces great enough to offset aggregation via hydrophobic interactions. However, when whey proteins are heat denatured, the pH-solubility profile shows a typical U-shaped profile with minimum solubility at pH 4.6 (Fig. 4) [39]. Obviously, heat denaturation alters surface hydrophobicity/hydrophilicity of the proteins and thereby the balance of protein-protein and protein-solvent interactions in favor of the former.

Salts affect the solubility of proteins in two different ways, depending on the physicochemical characteristics of the protein surface. Generally, at low ionic strength (<0.5 M), the solubility of proteins with a large number of exposed hydrophobic patches decreases (salting-out) and solubility of proteins with a high incidence of surface hydro-

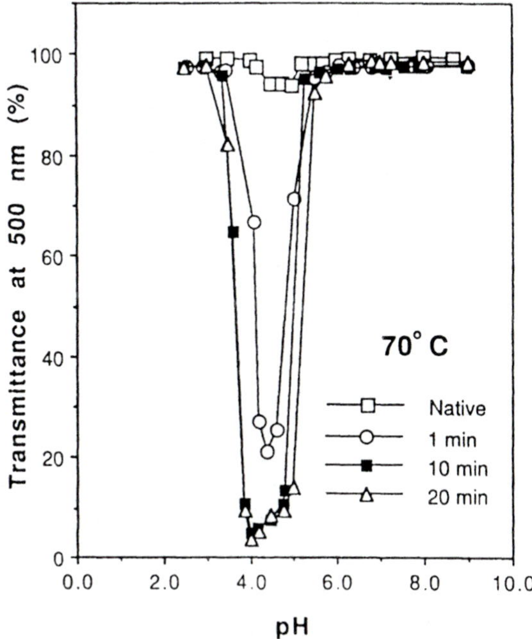

FIGURE 4 pH-solubility profile of whey protein isolate heat denatured at 70°C for various times. (From Ref. 39.)

philic groups increases (salting-in) with increase of ionic strength. That is, the salting-in and salting-out characteristics of a protein by salts is fundamentally related to the hydrophilicity-hydrophobicity characteristics of the protein surface.

The surface hydrophobicity of proteins is usually measured by the fluorescent probe (such as *cis*-parinaric acid and 8-anilino-1-naphthalenesulfonic acid)–binding method [40]. Although this approach provides some information about the nonpolarity of the protein surface, it is doubtful whether the measured value truly reflects the "hydrophobicity" of the protein surface. For instance, the surface hydrophobicity (as measured by the *cis*-parinaric acid–binding method) of WPI decreased with heating time at 70°C [39], whereas the solubility of the same heat-denatured WPI samples showed a decrease at pH 4.6 with increase in heating time (Fig. 4), indicating an increase in hydrophobicity of the protein surface with heating time. This contradiction between fluorescence probe–binding value and the solubility characteristics can be resolved only by recognizing that these two techniques probe two different properties of the protein surface. It has been shown that binding of fluorescent probes to proteins occurs only at well-defined hydrophobic cavities formed by grouping of nonpolar residues on the protein surface [38]. These cavities are not accessible to water but are accessible to nonpolar ligands. In this regard, these cavities do not affect the solubility characteristics of the protein. When these hydrophobic cavities are dissociated upon heating and the constituent hydrophobic residues are distributed randomly on the protein surface, then no significant binding of the fluorescent probe to the protein occurs. However, distribution of these hydrophobic residues on the protein surface exposed to solvent water would greatly enhance the hydrophobic character or the true surface hydrophobicity of the protein and would significantly alter its solubility characteristics.

The true relative surface hydrophobicity of proteins can be measured from their solubility behavior in salt solutions at isoelectric pH. For example, Figure 5 shows the effect of NaCl on the solubility at pH 4.6 of WPI that had been heat denatured at 70°C for various periods of time. The effectiveness of NaCl in solubilizing heat-denatured WPI decreases as the heating time is increased, showing that the hydrophobic character of the protein surface was enhanced as the heating time at 70°C was increased. The initial slope of these salt-solubility profiles can be used as an index of surface hydrophobicity of the proteins.

C. Viscosity and Thickening

The consumer acceptability of several liquid and semi-solid–type foods (e.g., gravies, soups, beverages) depends on the viscosity or consistency of the product. The viscosity or consistency of solutions is greatly influenced by solute type. Large molecular weight soluble polymers greatly increase viscosity even at low concentration. This again depends on several molecular properties such as size, shape, flexibility, and hydration. Solutions of randomly coiled polymers display greater viscosity than do solutions of compact folded polymers of same molecular weight. Thus, polysaccharides and gums, which are large, highly flexible and hydrophilic polymers, exhibit high viscosity even at low concentrations. These hydrocolloids are often used in food products as thickening agents. Certain proteins, such as gelatin, myosin, etc., that have large axial ratios exhibit high viscosity even at low concentrations.

The viscosity of protein solutions generally increases exponentially with protein concentration; this is attributable to increased interaction between the hydrated protein

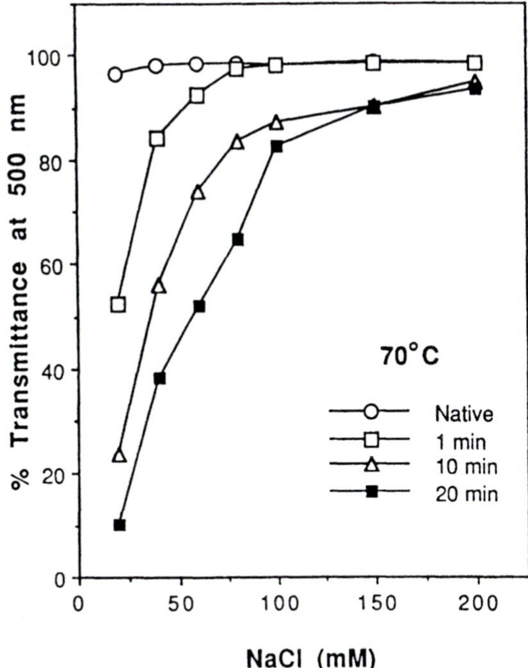

FIGURE 5 Solubility of heat-denatured (70°C for various times) whey protein isolate at pH 4.6 as a function of NaCl concentration.

TABLE 3 Effect of Heat on Specific Viscosity of WPI Solutions (5%, 25°C, pH 7.0)

	Specific viscosity[a]	
Heating time (min)	70°C, 5%	90°C, 9%
0	0.21	0.21
1	0.28	—
5	0.28	1.87
10	0.32	—
20	0.43	2.25
40		2.33
60		2.38
90		2.30

[a]Average of triplicate measurements.
Source: Ref. 39.

molecules. Also, the ability of the protein to absorb water and swell affects its viscosity. Partial denaturation and/or heat-induced polymerization, which causes an increase in the hydrodynamic size of proteins, increases viscosity. For example, Table 3 shows the effect of heat on specific viscosity of WPI solutions at two different heating conditions [39]. At 70°C, the viscosity of a 5% WPI solution increases with heating time; however, the increase was only about twofold after 20-minute heating time compared to that of the unheated control. On the other hand, when a 9% WPI solution was heated at 90°C for various periods of time and the diluted to 5% final concentration, the specific viscosity increased almost 10-fold after 20-minute heating. These differences are mainly due to the extent of polymerization of the proteins under these two heating conditions. The two most important factors that affect viscosity of protein solutions are the hydrodynamic shape and the size of the protein molecules, which follow the relationship:

$$\eta_{sp} = \beta C(\upsilon_2 + \delta_1 \upsilon_1)$$

where β is shape factor, C is the concentration, υ_1 and υ_2 are the partial specific volumes of the solvent and the protein, respectively, and δ_1 is the gram of water bound per gram of protein. Since heating of WPI under two different heating conditions does not significantly change its υ_2 and δ_1 values, the phenomenal increase in the specific viscosity of the WPI sample heated at 90°C must be due to an increase in its hydrodynamic size and shape. This has been shown to be related to formation of soluble polymers via sulfhydryl-disulfide interchange reactions [39].

At high protein concentrations, protein solutions display a pseudoplastic behavior [41,42]. The pseudoplastic behavior is the result of the tendency of protein molecules to orient their major axis in the direction of flow. When shearing is stopped, the oriented protein molecules, depending on their molecular properties and intermolecular interactions in the oriented state, may or may not return to the state of random orientation and dispersion in solution. Spontaneous return to random orientation will cause return of the viscosity to its original value. Solutions of fibrous proteins, e.g., gelatin and actomyosin, do not quickly relax to random orientation, and thus the viscosity remains low compared to its original value. On the other hand, solutions of globular proteins, e.g., whey proteins and soy proteins, rapidly regain their original viscosity when the flow is stopped [43].

The viscosity of some protein solutions, e.g., β-lactoglobulin, increases with time when the solution is sheared at a constant shear rate [41]. This is attributable to shear-induced partial denaturation and sulfhydryl disulfide–induced polymerization of β-lactoglobulin.

Solution conditions, such as pH, ionic strength, and temperature, affect viscosity of protein solutions. The viscosity of globular protein solutions generally decreases as the pH is decreased toward the protein's isoelectric pH. Partial heat denaturation generally increases the viscosity of commercial whey protein concentrates (WPC). For instance, heat denaturation of WPC at acidic pH caused a 10-fold increase in apparent viscosity compared to unheated WPC under comparable protein concentration [44]. Increase of ionic strength generally decreases the viscosity of protein solutions by affecting their hydration capacity. In this respect, the effects of divalent salts, e.g., calcium salts, is more pronounced than monovalent salts. For example, progressive replacement of calcium ions with sodium ions progressively increased the viscosity of a solution of commercial WPC.

D. Flavor Binding

The flavor-binding ability of proteins is related to interaction of small molecular weight flavorants with hydrophobic pockets or crevices on the protein's surface [45–47]. The extent of binding thus depends on the number of hydrophobic patches on the protein's surface. In addition to hydrophobic interactions, flavorants with polar groups may also bind to proteins via hydrogen bonding or electrostatic interactions. The mechanism of fat- and flavor-binding properties of proteins are discussed in Chapter 5.

The natural ability of proteins to bind flavor compounds has both desirable and undesirable implications. For instance, aldehydes and ketones generated from oxidation of unsaturated fatty acids in oilseeds, such as soybean, bind to proteins and resist their removal during solvent extraction of the meal. For example, the grassy and beany off-flavors of soy protein concentrates and isolates are mainly due to bound hexanal. Upon binding to surface hydrophobic cavities on the protein, hexanal may diffuse into the hydrophobic interior of the protein and thus resist its removal during extraction with organic solvents such as hexane.

Proteins can also be used as flavor carriers or flavor modifiers in foods, especially in plant protein–based meat analogs. In such applications, the proteins should be able to bind flavors reasonably tightly and retain them during processing. However, proteins do not bind all flavorants with equal affinity. Because of this, the relative concentrations of various flavorants in the final food product is often different from that added before processing; this affects the flavor profile of the final product.

Proteins can modify the flavor profile of a food product by selectively binding some flavorants more tightly than the others. An example of this is the flavor profile of yogurt containing whey proteins. Because whey proteins, especially β-lactoglobulin, strongly bind some of the flavor compounds in yogurt, they alter the aroma profile. In addition, during eating, poor release of these bound flavor components in the mouth also affects the flavor profile of yogurt.

Protein-bound flavorants do not contribute to taste and aroma unless they are released readily in the mouth during mastication. Therefore, knowledge of the mechanisms of interaction and relative binding affinity of various flavorants under various environmental and processing conditions are necessary both for preparing flavored protein-containing products and for devising effective strategies to remove off-flavors from protein preparations.

E. Gelation

Gelation refers to the transformation of a protein in the sol state into a gel-like structure by heat or other agents [38]. The mechanisms involved in protein gelation and characterization of the rheological properties of protein gels are described in Chapter 4. Therefore, only a few molecular properties of proteins affecting their gelation are discussed below.

Proteins form two types of gels: coagulum and transparent gels. The type of gel formed by a protein is primarily influenced by its amino acid composition, although solution conditions, such as pH and ionic strength, also play a role. Proteins containing a high frequency of nonpolar amino acid residues tend to form coagulum-type gels [48], whereas proteins that contain a high frequency of hydrophilic amino acids form transparent gels. Shimada and Matsushita [48] found that proteins form coagulum-type gels when the sum of Val, Pro, Leu, Ile, Phe, and Trp residues of the protein is above 31.5 mol%. At high levels of apolar amino acid content, the rate and extent of protein-protein interactions among thermally unfolded proteins seem to be greater than those of protein-water (solvent) interactions, leading to formation of a weak coagulum-type gel.

Like all polymer gels, including gels of synthetic polymers, the stability of a protein gel network against thermal and mechanical stresses depends on the number and nature of the cross-links formed per polymer chain. Theoretically, a gel network will be stable when the sum of the interaction energies of a polymer chain in the network is greater than its thermal kinetic energy at a given temperature. Since at a given temperature the thermal kinetic energy of a large polymer chain is lower than that of a small polymer chain, gel network structures formed with large polymers are more stable. Also, in a given mass of the gel network, the number of cross-links formed per polymer chain is usually greater in long-chain polymers than their short-chain counterparts; this is due to a decrease in the concentration of "ends" or "tails" of the polymer chains, which often do not take part in cross-linking. It has been shown that the square root of the hardness of gelatin gels is linearly related to molecular weight (i.e., chain length) [49]. Globular proteins also exhibit a similar relationship [50]. Furthermore, globular proteins with molecular weights less than 23,000 and devoid of either cystine or cysteine residues do not form heat-induced gels at any reasonable protein concentration [50]. Proteins that contain cysteine and cystine residues can form a gel even when the molecular weight is <23,000; this is because of the occurrence of sulfhydryl-disulfide interchange reaction during heating, which increases the molecular weight (chain length) to >23,000 [50].

Protein gels are highly hydrated structures containing anywhere from 85 to 98% water depending on the protein concentration used. Experimental evidence shows that the water in these gels has chemical potential (activity) close to that of liquid water. The mechanism by which the liquid water is transformed to a semi-solid–like state and held within the gel network in a nonflowable form is not properly understood. It is known that translucent gels, such as gelatin gels, hold more water than coagulum-type gels. Translucent gel networks are primarily held together by hydrogen bonding and electrostatic interactions and only to a limited extent by hydrophobic interactions. Thus, the mode of retention of water within the gel network may involve hydrogen bonding with the polar groups of the polypeptide backbone (i.e., NH and CO groups) and the side chain charged groups (i.e., COO^- and NH_3^+). However, these alone cannot account for the amount of water retained in the gel network. For instance, the number of moles of water bound per mole of each of these hydrogen bonding and ionic groups is known [51]. A simple estimation would suggest that, in a 10% protein gel, only a fraction of the total water in the gel is physically bound to these polar groups. Therefore, water binding alone cannot explain immobilization of a large amount of water within a gel network. A possible hypothetical explanation is as follows: A translucent gel can be considered as a two-phase system in which minute droplets of water are dispersed in a continuous network of polypeptide chains. In this confined environment, each water droplet (cell) is surrounded by a force field emanating from the charged side chains and the dipole groups of the polypeptide backbone (Fig. 6). Because of this force field, water molecules may reorient and form an extensively hydrogen-bonded water-water network, bridging the peptide polar groups across the cell. This extensively hydrogen-bonded water structure may resemble that of the icelike structure, thus transforming the liquid water into a semi-solid state. The rigidity of the hydrogen-bonded continuum structure may increase with a decrease in cell size.

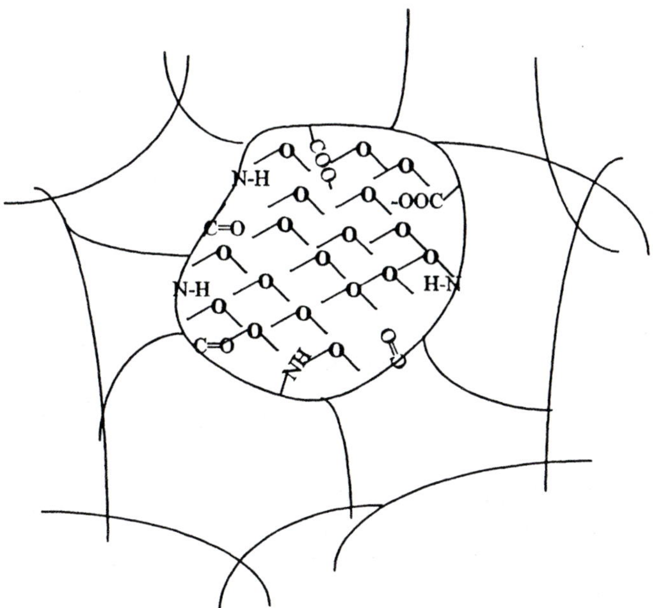

FIGURE 6 Hypothetical hydrogen-bonded state of water in a protein gel matrix.

V. CONCLUSIONS

The precise role of protein structure and how its structural transformation in a food milieu contributes to the expression of sensory properties is not well understood and is the topic of investigation in several laboratories. It must be emphasized that the sensory qualities of a food are not simply related to a single functional attribute of the protein ingredient; it is rather a manifestation of complex interactions among its various functional attributes. For example, for a protein to be functional in a cake, it should possess gelling/heat-setting, foaming, and emulsifying properties in the presence of the other ingredients used in cake mixes. In other words, it should be able to display multiple functionalities in a complex food milieu. Theoretically, a single protein cannot be expected to possess all the desirable multiple functionalities required in a particular food system. In practice, however, this can be achieved by using a mixture of different proteins, each one contributing to different functional attribute required in a particular food product. The proteins of animal origin, e.g., meat, milk, and egg proteins, which are mixtures of several proteins, are capable of performing multiple functions. For example, egg white possesses multiple functionalities such as gelation, emulsification, foaming, water binding, and heat coagulation, which makes it a highly desirable protein in many foods. The multiple functionalities of egg white are products of complex interactions among its protein constituents, namely, ovalbumin, conalbumin, lysozyme, ovomucin, globulins, and other proteins. The unique viscoelastic properties of wheat gluten arise from its protein constituents, namely, glutenins and gliadins. The elasticity of dough is due to intermolecular interactions between glutenins (disulfide cross-linking, hydrogen bonding, hydrophobic interactions), which results in the formation of threadlike polymers. These linear polymers subsequently interact with each other via hydrogen bonding, hydrophobic interactions, and disulfide cross-linking to form a sheetlike network with elastic properties. Gliadins, on the other hand, contribute to the viscosity of the dough. Although gliadins contain about 2–3% half-cystine residues, they do not undergo extensive polymerization via sulfhydryl-disulfide interchange reactions. These disulfide bonds remain as intramolecular disulfide bonds. This must be related to the unique structure of gliadins, which is different from that of glutenins. The glutenin-free dough is viscous but not viscoelastic; similarly, the gliadin-free dough is elastic but not viscoelastic. The viscoelasticity of wheat dough therefore depends on the ratio of glutenin and gliadin contents of wheat flour. Thus, attempts to inculcate the properties of egg white, wheat gluten, or other conventional food proteins in a novel protein, e.g., isolated β-lactoglobulin, α-lactalbumin, or oilseed proteins, for use in a particular food product may prove to be an exercise in futility. On the other hand, fundamental investigations on the functional properties of protein mixtures may prove to be a better approach. To achieve this, it is necessary to have a knowledge of the unique functional attributes of individual proteins in order to assess their potential role in a protein mixture.

The physicochemical properties that affect the functional properties of proteins are related to the size, charge distribution, hydrophobicity, hydrophilicity, molecular flexibility, and steric properties of the protein surface. Among these, the molecular flexibility and hydrophobicity are considered to be the most important factors influencing several functional properties. Genetic, chemical, and enzymatic modification approaches can be employed to alter these molecular properties. For example, polypeptides of legumins and vicilins proteins of legumes and oilseeds generally possess a β-barrel structure (see Chapter 9). The high heat stability, poor proteolytic digestibility, and probably less-than-desirable functionality of these proteins might be related to the structural stability of this β-barrel. Introduction of proline residues at critical locations in the β-barrel via site-directed mutagenesis might increase their molecular flexibility and thereby decrease their thermal and interfacial stability, increase their susceptibility to proteases, and possibly improve their functional properties. However, in order to apply such approaches to improve the functional properties of proteins, one needs to know how hydrophobic and how flexible certain regions of a protein should be in order for it to exhibit improved functionality. Thus, future research should focus on generating information on molecular architectures (or the descriptors) that define the functional properties of proteins.

REFERENCES

1. Y. Nozaki and C. Tanford, The solubility of amino acids and two glycine peptides in aqueous ethanol and dioxane solutions, *J. Biol. Chem. 246*:2211 (1972).
2. S. Kamtekar, J. Schiffer, H. Xiong, J. M. Babik, and M. H. Hecht, Protein design by binary patterning of polar and nonpolar amino acids, *Science 262*:1680 (1993).
3. A. Mak, L. B. Smille, and G. Stewart, A comparison of the amino acid sequences of rabbit skeletal muscle α- and β-tropomyosins, *J. Biol. Chem. 255*:3647 (1980).
4. S. Damodaran, Refolding of thermally unfolded soy proteins during the cooling regime of the gelation process: effect on gelation, *J. Agric. Food Chem. 36*:262 (1988).
5. S. Damodaran, Influence of protein conformation on its adaptability under chaotropic conditions, *Int. J. Biol. Macromol. 11*:2 (1989).
6. B. Lee and F. M. Richards, The interpretation of protein structure: Estimation of static accessibility, *J. Mol. Biol. 55*:379 (1971).

7. K. E. Van Holde, Effects of amino acid composition and microenvironment on protein structure, *Food Proteins* (J. R. Whitaker and S. R. Tannenbaum, eds.), AVI Publishing Co., Inc., Westport, CT, 1977, p. 1.
8. W. Bushuk and E. MacRitchie, Wheat proteins: aspects of structure that determine breadmaking quality, *Protein Quality and the Effects of Processing* (R. Dixon Phillips and J. W. Finley, eds.), Marcel Dekker, Inc., New York, 1989, p. 345.
9. N. C. Nielsen, Structure of soy proteins, *New Protein Foods: Seed Storage Proteins,* Vol. 5 (A. M. Altshul and H. L. Wilcke, eds.), Academic Press, New York, 1985, p. 27.
10. K. Kuwajima, A folding model of α-lactalbumin deduced from the three-state denaturation mechanism, *J. Mol. Biol. 114*:241 (1977).
11. D. A. Dolgisk, A. P. Kolomiets, I. A. Bolotina, and O. B. Ptitsyn, 'Molten-globule' state accumulates in carbonic anhydrase folding, *FEBS Lett. 165*:88 (1984).
12. F. M. Hughson, P. E. Wright, and R. L. Baldwin, Structural characterization of a partly folded apomyoglobin intermediate, *Science 249*:1544 (1990).
13. D. N. Brems and H. A. Havel, Folding of bovine growth hormone is consistent with the molten globule hypothesis, *Proteins: Struct. Funct. Genetics 5*:93 (1988).
14. K. Kuwajima, The molten globule state as a clue for understanding the folding and cooperativity of globular-protein structure, *Proteins: Struct. Funct. Genetics 6*:87 (1989).
15. K. Kuwajima, Y. Ogawa, and S. Sugai, Role of the interaction between ionizable groups in the folding of bovine α-lactalbumin, *J. Biochem. (Tokyo) 89*:759 (1981).
16. J. J. Ewbank and T. E. Creighton, The molten globule protein conformation probed by disulfide bonds, *Nature 350*:518 (1991).
17. J. L. Maubois, A. Pierre, J. Fauquant, and M. Piot, Industrial fractionation of main whey proteins, *Bull. Int. Dairy Fed. 212*:154 (1987).
18. P. M. Mulqueen and M. J. Kronman, Binding of naphthalene dyes to the N and A conformers of bovine α-lactalbumin, *Arch. Biochem. Biophys. 215*:28 (1982).
19. I. Hanssens, C. Houthuys, W. Herreman, and F. H. Van Cauwelaert, Interaction of α-lactalbumin with dimyristoyl phosphatidylcholine vesicels. I. A microcalorimetric and fluorescence study, *Biochim. Biophys. Acta 602*:539 (1980).
20. J. Kim and H. Kim, Fusion of phospholipid vesicels induced by α-lactalbumin at acidic pH, *Biochemistry 25*:7867 (1986).
21. J. M. Gonzalez and S. Damodaran, Sulfitolysis of disulfide bonds in proteins using a solid state copper carbonate catalyst, *J. Agric. Food Chem. 38*:149 (1990).
22. D. J. Merkler, G. K. Farrington, and F. C. Wedler, Protein thermostability, *Int. J. Peptide Protein Res. 18*:430 (1981).
23. R. Singleton, Jr., C. R. Middaugh, and R. D. MacElroy, Comparison of proteins from thermophilic and nonthermophilic sources in terms of structural parameters inferred from amino acid composition, *Int. J. Peptide Protein Res. 10*:39 (1977).
24. M. F. Perutz, Electrostatic effects in proteins, *Science 201*:1187 (1978).
25. G. Biesecker, J. I. Harris, J. C. Thierry, J. E. Walker, and A. J. Wonacott, Sequence and structure of D-glyceraldehyde 3-phosphate dehydrogenase from *Bacillus stearothermophilus, Nature 266*:328 (1977).
26. R. L. Baldwin, Temperature dependence of the hydrophobic interaction in protein folding, *Proc. Natl. Acad. Sci. USA 83*:8069 (1986).
27. K. A. Dill, D. O. V. Alonso, and K. Hutchinson, Thermal stabilities of globular proteins, *Biochemistry 28*:5439 (1989).
28. H. A. Scheraga, G. Nemethy, and I. Z. Steinberg, The contribution of hydrophobic bonds to the stability of protein conformation, *J. Biol. Chem. 237*:2506 (1962).
29. P. L. Privalov, Y. V. Griko, Y. S. Venyaminov, and V. P. Kutyshenko, Cold denaturation of myoglobin, *J. Mol. Biol. 190*:487 (1986).
30. H. B. Bull and K. Breese, Thermal stability of proteins, *Arch. Biochem. Biophys. 158*:681 (1973).

31. C. N. Pace, Protein conformations and their stability, *J. Am. Oil Chem. Soc. 60*:970 (1983).
32. P. Puyol, M. D. Perez, J. M. Peiro, and M. Calvo, Effect of binding of retinol and palmitic acid to bovine β-lactoglobulin on its resistance to thermal denaturation, *J. Dairy Sci. 77*:1494 (1994).
33. P. K. Ponnuswamy, R. Muthusamy, and P. Manavalan, Amino acid composition and thermal stability of proteins, *Int. J. Biol. Macromol. 4*:186 (1982).
34. J. E. Kinsella, Functional properties of food proteins: a review, *CRC Crit. Rev. Food Sci. Nutr. 7*:219 (1976).
35. I. D. Kuntz and W. Kauzman, Hydration of proteins and polypeptides, *Adv. Protein Chem. 28*:239 (1974).
36. J. A. Rupley, P.-H. Yang, and G. Tollin, Thermodynamic and related studies of water interacting with proteins, *Water in Polymers* (S. P. Rowland, ed.), ACS Symp. Ser. 127, American Chemical Society, Washington, DC, 1980, p. 91.
37. C. C. Bigelow, On the average hydrophobicity of proteins and the relation between it and protein structure, *J. Theor. Biol. 16*:187 (1967).
38. S. Damodaran, Interrelationship of molecular and functional properties of food proteins, in *Food Proteins* (J. E. Kinsella and W. G. Soucie, eds.), American Oil Chemists' Society, Champaign, IL, 1989, p. 21.
39. H. Zhu and S. Damodaran, Heat-induced conformational changes in whey protein isolate and its relation to foaming properties, *J. Agric. Food Chem. 42*:846 (1994).
40. A. Kato and S. Nakai, Hydrophobicity determined by a fluorescence probe method and its correlation with surface properties of proteins, *Biochim. Biophys. Acta 624*:13 (1980).
41. P. Pradispena and C. Rha, Pseudoplastic and rheopectic properties of globular protein (β-lactoglobulin) solution, *J. Text. Stud. 8*:311 (1977).
42. M. A. Tung, Rheology of protein dispersions, *J. Text. Stud. 9*:3 (1978).
43. A. M. Hermansson, Aspects of protein structure, rheology and texturization, in *Food Texture and Rheology* (P. Sherman, ed.), Academic Press, New York, 1979, p. 265.
44. W. Modler and D. Emmons, Properties of whey protein concentrate prepared by heating under acidic conditions, *J. Dairy Sci. 60*:177 (1977).
45. S. Damodaran and J. E. Kinsella, Flavor-protein interactions: binding of carbonyls to bovine serum albumin: thermodynamic and conformational effects, *J. Agric. Food Chem. 28*:567 (1980).
46. S. Damodaran and J. E. Kinsella, Interaction of carbonyls with soy protein: thermodynamic effect, *J. Agric. Food Chem. 29*:1249 (1981).
47. T. E. O'Neill and J. E. Kinsella, Binding of alkanone flavors to β-lactoglobulin: effects of conformational and chemical modifications, *J. Agric. Food Chem. 35*:770 (1987).
48. K. Shimada and S. Matsushita, Relationship between thermo-coagulation of proteins and amino acid compositions, *J. Agric. Food Chem. 28*:413 (1980).
49. J. D. Ferry, Mechanical properties of substances of high molecular weight. IV. Rigidities of gelatin gels; dependence on concentration, temperature and molecular weight, *J. Amer. Chem. Soc. 70*:2244 (1948).
50. C.-H. Wang and S. Damodaran, Thermal gelation of globular proteins: Weight average molecular weight dependence of gel strength, *J. Agric. Food Chem. 38*:1154 (1990).
51. I. D. Kuntz and W. Kauzmann, Hydration of proteins and polypeptides, *Adv. Protein Chem. 28*:239 (1974).
52. J. E. Kinsella, S. Damodaran, and J. B. German, Physicochemical and functional properties of oilseed proteins with emphasis on soy proteins, *New Protein Foods: Seed Storage Proteins* (A. M. Altschul and H. L. Wilke, eds.), Academic Press, London,1985, p. 107.

2

Thermal Denaturation and Coagulation of Proteins

J. I. Boye, C.-Y. Ma, and V. R. Harwalkar
Agriculture Canada
Ottawa, Ontario, Canada

I. INTRODUCTION

In recent years, the increased dependence of food-processing industries on the manufacture of fabricated foods has drawn much attention to the functional properties of individual ingredients used in food formulation and in product development. The term "functionality" as applied to food ingredients has been defined as any property aside from nutritional attributes that influences an ingredient's usefulness in foods [1]. Proteins are of particular significance, since they play a major role in determining the sensory, textural, as well as nutritional characteristics of various food products and have the ability to act as a matrix for holding water, lipids, sugars, flavors, and other ingredients during aggregate and gel formation [2].

Functional properties of a protein are related to the physical, chemical, and conformational properties, which include size, shape, amino acid composition and sequence, charge, and charge distribution. The properties of food proteins that affect functionality include hydrophilicity/hydrophobicity ratio, secondary structure content and their distribution (e.g., α-helix, β-sheet, and aperiodic structures), tertiary and quaternary arrangements of the polypeptide segments, inter- and intrasubunit cross-links (e.g., disulfide bonds), and the rigidity/flexibility of the protein in response to external conditions [3]. Most functional properties affect the textural qualities of a food and play an important role in determining the physical behavior of a food during preparation, processing, and storage.

Protein stability is of particular importance in determining their functionality in food systems. This is because a particular functional property is often governed by a specific conformational state of a protein and any alteration of that state affects its functionality [4]. The denaturation of food protein is a prerequisite in the exhibition of any functional property. The extent to which proteins unfold when denatured and the conformation they assume upon denaturation affect the functional and nutritional quality of foods during processing. It is important, therefore, that the interrelationship between protein denaturation and protein functionality be firmly established.

The most important food-processing operation that contributes to protein denaturation involves heat treatment. The mechanism of protein denaturation and protein interactions in foods can be described in terms of intermolecular forces, kinetics, and environmentally induced changes in protein structure during processing. Several reviews on protein denaturation and the structure-function properties of food proteins have been published [4–6]. This chapter is devoted to thermal denaturation and coagulation of food proteins and reviews the mechanisms of thermal denaturation and the factors that affect thermal denaturation as they relate to protein coagulation.

II. MECHANISM OF THERMAL DENATURATION

A. Definitions

Denaturation of food proteins has been defined as a process (or sequence of processes) in which the spatial arrangement of polypeptide chains within the molecule is changed from that typical of the native protein to a more disordered arrangement [7]. Cheftel et al. [8] defined protein denaturation more specifically as any modification in conformation (secondary, tertiary, or quartenary) not accompanied by the rupture of peptide bonds involved in primary structure.

The difficulty in defining denaturation is in deciding to what extent certain modifications of the protein molecule can be included in the concept of denaturation since a minute fraction of the protein conformation can undergo change without affecting the functional property of the protein. A more restrictive definition of denaturation specifies the loss of one or more of the most characteristic properties of the protein, e.g., loss of solubility in solvents in which the protein was previously soluble, loss of enzymatic activity, or changes in molecular weight of the protein. Depending on the protein and conditions, denaturation may be confined to a region of the protein or may involve the complete molecule in an "all-or-none" reaction, which reflects the cooperative nature of the transition from the native conformation to the least structured state [5].

One of the most common methods of denaturing proteins is to heat them in solution. Heat treatment of globular proteins in water or solvent increases their thermal motion, leading to the rupture of various intermolecular and intramolecular bonds stabilizing the native protein structure. This results in reorganization of both the secondary and tertiary configuration where previously inward-oriented "hydrophobic" amino acid residues become exposed to solvent. This process normally results in the formation of new short-lived intermediary conformations.

Thermal coagulation is the random interaction of protein molecules by heat treatment, leading to formation of aggregates that could be either soluble or insoluble (precipitates) [9]. Thermal gelation (to be covered in a separate chapter), on the other hand, is the formation of a three-dimensional network exhibiting certain degree of order [9]. For monomeric proteins, coagulation is normally preceded by denaturation. For oligomeric proteins with complex quaternary structures, heat may cause association/dissociation of the oligomers, and disruption of the quaternary structure itself may result in aggregation [10].

B. Structure of Protein Molecules

Proteins are polypeptide chains existing at four different structural levels: the primary, secondary, tertiary, and quartenary levels. The primary structure refers to the sequential assembly of amino acids linked together (in long chains) by peptide bonds joining α-NH_2 and α-COOH groups to form a linear series of tetrahedral carbon atoms. Proteins, however, rarely exist in linear chains. This is because the relative degree of interaction between various amino acid side chains and water generally precludes a linear arrangement of the polypeptide backbone [11]. Globular proteins, therefore, assume a secondary structure, which is a regular arrangement or orientation of the main-chain atoms and may be defined as the spatial structure the polypeptide chain assumes exclusively along the axis. Periodic secondary structures commonly encountered in proteins are α-helices (α, α_{11}, 3_{10}), β-sheets, and reverse turns, which include β-hairpins, β-bends, or β-turns; there is also one ill-defined structure without any plane or axis of symmetry—the random coil [8,12].

Under normal conditions of pH and temperature, each polypeptide assumes one specific conformation, called *native*. This corresponds to a thermodynamically stable and organized system with a minimal free energy, ΔG. The native conformation is closely related to the polarity, the hydrophobicity, and the steric hindrance of the side chains [8].

The tertiary structure relates to the three-dimensional organization of a polypeptide chain containing regions of well-defined (α-helix, β-bends, and β-pleated sheets) or ill-defined (random coil) structures, while the quartenary structures are produced by the association of separate folded polypeptide (subunits) into an aggregate multimeric structure under specified conditions of pH and temperature and without regard to the internal geometry of the subunits. The subunits may be identical polypeptide chains or chemically distinct species [11].

The properties of various noncovalent and covalent interactions that contribute to structural stability of proteins are described in Chapter 1.

C. Conformational Changes During Denaturation

Although each protein possesses a unique, well-defined structure in the native state, after denaturation, the same protein may have several nonspecific structures according to the type and extent of the denaturing treatment. These denatured structures are characterized by a degree of random configuration higher than that of the native molecule [13]. Thus, denaturation appears as an unfolding process, and, conversely, reversal occurs when refolding takes place. Various levels of denaturation can be distinguished according to whether the secondary, tertiary, or quartenary structure of the protein is involved in the process. Globular proteins can exist in a number of low-entropy but energetically equivalent states, seen as "quasi-native" states, and do not involve major changes in protein structure [4]. Depending on the protein and conditions, therefore, denaturation may be confined to a region of the protein or may involve the complete molecule in an "all-or-none" reaction, which reflects the cooperative nature of the transition from the native conformation to the least structured state [5]. This cooperativity arises because the transitions normally occur within a narrow range of temperature or concentration of denaturant [5]. The two-state coperative transition from native (N) to denatured (D) is temperature dependent with the degree of order of the protein molecule decreasing as the temperature increases [4].

Unfolded protein molecules associate through intermolecular interaction to form aggregates of irreversibly denatured molecules, which may lead to precipitation, coagulation, or gelation. The interactions involved in determining native protein structure are therefore similarly involved in protein coagulation and gelation. The formation of a thermally induced gel matrix or coagulum from proteins involves the following three sequential events [14]:

$$\text{denaturation} \longrightarrow \text{aggregation} \longrightarrow \text{cross-linking}$$

Protein *aggregation* involves the formation of higher molecular weight complexes from the denatured protein [15], which then *cross-link* by specific bonding at specific sites on the protein strands or by nonspecific bonding occurring along the protein strands. Denaturation is therefore a prerequisite for the formation of protein aggregates or gels. The initial steps in heat coagulation of an aqueous protein solution are:

1. Reversible dissociation of the quaternary structure into subunits or monomers (reversible dissociation of native polymers may also take place as the first step of denaturation).
2. Irreversible denaturation of secondary and tertiary structures (unfolding generally remains partial).

The final coagulum corresponds to aggregates of partly denatured proteins and involves the following mechanism:

$$nP_N \longrightarrow nP_D \longrightarrow [P_D]n$$

where n is the number of protein molecules, P_N is the native protein, and P_D is denatured protein.

In the absence of denaturation, as mentioned previously, the native state of proteins is maintained by a delicate balance of chain-interaction energies involving electrostatic interactions, hydrogen bonding, disulfide bonds, and hydrophobic interactions. Dissociation and/or unfolding of protein molecules generally increases the exposure of reactive groups especially the hydrophobic groups of globular proteins. Protein-protein hydrophobic interactions are, therefore, favored and are usually the main cause of subsequent aggregation [8] provided that the protein concentration, thermodynamic conditions, and other conditions are optimal for the formation of the tertiary matrix [15].

Cross-linking of protein aggregates, following denaturation, usually involves one or more of the following four mechanisms [16]:

1. Oxidative chemical reactions of proteins resulting from the covalent interaction of their functional groups
2. Cross-linking of proteins by polyfunctional agents (including metal ions) in solution
3. Physicochemical conditions resulting in limited solubility
4. Chemical modification of the proteins leading to limited protein solubility

Both disulfide and hydrogen bonding, as well as ionic interactions, are involved in the cross-linking of aggregates from denatured protein [17,18].

It has been suggested that the ability of certain proteins to form intermolecular disulfide bonds during heat treatment may be a prerequisite for their coagulation and gelation [19]. Heat treatment results in cleavage of existing disulfide bond structure or "activation" of buried sulfhydryl groups through unfolding of the protein. These newly formed or activated sulfhydryl groups can form new intermolecular disulfide bonds, which are essential for the formation of aggregate structures in some protein systems. The role of disulfide-sulfhydryl interchange reactions in the cross-linking of protein gels and aggregates is depicted in Figure 1.

Hydrogen bonding plays a major role in increased viscosity, which precedes the onset of coagulation. This type of cross-link allows for the open orientation necessary for water immobilization [15] and may be the most important type of cross-linking in reversible gels [21]. Ionic bonding has been suggested to be of primary importance at the protein/solvent interface and to solvent immobilization [8]. Protein solvation increases with increased salt addition due to decreased protein-protein attraction and increased protein-solvent attraction. At higher ionic strength, there is increased protein-protein attraction as the ions compete with the protein for solvent, which results in the formation of larger aggregates [9,15]. Hydrophobic interactions are also involved in protein aggregation and gelation. Nonspecific hydrophobic interactions have been implicated in the dissociative-associative reactions that initiate aggregation and gelation and also in improved strength and stability of protein gels on cooling [15,21].

D. Thermodynamic Considerations

Proteins assume a configuration that expends the least amount of free energy. This causes the peptide chain to orient its hydrophilic amino acid towards the outside of the molecule

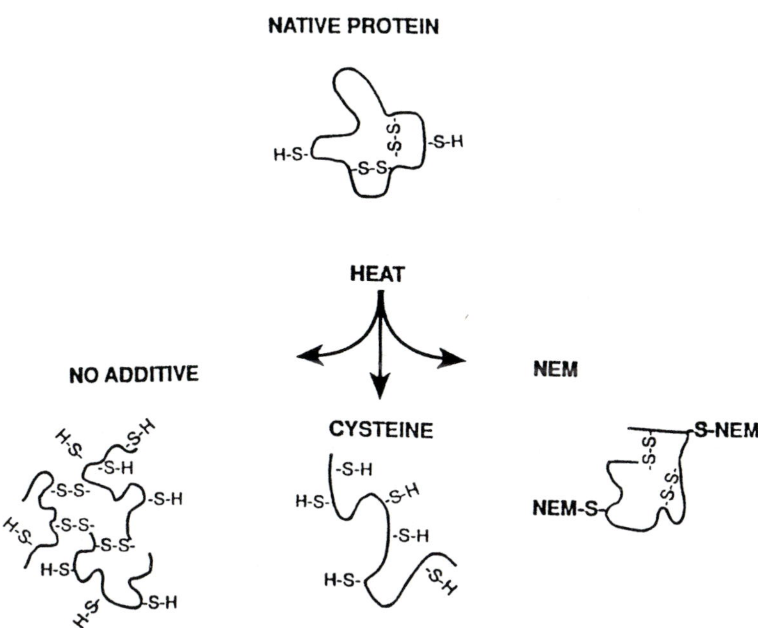

Figure 1 Role of sulfhydryl-disulfide interchange reactions during gelation in the presence and absence of *N*-ethylmaleimide (NEM), a sulfhydryl blocking agent (inhibits unfolding of the protein molecule), and cysteine, a reducing agent (prevents disulfide-sulfhydryl interchange reactions). (Adapted from Ref. 20.)

and to bury the hydrophobic amino acids in the interior of the molecule with the exclusion of water from this core. The conformation adopted by a given polypeptide sequence is that of minimum free energy for the system; that is, of all the minimal energy states, the native protein inhabits the lowest one, or "global minimum" (Fig. 2).

A common observation is that denatured proteins, upon restoration of the "native" environment, can regain precisely and spontaneously its native conformation, as depicted in Figure 2C [22]. However, if the random coiled peptide chain undergoes a random search for the global minimum, the number of conformations to be explored would require times of the order of 10^{26} years [23]; an alternative suggestion is that the folding and refolding of a protein chain occurs along certain kinetically accessible pathways that cover only a very small fraction of the potentially possible conformations.

The change of state of a substance is generally accompanied by a change in energy level and is manifested by the absorption or liberation of heat. The exclusion of water from contact with hydrophobic residues in the interior of the protein structure causes a large increase in entropy, while the close packing of protein interiors leads to low enthalpy. The net stability of the native protein is a result of the lowered enthalpy and the large increase in entropy.

The thermodynamic parameters for changing a protein from a native state (N) to a denatured state (D) can be developed as:

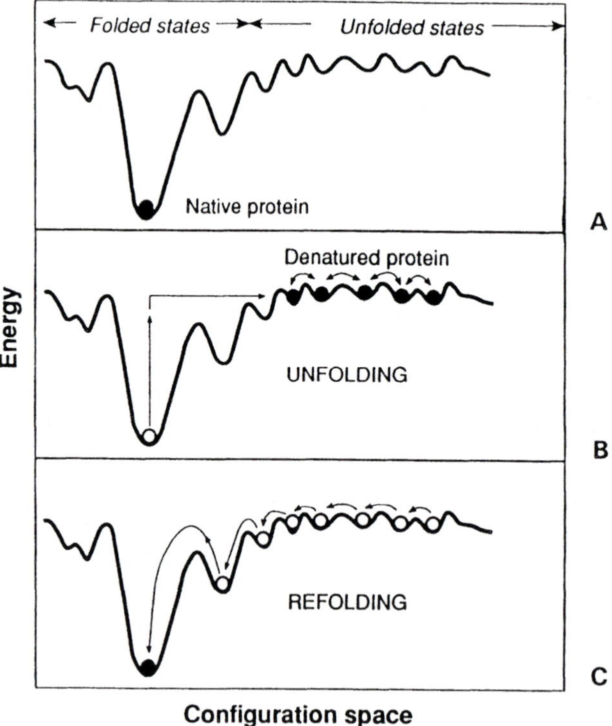

FIGURE 2 Diagrams of the energy of a protein molecule as a function of chain conformation. (Adapted from Ref. 22.)

$$N \underset{k}{\overset{heat}{\longleftrightarrow}} D$$

The equilibrium constant can be written as follows:

$$K = (D)/(N) \tag{1}$$

The thermodynamic parameters $\Delta G°$, $\Delta H°$, $\Delta S°$, and $\Delta Cp°$ can be obtained from the usual equations:

$$\Delta G° = -RT\ln K \tag{2}$$

$$\Delta H° = R\frac{\ln K}{1/T} \tag{3}$$

$$\Delta S° = \frac{\Delta H° - \Delta G°}{T} \tag{4}$$

$$\Delta Cp° = T(\Delta S°/\Delta T)_p \tag{5}$$

where $\Delta G°$ is the change in standard free energy (difference between the free energy of the native system containing all reactants at equimolar concentration and the free energy of the same system at equilibrium after denaturation), R is the gas constant, T is absolute temperature, $\Delta H°$ is the change in enthalpy at constant pressure, $\Delta S°$ is the change in entropy, and $\Delta Cp°$ is the change in heat capacity at constant pressure.

The value of ΔG is the fundamental measure of the stability of the protein. A central feature of the energetics of protein denaturation is that ΔH and ΔS are temperature dependent, as shown in the following equations:

$$\Delta H_T = \Delta H_o + \Delta C_p (T - T_o) \tag{6}$$

$$\Delta S_T = \Delta S_o + \Delta C_p \ln (T/T_o) \tag{7}$$

where ΔH_T and ΔS_T are at temperature T, while ΔH and ΔS are at a reference temperature T_o. ΔC_p is the difference in heat capacity between the N and D states [24].

ΔCp $(\Delta H/\delta T)$ gives the heat absorbed by the water of hydrophobic hydration and is about 1–2 kcal $mol^{-1}k^{-1}$ [25]. In the temperature range of 0–80°C, ΔCp is constant or nearly constant. Because of such large values of ΔCp, ΔH_T, and ΔS_T are strongly dependent on temperature. For example, a temperature change of 1°C causes changes in ΔH_T and in ΔS_T of about 1–2 kcal mol^{-1}. ΔH and ΔS have large compensating values at temperatures at which proteins denature, the mechanism of which, however, is not well understood [26].

From the equations above, it can be observed that δH (enthalpy) and δS (entropy) are dependent on temperature. Increase in entropy corresponds to the disorder accompanying unfolding, and the increase of $\delta Cp°$ is related to the transfer during denaturation of aliphatic or aromatic apolar groups to an aqueous environment and the energy associated with the disruption of water structure [8]. Increase in enthalpy upon denaturation is indicative of a much lower energy level for the native conformation of the protein [27].

In the denatured state and at low temperatures, water is oriented into structures of solvation around nonpolar groups, whereas at high temperatures a more random and disorganized solvation interaction occurs. The net result of increasing the temperature, therefore, leads to a more positive contribution to the entropy of unfolding caused by solvation effects [27].

E. Kinetic Considerations

The two-state nature of protein unfolding is a simplistic representation of the transitions occurring during protein denaturation. At any instant in time, each molecule in a population of unfolding protein has a unique conformation which is not static but is constantly being altered by rapid rotation about single bonds. The kinetic pathways of unfolding and refolding may be determined by which intermediate conformational state (I) the molecule passes through [12]. Thus, the denaturation reaction may be more accurately written as:

$$N \rightleftharpoons I_1 \rightleftharpoons I_2 \rightleftharpoons \ldots\ldots \rightleftharpoons I_{n-1} \rightleftharpoons I_n \rightleftharpoons D$$

where I_l represent molecules with structure intermediates between those of the native and fully denatured states.

The denaturation of β-lactoglobulin by urea, for example, conforms with:

$$N_s \rightleftharpoons N_u \rightleftharpoons D_r \longrightarrow D_i$$

where N_s and N_u are the stable and unstable forms of the native protein, respectively, and D_r and D_i are the reversible and irreversible states of the denatured protein, respectively [13]. The third step is more rapid than the fourth. In the thermal denaturation of bovine β-lactoglobulin solutions at low concentration, four periods have also been distinguished by light scattering measurements. In the initial period the molecules break into two fragments of about equal size, which then combine to form aggregates. In the second period the apparent activation energy is less than 30 kcal/mol. In the third and fourth periods, the reaction rate passes through a maximum at about 87°C [13]. For a simple two-state transition (N ⇔ D), the equilibrium constant (K) can be related to the average degree of unfolding (α) by the following equation:

$$K = \frac{[N]}{[D]} = \frac{1 - \alpha}{\alpha} \quad (8)$$

The kinetic equation for this transition may be written as

$$R = k\,(1 - \alpha)^n \quad (9)$$

where R is the rate of reaction in s^{-1}, k is the reaction rate constant in s^{-1}, α is the degree of conversion, and n is the order of reaction. When

$$n = 1; \quad \alpha = 1 - e^{-kt} \quad \text{and} \quad R = ke^{-kt}$$

$$n \neq 1; \quad \alpha = 1 - [kt(n - 1) + 1]^{1/(1-n)} \quad \text{and} \quad R = k[kt(n - 1) + 1]^{n/(1-n)}$$

where t is the time of reaction. In general, for the single-step reaction:

$$A \xrightarrow{k} B$$

kinetic parameters may be written as follows:

$$\frac{-dA}{dt} = \frac{dB}{dt} = kA \quad (10)$$

where A and B represent the concentrations of reactant and product, respectively, and t is time. The differential equation may be readily solved as:

$$\frac{\ln A}{A_o} = -kt \tag{11}$$

where A_o is the initial concentration. A logarithmic plot of the time-dependent change in A (or B) relative to A_o should be linear, with the slope giving the value of the rate constant (Fig. 3A). If the observed curve is not linear but has multiple phases (Fig. 3B), a corresponding greater number of reaction steps (and species) must be involved [12].

F. The Molten Globule State

The molten globule state has been described as a compact intermediate protein conformation that has a secondary structure content like that of the native state but with a poorly defined tertiary structure [28]. "Globule" refers to the native compactness, and "molten" refers to the increased enthalpy and the entropy on transition from the native structure to the new state [29]. The molten globule state may therefore be defined as a stable partially folded conformation that can be distinguished from either the native or the fully denatured forms [30].

α-Lactalbumin is a protein that exhibits the molten globule state. In solution, it undergoes intermolecular interactions leading to varying degrees of polymerization on both sides of its zone of insolubility (~pH 4.8) [31]. At acidic pH values, the protein undergoes a rapid reversible association and slow aggregation [32]. Between pH 6 and 8.5 there is very little association, and above pH 9.5 there is expansion without aggregation [33]. Denaturation of the protein involves dissociation of Ca^{2+}; this occurs regardless of whether the protein is at pH values below 4.0 or above 9.0, heated above 50°C, exposed to low concentrations of guanidine hydrochloride, or subjected to Ca^{2+} removal from the native form. These denaturing conditions result in a transition of the native state to a stable transient state designated as the "A" or the "molten globule" state that is different from the unfolded denatured (U) state [34].

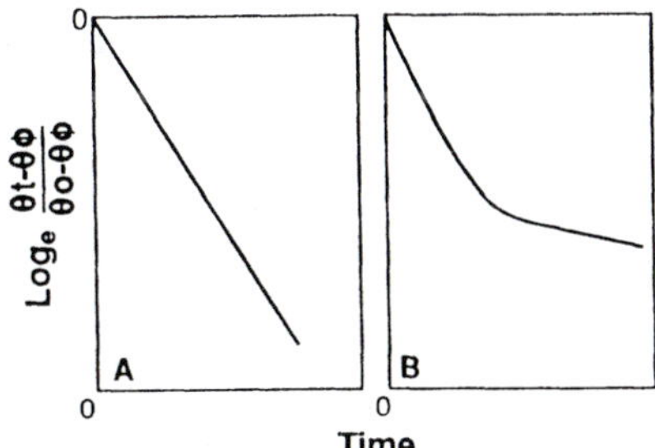

Figure 3 Example of one-step (A) aand two-step biphasic (B) reaction kinetics. A parameter of the extent of the reaction, θt, is plotted versus time; θo is the initial value of the parameter, θϕ, the final value. The slope of each curve gives the value of the apparent rate constant; if the reverse of a reaction is significant, the apparent rate constant is the sum of those of the forward and reverse steps. (Adapted from Ref. 12.)

Ovalbumin is another protein that has been shown to exist in the molten globule state. Koseki et al. [35] found that at low ionic strength and neutral pH, the protein forms soluble oligomers at high temperatures but does not gel. The far-UV CD spectrum indicated that only small changes in the secondary structure occurred on heating, compared with those induced by a high concentration of guanidine hydrochloride. The near-UV CD spectrum and difference UV absorption spectrum, however, indicated that the side chains of aromatic amino acids in the heated protein fluctuated as in the fully denatured state, suggesting that ovalbumin takes on a molten globule–like state at high temperatures.

The molten globule state is a strong candidate for explaining the molecular conformation of functional food proteins [30]. Kato et al. [36] found a significant improvement in functional properties of dried egg white by heating the proteins in a dry state at 80°C for several days. The gel strength, foam stability, and emulsion stability of the egg white proteins were increased almost fourfold by dry heating for 10 days. These increases in functionality have been attributed to the molten globule state [30].

G. Reversibility of Thermal Denaturation

Although most thermal denaturation is irreversible in nature, certain proteins have been observed to undergo reversible denaturation when the thermal influence is removed. When a protein has undergone a mild denaturation, a more or less complete reversal of some properties may occur on removing the denaturing agent, but generally the renatured protein is not completely identical with the native protein. The extent of reversibility is dependent on the criterion chosen to define denaturation and will depend on the protein and the severity of treatment.

Human serum albumin denatured by heating for 15 minutes at 70°C in 0.5 M phosphate buffer at pH 6 can be renatured by exposure to a pressure of 2000 kg/cm^2, even if left for a long time in the denatured condition. Solubility, viscosity, optical rotation, and tryptic digestion are identical in native and renatured protein [13]. Similarly, the heat-induced collagen-gelatin transition, which involves the conversion of highly asymmetrical, rigid, three-stranded collagen molecule to single-chain randomly coiled molecules of gelatin, may be largely reversed by cooling the solution to a temperature well below the melting temperature of collagen [37]. The reversibility of lysozyme denaturation has been demonstrated by differential scanning calorimetry and the complete recovery of enzymatic activity on cooling [38].

At pH values above 3.3, the thermal denaturation of α-lactalbumin is primarily a reversible process with 80–90% renaturation [39]. Below pH 3.3, a reduction in the denaturation temperature and enthalpy values are observed, which are accompanied by a reduction in the ability of the protein to return to the native conformation. The reversibility of α-lactalbumin is Ca^{2+} dependent. When a chelator (e.g., EGTA) is added to bind endogenous Ca^{2+}, only 2% or less of active α-lactalbumin is regenerated as compared with as much as 90–100% in the presence of 100 mM Ca^{2+} [40]. Several reviews have been written on the thermodynamics and physicochemical changes associated with the denaturation and refolding of α-lactalbumin [40]. α-Lactalbumin is a Ca^{2+}-binding protein with four disulfide bonds, which place considerable restraints on the polypeptide chain even in the presence of high concentrations of denaturants. Rao and Brew [41] proposed that in the absence of Ca^{2+}, the region of sequence in α-lactalbumin that constitutes the Ca^{2+}-binding site becomes a feature that blocks the formation of the correct three-dimensional structure and prevents renaturation.

III. FACTORS AFFECTING THERMAL DENATURATION

The susceptibility of proteins to denaturation by heat depends on other factors such as heat-treatment conditions, the presence of metallic ions, salt, urea, and detergents, among others.

A. Heat

Heat is the most common physical agent capable of denaturing proteins. The rate of denaturation depends on the temperature, and for protein denaturation the rate increases about 600 times with a 10°C increase in temperature in the range typical of denaturation. This results from the low energy involved in each of the interactions stabilizing the secondary, tertiary, and quartenary structure [8]. Mulvihill and Donovan [4] reported that a temperature increase of 7.5°C gives a 10-fold reduction in the time required for a fixed level of denaturation. In the temperature range 62–80°C, the relationship between temperature and time for a constant level of denaturation is semi-logarithmic, a plot of logarithm of heating time versus temperature yield a set of essentially parallel straight lines for denaturation levels in the range 5–40%. Harwalkar [42] reported that denaturation of β-lactoglobulin was detectable only at 75°C or above and appeared to proceed in two distinct stages: the first stage occurred in the first 5 minutes of heating and was faster than the subsequent stage. The author concluded that thermal denaturation leads to partial unfolding that increases progressively with severity of heat treatment. In reality, most "denatured" proteins are not completely unfolded and some ordered structure may exist after heating [5]. Byler and Purcell [43] observed that neither β-lactoglobulin nor bovine serum albumin was totally unfolded into a disordered state when heated above their transition temperature as determined by differential scanning calorimetry. When G-actin was heated more than 30°C above its denaturation temperature of 57.2°C at pH 8, the protein retained about 60% of its native helical structure [44].

The functional characteristics of food proteins are influenced by heating temperature as well as heating rate and heating time. Alteration of these conditions affects both the macroscopic and microscopic structural attributes of the protein. Depending upon the molecular properties of the protein in the unfolded state, it undergoes two types of interactions. Proteins that contain high levels of apolar amino acid residues undergo random aggregation via hydrophobic interaction, resulting in an opaque coagulum or precipitate. On the other hand, proteins that contain a below-critical level of apolar amino acid residues form soluble aggregates, which set into a thermally reversible, transparent gel network upon cooling [3]. When the heating conditions are extreme, protein molecules may not have time to align themselves in an ordered fashion. In these circumstances, poorly hydrated aggregates or precipitates that lack continuous matrix are formed [18]. Excessive heating of protein dispersions at temperatures far higher than the denaturation temperature can lead to a metasol state, which does not set into a gel upon cooling [17].

B. pH

In aqueous solutions, the denaturing effects of pH and temperature are so closely connected that denaturation processes can only rarely be considered as purely thermal. The pH of a protein solution affects both the denaturation temperature (Fig. 4) as well as functional characteristics such as gelation, foaming, and emulsification. Most proteins

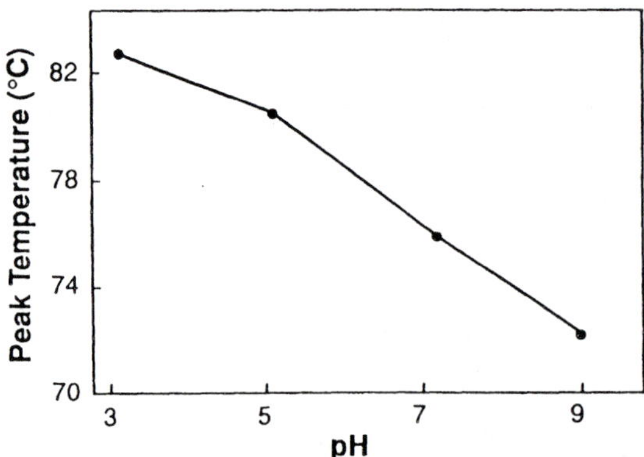

FIGURE 4 Plot of the denaturation temperature (peak temperature of DSC endotherm, T_d) of whey protein concentrate (30% w/v, pH adjusted by the use of phosphate buffer, ionic strength, 0.2) as a function of pH, showing a decrease in T_d with increasing pH. Heating rate was 5°C/min. (Adapted from Ref. 45.)

are stable over certain pH ranges, and at extremely high or low pH they usually denature [18]. Exposure to moderately high pH followed by readjustment to neutral pH has been shown to "activate" the protein molecules, thereby improving their functional properties [46]. This could be related to unfolding of the protein and/or activation of buried sulfhydryl groups.

At the isoelectric pH, the lack of repulsive forces inhibits unfolding of the protein molecules. At pH values well removed from the isoelectric points, repulsive electrostatic forces are induced by the large net charge of the proteins resulting in unfolding (denaturation). Prolonged exposure to extremely high pH, however, suppresses aggregate formation [47]. At strongly basic pH values, carboxylate-phenolic group and carboxylate-protonated amino (NH_3^+) group interactions are inhibited [17]; this interferes with the cross-linking of aggregates and affects gel formation.

C. Metallic Ions

Both pH and ionic strength of a solution determine the overall charge of the protein molecule and their susceptibility to thermal denaturation. The ability of electrolytes to influence the conformation and therefore stability of a protein depends on the concentration and/or ionic strength (μ) of the salt. At low salt concentration ($\mu < 0.5$), the stabilizing effect has been attributed to an electrostatic response [48]. At $\mu > 0.5$, the ability of salts to stabilize protein structure has been related to the preferential hydration of the protein molecule as a result of a salt induced alteration of the water structure in the vicinity of the protein [49].

At high concentrations of neutral salts (>1 M), proteins display decreased solubility, which may result in precipitation. This arises from the competition between the protein and the ions for the water molecules—a phenomenon called "salting-out." Protein-protein interactions are favored over protein-solvent interactions at high salt concentrations due to lack of enough water molecules, leading to aggregation and precipitation of the protein molecules [18]. Increased solubility of protein molecules increases their susceptibility to thermal denaturation; the denaturation temperature of most proteins is therefore generally higher in the presence of higher concentrations of stabilizing salts (Table 1).

The salting-out (stabilizing) action of different salts increases with their hydration energy and their steric hindrance. According to the Hofmeister series, ions may be ordered as follows: SO_4^{2-} < F^- < CH_3COO^- < Cl^- < Br^- < NO_3^- < I^- < ClO_4^- < SCN^-; NH_4^+ < K^+ < Na^+ < L^+ < I^+ < Mg^+ < Ca^{2+}. Ions to the left (SO_4^{2-}, F^-, CH_3COO^-, NH_4^+, K^+) promote salting-out, aggregation, and stabilization of the native conformation. Ions to the right (I^-, ClO_4^-, SCN^-) promote unfolding, dissociation, and salting-in. Cations at the high order of the series (e.g., Ca^{2+}) could reduce the free energy required to transfer the nonpolar protein groups into water and could weaken intramolecular hydrophobic interaction and enhance the unfolding tendency of the protein [50]. The ranking of anions and cations in the Hofmeister series are, however, based on molar concentrations rather than μ values, which influences the location of the divalent ions, in particular sulfate and calcium [51].

The effect of salts on the thermal properties of vicilin has been studied [51,52]. At salt concentrations of 0.5 M, the thermal stability of vicilin increased when mixed with salts of increasing molar surface tension (greater salting-out effect). The destabilizing ions with low molar surface tension (SCN^-, I^-), on the other hand, tended to denature the vicilin as evidenced by a decrease in the temperature of denaturation (Table 1). A similar effect was observed in oat globulin [53]. In the presence of Cl^- and Br^-, the temperature of denaturation increased with increasing ion concentration, whereas a gradual decrease in the denaturation temperature was observed with I^- and SCN^-. Increasing NaCl concentration has been shown to increase the stability of several proteins (e.g., vicilin, legumin, and β-lactoglobulin) [51,54]. K^+, Na^+, and Li^+ have also been observed to stabilize both vicilin and legumin at $\mu \geq 0.5$, while Ca^{2+} destabilized the protein isolate [51].

There are deviations in the effect of salts on protein denaturation. For example, low concentrations (<0.3 M) of K^+ and Ca^{2+} salts of Cl^-, F^-, and SCN^- decreased the

Table 1 Effect of Sodium Salts on Thermal Stability (T_d, °C) of Vicilin and Legumin in a Micelle Isolate of Faba Bean Protein

	Vicilin			Legumin		
Salt	0.15 M	0.5 M	2.0 M	0.15 M	0.5 M	2.0 M
NaSCN	88	84	69	95	94	91
NaI	87	89	81	95	97	99
$NaC_2H_3O_2$	90	95	106	96	104	119
NaCl	83	92	102	93	103	116
Na_2SO_4	89	96	104	96	103	108
$Na_3C_6H_5O_7$	90	92	98	89	93	108
$CaCl_2$	—	—	—	95	94	93
LiCl	87	95	96	97	101	104
NaCl	83	92	102	93	103	116
KCl	87	94	103	99	107	122

Source: Adapted from Ref. 51.

denaturation temperature of collagen; this has been attributed to swelling of the protein and a weakening of protein-protein interactions [55]. With KF and KCl, this trend is reversed at higher salt levels (0.3–2.0 M), giving a stabilizing effect similar to that observed for plant proteins.

Aggregation of globular proteins is also affected by ionic strength and the presence of certain salts. Hermansson [9] showed that the pH-dependent aggregation of 0.9% whey protein concentrate (WPC) was dependent on the ionic strength in the range of 0.05–0.2. Addition of 0.2 M NaCl was also shown to enhance protein-protein interactions in soy proteins [9]. Hickson et al. [56] reported that the viscosity index of 8.0% bovine plasma protein suspensions increased in the presence of 0.2 M Na^+ and 0.2 M Ca^{2+}; the increase was significantly higher (1.5–2 times) for Ca^{2+} than for Na^+. This effect of Ca^{2+} was attributed to the formation of bridges between molecules, generating a stronger gel network. Varunsatian et al. [57] showed that on the alkaline side of the isoelectric zone, aggregation of WPC was increased by the addition of $CaCl_2$, $MgCl_2$, or NaCl, of which $CaCl_2$ showed the greatest effect. The denaturation temperature of WPC as determined by differential scanning calorimetry significantly decreased in the presence of $CaCl_2$ or $MgCl_2$, but increased slightly in the presence of NaCl.

Depending on the individual properties of the ion and their concentrations, either the effect of ions can be just an ionic strength effect, or, when the concentration is higher, they can act as structure stabilizers (kosmotropic salts) or as destructing ions (chaotropic) [58]. The hydration of these ions results in a change in the water structure around the protein, and consequently hydrophobic interactions. The effect, therefore, depends on the extent of their binding to protein and the concomitant change in the water structure [58]. Certain proteins are metal binding, which implies that when specific ions are present, they bind to the ionic sites present on these proteins and alter their structure. α-Lactalbumin, for example, is considered as the most heat-stable whey protein. The binding of Ca^{2+} to apo-α-lactalbumin (Ca^{2+}-free) or decalcification of holo-α-lactalbumin (Ca^{2+}-bound) has pronounced effects on its stability to heat and chemical denaturants [40]. In the presence of Ca^{2+}, α-lactalbumin is stabilized against thermal denaturation, and any denaturation observed on heating is reversed on cooling. Removal of the bound Ca^{2+} reduces the thermal stability of the native tertiary structure and further decreases its ability to renature on cooling [59].

D. Sugars and Polyols

Sugars and polyols have also been shown to affect the thermal stability of food proteins. DSC studies of β-lactoglobulin, bovine serum albumin, ovalbumin, and conalbumin demonstrate an increase in denaturation temperature in the presence of both glucose and sucrose, with the greatest increase occurring with glucose [60–63]. In another study, Ball et al. [64] reported that sugars protect proteins against solubility losses during drying and may inhibit coagulation.

The influence of solvent components (such as sugars) on structure and conformational stability can be either by direct interaction of the protein or indirect interaction through modification of the solvent environment, or a combination of both [65]. As with stabilizing salts, sugars influence protein conformation primarily through their indirect effect on hydrophobic interactions [66]. Low levels of sugars increase the number of hydrophobic associations as evidenced by the decrease in surface hydrophobicity [67]; as sugar levels are further increased, these interactions are strengthened [61].

The extent of stabilization of sugars and polyols depends upon the type and concentration of the polyol as well as the nature of the protein [61,66]. The effect of some polyols on the thermal transition characteristics of β-lactoglobulin is given in Table 2. In the presence of sucrose, glucose, and glycerol, the denaturation temperature of β-lactoglobulin increased with increasing sugar concentration; in the presence of ethylene glycol, however, the temperature of denaturation decreased. Gerlsma and Stuur [69] suggested that ethylene glycol, although a polyol, lowers the dielectric constant of the medium and interacts with nonpolar side chains of proteins, weakening hydrophobic interaction and thereby lowering thermal stability.

Boye et al. [62] studied the denaturation and aggregation of β-lactoglobulin in the presence of increasing concentrations of glucose and sucrose using Fourier transform infrared (FTIR) spectroscopy. Their results showed that the sugars stabilized the partially denatured protein, inhibiting aggregation, with sucrose having a greater effect in inhibiting aggregation than glucose. Lee and Timasheff [70] proposed that as the level of sugar increases, there is an increase in solvent cohesive force, which increases the energy required for cavity formation for the associated structures in the solvent; the energy required for aggregate formation therefore becomes prohibitive.

E. Protein Modifiers

Addition of chemical agents such as urea, guanidine hydrochloride, and anionic detergents (e.g., sodium dodecyl sulfate) can modify protein structure and influence thermal behavior. Urea and guanidine hydrochloride can cause protein denaturation without heat treatment, but only at very high concentrations [71]. These reagents cause the breakdown of quartenary structure or subunit orientation of some proteins leading to aggregation [13].

The denaturation of egg albumin by urea proceeds in two stages: aggregation of the molecules, which makes the protein insoluble, and splitting of the molecules, which makes them soluble. These two reactions may occur simultaneously; the more the urea is diluted, the larger is the average size of the polydisperse particles. Both reactions have a positive temperature coefficient, but as the concentration of urea increases, the second reaction is accelerated more by increase in temperature than the first; consequently the opalescence of the solutions decreases with increasing temperature.

TABLE 2 Effect of Sugars and Polyols on DSC Characteristics of β-Lactoglobulin

Polyol conc.	Sucrose		Glucose		Glycerol		Ethylene glycol	
(%)	ΔT_d(°C)[a]	ΔH(J/g)[b]	ΔT_d(°C)[a]	ΔH(J/g)[b]	ΔT_d(°C)[a]	ΔH(J/g)[b]	ΔT_d(°C)[a]	ΔH(J/g)[b]
10	0.9	10.6	1.05	11.35	0.05	16.0	−2.1	13.5
20	1.15	12.3	1.55	7.9	1.1	16.6	−4.2	14.1
30	2.9	10.7	4.4	9.9	1.6	15.4	−7.6	10.9
40	3.9	11.8	4.0	11.5	2.8	14.7	−11.5	9.5
50	5.0	13.0	9.3	—	3.6	16.5	−14.9	11.0

[a] $\Delta T_d = T_d(\beta\text{-LG + polyols}) - T_d(\beta 0\text{-LG})$.
[b] Enthalpy.
Source: Adapted from Ref. 74.

Anionic detergents such as sodium dodecyl sulfate (SDS) also have marked effect on the thermal stability of food proteins. SDS has been reported to decrease the thermal stability of conalbumin and ovalbumin [72]. Low concentrations of SDS have, however, been shown to stabilize β-lactoglobulin by up to 7°C [73]. A similar effect has also been reported for oat globulin [53] and micellar isolate from faba bean [66]. In the presence of up to 0.75% SDS (26 mM), an increase in the denaturation temperature was observed. Beyond this, the denaturation temperature values were relatively stable but the size of the DSC endotherm was reduced, indicating that SDS induced protein denaturation. It has been suggested that SDS stabilizes proteins by forming bridges between positively charged groups in one loop of the polypeptide chain and hydrophobic regions in another [74]. Arntfield et al. [66] suggest that the critical factor is the specific conformation of the protein, and if this is inappropriate, no stability increase is observed, which would explain the varied response in the thermal stability of food proteins in the presence of SDS.

IV. METHODS OF MEASURING THERMAL DENATURATION AND COAGULATION

A. Solubility

Protein solubility is controlled by a delicate balance between repulsive and attractive intermolecular forces. These forces are dependent upon protein and water structures and are affected by solvent conditions [75]. One of the major consequences of thermal denaturation is a reduction in protein solubility. Proteins are soluble in water when electrostatic repulsive (and hydration repulsion) forces are greater than attractive hydrophobic interactions. Any condition that decreases the interactions between protein and water molecules decreases protein solubility. The tendency of a protein molecule to stay in solution is due to its interaction with the solvent molecules. Increasing the temperature progressively disrupts ionic bonding (including hydrogen bonds), which results in disordering of both protein and water structure. Protein-protein interactions are enhanced as a consequence of increased surface hydrophobicity, which effectively displaces water molecules from protein surfaces, thus promoting aggregation [76].

Several expressions have been used to define protein solubility. These include nitrogen solubility index (NSI), protein dispersibility index (PDI), water-soluble nitrogen, and water-dispersible protein (WDP). Betschart [77] defined protein solubility as the proportion of nitrogen of a protein concentrate that is determined as soluble after a specifically defined procedure. Various methods have been used to determine the extent of protein solubility after heat treatment. These methods are, however, prone to several errors, such as the prior processing history of the sample, extraction methods, ions, pH, solvents, etc. [77].

A standard procedure for determining protein solubility was developed using a micro-Kjeldahl procedure to measure the nitrogen solubility index [78]. The method involves dispersal of the dried protein in a solution of NaCl (0.1 M), pH adjustment to a desired value, centrifugation, filtration, and determination of nitrogen content of the supernatant by using the micro-Kjeldahl procedure. It was determined that although the biuret method and micro-Kjeldahl procedure generally provided comparable accuracy and precision for protein content and solubility of some proteins, the biuret procedure exhibited considerable error and variability for other proteins.

B. Ultraviolet/Visible and Fluorescence Spectrophotometry

Spectrophotometric measurements in the visible and ultraviolet region have been widely used to study proteins. The ultraviolet region extends from about 10 to 380 nm, but the most analytically useful region is from 200 to 380 nm. The visible region extends from the near-ultraviolet region (380 nm) to about 780 nm and is the region in which light appears as a color and is therefore useful for colorimetric analysis. Protein solutions characteristically show an absorbance at 280 nm. The absorption of electromagnetic radiation by proteins in the range of 250–300 nm is primarily due to the electronic excitation of tyrosine, tryptophan, and phenylalanine.

Denaturation is almost invariably accompanied by changes in molecular configuration, breaking and formation of side-chain hydrogen bonds involving tyrosyl and tryptophanyl residues, and ionization of tyrosyl residues. These contribute to alteration in spectral characteristics of proteins and can be used to monitor changes that occur in protein structure during denaturation. Furthermore, tryptophan and tyrosine residues buried within the core of the protein molecule may be exposed during heat denaturation of proteins. The exposure of these residues could result in an increase in absorbance, which can be related to the extent of denaturation. Protein denaturation could also lead to shifts in the peaks of absorption spectrum to shorter wavelength, or blue shift [79].

Aromatic amino acids also contribute to fluorescence, but fluorescence of most proteins is dominated by tryptophan residues with an emission maximum at 348 nm in water [80]. Denaturation is generally accompanied by a red shift in fluorescence emission [81].

UV and fluorescence spectrophotometries were used to monitor the thermal denaturation of dilute (<0.05%) oat globulin solutions at high ionic strength [82]. A significant red shift was observed in the absorption maximum of the UV spectra when the protein was heated at 110°C. Second derivative spectra suggested exposure of tryptophan residues in the heated samples. When 1% oat globulin was heat aggregated and fractionated into soluble and insoluble fractions, the UV and fluorescence spectra indicated no marked protein unfolding in the soluble fraction, but extensive denaturation was observed in the insoluble aggregates. In addition, the soluble fraction exhibited significantly more surface hydrophobicity than either the insoluble or the unheated protein samples.

Association of denatured molecules during heat treatment produces aggregates that increase the turbidity of protein solutions [83]. This change in the clarity of protein dispersion during heat treatment can be measured spectrophotometrically. Several studies have correlated an increase in absorbance measurements of protein solutions with an increase in aggregates formation [84,85]. In these studies, the absorbance of protein solutions (in the 500–660 nm region) are measured as the samples are heated either isothermally or as a function of temperature.

C. Infrared Spectroscopy

Infrared absorption spectroscopy in the 1400–1700 cm^{-1} spectral region has been used to study the secondary structure of polypeptides in the solid state [86] and in deuterium oxide solutions [87]. Deuterium oxide is usually employed instead of water because of its greater transparency in the region of interest [88].

The amide I′ band in the region of 1600–1700 cm^{-1} has been found to be the most useful for protein structure studies by infrared spectroscopy [89]. This region involves C=O stretching vibrations of peptide groups [86,87]. The broad infrared bands of proteins, which result from the vibrations of the peptide groups in this region, are composed of several overlapping bands due to the various protein segments with different secondary structures [90]. Each type of substructure, such as the α-helix, β-strands, and the various kinds of turns, gives rise to different C=O stretching bands as a result of differences in the orientation of molecular subgroups or changes in interchain and intrachain hydrogen-bonding interactions of the peptide groups. These subbands usually cannot be resolved by conventional spectroscopic techniques because their inherent widths are greater than the instrumental resolution. Resolution enhancement of the band, also known as Fourier self-deconvolution, results in a narrowing of the amide I′ band to reveal the hidden peaks. Figure 5 shows the infrared spectra of β-lactoglobulin in D_2O before and after deconvolution [62]. This technique not only enriches the qualitative interpretation of infrared spectra, but also provides a basis for the quantitative estimation of protein secondary structure. The principles and instrumentation of FTIR spectroscopy have been extensively reviewed [88–91]. Infrared studies of proteins to date have addressed the determination of secondary structure using the structure-sensitive amide I′ band in the infrared spectrum [92,93]. Some workers have examined the side-chain vibrations of proteins for additional information on interactions within the protein [94]. In addition, events leading to protein denaturation and reorganization [95] as well as intermolecular reactions not involving conformational changes have also been studied using the amide I′ region of the infrared spectrum [92].

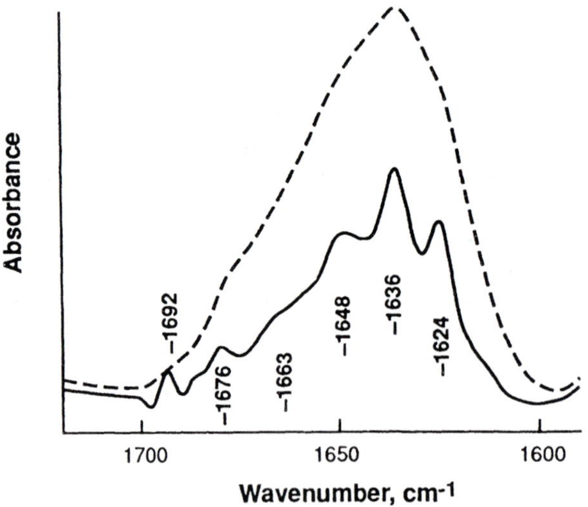

Figure 5 Infrared spectra of β-lactoglobulin in D_2O (200 mg/ml, pH unadjusted, 26°C) before (- - -) and after (—) deconvolution. The numbers refer to the wavenumber (cm^{-1}) of the six main bands assigned as follows: 1692—turns/β-sheet; 1676—β-sheet/turns; 1663—turns; 1648—overlap of α-helix and random coils; 1636—β-sheet; 1624—β-sheet. (Adapted from Ref. 62.)

Infrared spectroscopy can be used to monitor aggregation of protein during heat treatment. The particular changes that occur in the protein secondary structure and the manner in which the networks build from individual molecules when heated affect aggregate and gel formation [96]. For example, the formation of nonnative ordered structure of hydrogen-bonded β-sheets has been observed when gelling proteins were heated above their transition temperatures [44]. Clark et al. [97] suggested that the formation of a band at 1620 cm^{-1} correlated with protein aggregation and the formation of hydrogen-bonded β-sheet. These authors observed that formation of a second, less intense band at 1680 cm^{-1} together with the band at 1620 cm^{-1} indicated formation of antiparallel β-sheet. Figure 6 shows the FTIR spectra of β-lactoglobulin heated from 26 to 97°C at pH 3, 5, and 9. Denaturation of the protein (depicted by major changes in the bands between 1682 and 1626 cm^{-1}) was observed at 72°C at pH 3. At pH 9, however, denaturation was observed as early as 51°C, suggesting that β-lactoglobulin was more thermally stable at pH 3 than at pH 9. The two bands at 1684 and 1618 cm^{-1} associated with aggregate formation [97] were observed following denaturation at all the pH values studied.

D. Optical Methods: CD, ORD

Another spectroscopic technique used in the study of protein structure is circular dichroism (CD). This phenomenon is a result of the absorptions of visible or UV light by a chiral solute such as a protein. The most widely used absorptions in spectroscopic studies of proteins by CD are the electronic absorptions of the amides of the polypeptide backbone between wavelengths of 180 and 240 nm [98].

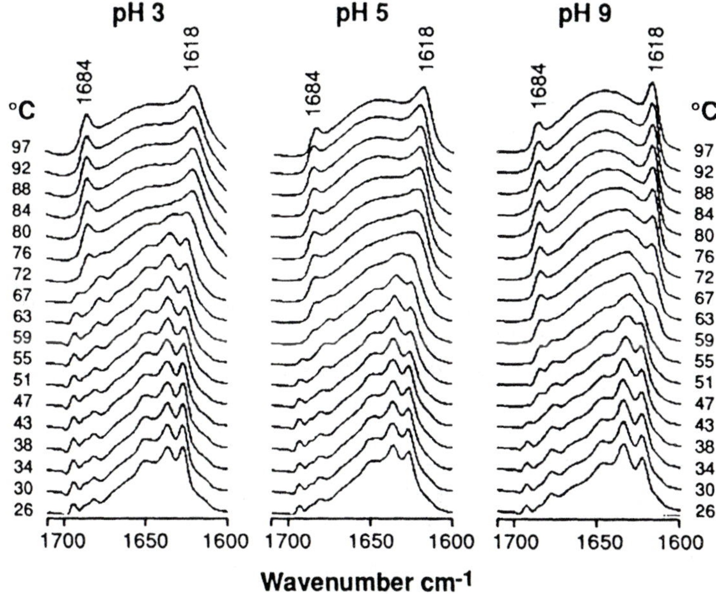

FIGURE 6 Stacked plot of deconvoluted infrared spectra of β-lactoglobulin (200 mg/ml) in deuterated phosphate buffer (ionic strength 0.2) as a function of temperature (1618 and 1684 cm^{-1} bands have been attributed to hydrogen-bonded antiparallel β-sheet structures associated with aggregate formation). (Adapted from Ref. 62.)

The optical activity of proteins results from their ability to rotate plane-polarized light. This light is characterized by an electric vector which oscillates within a plane and has two rotating components (clockwise, d, and counterclockwise, l) called circular polarizations. When these two components encounter a chiral object such as a protein in solution, they are absorbed and retarded unequally, resulting in either circular dichroism or optical rotation.

Optical rotation and dispersion (ORD) measures the wavelength dependence of the molecular rotation of a compound. Circular dichroism, on the other hand, is dependent on the fact that molar absorbtivity of an optically active compound is different for the two types of circularly polarized light. The extent of CD is usually expressed by ellipticity (θ), which is defined as:

$$\theta = \frac{(2303)9\ (\epsilon_l - \epsilon_d)}{2\pi}$$

where ϵ_l and ϵ_d are molar absorptivities in l- and d-polarized light, respectively.

A polypeptide folded entirely as an α-helix has a circular dichroic spectrum distinct from that of a polypeptide folded entirely in β-structure. Both of these spectra are distinct from that of a polypeptide unfolded as a random coil. One of the more important and informative uses of circular dichroism is to provide evidence that the structure of the protein has changed under particular circumstances. Damodaran [99] used CD to study the thermal denaturation and gelation of soy proteins in the presence and absence of NaCl and $NaClO_4$. Matsuura and Manning [100] studied the effect of pH, salts, and reducing agents (dithiothreitol and β-mercaptoethanol) on the secondary and tertiary structure of β-lactoglobulin during heat treatment.

There are, however, some limitations in the use of CD to study protein aggregation and gel formation. Cloudy aggregates or gels diffract light and produce artificial CD signals [101]. This greatly limits the conditions under which heated protein dispersions can be studied by CD. For further reading on the use of CD to study the denaturation of food proteins, the reader is referred to Woody [101].

E. Electrophoresis

In general, electrophoresis involves the separation of charged ions on the basis of their mobility under the influence of an externally applied electric field and can be used to evaluate the changes occurring in food proteins during heat treatment. When proteins aggregate, they form complex structures with large molecular weights that do not migrate at the same rate through the separation gel used in electrophoretic analysis. Differences between the electrophoretic patterns of proteins before and after heat treatment have been used to determine changes in the molecular size of the proteins [102,103] and to monitor denaturation and aggregation [45]. In Figure 7, the electrophoregram of native (unheated) whey protein concentrate is shown to have four main bands, identified as β-lactoglobulin A and B, α-lactalbumin, and bovine serum albumin (BSA), and traces of proteose-peptones and immunoglobulins. The electrophoretic pattern of these proteins heated at 50 and 60°C did not vary much from that of the native protein. At 65°C, however, only traces of the BSA and α-lactalbumin bands could be observed suggesting aggregation of these proteins. Above 70°C very little of the proteins were detected, suggesting extensive aggregation of all the proteins at higher temperatures [45].

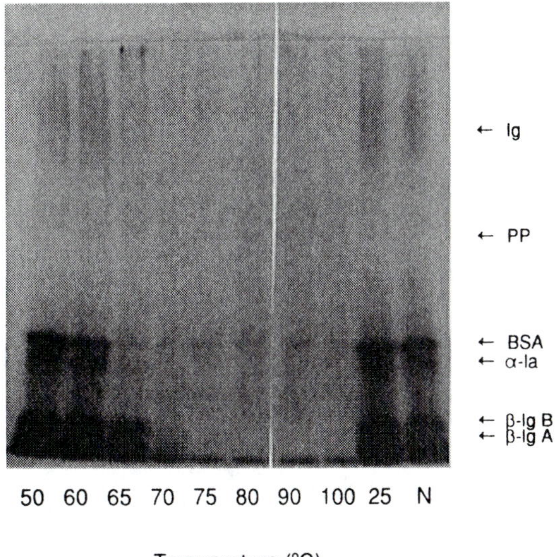

FIGURE 7 Electrophoregrams of whey protein concentrate (WPC) dispersions (150 mg WPC/g H_2O, pH unadjusted) heated for 60 minutes at various temperatures: α-la—α-lactalbumin); β-lg—β-lactoglobulin; BSA—bovine serum albumin; Ig—immunoglobulins; PP—proteose peptones; N—native WPC. (Adapted from Ref. 45.)

F. Reversed-Phase High-Performance Liquid Chromatography

Reversed-phase high-performance liquid chromatography (RP-HPLC) has been utilized to assess the changes that occur in food proteins during processing [104]. The separation of proteins by RP-HPLC is based on an adsorptive interaction between the hydrophobic side chains of the amino acid residues of proteins and the alkyl function of the stationary phase, as well as ionic and polar interactions between the hydrophilic side chains of the amino acid residues and the free silanols of the stationary phase [105]. The amino acid composition, the polypeptide chain length, and the net hydrophobicity of the protein molecules contribute to the retention of the proteins during the separation process [106]. Any heat-induced changes in these properties of the protein molecules can be easily monitored by HPLC. Parris et al. [103] monitored the extent of protein denaturation by comparing the sum of the normalized absorbance peaks of heated whey proteins eluted from the RP-HPLC column to those of an unheated control sample. In another study [102], integration of peak areas obtained from RP-HPLC and comparison of peak retention times with a calibration plot of log molecular weight versus retention time of known standards allowed for characterization of whey protein concentrate and analysis of heat treatment effects. Figure 8 shows the HPLC chromatogram of whey protein concentrate dispersion heated for 60 minutes at 60, 65, and 70°C. At 60°C, the peaks representing α-lactalbumin and β-lactoglobulin could be observed, suggesting that these proteins were not aggregated. At 65°C, the α-lactalbumin peak could not be detected, which suggests that it had aggregated. The β-lactoglobulin peak, however, was substantially reduced in

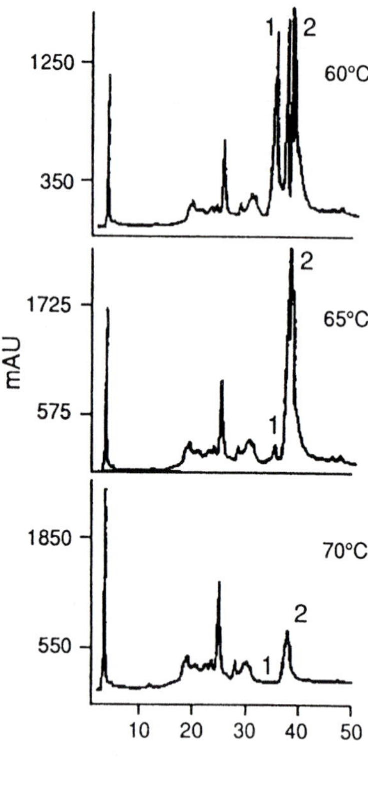

FIGURE 8 HPLC chromatogram of whey protein concentrate (WPC) dispersions (150 mg WPC/g H_2O, pH unadjusted) heated for 60 minutes at 60, 65, and 70°C. (1) α-lactalbumin; (2) β-lactoglobulin. (Adapted from Ref. 45.)

intensity only at 70°C, which indicates that it required higher temperatures to aggregate. For further reading on RP-HPLC, the reader is referred to Macrae [107], Newton [108], and Simpson [109].

G. Light Scattering

Although the changes in molecular size of proteins during thermal aggregation can be monitored by techniques such as SDS-PAGE, size-exclusion chromatography, and ultracentrifugation, a method for rapidly estimating the molecular weight of aggregates is desirable. Takagi [110] has developed a low-angle laser light scattering technique in combination with HPLC that can measure the molecular weight of giant molecules such as soluble aggregates of proteins.

In a typical setup, heat aggregated protein is fractionated by a HPLC column, and both light scattering (LS) and refractive index (RI) of the effluent are measured. Molecular weight (M_W) of the aggregates can be estimated from the ratio of the output of the scattering photometer, $(\text{Output})_{LS}$, to that of the refractometer, $(\text{Output})_{RI}$. M_W of protein can be estimated from the following equation:

$$M_W = \frac{(\text{Output})_{LS} \text{ of sample}}{(\text{Output})_{RI} \text{ of sample}} \cdot \frac{1}{K} \cdot \frac{(\text{Output})_{RI} \text{ of standard}}{(\text{Output})_{RI} \text{ of sample}} \cdot \frac{C_{\text{sample}}}{C_{\text{standard}}} \tag{12}$$

where C is the concentration of sample and standard proteins and K is an instrument constant, which can be determined from measurements of known molecular weight standard proteins.

Using this technique, the apparent M_W of soluble aggregates of ovalbumin was found to increase with heating temperature from 4 million at 76°C to 35 million at 100°C [111]. The M_W of heat-induced soluble aggregates of bovine globulin were also found to increase from 800,000 at 60°C to 1–100 million at >80°C [112].

H. Differential Scanning Calorimetry

Differential scanning calorimetry (DSC) has emerged as the technique of choice for the study of thermal transitions of food. In this technique, the substance of interest and an inert reference are heated at a programmed rate. Any thermally induced changes occurring in the sample are then recorded as a differential heat flow displayed normally as a peak on a thermogram.

The conversion of a protein from native to denatured state by heat is a cooperative phenomenon and is accompanied by a significant uptake of heat, seen as an endothermic peak in the DSC thermogram [113]. Parameters obtained from the DSC thermogram include enthalpy of denaturation (ΔH), the peak or denaturation temperature (T_d), and width of the calorimetric transition at half peak height (ΔT_W) (Fig. 9). ΔH is an estimate of the thermal energy required to denature the protein and is measured from the peak area of the thermogram. T_d is a measure of the thermal stability of the proteins, and at this temperature the change in Gibbs free energy (ΔG) between the folded and unfolded protein structure is zero. A list of T_d and ΔH values of several proteins is given in Table 3. T_W is an estimate of the cooperativity of protein unfolding [114]. If denaturation occurs

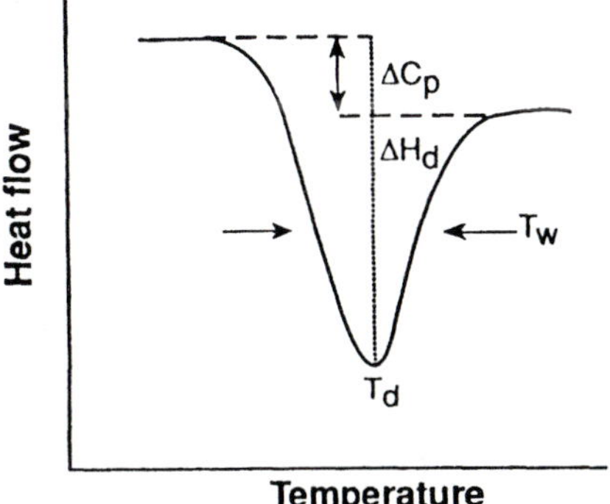

FIGURE 9 DSC thermogram showing change in heat capacity (ΔC_p), change in enthalpy, (ΔH_d), peak temperature of denaturation (T_d), and width of the peak at half-height (T_W).

TABLE 3 Enthalpy (ΔH) and Denaturation Temperatures (T^d) of Various Food Proteins

Protein	ΔT_d(°C)	ΔH(J/g)	Ref.
Oat	108.9	26.30	68
Faba bean	91.0	5.27	66
Canola	83.5	4.90	66
Cowpea (protein fraction)	85.5	7.74	66
Lentil (four)	82.5	8.37	66
Lima bean (flour)	98.0	7.11	66
Ovalbumin	84.5	15.12	38
Conalbumin	61.0	5.38	38
α-Lactalbumin	61.0	20.39	39
β-Lactoglobulin	81.5	14.40	66

within a narrow range of temperature (low T_W value), the transition is considered highly cooperative [113].

In addition, DSC can give information on the reversibility of the protein denaturation. The extent of reversibility is easily estimated by rescanning samples after heat denaturation and comparing relative areas of the transitions produced after the first and second heating [115]. Under conditions of reversible denaturation, the thermodynamics of protein stability can be determined [25], e.g., the thermodynamic characteristics of 11S globulins from some seeds have been determined by microcalorimetry [116].

Kinetic parameters of protein denaturation can also be determined by DSC using the same data as in the thermodynamic studies. In this case, emphasis is placed on the dynamics of the reaction as opposed to its energetics [6]. The basic equation for rate of change of a species (A) with time (t) can be written as:

$$\pm\frac{dA}{dt} = kA^n \tag{13}$$

where k is the rate constant (sec^{-1}) and n is the apparent reaction order. The temperature dependence of the rate constant can be determined from the Arrhenius equation:

$$k = Ze^{-Ea/RT} \tag{14}$$

where Z is the preexponential factor (sec^{-1}) and Ea is the activation energy (J/mol).

I. Rheological Measurements

Heat treatment of proteins often results in increased water uptake and swelling as proteins unfold. This results in an increase in hydrodynamic volume and increased resistance to flow [117]. The increase in viscosity also reflects intermolecular interactions resulting from attractions between adjacent molecules with the formation of weak transient networks. Where there is no interaction between particles, flow properties depend only on the volume fraction or concentration of the suspended material and show Newtonian behavior. In concentrated solutions the hydrodynamic domains of the protein molecules come into contact, resulting in interactions between suspended particles. Concentrated protein solutions, therefore, exhibit non-Newtonian behavior and show viscoelastic properties [118]. Several reviews on the rheology of foods have been published [118,119].

Dynamic rheological measurement in which sinusoidally oscillating stress or strain is applied to the sample is the preferred method for characterizing viscoelastic foods [6]. Dynamic measurements allow coagulation and gelation to be monitored since the induced deformations are usually so small that their effect on structure is negligible. Two independent parameters are obtained from dynamic measurements: the storage modulus (G′) describing the amount of energy that is stored elastically in the structure and the loss modulus (G″), which is a measure of the energy loss or the viscous response. The phase angle (δ) is a measure of how much the stress and strain are out of phase with each other. For a completely elastic material, the phase angle is 0° and for a purely viscous fluid δ is 90°. The ratio G″/G′, called the loss tangent, is equal to the tangent of the phase angle:

$$\tan \delta = G''/G'$$

and is proportional to the (energy dissipated)/(energy stored) per cycle.

Dynamic viscoelastic measurements have been used to monitor the properties of various food proteins that coagulate or gel upon thermal denaturation. These include β-lactoglobulin and muscle proteins [120,121].

J. Electron Microscopy

Electron microscopy is a technique that is increasingly being used to study the microstructure of protein aggregates and gels. Proteins can form either opaque aggregates, coarse coagulum, or soft or firm gels. The microstructures of these aggregates and gels differ markedly. The molecules of the denatured proteins in these structures are, however, not large enough to make the visualization of their unfolded nature possible. Using electron microscopy, the microstructure of the protein aggregates can be monitored and related to the extent of protein unfolding during heat treatment. Harwalkar and Kalab [122] observed that β-lactoglobulin heated at pH 2.5 and ionic strength of 0.2 formed uniform matrices composed of strands of protein particles and a few chainlike structures linked together. At higher ionic strength, severely aggregated and collapsed networks (precipitates) were observed. In another study, the microstructure of β-lactoglobulin dispersions heated at 90°C for 30 minutes at pH 2.5, 4.5, and 6.5 and precipitated at pH 4.5 was studied (Fig. 10) [123]. At pH 2.5 (Fig. 10A), compact aggregates were formed, a large proportion of which was fused, as well as some fibrous aggregates. At pH 4.5 (Fig. 10B), primary globular aggregates that were fused together were formed, and at pH 6.5 (Fig. 10C) both compact aggregates and fibrous structures were formed.

V. CONCLUSION

The relationship between protein structure and functionality continues to be an area of significant interest to food scientists and protein chemists. A detailed understanding of the mechanisms by which proteins denature and in so doing affect functional properties is imperative to the achievement of specific textural attributes in foods. Our current knowledge of the functional properties of proteins as they relate to protein denaturation during heat treatment is restricted to the behavior of individual proteins studied under univariate conditions, which fails to consider the extensive conformational changes that occur under multivariate systems and further ignores the complex interactions that occur between proteins and other food components during heat treatment. Future studies on

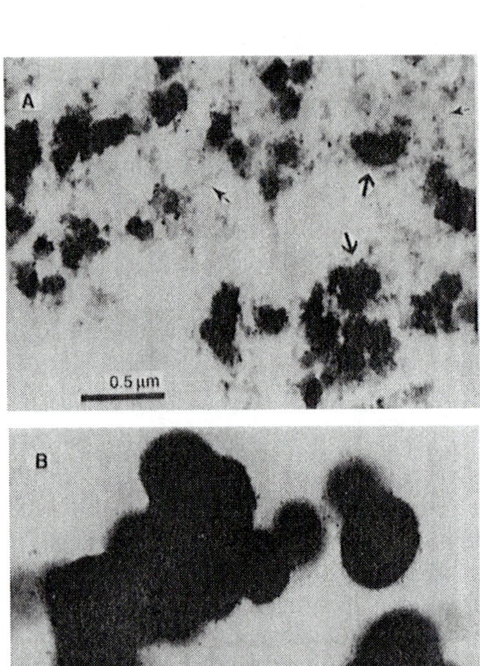

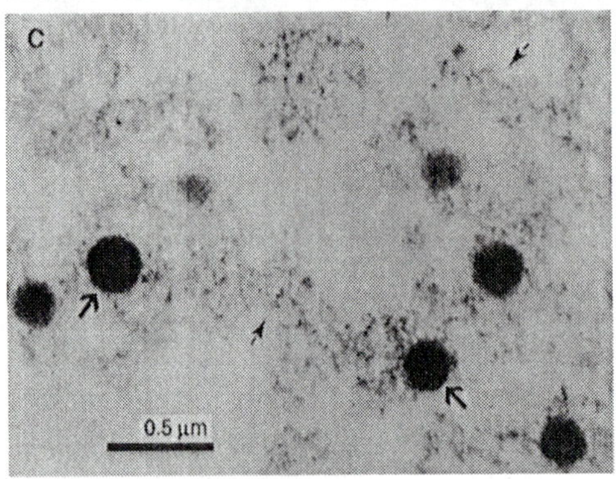

FIGURE 10 Isoelectric precipitate of β-lactoglobulin heated at 90°C for 30 minutes. (A) β-Lactoglobulin heated at pH 2.5 and precipitated at pH 4.5. Large arrows point to compact aggregates, a large proportion of which is fused (asterisks). Small arrows point to fibrous aggregates. (B) β-Lactoglobulin heated and precipitated at pH 4.5. Primary globular aggregates are fused together. (C) β-Lactoglobulin heated at pH 6.5 and precipitated at pH 4.5. (Adapted from Ref. 123.)

the denaturation and coagulation of proteins should include those in which the combined effect of factors that alter the molecular structure of proteins are analyzed simultaneously. This would be a better reflection of what actually occurs in real food systems.

Until recently, thermal treatment has been the most common processing technique used in denaturing food proteins. Most of our understanding of the functional behavior of food proteins has been based on their thermal properties. More recently, other processes such as high pressure have been used to denature proteins and to induce specific functional properties such as coagulation and gel formation. This is a growing area, which may hopefully shed more light into the mechanisms of protein denaturation and functionality.

One of the major problems in defining protein coagulation is the difficulty in clearly distinguishing aggregation, precipitation, and gel formation. This is primarily because, far from being sequential, these events tend to occur simultaneously under certain conditions. A clearer distinction of these processes will require that they be studied under conditions that allow them to occur individually. The application of novel techniques such as high-intensity ultrasound [124] and electron microscopy in combination with image analysis techniques [125] in the analysis of food structure should improve our ability to observe differences in these phenomena.

REFERENCES

1. A. Pour-El, Protein functionality: classification, definition and methodology, *Protein Functionality in Foods* (A. Pour-El, ed.), ACS Symp. Series, No. 147, American Chemical Society, Washington, DC, 1981, p. 5.
2. J. E. Kinsella, Functional properties of proteins in foods: A survey, *CRC Crit. Rev. Food Sci. Nutr. 7*:219 (1976).
3. S. Damodaran, Interrelationship of molecular and functional properties of food proteins, *Food Proteins* (J. E. Kinsella and W. G. Soucie, eds.), American Oil Chemists Society, Champaign, IL, 1989, p. 43.
4. D. M. Mulvihill and M. Donovan, Whey proteins and their thermal denaturation—a review *Irish J. Food Sci. Technol. 11*:43 (1987).
5. C. Tanford, Protein denaturation, *Adv. Protein Chem. 23*:121 (1968).
6. D. W. Stanley and R. Y. Yada, Physical consequences of thermal reactions in food protein systems, *Physical Chemistry of Foods* (H. G. Schwartzberg and R. W. Hartel, eds.), Marcel Dekker, New York, 1992, p. 669.
7. W. Kauzmann, Some factors in the interpretation of protein denaturation, *Adv. Protein Chem. 14*:1 (1959).
8. J. C. Cheftel, J.-L. Cuq, and D. Lorient, Amino acids, peptides and proteins, *Food Chemistry* (O. R. Fennema, ed.), Marcel Dekker, New York, 1985, p. 45.
9. A.-M. Hermansson, Aggregation and denaturation involved in gel formation, *Functionality and Protein Structure* (A. Pour-El, ed.), ACS Symp. Ser. 92, American Chemical Society, Washington, DC, 1979, p. 82.
10. B. German, S. Damodaran, and J. E. Kinsella, Thermal dissociation and association behaviour of soy proteins, *J. Agric. Food Chem. 30*:807 (1982).
11. A. R. Fersht, *Enzyme Structure and Mechanism*, 2nd ed., W. H. Freeman & Co., New York, 1985.
12. T. E. Creighton, *Proteins: Structure and Molecular Properties*, W. H. Freeman & Co., New York, 1984.
13. M. Joly, *A Physicochemical Approach to the Denaturation of Proteins*, Academic Press, New York, 1965.

14. S. F. Edwards, P. J. Lillford, and J. M. V. Blanshard, Gels and networks in practice and theory, *Food Structure and Behaviour* (J. M. V. Blanshard and P. J. Lillford, eds.), Academic Press, New York, 1987, p. 1.
15. R. H. Schmidt, Gelation and coagulation, *Protein Functionality in Foods* (A. Pour-El, ed.), ACS Symp. Ser. No. 147, American Chemical Society, Washington, DC, 1981, p. 131.
16. T. G. Parker and D. G. Dalgleish, The potential application of the theory of branching processes to the association of milk protein, *J. Dairy Res. 44*:79 (1977).
17. N. Catsimpoolas and E. W. Meyer, Gelation phenomena of soybean globulins: Protein-protein interactions, *Cereal Chem. 47*:559 (1970).
18. P. A. Morrissey, D. M. Mulvihill, and E. M. O'Weill, Functional properties of muscle proteins, *Development in Food Proteins*, Vol. 5 (B. J. F. Hudson, ed.), Elsevier Appl. Sci., London, 1987, p. 237.
19. S. Utsumi and J. E. Kinsella, Forces involved in soy protein gelation: Effects of various reagents on the formation, hardness and solubility of heat induced gels made from 7S, 11S and soy isolate, *J. Food Sci., 50*:1278 (1985).
20. C. H. Wang and S. Damodaran, Thermal gelation of globular proteins: Weight-average molecular weight dependence of gel strength, *J. Agric. Food Chem. 38*:1164 (1990).
21. G. Stainsby, The gelatin gel and the sol-gel transformation, *The Science and Technology of Gelatin* (A. G. Ward and A. Courts, eds.), Academic Press, New York, 1977, p. 179.
22. K. E. Van Holde, Effects of amino acid composition and microenvironment on protein structure, *Food Proteins* (J. R. Whitaker and S. R. Tannenbaum, eds.), AVI Publ. Inc., Westport, CT, 1977, p. 1.
23. C. B. Anfinsen and H. A. Scherga, Experimental and theoretical aspects of protein folding, *Adv. Prot. Chem. 29*:205 (1975).
24. R. Hawkes, M. G. Grutter, and J. Schellman, Thermodynamic stability and point mutations of bacteriophase t4 lysozyme, *J. Mol. Biol. 175*:195 (1984).
25. P. L. Privalov, Stability of proteins. Small globular proteins, *Adv. Prot. Chem. 33*:167 (1979).
26. N. Go, Theory of reversible denaturation of globular proteins, *Int. Pept. Prot. J. 7*:313 (1975).
27. A. Kilara and T. Y. Sharkasi, Effect of temperature on food proteins and its implications on functional properties, *CRC Crit. Rev. Food Sci. Nutr. 23*:323 (1986).
28. A. K. Lala and P. Kaul, Increased exposure of hydrophobic surface in molten globule state of α-lactalbumin, *J. Biol Chem. 267*:19914 (1992).
29. M. Ohgushi and A. Wada, 'Molten-globule state': a compact form of globular proteins with mobile side chains, *FEBS Lett. 164*:21 (1983).
30. Y. Mine, Recent advances in the understanding of egg white protein functionality, *Trends Food Sci. Technol. 6*:225 (1995).
31. T. P. Shukla, Chemistry and biological function of α-lactalbumin, *CRC Crit. Rev. Food Technol. 3*:41 (1973).
32. M. J. Kronman, R. Andreotti, and R. Vitols, Inter- and intramolecular interactions of α-lactalbumin. II. Aggregation reactions at acid pH, *Biochemistry 3*:1152 (1964).
33. M. J. Kronman, L. G. Holmes, and F. M. Robbins, Inter- and intramolecular interactions of α-lactalbumin. VIII. The alkaline conformational change, *Biochim. Biophys. Acta 133*:46 (1967).
34. O. B. Ptitsyn, Protein folding: hypothesis and experiments, *J. Prot. Chem. 6*:273 (1987).
35. T. Koseki, N. Kitabatake, and E. Doi, Irreversible thermal denaturation and formation of linear aggregates of ovalbumin, *Food Hydrocolloids 3*:123 (1989).
36. A. Kato, H. R. Ibrahim, H. Watanabe, K. Honma, and K. Kobayashi, New approach to improve the gelling and surface functional properties of dried egg white by heating in dry state, *J. Agric. Food Chem. 37*:433 (1989).
37. W. F. Harrington and P. H. von Hippel, Formation and stabilization of the collagen-fold, *Arch. Biochem. Biophys. 92*:100 (1961).

38. S. Barbut and C. J. Findlay, Thermal analysis of egg proteins, *Thermal Analysis of Foods* (V. R. Harwalkar and C.-Y. Ma, eds.), Elsevier Appl. Sci., London, 1990, p. 126.
39. V. Bernal and P. Jelen, Effect of calcium binding on thermal denaturation of bovine α-lactalbumin, *J. Dairy Sci. 67*:2452 (1984).
40. K. Brew and J. A. Grobler, α-Lactalbumin, *Advanced Dairy Chemistry*, Vol. 1, *Proteins* (P. F. Fox, ed.), Elsevier Appl. Sci., London, 1992, p. 191.
41. K. R. Rao and K. Brew, Calcium regulates folding and disulfide-bond formation in α-lactalbumin, *Biochem. Biophys. Res. Commun. 163*:1390 (1989).
42. V. R. Harwalkar, Kinetics of thermal denaturation of β-lactoglobulin at pH 2.5, *J. Dairy Sci. 63*:1052 (1980).
43. D. M. Byler and J. M. Purcell, FTIR examination of thermal denaturation and gel formation in whey proteins, *SPIE 1145*:415 (1989).
44. D. M. Smith, Protein interactions in gels: Protein-protein interactions, *Protein Functionality in Food Systems* (N. S. Hettiarachchy and G. R. Ziegler, eds.), Marcel Dekker, New York, 1994, p. 209.
45. J. I. Boye, I. Alli, A. I. Ismail, B. F. Gibbs, and Y. Konishi, Factors affecting molecular characteristics of whey protein gelation, *Int. Dairy J. 5*:337 (1995).
46. A. Pour-El and T. A. Swenson, Gelation parameters of enzymatically modified soy protein isolates, *Cereal Chem. 53*:438 (1976).
47. R. H. Schmidt and B. L. Illingworth, Gelation properties of whey protein and blended protein systems, *Food Prod. Dev. 12*:60 (1978).
48. P. H. von Hippel and T. Schleich, The effect of neutral salts on the structure and conformational stability of macromolecules in solution, *Structure and Stability of Biological Macromolecules* (S. N. Timasheff and G. D. Fasman, eds.), Marcel Dekker, New York, 1969, p. 417.
49. J. Arakawa and S. N. Timasheff, Preferential interactions of proteins with salt in concentrated solutions, *Biochemistry 24*:6545 (1982).
50. P. H. von Hippel and K. Y. Wong, On the conformational stability of globular proteins, *J. Biol. Chem. 240*:3909 (1965).
51. S. D. Arntfield, E. D. Murray, and M. A. H. Ismond, Effect of salt on the thermal stability of storage proteins from faba bean (*Vicia faba*), *J. Food Sci. 51*:371 (1986).
52. M. A. H. Ismond, E. D. Murray, and S. D. Arntfield, The role of non-covalent forces in micelle formation by vicilin from *Vicia faba*. II. The effect of stabilizing and destabilizing anions on protein interactions, *Food Chem. 21*:27 (1986).
53. V. R. Harwalkar and C.-Y. Ma, Study of thermal properties of oat globulin by differential scanning calorimetry, *J. Food Sci. 52*:396 (1987).
54. T. Itoh, Y. Wada, and T. I. Nakanishi, Differential thermal analysis of milk proteins, *Agric. Biol. Chem. 40*:1083 (1976).
55. A. Finch and D. A. Ledward, Differential scanning calorimetric study of collagen fibres swollen in aqueous neutral salt solutions, *Biochim. Biophys. Acta 295*:296 (1973).
56. D. W. Hickson, C. W. Dill, R. G. Morgan, D. A. Suter, and Z. L. Carpenter, A comparison of heat induced gel strengths of bovine plasma and egg albumin protein, *J. Anim. Sci. 51*: 69 (1980).
57. S. Varunsatian, K. Watanabe, S. Hayakawa, and R. Nakamura, Effects of Ca^{++}, Mg^{++} and Na^{+} on heat aggregation of whey protein concentrates, *J. Food Sci. 48*:42 (1983).
58. M. M. G. Koning and H. Visser, Protein interactions. An overview, *Protein Interactions* (H. Visser, ed.), VCH Publishers, New York, 1992, p. 1.
59. Y. Hiraoka, T. Segawa, K. Kuwajima, S. Sugai, and N. Moroi, β-Lactalbumin: a calcium metallo-protein, *Biochem. Biophys. Res. Commun. 95*:1098 (1980).
60. J. W. Donovan, C. J. Mapes, J. G. Davis, and J. A. Garibaldi, A differential scanning calorimetric study of the stability of egg white to heat denaturation, *J. Sci. Food Agric. 26*: 73 (1975).

61. J. F. Back, D. Oakenfull, and M. B. Smith, Increased thermal stability of proteins in the presence of sugars and polyols, *Biochemistry 18*:5191 (1979).
62. J. I. Boye, A. I. Ismail, and I. Alli, Effect of physico-chemical factors on the secondary structure of β-lactoglobulin. *J. Dairy Res. 63*:97 (1996).
63. J. I. Boye, I. Alli, and A. I. Ismail, Interactions involved in the gelation of bovine serum albumin. *J. Agric. Food Chem. 44*:996 (1996).
64. C. D. Ball, C. R. Hardt, and W. J. Duddles, The influence of sugars on the formation of sulfhydryl groups in heat denaturtion and coagulation of egg albumin, *J. Biol. Chem. 151*: 163 (1943).
65. D. Eagland, Nucleic acids, peptides and proteins, *Water: A Comprehensive Treatise* (F. Franks, ed.), Plenum Press, New York, 1975, p. 305.
66. S. D. Arntfield, M. A. H. Ismond, and E. D. Murray, Thermal analysis of food proteins in relation to processing effects, *Thermal Analysis of Foods* (V. R. Harwalkar and C.-Y. Ma, eds.), Elsevier Appl. Sci., London, 1990, p. 51.
67. M. A. H. Ismond, E. D. Murray, and S. D. Arntfield, The role of noncovalent forces in micelle formation by vicilin from *Vicia faba*. III. The effect of urea, guanidine hydrochloride and sucrose on protein interactions, *Food Chem. 29*:189 (1988).
68. V. R. Harwalkar and C.-Y. Ma, Effects of medium composition, preheating and chemical modification upon thermal behavior of oat globulin and β-lactglobulin, *Food Proteins* (J. E. Kinsella and W. G. Soucie, eds.), American Oil Chemists Society, Champaign, IL, 1989, p. 219.
69. S. Y. Gerlsma and E. J. Stuur, The effect of polyhydric and monohydric alcohols on the heat-induced reversible denaturation of lysozyme and ribonuclease, *Int. J. Peptide Protein Res. 4*:377 (1972).
70. J. C. Lee and S. N. Timasheff, The stabilization of proteins by sucrose, *J. Biol. Chem. 256*: 7193 (1981).
71. R. Lumry and H. Eyring, Conformation changes of proteins, *J. Phys. Chem. 58*:110 (1954).
72. P. O. Hegg, H. Martens, and B. Löfqvist, The protective effect of sodium dodecyl sulfate on the thermal precipitation of conalbumin. A study on thermal aggregation and denaturation, *J. Sci. Food Agric. 29*:245 (1978).
73. P. O. Hegg, Thermostability of β-lactoglobulin as a function of pH and concentration of sodium dodecyl sulfate, *Acta Agric. Scand. 30*:401 (1980).
74. P. O. Hegg and B. Löfqvist, The protective effect of small amounts of anionic detergents on the thermal aggregation of crude ovalbumin, *J. Food Sci. 39*:1231 (1974).
75. L. G. Phillips, D. M. Whitehead, and J. E. Kinsella, Protein stability, *Structure-Function Properties of Food Proteins*, Academic Press, New York, 1994, p. 25.
76. I. D. Kuntz and W. Kauzmann, Hydration of proteins and polypeptides, *Adv. Protein Chem. 28*:239 (1974).
77. A. A. Betschart, Nitrogen solubility of alfafa protein concentrate as influenced by various factors, *J. Food Sci. 38*:324 (1973).
78. C. V. Morr, B. German, J. E. Kinsella, J. M. Regenstein, J. P. van Buren, A. Kilara, B. A. Lewis, and M. E. Mangino, A collaborative study to develop a standardized food protein solubility procedure, *J. Food Sci. 50*:1715 (1985).
79. J. N. Solli and T. T. Herskovits, Solvent perturbation studies and analysis of protein and model compound data in denaturing organic solvents, *Anal. Biochem. 54*:370 (1973).
80. J. R. Lackowicz, Fluorescence of proteins, *Principles of Fluorescence Spectroscopy*, Plenum, New York, 1983, p. 1.
81. M. J. Kronman and L. G. Holmes, The fluorescence of native, denatured and reduced-denatured proteins, *Photochem. Photobiol. 14*:113 (1971).
82. C.-Y. Ma, and V. R. Harwalkar, Study of thermal denaturation of oat globulin by ultraviolet and fluorescence spectrophotometry, *J. Agric. Food Chem. 36*:155 (1988).

83. G. R. Ziegler and E. A. Foegeding, The gelation of proteins, *Adv. Food Nutr. Res. 34*:204 (1990).
84. G. R. Ziegler and J. C. Acton, Heat-induced transitions in the protein-protein interaction of bovine natural actomyosin, *J. Food Biochem. 8*:25 (1984).
85. T. Sano, S. F. Noguchi, J. J. Matsumoto, and T. Tsuchiya, Thermal gelation characteristics of myosin subfragments, *J. Food Sci. 55*:55 (1990).
86. S. Krimm, Infrared spectra and chain conformation of proteins, *J. Mol. Biol. 4*:528 (1962).
87. H. Susi, Infrared spectroscopy-conformation, *Meth. Enzymol. 26*:455 (1972).
88. H. Susi and D. M. Byler, Protein structure by Fourier transform infrared spectroscopy: Second derivative spectra, *Biochim. Biophys. Acta 115*:391 (1983).
89. H. Susi and D. M. Byler, Resolution-enhanced Fourier transform infrared spectroscopy of enzymes, *Meth. Enzymol. 130*:291 (1986).
90. W. K. Surewicz and H. H. Mantsch, New insight into protein secondary structure from resolution-enhanced infrared spectra. Review, *Biochim. Biophys. Acta 952*:115 (1988).
91. S. Krimm and J. Bandekar, Vibrational spectroscopy and conformation of potides, polypeptides and proteins, *Adv. Prot. Chem. 38*:181 (1986).
92. H. L. Casal, U. Kohler, and H. H. Mantsch, Structural and conformational changes of β-lactoglobulin B: an infrared spectroscopic study of the effect of pH and temperature, *Biochim. Biophys. Acta 957*:11 (1988).
93. D. M. Byler and H. M. Farrell, Jr., Infrared spectroscopic evidence for calcium ion interaction with carboxylate groups of casein, *J. Dairy Sci. 72*:1719 (1989).
94. P. W. Holloway and H. H. Mantsch, Infrared spectroscopic analysis of salt bridge formation between cytochrome b_5 and cytochrome *c*, *Biochemistry 27*:7991 (1988).
95. A. A. Ismail, H. H. Mantsch, and P. T. T. Wong, Aggregation of chymotrypsinogen: Portrait by infrared spectroscopy, *Biochim. Biophys. Acta 1121*:183 (1992).
96. J. E. Kinsella, and D. M. Whitehead, Proteins in whey: Chemical, physical and functional properties, *Adv. Food Nutr. Res. 33*:343 (1989).
97. A. H. Clark, D. H. P. Saunderson, and A. Suggett, Infrared and laser Raman spectroscopic studies of thermally-induced globular proteins gels, *Int. J. Pept. Prot. Res. 17*:353 (1981).
98. J. Kyte, Physical measurements of structure, *Structure in Protein Chemistry*, Garland Publishing Inc., New York, 1995, p. 393.
99. S. Damodaran, Refolding of thermally unfolded soy proteins during the cooling regime of the gelation process: Effect on gelation, *J. Agric. Food Chem. 136*:262 (1988).
100. H. E. Matsuura and M. C. Manning, Heat-induced gel formation of β-lactoglobulin: A study on the secondary and tertiary structure as followed by circular dichroism spectroscopy, *J. Agric. Food Chem. 42*:1650 (1994).
101. R. W. Woody, Circular dichroism of peptides, *The Peptides*, Vol. 7 (V. Hruby, ed.), Academic Press, San Diego, CA, 1985, p. 16.
102. E. Li-Chan, Heat-induced changes in the proteins of whey protein concentrate, *J. Food Sci. 48*:47 (1983).
103. N. Parris, J. M. Purcell, and S. M. Ptashkin, Thermal denaturation of whey proteins in skim milk, *J. Agric. Food Chem. 39*:2167 (1991).
104. R. J. Pearce, Analysis of whey proteins by high performance liquid chromatography, *Aust. J. Dairy. Technol. 38*:114 (1983).
105. L. Lemieux and J. Amiot, Application of reversed phase high-performance liquid chromatography to the separation of peptides from phosphorylated and dephosphorylated casein hydrolysates, *J. Chromatog. 473*:189 (1989).
106. M. P. Young and M. Merion, Capillary electrophoresis analysis of species variation in the tryptic maps of cytochrom C, *Current Research in Protein Chemistry: Techniques, Structure and Function* (J. J. Villafraca, ed.), Academic Press, New York, 1989, p. 217.
107. R. Macrae, Theory of liquid column chromatography, *HPLC in Food Analysis* (R. Macrae, ed.), Academic Press, New York, 1982, p. 141.

108. R. Newton, Instrumentation for HPLC, *HPLC in Food Analysis* (R. Macrae, ed.), Academic Press, New York, 1982, p. 28.
109. C. F. Simpson, Separation modes in HPLC, *HPLC in Food Analysis* (R. Macrae, ed.), Academic Press, New York, 1982, p. 80.
110. T. Takagi, Confirmation of molecular weight of *Aspergillus oryzae* α-amylase using the low angle laser light scattering technique in combination with high pressure silica gel chromatography, *J. Biochem. 89*:363 (1981).
111. A. Kato, Y. Nagase, N. Matsudomi, and K. Kobayashi, Determination of molecular weight of soluble ovalbumin aggregates during heat denaturation using low angle light scattering technique, *Agric. Biol. Chem. 47*:1829 (1983).
112. S. Hayakawa, Y. Suzuki, R. Nakamura, and Y. Sata, Physicochemical characterization of heat-induced soluble aggregates of bovine globulin, *Agric. Biol. Chem. 47*:395 (1983).
113. D. J. Wright, Scanning calorimetry in the study of protein behaviour, *Developments in Food Proteins*, Vol. 1 (B. J. F. Hudson, ed.), Applied Science Publ., London, 1986, p. 61.
114. P. L. Privalov, N. N. Khechinashvili, and B. P. Atanaasov, Thermodynamic analysis of thermal transitions in globular proteins. I. Calorimetric study of chymotrypsinogen, ribonuclease and myoglobin, *Biopolymers 10*:1865 (1971).
115. P. Relkin, B. Launay, and L. Eynard, Effect of sodium and calcium addition on thermal denaturation of apo-lactalbumin: A differential scanning calorimetry study, *J. Dairy Sci. 76*: 36 (1993).
116. V. B. Tolstoguzov, Some physico-chemical aspects of protein processing into foodstuffs, *Food Hydrocolloids* 2:339 (1988).
117. J. E. Kinsella, Milk proteins: Physicochemical and functional properties, *CRC Crit. Rev. Food Sci. Nutr. 21*:197 (1984).
118. C. Rha and P. Pradipasena, Viscosity of proteins, *Functional Properties of Food Macromolecules* (J. R. Mitchell and D. A. Ledward, eds.), Elsevier Appl. Sci., New York, 1986, p. 79.
119. C. Rha, Rheology of fluid foods, *Food Technol. 32*(7):77 (1978).
120. M. Paulsson, P. Dejmek, and T. Van Vliet, Rheological properties of heat induced β-lactoglobulin gels, *J. Dairy Sci. 73*:45 (1990).
121. A. P. Stone and D. W. Stanley, Muscle protein gelation at low ionic strength, *Food Res. Int. 27*:155 (1994).
122. V. R. Harwalkar and M. Kalab, Thermal denaturation and aggregation of β-lactoglobulin at pH 2.5. Effect of ionic strength and protein concentration, *Milchwissenschaft 40*:31 (1985).
123. V. R. Harwalkar and M. Kalab, Microstructure of isoelectric precipitates from β-lactoglobulin solution heated at various pH values, *Milchwissenschaft 40*:665 (1985).
124. D. J. McClements, Advances in the application of ultrasound in food analysis and processing, *Trends Food Sci. Technol. 6*:293 (1995).
125. R. Y. Yada, G. Harauz, M. F. Marcone, D. R. Beniac, and F. P. Ottensmeyer, Visions in the mist: The *Zeitgeist* of food protein imaging by electron microscopy, *Trends Food Sci. Technol. 6*:265 (1995).

3

Protein-Stabilized Foams and Emulsions

Srinivasan Damodaran
University of Wisconsin—Madison
Madison, Wisconsin

I. INTRODUCTION

Foods are multiphasic and multicomponent systems composed of proteins, polysaccharides, fats, water, and other minor nutrients and additives. Depending on their relative concentrations and solubility limits and limited thermodynamic compatibility in this complex milieu, the lipids and macromolecules generally exist as colloidally dispersed particles and aggregates. For example, most foods, such as milk, butter, margarine, spreads, salad dressings, frozen desserts, sausages, cakes, and ice cream, are emulsion and foam-type products in which apolar oil or air is dispersed as particles in an aqueous continuous phase containing soluble or dispersed macromolecule such as proteins and polysaccharide. Gels, such as sausages and frankfurters, may be defined as "solidified emulsions" in which the dispersed fat droplets are covered by a proteinaceous membrane in a continuous gel network of colloidal particles (proteins or polysaccharide) [1] in a semi-solid–like aqueous continuous phase. The semi-solid–like state of these meat emulsions is due primarily to the three-dimensional polymer network providing mechanical energy against deformation.

The textural properties of a food are determined by the size, shape, and distribution of these colloidal particles in the food. The creamy texture of ice cream, which is both a foam and an emulsion, is the result of micron-size droplets of fat, ice crystals, and air cells dispersed in an aqueous phase. Similarly, the textural characteristics of cakes depend on the size of air cells dispersed in a continuous polymeric network.

The multiphasic state of foods is mainly due to thermodynamic incompatibility among its major components. Fats, because of their energetically unfavorable interaction with water, tend to separate and exist as a separate phase. Although proteins and polysaccharide are soluble in water, because of their limited thermodynamic compatibility they tend to form protein-rich and polysaccharide-rich regions within the aqueous phase, a phenomenon known as water-in-water emulsion. Because of the tendency of the major food components to separate into distinct phases, especially in liquid-type products, food colloids are inherently unstable. Their stability against phase separation during storage is critically dependent on the type of surfactants present at the interfaces of the various dispersed phases.

A majority of foods are emulsions and foams, which are two-phase systems in which one of the phases (oil or gas or both) is dispersed in an aqueous continuous phase. Because oil/water and air/water interfaces are high-energy interfaces, emulsions and foams collapse as soon as they are created unless an emulsifier or a foaming agent is added to the system. The emulsifier or foaming agent (surfactant), because of its amphiphilic chemical nature (i.e., its affinity to both water and nonpolar phases), adsorbs and orients itself with the lipophilic group towards the nonpolar phase and the hydrophilic group towards the aqueous phase. This molecular ordering of and film formation by a surfactant decreases the interfacial tension; this facilitates continuous formation of new interface as the oil-water or gas-water system is continuously mixed either by homogenization or sparging.

Two types of emulsifiers or foaming agents are being used in foods: low molecular mass surfactants, such as phospholipids (lecithin), mono- and diglycerides, sorbitan monostearate, polyoxyethylene sorbitan monostearate, etc., and high molecular mass surfactants, such as proteins and certain gums. At similar bulk concentrations (w/v), low molecular mass surfactants decrease the surface or interfacial tension to a greater extent than the macromolecular surfactants. This difference is mainly related to differences in orientation and configuration of these surfactants at an interface. In the case of low molecular mass surfactants, the entire molecule adsorbs and instantaneously orients itself so that the hydrophilic head group is immersed in the aqueous phase and the lipophilic hydrocarbon chain is buried in the nonpolar phase (Fig. 1). This partitioning of the entire molecule between the two phases facilitates a maximum reduction of interfacial tension. Unlike low molecular mass surfactants, which have no conformational constraints to adsorption, adsorption of proteins at interfaces is a complex process owing to their complex conformational characteristics. Since there are no clearly defined hydrophilic

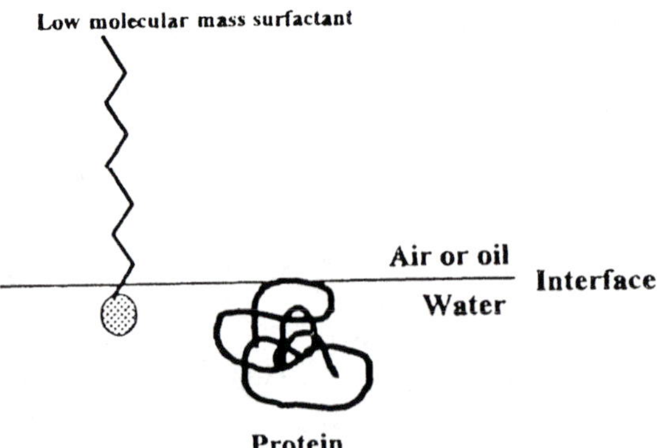

FIGURE 1 Orientation of low molecular weight and macromolecular surfactants at liquid interfaces.

head and hydrophobic tail regions in proteins, and since such groups are spread all over the protein surface, the relative orientation of these groups at the interface is not geometrically as precise as in the case of low molecular mass surfactants. Generally, protein adsorption proceeds through sequential attachment of polypeptide segments or nonpolar patches on the protein surface. However, because of conformational constraints (see Fig. 1) as well as the nature of distribution pattern of hydrophilic and hydrophobic residues in its primary structure, only a fraction of the peptide segments of the protein molecule lies in the interface ("trains"), and a great portion of the protein molecule remains suspended in the aqueous phase in the form of "loops" and "tails." A consequence of this configurational order of proteins at interfaces is that even at saturated monolayer coverage, only a fraction of the liquid "surface" or liquid-liquid "interface" is actually covered by the protein segments and the other regions of the surface or interface remain at high free energy levels. Thus, the surface tension of most protein solutions remains as high as 50 mN/m, whereas those of low molecular mass surfactants is in the range of 10–30 mN/m. In other words, proteins are generally less surface active than low molecular mass surfactants.

Although low molecular mass surfactants are more effective than proteins in reducing the interfacial tension, foams and emulsions formed by such surfactants are mostly unstable. This is because proteins, in addition to lowering interfacial tension, can form a continuous viscoelastic membrane–like film around oil droplets or air cells via noncovalent intermolecular interactions and via covalent disulfide cross-linking, whereas the low molecular mass surfactants cannot form such a viscoelastic film. Thus, in foods, which contain both low molecular and macromolecular surfactants, the stability of colloidally dispersed phases is primarily dependent on protein films adsorbed at the interfaces. However, practical observations indicate that all proteins are not equally surface active, even though all are amphiphilic and a majority of them contain similar percentages of polar and nonpolar amino acid residues. The wide differences in the surface activities of various proteins therefore must be related to differences in their conformation and the susceptibility of those conformations to unfolding at interfaces. Intuitively, the molecular factors that influence surface activity of proteins must be related to flexibility, conformational stability at interfaces, rapid adaptability of the conformation to changes in its environment, and the distribution pattern of hydrophilic and hydrophobic residues in its primary structure as well as on its folded surface. All of these molecular factors will collectively influence the surface activity of proteins [2,3].

Apart from the intrinsic molecular factors, the surface activity of a protein in a complex food system will be dictated by several other extrinsic factors such as pH, ionic strength, and temperature; apart from these environmental factors, interaction of the protein with other food components such as lipids, proteins, and polysaccharide will influence its surface activity in a food milieu. Food proteins are generally mixtures of several protein components. Thus, the surface activity of a protein source in foams and emulsions is innately dependent on the composition and relative surface activity of each protein component in the mixture. For example, the superior foaming and emulsifying properties of egg white, which contains several proteins, may be attributable to complex interactions among the protein constituents at the interface. Another phenomenon that could influence the surface activity of a protein in a food milieu is competitive adsorption of other proteins/peptides and other surface-active polar lipids at interfaces. In this respect, the surface activity of a protein will depend on its rate of adsorption, the rapidity of its conformational changes at the interface, the number of peptide segments bound to the interface, the irreversibility of its adsorption, and its resistance to displacement from the interface by other surface-active proteins/peptides or low molecular mass surfactants.

The physical principles governing the formation and stability of food colloids, e.g., foams and emulsions, are complex, especially if protein macromolecules are involved as surfactants. In this chapter, first the mechanism of adsorption and formation of protein films at interfaces will be examined, followed by analyses of the molecular factors affecting the stability of protein-stabilized foams and emulsions.

II. KINETICS OF ADSORPTION

A. Change in Surface Concentration

When a two-phase system (i.e., air-water or oil-water system) containing an aqueous protein solution is left to stand unperturbed, proteins from the aqueous phase migrate spontaneously to the interface. At equilibrium, the concentration of the protein in the interfacial region is always in excess of that in the bulk aqueous phase. This spontaneous migration towards, and accumulation at, liquid/liquid and liquid/gas interfaces indicates that the free energy of proteins is lower at interfaces than at either of the bulk phases. The free energy loss of proteins at interfaces is principally a result of two events: (a) removal of low-entropy water clathrates from the hydrophobic regions of protein surface to their high-entropy state with the result of transfer of the hydrophobic regions to the nonpolar phase, and (b) reorganization of the protein structure from a low-entropy native state to an high-entropy unfolded state. In other words, the adsorption of proteins at interfaces is purely an entropy-driven process, involving changes in the entropy of both water and protein.

The rate of accumulation of protein at an interface is generally assumed to be diffusion controlled [4] and follows the relation:

$$d\Gamma/dt = C_o(D/\pi t)^{1/2} \tag{1}$$

where Γ is the surface concentration, C_o is the bulk concentration, D is the diffusion coefficient, and t is the adsorption time. Integration of Eq. (1) gives:

$$\Gamma = 2C_o(Dt/\pi)^{1/2} \tag{2}$$

According to Eq. (2), a plot of Γ versus $t^{1/2}$ is expected to be a straight line passing through the origin, and the diffusion coefficient is calculated from the initial slope. However, several investigations have shown that the $\Gamma - t^{1/2}$ plot for many proteins is linear only at extremely low Γ values and the curve becomes nonlinear at higher Γ values [5–7]. Even in cases where the $\Gamma - t^{1/2}$ plots are linear, the diffusion coefficient values determined from the slopes are significantly lower than the values from solution studies [5,8]. The nonlinearity in the $\Gamma - t^{1/2}$ curve usually occurs when Γ is about 0.5–1 mg m^{-2} and the surface pressure is above 0.1 mN/m [7]. Ward and Tordai [4] proposed that the nonlinearity of the $\Gamma - t^{1/2}$ plot might be due to development of an energy barrier to adsorption once some protein molecules are adsorbed at the interface. The origin of this energy barrier to adsorption is debatable; MacRitchie and Alexaner [7] suggested that it may arise from the energy required to clear an area of ΔA at the interface against a surface pressure of Π in order to adsorb a protein molecule. That is, once a protein film with a surface pressure value of Π is formed at the interface, the subsequent rate of adsorption will be:

$$\frac{d\Gamma}{dt} = KC_0 \exp(-\Pi\Delta A/kT) \tag{3}$$

where K is the rate constant, k is the Boltzman constant, and T is the temperature. According to Eq. (3), a plot of $\ln(d\Gamma/dt)$ versus Π is linear and the value of ΔA can be calculated from the slope. Analyses of protein adsorption data for the air/water interface have shown that the ΔA value for most globular proteins is in the range of 50–150 Å^2, which is much smaller than the cross-sectional area of most proteins. Further, the ΔA values are apparently independent of the three-dimensional structure or other physicochemical properties of the proteins [9]. Since the cross-sectional area of an amino acid residue is about 15 Å^2, it is reasoned that a protein can adsorb to the air/water interface containing an unsaturated protein monolayer film by creating a ~100 Å^2 hole sufficient to present a hydrophobic patch consisting of about six hydrophobic residues to the interface.

The ability of Eq. (3) to explain the energy barrier to protein adsorption at high surface pressures has been questioned by several investigators [5,8,10]. A major argument against this theory is that, since the cross-sectional area of proteins is much larger than ~100 Å^2 (e.g., 900 and 10,000 Å^2 for lysozyme and bovine serum albumin, respectively), it would be physically impossible for a protein to bind to the interface through a 100 Å^2 hole against steric repulsions involving other protein molecules that are already adsorbed to the interface. In order for this to occur, the protein has to be extremely flexible and/or should be able to undergo extensive dynamic changes in its conformation as it approaches the interface from the subsurface, both of which are very unlikely in the case of globular proteins. Another fact that goes against the surface pressure barrier theory is that the $\Gamma - t^{1/2}$ plot for the adsorption of β-casein is fairly linear up to $\Pi = 10$ mN/m [8]. The exhibition of a $t^{1/2}$ kinetics despite a large build-up of surface pressure essentially suggests that the surface pressure does not function as an energy barrier for the adsorption of β-casein [10]. These inconsistencies apparently indicate that the contribution of surface pressure to the energy barrier should be minor, and that other forces, such as electrostatic and conformational constraints, of the protein may play a major role in the kinetics of protein adsorption at an interface.

The basic assumption involved in the diffusion-controlled adsorption theory is that when a fresh interface is created, the protein molecules at the subsurface (i.e., few molecular diameters from the interface) adsorb instantaneously to the interface. This sets up a concentration gradient between the bulk phase and the subsurface, which acts as the driving force for mass transport from the bulk phase to the subsurface and then on to the interface. Implicit in this theory is that, once a protein molecule reaches the subsurface, it invariably adsorbs to the interface. That is, each and every collision of the protein molecule with the interface leads to adsorption. The validity of this assumption is questionable. In order for the probability of every collision leading to adsorption to be unity, the entire protein surface must be nonpolar. However, this is not the case with globular proteins. Therefore, to a first approximation, the probability of a collision leading to adsorption should be a function of the hydrophobicity/hydrophilicity ratio of the protein surface. If the protein's surface is extremely hydrophilic, it may not adsorb to an interface. If the protein surface contains a few well-defined hydrophobic patches, then adsorption may take place when this hydrophobic patch collides with the interface. This is shown schematically in Figure 2. Thus, even in the absence of an energy barrier, the rate of adsorption of a protein at an interface should follow the relation [11]:

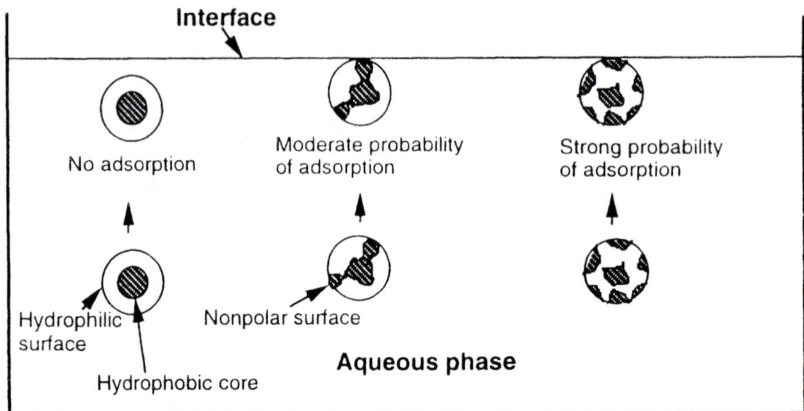

FIGURE 2 Dependence of probability of adsorption of a protein on the number of hydrophobic patches on its surface. (From Ref. 2.)

$$\Gamma = 2C_o (Dt/\pi)^{1/2} P_a \tag{4}$$

where P_a is the statistical probability of adsorption related to distribution of hydrophilic and hydrophobic patches on the protein surface. It should be emphasized, however, that single hydrophobic residues randomly distributed on the protein surface do not constitute a hydrophobic patch, nor do they possess sufficient interaction energy to strongly anchor the protein at an interface. Even though about 40% of a typical globular protein's overall solvent accessible surface is occupied by nonpolar amino acid residues [12], they will not enhance protein adsorption unless they exist as segregated regions or patches.

These considerations suggest that, at the molecular level, the mechanism of adsorption of proteins at interfaces is more complex than has been previously assumed. It is primarily governed by the dynamics of interaction of the interfacial force field with the molecular forces of the protein molecule as the protein approaches the interface. To understand the nature and influence of this interaction on the kinetics of adsorption, Damodaran et al. [8,13–16] studied adsorption behaviors of several structurally very different proteins, such as lysozyme, bovine serum albumin (BSA), β-casein, phosvitin, etc., using a surface radiotracer method. The square-root-of-time kinetics of adsorption of these proteins at the air/water interface under identical solution conditions are shown in Figure 3. It is apparent that these structurally and physicochemically very different proteins display remarkably different adsorption behaviors at the air/water interface. The most interesting behavior is that of lysozyme: during the first 60 minutes after forming a fresh air/solution interface, there is a gradual but significant decrease in the surface concentration (negative adsorption), implying that, rather than the lysozyme molecules at the subsurface instantaneously adsorbing to the interface as suggested by the diffusion-controlled adsorption theory, the lysozyme molecules that were originally at the surface or in the close vicinity of the surface migrated away from the interface and into the subsurface. After this initial desorption phase, the negative surface excess remained constant for a period of time (lag phase), followed by a rapid positive adsorption to the

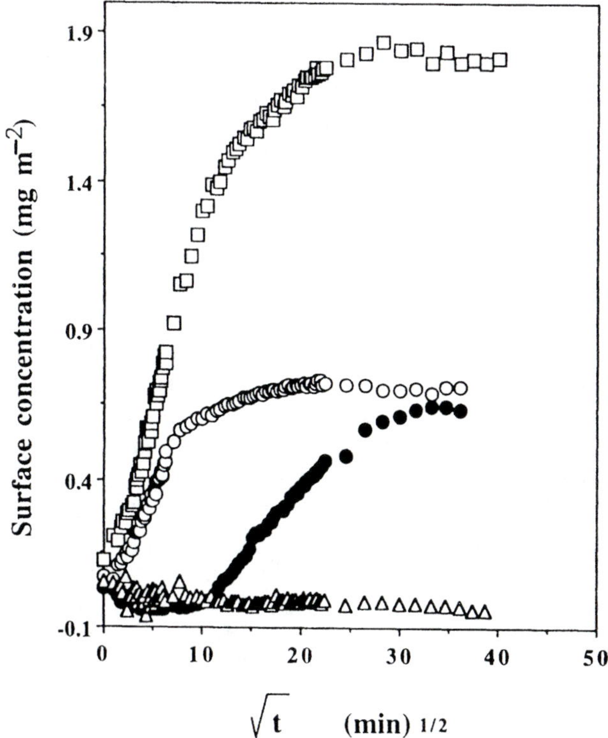

FIGURE 3 Variation of surface concentration with time during adsorption of β-casein (□), bovine serum albumin (○), lysozyme (●), and phosvitin (△) at the air/water interface from a 1.5 μg/ml protein solution at pH 7. (From Refs. 8, 15, and 24.)

interface. The diffusion coefficient value of lysozyme, calculated from the initial positive slope of its $\Gamma - t^{1/2}$ curve according to Eq. (3), is about one order of magnitude lower than its diffusion coefficient values from solution studies, indicating that it experiences an energy barrier for adsorption at the air/water interface. The equilibrium surface concentration of lysozyme only reaches to about 0.6 mg m^{-2} after about 900 minutes of adsorption.

Unlike lysozyme, β-casein and BSA show neither initial desorption nor a long lag period in their adsorption to the air/water interface. In both cases, positive adsorption commences soon after a fresh air/water interface is created. The diffusion coefficient value of β-casein, determined from the initial slopes of $\Gamma - t^{1/2}$ curves according to Eq. (3), is very close to its diffusion coefficient in solution, suggesting that every collision of β-casein with the interface indeed leads to adsorption. On the other hand, the diffusion coefficient of BSA is about two times lower than its diffusion coefficient in solution, indicating that there is an energy barrier to its adsorption; however, its energy barrier to adsorption appears to be lower than that of lysozyme. The equilibrium surface concentration of β-casein reaches a value of about 1.8 mg m^{-2}, compared to about 0.7 mg m^{-2} for BSA (Fig. 3) [16].

In contrast to the above three proteins, phosvitin, which is a phosphoglycoprotein, does not bind to the air/water interface at pH 7.0 (Fig. 3). The inability to adsorb to the air/water interface is due to its highly polyanionic character: Phosvitin has a molecular weight of about 35,000 and contains 217 amino acid residues, of which 123 residues are serine residues [17]. Of the 123 serine residues, 118 are phosphorylated [18]. The amino acid sequence (Fig. 4) indicates that only about 10% of the amino acid residues are nonpolar and the middle region of the sequence contains long stretches of phosphoserine residues. The net charge of phosvitin, calculated from its amino acid sequence, it about -179 at pH 7.0. Because of this high net charge, the physicochemical properties of phosvitin resemble that of a typical polyanionic polymer. This high net charge at pH 7.0 may also be the primary reason for its inability to adsorb to the air/water interface, because at pH 2.0, where its net charge is -15, it does adsorb to the air/water interface [15].

The remarkable differences in the kinetics of adsorption of the above four proteins under identical bulk concentration and other solution conditions are more complex than previously assumed. Neither simple diffusion-controlled adsorption nor surface pressure barrier-controlled adsorption theories can adequately explain the mechanism of adsorption of these proteins to the air/water interface. All these theories implicitly assume that the potential energy of a protein is *always* lower at the interface than at either the subsurface or the bulk phase. Such a simple assumption may not be true for all proteins. Phenomenologically, the tendency of a protein to adsorb to an interface, as it approaches the interface from the subsurface, should be related to its chemical potential, μ, at a distance ξ from the interface. The driving force for mass transfer either from the interface to the subsurface or from the subsurface to the interface must fundamentally be related to the chemical potential gradient $\delta\mu/\delta\xi$. This chemical potential gradient must include

1									10										20
Ala	Glu	Phe	Gly	Thr	Glu	Pro	Asp	Ala	Lys	Thr	Ser	Ser	Ser	Ser	Ser	Ser	Ala	Ser	Ser
21									30										40
Thr	Ala	Thr	Ser	Ser	Ser	Ser	Ser	Ser	Ala	Ser	Ser	Pro	Asn	Arg	Lys	Lys	Pro	Met	Asp
41									50										60
Glu	Glu	Glu	Asn	Asp	Gln	Val	Lys	Gln	Ala	Arg	Asn	Lys	Asp	Ala	Ser	Ser	Ser	Ser	Arg
61									70										80
Ser	Ser	Lys	Ser	Ser	Asn	Ser	Ser	Lys	Arg	Ser	Ser	Ser	Lys	Ser	Ser	Asn	Ser	Ser	Lys
81									90										100
Arg	Ser	Ser	Ser	Ser	Ser	Ser	Ser	Ser	Ser	Ser	Ser	Ser	Arg	Ser	Ser	Ser	Ser	Ser	Ser
101									110										120
Ser	Ser	Ser	Ser	Asn	Ser	Lys	Ser	Ser	Ser	Ser	Ser	Ser	Lys	Ser	Ser	Ser	Ser	Ser	Ser
121									130										140
Arg	Ser	Arg	Ser	Ser	Ser	Lys	Ser	Ser	Ser	Ser	Ser	Ser	Ser	Ser	Ser	Ser	Ser	Ser	Ser
141									150										160
Ser	Lys	Ser	Ser	Ser	Ser	Arg	Ser	Ser	Ser	Ser	Ser	Ser	Lys	Ser	Ser	Ser	His	His	Ser
161									170										180
His	Ser	His	His	Ser	Gly	His	Leu	Asn	Gly	Ser	Ser	Ser	Ser	Ser	Ser	Ser	Ser	Arg	Ser
181									190										200
Val	Ser	His	His	Ser	His	Glu	His	His	Ser	Gly	His	Leu	Glu	Asp	Asp	Ser	Ser	Ser	Ser
201									210										
Ser	Ser	Ser	Ser	Val	Leu	Ser	Lys	Ile	Trp	Gly	Arg	His	Glu	Ile	Tyr	Gln			

FIGURE 4 Amino acid sequence of hen egg phosvitin. (From Ref. 17.)

the energetics of interaction of the interfacial force field with the various molecular forces of the approaching protein molecules, not necessarily only the concentration gradient.

The chemical potential of an ideal solution is:

$$\mu_{ideal} = \mu^0 + RT \ln a \tag{5}$$

where μ^0 is the chemical potential of the solute in the solution at the standard state, a is the activity of the solute, R is the gas constant, and T is the temperature. Under ideal conditions, i.e., in the absence of an external force, mass transport within a solution is only dependent on concentration gradient, that is:

$$J_i = -L_i(\mu_{ideal}/\delta\xi)_T \tag{6}$$

where J_i is the flux and L_i is a phenomenological coefficient, which is a function of concentration, diffusion coefficient, and temperature. However, mass transport of proteins from the subsurface to an interface is far from ideal. If the subsurface is under the influence of the interfacial force field, then the chemical potential of molecules in the subsurface region cannot be same as that in the bulk phase. Also, the chemical potential of the molecules in this region will vary as a function of their distance from the interface. The change in the chemical potential of the protein molecule must arise from interaction of the interfacial force field with the electrostatic, hydrophobic, hydration, and conformational (entropic) forces of the protein molecule. If this is the case, then the mass transport from the subsurface to the interface should follow the linear relationship:

$$J_i = L_i\,(\delta\mu_{ideal}/\delta\xi)_T + L_{ele}\,(\delta\mu_{ele}/\delta\xi)_T + L_{H\phi}\,(\delta\mu_{H\phi}/\delta\mu)_T + L_{hyd}\,(\delta\mu_{hyd}/\delta\xi)_T + L_{conf}\,(\delta\mu_{conf}/\delta\xi)_T \tag{7}$$

The rate of mass transfer of the protein from the subsurface to the interface will depend on the sign and magnitude of the sum of the various chemical potentials. If the sum of the chemical potential gradients is negative at a given distance from the interface, then the protein will flow towards the interface; if positive, then the molecules should flow away from the vicinity of the interface until they reach a distance where the sum of the chemical potential gradients is zero. Among these chemical potential gradients, the electrostatic and hydration chemical potential gradients are expected to be positive, which should force the protein to flow away from the interface; the concentration gradient and hydrophobic and conformational (entropic) chemical potential gradients are expected to be negative, which should favor flow of the protein towards the interface.

The potential energy functions for the conformational and hydration forces are difficult to determine, whereas explicit functions for hydrophobic and electrostatic chemical potentials are available [19–21]. The hydrophobic chemical potential of a particle at a distance ξ from a planar nonpolar interface is [20]:

$$\mu_{H\phi} = -84R \exp(\xi/\xi_0)\ \text{kJ/mol} \tag{8}$$

where R is the radius of curvature of the particle and ξ_0 is the decay length (1 nm). If e is the net charge of a particle and ε_0 and ε are the dielectric constants of the aqueous and the nonpolar phases, then, according the electrostatic theory, as the particle moves towards the interface, an image charge, $e' = e(\varepsilon_0 - \varepsilon)/(\varepsilon_0 + \varepsilon)$, would appear in the nonpolar phase. The electrostatic repulsive potential energy between the real and image charges would be

$$\mu_{ele} = \frac{ee'}{2\xi\epsilon_0} = \frac{\epsilon^2}{2\xi\epsilon_0}\frac{(\epsilon_0 - \epsilon)}{(\epsilon_0 + \epsilon)} \tag{9}$$

where ξ is the distance of the charged particle from the interface. In the absence of any other attractive force, the particle would move toward the interface by Brownian motion only up to a distance where the electrostatic repulsive energy is numerically equal to the thermal energy, kT, of the particle. At distances closer than this critical distance from the interface, the electrostatic chemical potential of the particle would be greater than that in the bulk phase. Conversely, if the particle initially happens to be at the interface, because of its high electrochemical potential, it would desorb into the subsurface.

Another effect of electrostatic forces on protein adsorption is that once a certain amount of protein is adsorbed to the interface, the charged protein film will create an additional electrical potential energy barrier equivalent to:

$$\mu_{p,\ ele} = \int e\ d\psi = e\psi \tag{10}$$

The dynamics of adsorption of β-casein, BSA, lysozyme, and phosvitin at the air/water interface were explained in terms of interaction of their molecular forces with the interfacial force field [8,14,15]. In the cases of β-casein and BSA, their net chemical potential at the subsurface seems to be negative; thus, when a fresh air/solution interface is created, the molecules at the subsurface spontaneously flow toward the interface. However, in the case of lysozyme, which is a highly rigid, hydrophilic, and positively charged (net charge is +9 at pH 7.0) globular protein with no discernible hydrophobic patch on its surface, the sum of the positive hydration and electrostatic chemical potentials is larger than the sum of the negative hydrophobic and conformational chemical potentials; thus, the net positive chemical potential drives the lysozyme molecules at the interface or in the vicinity of the interface toward the subsurface phase. The initial desorption observed in the $\Gamma - t^{1/2}$ curve (Fig. 3) is the result of this process. In the subsurface, during the lag phase, lysozyme undergoes slow conformational change, which exposes previously buried hydrophobic groups; the increase (i.e., negative) in hydrophobic and conformational (entropic) chemical potentials alters the chemical potential balance at the subsurface in favor of adsorption to the interface. On the other hand, in the case of phosvitin, which is a polyanionic polymer in a random coil–like state, the electrochemical potential at pH 7.0 overrides all other chemical potentials and thus inhibits its adsorption to the air/water interface.

Thus, adsorption of proteins is fundamentally governed by the chemical potential gradient resulting from the interaction of the interfacial force field (the operator) with the various molecular forces of proteins. Theoretically, this basic mechanism should be true for both quiescent and dynamic flow systems (as in bubbling or stirring) as long as the conformation of the protein is not altered under turbulent flow conditions.

B. Change in Surface Pressure

The most critical requirement for the formation of foams and emulsions during whipping or homogenization is rapid reduction of the free energy (interfacial tension) of the newly created interface. Although rapid adsorption of protein is necessary to facilitate this reduction in surface tension, it is not a rate-limiting step under dynamic flow conditions. What is rate-limiting is the rapidity with which the protein undergoes conformational rearrangement and reorientation at the interface and the rate at which it decreases the interfacial tension.

The rate of change of surface pressure, Π (= $\gamma_0 - \gamma$, where γ_0 and γ are surface tensions in the absence and in the presence of adsorbed surfactant, respectively), varies greatly among proteins. Figure 5 illustrates the relative differences in the capabilities of β-casein, BSA, lysozyme, and phosvitin with regard to reducing the surface tension during adsorption at the air/water interface. Among these structurally very different proteins, the rate of increase of surface pressure (i.e., the rate of decrease of surface tension) parallels the rate of increase of surface concentration. Furthermore, both surface pressure and surface concentration reach equilibrium values simultaneously, suggesting that β-casein was able to completely unfold and rearrange/reorient and reduce the interfacial energy as soon as it arrives at the interface. However, this is not the case with BSA. In the initial stages of adsorption, the rate of increase of surface pressure almost parallels that of the surface concentration, indicating that it unfolds and rearranges its polypeptide segments when it arrives at the interface. However, it should be noted that the surface pressure did not reach a steady-state value and continues to increase even after the surface concentration has reached a steady-state value (see Figs. 3 and 5). This

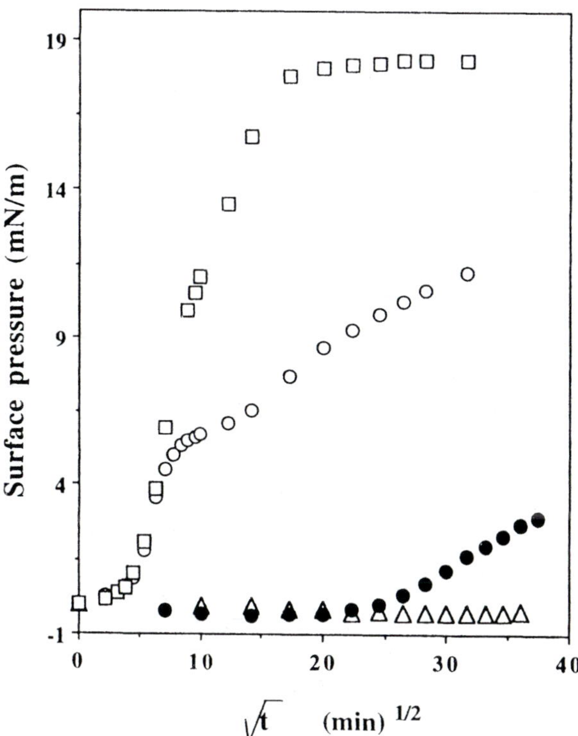

FIGURE 5 Variation of surface pressure with time during adsorption of β-casein (□), bovine serum albumin (○), lysozyme (●), and phosvitin (△) at the air/water interface from a 1.5 μg/ml protein solution at pH 7. (From Refs. 8, 15, and 24.)

suggests that the rates of unfolding and rearrangement processes for BSA are not as instantaneous as in the case of β-casein, and these processes continue even after a saturated monolayer film has been formed at the interface. The constraints for rapid conformational rearrangement at the interface are mainly due to its compact globular structure, which has 17 disulfide bonds and about 60% α-helix content.

The change in surface pressure of the lysozyme solution is far more slower than that of BSA (Fig. 5); the surface pressure did not increase for over 600 minutes, although during this time period the surface concentration reached a value of about 0.5 mg m^{-2} (Fig. 3). Furthermore, similar to BSA, the surface pressure of lysozyme solution did not reach a steady-state value even after 1500 minutes of adsorption, even though the surface concentration reached a steady-state value at about 1200 minutes, indicating that its rate of conformational change at the interface is very slow. It is notable that even though the surface concentrations of BSA and lysozyme at 1200 minutes of adsorption were very similar (about 0.7 mg m^{-2}), the surface pressure values were very different, that is, about 12 and 2 mN/m, respectively. This difference in the surface pressures of these two globular protein films reflects the differences in their ability to unfold and reorient at the interface and reduce the interfacial tension. In this regard, the conformation of lysozyme is much more rigid than the other proteins, and the relative rigidity follows the order lysozyme > BSA > β-casein.

The argument that the effectiveness of a protein film in reducing the interfacial energy is dependent on its conformational flexibility at the interface is further supported by the adsorption behavior of denatured and reduced lysozyme (Fig. 6) [8]. Unlike the native lysozyme, the fully denatured and S-S groups reduced and blocked lysozyme rapidly adsorbs as soon as a fresh air/water interface is created; it does not exhibit a lag time. At steady-state, the extent of adsorption of the denatured lysozyme is significantly higher than that of the native lysozyme. The rate of increase of surface pressure parallels that of surface concentration; both surface concentration and surface pressure reach steady-state values simultaneously, indicating that the denatured and reduced lysozyme rapidly attains an equilibrium unfolded structure as soon as it arrives at the interface. The steady-state surface pressure of the denatured and reduced lysozyme is about 18 mN/m, which is similar to that of β-casein. This suggests that, unlike the native lysozyme, the denatured lysozyme rapidly rearranges and reorients its hydrophobic and hydrophilic segments and thus effectively reduces the interfacial energy. Studies with structural intermediates of BSA also have shown that, under similar solution conditions, the rate of increase of surface pressure increased with increase of the extent of unfolded state of BSA in the bulk phase, suggesting that molecular flexibility at the interface is quintessential for a protein to effectively reduce the free energy of the interface [13].

In the case of phosvitin, there is no development of surface pressure with time (Fig. 5); this is not surprising, because phosvitin does not adsorb to the air/water interface at pH 7.0 owing to its high net negative charge at pH 7.0 (Fig. 3). However, even under conditions that promote adsorption of phosvitin to the air/water interface, no increase in surface pressure occurs [15]. This is illustrated in Figure 7, which shows time-dependent changes in surface concentration and surface pressure of a dilute solution of phosvitin (1.5 μg/ml) at pH 2.0 [15]. Although phosvitin is able to adsorb to an extent of about 1.25 mg m^{-2} at the air/water interface at pH 2.0, no increase in surface pressure occurs as a result of adsorption. In fact, the surface tension actually increases slightly as a result of adsorption. It should be noted that at comparable surface concentrations, other proteins generally cause significant reductions in surface tension. The

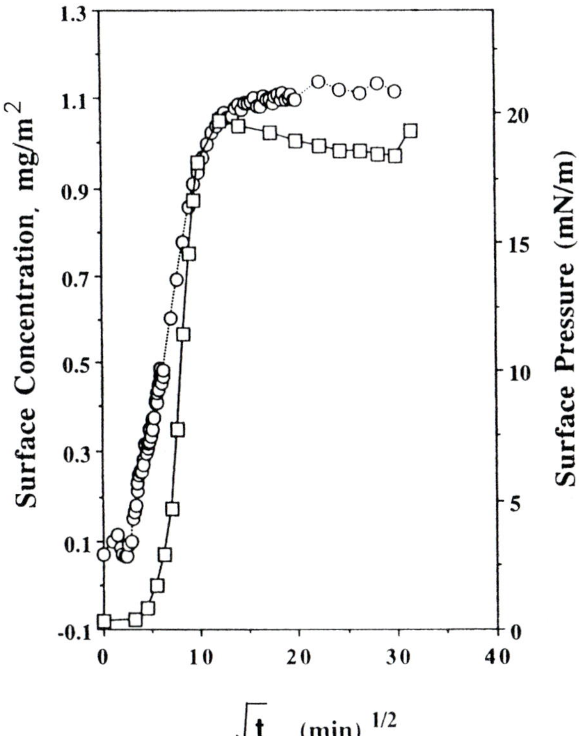

Figure 6 Variation of surface concentration (○) and surface pressure (□) with time during adsorption of reduced and heat- and urea-denatured egg white lysozyme at the air/water interface from 1.5 μg/ml protein solution. (From Ref. 8.)

slight increase of surface tension in spite of protein accumulation at the interface is incongruous and apparently violates the Gibbs adsorption equation:

$$\Gamma_2 = -(1/RT)\,(d\gamma/d\ln C) \tag{11}$$

where γ is the surface tension and C is the bulk concentration. This anomaly was explained based on the unique configuration of phosvitin at pH 2.0 [15]. At pH 2.0, the net charge of phosvitin is about −15 and its conformation assumes the β-sheet form as opposed to the random coil form at pH 7.0. Examination of its primary sequence shows that the only hydrophobic region in this protein is the peptide segment 205–216 at the C-terminus of the protein; the rest of the molecule has no discernible hydrophobic stretches. Thus, segment 205–216 ought to be the one that binds to the air/water interface. Since phosvitin seems to form about 10% α-helix below pH 2, and since only peptide segments with the binary code of -P-N-P-P-N-N-P- (where P and N are polar and nonpolar amino acid residues, respectively) can form α-helix, it can be speculated that segment 205–216 assumes α-helix structure at pH 2.0. Construction of an α-helical wheel of the segment 205–216 shows an amphiphilic helix with the hydrophobic residues Val, Tyr, Ile, His, Leu, and Trp on one half of the helix surface and the hydrophilic residues Glu, Ser, Gly, Ser, Lys, and Arg on the other half of the helix surface (Fig. 8).

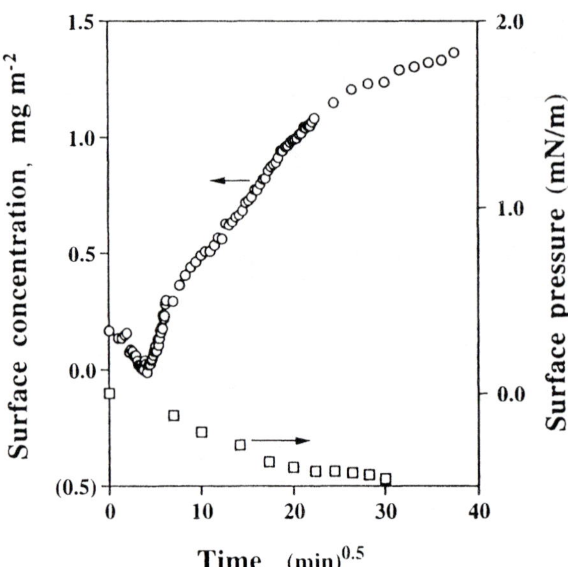

FIGURE 7 Variation of surface concentration (○) and surface pressure (□) with time during adsorption of phosvitin at the air/water interface from a 1.5 μg/ml protein solution at pH 2. (From Ref. 15.)

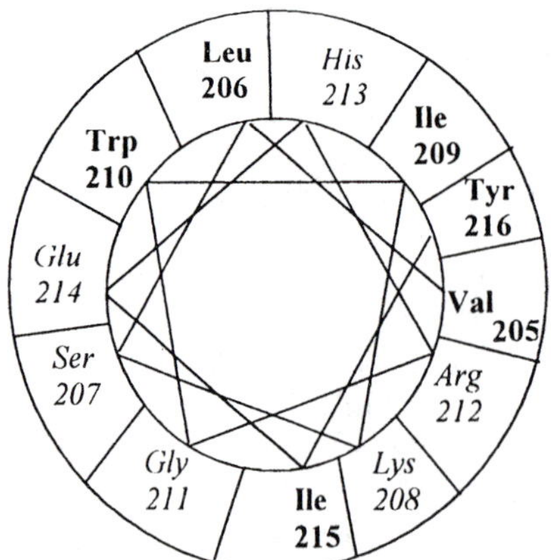

FIGURE 8 Cross-sectional view of the proposed helical structure of residues 205–216 of phosvitin. The amino acid residues in bold letters are hydrophobic and those in italics are hydrophilic. Note that the hydrophobic residues are segregated on one half of the helix surface and the hydrophilic residues on the other.

Since there are no other hydrophobic regions in the protein, the hydrophobic surface of this amphiphilic helix ought to be the one that attaches phosvitin to the air/water interface. The rest of the protein is suspended onto the subsurface, which, being extremely hydrophilic, exerts a pull on the surface water molecules, causing a slight increase in surface tension (about 0.5 erg cm^{-2}). What this means is that adsorption of about 1.2 mg m^{-2} phosvitin at the air/water interface increases the free energy of the interface by about 0.5 erg cm^{-2}. According to the Gibbs adsorption equation, this cannot occur. However, Damodaran and Xu [15] argued that the only thermodynamic requirement for adsorption is a reduction of free energy of the *system*, which includes the interface, the bulk phase, and the solute. Whether the reduction in free energy of the system occurs from a loss of free energy of the interface or the adsorbed solute or both is irrelevant. If adsorption occurs in spite of an increase in the free energy of the interface, then it must occur because of a larger reduction in the free energy of the adsorbed phosvitin. In fact, at 1.25 mg m^{-2} surface concentration, transfer of the hydrophobic surface of the α-helix from water to the air phase decreases the free energy of the system by -1.89 erg cm^{-2}, which is significantly higher than the increase of free energy of the interface ($\sim$0.5 erg cm^{-2}). Thus, although the adsorption behavior of phosvitin at the air-water interface apparently seems to violate the Gibbs adsorption equation, it does not violate the basic thermodynamic principle.

The remarkable differences in the $\Pi - t^{1/2}$ curves of the above four proteins (Fig. 5) clearly show that the rate of interfacial denaturation, reorientation, and the final configuration of proteins at the interface play a major role in their abilities to reduce the interfacial tension. Another important characteristic of the $\Pi - t^{1/2}$ curves is that the rate of surface pressure development generally lags behind the rate of accumulation of the proteins at the interface (see Figs. 3 and 5). For example, even for the highly flexible β-casein, a measurable increase in surface pressure occurs only at surface concentration of >0.4 mg m^{-2}. This also seems to be true for BSA and lysozyme. The surface pressure of protein film at any given surface concentration is sum total of three forces [22]:

$$\Pi_{a/w} = \Pi_{kin} + \Pi_{ele} + \Pi_{coh} \tag{12}$$

where Π_{kin}, Π_{ele}, and Π_{coh} are the contributions from kinetic, electrostatic, and cohesive forces, respectively. Since proteins are large, the contribution of Π_{kin} to the total surface pressure is negligible. In general, Π_{coh} is the major contributor to the surface pressure of protein films at interfaces [23]. It appears that below 0.4 mg m^{-2} surface concentration, most globular proteins, including β-casein, do not form a continuous cohesive film at the air/water interface. The saturated monolayer coverages at the air/water interface for β-casein, BSA, and lysozyme are 1.85, 1.0, and 0.7 mg m^{-2}, respectively [16,24]. Thus, the minimum surface coverages needed to develop measurable surface pressure for β-casein, BSA, and lysozyme films seem to be about 22, 40, and 70% of saturated monolayer coverage, respectively. What this means is that because β-casein is a flexible protein, even at 22% of saturated monolayer coverage it assumes an expanded configuration at the interface, interacts with adjacent protein molecules, and thus develops surface pressure. In contrast, since both BSA and lysozyme are rigid globular proteins with intramolecular disulfide bonds, they do not expand as well as β-casein, and thus require a higher percentage of monolayer coverage in order to interact with adjacent protein molecules and develop surface pressure. The percentages given above are truly a measure of the expandability of these proteins at the air/water interface; in this respect the expandability of these proteins follows the order β-casein > BSA > lysozyme.

C. Competitive Adsorption

Food proteins are generally mixtures of several protein components. Thus, the foaming and emulsifying properties of commercial food proteins, such as egg white, soy protein isolate, caseins, whey proteins, etc., are dependent on relative rates of binding and affinity of the protein components. Changes in composition during protein isolation or intentional manipulation of the composition of a protein mixture may alter the functional properties of the protein. Therefore, knowledge of the molecular factors that affect competitive adsorption of proteins at interfaces may be very useful in preparing protein ingredients that exhibit optimal functional properties in food systems.

To elucidate the influence of one protein on the adsorption of another protein in binary protein mixtures, Damodaran et al. [16,24–26] studied the kinetics of competitive adsorption of proteins from four binary protein systems—β-casein–lysozyme, lysozyme-BSA, BSA–β-casein, and α_{s1}-casein–β-casein—at the air/water interface. The rationale for selecting these binary systems was as follows: β-casein and α_{s1}-casein represent a random coil–type hydrophobic protein with a net negative charge, lysozyme represents a highly rigid hydrophilic globular protein with a net positive charge, and BSA represents a negatively charged protein with a molecular flexibility somewhere between those of β-casein and lysozyme. Thus, the β-casein–lysozyme pair would represent a random/globular and negatively/positively charged protein binary system; the lysozyme-BSA pair would represent a globular/globular and positively/negatively charged protein binary system; the BSA–β-casein would present a globular/random and negatively/negatively charged protein binary system; and the α_{s1}-casein–β-casein would represent a random/random and negatively/negatively charged protein binary system. Thus, a fundamental understanding of the adsorption behavior of these proteins in binary systems should provide the roles of charge-charge interactions, structural flexibility/rigidity, and hydrophilicity/hydrophobicity factors in competitive adsorption of proteins at interfaces.

A major conclusion of these studies is that, in the case of binary protein systems involving random coil/globular and globular/globular proteins, the competitive adsorption did not follow a Langmuir-type competitive adsorption mechanism, which states that the interfacial concentrations of two proteins A and B at any bulk protein ratio should be:

$$\Gamma_A = \frac{K_A C_A}{(1 + K_A a_A C_A + K_B a_B C_B)} \tag{13}$$

and:

$$\Gamma_B = \frac{K_B C_B}{(1 + K_A a_A C_A + K_B a_B C_B)} \tag{14}$$

where Γ_A and Γ_B are the surface concentrations of A and B, respectively; K_A and K_B are equilibrium constants; a_A and a_B are the average area occupied per molecule of A and B, respectively, at monolayer coverage in single-protein systems; and C_A and C_B are concentrations of A and B in the bulk phase at equilibrium. This Langmuir-type adsorption mechanism for a binary system is based on the assumption that the surface concentrations of A and B at equilibrium are related to their relative binding affinities to the interface, i.e., the composition of the binary film at the air/water interface at equilibrium is thermodynamically controlled. However, studies on β-casein–lysozyme, lysozyme-

BSA, and BSA–β-casein binary systems clearly showed this not to be the case [16,24,25]. It was shown that the interfacial composition of the mixed-protein film was primarily determined by the rate of arrival of each protein at the interface and the molecular area available at the interface at the time of arrival, i.e., the interfacial protein composition was kinetically controlled. The protein that arrived first at the interface adsorbed first and could not be displayed by the late-arriving protein component, even when the affinity of the latter to the interface was greater than that of the former in single-protein systems.

An example of the above phenomenon in the case of BSA-lysozyme binary system is shown in Figure 9. In single-protein systems, adsorption of lysozyme begins after about 110 minutes and reaches an apparent equilibrium surface concentration of about 0.67 mg m^{-2}, whereas BSA begins to adsorb immediately after creation of a fresh air/water interface and the surface concentration reaches a value of 0.9 mg m^{-2} within about 120 minutes. The important point to note here is that the adsorption of BSA is almost complete before lysozyme even begins to adsorb to the interface. In the 1:1 binary system, the equilibrium surface concentration of lysozyme is only about 0.06 mg m^{-2}, whereas that of BSA is about 0.82 mg m^{-2}. In other words, in the 1:1 binary system, BSA completely suppresses adsorption of lysozyme to the air/water interface. This is not because BSA has higher affinity than lysozyme to the interface, but because BSA arrives first at the interface, occupying most of the interfacial area before lysozyme

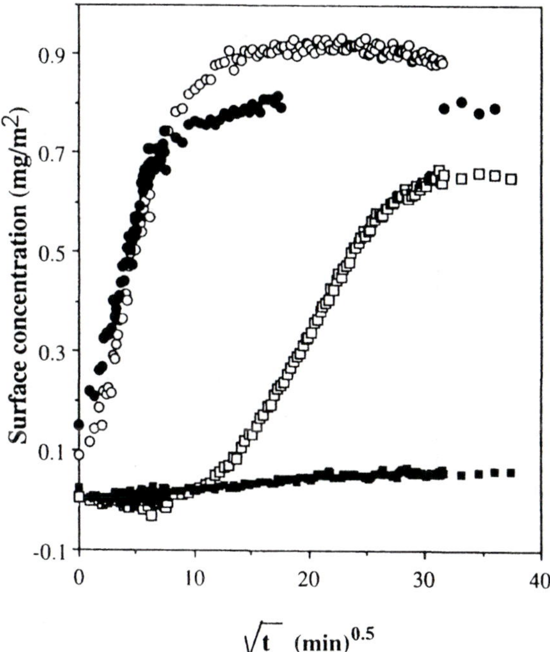

Figure 9 Kinetics of adsorption of lysozyme (squares) and bovine serum albumin (circles) in single-component (open symbols) and in 1:1 binary mixture (filled symbols) systems. Concentration of each protein in both systems was 1.5 μg/ml. (From Ref. 24.)

arrives at the interface. Since the adsorbed BSA molecules are incompressible, the extent of adsorption of lysozyme is then dependent only on the unoccupied area available at the time of its arrival at the interface.

The above interpretation is further confirmed by the data shown in Figure 10 [24]. In this case, ^{14}C-labeled lysozyme was first allowed to adsorb to the air/water interface; when unlabeled BSA is injected into the bulk phase at 250 minutes during the growth phase of adsorption of lysozyme, it abruptly stops further adsorption of lysozyme to the interface. However, more interestingly, no desorption of the already adsorbed lysozyme occurs. Similarly, when unlabeled BSA is injected into the bulk phase after lysozyme has reached equilibrium adsorption (1000 min), only a small amount of adsorbed lysozyme is desorbed by BSA from the interface. These results clearly indicate that once lysozyme is adsorbed to the interface, bulk phase BSA molecules cannot displace it.

Taken together, the results presented in Figures 9 and 10 clearly indicate that it is the kinetics of adsorption, not the thermodynamics of adsorption, that determines the composition of the interfacial protein film. This is true only for binary systems involving

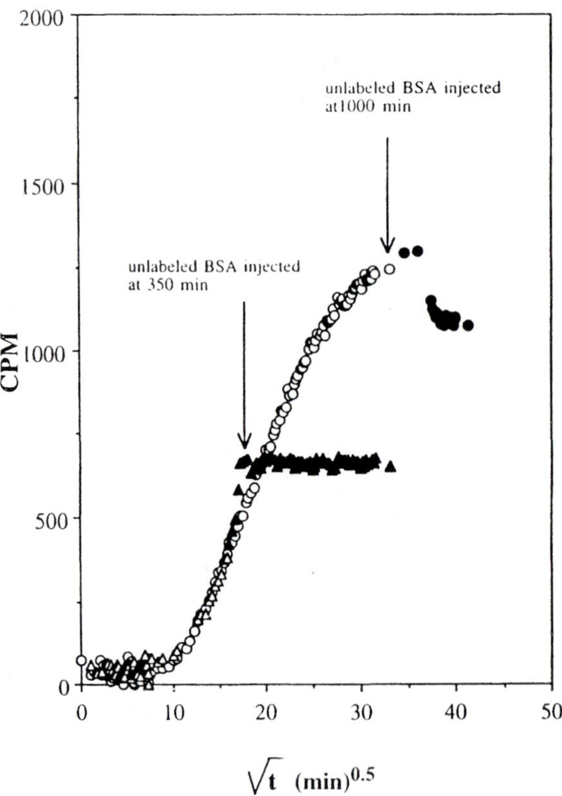

FIGURE 10 Displacement of adsorbed ^{14}C-lysozyme by unlabeled BSA. Symbols ○ and ● represent surface c.p.m of ^{14}C-lysozyme before and after injection, respectively, of unlabeled BSA at 1000 min of adsorption. Symbols △ and ▲ represent the same when unlabled BSA was injected at 350 min of adsorption. (From Ref. 24.)

random coil–globular and globular-globular protein binary systems, and seems to be not true for random coil–random coil protein binary systems. For example, Figure 11 shows the kinetics of adsorption of α_{s1}-casein and β-casein to the air/water interface in single and binary protein systems. In single-protein systems, for both α_{s1}- and β-caseins, adsorption commences immediately after a fresh air/water interface is created. The equilibrium surface concentration reaches a value of 1.66 mg m^{-2} for α_{s1}-casein and about 1.8 mg m^{-2} for β-casein. However, in the 1:1 binary system, the kinetics were more complex: The surface concentration of α_{s1}-casein increases first to a value of 1.0 mg m^{-2} within about 100 minutes and then decreases with time and reaches an equilibrium value of about 0.6 mg m^{-2}. In contrast, the surface concentration of β-casein increases continuously and reaches an equilibrium value of about 1.1 mg m^{-2}. It should be noted that the time at which the sum of the surface concentrations of α_{s1}- and β-caseins (Γ_{total}) reaches a steady-state value coincides with the time at which Γ of α_{s1}-casein reaches its maximum value. Beyond this value, even though the surface concentration of α_{s1}-casein decreases and that of β-casein increases with time, the Γ_{total} remains unchanged (Fig. 11B). This indicates that, in the time interval indicated by the vertical dotted lines in Figure 11B, β-casein adsorbs to the interface by displacing the already adsorbed α_{s1}-casein from the interface; indeed, for each β-casein molecule adsorbed, one molecule of α_{s1}-casein is displaced from the interface [24].

The adsorption behavior of the α_{s1}-casein–β-casein binary system is in stark constrast with those of the other binary systems discussed above. Whereas the composition of the interfacial protein film in the cases of the β-casein–lysozyme, lysozyme-BSA,

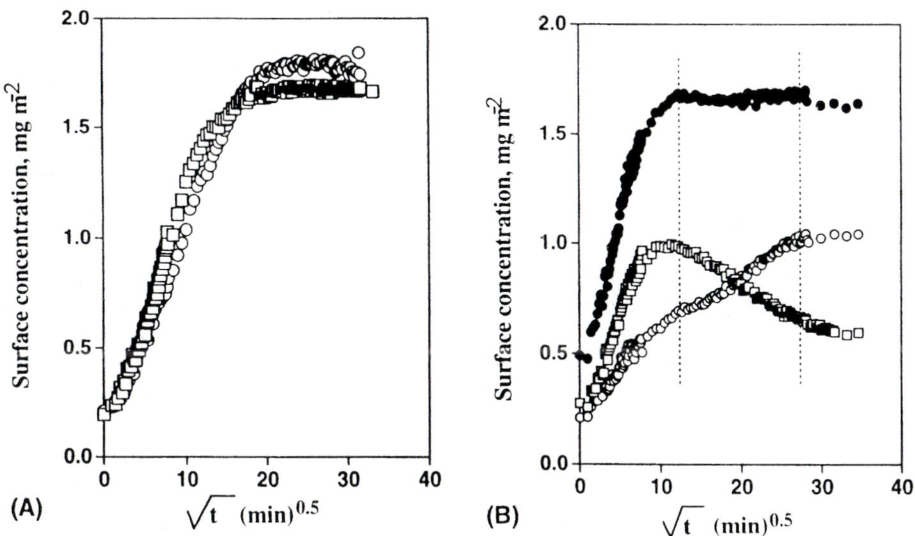

FIGURE 11 Kinetics of adsorption of α_{s1}-casein (□) and β-casein (○) in (A) single-component systems and (B) 1:1 binary mixture system. Concentration of each protein in both systems was 1.5 μg/ml. ● represents total surface concentration of α_{s1}-casein plus β-casein as function of adsorption time. The vertical dotted lines denote the time zone in which displacement of α_{s1}-casein by β-casein occurs. (From Ref. 26.)

and BSA–β-casein systems (i.e., systems involving random coil–globular and globular-globular proteins) is kinetically controlled, that of α_{s1}-casein/β-casein (that is, systems involving random coil–random coil proteins) is thermodynamically controlled.

1. Irreversibility and Displacement

Generally, proteins irreversibly adsorb to interfaces. One primary reason for this behavior is that, once adsorbed, proteins unfold and change conformation; this process makes it possible for the polypeptide chain to anchor itself to the interface with several peptide segments at the interface. Although the energy of interaction of each segment is very small, the sum of the energies of all the bound segments per molecule is much larger than the thermal energy, kT, of the molecule. To desorb the molecule from the interface, all bound segments must be simultaneously detached from the interface, which is statistically improbable. Desorption can occur when the protein film is compressed to high surface pressures [27–29]. The energy input in the form of surface pressure sequentially detaches adsorbed segments from the interface and eventually desorbs the protein. However, under normal adsorption conditions, the surface pressure of a protein film does not reach the levels reported by MacRitchie [29], and therefore it is questionable if reversible adsorption or exchange between surface and bulk phase molecules occurs under normal conditions.

To determine the molecular factors affecting exchange between adsorbed and bulk phase molecules or displacement of an adsorbed protein by another protein, sequential adsorption studies were performed on β-casein–lysozyme, lysozyme-BSA, BSA–β-casein, and α_{s1}-casein–β-casein binary systems [16,24–26]. Homo-displacement experiments showed that bulk phase lysozyme could not displace adsorbed lysozyme. This was also true for BSA. On the other hand, bulk phase β-casein could displace adsorbed β-casein, and so did α_{s1}-casein. In hetero-diplacement experiments, neither β-casein nor lysozyme was able to displace the other protein from the interface. Similarly, in the case of BSA–β-casein and BSA-lysozyme systems, neither of the proteins could displace the other from the air/water interface. In the case of α_{s1}-casein–β-casein, however, β-casein was able to displace α_{s1}-casein and α_{s1}-casein was able to displace β-casein from the air/water interface. This is interesting because, since β-casein is known to be more surface active than lysozyme and BSA, one would expect it to easily displace lysozyme and BSA from the interface.

The results of these studies can be summarized as follows. A random coil protein (such as α_{s1}-casein or β-casein) cannot displace an adsorbed globular protein, and a globular protein cannot displace or exchange with an adsorbed globular protein or displace a random coil protein from the air/water interface. In contrast, a random coil protein can exchange or displace another random coil protein. What this means in molecular terms is that, in the cases of globular-globular and globular–random coil protein binary systems, adsorption essentially follows a noncompetitive (in a thermodynamic sense), irreversible mechanism; in the case of random coil–random coil binary systems, adsorption follows a thermodynamically controlled competitive, reversible mechanism.

2. Thermodynamic Compatibility

The ability of β-casein to displace α_{s1}-casein from the interface, and vice versa, can be arbitrarily attributed to the molecular flexibility of α_{s1}-casein at the interface and therefore to its amenability to sequential detachment from the interface by β-casein. Similarly, the inability of β-casein to displace either lysozyme or BSA may be attributed a priori to the relatively rigid conformation of these proteins at the interface and therefore the necessity to simultaneously detach all of the adsorbed segments from the interface, which is not possible even for the highly surface-active β-casein. Although such simplistic explanations are quite satisfactory, it is likely that a more fundamental mechanism, namely, thermodynamic compatibility, may be involved in these processes. In order for a protein to displace another protein from an interface, it should be able to mix (or dissolve) in the film of the latter. Since the volume-concentration of a protein film is generally very high (20–30%, w/v), dissolution of one protein into the interfacial film of another protein can occur only if the two proteins are thermodynamically compatible. Since α_{s1}-casein and β-casein are flexible disordered polymers belonging to a similar class of proteins, they are thermodynamically compatible with each other [30,31]. This thermodynamic compatibility may allow β-casein to completely mix with α_{s1}-casein film (and vice versa) at the interface and thereby displace α_{s1}-casein purely on the basis of differences in affinities to the interface. However, lysozyme and β-casein belong to two different classes of proteins (i.e., globular vs. disordered). They are thermodynamically incompatible and therefore they cannot mix in each other's films and displace each other. It should be noted that bulk phase β-casein can displace or exchange with adsorbed β-casein [26], whereas bulk phase lysozyme cannot displace adsorbed lysozyme [24]. This can be explained as follows: since β-casein is a disordered protein, its conformation in the bulk phase and at the interface is very similar, that is, it goes from one disordered state to another disordered state. Because of structural similarities, its physicochemical properties, notably the surface hydrophobicity/hydrophilicity character, remain the same, and therefore they are still thermodynamically compatible. However, in the case of lysozyme, because of interfacial denaturation, the adsorbed lysozyme molecules are structurally different from those of the bulk phase molecules. Because of significant changes in the surface hydrophobic/hydrophilic character, probably the mixing of bulk phase lysozyme with the adsorbed (denatured) lysozyme in the film becomes thermodynamically incompatible. (Generally denatured proteins are thermodynamically incompatible with their native counterparts, which manifests itself in phase separation).

Thus, it appears that thermodynamic compatibility among proteins might be the fundamental mechanism by which displacement or exchange between proteins might occur at interfaces. This may also control the surface load and the composition of mixed-protein films [16].

III. FOAMS

Foams are colloidal systems in which tiny air bubbles are dispersed in an aqueous continuous phase. Many processed foods, such as whipped cream, ice cream, cakes, meringue, brad, souffles, mousses, and marshmallow, are foam-type products to start with. In all of these products, proteins are the main surface-active agents that help in the formation and stabilization of the dispersed gas phase.

The volume fraction, ϕ, of dispersed air is a critical factor in determining the size distribution, structure, and behavior of foams [32,33]. For foams containing close-packed spherical bubbles of uniform size, the upper limit of the volume fraction of liquid in the foam is about 0.26. If the liquid content is sufficiently high so that the bubbles are completely spherical and mobile, the dispersion is not considered a true foam but rather a "gas emulsion."

At the moment of creation, foam bubbles are spherical in shape. However, in a typical foam these bubbles aggregate to form a polyhedral foam with a thin, flat liquid film separating the bubbles. Some foams may never reach polyhedral forms due to their low stability. These foams are referred to as "wet" foams; the polyhedral foams are normally referred to as "dry" foams. The question of foam stability is meaningful only when the "wet" foam has a lifetime long enough to become a polyhedral "dry" foam.

Generally, protein-stabilized foams are formed by bubbling, whipping, or shaking a protein solution. The foaming properties of proteins encompass two aspects: (a) the ability to produce a large interfacial area so that a large volume of gas can be incorporated into the liquid (commonly referred to as foamability or foaming capacity) and (b) the ability to form a tenacious interfacial film that can withstand internal and external forces. The foaming capacity of a protein, which refers to the amount of interfacial area created by the protein, is often expressed as "overrun," which is defined as [34]:

$$\text{Overrun} = \frac{\text{Volume of foam} - \text{Volume of initial liquid}}{\text{Volume of initial liquid}} \times 100 \tag{15}$$

Foam stability, which refers to the ability of the protein to stabilize foam against gravitational and mechanical stresses, is often expressed as the time required for 50% of the liquid to drain from a foam or for a 50% reduction in foam volume [34,35]. These are very empirical methods, and they do not provide fundamental information about the factors that affect foamability and stability of protein-stabilized foams.

The most direct measures of foamability and stability are determination of the initial interfacial area of a foam and the rate of decay of the interfacial area. A more rigorous method based on the equation of state of foam has been described [36,37]. The equation of state of a spherical liquid surface of a foam bubble is given by the Laplace equation:

$$P_i - P_e = \frac{2\gamma}{r} \tag{16}$$

where P_i and P_e are the pressures inside and outside of the bubble, γ is the surface tension, and r is the radius of the bubble. Assuming that the foam bubble obeys the ideal gas law:

$$P_i V = nRT \tag{17}$$

where V is the volume, n is the number of moles of gas, and R and T have the usual meaning. Substitution of Eq. (16) in Eq. (17) gives:

$$\left(P_e + 2\frac{\gamma}{r}\right) V = nRT \tag{18}$$

or, since $V = Ar/3$:

$$P_e V + \frac{2}{3}\gamma A = nRT \tag{19}$$

where A is the surface area of the bubble. Equation (19) is the equation of the state of a foam. In a closed system with constant volume and number of moles of the gas, a change in surface area (ΔA) will cause a corresponding change in the external pressure (ΔP), and Eq. (19) takes the form:

$$3V\Delta P + 2\gamma\Delta A = 0 \tag{20}$$

Equation (20) describes the changes that take place during the collapse of a foam at constant volume and temperature. If A_0 is the interfacial area of a freshly formed foam, the interfacial area A_t at any time during the decay process is:

$$A_t = A_0 + \Delta A = A_0 - \frac{3V\Delta P_t}{2\gamma} \tag{21}$$

where ΔP_t is the net change in the external pressure. At infinite time, when the foam is completely collapsed, Eq. (21) reduces to:

$$A_\infty = 0 = A_0 - \frac{3V\Delta P_\infty}{2\gamma} \tag{22}$$

Therefore:

$$A_0 = \frac{3V\Delta P_\infty}{2\gamma} \tag{23}$$

The value of A_0, which is the interfacial area of a freshly formed foam, is the foamability of the protein under the experimental conditions used. Substitution of Eq. (23) in Eq. (21) gives:

$$A_t = \frac{3V}{2\gamma}(\Delta P_\infty - \Delta P_\tau) \tag{24}$$

and:

$$\frac{A_t}{A_0} = \frac{(\Delta P_\infty - \Delta P_t)}{\Delta P_\infty} \tag{25}$$

Equation (25) describes the kinetics of foam decay. Experimental determination of ΔP_t as a function of time provides the means to determine the total interfacial area of the foam as a function of time during the decay of the foam. This approach has recently been used to investigate the foaming properties of food proteins [38–42].

A. Foam Formation and Foam Stability

In order for a protein to be a good foaming agent, it should possess the following attributes [43]:

1. It should be able to rapidly adsorb at the air-water interface during whipping or bubbling.
2. It should undergo rapid conformational change and rearrangement at the air-water interface and rapidly reduce the surface tension.
3. It should be able to form a cohesive, viscoelastic film through intermolecular interactions.

The first two criteria are essential for better foamability, whereas the third criterion is important for the stability. Graham and Phillips [44] demonstrated that the single most important factor for foamability of a protein solution is the rate at which the protein can reduce the interfacial tension as new interfacial area is being continuously created during bubbling or whipping. Studies on the foaming properties of β-casein, BSA, and lysozyme showed that the relative foamability of these proteins followed the order β-casein > BSA > lysozyme, which is the same order in which they undergo conformational change and affect the rate of decay of surface tension in model systems (see Fig. 5). Thus, proteins that rapidly adsorb and undergo rapid conformational change at the freshly formed air/liquid interface during bubbling often exhibit better foamability than those that adsorb slowly and resist unfolding at the interface. Kitabatake and Doi [45] also reported that the foamability of proteins was not related to their equilibrium surface tension, but to the rate of surface tension decay (Fig. 12). The reasons for this can be easily explained as follows. Consider a situation when an air bubble is at an orifice (see Fig. 13). The escape of the bubble from the orifice is dependent on the magnitude of two forces, namely, the buoyancy force and the interfacial tension [32]. The buoyancy force is given by:

$$F_{buoy} = (4/3)\pi R^3 \rho g \tag{26}$$

where ρ is the density of the liquid, R is the radius of the bubble, and g is the acceleration due to gravity. The adhesion force that tends to retain the bubble at the orifice is:

$$F_{int} = 2\pi r \gamma \tag{27}$$

where r is the radius of the orifice and γ is the surface tension. At the moment of escape of the bubble from the orifice, F_{buoy} is equal to F_{int}, and therefore:

$$R = (3\gamma r/2\rho g)^{1/3} \tag{28}$$

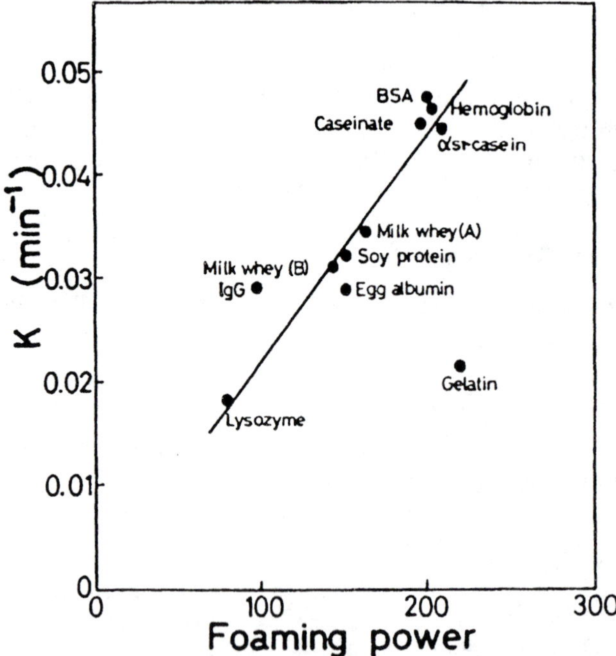

FIGURE 12 Relationship between foaming power (foamability) and the rate constant of surface tension decay of 1% protein solutions. (From Ref. 45.)

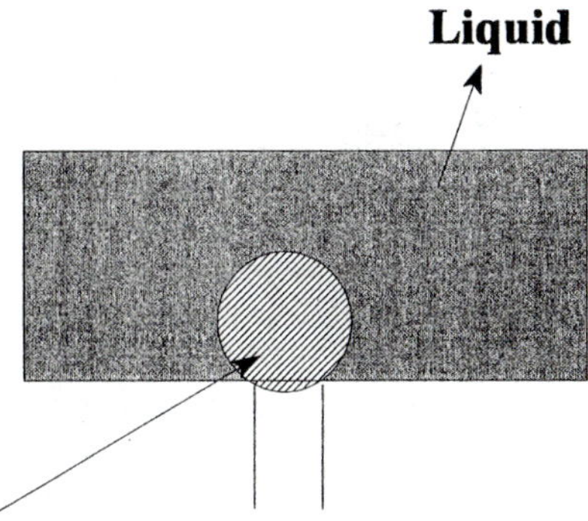

FIGURE 13 A gas bubble at the time of escape from an orifice.

According to Eq. (28), the larger the surface tension value, the larger will be the radius of the bubble formed, that is, the lower will be the total interfacial area created during foaming.

1. Disjoining Pressure

Two macroscopic processes in foams affect the kinetic stability of protein-stabilized foams. These are the rate of liquid drainage from the lamellae and film rupture. The rates of these two processes are dependent on the physical properties of the protein film and the physics of the lamella itself. These include film viscosity, shear resistance, elasticity, and disjoining pressure between the films [46]. The factors that increase liquid drainage from the lamella film decrease foam stability. The rate of film drainage can be described using the Reynolds equation [47]:

$$V = -\frac{dh}{dt} = \frac{2h^3}{3\mu R^2}\Delta p \tag{29}$$

where h is film thickness, t is time, μ is dynamic viscosity, R is the radius of the bubble, and $\Delta p = \pi_h - \pi_d$, where π_h is the capillary hydrostatic pressure and π_d is the disjoining pressure between the two protein films. In general, π_h is always greater than π_d, and therefore foams do not reach a thermodynamic equilibrium; the film drains continuously, leading to eventual thinning and collapse of the bubble [48].

According to Eq. (29), the stabilities of various protein-stabilized foams with similar bubble size distribution depend on the disjoining pressure between the two adsorbed protein layers of the lamella. The disjoining pressure is related to the physicochemical properties of proteins, because:

$$\pi_d = \pi_s + \pi_e + \pi_v + \pi_{hyd} \tag{30}$$

where π_s, π_e, π_v and π_{hyd} are the contributions from steric repulsion, electrostatic attraction/repulsion, van der Waals (including hydrophobic), and hydration repulsion forces between the protein layers. The π_v negatively contributes to the disjoining pressure, because the attractive van der Waals and hydrophobic interactions between the protein layers usually promote film thinning and film drainage. On the other hand, π_e positively contributes to the disjoining pressure at pH values away from the isoelectric pH of the protein, where proteins carry a net positive or negative charge. The charge repulsion between the protein layers tends to oppose thinning of the lamella. However, in the absence of a overwhelming cohesive interactions within each protein layer, excessive electrostatic repulsion between protein molecules with the layer will impair the integrity of the protein layer and thus may cause breakage of the film. The contribution of π_s to the disjoining pressure is always positive. The steric effects arise mainly from the loops and tails of the adsorbed protein molecules. This again depends on the structural rigidity of proteins at the air/water interface. Proteins that are very flexible tend to form dilute monolayers, whereas rigid globular proteins form concentrated thick monolayers with relatively rigid loops protruding into the liquid phase of the lamella.

A case in point is the difference in the stabilities of foams stabilized by β-casein, BSA, and lysozyme [44]. Even though the foamability of β-casein is better than that of the other proteins, its foam has the fastest drainage rate for the following reasons. Since β-casein is highly hydrophobic, the magnitude of negative contribution of π_v to the disjoining pressure is significant. Moreover, since β-casein is highly flexible, most of its peptide backbone assumes the train configuration at the air/water interface; thus, because of the limited number of loops, the magnitude of steric repulsion contributing positively to the disjoining pressure is marginal. On the other hand, the slow rate of drainage of lysozyme foams is attributable to a significant amount of steric, hydration, and electrostatic repulsions and a lack of attractive hydrophobic interaction between the protein layers.

2. Marangoni Effect

Another important factor that retards film drainage is the Marangoni effect. When adsorbed protein layers are present on either side of the film (Fig. 14), bursting of the lamella will occur only by stretching and local thinning. When the lamella is stretched, the increase in surface area of the film causes a decrease of surfactant concentration at

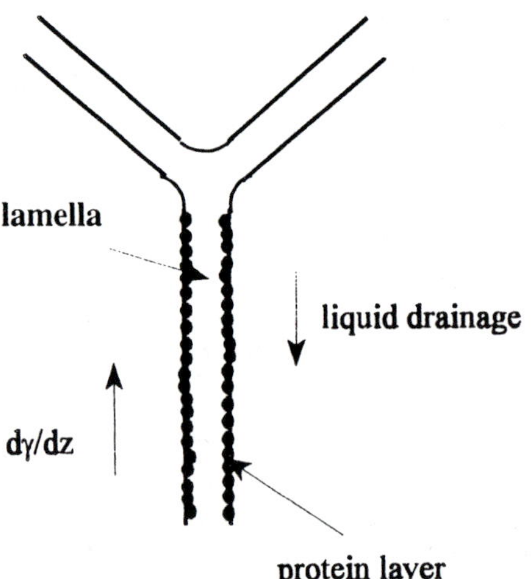

FIGURE 14 Schematic representation of the junction of three lamella films at a Plateau border. $d\gamma/dz$ is surface tension gradient created at the interface due to flow of the liquid through the lamella.

the surface, and consequently an increase in surface tension. At this moment in time, if the adsorption of surfactant molecules from the continuous phase is not quick enough to restore the surface tension (or surface pressure), then the local increase in surface tension would drag the surfactant layer from the adjacent low-surface-tension regions. As the film moves toward the high-surface-tension regions, it carries an appreciable volume of underlying liquid along with it, which restores surface tension and the thickness of the lamella. This process is known as the Marangoni effect. This occurs in foams under dynamic conditions as a consequence of drainage. In foams, the streaming of liquid through the lamella exerts a shearing stress on the adsorbed protein layer, creating a surface tension gradient along the height of the lamella (Fig. 14). As a result, the protein layers on both sides of the lamella tend to move in the direction opposite to the liquid flow. The movement of the protein layer tends to retard the rate of liquid drainage.

The magnitude of the Marangoni effect is a function of the surface dilational viscosity of the protein film. The surface dilational viscosity under dynamic conditions is given by:

$$\eta_d = \frac{\Delta\gamma}{(\mathrm{dlnA/dt})} \tag{31}$$

where $\Delta\gamma$ is the change of surface tension with respect to the equilibrium value, t is time, and A is the area of the lamella surface. Under equilibrium conditions, η_d is defined as surface elasticity, ε, which is:

$$\varepsilon = 2(\mathrm{d}\gamma/\mathrm{dlnA}) \tag{32}$$

Equation (32) suggests that for ε to be large, the surface tension gradient must be large. This can occur only if the rate of adsorption of the protein from the bulk phase to the surface is slow so that the surface tension gradient can be maintained long enough for the operation of the Marangoni effect. If the protein adsorbs to the surface as soon as the film is stretched, the surface tension gradient will be very low, and the lamella film will expand continuously to a breaking point. The Marangoni effect plays a vital role during foam formation, where constant stretching of foam lamellae occur.

Another molecular factor that affects the elasticity of protein films is the rate of configurational transitions of the polypeptide chain from trains to loops, and vice versa, during compression and expansion of the film. For instance, the elasticity modulus of β-casein at the air/water interface is about 5–30 mM/m, and the relaxation time for train-to-loop configurational change is about 10^{-8} s [49]. Because of the rapid configurational changes in β-casein, when the lamella is stretched, instead of the β-casein film physically moving toward the low-surface-tension regions, it expands instantaneously and prevents formation of a surface tension gradient. Because of the lack of viscous drag of the lamella fluid (Marangoni effect), rapid drainage and thinning occur. In contrast, films of globular proteins exhibit higher elasticity moduli. For instance, the elasticity moduli of lysozyme and BSA are in the range of 200–400 and 60–400 mM/m, respectively [49]. In these globular proteins, the relaxation rate for train-to-loop configurational fluctuations is very slow; therefore, when the lamella film is stretched, instead of the protein layer expanding toward the low-surface-tension regions, the entire protein layer moves towards the low-surface-tension regions. The consequent viscous drag of the liquid beneath the protein layer retards the drainage. Thus, elasticity of protein films is fundamentally related to their configurational flexibility at the interface. The greater the molecular inflexibility, the greater is the elasticity.

It should be pointed out that while the flexibility of the polypeptide chain seems to be essential for foamability, it is detrimental to foam stability. Conversely, proteins that foam very well may not be able to impart stability to the foam, whereas the proteins that foam very poorly may have the molecular characteristics that impart stability. Thus, for a protein to foam well and stabilize the foam, it should display a proper balance of flexibility and rigidity at the air/water interface.

3. Surface Film Rigidity/Shear Viscosity

Another factor that influences the stability of the lamella is the rigidity of the protein layers on either side of the lamella film. In most cases, the viscosity of these layers is much larger than that of the interstitial liquid. Often, the viscosity is not Newtonian, but displays a yield stress σ_y. The rigidity of the protein layer is proportional to its yield stress, and if it has very high yield stress, it may behave like a solid. Although a highly rigid protein film may interfere with the Marangoni effect, it may stabilize the lamella against mechanical disturbances and, more importantly, retard the liquid drainage.

The rheological properties of protein films are dependent on the extent of intermolecular interactions within the film. Proteins that exhibit optimum intermolecular interactions and form a cohesive continuous network can create very stable foams. For example, the shear viscosity of lysozyme film at the air/water interface is about 1000 mN m^{-1} s [49], indicating that lysozyme forms a highly cohesive film, whereas the shear viscosity of β-casein is less than 1.0 mN m^{-1} s, even though its surface concentration at equilibrium is higher than that of lysozyme.

4. Film Rupture

The mechanism of film rupture is believed to follow a nucleation process, although the forces that initiate this nucleation are not well studied. However, the factors that influence retardation of the growth of a hole in the film are film thickness, interfacial tension, and the elasticity of the protein [49]. Of these, film thickness is the most critical parameter. Below a critical thickness of about 500 Å, the growth of the hydrodynamic surface waves accelerates thinning of the lamella [50,51]. However, Dickinson [52] argued that the probability of spontaneous hole formation due to thermal fluctuations is negligible as long as the local film thickness exceeds 100 Å. Further, the Marangoni effect also would resist inhomogeneous drainage and local thinning when a surfactant is present.

Prins [32] suggested that the presence of hydrophobic particles in the foam can cause film rupture via two different mechanisms: If the hydrophobic particle is large and nonspreading, contact of the particle with both surfaces of a lamella will increase the Laplace pressure at the location of the contact because of formation of a convex surface (Fig. 15A). This causes flow of the film away from the hydrophobic particle and eventual loss of contact with the particle and formation of a hole. If the hydrophobic particle is of a spreading type, spreading of the particle on the film surface will displace the original surfactant; if the spreading particle is not able to stabilize the film, then the foam will collapse. If the spreading causes flow of the original film surface away from the particle, local thinning and eventual breakage of the film will occur (Fig. 15B).

Studies have shown that certain rigid, nonspreading hydrophobic particles, such as hydrophobic SiO_2 particles and polystyrene particles, stabilize protein foams [53,54], whereas Teflon particles destabilize foams [55]. This is probably because the size of the hydrophobic SiO_2 particles and polystyrene particles (0.7 and 6.0 μM, respectively) were not large enough to make contact with both surfaces of the film, whereas the Teflon

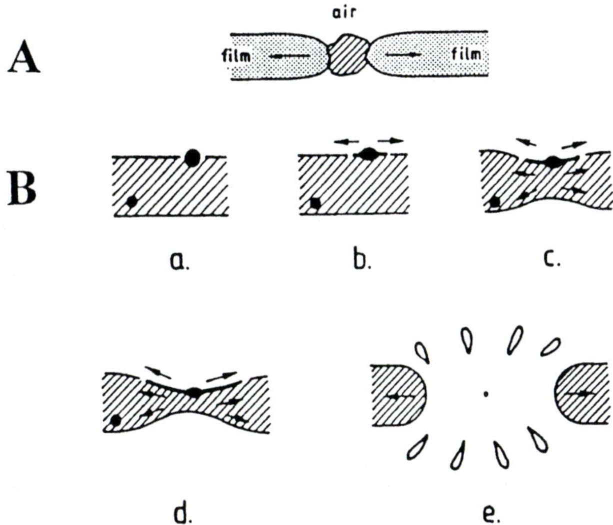

FIGURE 15 (A) Mechanism of film breakage by a rigid hydrophobic particle larger in diameter than the thickness of the lamella film. (B) Sequence of events leading to film breakage by a spreading hydrophobic particle. (From Ref. 32.)

particles were large enough to make contact with both surfaces of the film. This also indicates that while rigid hydrophobic particles larger than the thickness of the lamella cause film rupture, rigid hydrophobic particles smaller than the thickness of the lamella may actually stabilize foams. This stabilizing effect of small hydrophobic particles on foams is probably due to a decrease in the pressure difference between gas pressures at the lamella film and the Plateau border, as depicted in Figure 16. In a polyhedral foam, the pressure difference between the lamella film and the Plateau border is:

$$P_{film} - P_{PB} = \Delta P = (P_{gas} - \gamma/r_{film}) - (P_{gas} - \gamma/r_{PB}) \tag{33}$$

Since r_{film} is infinity in a polyhedral foam, Eq. (33) reduces to:

$$\Delta P = \frac{\gamma}{r_{RB}} \tag{34}$$

where γ is the surface tension and r_{PB} is the radius of curvature of the Plateau border. However, when a hydrophobic particle is present at the surface of the lamella film, r_{film} is no longer infinite and the ΔP becomes smaller. When more hydrophobic particles are bound to the film, its radius of curvature approaches that of the Plateau border and the pressure difference approaches zero. The decrease in the pressure difference decreases the driving force for liquid drainage and thus stabilizes the foam.

Another phenomenon known as Ostwald ripening or disproportionation also contributes to foam breakage. This is related to the Laplace pressure difference between small and large bubbles in a foam. According to Eq. (16), the Laplace pressure inside a small bubble must be greater than that in a large bubble. Because of this pressure difference, the air (or gas) from small bubbles diffuse into adjoining large bubbles, and eventually the small bubbles collapse and disappear. The rate of diffusion of gas from

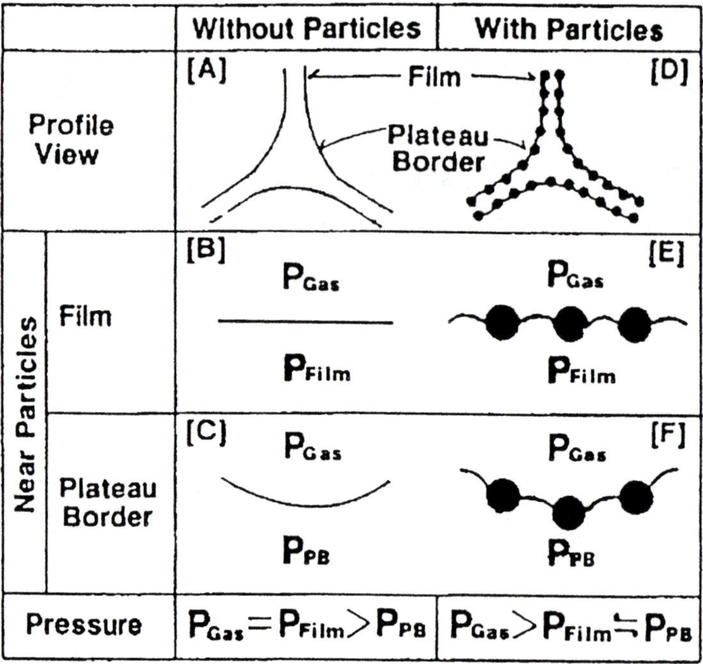

Figure 16 Mechanism of stabilization of foams by small hydrophobic particles via their effect on the pressure difference between the foam film and plateau border. ●, Hydrophobic particles.

small to large bubbles depends on the barrier properties of the protein layer, the surface dilational modulus [see Eq. (32)], the solubility of the gas in the lamella fluid, and the thickness of the lamella.

B. Molecular Properties Affecting Foam Formation and Stability

Apart from molecular flexibility, the foaming properties are affected by several physicochemical properties of proteins, including surface and molecular hydrophobicity, net charge and charge distribution, hydrodynamic properties, and other environmental factors.

Protein hydrophobicity can be classified in two ways: as surface hydrophobicity of the protein's exterior and as the average hydrophobicity of the amino acid residues of the protein, calculated from transfer free energy of amino acid side chains from ethanol to water [56]. The surface hydrophobicity is determined from the extent of binding of hydrophobic fluorescent probes, such as *cis*-parinaric acid or 1-anilino-8-naphthalenesulfonate [57]. Since the initial adsorption of a protein molecule to an interface involves the anchoring of a hydrophobic patch on the protein surface to the interface, one should expect a correlation between surface hydrophobicity and foamability of proteins. On the contrary, no direct correlation has been found (Fig. 17) [58]. Instead, the surface hydrophobicity-foamability plot exhibits a curvilinear relationship. On the other hand, the

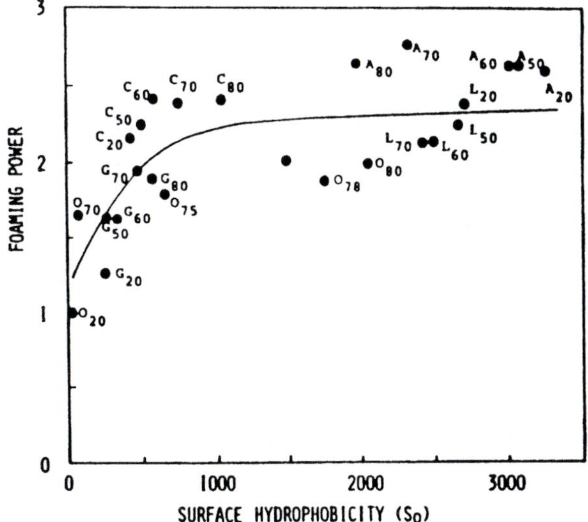

Figure 17 Correlation between foaming power and mean hydrophobicity of proteins. (From Ref. 61.)

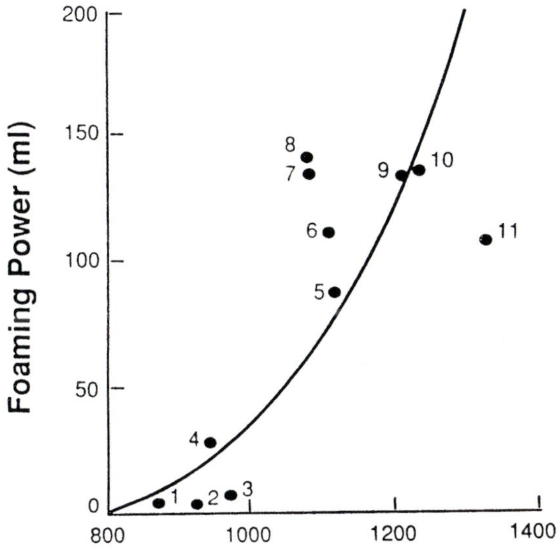

Figure 18 Correlation between foaming power and surface hydrophobicity of proteins. (From Ref. 58.)

foamability of proteins shows a positive correlation with the average hydrophobicity (Fig. 18). Taken together, the results shown in Figures 17 and 18 suggest that a surface hydrophobicity value of about 1000 is sufficient to successfully anchor proteins to the air/water interface during bubbling (i.e., the probability of each collision leading to adsorption may be close to unity above this value). However, once the protein is adsorbed to the interface, its ability to rapidly decrease the surface tension depends on its average hydrophobicity and its ability to expose all the hydrophobic residues to the interface. Experimental evidence indicates that proteins indeed undergo substantial conformational change at the air/water interface [59,60]. Thus, the rate of reduction in surface tension and the expansion of the interfacial area during foaming are only limited by the total number of hydrophobic groups available in the protein, not by the number of hydrophobic patches on the protein surface.

Although foamability exhibits a positive correlation with average hydrophobicity, it is only conceivable that proteins that are highly hydrophobic may coagulate via hydrophobic interactions at the interface, where the concentration is much higher than in the bulk phase. If this occurs, it would make holes in the protein layer and thus would destabilize the foam. If a protein is highly charged, such as phosvitin, because of strong electrostatic repulsion at the interface, the protein may not be able to form a cohesive continuous film necessary for the stability. Indeed a negative correlation exists between foam stability and charge frequency (i.e., the average number of charged residues per residue in a protein) (Fig. 19) [61]. Thus, it appears that, in order to form a cohesive film with optimal rheological properties, the protein should have a proper balance of hydrophobic and hydrophilic characteristics.

Although flexibility, which is defined as the relative movement of various domains in a protein or the reorientational relaxation rates of amino acid residues in a polypeptide

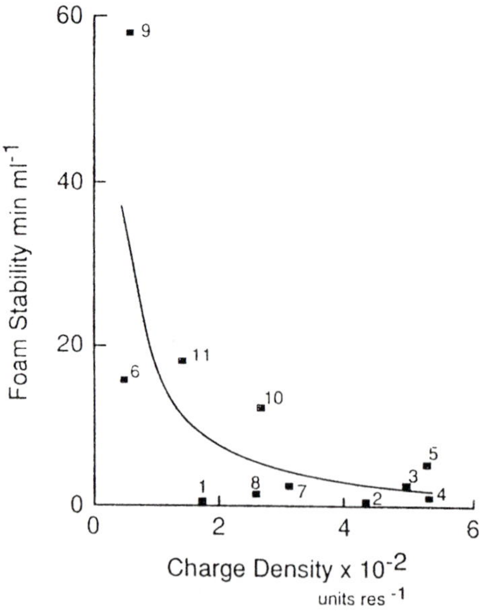

FIGURE 19 Correlation between foam stability and charge density of proteins. (From Ref. 61.)

chain, has been recognized as the most important criterion for functionality (especially the foaming properties), the relationship between initial conformation of a protein in solution and its foaming properties is not fully understood. This is important, especially in technological applications, because if such a relationship exists, proteins can be physically or chemically modified to assume a certain structure prior to foaming.

In an attempt to understand the role of protein structure in solution on its foaming properties, Zhu and Damodaran [40] studied heat-induced conformational changes in whey protein isolate and studied its foaming properties by measuring the pressure change above a foam in a specially designed foaming apparatus according to Eq. (25) [38]. Figure 20 shows the kinetics of interfacial area decay of foams of 5% native WPI and 5% WPI that was heat denatured for various times at 70°C. The decay of WPI foams exhibit a nonlinear first-order kinetics, indicating involvement of at least two microscopic processes in foam decay [38]. These two processes are gravitational liquid drainage and disproportionation of gas bubbles due to gas diffusion from smaller bubbles to larger bubbles (disproportionation). Thus, since drainage and disproportionation are the two major processes contributing to foam collapse, the rate of decay of a protein foam must follow at the least a biphasic first-order kinetics:

$$\frac{A_t}{A_o} = Q_g \exp(-k_g t) + Q_d \exp(-k_d t) \tag{35}$$

where k_g and k_d are first-order rate constants for the drainage and disproportionation processes and Q_g and Q_d are the amplitude parameters of the two kinetic processes. Among these processes, the first kinetic phase ought to be the gravitational drainage and the second the disproportionation [38]. When WPI was progressively heat denatured at 70°C, the sample heated for 1 minute exhibited more foam stability than either the unheated control or the other heated samples (Fig. 20). The foamability, i.e., the total interfacial area A_o, also was highest for this sample [40]. The shapes of the decay curves were also qualitatively different: while the foams of samples heated 1, 5, and 10 minutes exhibited a concave-type decay curve, the native and 20-minute heated samples exhibited a convex-type decay curve. In a convex-type decay curve, drainage is the rate-limiting step, and in a concave-type decay curve breakage due to disproportionation is the rate-limiting step. The transformation from convex to concave and from concave to convex biphasic first-order decay as a function of heating time at 70°C reflects some fundamental changes in the physical properties of whey proteins. This may be related to structural changes in the protein. The changes in the secondary structure of the native and heated WPI, as determined from circular dichroic measurements, are shown in Figure 21. The sample heated 1 minute (70°C, 5% solution), which showed better foam stability, had a gross secondary structure content of about 28.5% β-sheet, 16.5% α-helix, and 55% aperiodic structures. It should be emphasized, however, that changes in the tertiary structure, and the consequent changes in the physicochemical properties of the protein surface and its conformational stability at the interface, are more important than changes in the secondary structure per se. In this regard, it should be noted that the sample heated 20 minutes, which had higher aperiodic structure content, in fact had lower foamability and foam stability than the sample heated for 1 minute. This might be due to differences in their tertiary structures.

In contrast to the foaming properties of the 70°C heated samples, WPI solutions (9%) heated at 90°C exhibited poor foamability and foam stability (Fig. 22). There was no remarkable change in the secondary structure content of these samples (Fig. 21).

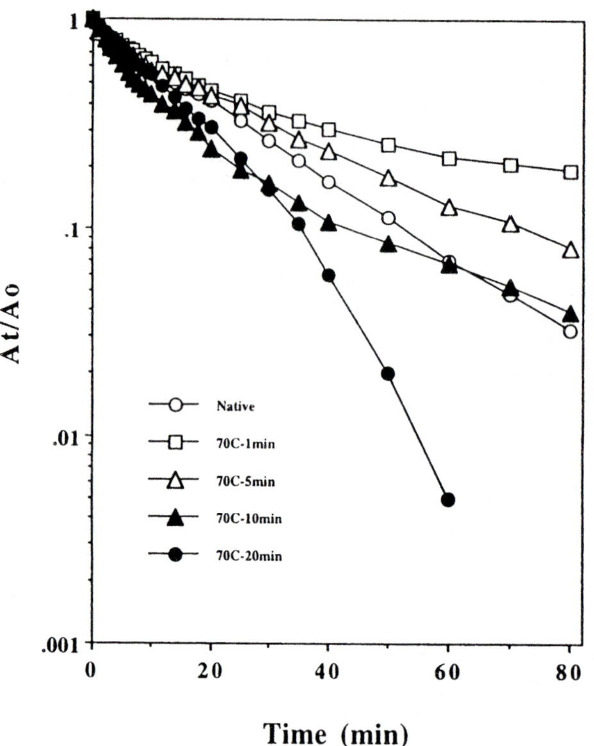

FIGURE 20 Surface area decay of foams of whey protein isolate (WPI) heated at 70°C and 5% protein concentration conditions for various times. Foams were formed using 5% WPI in water at pH 6.8, 25°C. (From Ref. 40.)

However, more than 98% of the protein in these heated samples took the form of very high molecular weight polymers, and the specific viscosities of these solutions were an order of magnitude greater than that of the native WPI [40]. The poor foamability and foam stability of these severely heat-treated samples is probably due to excessive polymerization via sulfhydryl-disulfide exchange reactions. When native WPI was mixed with WPI heated at 90°C for 20 minutes, the mixture exhibited maximum foam stability when the ratio of native to polymerized WPI was about 40:60 (Fig. 23) and maximum foamability at about a 60:40 ratio (Fig. 24). These results indicated that, in addition to structural changes at the secondary and tertiary levels, the ratio of monomeric to polymeric proteins in heat-denatured WPI affected its foaming properties. Although the fundamental reasons for this is not clear, it appears that the monomeric species contribute to foam generation and the polymeric species contribute to foam stability. Since the diffusivity of monomeric proteins is greater than that of polymerized species, they rapidly adsorb and form a film at the interface during bubbling. The protein films formed by the monomeric species do not appear to have the required viscoelastic properties to stabilize the foam. When the late-arriving polymerized species adsorbs to the preformed protein layer, it increases the viscoelastic properties of the film and thus stabilizes the foam against gravitational drainage and interbubble gas diffusion. Recently, it has been shown that high molecular weight polymers formed during whipping of egg white af-

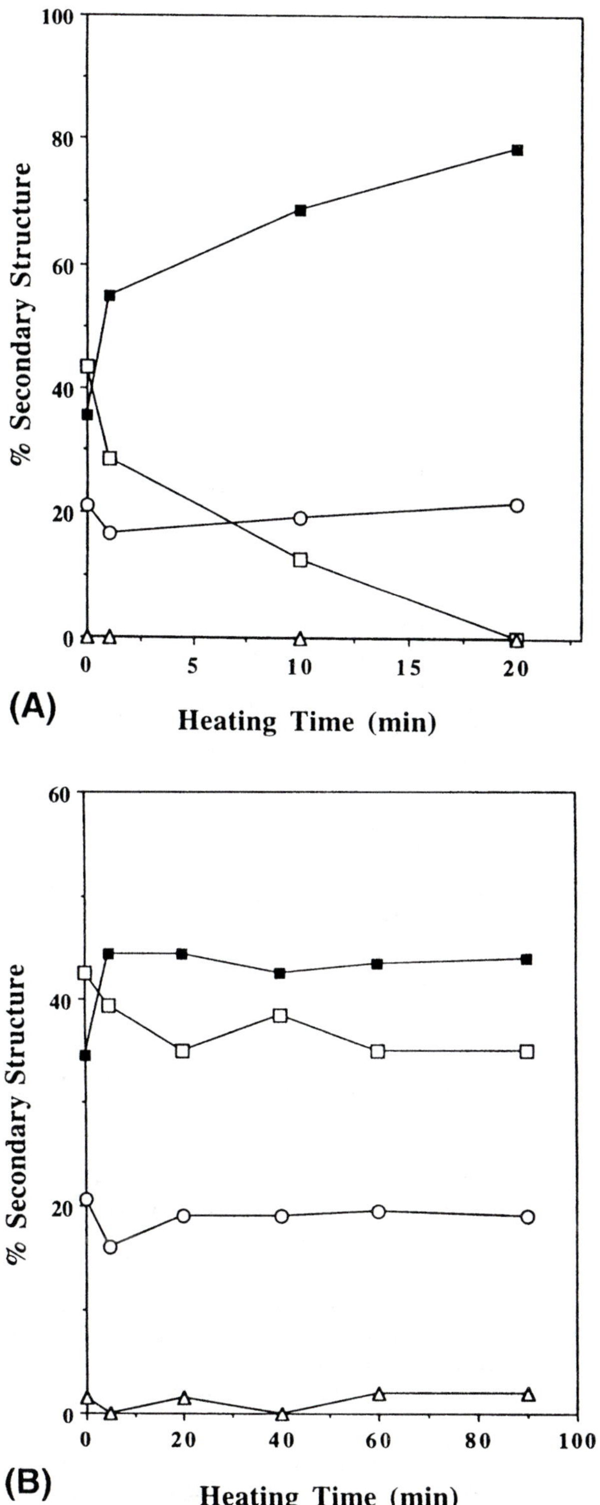

FIGURE 21 Changes in secondary structure content of WPI as a function of time at (A) 70°C and (B) 90°C heating conditions. ○, α-helix; □, β-sheet; △, β-turns; ■, aperiodic structures. (From Ref. 40.)

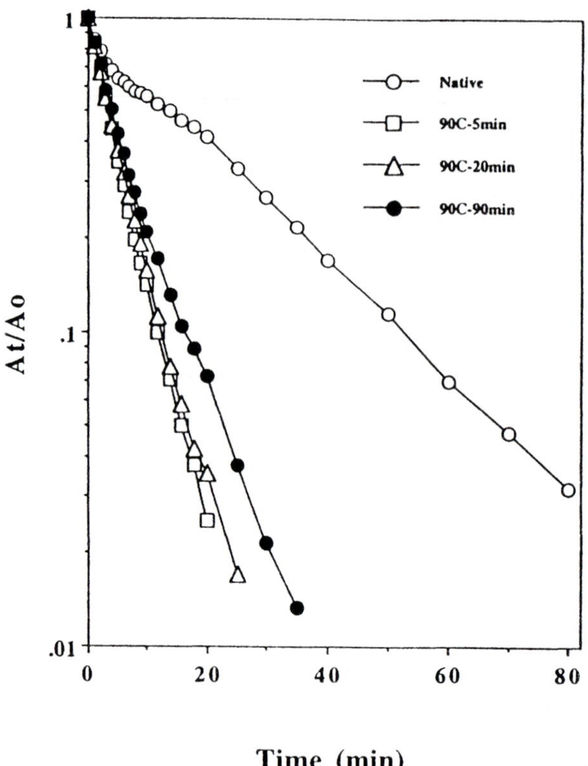

Figure 22 Surface area decay of foams of WPI samples heated at 90°C and 9% protein concentration conditions for various times. Foams were formed using 5% WPI in water at pH 6.8, 25°C. (From Ref. 40.)

fected its foaming properties [62]. Whether or not this is also true for other proteins remains to be investigated.

Cleavage of intramolecular disulfide bonds, which removes constraints for molecular flexibility at the interface, also influences the foamability and foam stability of proteins. When disulfide bonds in bovine serum albumin were progressively cleaved using increasing concentrations of dithiothreitol, the stability of BSA foam increased with the amount of dithiothreitol (Fig. 25) [38]. It is notable that the shape of the decay curve changed from a monophasic first order to a convex-type biphasic first order as the dithiothreitol was increased from 0 to 10 mM, indicating that cleavage of disulfide bonds facilitates formation of a highly viscoelastic protein film that is able to retard the liquid drainage. Similar results also were obtained with foams of soy glycinin and whey proteins [63,64].

Proteins that possess similar structures in solution have been found to display different foaming properties [65]. Although human, turkey egg, and hen egg lysozymes have very similar tertiary and secondary structures, their foaming properties were different. The order of increasing foam stability and foamability followed the order turkey

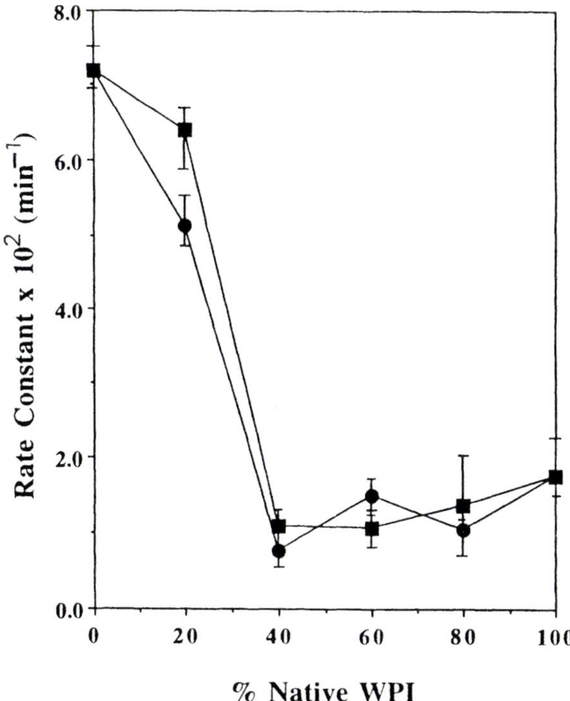

FIGURE 23 Effect of the ratio of native and polymerized proteins in WPI on the gravitational (●) and disproportionation (■) rate constants of decay of WPI foams. The bars represent standard error. (From Ref. 40.)

< hen < human [65]. This suggests that slight variations in amino acid sequence, and probably in the tertiary structure, have significant effects on the foaming properties of proteins.

Several nonprotein additives, such as phospholipids, polyphenols, and ethanol, affect the foaming properties. Phospholipids and other low molecular weight emulsifiers, such as Tween 20, destabilize protein-stabilized foams. This is principally because of competition of these surface-active molecules with proteins for adsorption at interfaces, which causes partial displacement of proteins from the interface [66,67]. Addition of low concentrations of polyphenols, such as (+)-catechin, to a Tween 20–β-lactoglobulin foaming system countered the destabilizing effect of Tween 20 and improved the foamability and foam stability [67]. This was attributed to catechin-induced cross-linking of proteins in the adsorbed layer, which prevented its displacement by the low molecular weight surfactant and probably also improved the viscoelastic properties of the film.

Addition of a small amount of ethanol to protein solutions prior to foaming has been shown to improve the stability of several proteins foams. The maximum effect was observed at 0.2–0.3 wt% ethanol [68]. At higher concentrations, the foam stability decreased gradually. The positive effect of ethanol on foam stability might be related to partial unfolding of the proteins, which may facilitate formation of films with better

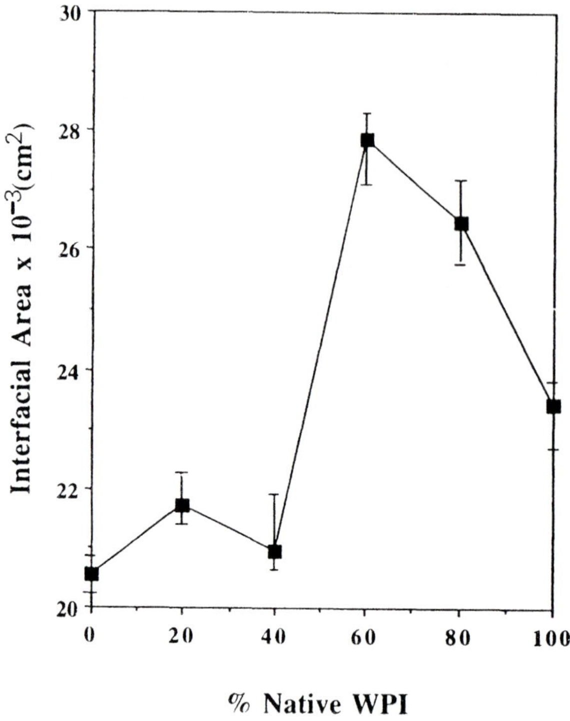

Figure 24 Effect of the ratio of native and polymerized proteins in WPI on the initial interfacial area of WPI foams. The bars represent standard error. (From Ref. 40.)

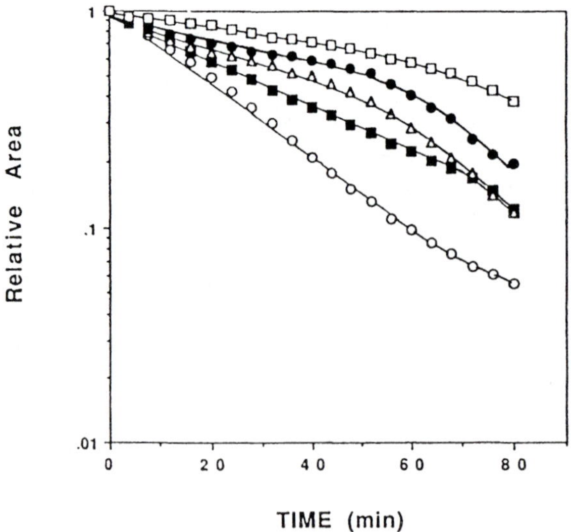

Figure 25 Effect of dithiothreitol concentration on surface area decay of BSA foam (1%, pH, 7) at 25°C. (○) 0 mM; (■) 0.5 mM; (△) 1.0 mM; (●) 5.0 mM; (□) 10 mM. (From Ref. 38.)

rheological properties. Among the proteins studied, the effect of ethanol on the stabilities of protein foams followed the order gelatin > β-casein > β-lactoglobulin > lysozyme.

Since commercial proteins are mixtures of several proteins, the effects of interactions among protein components in a mixture on the foaming properties have been studied in model systems [69–71]. Most notably, studies have shown that the foaming properties of acidic proteins can be improved by mixing with basic proteins, such as lysozyme and clupeine [71]. For instance, addition of lysozyme up to a level of 0.1% to a solution of 0.5% BSA dramatically improved the foaming properties of BSA at pH 8.0. The pH and ionic strength dependability of this effect suggested that electrostatic complex formation between the acidic and basic proteins was responsible for foam enhancement by basic proteins. However, it is debatable whether or not lysozyme exists as a part of the adsorbed primary layer along with the acidic protein. It has been shown in model adsorption systems that when the concentration ratio of BSA to lysozyme or β-casein to lysozyme is above 2.8, the primary adsorbed layer is entirely composed of BSA (or β-casein) only, and lysozyme formed a secondary layer beneath the primary layer of the acidic protein (see Fig. 9) [24,25]. This may occur in foams as well. If so, then the stabilizing effect of basic proteins on acidic protein foams might be related to the formation of an electrical double layer of acidic and basic proteins at the interface, which may improve the viscoelastic properties of the film and increase the disjoining pressure as well.

IV. EMULSIONS

Food emulsions can be divided into two types: oil-in-water emulsions, in which an oil is dispersed in an aqueous continuous phase, and water-in-oil emulsions, in which water is dispersed in a continuous oil phase. These colloidal dispersions are thermodynamically unstable, and they immediately separate into two phases unless a surfactant is present at the interface. Even in the presence of an adsorbed surfactant at the interface, these dispersed systems are only kinetically stable; they flocculate and coalesce and eventually separate into two phases after a period of time. The kinetic stability of an emulsion, then, depends on the physical and chemical properties of the adsorbed surfactant layer and its ability to prevent flocculation and coalescence of oil droplets. Proteins, being amphiphilic and able to form cohesive viscoelastic films at air/water and oil/water interfaces, are preferred over low molecular weight surfactants as emulsifiers in food applications. However, although all proteins are amphiphilic, they display considerable differences in emulsifying properties. This chapter will present a general discussion on emulsion stability and the molecular properties of proteins affecting their emulsifying properties.

A. Emulsion Stability

Several physical processes in an emulsion affect its kinetic stability. The principal ones are creaming, flocculation/aggregation, and coalescence. The sequence of occurrence of these processes leading to phase separation is depicted in Figure 26 [72]. All of these processes are interdependent. For example, coalescence increases creaming, flocculation/aggregation increases both coalescence and creaming, and creaming enhances flocculation/aggregation. A general description of these physical processes is

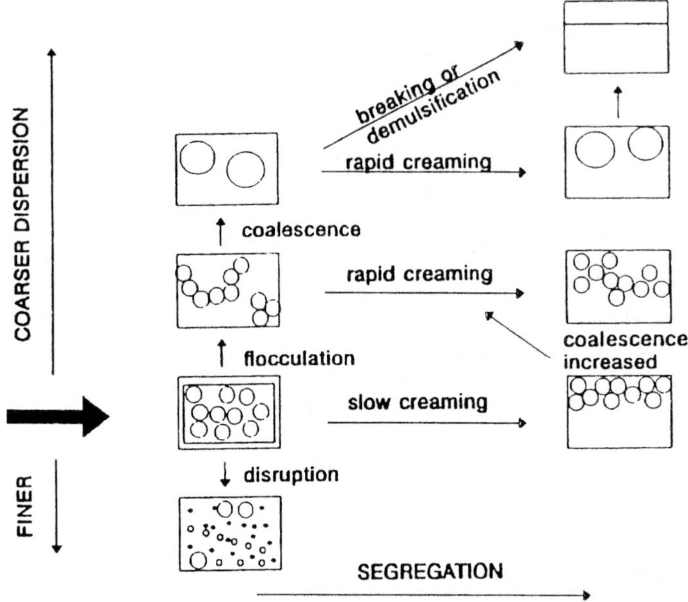

FIGURE 26 Various types of instability of an oil-in-water emulsion. (From Ref. 72.)

given below. More detailed analyses of the theories describing these phenomena can be found elsewhere [52,72,73].

1. Creaming

Creaming, which is a reversible process, occurs because of density differences between the oil and the aqueous phases. The oil droplets, being of lower density than water, tend to move to the top of an emulsion sample. The velocity of rise of a single droplet, based on Stokes law, is:

$$\nu = \frac{2r^2 \Delta\rho g}{9\eta_o} \tag{36}$$

where r is the radius of the oil droplet, $\Delta\rho$ is the density difference between the aqueous and the oil phases, g is acceleration due to gravity, and η_o is the viscosity of the aqueous phase. The velocity of creaming is directly proportional to the square of the radius and density difference and inversely proportional to the viscosity of the continuous phase. For a dilute emulsion with polydispersity, the creaming rate is [74]:

$$\nu = \frac{(d^2 + \sigma^2)\Delta\rho g}{18\eta_o} \tag{37}$$

where d and σ are the weight average and standard deviation of the droplet diameter distribution. Increase of viscosity of the aqueous phase by the addition of hydrocolloids, such as various polysaccharides and gums, can effectively slow down the rate of creaming [74,75].

The rate of creaming also can be decreased by reducing the size of the oil droplets. Under turbulent conditions, as it is encountered in a homogenizer, the average diameter of droplets formed is empirically related to [76]:

$$d_{ave} \propto E^{-2/5}\,\gamma^{3/5}\,\rho^{-1/5} \tag{38}$$

where E is the energy density, i.e., the amount of energy input per unit volume per unit time, γ is the interfacial tension, and ρ is the density of the continuous phase. At sufficiently high protein concentration, the equilibrium interfacial tension of a protein solution/oil interface is similar for most food proteins, although there are exceptions, such as lysozyme (which is not a typical food protein). If this is the case, then at a given protein concentration the size of the oil droplet formed should be dependent only on the energy density. Indeed, above 3% concentration and under identical homogenization conditions, the specific surface area of various food protein emulsions has been shown to be the same (see Ref. 77). At 1% protein concentration, however, the specific interfacial areas of emulsions formed under identical emulsification conditions were different for various proteins. For the proteins studied, it followed the order soy proteins < whey proteins < Na-caseinate < soluble wheat protein < blood plasma proteins [77]. The low specific surface area for soy proteins is not due to protein depletion in the continuous phase. It should be pointed out that even at 1% concentration, the static oil/water interfacial tension for these proteins is very similar. Thus, the differences in the abilities of these proteins to create interfacial area must be related to dynamic interfacial gradients formed in oil droplets during homogenization. When interfacial gradients are formed around the oil droplets, the stability of the droplets against coalescence will depend on the rate of flow of the adsorbed protein layer from low-surface-tension regions to high-surface-tension regions (Marangoni effect) and/or the rate at which protein molecules from the bulk phase adsorb to the interface. Walstra [76] has argued that under typical emulsification conditions in a homogenizer, the time needed for a protein to cover a droplet is in the range of 10^{-5}–10^{-7} s, and therefore the rate of protein adsorption has no bearing on emulsification. It should be recognized, however, that, as it is true for all turbulent flow systems, around each oil droplet there is a stationary aqueous layer, and the thickness of this boundary layer will affect mass transport of the protein from the continuous phase to the oil/water interface. Thus, if the rate of formation of interfacial tension gradients around droplets under highly turbulent conditions in a high-pressure homogenizer is faster than either the rate of adsorption of the protein or the rate of flow of the adsorbed protein layer from low- to high-interfacial-tension regions, then collisions between droplets at the high-interfacial-tension sites would inevitably lead to coalescence during emulsification, resulting in large droplet size and low-specific-interfacial area. Thus, the physicochemical properties of proteins and the viscoelastic properties of the protein film formed at the oil/water interface will also influence the droplet size.

2. Aggregation/Flocculation

Flocculation occurs when two or more droplets collide leading to formation of an aggregate. In this aggregate, each droplet maintains its shape and identity. The formation of a stable aggregate occurs when the attractive forces between the droplets are greater than repulsive forces. The attractive interactions arise from van der Waals and hydrophobic interactions between the droplets, and repulsion arises from electrostatic repulsive interactions and steric repulsion between the adsorbed layers. For two droplets of radius r in water at room temperature (~25°C), the attractive and repulsive potential energies of interaction, based on the Deryagin-Landau-Verwey-Overbeek (DLVO) theory, are [78]:

$$G_A = \frac{-Aa}{12D} \tag{39}$$

$$G_R = 4.3 \times 10^{-9} \text{ a } \psi_o^2 \ln (1 + e^{-\kappa D}) \tag{40}$$

where A is the Hamaker constant (in Joules), a is the radius of the droplets (in m), D is the distance of separation ($D \ll a$), ψ_o is the surface potential (in V), and κ is the Debye-Huckel parameter (in m^{-1}). For a triglyceride with 10- to 18-carbon acyl chains, the Hamaker constant in water is about 5×10^{-21} J [79]. The general profiles of van der Waals attractive and electrostatic repulsive interaction free energies as a function of surface separation distance between two oil droplets of 4 μm diameter coated with a 2 mg m^{-2} β-casein film are shown in Figure 27. When only van der Waals and electrostatic interactions are considered, then flocculation occurs at the secondary minimum of the

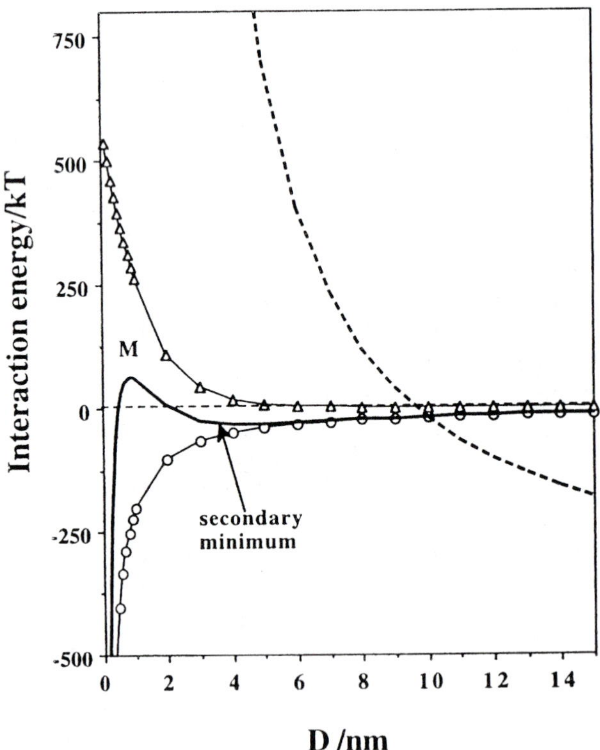

Figure 27 Calculated energy profiles of attractive van der Waals (○) and repulsive electrostatic (△) interactions between two oil droplets coated with a β-casein–adsorbed layer, based on Eqs. (39) and (40). The diameter of each droplet was assumed to be 4 μm. The surface potential ψ_o for the β-casein layer was assumed to be −20 mV. The thick solid line represents the sum of attractive van der Waals and repulsive electrostatic interactions. The thick broken line represents the net free energy profile as a function of distance if steric repulsion is also taken into account (see text and Fig. 29).

net free energy curve. The strength of the flocs is dependent on the depth of this secondary minimum: If the minimum is deep, the flocs tend to be strong aggregates. If the positive potential at location M is not too large compared to the thermal energy kT, the particles may climb over this barrier and form strong aggregates in the primary minimum.

One of the most important forces that opposes flocculation/aggregation and coalescence of emulsion droplets is the steric repulsion between the adsorbed protein layers (Fig. 28). When two droplets covered with protein layers approach each other, the overlap of the protruding polymer chains produces a net repulsion between the adsorbed layers. This interaction has two components. The first is the osmotic repulsion between the overlapping segments, which favors stretching of the chains. The second is related to the elastic energy of the chains, which opposes stretching. The net steric repulsive energy per unit area of the interacting surfaces is given by [80]:

$$G_{SR} = (kT\Gamma L/s)[(2L/D)^{2.25} - (D/2L)^{0.75}] \quad (41)$$

where k is the Boltzman constant, T is the temperature, Γ is the number of adsorbed polymer molecules per unit area of the surface, s is the mean distance between the polymers [$s = \sqrt{(1/\Gamma)}$], L is the thickness of the protruding chains, and D is the distance between the droplet surface. Recently, neutron reflectivity measurements have shown that the thickness of the β-casein layer adsorbed at the air/water interface is about 5–10 nm [81]. Assuming that these values are also true for the oil/water interface, the potential energy profile for steric repulsion, calculated using Eq. (41), between two adsorbed β-casein layers (for $\Gamma = 2$ mg.m^{-2}) is as shown in Figure 29. The steric repulsion between the layers increases very steeply when $D < 2L$. The magnitude of this steric repulsive free energy is far greater than the attractive van der Waals forces. Indeed, when steric repulsion is included in the net free energy calculation (Fig. 27), the secondary minimum disappears and the free energy of interaction increases very steeply once the brushes of the protein layers come into contact. Hence, steric repulsion between the adsorbed protein layers appear to be the dominant force controlling flocculation and coalescence of protein-coated droplets.

If the protruding segments are highly hydrated, then, in addition to the steric repulsion, hydration repulsion also would inhibit flocculation of the droplets. In the absence of attractive intersegment interactions between the chains, for example, hydrophobic interactions, the steric and hydration repulsions will act against flocculation/aggregation

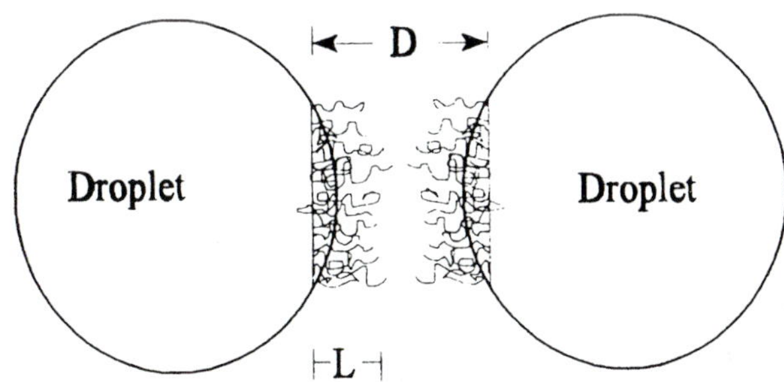

FIGURE 28 Schematic representation of the brushes of adsorbed protein layers on emulsion droplets exerting steric repulsion.

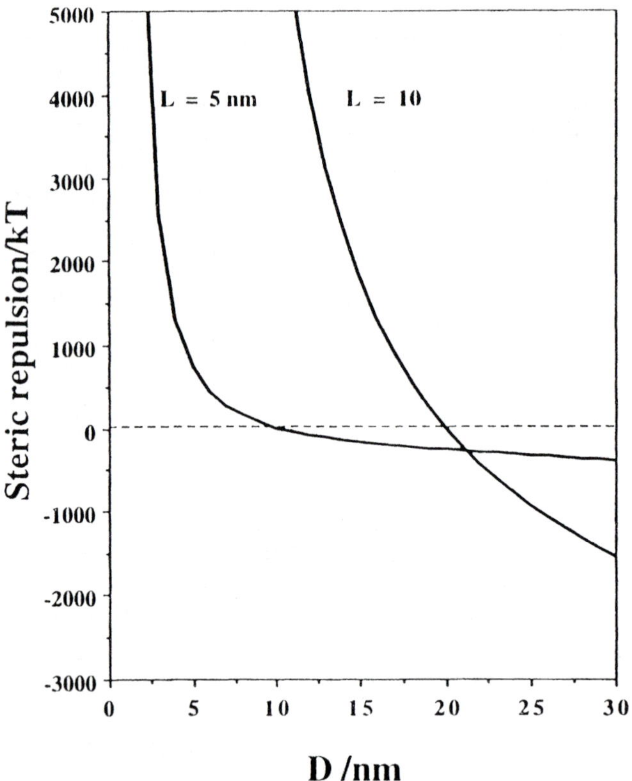

FIGURE 29 Calculated steric repulsive interaction free energy between two adsorbed β-casein layers on emulsion droplets based on Eq. (41). The following assumptions were used: $\Gamma = 2$ mg m^{-2} (i.e., 5.02×10^{16} β-casein molecules per m^2); surface area of contact between the interacting surfaces $= 3.14 \times 10^{-15}$ m^2 (1/16000 of total surface area of a 4-μm droplet); s $= 4.464 \times 10^{-9}$ m. The two curves are for protruding protein layer thicknesses of 5 and 10 nm.

of the droplets. The net potential energy of interaction between two droplets can then be given as:

$$U_{net} = U_{vdW} + U_{ele} + U_{steric} + U_{hyd} \tag{42}$$

Flocculation/aggregation of emulsion droplets is affected by solution conditions, such as pH, temperature, ionic strength, and macromolecular stabilizers.

3. Coalescence

Coalescence between particles occurs when collision between them leads to rupture of the protein film. This processes is irreversible. The probability of coalescence increases when the collisional encounters between particles is high as in highly concentrated emulsions. Flocculation and creaming also enhances the likelihood of coalescence.

The coalescence rate is fundamentally related to the particle size distribution and the mechanical/viscoelastic properties of the interfacial film. The rate of coalescence is slow when the droplet size is small. When the droplet size is larger than 5 μm, the coalescence rate is generally independent of the viscoelastic properties or the thickness of the protein layer. The droplet size can be reduced by using a high protein concentration and a high energy density during emulsification. However, even under the best of conditions, the minimum average droplet diameter of protein-stabilized emulsions, e.g., β-casein emulsion, is about 0.3–0.4 μm [82]. The droplet size distribution in commercially prepared emulsions is typically in the range of 1–10 μm.

The thicker the adsorbed protein layer, the lower is the coalescence rate. A thick absorbed layer has high surface viscosity and elasticity and provides steric stability against flocculation and coalescence. The thickness of the absorbed layer is related to surface load, Γ. Under turbulent conditions in a high-pressure homogenizer, the amount of protein adsorbed to oil droplets is given by [83]:

$$\Gamma(t) = KCd\,(1 + x/d)^3\,t \tag{43}$$

where K is a constant, C is the concentration of protein in the bulk phase, x is the diameter of the adsorbing protein particles, d is the diameter of the emulsion droplets, and t is the time. Equation (43) suggests that the surface load will be higher if the size of the protein particle is larger.

The stability against coalescence is also dependent on the integrity of the protein film, which can be affected by solution conditions, such as pH, temperature, and ionic strength. Conditions that lead to aggregation or coagulation of the protein in the adsorbed layer can create "holes" in the film, causing coalescence. The coalescence rate generally increases as the pH of the emulsion approaches the isoelectric point of the protein [84], primarily because of precipitation of the protein in the adsorbed layer. The decrease in the electrostatic repulsion between droplets at the isoelectric pH of the protein also contributes to increased flocculation and coalescence.

The presence of low molecular weight surfactants, such as phospholipids, often promotes coalescence. This is principally because of displacement of proteins from the interface by the low molecular weight surfactants. This results in a decrease of surface viscosity, elasticity, and steric and electrostatic repulsions. Formation of fat crystals at the oil/water interface also causes partial coalescence [85].

B. Molecular Factors Affecting Emulsifying Properties

1. Solubility

Solubility of proteins plays a role in their emulsifying properties; however, 100% solubility is not an absolute requirement. While highly insoluble proteins do not perform well as emulsifiers, no reliable relationship exists between solubility and emulsifying properties in the 25–80% solubility range [86]. However, since highly insoluble proteins may precipitate at the oil/water interface, which may promote coalescence, some degree of solubility may be necessary in order to form a stable elastic film at the interface.

2. Surface Hydrophobicity

The emulsifying properties of proteins show a positive, albeit a weak, correlation with surface hydrophobicity (as measured by the *cis*-parinaric acid–binding method) [57], but not with mean hydrophobicity of amino acid residues. It appears that this relationship may result from the fact that proteins having many hydrophobic patches or cavities on their surface are potentially unstable at the oil/water interface. On the other hand, the poor correlation between the emulsifying properties and mean hydrophobicity indicates that although proteins are unstable and undergo conformational change at the oil/water interface, they are not extensively denatured as it occurs at the air/water interface. This is reasonable because the interfacial free energy at the oil/water interface is considerably lower than that at the air/water interface; this lower interfacial energy probably is insufficient to overcome the activation energy barrier for complete unfolding of proteins at the oil/water interface. In any case, the positive correlation between surface hydrophobicity and emulsifying properties is by no means perfect, because the emulsifying properties of certain proteins, such as β-lactoglobulin and α-lactalbumin, do not show strong correlation with surface hydrophobicity [87]. This may be related to differences in the molecular flexibility of these proteins. It should be noted that while *cis*-parinaric acid binding may provide information on the number of hydrophobic cavities on a protein's surface, it provides no direct information on molecular flexibility, which may be the most important factor affecting the emulsifying properties.

3. Heat Denaturation

Controlled heat denaturation of proteins prior to emulsification, which does not result in insolubilization, can improve the emulsifying properties of proteins [88,89]. This is probably due to an increase of surface hydrophobicity and flexibility of heat-treated proteins. Matsumura et al. [90] showed that α-lactalbumin in the molten globule state (see Chapter 1) unfolded more readily than the native protein at interfaces. Moreover, in competitive adsorption experiments involving native β-lactoglobulin and the molten globule state α-lactalbumin, the latter preferentially adsorbed to the oil/water interface than the former. Excessive heat treatment that leads to aggregation and/or insolubilization of proteins impairs their emulsifying properties.

4. Competitive Adsorption

Typical food proteins are mixtures of several proteins. Thus, competitive adsorption among the proteins is to be expected, and hence the composition of the interfacial film and its stability will be dependent on the initial ratio of protein components in the bulk phase. Shimizu et al. [87] showed that the proteins of whey exhibited selective adsorption at the oil/water interface. This selectivity was affected by the pH. At pH 3, 48% of the proteins in the film at the oil/water interface was α-lactalbumin, which decreased to about 10% at pH 9; in contrast, the surface composition of β-lactoglobulin increased from 13% at pH 3 to about 62% at pH 9. In the case of caseinate-stabilized emulsions, the composition of α_{s1}- and β-caseins at the oil/water interface immediate after formation of the emulsion was almost the same as that of the initial bulk solution; however, during aging, the ratio of the β-casein to α_{s1}-casein in the adsorbed film increased, presumably because of displacement of α_{s1}-casein by β-casein [91]. Generally, highly hydrophobic proteins displace less hydrophobic protein from liquid interfaces, provided the proteins under investigation are thermodynamically compatible [26]. Because α_{s1}-casein and β-casein are similar proteins and thermodynamically compatible, they can displace each other from the oil/water interface. However, this may not be the case with other protein pairs. For instance, investigations on gelatin-caseinate mixtures have shown that at a gelatin-to-caseinate weight ratio of less than 2:1, the interfacial film of freshly formed emulsion contained only caseinate [92]. However, in the case of gelatin–β-lactoglobulin mixtures, a significant amount of gelatin was able to adsorb to the oil/water interface even at high ratios of β-lactalbumin to gelatin in the bulk phase. This must be related to differences in thermodynamic compatibility of mixing of gelatin with β-casein and β-lactoglobulin at the interface.

Information on competitive adsorption of proteins at the oil/water interface is limited. More experimental studies are needed to elucidate the fundamental mechanism affecting competitive adsorption of globular proteins and its effect on the formation and stability of emulsions.

5. Low Molecular Weight Surfactants

Typical food emulsions contain significant amounts of low molecular weight surfactants, such as phospholipids and mono- and diglycerides. Competitive adsorption between proteins and these low molecular weight surfactants can influence the stability of food emulsions. Because low molecular weight surfactants are more surface-active than the macromolecular protein surfactants, they effectively compete for adsorption at interfaces. When added to protein-stabilized emulsions, they displace proteins from the interface [93–96]. The displacement of the protein from the interface decreases interparticle electrostatic and steric repulsions and reduces the thickness of the protein layer and the viscoelastic properties of the protein film. These adverse changes induce instability and rapid coalescence of emulsion particles. Notably, the low molecular weight surfactants affect the stability but not the droplet size [95,97]. The time of addition of the low molecular weight surfactant also has an effect on the stability of the emulsion. Generally, when a low molecular weight surfactant is added prior to emulsification, only partial displacement of the protein occurs. When it is added after emulsification, complete displacement occurs, depending on the amount added.

Unfortunately, a majority of the studies on competitive adsorption of low molecular weight surfactants and proteins were done on dairy proteins, especially β-casein. Since β-casein is a flexible disordered protein, it is not possible to determine what role, if any, folded conformations of globular proteins play in their displacement from the oil/water interface by low molecular weight surfactants.

6. Conformation of Adsorbed Protein

Investigation of the secondary and tertiary conformations of proteins in situ at liquid interfaces is difficult. However, indirect evidence suggests that proteins undergo significant conformational changes at interfaces. For example, two of the four nonreactive cysteine residues in native ovalbumin become reactive in foamed ovalbumin, indicating denaturation of ovalbumin at the air/water interface [59]. Conformational changes in proteins at interfaces are to be expected, because, since the conformation of a protein is the manifestation of its interaction within itself and with the environment, any change in a protein's environment must cause conformational change. This must involve a net decrease in its free energy compared to that in the bulk phase. Since the exact secondary and tertiary conformation of a protein at an interface is difficult to decipher, its configuration is usually depicted in terms of "trains," "loops," and "tails" (Fig. 30). The trains are the hydrophobic segments that lie flat on the interface, making contact with both the aqueous and oil phases (only the nonpolar side chains of amino acids are oriented towards the oil phase); the loops are the polypeptide segments between the trains that are suspended into the aqueous continuous phase; and the tails are the N- and C-terminal segments invariably suspended into the aqueous phase. The relative distribution of trains, loops, and tails in the adsorbed molecule determines the properties of emulsions. If trains

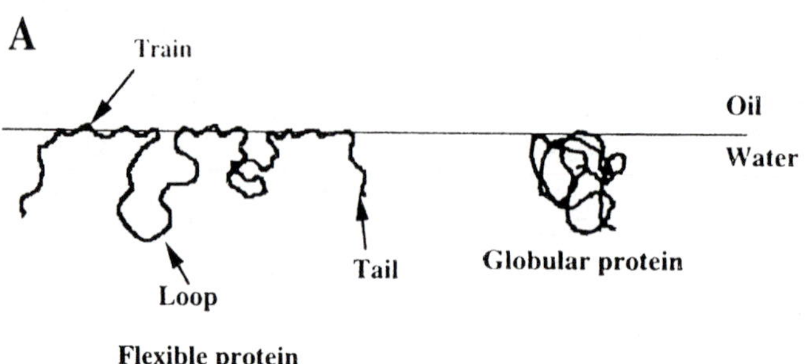

FIGURE 30 Various protein configurations at liquid interfaces. B is the "two pancakes" model proposed by Douillard [99].

are the predominant configuration, then the protein will exert a high surface pressure. For this to happen, the protein should be highly flexible and highly hydrophobic (e.g., β-casein). The stability of the emulsions of such proteins, however, may be no better than those that form more loops, e.g., globular proteins. The presence of loops extended into the continuous phase will exert steric repulsion against flocculation and coalescence.

The relative amounts of trains and loops in an adsorbed protein depends at least on two factors: the rate of adsorption and the rate of unfolding at the interface. If the rate of adsorption is faster than the rate of unfolding of the protein at the interface, as it is at high protein concentration, then because of the limited availability of unoccupied area at the interface, the protein may only partially unfold or only a limited number of trains may form; a majority of the polypeptide may assume the loop configuration. Thus, in a typical emulsion formed with 3–5% protein solution, a major fraction of the protein at the interface may be in the loop configuration.

On thermodynamic grounds, it is reasonable to argue that the loops formed at the interface must be oriented only towards the aqueous continuous phase and not into the oil phase. For a polypeptide segment to form a loop into the oil phase, all the amino acid residues in that loop should be nonpolar. Also, since the formation of even a tight bend, such as β-bends in proteins, requires at least five amino acid residues, a segment containing more than five nonpolar amino acid residues will be required to form a loop into the oil phase. Such a long stretch of nonpolar segment is not common in proteins. Even if one of the amino acid residues in the loop-forming segment is a charged polar residue, the loop formation in the oil phase will not occur because of the free energy demand for transferring this polar residue from the aqueous phase to the oil phase. Moreover, since the polar peptide bonds of the segment also would prefer to be in contact with the aqueous phase, it is unlikely that typical proteins can form loops into the oil phase. The "two-pancakes" model proposed by Douillard (Fig. 30B) [98,99] for protein configuration, in which one of the pancakes (loops) is extended into the oil phase and the other into the aqueous phase, is therefore untenable. This type of configuration may be possible in synthetic block copolymers containing alternating stretches of hydrophilic and hydrophobic segments, but not with water-soluble globular proteins. Another argument against this model is that if a protein can indeed form several loops into the oil phase or undergo molecular partitioning at the interface, then it should be possible to make both oil-in-water and water-in-oil emulsions using such a protein. However, it is known that a water-in-oil emulsion cannot be made with a typical water-soluble protein. In this regard, proteins strictly follow Bancroft's rule, which stipulates that the continuous phase of an emulsion is the one in which the surfactant has good solubility.

V. CONCLUSION

The physics of foams and emulsions is well understood. Also, considerable progress has been made in understanding the formation and stability of protein-stabilized foams and emulsions. Kinetic studies on the adsorption of proteins at planar interfaces have shed light on the molecular factors affecting adsorptivity of proteins at liquid interfaces. It has often been said that adsorption studies with very dilute protein solutions have no relevance to emulsion formation under turbulent conditions, where the time of adsorption is of the order of less than a millisecond instead of minutes. However, it should be recognized that although the collisional frequency of proteins with an interface is dependent on the concentration and the degree of turbulence, whether or not the collision will lead to adsorption depends on the molecular properties of proteins. The elucidation of the molecular factors that affect surface activity of proteins can be assessed only in dilute systems, not in emulsions. Indeed the results of model dilute systems relate very well with those of emulsion systems. For example, the competitive adsorption and displacement/exchange behaviors of α_{s1}-casein and β-casein at the planar air/water interface in dilute systems [26] has been shown to be exactly similar to those that occur in the oil-in-water emulsion system [91].

Although there is a copious amount of information on the factors affecting protein adsorption at the planar air/water interface, no such studies have been reported for the planar oil/water interface. Kinetics and thermodynamics of adsorption of different proteins under various solution conditions and conformational states in the bulk phase may provide useful information on the molecular factors affecting the emulsifying properties of proteins.

ACKNOWLEDGMENT

Financial support from the National Science Foundation (Grant #BES 9315123), which was responsible for some of the research results presented in this chapter, is acknowledged.

REFERENCES

1. K. W. Jones and R. W. Mandigo, Effects of chopping temperature on the microstructure of meat emulsions, *J. Food Sci. 47*: 1930 (1982).
2. S. Damodaran, Amino acids, peptides, and proteins, *Food Chemistry* (O. Fennema, ed.), Marcel Dekker, New York, 1996, pp. 321–429.
3. S. Damodaran, Functional properties, *Food Proteins: Properties and Characterization* (S. Nakai and H. W. Modler, eds.), VCH Publishers, Inc., New York, 1996, pp. 167–234.

4. A. F. H. Ward and L. Tordai, Time-dependence of boundary tensions of solutions: 1. The role of diffusion in time effects, *J. Chem. Phys. 39*: 453 (1946).
5. J. A. De Feijter and J. Benjamins, Adsorption kinetics of proteins at the air-water interface, *Food Emulsions and Foams* (E. Dickinson, ed.), Royal Society of Chemistry, London, 1987, pp. 72–85.
6. D. E. Graham and M. C. Phillips, Proteins at liquid interfaces: 1. Kinetics of adsorption and surface denaturation. *J. Colloid Interface Sci. 70*: 403 (1979).
7. F. MacRitchie and A. E. Alexander, Kinetics of adsorption of proteins at interfaces: II. The role of pressure barriers in adsorption, *J. Colloid Sci. 18*: 458 (1963).
8. S. Xu and S. Damodaran, Comparative adsorption of native and denatured egg-white, human, and T_4 phage lysozyme at the air-water interface, *J. Colloid Interface Sci. 159*: 124 (1993).
9. L. Ter-Minassian-Saraga, Protein denaturation on adsorption and water activity at interfaces: an analysis and suggestion, *J. Colloid Interface Sci. 80*: 393 (1981).
10. S. Damodaran and K. B. Song, diffusion and energy barrier controlled adsorption of proteins at the air-water interface, *Interactions of Food Proteins* (N. Parris and R. Barford, eds.), ACS Symposium Series 454, American Chemical Society, Washington, DC, 1991, pp. 104–121.
11. S. Damodaran, Interrelationship of molecular and functional properties of food proteins, *Food Proteins* (J. E. Kinsella and W. G. Soucie, eds.), The American Oil Chemists' Society, Champaign, IL, 1989, pp. 21–51.
12. B. Lee and F. M. Richards, The interpretation of protein structure: estimation of static accessibility, *J. Mol. Biol. 55*: 379 (1971).
13. S. Damodaran and K. B. Song, Kinetics of adsorption of proteins at interfaces: role of protein conformation in diffusional adsorption, *Biochim. Biophys. Acta 954*: 253 (1988).
14. S. Xu and S. Damodaran, The role of chemical potential in the adsorption of lysozyme at the air-water interface, *Langmuir 8*: 2021 (1992).
15. S. Damodaran and S. Xu, The role of electrostatic forces in anomalous adsorption behavior of phosvitin at the air/water interface, *J. Colloid Interface Sci. 178*: 426 (1996).
16. Y. Cao and S. Damodaran, Coadsorption of β-casein and bovine serum albumin at the air-water interface from a binary mixture, *J. Agric. Food Chem. 43*: 2567 (1995).
17. B. M. Byrne, A. D. van het Schip, J. A. M. van de Klundert, A. C. Arnberg, M. Gruber, and A. B. Greet, Amino acid sequence of phosvitin derived from the nucleotide sequence of part of the chicken vitellogenin gene, *Biochemistry 23*: 4275 (1984).
18. R. C. Clark, The primary structure of avian phosvitins, *Int. J. Biochem. 17*: 983 (1985).
19. M. F. Perutz, Electrostatic effects in proteins, *Science 201*: 1187 (1978).
20. J. N. Israelachvili and R. M. Pashley, The hydrophobic interaction is long range, decaying exponentially with distance, *Nature 300*: 341 (1982).
21. J. N. Isrealachvilli and P. M. McGuiggan, Forces between surfaces of liquids, *Science 241*: 795 (1988).
22. J. T. Davies and E. K. Rideal, *Interfacial Phenomena*, 2nd ed., Academic Press, New York, 1963.
23. K. B. Song and S. Damodaran, Influence of electrostatic forces on the adsorption of succinylated β-lactoglobulin at the air-water interface, *Langmuir 7*: 2736 (1991).
24. K. Anand and S. Damodaran, Kinetics of adsorption of lysozyme and bovine serum albumin at the air-water interface from a binary mixture, *J. Colloid Interface Sci. 176*: 63 (1995).
25. S. Xu and S. Damodaran, Kinetics of adsorption of proteins at the air-water interface from a binary mixture, *Langmuir 10*: 472 (1994).
26. K. Anand and S. Damodaran, Dynamics of exchange between α_{s1}-casein and β-casein during adsorption at air-water interface, *J. Agric. Food Chem. 44*: 1022 (1996).
27. F. MacRitchie and L. Ter-Minnassian Saraga, Stability of highly compressed monolayers of I-labeled and cold BSA, *Prog. Colloid Polymer Sci. 68*: 14 (1983).
28. F. MacRitchie and L. Ter-Minnassian Saraga, Concentrated protein monolayers: Desorption studies with radiolabeled bovine serum albumin, *Colloid Surf. 10*: 53 (1984).

29. F. MacRitchie, Desorption of proteins from the air/water interface, *J. Colloid Interface Sci. 105*: 119 (1985).
30. V. I. Polyakov, V. Y. Grinberg, Y. A. Antonov, and V. B. Tolstoguzov, Limited thermodynamic compatibility of proteins in aqueous solutions, *Polym. Bull. 1*: 593 (1979).
31. V. I. Polyakov, I. A. Popello, V. Y. Grinberg, and V. B. Tolstoguzov, Thermodynamic compatibility of proteins in aqueous medium, *Nahrung 30*: 365 (1986).
32. A. Prins, Principles of foam stability, *Advances in Food Emulsions and Foams* (E. Dickinson and G. Stainsby, eds.), Elsevier Applied Sci. Publishers, London, 1988, pp. 91–122.
33. M. Yasukawa, K. Yamagiwa, and A. Ohkawa, Foam ratio as a measure for mechanical foam-breaking in a gas bubbling system, *Bull. Chem. Soc. Jpn. 63*: 3307 (1990).
34. P. J. Halling, Protein-stabilized foams and emulsions, *CRC Crit. Rev. Food Sci. Nutr. 13*: 155 (1981).
35. J. B. German, T. E. O'Neill, and J. E. Kinsella, Film forming foaming behavior of food proteins, *J. Am. Oil Chem. Soc. 62*: 1358 (1985).
36. G. Nishioka and S. Ross, A new method and apparatus for measuring foam stability, *J. Colloid Interface Sci. 81*: 1 (1981).
37. S. Ross, Bubbles and foam, *Ind. Eng. Chem. 61*: 48 (1969).
38. M.-A. Yu and S. Damodaran, Kinetics of protein foam destabilization: evaluation of a method using bovine serum albumin, *J. Agric. Food Chem. 39*: 1555 (1991).
39. M.-A. Yu and S. Damodaran, Kinetics of destabilization of soy protein foams, *J. Agric. Food Chem. 39*: 1563 (1991).
40. H. Zhu and S. Damodaran, Heat-induced conformational changes in whey protein isolate and its relation to foaming properties, *J. Agric. Food Chem. 42*: 846 (1994).
41. H. Zhu and S. Damodaran, Effects of calcium and magnesium ions on aggregation of whey protein isolate and its effect on foaming properties, *J. Agric. Food Chem. 42*: 856 (1994).
42. H. Zhu and S. Damodaran, Proteose peptones and physical factors affect foaming properties of whey protein isolate, *J. Food Sci. 59*: 554 (1994).
43. S. Damodaran, Structure-function relationship of food proteins, *Protein Functionality in Food Systems* (N. S. Hettiarachchy and G. R. Ziegler, eds.), Marcel Dekker, New York, 1994, pp. 1–38.
44. D. E. Graham and M. C. Phillips, The conformation of proteins at the air-water interface and their role in stabilizing foams, *Foams* (R. J. Akers, ed.), Academic Press, New York, 1976, pp. 237–255.
45. N. Kitabatake and E. Doi, Surface tension and foamability of protein and surfactant solutions, *J. Food Sci. 53*: 1542, 1569 (1988).
46. S. Ross and I. D. Morrison, Foams, *Colloidal Systems and Interfaces*, Wiley & Sons, New York, 1988, pp. 294–326.
47. E. D. Manev, S. V. Sazdanova, and D. T. Wasan, Emulsion and foam stability: the effect of film size and film drainage, *J. Colloid Interface Sci. 97*: 591 (1984).
48. A. Marmur, The effect of gravity on thin fluid films, *J. Colloid Interface Sci. 100*: 407 (1984).
49. M. C. Phillips, Protein conformation at liquid interfaces and its role in stabilizing emulsions and foams, *Food Technol. (Chicago) 35*: 50 (1981).
50. E. Ruckenstein and A. Sharma, A new mechanism of film thinning: enhancement of Reynolds velocity by surface waves, *J. Colloid Interface Sci. 119*: 1 (1987).
51. A. Sharma and E. Ruckenstein, Critical thickness and lifetimes of foams and emulsions: role of surface wave-induced thinning, *J. Colloid Interface Sci. 119*: 14 (1987).
52. E. Dickinson, *An Introduction to Food Colloids*, Oxford University Press, Oxford, 1992, p. 129.
53. F.-Q. Tang, Z. Xiao, J.-A. Tand, and L. Jiang, *J. Colloid Interface Sci. 131*: 498 (1989).
54. H. Kumagai, Y. Torikata, H. Yoshimura, M. Kato, and T. Yano, Estimation of the stability of foam containing hydrophobic particles by parameters in the capillary model. *Agric. Biol. Chem. 55*: 1823 (1991).

55. G. C. Frye and J. C. Berg, Antifoam action by solid particles, *J. Colloid Interface Sci. 127*: 222 (1989).
56. C. C. Bigelow, On the average hydrophobicity of proteins and the relation between it and protein structure, *J. Theor. Biol. 16*: 187 (1967).
57. A. Kato and S. Nakai, Hydrophobicity determined by a fluorescence probe method and its correlation with surface properties of proteins, *Biochim. Biophys. Acta 624*: 13 (1980).
58. A. Kato, Y. Osako, N. Matsudomi, and K. Kobayashi, Changes in emulsifying and foaming properties of proteins during heat denaturation, *Agric. Biol. Chem. 47*: 33 (1983).
59. N. Kitabatake and E. Doi, Conformational change of hen egg ovalbumin during foam formation detected by 5,5′-dithiobis(2-nitrobenzoic acid), *J. Agric. Food Chem. 35*: 953 (1987).
60. L. G. Phillips, S. E. Hawks, and J. B. German, Structural characteristics and foaming properties of β-lactoglobulin: effects of shear rate and temperature, *J. Agric. Food Chem. 43*: 613 (1995).
61. A. Townsend and S. Nakai, Relationship between hydrophobicity and foaming capacity of proteins, *J. Food Sci. 48*: 588 (1983).
62. T. Trziszka, Protein aggregation during whipping of egg-white and its effect on the structure and mechanical properties of foams, *Archiv. Geflugelkunde 57*: 27 (1993).
63. S. H. Kim and J. E. Kinsella, Surface active properties of food proteins: effects of reduction of disulfide bonds on film properties and foam stability of glycinin, *J. Food Sci. 52*: 128 (1987).
64. N. K. D. Kella, S. T. Yang, and J. E. Kinsella, Effect of disulfide bond cleavage on structural and interfacial properties of whey proteins. *J. Agric. Food Chem. 37*: 1203 (1989).
65. J. R. Bacon, J. W. Hemmant, N. Lambert, R. Moore, and D. J. Wright, Characterization of the foaming properties of lysozymes and α-lactalbumins: a structural evaluation. *Food Hydrocolloids 2*: 225 (1988).
66. D. C. Clark, A. R. Mackie, P. J. Wilde, and D. R. Wilson, Differences in the structure and dynamics of the adsorbed layers in protein-stabilized model foams and emulsions. *Farady Discuss. 98*: 253 (1994).
67. D. K. Sarker, P. J. Wilde, and D. C. Clark, control of surfactant-induced destabilization of foams through polyphenol-mediated protein-protein interactions, *J. Agric. Food Chem. 43*: 295 (1995).
68. M. Ahmed and E. Dickinson, Effect of ethanol on the foaming of aqueous protein solutions, *Colloids Sufaces 47*: 353 (1990).
69. L. G. Phillips, M. J. Davis, and J. E. Kinsella, The effects of various milk proteins on the foaming properties of egg-white, *Food Hydrocolloids 3*: 163 (1989).
70. D. C. Clark, A. R. Mackie, L. J. Smith, and D. R. Wilson, The interaction of bovine serum albumin and lysozyme and its effect on foam composition, *Food Hydrocolloids 2*: 209 (1988).
71. S. Poole, S. I. West, and C. L. Walters, Protein-protein interactions: Their importance in the foaming of heterogeneous protein systems, *J. Sci. Food Agric. 35*: 701 (1984).
72. H. Mulder and P. Walstra, *The Milk Fat Globule: Emulsion Science as Applied to Milk Products and Comparable Foods*, CAB, Farnham Royal, Pudoc, Wageningen, 1974.
73. P. Walstra, Introduction to aggregation phenomena in food colloids, *Food Colloids and Polymers: Stability and Mechanical Properties* (E. Dickinson, and P. Walstra, eds.), Royal Society of Chemistry, Cambridge, 1993, pp. 3–15.
74. A. M. Howe, A. R. Mackie, P. Richmond, and M. M. Robins, Creaming of oil-in-water emulsions containing polysaccharides, *Gums and Stabilizers for the Food Industry 3* (G. O. Phillips, D. J. Wedlock, and P. A. Williams, eds.), Elsevier Applied Science, London, 1985, pp. 295–309.
75. R. G. Morley, Utilization of hydrocolloids in formulated foods, *Gums and Stabilizers for the Food Industry 2* (G. O. Phillips, D. J. Wedlock, and P. A. Williams, eds.), Pergamon, Oxford, 1984, pp. 211–239.

76. P. Walstra, The role of proteins in stabilization of emulsions, *Gums and Stabilizers for the Food Industry 4* (G. O. Phillips, D. J. Wedlock, and P. A. Williams, eds.), IRL Press, Oxford, 1988, pp. 323–336.
77. P. Walstra and A. L. De Roos, Proteins at air-water and oil-water interfaces: static and dynamic aspects, *Food Rev. Int. 9*: 503 (1993).
78. P. Walstra, Introduction to aggregation phenomena in food colloids, *Food Colloids and Polymers: Stability and Mechanical Properties* (E. Dickinson and P. Walstra, eds.), Royal Society of Chemistry, Cambridge, 1993, pp. 3–15.
79. J. Israelachvili, *Intermolecular and Surface Forces*, Academic Press, London, 1992, pp. 190–191.
80. P. G. de Gennes, Polymers at an interface: a simplified view, *Adv. Colloid Interface Sci. 27*: 189 (1987).
81. P. J. Atkinson, E. Dickinson, D. S. Horne, and R. M. Richardson, A neutron reflectivity study of the adsorption of β-casein at the air-water interface, *Food Macromolecules and Colloids* (E. Dickinson and D. Lorient, eds.), Royal Society of Chemistry, Cambridge, 1995, pp. 77–80.
82. J. A. Hunt and D. G. Dalgleish, Adsorption behavior of whey protein isolate and caseinate in soy oil-in-water emulsions, *Food Hydrocolloids 8*: 175 (1994).
83. P. Walstra and H. Oortwijn, The membranes of recombined fat globules. 3. Mode of formation, *Neth. Milk Dairy J. 36*: 103 (1982).
84. K. P. Das and J. E. Kinsella, pH dependent emulsifying properties of β-lactoglobulin, *J. Dispersion Sci. Technol. 10*: 77 (1989).
85. K. Boode, P. Walstra, and A. E. A. Degrootmostert, Partial coalescence in oil-in-water emulsions. 2. Influence of the properties of the fat, *Colloids Surfaces A Physicochem. Eng. Aspects 81*: 139 (1993).
86. S. Y. Liao and M. E. Mangino, Characterization of the composition, physicochemical and functional properties of acid whey protein concentrates, *J. Food Sci. 52*: 1033 (1987).
87. M. Shimizu, M. Saito, and K. Yamauchi, The adsorptivity of whey proteins on the surface of emulsified fat, *Agric. Biol. Chem. 45*: 2491 (1981).
88. A. Kato, Y. Osako, N. Matsudomi, and K. Kobayashi, Changes in emulsifying and foaming properties of proteins during heat denaturation, *Agric. Biol. Chem. 47*: 33 (1983).
89. E. Dickinson and S. T. Hong, Surface coverage of β-lactoglobulin at the oil-water interface: influence of protein heat-treatment and various emulsifiers, *J. Agric. Food Chem. 42*: 1602 (1994).
90. Y. Matsumura, S. Mitsui, E. Dickinson, and T. Mori, Competitive adsorption of α-lactalbumin in the molten globule state, *Food Hydrocolloids 8*: 555 (1994).
91. E. Dickinson, S. E. Rolfe, and D. G. Dalgleish, Competitive adsorption of α_{s1}-casein and β-casein in oil-in-water emulsions, *Food Hydrocolloids 2*: 397 (1988).
92. J. Castle, E. Dickinson, A. Murray, B. S. Murray, and G. Stainsby, Surface behavior of adsorbed films of food proteins, *Gums and Stabilizers for the Food Industry* (G. O. Phillips, D. J. Wedlock, and P. A. Williams, eds.), Elsevier Applied Sci., London, 1986, p. 409.
93. J. L. Courthaudon, E. Dickinson, and D. G. Dalgleish, Competitive adsorption of lecithin and β-casein in oil-in-water emulsions, *J. Agric. Food Chem. 39*: 1365 (1991).
94. J. L. Courthaudon, E. Dickinson, Y. Matsumura, and D. C. Clark, Competitive adsorption of β-lactoglobulin + Tween-20 at the oil-water interface, *Colloids Surf. 56*: 293 (1991).
95. Y. Fang and D. G. Dalgleish, Competitive adsorption between dioleoylphosphatidylcholine and sodium caseinate on oil-water interfaces, *J. Agric. Food Chem. 44*: 59 (1996).
96. D. G. Dalgleish, M. Srinivasan, and H. Singh, Surface properties of oil-in-water emulsion droplets containing casein and Tween 20, *J. Agric. Food Chem. 43*: 2351 (1995).
97. A. Tomas, J. L. Courthaudon, D. Paquet, and D. Lorient, Effect of surfactant on some physicochemical properties of dairy oil-in-water emulsions, *Food Hydrocolloids 8*: 543 (1994).

98. R. Douillard, Adsorption of serum albumin at the oil/water interface, *Colloids Surf. 91*: 113 (1994).
99. J. Hargreaves, R. Douillard, and Y. Popineau, Application of polymer scaling concepts to purified gliadins at the air-water interface, *Food Macromolecules and Colloids* (E. Dickinson and D. Lorient, eds.), Royal Society of Chemistry, Cambridge, 1995, pp. 71–76.

4

Protein Gelation

David Oakenfull, John Pearce, and Ralph W. Burley
CSIRO Division of Food Science and Technology
North Ryde, New South Wales, Australia

I. INTRODUCTION

Protein gelation might first have been experienced by a slack Bronze Age cook who put off doing the dishes and found the gravy congealed in the pot next morning. Gels, such as those from milk and egg, must have been part of human food for a very long time, and until recently they must have seemed very mysterious. The word "jelly" first appeared in the fourteenth century, derived from the Latin *gelare*, meaning to freeze, via the French *gelée*, meaning frost. A book from that time mentions: "Of the shepe . . . Of whose hede boylled . . . Ther cometh a gelly" [1]. By the eighteenth century, protein gelation had reached multicolored sophistication: "To make Riben Jelly . . . run the Jelly into little high Glasses . . . one Colour must be thorough cold before you put another Colour on . . . colour red with Cochineal, green with Spinage . . . and sometimes the Jelly by itself" [1].

The scientific term "gel" was probably invented by Thomas Graham, the founding father of colloid chemistry, who died in 1869. Graham made an important advance in the understanding of gels in 1861 when he recognized "colloids" as nonsettling dispersions in a liquid of particles intermediate in size between small molecules and microscopic particles [2]. "Gel" was originally a suffix derived from the first syllable of gelatin and applied by Graham to terms such as *alcogel* and *hydrogel* [1]. Formal definitions applicable to all gels are given in Section II, where it is concluded that the most useful definition is in terms of rheology.

We now know that food gels are aqueous solutions or dispersions of high molecular weight carbohydrates or proteins, cross-linked so as to form an interconnected molecular network that spans the volume of the liquid medium. A feature of such gels is that the properties of the liquid, including colligative properties, are not altered much by the presence of the stabilizing network. The chemical structure of proteins allows a range of properties unequalled by other polymers. It is therefore not surprising that in biology—especially in animals—natural gels often consist of protein or protein attached to carbohydrate. Blood clots are made by the action of the enzyme thrombin on the protein fibrinogen, causing end-to-end aggregation and formation of a gel network structure. Collagen appears to be the gel-forming protein in jellyfish [3]. Another obvious example of a natural gel is the albumen of (fresh) birds' eggs. Thus, it is not surprising that food proteins can be modified by processing or cooking to give useful gels—boiled eggs, yogurt, surimi, tofu, and, especially, gelatin with its multitudinous food and non-food uses.

Unfortunately the characteristics of proteins responsible for their usefulness also make quantitative treatment of their physical properties difficult, even for pure proteins. Food proteins, which are rarely pure and are often used with other ingredients, are even more difficult to deal with quantitatively, so in general their physical properties must be treated empirically. Therefore, although this chapter is aimed primarily at explaining current theories of gelation by macromolecules and at pointing out the properties of proteins likely to contribute to gelation, it deals largely with specific examples and with experimental results. Gelation in food systems in general has been recently reviewed by Oakenfull [4] and Clark and Ross-Murphy [5]; gelation of proteins has been reviewed by Ziegler and Foegeding [6].

II. THE THEORY OF GELATION

A. Basic Concepts

Gels are a form of matter intermediate between a solid and a liquid. About 70 years ago, Dorothy Jordan Lloyd commented: "The colloidal condition, the gel, is one that is easier to recognize than to define" [7]. But 20 years or so later, Bungenberg de Jong [8] was able to define a gel as a "system of a solid character, in which the colloidal particles somehow constitute a coherent structure." It was known from light scattering studies and the ultramicroscope that gelation involved (very much as Thomas Graham had suspected) aggregation of colloidal particles [2], but beyond that, gelation was one of nature's mysteries. The work of Flory [9], Treloar [10], and many others in the 1940s on cross-linked polymers (such as rubber) led to the suggestion that gels consist of polymeric molecules, or submicroscopic particles, cross-linked to form an interconnected molecular network immersed in a liquid medium [11]. By 1951, Williams and Alberty [12] could report that "there is at present almost universal acceptance of the view that gels possess a ramifying, more or less coherent, framework that retains the liquid component and confers elasticity and rigidity on the system as a whole."

In food gels this liquid is invariably water; the molecular network consists of proteins or polysaccharides or a combination of the two. The properties of the gel are the net results of the complex interactions between the solvent and the molecular network (Fig. 1). The water, as a solvent, influences the nature and magnitude of the intermolecular forces that maintain the integrity of the polymer network; the polymer network holds the water, preventing it from flowing away. In effect, the polymer molecules are aggregated into one immense molecule with a three-dimensional structure that "fills" the liquid. We now think of gels in terms of this three-dimensional structure and the rheological properties it confers on the gelled material.

FIGURE 1 Schematic diagram of a gel network. The hatched areas represent junction zones where the polymer molecules are cross-linked.

B. Rheological Properties

Rheology is the science of deformation and flow of matter [13]. When we refer to a material as being gelatinous, we are thinking of its rheology. A gelatinous material is semi-solid; it has rigidity but readily deforms under stress. When we apply a force to an elastic solid, the shape changes (e.g., stretching a rubber band) and the deformation is proportional to the applied force. When this force is removed, the material springs back to its original shape and most of the energy used in producing the deformation is recovered. In contrast, when we apply a force to a liquid, it responds by flowing, and the rate of flow is proportional to the applied force. A gel is viscoelastic—it has both solidlike and liquidlike rheological properties. It is this specific rheological behavior that is most characteristic of the gel state.

There are two broad classes of rheological measurement: those made with "small deformation" and those made with "large deformation." This is an important distinction, because they provide very different classes of information about the nature of the gel. Open a carton of yogurt and touch the surface of the gel. Gentle pressure deforms the surface. Push harder and the gel ruptures. These are small-deformation and large-deformation "measurements," respectively. More quantitatively, some of the measurements that can be made are illustrated in Figure 2, which shows the relationship between load (stress, σ) and deformation (strain, ε) for a slab of gel under compression. The initial slope, where there is only very small deformation of the sample, gives the rigidity or shear modulus (G = stress/strain); the maximum load that the gel can sustain, typically under high deformation, is its rupture strength (RS). Another quantity sometimes measured is the cohesiveness, which is the stress produced by a strain of half the RS.

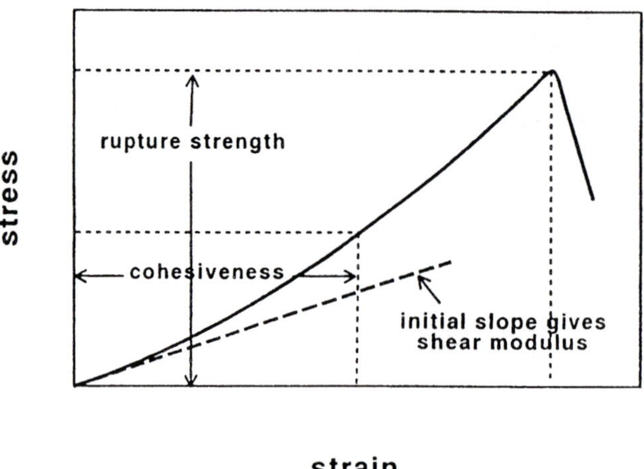

FIGURE 2 The relationship between stress and strain for a typical gel under compression.

Measurements with large deformation are easy to make with relatively simple apparatus, so it is not surprising that they are the most frequently reported type of gel measurement. Typically, a probe is inserted into the gel with constant velocity and the peak force before the gel fractures is recorded as the "gel strength." Unfortunately, this type of measurement is the least informative and the most susceptible to random experimental error. Values obtained for the stress-to-break (σ_B) and the strain-to-break (ε_B) both depend on the rate at which the probe is inserted. Moreover, the molecular events preceding fracture of the gel are inherently random. Even for carefully controlled replicate experiments there will be a distribution of values. In the classical (Griffith's) theory of fracture of linear elastic materials, failure is related to the concentration of elastically stored energy at the tip of a macroscopic crack [14]. In elastic materials, rupture is said to occur because there is a distribution of lengths of stress-bearing (elastically active) network chains [15]. Each of these chains has a limit to the extent to which it can be stretched, and as this is exceeded for the shortest chains, the extra stress is "concentrated" on the next shortest chains. In this way the material ruptures by a series of very fast, but apparently random, events. The network chains in a protein gel have very limited capacity to accept molecular deformation, so failure occurs at much smaller strains than for materials like rubber. The distribution of pore sizes within the gel may have a profound impact on its fracture properties. Cracks may propagate more easily through areas of lower protein concentration [15]. Despite these inherent difficulties, it is possible to obtain a set of data characteristic of the gel by measuring σ_B and ε_B at different extension rates and for different replicates [16]. This gives the "failure envelope" as shown in Figure 3.

Until recently, measurements with small deformation were much more difficult to carry out. Curves of stress versus strain, as shown in Figure 2, are rarely linear, and it is necessary to estimate the initial slope. Modern instruments linked to computers make this quicker and more reliable than in the past. It has also recently been found possible to measure the absolute shear modulus of a gel from a simple compression test with

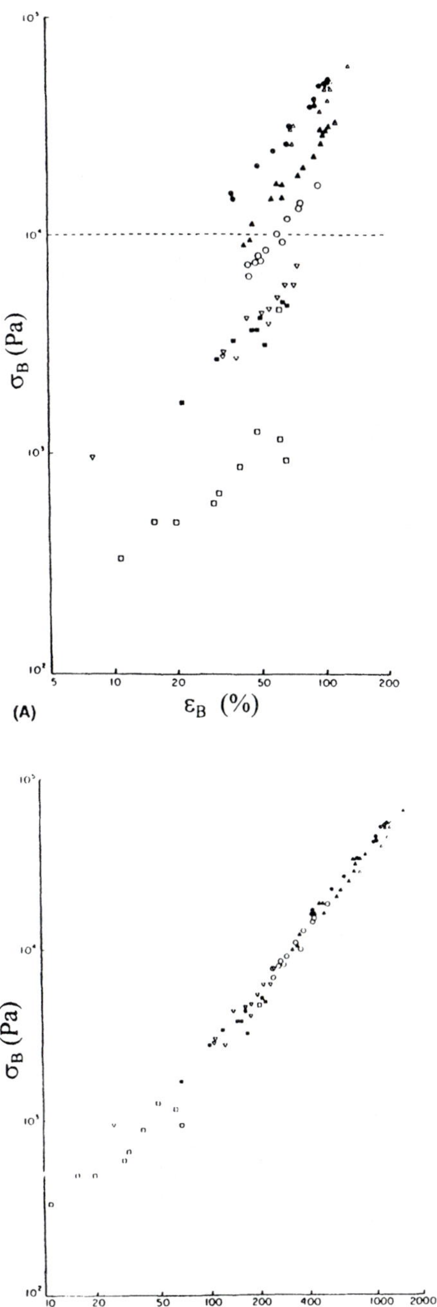

FIGURE 3 (A) Failure envelope for gelatin gels, concentrations □ 5%, ■▽ 10%, ○ 15%, ▲ 20%, ● 25%. The dotted line at 10^4 Pa is used to illustrate the horizontal shift of the 15 and 20% data by b_c (see text) to give (B) the superposed envelope. (From Ref. 16.)

insertion of a cylindrical probe [17]. In contrast to the situation for measurements with large deformation, the theory of small-deformation rheology is much more advanced and the measurements are correspondingly more informative.

But these types of measurements have the disadvantage of being essentially static, neglecting viscoelasticity. Because gels are viscoelastic, stress in a gel takes a finite time to respond to an applied strain and much more information can be obtained from dynamic measurements than is available from static (equilibrium) measurements of deformation. During the last 10 years improved instrumentation (see Sec. IV) has greatly facilitated dynamic testing of gels, and dynamic measurements are providing information about gels and gel networks not previously accessible. The viscoelastic behavior is measured by applying either a constant or a sinusoidally oscillating stress or strain. Time dependence of stress and strain are shown in Figure 4. In a perfectly elastic solid, strain responds immediately to stress, as shown in Figure 4A. In a gel, the material takes a finite time to respond (Fig. 4B). Measurements can be carried out with constant stress, expressing the results in terms of the creep compliance function:

$$J(t) = \frac{\text{strain (t)}}{\text{stress}} \tag{1}$$

or at constant strain, expressing the results as the stress relaxation function:

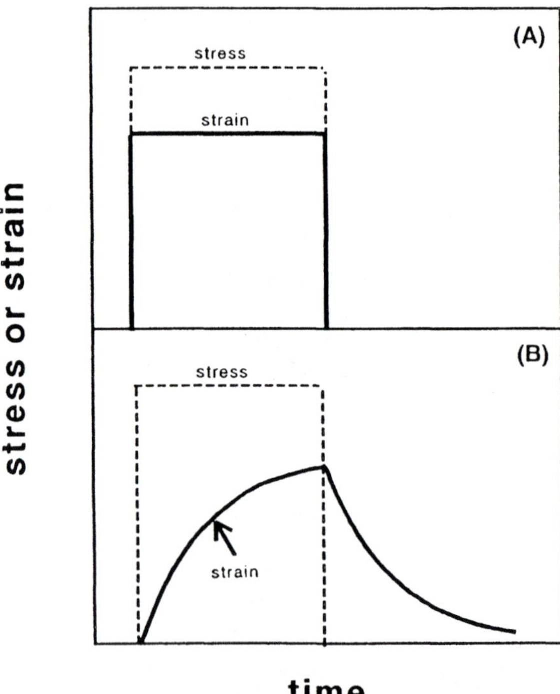

FIGURE 4 Response to a step in the applied stress of (A) a perfectly elastic solid and (B) a viscoelastic material.

$$G\ (t) = \frac{\text{stress (t)}}{\text{strain}} \tag{2}$$

A sinusoidally oscillating stress or strain is illustrated in Figure 5. When the material is an ideal elastic solid the stress is in phase with the strain, when it is an ideal liquid there is a phase difference of 90°, and when the material is a viscoelastic gel the phase angle is somewhere between these two extremes. The results of such measurements are usually expressed as the storage modulus G′, which is the ratio:

$$G' = \frac{\text{stress component in phase with strain}}{\text{strain}} \tag{3}$$

and the loss modulus G″, which is the ratio:

$$G'' = \frac{\text{stress component 90° out of phase with strain}}{\text{strain}} \tag{4}$$

The storage modulus, G′, characterizes the rigidity of the sample (the energy expended in deforming an elastic solid is stored and is recoverable when the stress is released); the loss modulus, G″, characterizes the resistance of the sample to flow (energy expended in inducing flow is dissipated as heat and therefore lost). The stress wave that results from the sinusoidal applied strain is given by:

$$\sigma_t = \varepsilon_0(G'\sin \omega t + G''\cos \omega t) \tag{5}$$

where σ_t is the stress at time t, ε_0 is the maximum strain amplitude, and ω is the frequency of the oscillation. Another quantity often calculated is the loss angle (δ), which is given by:

$$\tan \delta = \frac{G''}{G'} \tag{6}$$

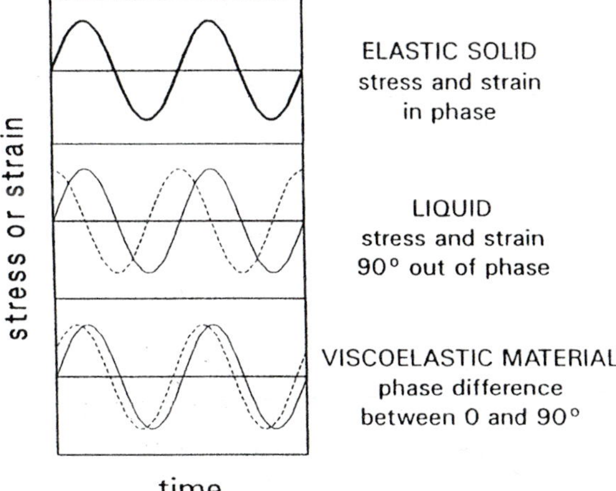

FIGURE 5 Response of different materials to a sinusoidally oscillating stress.

Thus, tan δ is a measure of the ratio of the energy lost to energy stored in a cyclic deformation. A small tan δ means that the material is highly elastic.

The frequency dependence of G′ and G″ is the "mechanical spectrum" [18]. For a gel, a typical mechanical spectrum, e.g., over the frequency range of 10^{-2} to 10^{2} rad s^{-1}, consists of two nearly horizontal straight lines (Fig. 6). G′ is typically 1–2 orders of

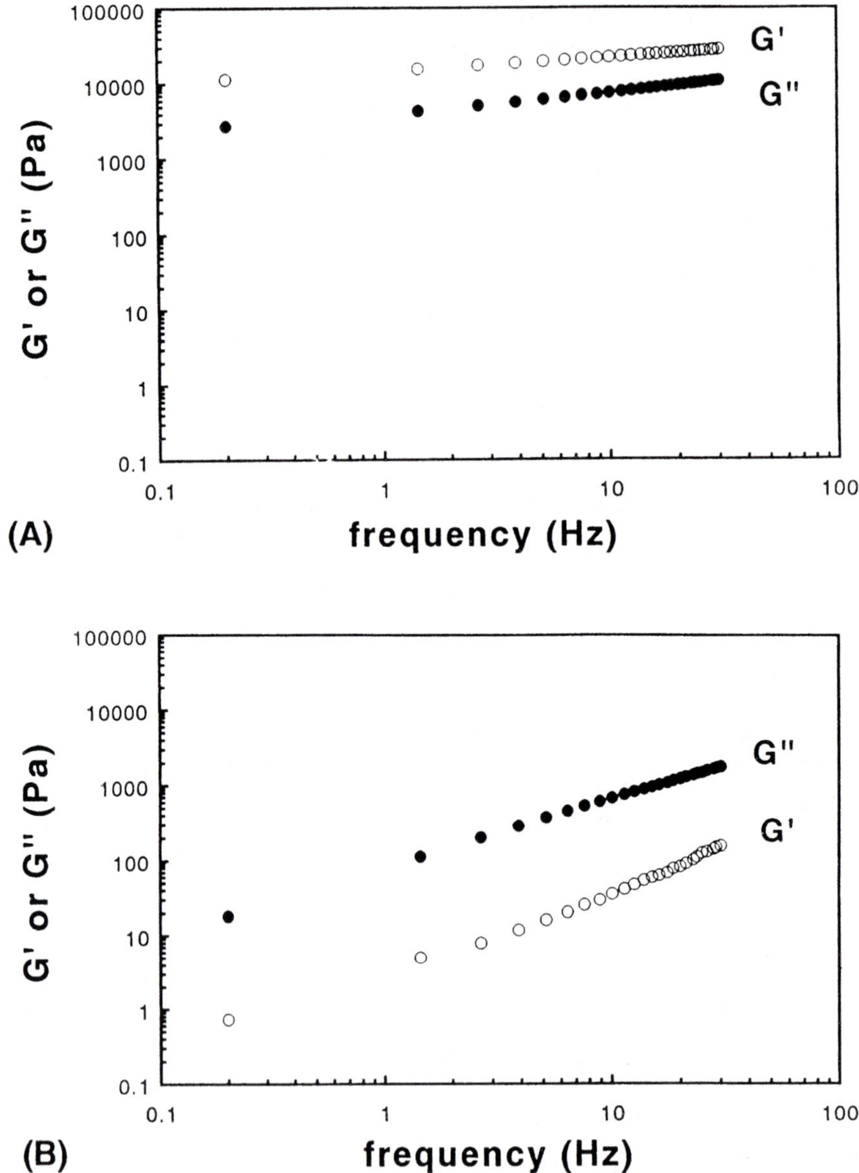

FIGURE 6 Mechanical spectrum of (a) freeze-thawed egg yolk compared with (B) unfrozen yolk, (C) a viscous solution (locust bean gum), and (D) a gel (1% ι-carrageenan). All measurements were made at 10°C.

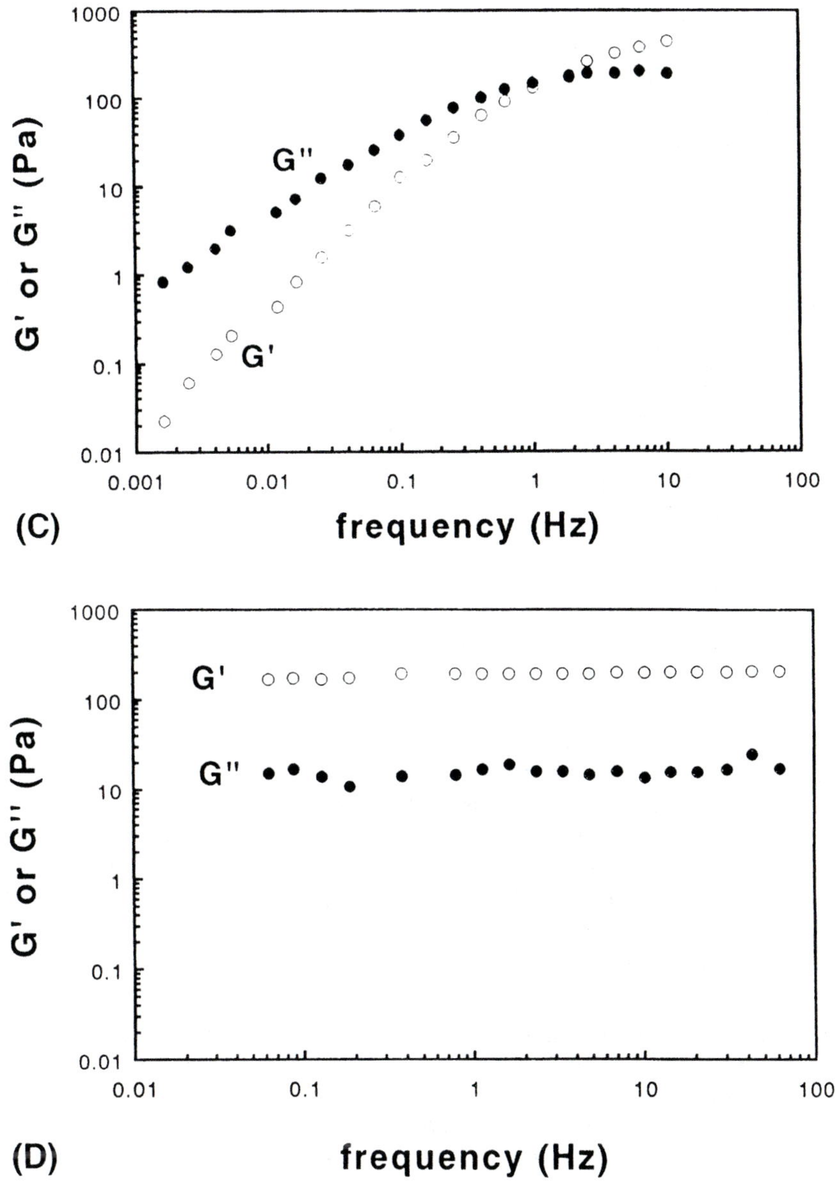

Figure 6 Continued.

magnitude greater than G″, and both may show some slight increase at higher frequencies. This behavior is quite different from that seen in the mechanical spectrum of a viscous solution, such as a solution of locust bean gum, which is not a true gel (Fig. 6). At low frequency, the G″ is greater than G′ and both parameters vary with frequency. As frequency increases, G′ crosses G″. This behavior is typical of macromolecular solutions in which the polymer molecules are mutually entangled [19]. At low frequencies there is sufficient time for the entanglements to make and break. Entanglement causes the high viscosity but there are no intermolecular cross-linkages, as in a gel. The mechanical spectrum thus provides a useful rheological definition of a gel: "A gel is a viscoelastic material for which G′ is greater than G″ and for which both G′ and G″ are almost independent of frequency" [18].* This is illustrated in Figure 6, which shows the mechanical spectrum of egg yolk that has been frozen and then thawed. When yolk is cooled to −6°C or below, the viscosity is permanently increased so that it remains pasty on rewarming. This change is called "yolk gelling," though whether or not the material is a true gel is uncertain [20]. The results shown in Figure 6 make it clear that frozen/thawed yolk is indeed gel-like. G′ is greater than G″ for the frozen/thawed yolk; G″ is greater for liquid yolk.

C. Networks and Junction Zones

The rheological theory of small deformations is now well established, following the development of the theory of rubber elasticity in the 1940s. The classical theory of rubber elasticity gives the following relationship between the shear modulus (G) and the concentration of the gel-forming polymer (c) [10]:

$$G = \frac{RTc}{M_c} \tag{7}$$

where R is the gas constant, T the absolute temperature, and M_c the number average molecular weight of the "active chains," that is, the segments of polymer chain joining adjacent cross-links (Fig. 1). This equation is the entry point for several different ways of using the concentration dependence of shear modulus to derive information about the gel network and the junction zones where the polymer molecules are cross-linked.

One of the earliest historically important studies was that by Hermans [22], who assumed that the association of reactive sites on the polymer to form a junction zone is effectively described by a monomer-dimer equilibrium. The association constant (K_J) is then given by an equation analogous to the Ostwald dilution law:

$$K_J = N_0 f \cdot \frac{(1 - \alpha)^2}{\alpha} \tag{8}$$

where N_0 is the number of polymer chains per unit volume, f is the functionality, defined as the number of functional groups or attachment sites available to form junction zones, and α is the fraction of these that have reacted.

More recently, Oakenfull [23] extended and generalized this treatment, allowing for greater than one-to-one association. K_J is then given by:

$$K_J = \frac{[J]M_J^n}{[n(c - M_J[J])]^n} \tag{9}$$

where n is the number of cross-linking sites forming a junction, c is the weight concentration of the polymer, [J] is the molar concentration of junction zones, and M_J is their

*In practice, caution is needed in using this definition, because it is possible, as Doublier has shown for solutions of xanthan gum, that the cross-over point for *G′* and *G″* can be at frequencies too low to be accessible experimentally [21].

number average molecular weight. In addition, from Eq. (7), the shear modulus is given by:

$$G = -\frac{RTc}{M} \cdot \frac{M\,[J] - c}{M_J\,[J] - c} \tag{10}$$

where M is the number average molecular weight of the polymer.

The derivation of Eq. (7) relies on the assumption that the "active chains," the segments of polymer linking junction zones, are Gaussian, i.e., that they are perfectly flexible and free to adopt all possible conformations. This appears to be true to a good enough approximation for weak, dilute, gelatin gels where the "active chains" are relatively long. It is unlikely, though, to be true for many protein gels where the active chains are composite structures formed from globular subunits. Within this limitation, Eqs. (9) and (10) can be solved numerically to give:

M—the number average molecular weight of the polymer
M_J—the number average molecular weight of the junction zones
n—the number of cross-linking loci that associate in forming a junction zone
K_J—the association constant for formation of junction zones

Use of this method is illustrated by data for gelatin gels, shown in Figure 7 [24]. Gelatin gels are stabilized by sucrose that causes a large increase in the elasticity modulus. The size of the junction zones decreases with added sucrose. The formation of

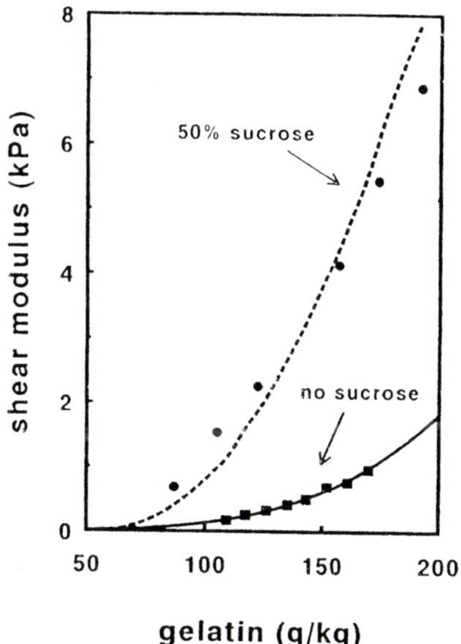

FIGURE 7 Comparison of experimental data with theoretical curves calculated from Eqs. (9) and (10). The results are for gelatin gels measured with and without addition of sucrose. The parameters used to generate the curves were 50% sucrose ($M = 48{,}000$, $M_J = 683$, $n = 3.18$) and no sucrose ($M = 50{,}000$, $M_J = 12{,}800$, $n = 3.09$). (Adapted from Ref. 24.)

junction zones is a cooperative process with many individually weak forces contributing to stability. Added sucrose enhances these intermolecular forces. Consequently, smaller junction zones become thermodynamically stable, extended ordered structures being inherently opposed by entropy. Smaller but more numerous junction zones produce a more extensive gel network and thereby increase the rigidity of the gel. As an indication of the general validity of the approach, the values calculated for M, the number average molecular weight of the polymer, were internally consistent and agreed with the value found by viscometry.

Clark and Ross-Murphy [25] have followed a different path and obtained the expression:

$$G = [Nf\,\alpha(1 - \nu)^2(1 - \beta)/2]aRT \tag{11}$$

In this equation the quantity in square brackets is a measure of the number of moles of active chain per unit volume. The factor aRT is a measure of the average contribution per mole of active chain to the increase in the free energy of the system per unit increase in strain, and hence to the modulus. The term f is the "functionality," the number of attachment sites or functional groups available to form cross-links; α is the proportion of these functional groups that have reacted. They used an extension of the Case-Scanlan network model [26] to obtain N_e, the number of elastically active junction zones as a function of f and α:

$$N_e = f\,\alpha(1 - \nu)^2(1 - \beta)/2 \tag{12}$$

where

$$\beta = \frac{(f - 1)\alpha\nu}{(1 - \alpha + \alpha\nu)} \tag{13}$$

and ν, the "extinction probability," is given by $\nu = R(1 - \alpha + \alpha\nu)^{f-1}$, where R implies that we take the lowest positive root of the expression within the parentheses. Compared with Eqs. (9) and (10), there are fewer assumptions implicit in the derivation of Eq. (11). It is valid over a wide range of concentrations and gives good fits to experimental data (Fig. 8). It seems, though, less useful than Eqs. (9) and (10), in that it provides less information about the size and thermodynamic stability of the junction zones.

III. PROPERTIES OF PROTEINS RELEVANT TO GELATION

To form a gel, the food material must contain large molecules capable of forming cross-links in three dimensions (see Sec. II). Because of their chemistry, proteins have many advantages for this purpose. Compared with carbohydrates, they are able to form a wider range of cross-links and they have a higher nutritional value. Furthermore, proteins are potentially more uniform than carbohydrates. However, these advantages have to be balanced against the increased cost of some proteins and the fact that generally much higher concentrations are required to form a protein gel than a carbohydrate gel.

For the production of gels, the important properties of proteins are their flexibility, including their ability to denature and give extended chains, and, especially, their ability to form extensive networks by cross-linking. Cross-linking and denaturation are discussed separately below.

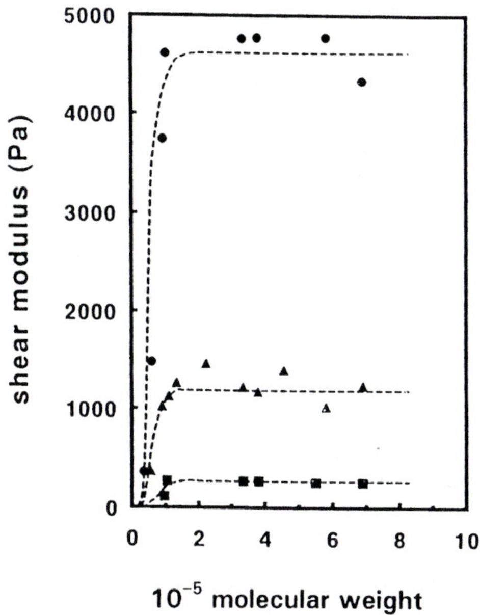

FIGURE 8 Best fits from Eq. (11) for elastic modulus (E = 3G; 10^5 dyne/cm^2) versus molecular weight data for κ-carrageenan in 0.1 M KCl: ●, ■, and ▲ represent polymer concentrations of 5, 10, and 20 g/liter, respectively. (Adapted from Ref. 25.)

A. Protein Cross-Linking

As we have seen, a gel's characteristic rheological properties result directly from a continuous molecular network spanning the volume of the gel. Gelation occurs when the conditions are such that the molecules are induced to aggregate in some way. Usually protein gels are made by heating. As the temperature increases, the molecules partly unfold, exposing previously hidden reactive groups. These groups can then react intermolecularly to form a continuous network. Gelatin is a notable exception; its gels are made by cooling. Gelatin is partly degraded collagen (see Sec. V. A). It readily dissolves in hot water as random coils; when the solution is cooled, junction zones are formed by small segments from two or three polypeptide chains reverting to the collagen triple helix structure.

The stability of protein structures arises from a combination of *covalent* bonds, particularly disulfide bones, with *noncovalent*, intermolecular connections provided by hydrogen and electrostatic bonds and hydrophobic effects. Their characteristics are summarized in Table 1.

Some of the noncovalent bonds listed in Table 1 are relatively weak although collectively they confer great strength; we can assume that for maximum stability all possible bonds will be formed. Electrostatic bonds may be stronger than other noncovalent bonds, but their existence is determined by the pH and salt concentration (i.e., ionic strength). In an important type of electrostatic bond, a metal ion bridges two anions.

Hydrogen bonds and hydrophobic forces depend very much on the unique nature of water as a solvent [27,28]. Because water is both a hydrogen bond donor and a

TABLE 1 Intermolecular Forces Stabilizing Protein and Polysaccharide Structures

Bond type	Example
Noncovalent	
Hydrogen bonds	-H···O···H- -H···N···H-
Hydrophobic interactions	$-CH_3\cdots CH_3-$
Electrostatic—salt links	$-COO^-\cdots{}^+H_3N-$
Electrostatic—metal ion bridges	$-COO^-\cdots Ca^{2+}\cdots{}^-OOC-$
Covalent	
Disulfide	-S-S-
Gamma glutamyl	$-CH_2-CH_2-CO\cdots NH-CH_2-CH_2-$

hydrogen bond acceptor, liquid water itself is a continuous molecular network—but an exceedingly transient one, continuously reforming and regrouping. No other liquid is held together by such strong and directional intermolecular forces. In ice, every water molecule is bonded to its four nearest neighbors. Its O-H bonds are directed towards lone pairs of electrons on two of these nearest neighbors, forming two O-H···O bonds; in turn, each of its lone pairs is directed towards an O-H bond of one of the neighbors, forming two O···H-O bonds. In liquid water, this structure is incomplete and transient, and is still not fully understood. In aqueous solution, nonpolar molecules, or molecules carrying nonpolar groups, are surrounded by an ordered or "structured" layer of water molecules. When these nonpolar molecules approach each other some of the ordered water molecules are "squeezed out," and the molecular rearrangements that this entails provide the thermodynamic driving force for hydrophobic interaction. Similarly, in electrostatic interactions the hydration shells of ordered water molecules surrounding the ions are important. Association of ions involves interpenetration of these hydration shells with consequent rearrangement of water molecules. The importance of noncovalent bonds for most of the functional properties of proteins cannot be overemphasized, although, as explained later, they often inhibit gelation.

The main covalent cross-links joining protein chains are the disulfide bonds of cystine residues (Table 1). These stabilize the structure of many proteins. They have great strength but are broken by sulfhydryl-containing compounds and some other reducing agents and by oxidizing agents such as peroxides. They are formed readily by the mild oxidation of sulfhydryl groups, for example:

$$\cdots SH + HS\cdots \xrightarrow{\text{oxidation}} \cdots S\text{-}S\cdots + H_2O$$

The ability of disulfide bonds to undergo sulfhydryl-disulfide interchanges confers flexibility on protein structures:

$$R_1\text{-}SH + R_2\text{-}S\text{-}S\text{-}R_3 \longrightarrow R_1\text{-}S\text{-}S\text{-}R_2 + H\text{-}S\text{-}R_3$$

where the Rs represent parts of a polypeptide chain.

Covalent linkages between proteins can also be introduced by the Maillard reaction [29]. Maillard reactions are a complex series of reactions between amino and carbonyl compounds. These reactions are of enormous practical importance in the development of colors and flavors. (Familiar examples are caramelization and the characteristic flavors of roast meat.) When proteins are heated with reducing sugars, covalent cross-linkages are formed, which contribute to the gel network. At the same time, there is a fall in pH that may also contribute to gelation. That there can be significant contributions from Maillard cross-linkages is shown by the fact that Maillard gels are not completely solubilized by sodium dodecyl sulfate or mercaptoethanol [29].

Other covalent cross-links are less common in proteins. They include cross-links between glutamyl and lysyl residues, sometimes known as "isopeptide" links, which are produced by the enzyme transglutaminase, and also other cross-links involving lysine residues, produced by the enzyme lysyl oxidase. In addition, glycoproteins are common in biology and some are notable gel formers, for example, ovomucin. It is not entirely clear how carbohydrate contributes to gelation. The sugar residues would be expected to increase the protein's solubility, and large carbohydrate attachments would be expected to increase the potential for cross-linking.

B. Denaturation and the Formation of Extended Protein Chains

For most food proteins, partial or complete denaturation is necessary for gelation. During denaturation, proteins undergo unfolding of their three-dimensional structures to give extended chains but without rupture of covalent bonds (see Chapter 2). Fully denatured proteins do not contain the stable secondary structures present in native proteins, such as α-helices or β-sheets. Proteins may be denatured in several ways, depending on the strength of the cross-links holding the native structure in place. Many proteins are denatured by heating, but a few resist denaturation in this way. Some proteins can be denatured by high pressure. Most soluble proteins can be denatured by mechanical agitation and by shear forces at interfaces. Some proteins are denatured or partially denatured by organic solvents such as alcohol. Disaggregating solvents, such as concentrated aqueous urea, very effectively denature most proteins, although such solvents are not of interest for food processing.

Denaturation, or partial denaturation, exposes at least some of the hydrophobic parts of the molecule to the solvent. Thus, regions of the molecule originally involved in maintaining the stability of the native form become available for intermolecular bonding and a network will form (provided that there are, on average, at least two attractive sites per molecule). Complete denaturation of the protein is not often necessary for gelation; in fact, a completely denatured protein usually forms an insoluble precipitate because of extensive hydrogen bonding or hydrophobic interactions between unfolded chains. Completely denatured proteins are likely to give gels only in solvents such as concentrated urea that prevent extensive noncovalent interchain cross-linking. During denaturation by heat, the protein concentration may be important; high concentrations are more likely to give precipitates, and low concentrations are likely to give gels. Gelling occurs especially at the boundary between aggregation and solubility [30].

C. Gel Networks Formed by Proteins

Generation of gel networks from globular proteins often requires a degree of unfolding of the protein as a first step. This can be brought about by increasing the temperature or changing the pH or ionic strength, in which case gelation is irreversible. (A familiar example is the gelled white in a hard-boiled egg.)

The network structures formed by globular proteins can have many forms. The active sites formed on partial denaturation of the protein can vary in number and relative reactivity; for example, electrostatic forces depend on exposure of charged groups and on the pH and ionic strength. Essentially, though, globular proteins can aggregate and form a gel network in only two ways. One is by random aggregation of the globular protein units that can give a network structure purely statistically [31]; the other is by a more ordered aggregation that gives rise to a "string-of-beads" network structure (Fig. 9). Random aggregates have been observed in heat-set gels of β-lactoglobulin and acid-induced gels of casein; string-of-beads structures have been observed in heat-set gels of serum albumin, lysozyme, and ribonuclease [32]. Some proteins (e.g., β-lactoglobulin and ovalbumin) can form a gel of either type, depending on the pH and ionic strength.

Clark and coworkers [33] used computer modeling to study these two types of aggregation in detail. They assumed that there was very little loss of form by the globular protein during aggregation and, proceeding on this basis, generated simulated two-dimensional sections of gel. They subdivided the string-of-beads model according to whether (a) the chains of spheres were absolutely straight with angles between successive protein-protein links all equal to 180°, (b) the chains of spheres had high persistence, where the angles between successive links were between 150 and 180°, or (c) the chains of spheres had low persistence, where the angles between successive links varied between 90 and 180°. Computer images derived from the linear model (a) were similar to electron micrographs of heat-set lysozyme gels at pH 2.0, while the more random models produced images similar to the microstructure of ribonuclease gels. However, the "high-persistence" model gave the type of image seen most often experimentally, suggesting that the surface of the reactive monomer is generally not uniform and that bonding between beads occurs at specific sites. Clark and coworkers concluded that the difference between random and ordered aggregation is not clear-cut. No gel network is precisely defined in a geometric sense, and the differences between random and more ordered aggregation depends on the number of binding sites available per unfolded protein molecule, their spatial distribution, and their relative bonding strengths.

Where the natural function of the protein depends on aggregation (e.g., actin and tubulin), highly organized rodlike structures can arise even from globular monomers. These can gel by forming entanglement networks ("lattice gels") with no clearly defined junction zones. Blood clots form when fibrinogen polymerizes to fibrin [6].

D. Opacity

In contrast to polysaccharide gels, which are almost invariably clear, protein gels are often opaque. Opaque gels are formed when fluctuations in polymer density approach macroscopic size and effectively scatter light [34]. Such networks are characterized by regions of high polymer concentration separated by regions nearly devoid of polymer. Transparent gels, on the other hand, have an almost homogeneous network. In some protein gel systems there can be a transition from transparency to opacity, or vice versa. Ovalbumin gels formed at pH 10 are transparent, but those formed at pH 5 are opaque. Electron micrographs show that the gels formed at pH 10 are almost homogeneous, whereas those formed at a pH 5 have localized regions of high protein concentration (Fig. 10).

Doi [32] has developed a two-step heating method as another way of getting transparent gels. Heated 5% ovalbumin solutions (pH 7.5) give transparent solutions in the

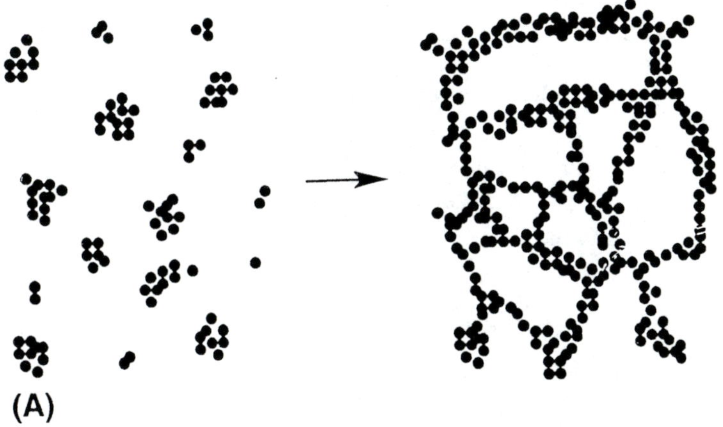

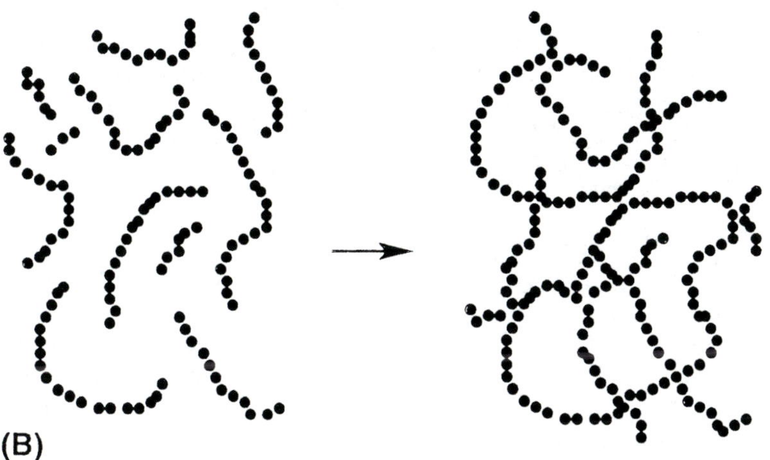

Figure 9 Two types of gel network formed by the aggregation of globular proteins. (A) Random aggregation of molecules; (B) aggregation of "string-of-beads" polymers. (Adapted from Ref. 30.)

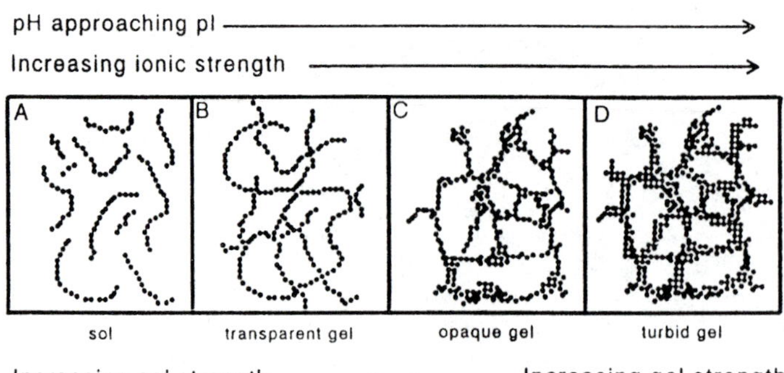

FIGURE 10 Model for the formation of transparent and opaque gels by ovalbumin. (A) At pH values far from the pI and at low ionic strength, electrostatic repulsive forces hinder the formation of random aggregates and linear polymers are formed. (B) With decreasing electrostatic forces (increasing ionic strength or change of pH from pI), three-dimensional networks form a transparent gel. (C) At intermediate ionic strength or pH, linear polymers and random aggregates are mixed together in the same gel, forming a translucent or opaque gel. (D) At high ionic strength or at pH values near the pI, proteins polymerize to form a turbid gel composed of random aggregates. (Adapted from Ref. 30.)

absence of NaCl, transparent gels at 10–20 mM NaCl, translucent gels at 30–40 mM NaCl, and turbid gels at higher NaCl concentrations. The transparent solution obtained by heating ovalbumin without added NaCl contains heat-denatured ovalbumin (its viscosity is higher than that of the unheated solution); if NaCl is now added (within the concentration range 10–200 mM) and the solution reheated, a transparent gel is formed [32]. Doi's explanation is that without NaCl, electrostatic forces restrict aggregation to linear strings-of-beads, and once these have formed, addition of NaCl and reheating gives a "fine" gel network. If ovalbumin is simply heated with NaCl, aggregation is more random and an opaque gel is formed.

E. Thermal Reversibility

Protein gels are almost invariably thermally irreversible (in contrast to many polysaccharide gelling systems, where gels often revert to the liquid state on heating). If a gel forms when the protein solution is heated, cooling or further heating does not reverse the process. Gelatin is the most notable exception. Its gelation mechanism is different from other proteins (see Sec. V.A). It consists of heavily degraded and denatured segments of collagen. Gelation occurs when the solution is cooled to below the helix-coil transition temperature and the protein chains can cross-link by forming small regions of collagen triple helix structure. This process is perfectly reversible. If you warm the gel to above the transition temperature, the helices revert to random coils and the gel liquefies [34].

It has also been claimed that the gelation of soy proteins can be thermoreversible. Here the situation is very different from gelatin, as the soy proteins are globular and the question of thermoreversibility remains a controversial one. Catsimpoolas and Meyer [35] proposed the following scheme for the gelation of soybean globulins:

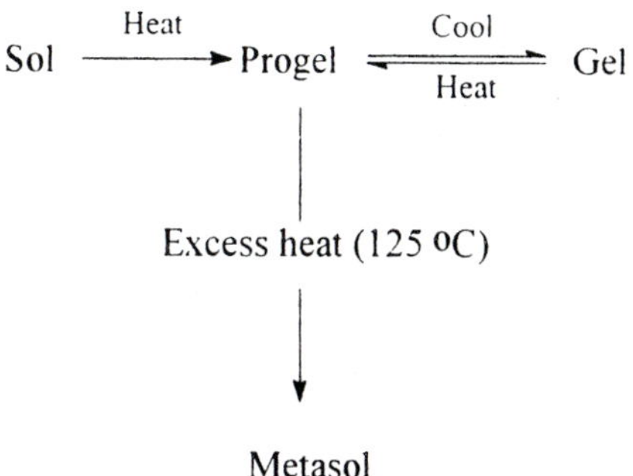

According to this scheme, the sol → progel conversion is irreversible but the progel ⇌ gel transition is reversible because it involves mainly noncovalent bonds. The question of thermoreversibility is thus partly a matter of semantics. Van Kleef [36] has argued that soybean protein gels are "clearly irreversible" because the native state of the protein is not restored when the solution is cooled. However, if we look only at the formation of the gel—the conversion of the system from a liquid to a solid state—the system is indeed reversible. An additional complication is that soybean protein contains several different fractions—2S, 7S, 11S. It has been suggested that the 11S fraction alone accounts for thermoreversibility and that the physical state of the 11S protein before aggregation determines the extent of reversibility [37].

F. Mixed Systems

Almost all gelled foods contain complex mixtures of different proteins or proteins in combination with polysaccharides that can also form gels. In these mixtures, molecular interactions occur which powerfully influence the gelation characteristics of the individual components.

Interaction between proteins and polysaccharides can result in one of three consequences [38,39]:

1. Co-solubility—in which there is no significant interaction between the two classes of polymer molecule and both coexist in solution
2. Incompatibility—where repulsion between the two types of polymer causes them to form separate phases
3. Complexing—where attraction between the two polymers causes them to form a single phase or precipitate

Polymers behave very differently from small molecules in solution because the entropy of mixing (which favors single-phase systems) is dependent on the number of molecules present, rather than on their concentration by weight. Because polymers are of high molecular weight, their solutions contain relatively fewer molecules than solutions of the same concentration by weight of a low molecular weight solute. The entropy of mixing therefore becomes insignificant for polymers, and the behavior of mixed polymer solutions is decided by the energies of interaction between the chains. When this is favorable, for example, in polyanion-polycation systems, the two polymers may associate into a single gel-like phase or form a precipitate. More commonly, the interactions between the two polymers are less favorable than between like segments of each type. There is therefore a tendency for each to exclude the other from its polymer domain, so that the effective concentration of both is raised in their respective domains. At sufficiently high concentrations, the system can separate into two liquid phases, or one component may be driven out of solution by the other.

When a polysaccharide is used to gel a meat or dairy product, there is almost always a strong interaction between the polymers, and the texture of the mixed gel can be very different from those of the gels formed by the components singly. The physical chemistry is very complex—we are confronted with a mass of empirical observations that are difficult to interpret. But there has been some progress towards setting up a theoretical framework of understanding.

When both polymers can independently form a network, three extremes of structure may be formed [40] (Fig. 11):

1. Interpenetrating networks, where the two types of polymer (A and B) are mutually entangled
2. A phase-separated network, in which domains of pure A are interspersed with domains of pure B (e.g., the agar/gelatin system)
3. A coupled network, in which at least some junction zones involve both polymers (e.g., the alginate/gelatin system)

Figure 12 shows how this theoretical framework can be applied to mixed gels of protein and polysaccharide. When the two polymers form a complex, they may gel as a coupled network; when they are incompatible, they may gel as two separate but interpenetrating networks or phase-separated networks, with isolated domains of each gel type.

Protein-polysaccharide complexes might be regarded as a new type of gelling agent where the formation conditions, as well as the rheological and other physicochemi-

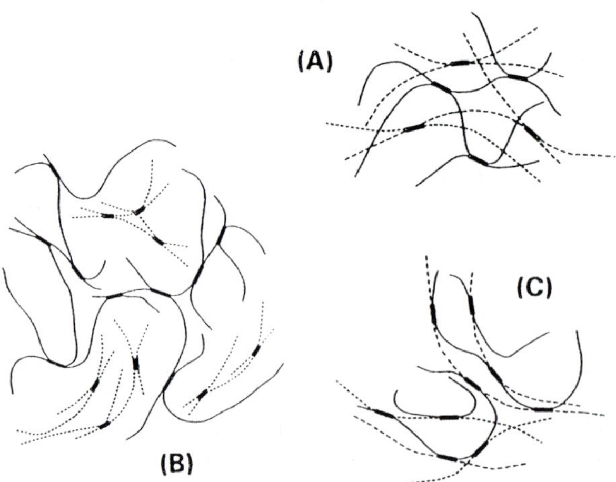

FIGURE 11 Mixed gel networks. (A) Interpenetrating networks—the two polymers interact only by mutual entanglements. (B) Phase separated network—pure gel A is interspersed with pure gel B. (C) Coupled networks—at least some of the junction zones involve both types of polymer.

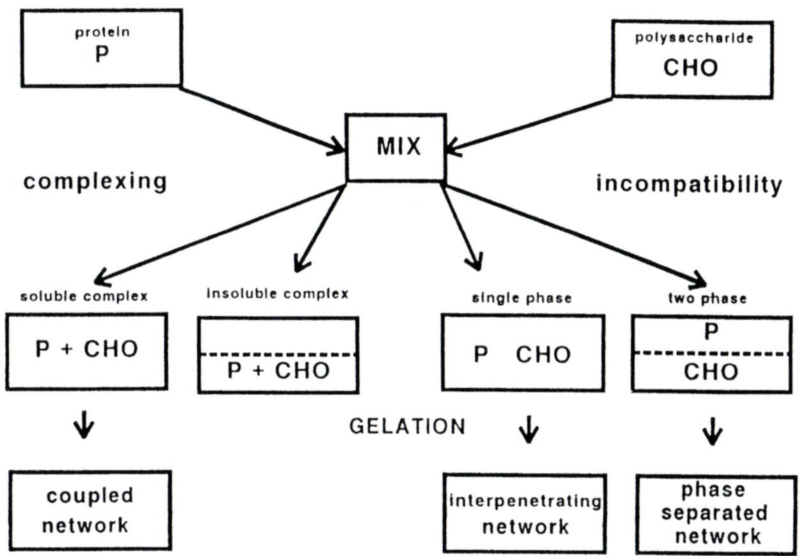

FIGURE 12 The different possible interactions of proteins and polysaccharides and their consequences in the formation of mixed gels.

cal properties, could be controlled to give a product of the desired textural characteristics [41].

IV. PRACTICAL CONSIDERATIONS

A. Measurement of Gel Strength

Many different instruments are available for measuring gel strength, some of which have a long history. Some simply provide empirical tests that cannot be related to fundamental rheological quantities; others are more fundamental and measure well-defined parameters such as shear modulus or rupture strength [42]. A survey of recent research papers on gelation showed that the Instron Universal Testing Machine is the most commonly used instrument for studying the rheology of gels. This instrument measures force versus displacement (under either extension or compression), as shown in Figure 2. A slab of free-standing gel can be compressed between parallel plates or a probe can be inserted into the gel. Using extension, a strip of gel can be stretched for measurement of rupture strength.

A major recent advance has been the development of relatively inexpensive instruments for dynamic testing that rely heavily on PCs for recording, storing, and handling the data. Typically the sample is subjected to rotational strain between parallel plates or between a cone and plate (Fig. 13). The moving cone or plate has an air bearing with very low friction and is rotated by a computer-controlled induction motor; movement is measured optically. Such instruments can measure viscosity versus rate of shear, creep, stress relaxation, and, using oscillation, the bulk modulus and storage modulus, etc. The complex calculations required to produce, for example, mechanical spectra, as shown in

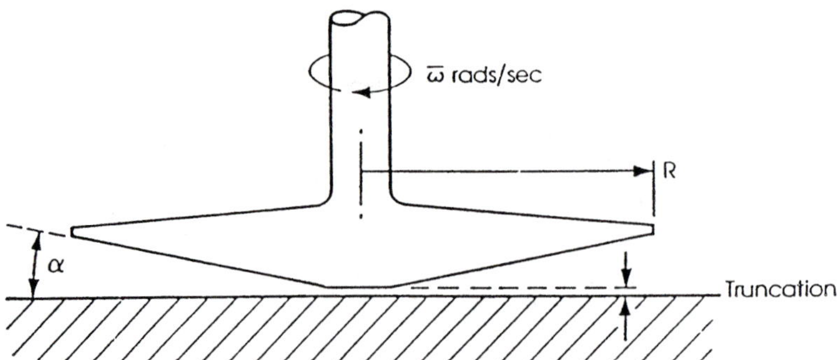

FIGURE 13 Cone and plate system used for dynamic rheometry.

Figure 6, are very conveniently done by the computer. But the ease with which such measurements can be made is in itself a hazard. Caution is needed, particularly in oscillatory rheology. Oscillatory measurement of viscoelastic properties is based on the assumption that the material's behavior is linear—i.e., the viscosity and modulus are constant with respect to stress or strain. Most materials are only linear at relatively small strains and therefore low stresses. In addition, the extent of linearity can depend on the frequency of the measurement. The computer can produce plots of G', G'', tan δ, etc. with no indication that the material's viscoelastic properties are nonlinear and the results therefore meaningless. And another word of warning: the fundamental rheological properties measured at low strain may bear no relation to the behavior and texture at high strains, including rupture of the gel.

B. Kinetics of Gelation

Measurement of gelation kinetics requires a method for following the development of the gel network without significantly affecting the process by mechanical disturbance. Oakenfull and Scott [43] used a simple procedure that measures the time required for the solution to form a gel of very small but predetermined rigidity. A fixed weight of solution is placed in each of a series of vials held thermostated at the gelation temperature and inverted sequentially. The time required to form a gel just firm enough to remain held in position is recorded and the reciprocal of the setting time is proportional to the rate of gelation. Schweid and Toledo [44] used the small volume changes that accompany denaturation and aggregation to measure rates of gelation of meat batters. They sealed the meat batters in a piston and measured the pressure change with increasing temperature (using an Instron Universal Testing Machine to measure the force exerted by the piston).

Other methods that have been used seem to involve more significant mechanical disturbance to the sample. For example, Gossett and colleagues [45] used an electrobalance to follow the time course of the small force exerted by the gel on a small hook-shaped probe. There must be some movement of the probe before the balance can measure the force, and, as they used only a single sample, the probe would be subjected to numerous tweaks in the course of a run. Marrs and Steele [46] described a system that follows the increase of viscosity with time by measuring the free fall of small droplets of carbon tetrachloride. The carbon tetrachloride is injected via a hypodermic needle and the velocity of the droplets measured electronically using fiber optics. More recently kinetic measurements have been made by forming the gel between the plates of an oscillatory rheometer [47]. With this type of instrument, provided that conditions are chosen appropriately, the movement can be so small as to be insignificant.

V. FOOD PROTEIN SYSTEMS

This section presents a brief summary of the role of gelation in some common protein food materials.

A. Collagen-Gelatin

A degraded form of collagen, gelatin, is the main protein gelling agent for food uses, where it is an ingredient of jellies, mousses, marshmallows, fruit gums, etc. As a protein it has a nutritional role as well. A notable feature of gelatin, when compared with many other protein gels, is that it forms gels over a wide range of concentrations so that a wide range of products, from soft jellies to firm confectionery, can be made.

Collagen is the most plentiful protein in vertebrates. At least 13 types of collagen (Types I–XIII) are known. They differ slightly in amino acid composition, in molecular size, and in their location in the animal. The essential structure of collagen is a rod-shaped triple helix of three polypeptide chains. The chains are not necessarily the same, although each contains large regions with the sequence Gly-X-Y, frequently Gly-Pro-Hypro. During collagen biosynthesis, the first 100 or so amino acid residues of each molecule do not have the Gly-X-Y sequence. Instead they have the important role of aligning the three chains precisely so that the triple helix will result. In vivo, the collagen helices are arranged in small groups (fibrils), which in turn are arranged in bundles. These structures are stabilized against enzyme degradation by unusual covalent cross-links. Such cross-links are formed over several years mainly from lysine and hydroxylysine side chains. The nature of the final cross-links has not yet been elucidated [48]. As a result of this increased cross-linking, collagen is easier to extract from the tissues of young animals than from old animals.

The collagen main chains are very long—more than 1000 residues for Type I, the molecular weight being more than 100,000 daltons. With this structure it almost seems as if collagen had been designed to form gels. Nevertheless, most collagen has a structural role (as tendons, ligaments, bone matrix, sheaths, etc.) rather than a gel-forming role. An interesting exception occurs in the coelenterates, because the jelly of jellyfish contains collagen that is responsible for the gel structure [3].

Gelatin is made by degrading and solubilizing the collagen in bones or skin. For the commercial production of gelatin, only Type I (found in skin and bones) and Type III collagens (found only in skin) are used. Most food gelatins are from skin—usually pig skin or cattle hides. Production of gelatin takes place in several stages. The skins are first pretreated with acid (diluted sulfuric, hydrochloric, or phosphoric acids) or alkali (lime, sodium hydroxide, or mixtures). A considerable time may be needed for pretreatment depending on the age of the skin—young skins with relatively few covalent cross-links require shorter times than old skins. During pretreatment, cross-links are broken and there may be extensive nonspecific main-chain hydrolysis. Extraneous material is also removed. Next, the soluble gelatin is extracted from the remaining skins by dissolving it in water at temperatures up to about 90°C. The gelatin solution is then filtered, decolorized if necessary, dried, and granulated or powdered. Solid gelatin can be dissolved in warm water, and it forms a gel on cooling.

As a result of the damage during production, gelatin is not a well-defined product. Although the processes for the production of gelatin are apparently simple, they have been the subject of a vast amount of research. In spite of this, the whole subject of gelatin production and use is still more of an art than a science. The following conclusions can be drawn: Gelatin is not a homogeneous material. The extraction procedures can cause hydrolysis of main chains, most likely the excision of nonhelical regions. At the same time there is loss of ammonia from asparagine and glutamine residues, untangling of the triple helices, and aggregation of the modified chains to give larger units. There is also rupture of some covalent cross-links. Consequently, it is not easy to define the nature of a sample of gelatin from a particular source or to predict the properties of the gels it might form.

Tests carried out on gelatin and its gels to determine if it has suitable properties include investigations into molecular weight distribution, solution opacity, gel strength, and rheological behavior. The rheology of gelatin gels was described in Sections II and III, where it was pointed out that the key to understanding the nature of gelation by gelatin is the triple helical structure of collagen. In a heated solution of gelatin, the triple helices are largely unraveled. On cooling, small regions of helix re-form and provide very effective junctions for a three-dimensional network.

B. Milk Proteins

Milk proteins are classified as caseins and whey proteins. Whey proteins are now an important functional ingredient for the food industry and are of great importance for their gelation potential. There are a number of casein proteins, but they are similar in aspects of their structure and properties and also in their collective assembly into casein micelles. On the other hand, the whey proteins are more diverse in size, structure, properties, and functions, and hence are frequently defined as those proteins remaining after removal of the caseins from milk [49,50].

1. Caseins

Caseins are a group of phosphoproteins insoluble at pH 4.6 (the isoelectric point), where the whey proteins remain soluble. There are four primary proteins, α_{s1}, α_{s2}, β, and κ, in the approximate ratio 40:10:35:12, all relatively small molecules of about 20,000–24,000 daltons. Because of their high content of phosphoseryl residues, caseins bind polyvalent ions strongly, principally Ca^{2+}, which promotes aggregation. In normal milk, about 95% of the casein exists as casein micelles. These are coarse colloidal particles with molecular weights of about 10^8 and mean diameters of about 100 nm. While the structure of casein micelles is not known in detail, it is widely (although not universally) accepted that the micelles are composed of spherical submicelles, 10–15 nm in diameter, which have porous, open structures. There are about 25,000 organic phosphate residues per micelle, about 40% of which are linked in pairs via $Ca_9(PO_4)_6$, with Ca^{2+} acting as counterions for most of the remainder. The surface of the micelle is rich in κ-casein, which appears to be a critical factor in determining micelle stability [50].

Casein gels can be formed by treatment with rennet or by acidification [51]. Coagulation of milk by rennet is a two-stage process. The first involves enzymatic cleavage of protein, and the second involves precipitation of the modified casein at temperatures above 20°C. Only κ-casein is hydrolyzed, and only its Phe_{105}-Met_{106} bond is cleaved. The peptide sequence containing residues 1–105 is called *para*-κ-casein, and the remaining sequence [106–169] is known as the "macropeptide." The macropeptides released from the casein micelles are highly charged and hydrophilic; hence hydrolysis of κ-casein during the primary phase of rennet action reduces the zeta potential of the casein micelles from −10/−20 to −5/−7 mV and at the same time removes protruding peptides from the surface. This has the effect of reducing the repulsive effects—electrostatic and steric—that prevent aggregation by keeping the casein micelles apart. When about 85% of the κ-casein has been hydrolyzed, the casein micelles begin to aggregate into a gel. Chemical modification of histidine, lysine, or arginine residues inhibits coagulation, suggesting that electrostatic interactions are important in this process. Aggregation of rennet-altered micelles also depends on Ca^{2+}, which may act by cross-linking micelles via serine phosphate residues or by charge neutralization. Small, chainlike aggregates form initially, followed by short discrete chains, which, as the reaction proceeds, grow, branch, and form a gel network [50].

When milk is acidified, colloidal calcium phosphate in the casein micelles progressively solubilizes and aggregation of the casein micelles occurs as the isoelectric point (pH 4.6) is approached. To form a gel (as opposed to a curd), the pH must fall slowly without stirring. This occurs during the production of fermented milk products, such as yogurt, as lactose is converted into lactic acid. The same effect can be achieved artificially by adding an acidogen such as glucono-δ-lactone (GDL). The mechanism of gelation is different from rennet-induced gels. It seems that as the pH drops to 5.2, weakly bound β- and κ-casein molecules are released from the micelles, leaving an otherwise intact size-determining framework of α_s-casein. After this partial disintegration at pH 5.5–5.2, small aggregated structures become evident, indicating interaction of free casein molecules. As the pH drops further, to 4.9, the aggregated structures contract to form discrete particles that are larger than the original micelles. These particles then rearrange and aggregate to form the final casein network [50]. The mechanism appears to be somewhat different if the milk is preheated to temperatures high enough to denature the whey proteins (as is standard practice in the production of fermented products such as yogurt and quarg). Beta-lactoblobulin then interacts with κ-casein on the micelle surface. The casein particles in heated milk coalesce less extensively than in unheated milk, and it appears that the presence of denatured β-lactoglobulin inhibits micelle contact and fusion [51].

Casein can also form gels spontaneously during storage of UHT-sterilized milk (age-gelation). The mechanism of age-gelation has not been definitely established [52], but there are probably two separate causes: microbiological and physicochemical. Extracellular proteases secreted by psychotrophic bacteria appear to be a major causative factor in UHT milk produced from raw milk of low microbiological quality. Nevertheless, perfectly sterile UHT milk will eventually gel. Whey proteins dissociate from the casein micelles in the UHT milk during storage; it is possible that this promotes dissociation of the micellar κ-casein and gelation occurs by much the same mechanism as acid-induced gels.

2. Whey Proteins

The composition of whey, and hence of the derived products, depends on the method of removal of the casein. Most whey-protein products are derived from cheese whey, a co-product of the manufacture of cheese. In current commercial practice, whey is mostly obtained either after rennet coagulation of milk or after direct or indirect acidification of milk, as in the preparation of acid casein and cottage cheese [53].

The simplest whey product, whey powder, is made by the evaporation, crystallization, and spray-drying of whey. Whey protein concentrate (WPC) is also widely used commercially. Ultrafiltration is used to concentrate the proteins relative to the low molecular weight constituents of whey (mainly lactose and mineral salts). WPC consists of about 55% protein. Higher concentrations of protein can be achieved by using the same membrane systems but applying additional stages of ultrafiltration that allow a further decrease in the concentration of lactose and minerals. Concentrations of up to 85% can be achieved from certain whey sources, especially those with low lipid contents.

Whey protein isolates (WPI), which have protein concentrations in excess of 90%, are manufactured using ion-exchange adsorption/desorption processes. Two systems are operated commercially, one based on a modified cellulose ion exchanger in a stirred tank (the Vistec process), the other based on silicate ion exchangers in fixed-bed columns (the Spherosil process). The major whey proteins are recovered by both processes, but minor proteins, particularly the immunoglobulins, are lost.

Typical compositions of the two major whey types processed commercially (cheese whey and acid whey) are given in Tables 2 and 3. There are clear differences in the protein content and composition and also in the proportion of lipid and ash. The principal protein in all types of whey is β-lactoglobulin, which accounts for 50–65% of the dry weight. Because this is the whey protein that most readily forms thermal gels, its presence in whey is especially important. β-Lactoglobulin is a globular protein with a single polypeptide chain of 162 residues and a monomeric molecular weight of 18,363 daltons for the most abundant (A) variant. A second variant (B), differing only in two amino acid residues, is also common. Both structures are stabilized by two disulfide cross-links and also contain a sulfhydryl group. β-Lactoglobulin is sensitive to heat denaturation at close to 72°C at the natural pH and ionic strength of cows' milk. At higher pH, structural changes occur and the protein becomes unstable and forms aggregates.

α-Lactalbumin is the second most plentiful protein in whey. In the milk of western cattle there is only a single variant with a molecular weight of 14,175 daltons (although other variants are known): a globular protein stabilized by four disulfide bonds. At neutral

Table 2 Protein Composition of Cheese and Acid Wheys

	Average composition (g/liter)	
	Cheese whey	Acid whey
Whole whey protein	6.7	5.8
β-Lactoglobulin	3.5	3.5
α-Lactalbumin	1.3	1.3
Serum albumin	0.1	0.1
Immunoglobulins	0.4	0.4
Lipoprotein	0.2	—
Proteose peptone	0.2	0.2
Glycomacropeptide	1.0	—

Table 3 General Composition of Cheese and Acid Whey

	Average composition (mg% w/w)	
	Cheese whey	Acid whey
Lactose	5000	4400
Lipid	600	100
Ash	520	590
Na	35	40
K	109	133
Ca	22.5	86
Mg	5.8	8.9
P	42.5	63.0

pH, α-lactalbumin is stable in milk or buffered solutions to temperatures exceeding 80°C. It was, until recently, considered the most stable of whey proteins. It has now, however, been found that α-lactalbumin exists naturally as a calcium metalloprotein and that large structural changes, accompanied by apparent denaturation, occur at approximately 65°C, associated with the loss of bound calcium ions. Similar structural changes occur below pH 4.0 and above pH 9.0 at ambient temperatures.

Gelation of whey proteins by heating at a concentration above a critical point occurs by a mechanism similar to that of other globular proteins. Initial denaturation/perturbation of the protein structure is followed by intermolecular interactions that form a cross-linked matrix. Thus, a complete understanding of the mechanism of formation of heat-set whey protein gels demands knowledge of denaturation processes and factors influencing subsequent interactions. Environmental factors that influence these intermolecular interactions (particularly pH and ionic strength) are of enormous importance [54].

Gelation of WPC requires a minimum concentration of about 8% protein, depending on the preparative technique and the heating temperature. The gelation temperature, usually in the range of 85–100°C, is itself concentration dependant. Exhaustive removal of ions from WPI, as occurs in its preparation by ion-exchange adsorption, results in an inability to form gels, which is overcome by addition of salts. In particular, addition of calcium chloride results in clear, elastic gels. In fact, a minimum concentration of calcium appears to be required for gelation to proceed at all [55]. The presence of nonionic materials also affects the gelation of whey proteins. Thus lipids and lactose may adversely effect gelation.

Gelation of isolated β-lactoglobulin has also been extensively studied. In solution at pHs from 2.5 to 8, it readily forms gels on heating. As with WPI, the appearance and viscoelastic behavior of these gels are highly sensitive to pH. Gels formed within the pH range of 4–6 are opaque and appear to be "aggregate" or "particle" gels. Outside this range, the gels are more transparent and appear to be "fine-stranded" gels. The critical concentration for gel formation is lower for the aggregate gels than for the fine-stranded gels and can be as low as 1% in the pH range of 4–5. The onset of gelation of 12% β-lactoglobulin solutions takes place at temperatures far below the denaturation temperature. At higher and lower pH the onset temperature of gelation is greater than the denaturation temperature. Gels formed at low pH (3.0–4.0) are brittle compared with those formed at higher pH (6–7.5), which are rubberlike, despite their superficial similarity in appearance [56].

The mechanism of gelation of β-lactoglobulin remains poorly understood. From studies on the heat gelation of the A and B variants of β-lactoglobulin, it is clear that small changes in structure can lead to significant changes in functionality. Thus, variant A had a lower gelation point and a higher initial gelling rate than the variant B [57]. An interesting attempt has been made to test the importance of disulfide cross-linking by using a genetically engineered derivative of variant A in which cysteine residues were substituted for an arginine (residue 40) or a phenylalanine (residue 82) or both [58]. Although no additional disulfides were present in the unheated modified proteins, the higher gel strength and lower critical gel concentration of heat-set gels were interpreted as evidence for the importance of disulfides as cross-links in gelation. (There is some doubt about this conclusion because the effect of arginine and phenylalanine residues was not allowed for.)

The gelation properties of egg proteins, soy proteins, and muscle proteins are discussed in other chapters.

G. Mixed Systems—Some Specific Examples

In most countries the range of gelling agents available for food use is limited by regulation. In the development of new textures and in the formulation of new products, we have to rely on combinations of two or more of these permitted gelling agents. Mixed systems are therefore of great practical importance and there is much research in this area, although most of it is very empirical and unhelpful in understanding or illustrating general principles. We give here some more informative recent examples.

1. Casein/κ-Carrageenan

In chocolate milk, the sedimentation of cocoa particles is prevented by a weak gel network provided by the addition of a small amount of κ-carrageenan. Carrageenan reacts with κ-casein, forming a strong complex [59]. Only part of the carrageenan chain is adsorbed by the casein micelle, most of it is free in solution in the form of loops or tails. As the solution is cooled to below the carageenan's helix-coil transition temperature, it appears that the free κ-carrageenan chains link by forming double helices (Fig. 14) [60].

2. Muscle Protein/Alginate or Xanthan

The effects of xanthan and alginate on the gelation of myofibrilar protein have recently been studied [61]. Gels were prepared by heating the mixture at pH 6 in 0.6 M NaCl. The polysaccharides appear to interact with myofibrillar protein because they markedly change the rheological characteristics of the sol. But, in contrast to the effect of κ-carrageenan on casein, they reduce the gelling ability of the protein.

3. Production of Meat Analogs

Phase separation (see Sec. III.F) has been deliberately exploited to generate gelled products with, it is claimed, the anisotropic, fibrillar character of materials like meat [62]. The general principles involved are outlined in Figure 15. The components are mixed so as to form a water-in-water emulsion and gelled while flowing. By subjecting it to

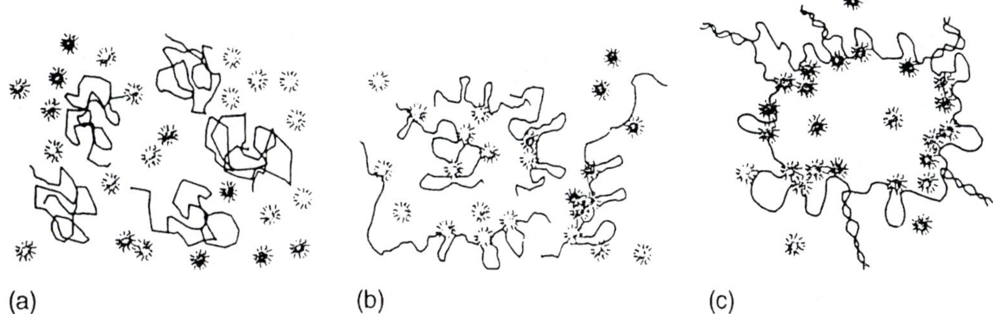

FIGURE 14 Schematic diagram of the mechanism of gelation of casein/κ-carrageenan mixed gels. (a) Casein micelles and random coil κ-carrageenan in solution. (b) Casein micelle/κ-carreenan interactions in solution. (c) Gel formed from casein and κ-carrageenan. (From Ref. 60.)

flow, the emulsion can be made anisotropic due to deformation of the dispersed droplets. If the dispersed phase is gelled under these conditions, fibers are produced; if the continuous phase is gelled, the result is a gel filled with liquid capillaries; gelation of both phases gives gels filled with orientated fibers. A gel with liquid channels also has advantages for the production of flavored products because liquids have better flavor release characteristics than gels.

VI. CONCLUSIONS

Steady but slow progress is being made in studies of the theory of gelation. Studies on proteins are lagging, largely because of the complexity of protein systems and the present

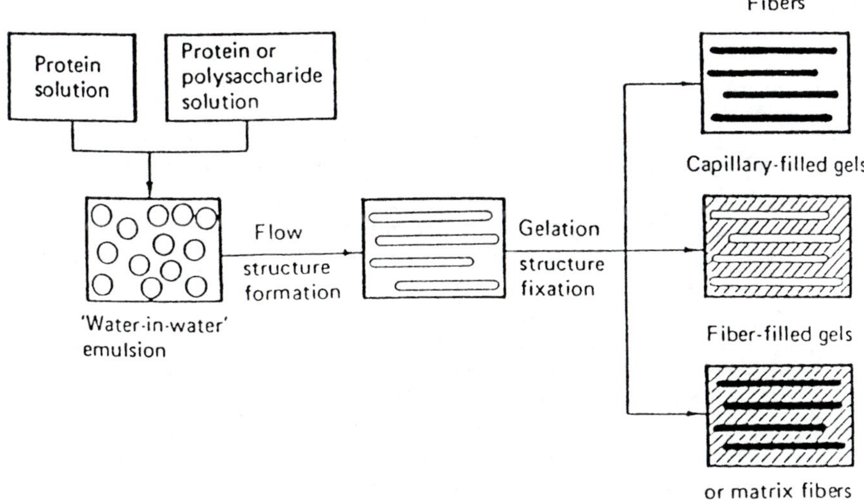

FIGURE 15 General scheme for the processing of two-phase liquid systems by spinneretless spinning technology. (From Ref. 62.)

high cost of research; nevertheless progress is continuing. Questions such as the contributions of various cross-links and their separation distances to properties of a gel could in theory be answered by constructing synthetic proteins with known sequences. At present this is not feasible, but a tailor-made gelling protein for food use is not an impossibility for the future.

Almost all gelled foods contain complex mixtures of proteins and polysaccharides. These mixed systems are of great practical importance and present a great theoretical challenge. In most countries the range of gelling agents available for food use is limited by regulation, consequently development of new textures and formulation of new products often relies on use of combinations of two or more of these permitted gelling agents. How these complex macromolecules interact is still only partly understood, but protein-polysaccharide complexes might be regarded as a new type of gelling agent where the formation conditions, as well as the rheological and other physicochemical properties, could be controlled to give a product with the desired textural characteristics.

REFERENCES

1. *The Oxford English Dictionary*, 2nd ed., Oxford University Press, London, 1992.
2. E. Heymann, *The Sol-Gel Transformation*, Hermann et Cie, Paris, 1936.
3. S. Kimura, S. Miura, and Y.-H. Park, Collagen as the major edible component of jellyfish (*Stomolophus nomural*), *J. Food Sci. 48*:1758 (1983).
4. D. Oakenfull, Gelling agents, *CRC Crit. Rev. Food Sci. Nutr. 26*:1 (1987).
5. A. H. Clark and S. B. Ross-Murphy, Structural and mechanical properties of biopolymer gels, *Adv. Polymer Sci. 83*:57 (1987).
6. G. R. Ziegler and E. A. Foegeding, The gelation of proteins, *Adv. Food Nutri. Res. 34*:203 (1990).
7. P. J. Flory, *Disc. Faraday Soc. 57*:7 (1974).
8. H. G. Bungenberg de Jong, Gels, *Colloid Science II* (H. R. Kruyt, ed.), Elsevier, Amsterdam, 1949, p. 2.
9. P. J. Flory, *Principles of Polymer Chemistry*, Cornell University Press, Ithaca, NY, 1953, p. 432.
10. L. R. G. Treloar, *The Physics of Rubber Elasticity*, Clarendon Press, Oxford, 1975, p. 59.
11. P. H. Hermans, Reversible systems, *Colloid Science II* (H. R. Kruyt, ed.), Elsevier, Amsterdam, 1949, p. 483.
12. J. W. Williams and R. A. Alberty, Colloids, *Treatise on Physical Chemistry*, Vol. 2, 3rd ed. (H. S. Taylor and S. Glasstone, eds.), Van Nostrand, New York, 1942, p. 675.
13. P. Sherman, *Industrial Rheology*, Academic Press, London, 1970.
14. T. L. Smith, Rupture processs in polymers, *Pure Appl. Chem. 23*:235 (1970).
15. H. McEvoy, S. B. Ross-Murphy, and A. H. Clark, Large deformation and ultimate properties of biopolymer gels. I Single biopolymer component systems, *Polymer 26*:1483 (1985).
16. S. B. Ross-Murphy, Rheological methods, *Biophysical Methods in Food Research* (H. W.-S. Chan, ed.), Blackwell Scientific Publications, Oxford, 1984, p. 138.
17. D. G. Oakenfull, N. S. Parker, and R. I. Tanner, Method for determining absolute shear modulus of gels from compression tests, *J. Texture Stud. 19*:407 (1989).
18. E. R. Morris, Polysaccharide solution properties: origin, rheological characterization and implications for food systems, *Frontiers in Carbohydrate Research 1: Food Applications* (R. P. Millane, J. N. BeMiller, and R. Chandrasekaran, eds.), Elsevier, London, 1989, p. 132.
19. J. D. Ferry, *Viscoelastic Properties of Polymers*, J. Wiley and Sons, New York, 1980.
20. R. W. Burley and D. V. Vadehra. *The Avian Egg; Chemistry and Biology*, John Wiley and Sons, New York, 1989.

21. J. L. Doublier., Rheological investigation of polysaccharide interactions in mixed systems, *Gums and Stabilisers for the Food Industry—7* (G. O. Phillips, P. A. Williams, and D. J. Wedlock, eds.), IRL Press, Oxford, 1994, p. 257.
22. J. R. Hermans, Elastic properties of the particle network in gelled solutions of hydrocolloids. I Carboxymethylcellulose, *J. Polymer Sci. Part A 3*:1859 (1965).
23. D. G. Oakenfull, A method for using measurements of shear modulus to estimate the size and thermodynamic stability of junction zones in noncovalently crosslinked gels. *J. Food Sci. 49*:1103 (1984).
24. D. G. Oakenfull and A. G. Scott, Stabilization of gelatin gels by sugars and polyols, *Food Hydrocolloids 1*:163 (1986).
25. A. H. Clark and S. B. Ross-Murphy, Concentration dependence of gel modulus, *Br. Polymer J. 17*:164 (1985).
26. M. Gordon and S. B. Ross-Murphy, The structure and properties of molecular trees and networks, *Pure Appl. Chem. 43*:1 (1975).
27. C. Tanford, *The Hydrophobic Effect*, John Wiley and Sons, New York, 1973.
28. A. Ben Naim, *Hydrophobic Interactions*, Plenum Press, New York, 1973.
29. S. E. Hill, J. R. Mitchell, and H. J. Armstrong, The production of heat stable gels at low protein concentration by use of the Maillard reaction, *Gums and Stabilisers for the Food Industry—6* (G. O. Phillips, P. A. Williams, and D. J. Wedlock, eds.), IRL Press, Oxford, 1992, p. 471.
30. P. Hegg, Conditions for the formation of heat-induced gels of some globular food proteins, *J. Food Sci. 47*:1241 (1982).
31. M. P. Tombs, Gelation of globular proteins, *Disc. Faraday Soc. 57*:158 (1974).
32. E. Doi, Gels and gelling of globular proteins, *Trends Food Sci. Technol. 4*:1 (1993).
33. A. H. Clark, F. J. Judge, J. B. Richards, J. M. Stubbs, and A. Suggett, Electron microscopy of network structures in thermally-induced globular protein gels, *Int. J. Peptide Protein Res. 17*:380 (1981).
34. T. Tanaka, Gels, *Sci. Am. 244*:124 (1981).
35. N. Catsimpoolas and E. W. Meyer, Gelation phenomena of soybean globulins. I Protein-protein interaction, *Cereal Chem. 47*:559 (1970).
36. F. S. M. van Kleef, Thermally induced protein gelation; gelation and rheological characterization of highly concentrated ovalbumin and soybean protein gels, *Biopolymers 25*:31 (1986).
37. A.-M. Hermannsson, Aggregation and denaturation involved in gel formation, *ACS Symp. Ser. 92*:81 (1979).
38. G. Stainsby, Proteinaceous gelling systems and their complexes with polysaccharides, *Food Chem. 6*:3 (1980).
39. V. B. Tolstoguzov, Functional properties of food proteins and the role of protein-polysaccharide interaction, *Food Hydrocolloids 4*:429 (1991).
40. V. J. Morris, Multicomponent gels, *Gums and Stabilisers for the Food Industry—3* (G. O. Phillips, D. J. Wedlock, and P. A. Williams, eds.), Elsevier, London, 1986, p. 87.
41. S. K. Samant, R. S. Singhal, P. R. Kulkarni, and D. V. Rege, Protein-polysaccharide interactions: a new approach in food formulations, *Int. J. Food Sci. Technol. 28*:547 (1993).
42. J. R. Mitchell, Rheology of gels, *J. Texture Stud. 7*:313 (1976).
43. D. G. Oakenfull and A. G. Scott, New approaches to the study of food gels, *Gums and Stabilisers for the Food Industry 3* (G. O. Phillips, D. J. Wedlock, and P. A. Williams, eds.), Elsevier, London, 1986, p. 465.
44. J. M. Schweid and R. T. Toledo, Changes in the physical properties of meat batters during heating, *J. Food Sci. 46*:850 (1981).
45. P. W. Gossett, S. S. H. Rizvi, and R. C. Baker, A new method to quantitate the coagulation process, *J. Food Sci. 48*:1400 (1983).
46. W. M. Marrs and D. J. Steele, An instrument for the measurement of gelation time, *J. Phys. E: Sci. Instruments 8*:270 (1975).

47. A.-M. Hermansson, Soy protein gelation, *J. Am. Oil Chem. Soc. 63*:658 (1986).
48. K. Barnard, N. D. Light, T. J. Sims, and A. J. Bailey, Chemistry of collagen crosslinks, *Biochem. J. 244*:303 (1987).
49. R. C. Bottomley, M. T. A. Evans, and C. J. Parkinson, Whey proteins, *Food Gels* (P. Harris, ed.), Elsevier Applied Science, London, 1990, p. 435.
50. P. F. Fox and D. M. Mulvihill, Casein, *Food Gels* (P. Harris, ed.), Elsevier, London, 1990, p. 121.
51. I. Heertje, J. Visser, and P. Smits, *Food Microstructure 4*:267 (1985).
52. V. R. Harwalkar, Age-gelation of sterilized milks, *Developments in Dairy Chemistry - 1 - Proteins* (P. F. Fox, ed.), Applied Science Publishers, 1982, p. 229.
53. R. Jost, Functional characteristics of dairy proteins, *Trends Food Sci. Technol. 4*:283 (1993).
54. C. V. Morr, Whey protein functionality: Current status and the need for improved quality and functionality, Proceedings of Dairy Products Technical Conference, Chicago, 1990, p. 69.
55. P. R. Kuhn and E. A. Foegeding, Factors influencing whey protein gel rheology: dialysis and calcium chelation, *J. Food Sci. 56*:789 (1991).
56. M. Stading and A. M. Hermansson, Large deformation properties of β-lactoglobulin gel structures, *Food Hydrocolloids 5*:339 (1991).
57. X. L. Huang, G. L. Catignani, E. A. Foegeding, and H. E. Swaisgood, Comparison of the gelation properties of β-lactoglobulin genetic variants A and B, *J. Agric. Food Chem. 42*: 1064 (1994).
58. S. P. Lee and C. A. Batt, Enhancing the gelation of beta-lactoglobulin, *J. Agric. Food Chem. 41*:1343 (1993).
59. T. H. M. Snoeren, P. Both, and D. G. Schmidt, An electron microscopic study of carrageenan and its interaction with kappa-casein, *Neth. Milk Dairy J. 30*:132 (1976).
60. S. Y. Xu, D. W. Stanley, H. D. Goff, V. J. Davidson, and M. Le Maguer, Hydrocolloid/milk gel formation and properties, *J. Food Sci. 57*:96 (1992).
61. Y. L. Xiong and S. P. Blanchard, Viscoelastic properties of myofibrillar protein-polysaccharide composite gels. *J. Food Sci. 58*:164 (1993).
62. V. B. Tolstoguzov, Creation of fibrous structures by spinneretteless spinning, *Food Structure—Its Creation and Evaluation* (J. M. V. Blanshard and J. R. Mitchell, eds.), Butterworths, London, 1988, p. 181.

5

Protein-Lipid and Protein-Flavor Interactions

JEAN-MARC CHOBERT AND THOMAS HAERTLÉ
Institut National de la Recherche Agronomique, Nantes, France

I. INTRODUCTION

Protein-lipid and protein-flavor interactions play several important roles in biological and technological processes. On the one hand, many interactions of lipids with proteins and peptides are responsible for the formation of the secondary and tertiary structure of several proteins and peptides, such as membrane proteins and several membrane-bound receptors. Leader peptides that interact with cellular membranes also may be placed in this category. On the other hand, many of these interactions, which often shape the conformations and the functionality of proteins, are responsible for signal transduction and also for the trafficking of many secreted or adsorbed proteins and other compounds across cells. The appropriate match of amphiphilic lipids and amphiphilic fragments of the proteins can create many of the desired biological and technological interfaces.

In more general and almost philosophical terms, protein-lipid interactions are responsible for the correct organization of a large number of crucial biological structures, such as membranes, organelles, cells, tissues, and entire organisms. One can say with certitude that both the morphology and physiology of whole living world are tributary to these interactions and they are largely determined by protein-lipid interactions.

In food technology, the interactions between proteins and lipids may present a major gambit, as can be seen in the case of formulations of cheeses, milk creams, mayonnaises, doughs, bakery, and several meat products. The structural organizations of lipid-protein complexes in biological systems are quite different from those that are formed in processed food products. However, the physicochemical characteristics of protein-lipid interactions that occur in foods as a result of heating, mixing, shearing, etc., are of the same kind as those responsible for the stability and functioning of living systems.

II. MAJOR FACTORS DETERMINING PROTEIN-LIPID AND PROTEIN-FLAVOR INTERACTIONS

The major forces involved in protein-lipid and protein-flavor interactions are similar to those that are responsible for protein folding. Thus, interactions between lipids and proteins can modify or interfere with the formation of native protein structures, as occurs when proteins are transferred from the aqueous phase to a lipid phase, e.g., lipid bilayers in biological systems.

In some cases, the protein-lipid interactions can also play an important role in the folding of a protein during its synthesis in vivo into its final native functional conformation. This is true of all membrane proteins. A good example is colicin A, an antibiotic protein that kills *Escherichia coli* cells by forming channels in their membranes. The bacterial toxin colicin A binds spontaneously to the surfaces of negatively charged membranes. The surface-bound toxin must subsequently, however, adapt an acidic "molten globule" conformation prior to fully inserting itself into the lipid bilayer. Clearly, electrostatic interactions must play a significant role in both events. A large positively charged surface exists in this protein, which is involved in its binding to negatively charged membranes. Surprisingly, colicin N, another similar bacterial toxin, also has a similarly positively charged surface, even though its overall amino acid composition is quite different from that of colicin A; for example, its isoelectric point is 10.20, whereas that of colicin A is 5.44. There is a single highly conserved aspartate residue (Asp78) on the positively charged face in all colicins, which plays an important role in the insertion of the toxin into the bacterial membrane. When this residue is replaced by asparagine in the mutant D78N, the mutant binds rapidly to negatively charged vesicles but inserts only half as fast as the wild-type protein into the membrane core.

The mechanism of incorporation of the ion channel forming C-terminal fragment of colicin A into the negatively charged lipid vesicles provides an example of insertion of a soluble protein into a lipid bilayer. The N-terminus portion of this protein is known; it consists of a 10-helix bundle containing a hydrophobic helical hairpin. This fragment forms a well-defined complex with dimyristoylphosphatidyl-glycerol [1].

Comparison of the amino acid sequences of transmembrane bacterial proteins shows that the residues fully exposed to the membrane lipids do not show significant conservation, indicating that few restrictions are placed on residues that are exposed to the membrane lipids. This implies that there are relatively few specific interactions between the transmembrane helices and the fatty acid chains of the membrane that require the presence of the specific residues. This is also consistent with the observation that membrane proteins can move within the plane of the membrane, by lateral diffusion, and are not at fixed positions.

Almost all membrane proteins lose their folds, activities, or structural functions once they are displaced from the lipid bilayer and into the aqueous phase. They usually become irreversibly denatured once removed from their native environment.

Evolution of biological systems has worked out very complicated and ingenious mechanisms to facilitate proper folding of proteins and interactions between proteins and lipids in order to form functional protein-lipid supramolecular assemblies such as membranes, membrane interfaces, membrane channels, receptors, etc. Any perturbation of this fine-tuning brings the whole system of delicate equilibria to a standstill, as can be seen in case of genetic diseases involving faulty membrane transport and signal transduction, many of which are lethal.

III. PROTEIN-LIPID INTERACTIONS IN FOOD SYSTEMS

A. Interaction of Animal Proteins with Lipids

1. Model Systems

Emulsion formation and stability in most food products are greatly aided by the presence of proteins, which, during the process of emulsification and new-surface creation, tend to adsorb at the oil-water (o/w) interface.

Once adsorbed, proteins undergo unfolding and rearrangement at the o/w interface; the unfolded molecules interact with each other via hydrophobic, hydrogen bonding, van der Waals, and ionic interactions and form a viscoelastic film. Proteins, however, are not the only surface-active components in real food emulsion systems. Typically, foods contain several low molecular weight surface-active agents, which will also compete for adsorption at the o/w interface. This may result either in cooperative interactions with adsorbed protein molecules or displacement of protein molecules from the interface; the latter might cause destabilization of the emulsion. Surface-active constituents naturally present in oils may also influence interfacial adsorption properties of proteins as observed in the case of olive oil, which contains diglycerides at a 2–2.5% level. Monoglycerides, on the other hand, are found at a much lower concentration. Addition of diolein, monoolein, or oleic acid to purified olive oil at concentration levels of diglycerides, monoglycerides, and free fatty acids found in the natural olive oil showed no cooperative interactions with bovine serum albumin (BSA) and sodium caseinate at the o/w interface; in fact they decreased the viscoelastic properties of adsorbed BSA and sodium caseinate films, although the sodium caseinate film was viscous [2]. It can therefore be concluded that diglycerides or free fatty acids, on their own, antagonize rather than aid the adsorption of proteins. On the other hand, in the case of natural olive oil–protein solution interfaces, all these natural olive oil constituents together with others (such as sterols, phospholipids, and oxidation products of lipids, which occur at a much lower concentration in olive oil) adsorb at the interface, cooperatively interact with the adsorbed protein molecules, and form mixed films exhibiting high viscoelasticity. Thus, whereas complex interactions of various lipid constituents of natural oils with proteins at the oil-water interface seem to create a viscoelastic mixed film, individually they cause destabilization of a protein at the interface.

One of the most common methods of improving the stability of dairy emulsions is the addition of various types of surfactants, such as mono- and diglycerides. The glycerides adsorb at the fat globule surface to lower its interfacial free energy and, consequently, improve stability against coalescence. Since milk proteins can also lower the interfacial free energy [3], it might be interesting to study the interactions between the glycerides and the milk proteins when used together in dairy emulsion formulations.

Monolayer techniques have been employed successfully to study the properties of milk proteins [4–10], but very few investigations have been undertaken to study the interactions between milk proteins and lipids [11,12]. Rahman and Sherman [13] reported a study on the interactions of individual caseins, α-lactalbumin, and β-lactoglobulin with mono- and diglycerides at the air/water interface.

The behavior of a model system—a mixture of sodium caseinate (NaCas)/glycerol monostearate (GMS)—spread at the air–phosphate buffer interface was studied to obtain information on the interactions taking place between lipids and proteins at the fat globule surface [14]. A Wilhelmy-type compression system was used to record the pressure-area

(π-A) isotherms of mixed films at four different GMS:NaCas surface ratios (0.2:0.8, 0.4:0.6, 0.6:0.4, and 0.8:0.2) and at three pH values (5.9, 6.8, and 7.7). Results showed no pH effect over the range tested. Condensation of the mixed films was detected in the low-pressure region of the isotherms (0–20 mN/m). Maximum condensation was encountered at an initial surface ratio of 0.6:0.4. In the high-pressure region, a portion of sodium caseinate remained at the interface and increased the collapse pressure of the film.

Oil-in-water food emulsions are often stabilized by proteins. After diffusion towards the o/w interface and adsorption, proteins constitute an interfacial film. Because of its electrical and rheological properties, this film prevents coalescence and flocculation. The amphipathic nature of proteins, i.e., their ability to interact with both water and with lipids of fat droplets, is necessary to form such films.

Interactions between proteins (bovine serum albumin, whole bovine casein or its pure fractions, lysozyme, fatty acid–binding protein of rat liver, whey proteins, β-lactoglobulin, α-lactalbumin) and lipids in model systems (fatty acids, triglycerides, or phospholipids) have been reported. Such studies have used spectroscopic (fluorescence, UV), resonance (nuclear magnetic resonance, electron spin resonance), or chromatographic methods (gas-liquid chromatography). Some of these studies have focused particularly on the influence of the structure of lipids and proteins on interactions. Proteins with hydrophobic regions embedded in the interior of their three-dimensional structures were less lipid-binding than those with more accessible hydrophobic regions [15]. Increasing the length of the hydrocarbon chain [16,17] or decreasing its degree of unsaturation improved interactions between lipids and proteins. Also, binding of retinol on the retinol-binding protein [18] or on β-lactoglobulin [19] suggested the presence of a nonpolar binding site. Such results are in accordance with the hypothesis that interactions between lipids and proteins are of a hydrophobic nature.

On the other hand, β-lactoglobulin seems to bind dipalmitoylphosphatidylcholine like a peripheral protein does to membrane phospholipids. Interactions may occur through lysyl or arginyl residues of the protein and negatively charged groups of lipids [20,21]. Studies about interactions between pure fractions of milk proteins and fatty acids at pH 7 stressed the importance of the polar head of lipids [22]. These studies seem to indicate an electrostatic mechanism for fatty acid binding to milk proteins.

The convincing electron spin resonance study of the interactions between nitroxide homologs of fatty acids and purified components of milk proteins in a model food emulsion was performed by Aynié et al. [23]. They showed that with or without the protein, the mobility of nitroxide moiety was dependent on its position on the fatty acid chain. The extent of its interactions with proteins in emulsions decreased in the order α_{s1}-casein > β-lactoglobulin > β-casein. α_{s1}-Casein induced a better organization of the lipid monolayer. Neither differences in flexibility nor tertiary structure of the proteins explained this result. It appeared that interactions occurred through lipid polar heads and protein polar side chains due to hydrogen bonds and/or electrostatic interactions. The authors have demonstrated that the affinity of milk proteins for fatty acids in solutions increased in the order β-lactoglobulin > α_{s1}-casein > β-casein. They concluded that most of these interactions are polar interactions and were unable to detect any hydrophobic interactions. The milk proteins seem to interact preferentially with polar heads of nitroxide-labeled fatty acids as was reported before for the same protein in solution [21].

It is well known that β-lactoglobulin, which is found in the milk of several mammal species, binds fatty acids and sodium dodecyl sulfate. Recent studies on β-lactoglobulin–binding properties show that it can also bind with high specificity small hydrophobic ligands such as retinoids [18,24], alkanone flavors [16], polyoxyethylene sorbitan monolaurate [25], protoporphyrin IX [26], and ellipticine [27]. X-ray crystal structure of β-lactoglobulin [19,28] suggests that this protein may belong to the superfamily of hydrophobic molecule transporters termed lipocalicins. Retinol-binding protein [29], bilin-binding protein [30], insecticyanin [31], and β-lactoglobulin [19,28] are the best known proteins of this class. All of these proteins share a common three-dimensional structural pattern, i.e., eight-stranded antiparallel β-sheet flanked on one side by an α-helix constituting a hydrophobic pocket.

The biological function of β-lactoglobulin still remains unclear. Since it is able to bind fatty acids and to increase the activity of pregastric lipase, it has been recently claimed that ruminant β-lactoglobulin could participate in fat digestion during the neonatal period. Bovine β-lactoglobulin, purified by nondenaturing methods at neutral pH, contains bound fatty acids [32]. It has also been postulated that classical isolation procedures of β-lactoglobulin involving exposure to low pH and/or drastic modifications of the ionic strength of milk caused conformational changes, which decrease the affinity of this protein for fatty acids [33]. Studies of thermodynamic stability of β-lactoglobulin at various pH values show, however, that the temperature of irreversible denaturation increases with a decrease in pH from 7.5 to 1.5, suggesting a net increase in protein stability. Similar behavior is also observed during β-lactoglobulin unfolding induced by urea at different pH.

Experiments aimed at elucidation of interactive properties of bovine and porcine β-lactoglobulin, isolated at acid pH, with 16 saturated and unsaturated fatty acids varying in aliphatic chain length have been conducted by Frapin et al. [34].

β-Lactoglobulin fatty acid–binding site was studied using fatty acids of different chain length and containing from zero to four double bonds. Figure 1 shows the binding of palmitic and arachidic acids to native β-lactoglobulin at neutral pH. The extent of enhancement of β-lactoglobulin tryptophan fluorescence was dependent on the fatty acid used. In all cases, the maximum fluorescence occurred at 1 : 1 stoichiometries of protein to fatty acid, suggesting that β-lactoglobulin has only one high-affinity fatty acid–binding site. It has been previously reported that bovine β-lactoglobulin has additional low-affinity fatty acid–binding sites [33]. The apparent binding constants and the apparent mole ratios of binding of various saturated and unsaturated fatty acids to β-lactoglobulin are presented in Tables 1 and 2. Among the fatty acids studied, palmitic acid shows the highest affinity ($K_d' = 1 \cdot 10^{-7}$ M, $n = 0.93$) for β-lactoglobulin.

The addition of caprylic or capric acid to β-lactoglobulin solution does not produce any fluorescence enhancement (Fig. 1), suggesting that, under the conditions used, these compounds do not bind to the protein. Lauric acid binds to β-lactoglobulin and the affinity of β-lactoglobulin for saturated fatty acids (Table 1) increases gradually from lauric acid ($K_d' = 7 \cdot 10^{-7}$ M) to palmitic acid ($K_d' = 1 \cdot 10^{-7}$ M) and then decreases for longer aliphatic chains, as shown in Table 1. It seems that the fatty acid–binding pocket of β-lactoglobulin can only accommodate the 16-carbon aliphatic chain with high steric precision. Apparently, the nature of fatty acid/β-lactoglobulin interactions is mainly hydrophobic since its affinity for fatty acids increases with an increase of chain length in the C10–C14 range. It should be emphasized that sodium dodecyl sulfate ($K_d' = 2.3 \cdot 10^{-7}$ M) and dodecanyl acetate ($K_d' = 2.6 \cdot 10^{-7}$ M) show similar affinities for β-lactoglobulin. Apparently, however, the binding of fatty acids to β-lactoglobulin

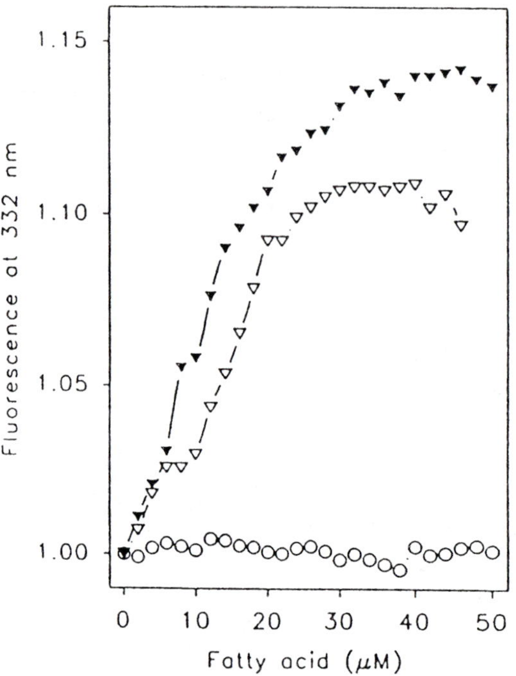

FIGURE 1 Corrected β-lactoglobulin (BLG) tryptophan fluorescence titration curves with caprylic acid (○), palmitic acid (▽), and arachidic acid (▼). During the titrations with caprylic acid, palmitic acid, and arachidic acid, BLG concentrations were 32, 25, and 35 μM, respectively. (From Ref. 34.)

TABLE 1 Apparent Dissociation Constants (K_d') and Apparent Molar Ratio (n) of Saturated Fatty Acid/β-Lactoglobulin Complexes

	K_d' (x 10^7 M)[a]	n
Native β-lactoblobulin		
Caprylic acid (C8:0)	—	—
Capric acid (C10:0)	—	—
Lauric acid (C12:0)	7.0 ± 1.6	0.86 ± 0.05
Myristic acid (C14:0)	3.3 ± 1.0	0.83 ± 0.01
Palmitic acid (C16:0)	1.0 ± 0.05	0.93 ± 0.11
Stearic acid (C18:0)	1.2 ± 0.07	0.89 ± 0.03
Arachidic acid (C20:0)	4.0 ± 0.03	0.85 ± 0.04
Delipidated β-lactoglobulin		
Myristic acid	1.8 ± 0.05	0.79 ± 0.05
Palmitic acid	1.3 ± 0.3	0.95 ± 0.2

[a]Each value is an average of at least three determinations.
Source: Ref. 34.

TABLE 2 Apparent Dissociation Constants (K'_d) and Apparent Molar Ratio (n) of Monounsaturated Fatty Acid/β-Lactoglobulin Complexes

	K'_d (x 10^7 M)[a]	n
Myristoleic acid (C14:1Δ9)	1.6 ± 0.05	0.83 ± 0.06
Palmitoleic acid (C16:1Δ9)	2.6 ± 0.1	0.82 ± 0.02
Oleic acid (C18:1Δ9)	1.3 ± 0.05	0.82 ± 0.05
Elaidic acid [C18:1Δ9 (*trans*)]	1.5 ± 0.02	0.82 ± 0.06
Linoleic acid (C18:2Δ9, 12)	1.9 ± 0.01	0.83 ± 0.08
Linolelaidic acid [C18:2Δ9, 12 (*trans, trans*)]	3.0 ± 0.7	0.86 ± 0.07
Linolenic acid (C18:3Δ9, 12, 15)	1.7 ± 0.06	0.91 ± 0.1
γ-Linolenic acid (C18:3Δ6, 9, 12)	1.3 ± 0.01	0.93 ± 0.04
Arachidonic acid (C20:4Δ5, 8, 11, 14)	3.3 ± 0.13	0.83 ± 0.07

[a]Each value is an average of at least three determinations.
Source: Ref. 34.

requires at least one hydrophobic chain end since hexadecanedioic acid does not bind to β-lactoglobulin. It should be mentioned that, in contrast to β-lactoglobulin, dicarboxylic acids formed during ω-oxidation of monocarboxylic acids bind to bovine serum albumin [35].

As shown in Table 2, unsaturated fatty acids also bind to native β-lactoglobulin. In most cases, except for myristoleic acid and arachidonic acid, the binding of unsaturated fatty acids to β-lactoglobulin is slightly weaker than the saturated molecules. The apparent dissociation constants of oleic acid/β-lactoglobulin, linoleic acid/β-lactoglobulin, and linolenic acid/β-lactoglobulin complexes are $1.3 \cdot 10^{-7}$, $1.9 \cdot 10^{-7}$, and $1.7 \cdot 10^{-7}$ M (Table 2), respectively, compared to $K'_d = 1.2 \cdot 10^{-7}$ M for the stearic acid/β-lactoglobulin complex (Table 1). β-Lactoglobulin binds natural fatty acids (oleic acid, linoleic acid) with slightly higher affinity (Table 2) than their *trans* isomers (elaidic acid, linolelaidic acid). Based on these results, it is apparent that structural constraints imposed by C=C double bonds in the aliphatic chain of fatty acids affect only slightly their binding to β-lactoglobulin. Surprisingly, prostaglandin A_1, a physiologically important derivative of dihomo-γ-linolenic acid, does not bind to β-lactoglobulin under the conditions studied. The binding of various saturated and unsaturated fatty acids by β-lactoglobulin compared with its relatively stringent binding specificity toward the β-isomer of ionones [36] and retinoids [24] suggests that these two classes of ligands are bound at two distinct binding sites. The characterization of retinol and fatty acid–binding sites of β-lactoglobulin may be achieved by using fluorescence spectroscopy, provided the ligand is a fluorophore [24,36]. Retinol fluorescence is weak in the unbound state in solution; however, it is greatly enhanced upon binding to β-lactoglobulin ($K'_d = 2 \cdot 10^{-8}$ M). Its complex with β-lactoglobulin has a fluorescence spectrum (excitation at 342 nm) with a maximum emission at 480 nm [37]. After excitation at 342 nm, palmitic acid-retinol-β-lactoglobulin complex (1 : 1 : 1) exhibits a typical fluorescence at 480 nm. In addition, the titration of retinol/β-lactoglobulin complex, as well as β-lactoglobulin, with palmitic acid shows a typical fluorescence enhancement of tryptophan fluorescence (data not shown). Derived apparent dissociation constants are similar in the two cases: $K'_d = 1 \cdot 10^{-7}$ M ($n = 0.93$) and $K'_d = 2.6 \cdot 10^{-7}$ M ($n = 0.9$)

for palmitic acid bound to β-lactoglobulin and retinol/β-lactoglobulin complex, respectively. These results indicate that retinol and palmitic acid are bound to β-lactoglobulin monomer at two different sites.

Alkylated and esterified β-lactoglobulin do not bind palmitic acid at neutral pH. However, it has been reported previously that chemical modifications of β-lactoglobulin enhance its affinity for retinol [24]. Apparently, chemical modification induces changes in the secondary structure [36], especially in the case of methylated β-lactoglobulin (30% β-sheet instead of 52% for native β-lactoglobulin) and thus destabilize the fatty acid–binding site. This observation gives additional evidence that retinol and fatty acids bind to β-lactoglobulin at two nonoverlapping binding sites.

Porcine β-lactoglobulin does not bind palmitic acid at neutral pH, whereas it binds retinol with an apparent dissociation constant of $K'_d = 1.2 \cdot 10^{-7}$ M. At a first sight, the differences in binding specificities of porcine and bovine β-lactoglobulins are surprising. Porcine and bovine β-lactoglobulins display 66% of sequence homology. Porcine β-lactoglobulin is monomeric at neutral pH, in contrast to ruminant oligomeric β-lactoglobulin [38]. For example, bovine β-lactoglobulin exists in various oligomeric states in function of pH, temperature, and concentration. At neutral pH, this protein is mainly dimeric, whereas it is monomeric at pH 3.0. It was proposed that the strand I (residues 145–150) is involved in the formation of the dimer by making antiparallel interactions with the diad-related strand [19]. Comparison of porcine and bovine β-lactoglobulin sequences shows major differences of their C-termini, while this region is highly conserved in all ruminant β-lactoglobulin sequences [38]. The deletion of residues 149–152 in porcine β-lactoglobulin, as well as the replacement of His-146 by Arg-146, may explain the monomeric state of porcine β-lactoglobulin. It may also suggest that the fatty acid–binding site of bovine β-lactoglobulin is close to the region involved in dimer formation, i.e., strand I, helix 130–140, and strand A [19]. Prevailing evidence indicates that β-lactoglobulin may bind retinol at the highly conserved area of this protein superfamily [19,39], i.e., inside the main hydrophobic pocket. Fatty acids could be bound elsewhere, may be in the external hydrophobic site described by Monaco et al. [28] or at the interface between monomers since both monomeric bovine β-lactoglobulin (acid pH) and monomeric porcine β-lactoglobulin do not bind fatty acids.

2. Food Systems

Protein-lipid interactions occur in foods during processing, especially in extruded foods. The effects of batch and extrusion cooking on lipid-protein interactions of processed cheese have been studied by Blond et al. [40]. In this particular study, samples of processed cheese were prepared (a) by a batch process in a mixer or (b) in an extrusion cooker. Effects of recipe variations and process conditions on protein-lipid interactions were evaluated. The addition of salts or premelted cheese mix increased lipid binding. The proportion of bound lipid did not increase with increasing fat content. Final cooling with slow mixing increased lipid binding in (b) but not in (a). Proteolysis was greater in (b) than in (a). Bound lipid content decreased and proteolysis increased with increasing temperature and rotation rate in (b). A new index for evaluation of protein-lipid binding in processed cheese, based on the ratio r = MGS/NSN, where MGS is lipid fraction bound to nonsedimentable nitrogen (NSN), has been proposed.

Lipid-protein interactions in concentrated infant formula milk have been studied [41]. In this case, ^{14}C-labeled κ-casein or β-lactoglobulin was added as a tracer protein to raw skim milk used for making the infant formula [41]. Ultracentrifugation of the sterilized product resulted in three fractions: a lipid phase with associated proteins, free casein micelles and other dense particles, and a fluid phase. Distribution of the ^{14}C-tracer protein and the protein content (measured by chemical methods) of the three phases varied significantly, and this depended on the processing conditions used (time and temperature of sterilization) and also on the amount of KOH and/or urea added to the formula. When KOH was added to the formula at 0–8 mEq/liter (after homogenization but before sterilization), the amount of ^{14}C-κ-casein in the lipid layer of the sterilized product decreased by 4.7% for each mEq/liter of added KOH; the corresponding reduction in the protein content of the lipid layer was 2 g/liter. These differences in the structure of the product, related to interactions of protein with lipid, protein, or calcium phosphate, may affect the physical properties and stability of milk-based lipid-rich products.

Gomes-Areas and Lawrie [42] have demonstrated that lipid-protein interactions played an important role in texture formation during thermoplastic extrusion of lung protein isolates. The texture of the extruded products was dependent on selective defatting of the isolate with solvents of increasing polarity. The use of organic solvents to remove lipids from lipid-protein complexes is widespread in the literature [43,44], and by careful choice of solvents, it is possible to disrupt the majority of lipid-protein complexes.

The effect of solvents of increasing polarity on lipid extraction of bovine lung and rumen proteins, isolated by two different procedures, namely, alkaline solubilization of the protein followed by its isoelectric precipitation and sodium dodecyl sulfate solubilization of the protein and its subsequent precipitation by ferric chloride was studied. A hyperbolic relationship was observed between the amount of lipid extracted and the dielectric constant of the solvent used. This relationship was described by an equation, which allowed the calculation of the total amount of lipids present in the isolates and the average intensity of the interaction between lipids and proteins in each system. Relative composition of fatty acids in lung isolates showed marked dependence on the polarity of the solvent employed.

In order to better understand the texturization process of the defatted offal isolates, it is essential to know the influence of lipid-protein association on the interaction of the isolates with water. Some aspects of these interactions have been investigated by Gomes-Areas [45]. Extensive data on water-holding capacity and solubility of beef tissues are found in the literature. However, data on the effect of lipid-protein interactions on these properties are scarce, and, due to the diversity of methods used, the results are not always comparable.

The simplest method to assess the thermodynamic properties of water in any system is to determine its equilibrium vapor pressure. The isotherms thus obtained provide several helpful parameters for the interpretation of the status of water in the system [46–48]. Iglesias and Chrife [49] studied the effect of fat on the water vapor isotherms of air-dried minced beef and found that variation in the concentration of added fat did not alter the pattern of water vapor isotherms. The amount of fat used by Iglesias and Chrife [49] (50–55%) was far beyond the normal content of lipids in offal protein isolates (12–20%). Thus, it was deemed necessary to study more carefully the effect of the naturally occurring lipids on the hydration characteristics of the offal protein isolates.

Water-protein interactions of rumen and lung protein isolates defatted by different solvents were studied using water sorption isotherms [50]. Two isolation procedures were employed to obtain the isolates: alkaline solubilization and isoelectric precipitation of

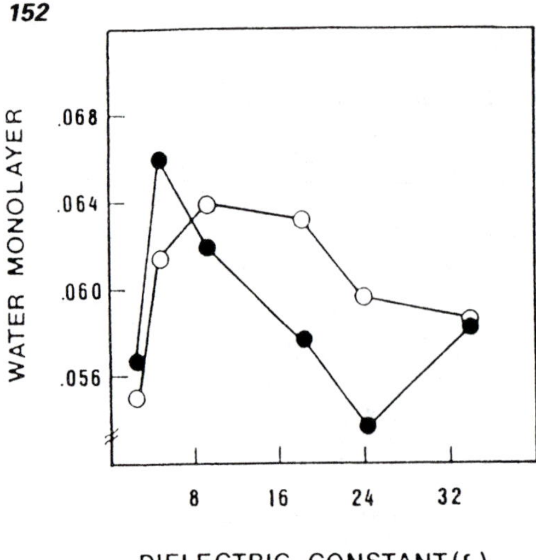

FIGURE 2 Lung isolates–Monolayer values for water adsorption (m_o) as a function of the dielectric constant of the solvent used in lipid extraction. (○) pH 10.5 isolate; (●) SDS isolate. (From Ref. 50.)

proteins, and SDS solubilization and $FeCl_3$ precipitation of proteins. Water monolayer values of the protein fraction showed a marked dependence on the dielectric constant of the solvent used for lipid extraction. In lung isolates, as seen in Figure 2, values increased up to a peak when solvents of intermediate polarity were employed, whereas in rumen isolates, as shown in Figure 3, values were initially constant and then decreased with increase of polarity of the solvent. Affinity of water to protein support showed a more complex pattern and was dramatically affected by the protein isolation procedure used.

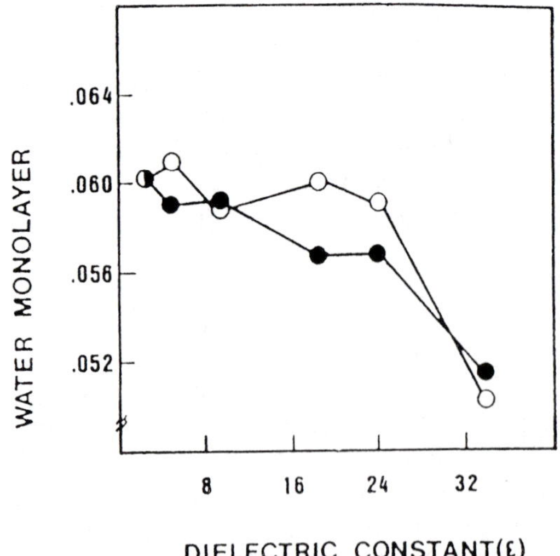

FIGURE 3 Rumen isolates–Monolayer values for water adsorption (m_o) as a function of the dielectric constant of the solvent used in lipid extraction: (○) pH 10.5 isolates; (●) SDS isolate. (From Ref. 50.)

The interaction of fish proteins (carp myofibrils) with lipids (fish oil) during freeze-drying and storage has been investigated by Kunimoto et al. [51]. Deterioration of fish products caused by autoxidation of highly unsaturated fatty acids is a particular problem in the utilization of dark-fleshed fish. Studies of changes in lipid and protein from a mixture of carp myofibrils and sardine lipids, freeze-dried and stored under various relative humidity conditions, demonstrated that autoxidation of lipids occurred during freeze-drying and storage. The relative humidity affected the autoxidation of lipids and the browning of protein. In a high humidity environment, peroxide value and thiobarbituric acid value were low, whereas dimers and trimers of lipid increased. Browning of protein moiety occurred more rapidly in high-humidity environments. Loss of available lysine was proportional to the browning of protein moiety, but little change in protein digestibility was observed during storage.

B. Interaction of Plant Proteins with Lipids

Under the high temperature, pressure, and mixing encountered in extrusion processing, oxidation of unsaturated lipids may occur [52]. Lipid hydroperoxides or their breakdown products may attack proteins to form covalent bonds. Funes and Karel [53] reported that linoleic acid incubated with lysozyme under oxidizing conditions produced small quantities of lysozyme-lipid and lysozyme-lysozyme-lipid. Investigation of protein-lipid interaction during extrusion was performed by Izzo and Ho [54], who used corn endosperm storage protein zein and corn oil. Zein was extruded with and without 5% added corn oil at 120 and 165°C. The temperatures were chosen over a range that would induce protein denaturation and would yield technologically easily extrudable products with a favorable snack stick appearance. Two solvent extraction methods of different degrees of severity were applied in order to remove lipid from protein. Effects of moisture content (20–30%), temperature (100–200°C), and screw speed (100–300 rev/min) on lipid-macromolecule interaction in corn meal was investigated. Interaction of native corn meal lipids with carbohydrates and proteins was evaluated by the amount of lipid available to hexane extraction. Extractable lipid decreased by 66.0–88.5% after extrusion, the magnitude of decrease mainly being affected by moisture content and the interaction of moisture content with barrel temperature. Maximal lipid-macromolecule interaction was observed under low-moisture and high-temperature conditions; minimal interactions were observed under high-moisture and high-temperature conditions. Lipids involved in the interactions were identified following extraction with chloroform-methanol (2:1) and enzymatic digestion of the starch and protein. During extrusion, lipids appear to be entrapped within the carbohydrate-protein network, decreasing the amount of available lipid. A preferential decrease of fatty acid, lysoPC, and glycolipid was observed. Both fatty acid and lysoPC were released by amylase digestion, while glycolipid could be recovered by amylase and proteinase digestion. Results suggest the formation of fatty acid–carbohydrate and lysoPC-carbohydrate inclusion complexes.

The model system of the corn endosperm storage protein zein and corn oil was used by Izzo and Ho [54]. Protein-lipid interaction in extruded model systems of zein and corn oil was investigated by analysis of lipids available to solvent extraction. Extracts were subjected to HPLC lipid class analysis (Table 3). Both methods indicated a decrease in extractable lipid on extrusion cooking. Slurry extraction revealed triglyceride and diglyceride binding, whereas the more harsh extraction by precipitation detected binding of fatty acid. Both methods showed that protein-lipid interaction was not increased by

TABLE 3 Quantitative Lipid Class Analysis of Extracts Obtained by the Precipitation Extraction Method

Extract source	% Triglyceride	% Fatty acid	% Diglyceride	% Peak B	% Unidentified
Unextruded zein	2.22	3.62	0.15	0.37	10.90
Extruded zein, 120°C	2.01	2.80	0.23	0.29	5.57
Extruded zein, 165°C	2.21	3.30	0.28	0.40	4.81
Extruded zein + corn oil, 120°C	6.18	2.75	0.21	0.36	6.00
Extruded zein + corn oil, 165°C	4.90	2.51	0.22	0.37	6.80

Source: Ref. 54.

elevating the temperature to 165°C, suggesting that heat-favored reactions did not predominate. Noncovalent interactions appeared to cause the decrease in lipid extractability.

However, this conclusion of Izzo and Ho [54] is contradicted by the results of Zamora et al. [55]. These authors performed the incubation experiments at 37°C in order to study the influence of temperature and mixture composition on the formation of fluorescent compounds, using glutathione (GSH) and 13-linoleate hydroperoxide (13-LAOOH) as model compounds for lipid-protein interaction. Highest yields of fluorescent compounds were obtained with a GSH:13-LAOOH ratio of 1:2 at 37°C. Similar kinetics of formation were observed with aldehydes (pentanal, butanal, propanal) in place of 13-LAOOH, but the formation was much slower.

Lipids in beer can affect a number of qualities of the product, one of which is the foam stability. The effect of lipids on beer foam stability has been a topic of many investigations. It is generally known that lipids destabilize beer foam stability by reducing the surface tension of the liquid film around the gas bubbles. It is thought that the endogenous levels of different classes of lipids (triglycerides or fatty acids) in beer are usually too low to have any significant effect. It has been demonstrated, however, that addition of fractions (combination of triglycerides and phospholipids) of malt lipid extract at levels of 1 mg/liter reduces foam stability by 62%. The effects of triolein and palmitic acid on head retention values (HRV) and foam proteins in beers brewed from (a) 100% barley malt or (b) grist containing 15% wheat flour were investigated [56]. Triolein and palmitic acid both reduced HRV. Experiments using ^{14}C triolein to assess lipid-protein interactions in beer brewed from (a) or (b) were also carried out. Results showed that the stability of triolein-protein complexes in (a) and (b) was very low and their effect on beer foam stability was minimal. Although acetic acid–soluble proteins of wheat flour showed slight affinity for triolein, there was no evidence that they were active in (b). The authors concluded that there is no significant difference in affinity of triolein for protein derived from beers brewed either an all-barley malt grist or one containing wheat flour. Apparently, lipid-protein interaction exists in beer, but it was impossible to identify the proteins involved. The involvement of lipids in wheat gluten has been observed for the first time more than 100 years ago by Osborne and Vorhees [57], who reported the presence of lecithin in the gluten washed from wheat flour.

The significance of interactions among wheat proteins, carbohydrates, and lipids has also been widely recognized in processed baked products. On the moisture-free basis, wheat flour contains approximately 80% starch, 14% proteins, 4–5% lipids, and 2% pentosans. Wheat proteins are broadly characterized, based on their solubility, into four classes: water-soluble albumins, salt-soluble globulins, alcohol-soluble gliadins, and acid- and alkali-soluble glutenins. Various proteins are distributed differently in the kernel. Inner endosperm consists of approximately equal amounts of gliadin and glutenin proteins. When water is added, wheat endosperm proteins, gliadin, and glutenin form a tenacious colloid complex, known as gluten, which consists of about 85% flour protein. Other 15% are non–dough-forming nongluten proteins. Lipids are minor component in wheat (3–5%). Thirty percent of the wheat lipids are located in germ, 25% in aleurone, and 45% in the starchy endosperm. Two thirds of the endosperm lipids are nonstarch lipids, and one third are starch lipids. Starch-associated lipids do not bind to proteins during dough mixing; they are bound tightly as inclusion bodies with amylose and consequently are not available for reaction with gluten proteins at least during gluten formation. Nonstarch lipids constitute 1.4–2% of starch weight. Their differences in solubility give a good means of separating them into two categories, free and bound. Free lipids can be extracted with nonpolar solvents such as ether or petroleum ether. Polar solvents such as water-saturated butanol are required for the separation of bound lipids. Free lipids constitute 0.8–1% and bound lipids 0.6–1% of flour weight. About two thirds of free lipids are nonpolar and the same proportion of bound lipids is polar. Polar lipids are the mixture of glycolipids and phospholipids. The very simple operations, such as addition of water and mixing, cause binding of lipids by the flour. The dependence of lipid binding on dough mixing is shown in Figure 4. For both flours the quantity of extractable free lipids decreased significantly at the first stage of mixing. It was also reported [58] that poor-quality flour may yield more extractable protein (Fig. 4, bottom). At any time of mixing, more protein can be extracted from poor- than from good-quality flour, and there is linear relationship between the mixing time and extractable protein yield. In good-quality flour, lipid binding reaches its maximum after approximately 4 minutes of mixing.

In the past, many researchers experienced difficulties in extracting lipids from dried gluten with ether. However, when determined by acid hydrolysis, gluten lipid content was as high as 11%. Consequently, the presence of lipoprotein complexes in gluten has been recognized. Figure 5 shows how flour components are involved in gluten formation.

Apparently it is hard to specify which of the proteins of the gluten system binds the most of lipids since the content of the bound lipids in the given protein fraction depends heavily on the method of protein fractionation used. However, two important technologically observations can be made:

1. The protein extractability from the dough decreases (Fig. 6) when the lipid content has been lowered by the preceding extraction. The protein solubilities of defatted flours were restored to a large extent when the extracted lipids were added to the flour.
2. There is an almost linear relationship between the lipid content and the loaf volume (Fig. 7).

In baked breads multiple interactions occur among wheat components and/or added components. The protein/starch interactions mediated by lipids have important functional roles in baked goods. At elevated oven temperature, starch is gelatinized and proteins are denatured. Bread proteins differ from flour and dough gluten proteins, and bread starch differs from the native flour and dough starch. Therefore, it is impossible to predict

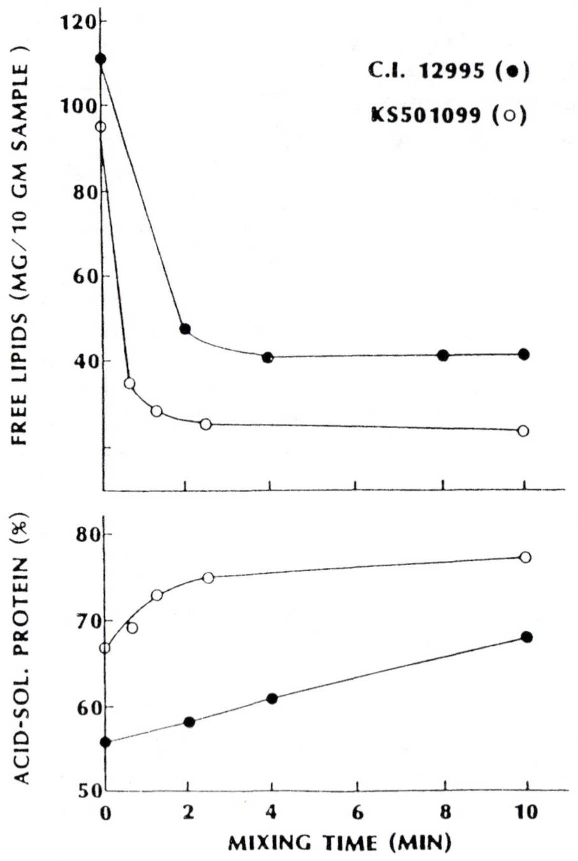

Figure 4 Effects of mixing time on free lipid extractability (top) and 0.05 N HOAc-soluble protein extractability (bottom) of good-quality flour (C.I. 12995, 1979 crop) and poor-quality flour (KS501099, 1979 crop). (From Ref. 58.)

the structure of protein/lipid complexes in the bread on the basis of their original configuration in the flour and dough system.

Several models of protein-lipid interactions in breadmaking have been proposed. In Hess and Mahl's model [59] (Fig. 8a), the adhesive protein is bound through a lecithin layer. The lipoprotein model (Fig. 8b) involving a bimolecular lipid layer is based on a structural study by electron microscopy. From x-ray evidence, Grosskreutz [60] deduced that lipids form well-oriented bilayers in gluten and that protein chains are bound to the outer edges of the phospholipids probably by electrostatic interactions. Figure 8c shows a model of starch-glycolipid-gluten complex proposed by Hoseney et al. [61]. They proposed that lipids are bound to gliadins via electrostatic interactions, whereas they bind to glutenins via hydrophobic interactions. Figure 8d depicts the model presented by Wehrli and Pomerantz [62] for a starch-glycolipid-gluten complex. This model puts more emphasis on the van der Waals and hydrophobic interactions.

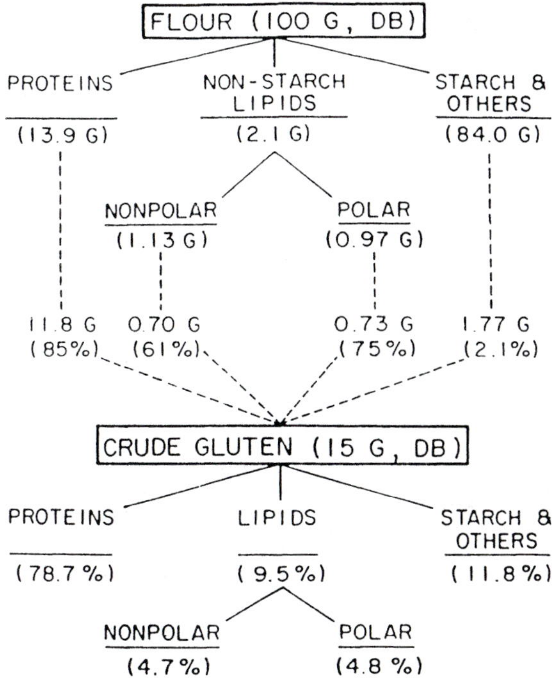

FIGURE 5 Gluten washing from 100 g (dry basis) flour. Involvement of flour components. (From Ref. 58.)

Interesting results from a number of published studies on lipid-protein interactions in wheat flour and gluten are reported by Laignelet [63]. They cover studies on free and bound lipids in flour, effects of soaps on flour proteins and lipids, biochemical study of the molecular species of lipids in flour and gluten, biophysical studies on gluten lipids (NMR, freeze-fracture/electron microscopy, photochemistry, spectrofluorometry), relationships between gluten lipids and baking quality, and study of lipid-binding proteins in wheat flour. Major changes occur in molecular species of lipids during flour mixing or gluten processing. Gluten behaves like a microemulsion stabilized by the protein network, the lipids of major technological importance are those bound to gluten, and flour contains at least two proteins with affinity for lipids (like transfer proteins), the origin and technological significance of which is yet to be clarified.

^{31}P-NMR spectra of gluten from undefatted and defatted wheat flour and freeze-fracture electron micrographs of gluten from undefatted and defatted wheat flour and gluten that had been hand-washed for 60 minutes have been described by Marion et al. [64]. The gluten used had been subjected to a heating/cooling cycle over the temperature range 25–70–25°C. Results indicate that complexes between lipids and proteins (such as those found in membranes) do not occur in the gluten, and that gluten should be regarded as a system containing stabilized microemulsions.

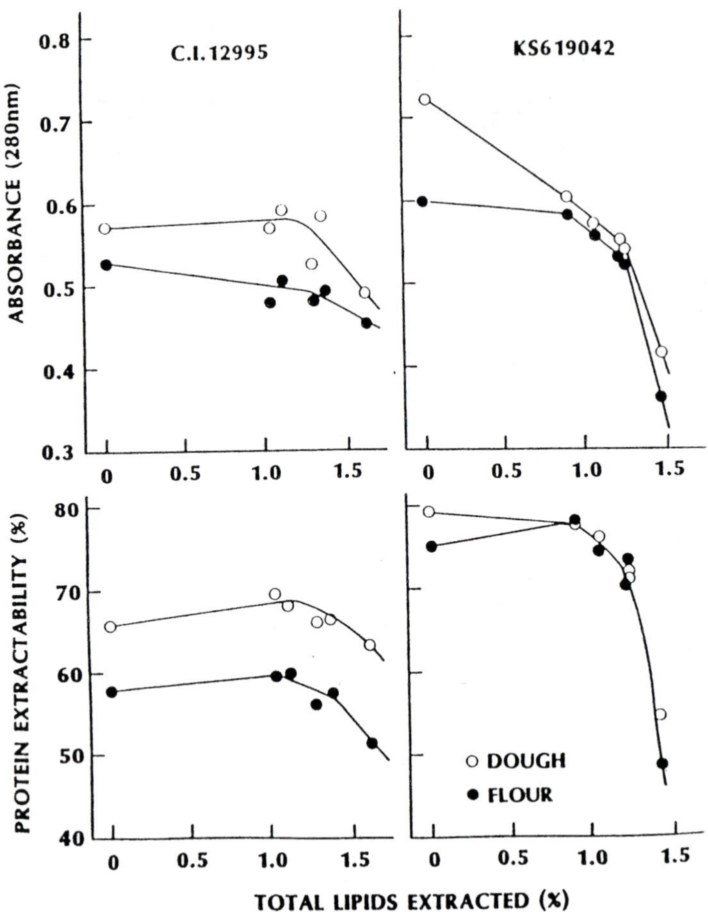

FIGURE 6 Effects of lipid removal (percentage of flour weight, dry basis), on absorbance (at 280 nm) of proteins extracted from two flours (good-quality on the left and poor-quality on the right) and their lyophilized doughs with 0.05 N HOAc (top) and protein extractability (percentage of total proteins) determined by Kjeldahl method (bottom). (From Ref. 58.)

Lipid oxidation is the major problem in storage of fresh and processed foods. The oxidation can adversely affect not only flavor, odor, and color qualities but also nutritive value. Defatted peanut flour contains residual lipids, which, because of their various reactive groups, can affect the ultimate quality of flour destined for human consumption. Involvement of these lipids in protein-lipid interaction was studied by St. Angelo and Graves [65]. Polar and nonpolar bound lipids were extracted from proteins with chloroform-methanol-concentrated HCl (2:1:0.03). The neutral fraction contained sterols, triglycerides, and esterified and free fatty acids. The polar fraction contained phospho- and glycolipids. The electrophoretic mobilities of the proteins that contained bound lipids changed after removal of the lipids by solvent extractions. The Oil Red O used to

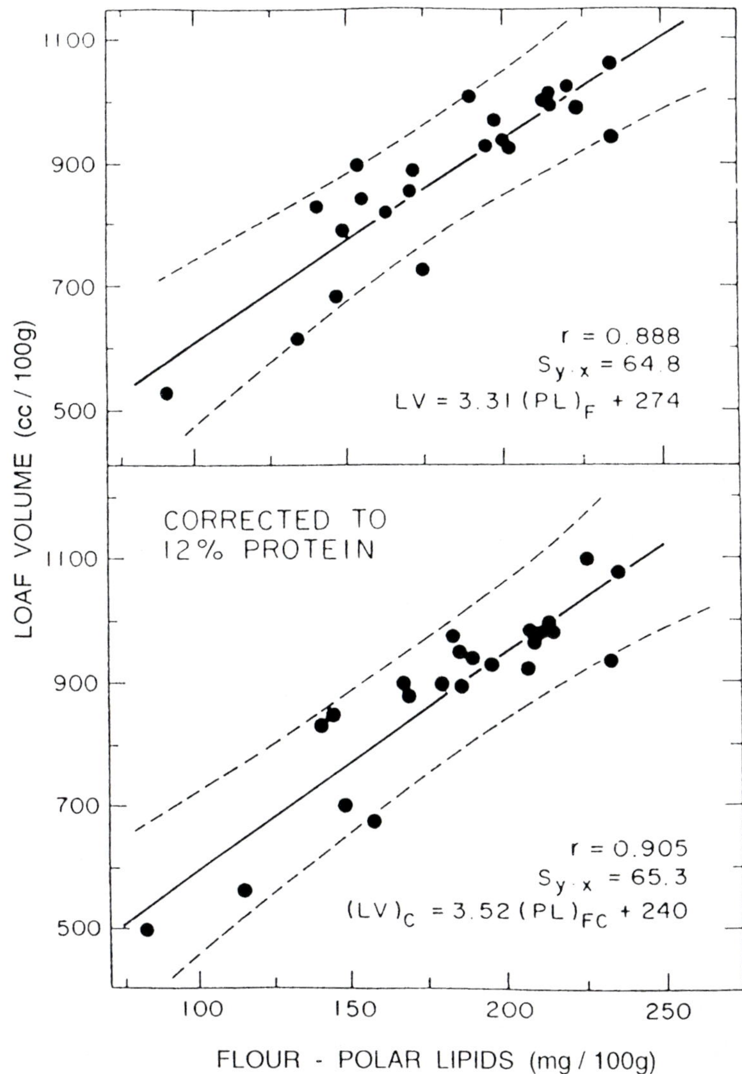

Figure 7 Relationship between loaf volume (LV) of bread baked from 100 g of flour and petroleum ether–extractable polar lipids of flour $(PL)_F$. Top, LV and $(PV)_F$ on an as-received protein basis; bottom, LV and $(PL)_F$ corrected to 12% protein content. C = Corrected; $S_{y \cdot x}$ = standard deviation. (From Ref. 58.)

stain gels for lipid was shown to be incompatible with SDS, and they should not be used simultaneously. Results showed that extraction of lipids from peanuts with hexane or chloroform-methanol was incomplete (giving a potential source of rancidity) unless acid was added to the solvent system. Results of this study may be useful to understanding an important factor that affects quality and stability of peanuts products and flours stored for long periods before their utilization.

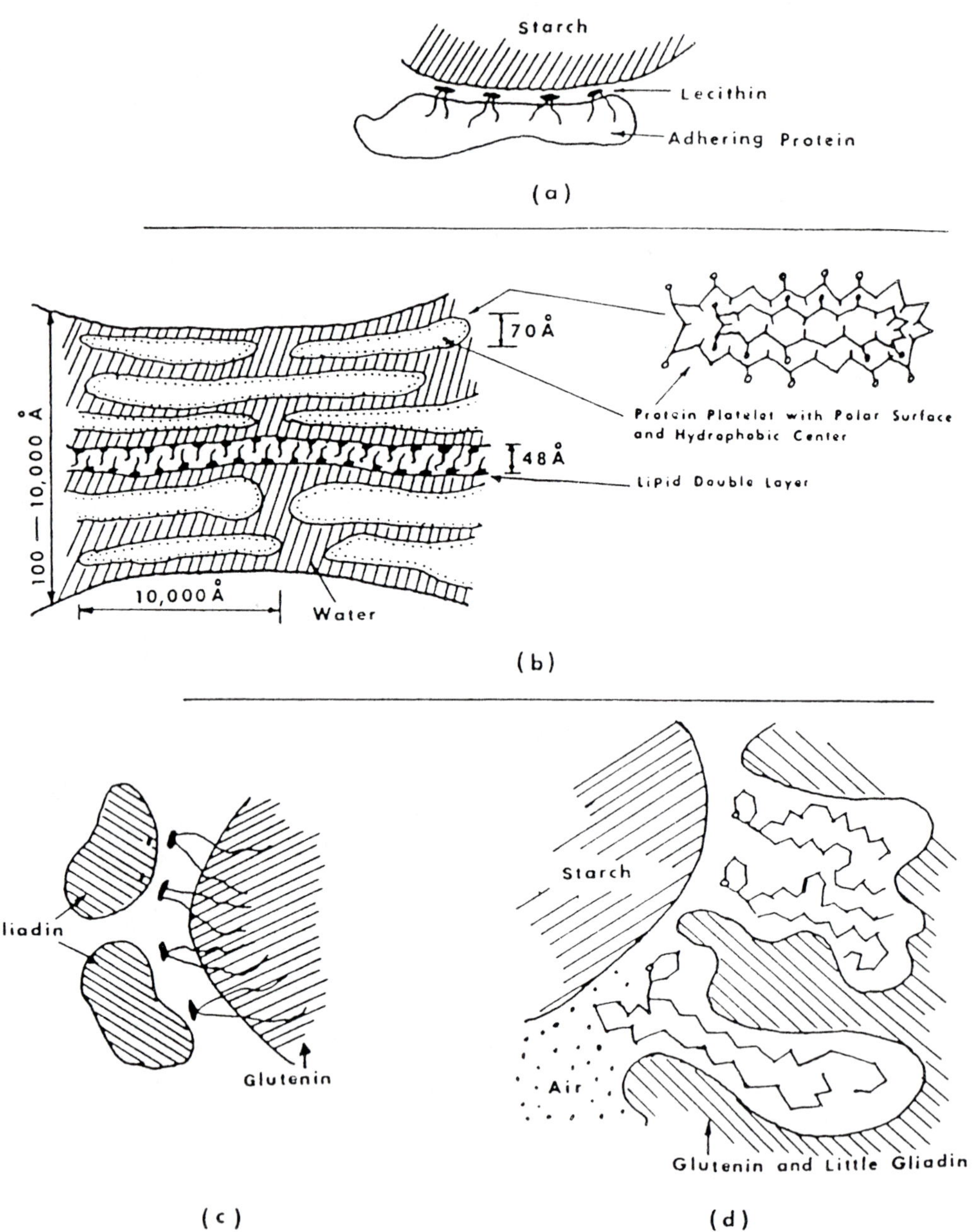

FIGURE 8 Models of the complexes formed in breadmaking. (a) Starch-lipid adhesive protein complex in flour, (b) lipoprotein model in gluten, (c) gliadin-glycolipid-glutenin complex, and (d) starch-glycolipid-gluten complex in bread. (From Ref. 58.)

IV. GENERATION OF FLAVORS DURING FOOD PROCESSING AND THEIR INTERACTIONS WITH PROTEINS

As described by Ho et al. [66], lipid-protein interactions can facilitate generation of volatiles in deep-fried foods. Due to the complexity of proteins and lipids found in foods, model systems were used to study these interactions.

Mixtures of 2,4-decadienal with either cysteine or glutathione (as model systems to study volatiles formed from lipid-protein interaction in deep-fat fried foods) were reacted in a closed sample cylinder in an aqueous medium [67]. Both cysteine and glutathione are sulfur-containing components found in natural food materials. They were selected as models for protein-lipid interactions and to study flavor generation in deep-fried foods. Each solution was adjusted to pH 7.5 and heated for 1 hour at 180°C, a representative frying temperature. The volatiles produced by degradations and interactions were isolated by simultaneous solvent-steam distillation and analyzed by gas chromatography (GC) and coupled GC-MS. Multiple compounds are produced by this heat-driven reaction of 2,4-decadienal with cysteine. Forty-five compounds were identified, and 2,4,6-trimethylperhydro-1,3,5-dithiazine was the major component of reaction products. Forty-two volatiles were determined from the thermal interaction of 2,4-decadienal and glutathione. 2-Pentylpyridine was the major component. A large number of long-chain alkyl-substituted heterocyclic compounds including thiophene, pyridine, thialdine, and many other sulfur-containing compounds were detected in the interacting systems. Most of the identified products are produced by well-known chemical pathways.

The binding of various nonpolar compounds to soy proteins has been studied in solutions and in the dry powder state. Damodaran and Kinsella [68,69] concluded, using an equilibrium dialysis method, that in aqueous systems the binding of aliphatic carbonyls to soy proteins is hydrophobic in nature and that the β-conglycinin fraction of soy protein may be responsible for the off-flavor binding by soy protein. The absorption coefficient at 280 nm of 1% solutions of pure soy protein, β-conglycinin, glycinin, and the acidic and basic subunits of glycinin were 6.04, 4.4, 8.04, 7.18, and 8.8, respectively. Using equilibrium dialysis, the binding affinities of these proteins for the model flavor compound 2-nonanone were determined [70]. On an equivalent weight basis, soy protein, β-conglycinin, and glycinin had approximately 5, 2, and 3 primary binding sites per 100,000 daltons and affinity constants of 570, 3050, and 540 M^{-1}, respectively, i.e. β-conglycinin showed a fivefold greater affinity for nonanone than the other soy protein. The acidic and basic subunits showed binding behavior similar to that of glycinin.

Interaction of pea protein isolates with diacetyl was studied by Dumont [71]. Effects of heat treatment of the protein (80 min in a boiling water bath) or use of the protein as an emulsifier in o/w systems were studied. Diacetyl concentrations of 2–5% were used in the trials; retention of diacetyl on the pea protein was evaluated by headspace GLC. Diacetyl retention was increased by heat treatment before diacetyl addition and was generally increased by emulsion preparation. Possible effects of processing on the structure and flavor compound retention capacity of the protein may include unmasking of the potential diacetyl-binding sites. Retention of diacetyl differed between two pea protein preparations tested. Problems linked with the accurate prediction of binding of flavor compounds to proteins are found by this author to be complex.

The foods containing plant or other nonconventional proteins may acquire unfamiliar and even undesirable flavor. The potential of many proteins to bind flavors makes it difficult to develop products with acceptable and desired flavor. Numerous studies on flavoring of soy proteins have been reported and the experience with this protein may illustrate the problems encountered when novel proteins are used in conventional foods. The flavor of soy flours was described by Covan et al. [72] as beany and bitter. Studies by Honig et al. [73] suggest that some of the compounds responsible for the objectionable flavor could be removed if appropriate processing techniques are developed.

The interaction between soy protein and chicken flavor on the perception of flavor was studied using a trained sensory panel [74]. Four levels of soy protein (0, 4, 8, and 16%) were hydrated in four concentration of chicken flavor (0, 0.1, 0.2, and 0.4%) before adding to a formulated soup. Eight panelists evaluated all 16 samples for pleasantness. Suppression of chicken flavor and pleasantness was observed at high levels of soy protein indicating masking or retention of the chicken flavor by the soy protein. The perception of soy flavor was not altered by increasing levels of chicken flavor. Soy protein and chicken flavor were found to contribute equally to the overall flavor intensity of the soup.

Fat substitutes constitute an interesting area of food research. Replacement of fats in foods requires overcoming new and unique flavor challenges. Fats assume multiple functions in food products, such as mouthfeel and richness, interaction with flavor components for sensory balance, and interaction with precursors of flavor formation as in cheeses and deep-fried foods. Approved fat substitutes are mainly made out of carbohydrates and proteins. Lipophilic particles may be produced from carbohydrates or proteins achieving quite satisfactory fatlike texture and mouthfeel. None of these products can be transformed to the same flavor derivatives as lipids during processing, nor will their interactions with the added flavors be comparable to fats. The same remains true for their sensory effects. Interaction of flavor compounds with microparticulated protein- or carbohydrate-based fat replacers was studied by Schirle-Keller et al. [75]. The vapor pressures of acetaldehyde, diacetyl, ethyl benzene, ethyl caproate, ethyl heptanoate, ethyl sulfide, hexanal, *trans*-2-hexenal, D-limonene, octanone, 1-pentanol, 2-pentanone, and styrene (50 ppm each) over a vegetable oil emulsion, water, or selected fat replacer (aqueous solutions, 37°C) was monitored. The five fat replacers were a whey-based and an egg-based microparticulated protein, a pregelatinized tapioca maltodextrin, microcrystalline cellulose (MCC), and a MCC/carboxymethylcellulose blend. The flavor compounds interacted with oil in a predictable manner following Raoult's law. The fat replacers generally had little influence on vapor pressure of the flavor compounds behaving in a way comparable to water system except for the Simplesse products, which had substantial interaction with aldehydes.

Recent developments in structural studies of small proteins interacting with hydrophobic ligands have shed new light on lipid-binding properties of β-lactoglobulin. It has been postulated recently that β-lactoglobulin belongs to the superfamily of proteins [19,38,76] involved in the strong interactions with small hydrophobic molecules, such as retinol and its derivatives, pheromones, biliverdines, pyrazines, etc. Retinol-binding protein [29], bilin-binding protein [30], insecticyanin [31], and β-lactoglobulin [19,28] are the best known proteins of this class. It is not known, however, how these structurally very similar proteins can exhibit binding specificities for a variety of chemically and structurally different ligands.

Proper understanding of the nature of the factors contributing to β-lactoglobulin-binding properties may generate ideas for applications of this abundant whey protein or of its nontoxic derivatives as food additives. For instance, β-lactoglobulin could be engineered (a) to bind and protect a wide range of volatile and unstable flavors during food manufacturing or to release them in more or less controlled way by a simple treatment, or (b) to trap undesirable compounds.

Modification of β-lactoglobulin by enzymatic or chemical treatment may be one of the ways to induce conformational changes in this protein. Consequently, it may alter or broaden its binding properties. Esterification of carboxyl groups [77] or alkylation of lysine residues [78] is one of the simplest, cheapest, and most nontoxic (if done properly) chemical modifications of proteins.

Dufour and Haertlé [36] have investigated by fluorescence the interactive properties of native, alkylated, or esterified β-lactoglobulin with several flavor compounds: R(+) and S(−) limonene, geraniol, and α- and β-ionone, which are terpenes isolated from the volatile oils of different plants. On the one hand, the monoterpenes constitute a large group of natural products with frequent aromatic properties. Geraniol plays a key role in isoprenoid metabolism, and R(+) and S(−) limonene are obtained by its cyclization [79]. On the other hand, α- and β-ionone are the degradation products of carotene with an aroma similar to violets. According to Weeks [80], β-carotene, the most widely known tetraterpene in plants, is the direct precursor of β-ionone and retinol.

The fluorescence emission spectra of BLG or its derivatives were studied as a function of added compounds, and the observed tryptophan fluorescence quenching, due to changes of the polarity in the neighborhood of indole [81], is indicative of the formation of a complex. Fugate and Song [37] determined that only one of two BLG tryptophan residues is involved in the binding of retinol to BLG.

The addition of geraniol or R(+) and S(−) limonene to protein solutions doesn't produce any fluorescence quenching, suggesting that these monoterpenes don't bind to BLG or to its derivatives, or that they don't interfere with BLG tryptophans. In the same way, α-ionone binding to BLG or its derivatives cannot be demonstrated by fluorometry (Figs. 9 and 10). The addition of β-ionone induces a significant quenching of BLG fluorescence. Corrected titration curves for various derivatives of BLG are shown in Figures 9 and 10, and the maximum fluorescence quenching is obtained at a β-ionone-protein ratio of 1:1. It should be pointed out that the decrease in the fluorescence intensity of the blank *N*-acetyl-L-tryptophanamide solution is not due, apparently, to the interaction between β-ionone and *N*-acetyl-L-tryptophanamide, but rather to the inner filter effect as a result of the absorbance of β-ionone at 290 nm.

It may be concluded on the basis of these results that, from the tested terpenes, BLG and its derivatives bind only β-ionone. β-Ionone and retinol can be derived from β-carotene, and they share the same structure (Fig. 11), differing only in the length of the isoprenoid chain. Two ionone isomers (α and β) differ in the placement of the double bond, which is conjugated with the one in the isoprenoid chain in the case of the β-isomer. The conjugation of the two double bonds imposes conformational constraints; β-ionone cycle and all the conjugated double-bonds are placed in the same plane, as shown by the structure energy minimization of β-ionone and retinol using the molecular modeling software Sybyl (Fig. 11). Prediction of the α-ionone conformation at the energy minimum yields a very different structure. When their binding to BLG is studied, these two isomers produce totally different results, suggesting that the binding site of BLG, besides its specificity to certain hydrophobic features, is also highly structure specific. It may be assumed that the structural constraints brought about by the conjugation of the cyclohexenyl ring double bond through the vicinal C=C with the double bond system of the isoprenoid tail are essential for the recognition by BLG and a complex

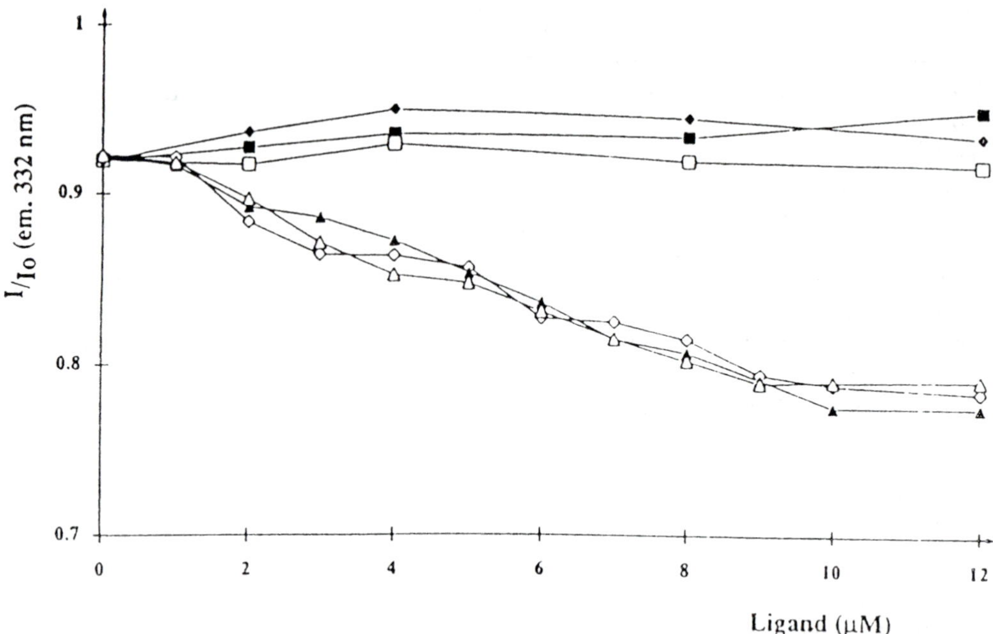

FIGURE 9 Corrected fluorescence titration curves of β-lactoglobulin and its esterified derivatives with α- and β-ionone. Excitation wavelength: 290 nm; emission wavelength: 332 nm. Binding of α-ionone to β-lactoglobulin: 10.6 μM (■), methyl-esterified BLG, 11.2 μM (□), and ethyl-esterified BLG, 11 μM (♦). Binding of β-ionone to β-lactoglobulin: 10.6 μM (◇), methyl-esterified BLG, 11.2 μM (▲), and ethyl-esterified BLG, 11 μM (△). (From Ref. 36.)

formation. As may be seen from the comparison of the apparent dissociation constants of retinol-BLG (2×10^{-8} M) and β-ionone–BLG (6×10^{-7} M) complexes, the length of the isoprenoid tail seems to be less important in the determination of the binding specificity. BLG and retinol-binding protein [29] display similar three-dimensional structures, overlapping more than 95% [19,39]. The binding specificity is certainly defined by the amino acid side chains lining the inner side of the β-barrel calyx.

Thus, in the case of BLG, we face an intriguing situation of a protein with a narrow specificity for the structural motif formed by the conjugated double bonds of β-ionone ring and isoprenoid chain. But BLG is also known to bind to structurally different molecules such as free fatty acids and triglycerides [32], alkanone flavors [16], etc. The above-mentioned BLG capacity to bind chemically and structurally miscellaneous ligands could mean that BLG, in addition to a presumably deep central pocket site, may potentially bind these ligands elsewhere in the outer surface site framed by hydrophobic residues.

Generally, esterification and alkylation of BLG enhance its binding properties. The apparent dissociation constant of the β-ionone–EtBLG complex is 3.5 times smaller than that of β-ionone–BLG. In addition, the study of the binding of the two ionone isomers demonstrates that BLG, which binds β-ionone but no α-ionone, has a narrow specificity

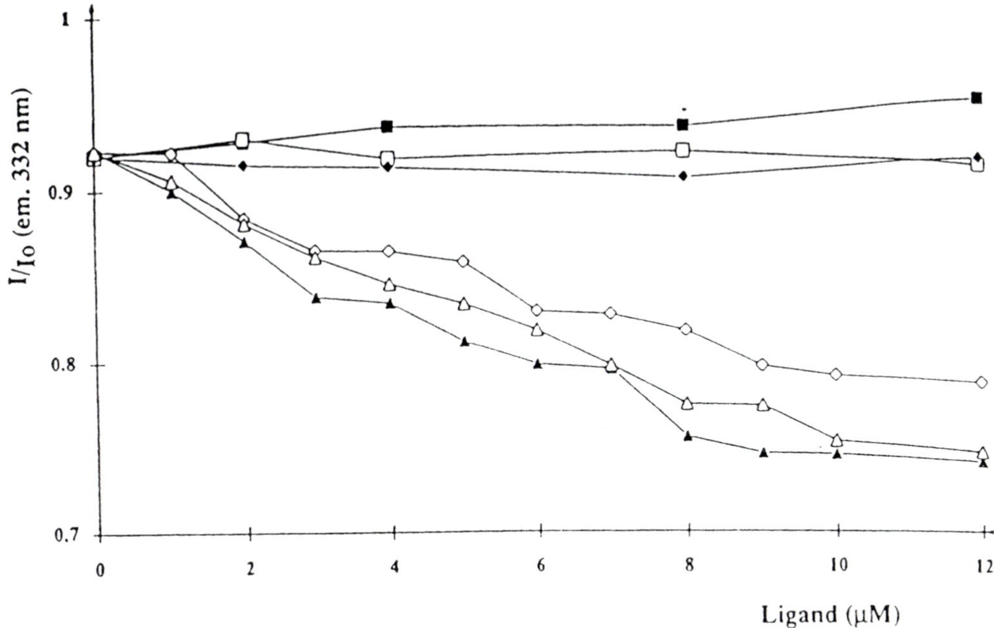

FIGURE 10 Corrected fluorescence titration curves of β-lactoglobulin and its alkylated derivatives with α- and β-ionone. Excitation wavelength: 290 nm; emission wavelength: 332 nm. Binding of α-ionone to β-lactoglobulin: 10.6 μM (■), *N*-methyllysyl-BLG, 11.5 μM (□), and *N*-ethyllysyl-BLG, 7.5 μM (♦). Binding of β-ionone to β-lactoglobulin: 10.6 μM (◇), *N*-methyllysyl–esterified BLG, 11.5 μM (▲), and *N*-ethyllylsyl-BLG, 7.5 μM (△). (From Ref. 36.)

for the pattern formed by the β-ionone ring and the double bonds of the isoprenoid chain. In the context of this relatively stringent binding specificity towards so chemically similar terpenes (e.g., the α and β isomers of ionone) and the binding of other very chemically different ligands by BLG, it may be assumed that a protein molecule displaying such varied specificities also has several binding sites.

V. CONCLUSIONS

It is evident that the interaction of lipids with proteins constitutes one of the major factors affecting food and cellular microstructures. The appropriate organization of the synthesized or produced (in the case of food technologies) interfaces determines the biological and technological functions. The proper arrangement of the interfaces, as can be observed in the case of microparticulated or extruded proteins, can profoundly modify the character of the interactions of proteins with lipids, aromas, and flavors. These interactions can also modify the aroma profile generated during heat processing of food products. This can occur when plant proteins are used as substitutes for animal proteins. Consequently, when the protein-flavor interactions change, alterations in flavor and aroma profile and perception will necessitate reformulation of the flavor and aroma components added to foods. Hence, it is clear that alterations in protein interfaces in foods, whether

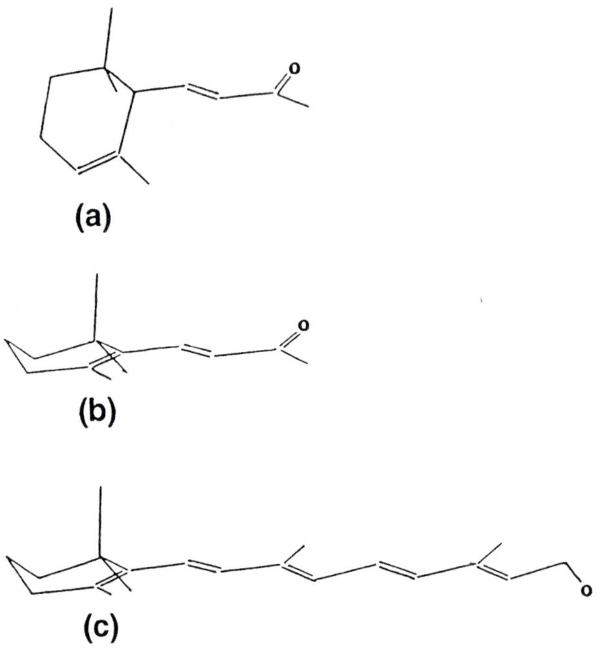

FIGURE 11 Optimized structures of (a) α-ionone, (b) β-ionone, and (c) retinol. The molecules were built and their geometries minimized with the help of the molecular modeling software Sybyl (Tripos Associates, St. Louis, MO). (From Ref. 36.)

induced by technological treatment or by the use of novel proteins, would radically change the nature and extent of interactions between proteins, lipids, and aroma components. Some of these changes were discovered and used successfully in several food products, for example, in baking and dairy technologies. However, several aspects of protein-lipid and protein-flavor interactions in other novel processed foods are yet to be addressed.

REFERENCES

1. J. H. Lakey, M. W. Parker, J. M. Gonzales-Manas, D. Duche, G. Vriend, D. Baty, and F. Pattus, The role of electrostatic charge in the membrane insertion of colicin A. Calculation and mutation, *Eur. J. Biochem. 220*:155 (1994).
2. V. Kiosseoglou and P. Kouzounas, The role of diglycerides, monoglycerides and free fatty acids in olive oil minor surface-active lipid interaction with proteins at oil-water interfaces, *J. Dispersion Sci. Technol. 14*:527 (1993).
3. C. V. Morr, Emulsifiers: milk, proteins, *Proteins Functionality in Foods* (J. P. Cherry, ed.), American Chemical Society, Washington, DC, 1981, p. 205.
4. J. Mitchell, L. Irons, and G. J. Palmer, A study of the adsorbed films of milk proteins. *Biochim. Biophys. Acta 200*:138 (1970).
5. J. V. Boyd, J. Mitchell, L. Irons, P. R. Musselwhite, and P. Sherman. The mechanical properties of milk proteins films spread at the air/water interface, *J. Colloid Interface Sci. 15*: 478 (1973).

6. D. E. Graham and M. C. Phillips, Proteins at liquid interfaces. I. Kinetics of adsorption and surface denaturation, *J. Colloid Interface Sci. 70*:403 (1979).
7. D. E. Graham and M. C. Phillips, Proteins at liquid interfaces. II. Adsorption isotherms, *J. Colloid Interface Sci. 70*:415 (1979).
8. D. E. Graham and M. C. Phillips, Proteins at liquid interfaces. III. Molecular structures of adsorbed films, *J. Colloid Interface Sci. 70*:427 (1979).
9. D. E. Graham and M. C. Phillips, Proteins at liquid interfaces. IV. Dilational properties, *J. Colloid Interface Sci. 76*:227 (1980).
10. D. E. Graham and M. C. Phillips, Proteins at liquid interfaces. V. Shear properties, *J. Colloid Interface Sci. 76*:240 (1980).
11. R. H. Jackson and M. J. Pallansch, Influence of milk proteins on interfacial tension between butter-oil and various aqueous phases, *J. Agric. Food Chem. 9*:424 (1961).
12. M. C. A. Griffin, R. B. Infanti, and B. A. Klein, Structural domains of κ-casein show different interaction with dimiristoyl phosphatidylcholine monolayers, *Chem. Phys. Lipids 36*:91 (1984).
13. A. Rahman and P. Sherman, Interaction of milk proteins with monoglycerides and diglycerides, *Colloid Polymer Sci. 260*:1035 (1982).
14. M. F. Laliberté, M. Britten, and P. Paquin, Interfacial properties of sodium caseinate/monoglyceride mixture; combined effects of pH and surface area ratios, *Can. Inst. Food Sci. Technol. J. 21*:251 (1988).
15. L. M. Smith, P. Fantozzi, and R. K. Creveling, Study of triglyceride-protein interactions using a microemulsion filtration method, *J. Am. Oil Chem. Soc. 60*:960 (1983).
16. T. E. O'Neill and J. E. Kinsella, Binding of alkanone flavors to β-lactoglobulin: effect of conformational and chemical modification, *J. Agric. Food Chem. 35*:770 (1987).
17. T. C. I. Wilkinson and D. C. Wilton, Studies on fatty acids binding protein: the binding properties of rat liver fatty acid binding protein, *Biochem. J. 247*:485 (1987).
18. S. Futterman and J. Heller, The enhancement of fluorescence and the decreased susceptibility to enzymatic oxidation of retinol complexed with bovine serum albumin, β-lactoglobulin and the retinol-binding protein of human plasma, *J. Biol. Chem. 247*:5168 (1972).
19. M. Z. Papiz, L. Sawyer, E. E. Eliopoulos, A. L. T. North, J. B. C. Findlay, R. Sivraprasadarao, T. A. Jones, M. E. Newcomer, and P. J. Kraulis, The structure of β-lactoglobulin: its similarity to plasma retinol-binding protein, *Nature 324*:383 (1986).
20. E. M. Brown, P. E. Pfeffer, T. F. Kumosinsky, and R. Greeberg, Accessibility and mobility of lysine residues in β-lactoglobulin, *Biochemistry 27*:5601 (1988).
21. D. G. Cornell and D. L. Patterson, Interactions of phospholipids in monolayer with β-lactoglobulin adsorbed from solution, *J. Agric. Food Chem. 37*:1455 (1989).
22. M. Le Meste, B. Closs, J. L. Courthaudon, and B. Colas, Interactions between milk proteins and lipids. A mobility study, *Interactions of Food Proteins.* (N. Parris and R. Barford, eds.), American Chemical Society Symposium Series, Washington, DC, 1991, p. 137.
23. S. Aynié, M. Le Meste, B. Colas, and D. Lorient, Interactions between lipids and milk proteins in emulsion, *J. Food Sci. 57*:883 (1992).
24. E. Dufour and T. Haertlé, Binding of retinoids and β-carotene to β-lactoglobulin. Influence of protein modifications, *Biochim. Biophys. Acta 1079*:316 (1991).
25. M. Coke, P. J. Wilde, E. J. Russel, and D. C. Clark, The influence of surface composition and molecular diffusion on the stability of foams formed from protein/surfactant mixtures, *J. Colloid Interface Sci. 138*:489 (1990).
26. E. Dufour, M. C. Marden, and T. Haertlé, β-Lactoglobulin binds retinol and protoporphyrin IX at two different binding sites, *FEBS Lett. 277*:223 (1990).
27. G. Dodin, M. Andrieux, and H. Al Kabbani, Binding of ellipticin to β-lactoglobulin. A physico-chemical study of the specific interaction of an antitumor drug with a transport protein, *Eur. J. Biochem. 193*:697 (1990).

28. H. L. Monaco, G. Zanotti, P. Spadon, M. Bolognesi, L. Sawyer, and E. E. Eliopoulos, Crystal structure of the trigonal form of bovine β-lactoglobulin and of its complex with retinol at 2.5 Å resolution, *J. Mol. Biol. 197*:695 (1987).
29. M. E. Newcomer, T. A. Jones, J. Aqvist, J. Sundelin, U. Eriksson, L. Rask, and P. Peterson, The three dimensional structure of retinol binding protein, *EMBO J. 3*:1451 (1984).
30. R. Huber, M. Schneider, O. Epp, I. Mayr, A. Messerschmidt, J. Plugrath, and H. Kayser, Crystallisation, crystal structure analysis and preliminary molecular model of the bilin binding protein from the insect *Pieris brassicae*, *J. Mol. Biol. 195*:423 (1987).
31. H. M. Holden, W. R. Rypniewski, J. H. Law, and I. Rayment, The molecular structure of insecticyanin from the tabacco hornworm *Manduca sexta* L. at 2.6 Å resolution, *EMBO J. 6*:1565 (1987).
32. M. C. Diaz de Villegas, R. Oria, F. J. Sala, and M. Calvo, Lipid binding by β-lactoglobulin of cow milk, *Milchwissenshaft 42*:357 (1987).
33. A. A. Spector and J. E. Fletcher, Binding of long chain fatty acids to β-lactoglobulin, *Lipids 5*:403 (1970).
34. D. Frapin, E. Dufour, and T. Haertlé, Probing the fatty acids binding site of β-lactoglobulins, *J. Prot. Chem. 12*:443 (1993).
35. J. H. Tonsgard and S. C. Meredith, Characterization of the binding sites for dicarboxylic acids on bovine serum albumin, *Biochem. J. 276*:569 (1991).
36. E. Dufour and T. Haertlé, Binding affinities of β-ionone and related flavor compounds to β-lactoglobulin: effects of chemical modifications, *J. Agric. Food Chem. 38*:1691 (1990).
37. R. D. Fugate and P. S. Song, Spectroscopic characterisation of β-lactoglobulin-retinol complex, *Biochim. Biophys. Acta 625*:28 (1980).
38. J. Godovac-Zimmerman, The structural motif of β-lactoglobulin and retinol binding protein: a basic framework for binding and transport of small hydrophobic molecules, *Trends Biochem. Sci. 13*:64 (1988).
39. A. C. T. North, Three-dimensional arrangement of conserved amino acid residues in a superfamily of specific ligand-binding protein, *Int. J. Biol. Macromol. 11*:56 (1989).
40. G. Blond, E. Haury, and D. Lorient, Effects of batch- and extrusion-cooking on lipids-proteins interactions of processed cheese, *Sci. Aliments 8*:325 (1988).
41. B. O. Rowley and T. Richardson, Protein-lipid interactions in concentrated infant formula, *J. Dairy Sci. 68*:3180 (1985).
42. J. A. Gomes-Areas and R. A. Lawrie, Effect of lipid-protein interactions on extrusion of offal protein isolates, *Meat Sci. 11*:275 (1984).
43. W. Christie, *Lipid Analysis*, Pergamon Press, Oxford, 1982, p. 17.
44. P. Zahler and V. F. Niggli, The use of organic solvents in membrane research, *Methods in Membrane Biology* (E. D. Korn, ed.), Plenum, New York, 1977, p. 1.
45. J. A. F. Gomes-Areas, Lipid protein interactions in offal protein isolates: effect of several solvents on lipid extraction, *J. Food Sci. 50*:1392 (1985).
46. I. D. Kuntz and W. Kauzmann, Hydration of proteins, *Adv. Prot. Chem. 28*:239 (1974).
47. M. Lüscher-Mattli and M. Rüegg, Thermodynamic functions of biopolymer hydration. I. Their determination by vapor pressure studies, discussed in an analysis of the primary hydration process, *Biopolymers 21*:403 (1982).
48. A. C. Zettlemoyer, F. J. Micale, and K. Klier, Adsorption of water on well-characterized solid surfaces, *Water-A Comprehensive Treatise* (F. Franks, ed.), Plenum Press, New York, 1975, p. 249.
49. H. A. Iglesias and J. Chrife, Effect of fat content on water sorption isotherm of air dried minced beef, *Lebensm. Wiss. Technol. 10*:151 (1977).
50. J. A. Gomes-Areas, Effect of lipid-protein interactions on hydration characteristics of defatted offal protein isolates, *J. Food Sci. 51*:880 (1986).

51. M. Kunimoto, K. Matsumoto, and K. Zama, The interaction of fish proteins and lipids during freeze-drying and storage. I. The interaction of carp myofibrils and fish oil. *Bull. Fac. Fisheries, Hokkaido Univ.* [Hokkaido Daigaku Suisangakubu Kenkyu Iho] *36*:50 (1985).
52. M. T. Izzo, Lipid-protein and lipid-carbohydrate interactions during twin extrusion of corn meal, *Diss. Abstr. Int.* 52 3363ISSN: 0419 (1992).
53. J. Funes and M. Karel, Free radical polymerisation and lipid binding of lysozyme reacted with peroxidizing linoleic acid, *Lipids 16*:347 (1981).
54. M. T. Izzo and C. T. Ho, Protein-lipid interaction during single-screw extrusion of zein and corn oil, *Cereal Chem. 66*:47 (1989).
55. R. Zamora, F. Millan, F. J. Hidalgo, M. Alaiz, M. P. Maza, and E. Vioque, Application of fluorescence techniques to the study of lipid-protein interactions. Preliminary trials, *Grasas Aceites 37*:317 (1986).
56. K. S. Morris and J. S. Hough, Lipid-protein interactions in beer and beer foam brewed with wheat flour, *J. Am. Soc. Brewing Chem. 45*:43 (1987).
57. T. B. Osborne and G. C. Voorhees, The proteins of the wheat kernel, *Am. Chem. J. 15*:392 (1893).
58. O. K. Chung, Lipid-protein interactions in wheat flour, dough, gluten, and protein fractions, *Cereal Foods World 31*:242 (1986).
59. K. Hess and H. Mahl, Elektronmikroskopische Beobachtungen an Mehl und Mehlpreparaten von Weizen, *Mikroskopie* 9:81 (1954).
60. J. C. Grosskreutz, A lipoprotein model of wheat gluten structure, *Cereal Chem. 38*:336 (1961).
61. R. C. Hoseney, Y. Pomerantz, and K. F. Finney, Functional (breadmaking) and biochemical properties of wheat flour components. VI. Gliadin-lipid-glutenin interactions in wheat gluten. *Cereal Chem. 47*:135 (1970).
62. H. P. Wherli and Y. Pomerantz, A note on the interactions between glycolipids and wheat flour macromolecules, *Cereal Chem. 47*:160 (1970).
63. B. Laignelet, French research on lipid-protein interactions in wheat flour and gluten, *Ind. Céréales 52*:13 (1988).
64. D. Marion, C. LeRoux, S. Akoka, C. Tellier, and D. Gallant, Lipid-protein interactions in wheat gluten: a phosphorus nuclear magnetic resonance spectroscopy and freeze-fracture electron microscopy study, *J. Cereal Sci. 5*:101 (1987).
65. A. J. St Angelo and E. E. Graves, Studies of lipid-protein interaction in stored raw peanuts and peanut flours, *J. Agric. Food Chem. 34*:643 (1986).
66. T. C. Ho, J. T. Carlin, T. C. Huang, L. S. Hwang, and L. B. Hau. Flavour development in deep fat fried foods, *Flavour Science and Technology* (M. Martens, G. A. Dalen, and G. H. Jr. Russwurm, eds.) Wiley, Chichester, 1987, p. 35.
67. Y. Zhang and C. T. Ho, Volatile compounds formed from thermal interaction of 2,4-decadienal with cysteine and glutathione, *J. Agric. Food Chem. 37*:1016 (1989).
68. S. Damodaran and J. E. Kinsella, Interaction of carbonyls with soy proteins. Thermodynamic effects, *J. Agric. Food Chem. 29*:675 (1981).
69. S. Damodaran and J. E. Kinsella, Interactions of carbonyls with soy proteins. Conformational effects, *J. Agric. Food Chem. 29*:1253 (1981).
70. T. E. O'Neill and J. E. Kinsella, Flavor protein interactions: characteristics of 2-nonanone binding to isolated soy protein fractions, *J. Food Sci. 52*:98 (1987).
71. J. P. Dumont, Diacetyl retention by pea proteins: effect of physical treatments, *Sci. Aliments 5 (Hors Série V)*:85 (1985).
72. J. C. Covan, J. J. Rackis, and W. J. Wolf, Soybean protein-flavour components: a review, *J. Am. Oil Chem. Soc. 50*:426 (1973).
73. D. H. Honig, K. A. Warner, E. Selke, and J. J. Rackis, Effects of residual solvents and storage on flavour of hexane/ethanol azeotrope extracted soy products, *J. Agric. Food Chem. 27*: 1383 (1979).

74. L. J. Malcolmson, M. R. McDaniel, and E. Hoehn, Flavor protein interactions in a formulated soup containing flavored soy protein, *Can. Inst. Food Sci. Technol. 20*:229 (1987).
75. J. P. Schirle-Keller, H. H. Chang, and G. A. Reineccius, Interaction of flavor compounds with microparticulated proteins, *J. Food Sci. 57*:1448 (1992).
76. S. Pervaiz and K. Brew, Homology of β-lactoglobulin, serum retinol-binding protein and protein HC, *Science 228*:335 (1985).
77. H. Fraenkel-Conrat and H. S. Olcott, Esterification of proteins with alcohols of low molecular weight, *J. Biol. Chem. 161*:259 (1945).
78. G. E. Means and R. E. Feeney, Reductive alkylation of amino groups in proteins, *Biochemistry 7*:2192 (1968).
79. R. Croteau, Biosynthesis of cyclic monoterpenes, *Biogeneration of Aromas* (T. H. Parliment, and R. Croteau, eds.), American Chemical Society, Chicago, 1986, p. 134.
80. W. W. Weeks, Carotenoids: a source of flavor and aromas, *Biogeneration of Aromas* (T. H. Parliment and R. Croteau, eds.), American Chemical Society, Chicago, 1986, p. 157.
81. R. J. Lakowicz, Protein fluorescence, *Principle of Fluorescence Spectroscopy* (R. J. Lakowicz, ed.), Plenum Press, New York, 1983, p. 342.

6

Protein-Polysaccharide Interactions

VLADIMIR B. TOLSTOGUZOV
Nestec Ltd. Research Center
Lausanne, Switzerland

I. INTRODUCTION

Protein-polysaccharide interactions play a significant role in the structure and stability of many processed foods. The control or manipulation of these macromolecular interactions is a key factor in the development of novel food processes and products as well as in the formulation of fabricated food products.

Functional properties of food proteins, such as solubility, surface activity, conformational stability, gel-forming ability, and emulsifying and foaming properties, are affected by their interactions with polysaccharides. Interactions of these biopolymers with each other and their competitive interactions with other system components (water, lipids, sugars, metal ions, surfactants, etc.) determine structure-property relationships in foods.

Nonspecific protein-polysaccharide interactions can be subdivided into two groups: attraction and repulsion between unlike macromolecules. These two types of interbiopolymer interactions are responsible for complex formation and the immiscibility of biopolymers, respectively. Since this is an example of polyelectrolyte interactions in solution, complex formation and thermodynamic incompatibility are primarily influenced by pH, ionic strength, conformation, charge density, and concentration of the biopolymers.

In this chapter, the mechanisms involved in protein-protein and protein-polysaccharide interactions that result in either complex formation or thermodynamic incompatibility will be discussed. The influences of these two phenomena on the functional properties of food proteins will be examined.

A. History

Investigations into protein-polysaccharide interactions began 100 years ago. In 1896, Beijerinck [1,2] published the first study on the phase behavior of mixed solutions of gelatin and soluble starch. Beijerinck's last study, published in 1910, was entitled, "The formation of emulsions by mixing the solutions of certain gel-forming colloids." Beijerinck discovered the impossibility of mixing aqueous solutions of 10% gelatin with 2% agar-agar and 10% gelatin with 10% soluble potato starch. Droplets of gelatin solution were dispersed in the bulk phase of the starch solution. The structure of this water-in-water emulsion remained unchanged after heating and extensive mixing. Beijerinck also noted that an osmotic equilibrium between the emulsion phases was mainly established by water redistribution between these phases. Owing to gelation of both phases, the emulsion structure was solidified by cooling.

The first study on the formation of complexes by oppositely charged biopolymers was carried out by Tiebackx [3] in 1911. He found that adding an acid to a mixed solution of gelatin and gum arabic broke it down into two phases. The highly concentrated precipitate contained both biopolymers. Its yield was dependent on the ratio of the macromolecular reactants.

The phenomena described by Beijerinck and Tiebackx were quantitatively studied by Ostwald and Hertel [4]. It was experimentally shown that both types of phase separation are typical of polymers [5–9]. In the 1950s and 1960s Albertsson [8] carried out the first systematic investigation of the incompatibility of polysaccharides and water-soluble polymers. He showed that aqueous two-phase polymer systems are of great importance for separation of biopolymers under mild conditions [8]. At the same time the theory of the thermodynamic incompatibility of polymers was developed by Flory [10]. It was a time of much intense development of the physical chemistry of synthetic polymers and their blends. The scientific aspects of incompatibility and its applications in polymer processing have been considered in many review papers (e.g., Ref. 11).

Pioneering work in the systematic investigation of the interaction between proteins and anionic polysaccharides was performed by Bungenberg de Jong and colleagues [5]. They showed that mixing aqueous solutions of gelatin and gum arabic resulted in two kinds of two-phase systems. The gelatin and gum arabic were mainly concentrated either in the same single bottom phase or in different phases of a two-phase system. These two phenomena of phase separation in biopolymer solutions have been called complex and single coacervation, respectively. Bungenberg de Jong [5] studied these phenomena only in systems containing gelatin and gum arabic.

Gelatin is, however, not a typical protein. Normally, proteins have a compact and rigid molecular structure, whereas gelatin behaves like a flexible chain polymer in solution. It is notable for its random coil conformation at temperatures above 35°C and for its high excluded volume compared to that of globular proteins. Food systems usually contain a heterogeneous mixture of proteins and polysaccharides differing in chemical nature, conformation, chain rigidity, size and shape of molecules, degrees of hydrolysis, denaturation, dissociation, and aggregation. This means that information on the phase behavior of polymer blends and mixed gelatin–gum arabic systems is not sufficient for understanding the functional properties of proteins and polysaccharides in real food systems.

Basic information concerning the main classes of food proteins and polysaccharides has been obtained during the last two decades [12–20]. It has been shown that interactions of proteins with polysaccharides and of various proteins between each other and with water govern the solubility and co-solubility of biopolymers, their ability to form viscous and viscoelastic solutions and gels, and their behavior at interfaces. It has been shown that nonspecific interpolymer interactions can be reversible and irreversible and cooperative and noncooperative in nature. Even a small alteration of the interactions between macromolecules may result in a change in food texture. This can lead to complex formation and thermodynamic incompatibility between various food hydrocolloids. A surprising finding was that biopolymers behave similarly to classical synthetic polymers, i.e., the main structure-property relationships for protein-polysaccharide mixtures are similar to those of synthetic polymers. It is a specific example of interpolymer interactions that are mainly influenced by specific conformations and the polyelectrolyte nature of biopolymers [12,14,21–25].

B. Basic Principles

Figure 1 shows that by mixing a protein and a polysaccharide, two kinds of single-phase systems (1 and 2) and two types of two-phase systems (3 and 4) can be produced. In other words, mixed solutions of biopolymers can exist either in stable or phase-separated states. The difference between the two-phase liquid systems 3 and 4 is that biopolymers are concentrated either in the same concentrated bottom phase of system 3 or within different phases of system 4. All four types of systems differ strongly in structure and properties. Functional properties of a macromolecular food component also vary in these systems [12–14,23,24].

Systems 1 and 3 contain soluble and insoluble protein-polysaccharide complexes, respectively. Interbiopolymer complex formation occurs at pH values below the isoelectric point (IEP) of proteins and at low ionic strengths, usually <0.3. At pH values below the IEP, protein molecules have a net positive charge and behave as polycations. They behave as polyanions at pH values above the IEP. The IEPs of most food proteins are between 4 and 7. However, some proteins and subunits of oligomeric proteins have IEPs in the basic pH range—for instance, lysozyme from egg white, albumin from rapeseeds, soybean trypsin inhibitors, protamines, and histones. When the pH is intermediate between the IEPs of two proteins, the proteins will be oppositely charged and will form an electrostatic protein-protein complex.

At mild acidic and neutral pH values, which are typical of most foods, carboxyl-containing polysaccharides behave as polyanions. Electrostatic complex formation between proteins and anionic polysaccharides generally occurs in the pH range between

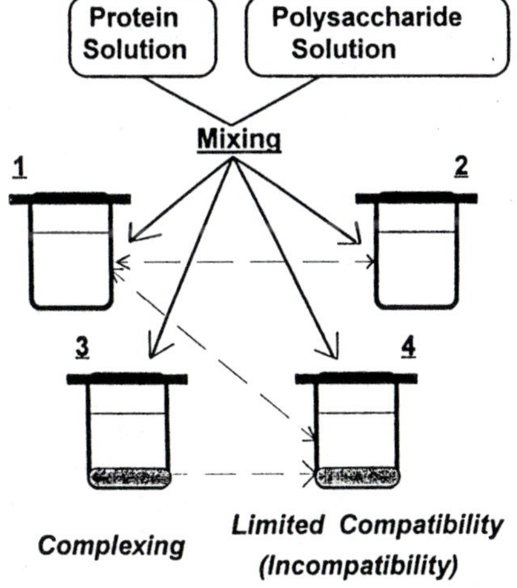

Figure 1 Schematic representation of the four possible systems obtained by mixing solutions of a protein and a polysaccharide. (From Refs. 23 and 24.)

the pK value of the anionic groups (carboxyl groups) on the polysaccharide and the protein's IEP.

The formation of an electrostatic complex is usually a reversible process depending on such variables as pH and ionic strength. Generally, electrostatic complexes dissociate when the ionic strength exceeds 0.2–0.3, or when the pH value is above the protein's IEP. At pH values above the IEP, e.g., at neutral pH, electrostatic interactions can still occur between anionic carboxyl-containing polysaccharides and positively charged subunits of oligomeric proteins. Sulfated polysaccharides are capable of forming soluble complexes at pH values above the protein's IEP. Nonelectrostatic macromolecular interactions can lead to formation of an irreversible complex. Interbiopolymer complexes can be regarded as a new type of food biopolymer whose functional properties differ strongly from those of the macromolecular reactants.

The interaction between oppositely charged biopolymers is enhanced when the net opposite charges of biopolymers are increased and the ratio of net charges of the polymer reactants approaches unity. Proteins and polysaccharides possess a large number of ionizable and other functional side chain groups with different pK values. They differ in shape, size, conformation, flexibility, and net charge to a given pH and ionic strength. Therefore, the formation of structurally regular junction zones in these complexes is very unlikely. The size and stability of junction zones are of importance for the stability and functional properties of interbiopolymer complexes (see Sec. II).

Systems 2 and 4 (Fig. 1) are the single-phase and two-phase mixtures of limitedly compatible biopolymers in solution. Thermodynamic incompatibility in solutions of a mixture of a protein and a polysaccharide usually takes place at a high ionic strength and at pH values above the protein's IEP, where the biopolymers have like charges. These conditions are typical of food systems. System 2 is formed when the bulk concentration of the biopolymers is below the co-solubility threshold. However, when the bulk biopolymer concentration is increased above this critical level, the mixed solution breaks down into two liquid phases (system 4). The term "biopolymer compatibility" implies miscibility of different biopolymers on a molecular level. This means that "compatibility" (system 1, Fig. 1) is not synonymous with the terms "miscibility" and "co-solubility" (system 2). The term "incompatibility" is illustrated by systems 2 and 4. This term is also not a synonym for such terms as "immiscibility," "demixing," and "phase separation," which correspond to system 4. The term "incompatibility" or "limited thermodynamic compatibility" covers both limited miscibility or limited co-solubility of biopolymers (i.e., system 2) and demixing or phase separation of mixed solutions (system 4).

The mixing process is spontaneous when changes in Gibbs free energy ($\Delta G = \Delta H - T\Delta S$) is negative. The mixing process can only give rise to complete compatibility (dissolution of different compounds in each other) when the entropy difference ($T\Delta S$) between the two-phase and single-phase states is larger than the mixing enthalpy (ΔH).

The entropy of mixing is the driving force for mixing two liquids to give an ideal solution. This means that the entropy of mixing low molecular weight substances in solution usually contributes to minimize the free energy of mixing more than the enthalpy of interactions of single-phase system. The entropy of mixing is positive and the enthalpy of mixing is zero when the molecules spread throughout the bulk of the mixture and there are the same molecules interactions in the mixture and in the initial liquids. This is true for low molecular weight compounds, but not for polymers.

Various intermolecular interactions, namely, polymer1-solvent, polymer2-solvent, polymer1-polymer1, polymer2-polymer2, polymer1-polymer2, and solvent-solvent interactions, exist in mixed biopolymer solutions. The entropy of mixing significantly decreases when monomers are transformed into (bio)polymers. Because of the large size and rigidity of macromolecules typical of biopolymers (e.g., highly expanded polysaccharide coils and highly compact nonpenetrable globular proteins), biopolymer solutions contain less independently moving particles (segments of macromolecules). Since the entropy of mixing is a function of the number of individual particles being mixed, the value of the entropy of mixing (ΔS) of biopolymers is several orders of magnitude smaller than that corresponding to monomers. Accordingly, molecularly homogeneous mixtures of biopolymers could be prepared if ΔH is negative. This means that the attractive forces between different macromolecules are equal to or greater than those between the same type of macromolecules. This in turn means that biopolymer compatibility is related to the ability to form soluble interbiopolymer complexes. Consequently, the incompatibility of biopolymers occurs when interbiopolymer complex formation is inhibited, i.e., under conditions that inhibit interactions between macromolecules of different types and promote association between macromolecules of the same type.

The thermodynamic incompatibility of biopolymers takes place when the Gibbs free energy of mixing is positive. Sufficiently concentrated solutions of biopolymers differing in chemical composition, conformation, chain rigidity, and affinity for solvent are usually immiscible. Phase separation in mixed solutions of a large number of biopolymers studied [24,26] is very sensitive to entropy factors given by the excluded volume of the macromolecules. The phenomenon of incompatibility relates to the occupation of a volume of the solution by macromolecules and the repulsion between unlike macromolecules (see Secs. III.A–D). Therefore, phase behavior of a mixed biopolymer solution strongly depends on the molecular weight and the conformation of the macromolecules. Phase separation usually occurs at biopolymer concentrations exceeding 2–20% (depending on the size and shape of the macromolecules). In contrast, since attraction between dissimilar macromolecules leads to formation of interbiopolymer complexes, insoluble complexes can form even when very dilute biopolymer solutions (bulk concentration less than 0.01%) are mixed. From food protein functionality viewpoint, the formation of a complex means a change in the composition, structure, and functional properties of the protein particles. Thus, to control the compatibility of functional food components, their environment must be manipulated. By manipulating the environmental factors, the effective concentration of functional macromolecules in the phases, the phase behavior and structure/texture of a food system can be controlled.

Since colloidal systems are only kinetically (not thermodynamically) stable, the critical concentration of a polysaccharide required to change the stability of a protein dispersion is significantly lower for protein suspensions and emulsions stabilized by proteins than for protein solutions. Owing to complex formation, an anionic polysaccharide can act either as a flocculating agent (bridging flocculation) or as a protective colloid.

Protein-polysaccharide incompatibility in solution may result in (depletion) flocculation of a protein suspension or an emulsion [24–26]. Depletion flocculation occurs when a disperse system contains two incompatible biopolymers, one of which is adsorbed on the dispersed particles and the other of which is dissolved in the dispersion medium. Because of biopolymer incompatibility, the concentration of the dissolved biopolymer is strongly reduced near the surface of dispersed particles consisting of, or covered by, the other incompatible biopolymer. The negative adsorption of the dissolved biopolymer by a colloidal particle corresponds to formation of a low-biopolymer concentration (depletion) layer around this colloid particle. If two dispersed particles surrounded by depletion layers come into close proximity due to Brownian motion, overlapping of depletion layers can take place within the gap between these particles. This results in formation of a microvolume with a very much reduced concentration of the dissolved biopolymer between the two particles. This microvolume contains pure solvent (or a solution with a very much reduced concentration of the dissolved biopolymer) and is surrounded by a biopolymer solution. This results in formation of a solvent chemical potential gradient. The solvent is transferred by diffusion from the gap between the particles to the dispersion medium, which in turn causes aggregation of the particles. Osmotic pressure favors the flow of the solvent from the gap between the particles, which also enables the particles to overcome the repulsive energy barrier between them; this facilitates aggregation via Brownian motion.

The phenomenon of depletion flocculation is similar to the phenomenon of limited thermodynamic incompatibility of biopolymers in solutions. The basic difference between the two phenomena is that depletion flocculation is of a nonequilibrium nature. Unlike biopolymer solutions, colloidal dispersions are thermodynamically unstable because of an inherent excess surface free energy. Therefore, in the case of depletion flocculation, the effect of an added polysaccharide on the stability of protein-stabilized emulsions or protein suspensions is more strongly pronounced than on phase separation of its mixed solutions with proteins. For instance, phase separation of a mixed protein-polysaccharide solution usually takes place at a total biopolymer concentration exceeding 4%, whereas depletion flocculation occurs at concentrations less than 1%.

Depletion flocculation can also be treated as a particular case of flocculation and coagulation of colloidal particles in a nonwettable medium. The poorer the quality of the dispersion medium as a solvent for the adsorbed biopolymer (i.e., the lower the compatibility of the dissolved and adsorbed biopolymers), the more pronounced will be the depletion flocculation.

II. ELECTROSTATIC INTERBIOPOLYMER COMPLEXES

Macromolecular interactions responsible for complex formation may be divided into three types: (a) between macroions, (b) between oppositely charged (acidic and basic) side groups, and (c) between other available side groups of macroions. In the first case, the net opposite charge, shape, size and flexibility of macroions are the important factors. In the other cases, reactivity of side chain groups of amino acid residues available on the exterior surface of the protein molecules and on the sugar units of polysaccharide backbone are important. The spacing, geometrical arrangements of the side chains, and the ability of side chain groups to mutually orient themselves also contribute to the efficiency of interbiopolymer complex formation.

During formation of electrostatic complexes, the overall net charge of anionic polysaccharides decreases with gradual attachment of each successive protein macroion. Diminishing net opposite charges on macromolecular reactants reduces both the hydrophilicity and the solubility of the resultant complex. It also leads to a decrease in the IEP of the complex compared to that of the initial protein [12]. The higher the relative content of polysaccharide, the lower the pH at which the complex precipitates. This process of mutual neutralization of macroreactants leads to an electrostatically neutral insoluble complex. Soluble complexes are formed when the ratio of biopolymer reactants is far from equivalent. When the ratio of polyacid and polybase is equivalent, a maximum yield of insoluble complex is produced, i.e., when the net charge ratio of biopolymer reactants is close to unity [12–14].

Dispersed particles of an insoluble complex are aggregated and precipitated. They form a dispersed phase of complex coacervate. Aggregation of individual complex particles is mainly due to hydrogen bond formation, hydrophobic interactions, as well as dipole-dipole and charge-dipole interactions. Formation of the concentrated phase of a complex is favorable for electrostatic interactions of side-chain groups bearing opposite charges. This leads to a decrease in electrostatic free energy of the system. The loss of entropy of rigid biopolymers upon complex formation may be compensated (counterbalanced) by the enthalpy contribution from interactions between the macroions as well as by liberation of counterions and water molecules from the interacting polymer surfaces. Nonelectrostatic interactions play an important role in composition-property relationships of complex coacervates. In particular, nonelectrostatic interactions can cause nonequilibrium effects in the complexes and thus affect gel-forming conditions and rheological properties of gels made from complexes [12,14,23,24].

A. Composition-Structure-Property Relationships

Biopolymers come into contact upon complexing and form junction zones. The junction zones of a complex are regions where segments of two or more molecules of different biopolymers are joined together. The size and thermodynamic stability of these junction zones depend on the number of segments involved and the type of interactions between them. The heterogeneity of junction zones in terms of size and stability is due to differences in chemical structure, size, and shape of interacting chains, as well as to the fact that the centers of mass of randomly interacting polymers do not generally coincide [27,28].

The contact area between interacting globular protein molecules is limited by their topography. The rigid and compact conformation of proteins restrict the number of oppositely charged groups capable of interacting with each other. If protein molecules associate as spherical particles, the junction zones can be expected to be small compared to those of proteins with disordered conformation. Linear flexible macromolecules are able to form more extended interchain junction zones. For instance, the random coil conformation of linear anionic polysaccharides and disordered structures of some proteins, such as caseins, gelatin (at temperatures above 35°C), or acid-denatured edestin, are favorable for steric adjustments between the oppositely charged interacting groups.

Junction zones formed by a side-by-side packing of oppositely charged linear polymers may contain many individual chains, especially in the case of insoluble complexes. Electrostatically associated insoluble complexes aggregate further due to ionic bonds as well as hydrophobic interactions and hydrogen bonding. A three-dimensional network of junction zones can be formed throughout the volume of the dispersed particles of the insoluble complex. As a result, all microions are incorporated into junction zones of the separated phase of a complex coacervate. This can lead to formation of complex gels (e.g., gelatin-alginate gels), which can be thermoirreversible in concentrated urea solutions and dissociated in salt solutions.

Macromolecules always move, if possible, to occupy a position of minimal energy. This corresponds to mutual attraction of oppositely charged chains. For instance, this happens when protein, polysaccharide macroions, and other counterions are incorporated into the separated phase of a complex coacervate. This concentrated phase contains electrically equivalent amounts of the randomly mixed macroions. This mutual neutralization of opposite charges minimizes the electrostatic free energy. Junction zone formation is a time-consuming process since it requires mutual adjustment of macroions and changes in local macroion configuration. For instance, the aging of gelatin-alginate (or gelatin–low methoxy pectin) complex gels is accompanied by the formation of electrostatically stabilized junction zones and the conversion of these thermoreversible gelatinlike gels into thermally irreversible gels [23,27].

1. Role of Conformations of Interacting Polymers

The composition-property relationships of complexes are affected by conformations of the interacting macromolecules. Proteins with an unfolded structure, such as gelatin and casein, are able to form a maximum number of contacts with an oppositely charged polysaccharide. It is, therefore, typical of gelatin, as exemplified in Figure 2, to form electrically neutral insoluble complexes with anionic polysaccharides over a wide range of system compositions.

Figure 2 shows that the composition of an insoluble complex formed between gelatin and sodium alginate depends on the pH of the system, but not on the ratio of their initial concentrations. At a given pH, the ratio of weight fraction of gelatin to anionic polysaccharide in an insoluble complex is equal to the ratio of their charges:

$$n = \frac{C_{\text{gelatin}}}{C_{\text{polysaccharide}}} = \frac{Z_{\text{polysaccharide}}}{Z_{\text{gelatin}}}$$

where n is the ratio of weight fractions (C) of gelatin to anionic polysaccharide in an insoluble complex and $Z_{\text{polysaccharide}}/Z_{\text{gelatin}}$ is the ratio of net charge of the macroions at a given pH. This shows that the stoichiometry is strongly affected by pH, since the net charges of proteins and anionic polysaccharides are oppositely changed with pH. After the precipitation of an insoluble complex of a constant composition, an excess of one of the polymer components remains in solution. When the pH is decreased below the protein's IEP, the net charge of the protein increases, while that of an anionic polysac-

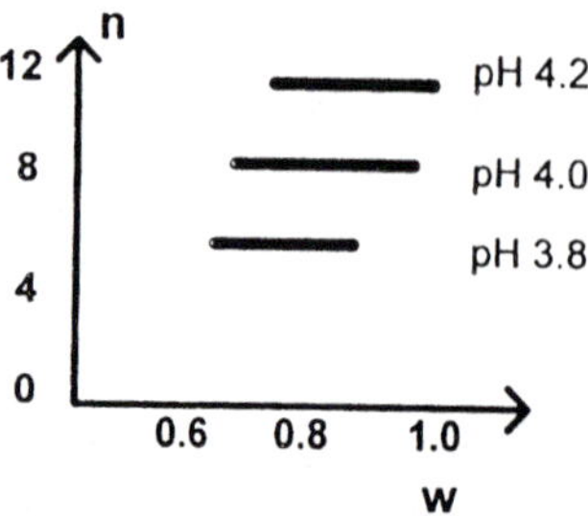

FIGURE 2 Weight ratio (n) of gelatin and sodium alginate in an insoluble complex versus gelatin weight fraction ($W = C_{\text{gelatin}}/C_{\text{gelatin}} + C_{\text{alginate}}$) in the initial mixed solution. (From Ref. 23.)

charide macroions decreases. As a result, the insoluble complex is enriched with polysaccharide [12,13,25].

In contrast, the composition of soluble complexes depends on the initial ratio of protein and polysaccharide concentrations. Usually, soluble complexes are formed in mixed gelatin solutions containing an excess of anionic polysaccharide. However, globular proteins interacting with anionic polysaccharides more readily form charged insoluble complexes.

Normally, soluble electrostatic complexes form at low bulk concentrations. These soluble complexes can be regarded as metastable and generally are able to aggregate when the concentration is increased. The aggregation of soluble complexes can be induced by the addition of a small amount of salt. At high salt concentration these complexes are dissociated [14].

Thus, in all cases, the composition of the phase formed by insoluble interbiopolymer complexes tends to satisfies the condition of electrical neutrality [23].

2. Nonequilibrium Effects of Complex Formation

The composition and properties of an electrostatic interpolymer complex depend on the pH, ionic strength, and the nature and ratio of the biopolymers. The composition of a complex can also depend on its method of preparation. At low pH and at low ionic strengths electrostatic complexes are not at equilibrium; thus, the composition and properties of a complex depend on the conditions of its formation. For instance, Figure 3 shows that the properties (such as dispersibility) of complexes are dependent on the method of their preparation, i.e., on the sequence in which the system components, namely, solutions of the protein, the polysaccharide, and an acid, are mixed. Nonequilibrium complexes are especially typical of polyelectrolytes with a high charge density. Nonequilibrium complex formation has been studied in detail for mixtures of bovine serum albumin and dextran sulfate [12,14]. Figure 3 schematically shows a nephelometric base (A-B)–acid (B-C) titration curve for this system. The intensity of light scattering

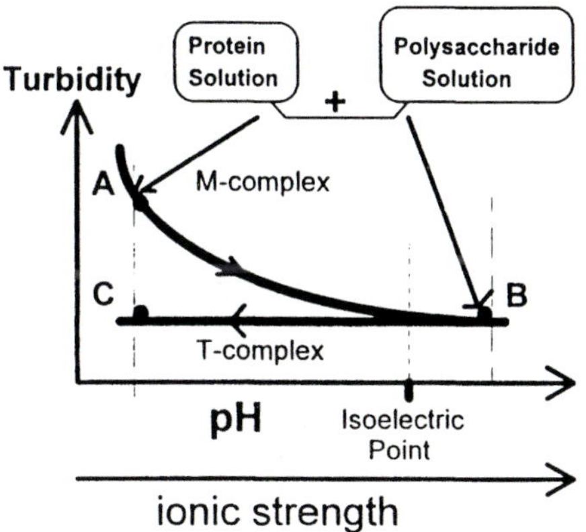

Figure 3 Changes in turbidity (arbitrary units) of aqueous mixed protein-polysaccharide (e.g., bovine serum albumin–dextran sulfate) solutions during nephelometric acid-basic titration (A-B-C). Turbidity (or an aggregation) hysteresis during alkali and acid titration of a mixture solution shows that the structure and properties of complexes depend on the way in which they are produced. (From Refs. 12,14.)

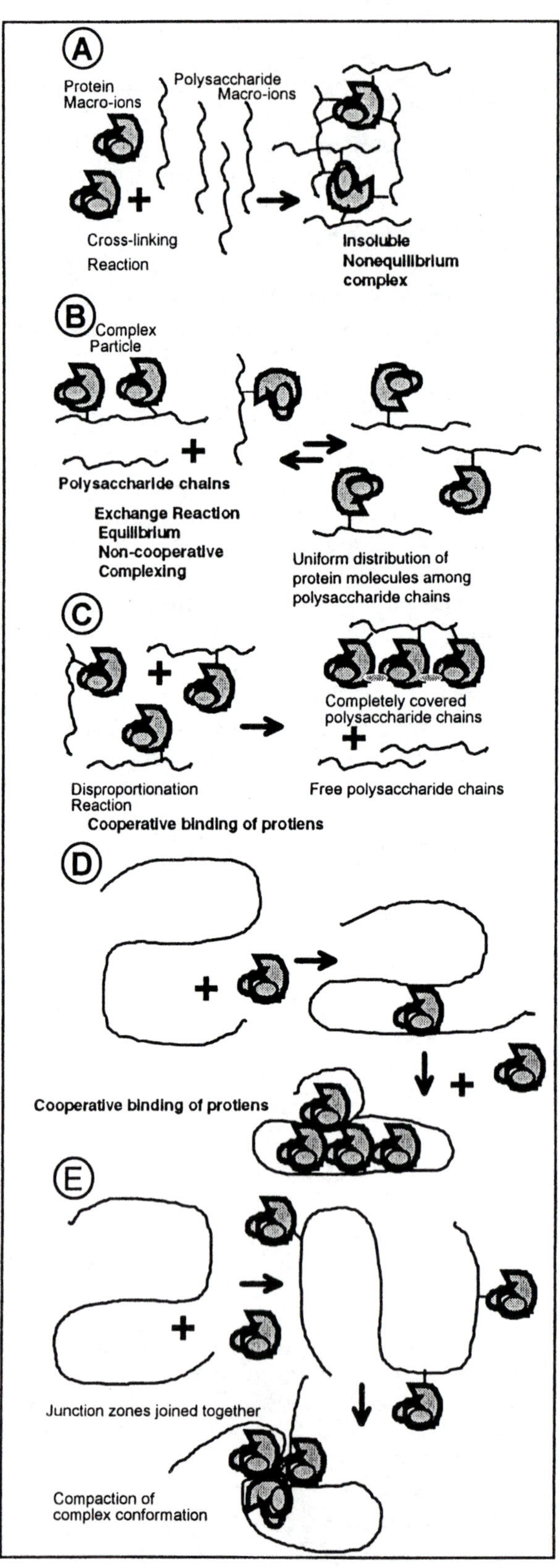
A
Protein
Macro-ions
Polysaccharide
Macro-ions
Cross-linking
Reaction
Insoluble
Nonequilibrium
complex
B
Complex
Particle
Polysaccharide chains
Exchange Reaction
Equilibrium
Non-cooperative
Complexing
Uniform distribution of
protein molecules among
polysaccharide chains
C
Completely covered
polysaccharide chains
Disproportionation
Reaction
Free polysaccharide chains
Cooperative binding of protiens
D
Cooperative binding of protiens
E
Junction zones joined together
Compaction of
complex conformation

serves here as a measure of the yield of insoluble complex (A) in a dilute mixed solution [14].

The mixed solution prepared by mixing polymer solutions at a pH below the protein's IEP separates into two phases (A). The insoluble complex (A) forms a concentrated bottom phase. The dissociation of complex (A) can be achieved by taking the system to alkaline conditions (B). A stable system (B), which does not contain electrostatic complexes, can also be prepared by mixing the polymer solutions at alkaline pH (B). Lowering of the pH from point (B) to acidic conditions leads to system (C) containing soluble complexes.

Of special concern is the fact that aggregation hysteresis (Fig. 3) is also observed (e.g., for protein-dextran sulfate complexes) in concentrated urea solutions. This means that strong electrostatic interaction between macromolecules is mainly responsible for nonequilibrium complex formation. Similar hysteresis (Fig. 3) also can be obtained by varying the ionic strength of the system [12]. Insoluble complexes can be produced by mixing protein and polysaccharide solutions at a low ionic strength and at pH values below the protein's IEP. For instance, this occurs at point A if the abscissa now represents the ionic strengths, not pH values. At higher ionic strengths, complex formation is inhibited, for instance, at point B. Subsequent dialysis of B-C of the system B against an acid solution at the same pH value as A produces a soluble T-complex (C). Complexes of the two types A and C have been given the names "mixing"and "titration" complexes, respectively, according to their method of preparation [14].

Figure 3 shows that there is a great difference in solubility behavior of mixing and titration (M- and T-) complexes (A and C) at a given pH and system composition. At low ionic strength and pHs below the protein's IEP (e.g., point A) the formation of insoluble M-complexes is a nonequilibrium process due to the strong electrostatic interaction between a positively charged protein and an anionic polysaccharide. Figure 4A shows that protein molecules act as cross-linking agents between polysaccharide chains. This forms a three-dimensional network of complex gel particles. The concentration of precipitated gel-like phase depends on the ratio of protein to polysaccharide, pH, and ionic strength. A protein can be quantitatively precipitated in the form of a M-complex by adding an equal amount of an anionic polysaccharide. However, the protein content of the M-complex (e.g., of serum albumin with sodium alginate or dextran sulfate) may be lower than that needed to satisfy electrical neutrality conditions [14].

FIGURE 4 Nonspecific attractive protein-polysaccharide interactions. Nonequilibrium (A) and equilibrium (B) protein-polysaccharide complexes, noncooperative complexing (A, B) and cooperative complexing (C, D, and E). (A) Strong electrostatic interbiopolymer interaction. The equilibrium is "frozen." Irreversible complexing. pH is well away from the protein IEP (typically pH $<$ 4.5 for carboxyl-containing and pH $<$ 5 for sulfated anionic polysaccharides), low ionic strength. Structure is typical of M-complexes. (B–E) Weak electrostatic interbiopolymer interaction. pH is around and above the protein IEP (typically pH $>$ 4.5 for carboxyl-containing polysaccharides and pH $>$ 5 for sulfated polysaccharides), low or moderate ionic strengths. Structure is typical of T-complexes. (C) Weak and comparable protein-polysaccharide and protein-protein interactions. pH is near the protein IEP, for instance, within the range 5.0 $<$ pH $>$ 5.6, and ionic strength is 0.3 for complexing bovine serum albumin–dextran sulfate.

A gradual decrease of pH (B-C) of a mixed protein-polysaccharide solution is accompanied by a gradual intensification of electrostatic interaction between the protein and the polysaccharide. This leads to formation of soluble equilibrium T-complexes. Figure 4B shows that the reversibility of complex formation results in uniform distribution of protein molecules among polysaccharide chains [14,28]. When there is a relative excess of an anionic polysaccharide the acid titration (B-C) does not lead to aggregation of the complex. Under these conditions a metastable concentrated dispersion of protein-polysaccharide complexes can be prepared [12].

Thus, when solutions of oppositely charged biopolymers are mixed under conditions of strong interbiopolymer interaction, i.e., at pH values below the protein's IEP, the aggregated M-complexes obtained are hardly soluble in water (Fig. 4A). In contrast, when the interaction between microions is gradually increased by titrating the mixed solution from a neutral to an acid pH values around the protein's IEP, the weak T-complexes obtained become quite dispersible in water. The properties of protein-polysaccharide complexes formed near the protein's IEP are independent of the way in which they are produced. This indicates that complex formation under weak protein-polysaccharide interaction conditions is an equilibrium-reversible process (Fig. 4B).

In the pH range close to the protein's IEP, an interesting phenomenon (Fig. 4C) of nonuniform redistribution of protein molecules among polysaccharide chains occurs [12,14]. This leads to formation of low-protein and high-protein fractions of the T-protein-polysaccharide complexes. The reason is that in the vicinity of the protein's IEP, hydrophobic protein-protein and electrostatic protein-polysaccharide interactions can be energetically comparable to each other. Protein-protein association (or self-association of proteins), which is mainly due to hydrophobic interactions, is usually enhanced when the pH of a solution approaches the IEP. Accordingly, near the IEP of the protein, where the protein-polysaccharide interaction (attraction) is relatively weak and protein-protein interaction (attraction) between the protein molecules bound to a polysaccharide chain is strong, each free site situated near the site already occupied by a protein molecule becomes thermodynamically preferable for further binding of protein molecules. This leads to cooperative protein adsorption on an anionic polysaccharide. Because of this, some parts of the polysaccharide chain tend to be completely covered by protein molecules while the other parts are completely free of protein.

Thus, protein sorption by anionic polysaccharides is a cooperative process over a narrow range of pH values near the IEP of the protein. For instance, in the range of $pH > 5.0$ and < 5.6, T-complexes of bovine serum albumin with dextran sulfate go over from a uniform to a nonuniform distribution of protein molecules among the polysaccharide chains. At higher pH values ($pH > 5.6$) the protein adsorption is a reversible, noncooperative process (Fig. 4B), and no remarkable effect on the protein distribution among and along dextran sulfate chains is observed; this is because of relatively weak protein-protein interaction above pH 5.6. At lower pH values ($pH < 5$), i.e., under conditions of strong interbiopolymer interaction, the equilibrium of complex formation is frozen. This means that the T-complex preserves its nonaggregated structure (each particle of the complex contains a single polysaccharide chain) below $pH < 5$ and that the composition of neither T- or M-complexes changes with pH during titration. At pH values well below the IEP or the protein, both T- and M-complexes are of nonequilibrium nature and complexing is noncooperative.

Cooperative protein adsorption can also result from conformational changes in polysaccharide induced by binding of protein molecules. The bound protein molecules can cross-link segments of an anionic polysaccharide chain (Fig. 4D). This can result in an increase in local concentration of polysaccharide segments and consequently increase the local charge density, which in turn may promote further protein adsorption and compact conformation (E) of protein-polysaccharide complexes [12,28]. The junction zones formed by the association of oppositely charged segments of proteins and polysaccharides have a lower hydrophilicity than that in the dissociated state. The compact structure of a complex reflects a less hydrophilic nature of its junction zones, which can form a "hydrophobic interior" (E). Thus, nonspecific complex formation between biopolymers can be either reversible or irreversible and cooperative or noncooperative.

3. Stability of Complexes

The stability of junction zones mainly depends on their length, the nature of interactions, and the number of interacting groups (i.e., cohesion energy density). Thus the melting temperature of protein-polysaccharide complex gels and their stability against ionic strength–induced dissociation are governed by the above factors. Freshly prepared gelatin-alginate complex gels can be dissolved in hot water or in urea solution in a manner similar to that of a gelatin gel. However, upon aging, the gel becomes thermoirreversible and can only dissolve in salt solutions. Upon addition of salt, the gelatin-alginate complex gel becomes thermoreversible again [12,14]. The thermoirreversibility of aged gelatin-alginate or gelatin-pectin mixed gels is attributable to a transition from collagen-type triple-helix junction zones of gelatin to other types of junction zones involving electrostatic interaction with alginate (or pectin) chains.

The stability of protein-anionic polysaccharide complexes against high ionic strength and pH values above the IEP of the protein may be increased by thermal denaturation of the bound globular protein and by adding divalent cations (such as Ca, Fe, Cu, etc.). The stabilizing effect of divalent cations may be attributable to their ability to cross-link proteins and anionic polysaccharides. The increase in stability caused by thermal denaturation is attributable to an increase in the number of hydrophobic interactions, hydrogen bonding, and coordinate bonds within the complex. These interactions, in addition to changes induced by precipitation, cause irreversible changes in the functional properties protein-polysaccharide electrostatic complexes [24,29].

Ternary complexes made of protein, multivalent cation, and anionic polysaccharide are usually of nonequilibrium nature. Their composition and properties are determined by the conditions during their formation [12,24,29,35]. For instance, the influence of divalent cations (such as Ca and Cu ions) on the properties of an oil-in-water emulsion stabilized by a protein-polysaccharide mixture is remarkably different depending on whether the cation was added either before or after emulsification. When the cation is added before emulsification, the ternary complex precipitates in the form of gel-like particles. On the other hand, when the cation is added after emulsification, there is no precipitation. Instead, an insoluble ternary complex is formed on the surface of the oil droplets. As a result of this encapsulation of the dispersed oil droplets, a highly stable oil-in-water emulsion with controllable rheological properties is obtained. For instance, the consistency of emulsions stabilized by the ternary complex can be varied from a liquid to a thixotropic solid state depending on the concentration of the divalent cation.

The formation of protein-anionic polysaccharide and protein-protein complexes also causes a localized increase in protein concentration. The denaturation of the bound protein molecules can promote protein-protein interactions (typical of protein gels) via hydrogen bonding and hydrophobic and electrostatic interactions. Therefore, protein denaturation generally leads to formation of stable complexes at higher yields. In the complexed state, renaturation or refolding of the denatured protein is inhibited [24,27,28]. An example of this is the irreversibility of thermal denaturation of the Kunitz trypsin inhibitor bound to anionic polysaccharides and other proteins. It has also been shown that the enthalpy and the temperature of denaturation of the Kunitz trypsin inhibitor bound to pectin are affected by the degree of esterification of the pectin and by the protein:pectin ratio [24,28].

B. Functional Properties of Electrostatic Complexes

The formation of junction zones affects both particle-solvent and particle-particle interactions. The solubility of proteins may be increased by electrostatically complexing them with anionic polysaccharides. Formation of T-complexes may increase protein solubility and inhibit protein precipitation at the IEP. Anionic polysaccharides can act as protective hydrocolloids inhibiting aggregation and precipitation of like-charged dispersed protein particles, e.g., of denatured proteins. This protective action also can increase the stability of protein suspensions and oil-in-water emulsions stabilized by soluble protein-anionic polysaccharide complexes [12,14,30].

The hydrophilic-hydrophobic character of a complex particle is controlled by the junction zone:polysaccharide chain ratio. An increase in junction zone content can improve surface properties of functional protein complexes. Formation of junction zones with decreased hydrophilicity also provides a compact conformation to the complex, increases the least concentration end point (LCE) for gelation, and decreases the viscosity of soluble complexes [12,23]. Macromolecular segments that are not incorporated into the junction zones play a key role in dictating the solubility and gelation of the complex. These segments also play important roles in the expression of several other functional properties of the complex.

An increase in junction zone size (especially multichain zones) decreases the solubility and causes precipitation of the complex. The resulting insoluble complex may be dissociated to recover and reuse the polysaccharide. Thus, protein-polysaccharide interactions can be used as a method to fractionate proteins. The difference between IEP and the net charge at a given pH of different proteins is the main factor enabling selective precipitation of a protein in a mixture by anionic polysaccharides. This technique has been applied to the fractionation of milk and yeast proteins [12,15–19].

At pH values below the IEP, anionic polysaccharides can also act as flocculents for the precipitation of protein suspensions or emulsions stabilized by proteins. The complexation with dispersed protein particles decreases interparticle repulsive forces causes cross-linking and flocculates by bridging-flocculation. The rate and extent of flocculation depend on the composition of the particle surface, the molecular weight, and conformation of the anionic polysaccharide. This approach can be used to recover proteins from dilute dispersions, including different effluents and waste streams produced by the food industry as well as for clarification of different liquid food systems.

It is known that protein-stabilized foams and emulsions are more stable at the protein's IEP than at other pH, where electrostatic repulsion between protein molecules is minimized. Also, in a binary protein system, the stability of foams and emulsions is greater at pHs between the IEPs of the proteins, where the electrostatic attractive forces are maximized.

Interactions between neighboring protein molecules in the adsorbed surface layer can be maximized by complexing with an anionic polysaccharide or other oppositely charged proteins [32–36]. Films of protein-polysaccharide complexes at oil/water and air/water interfaces, formed either by direct adsorption of the complex or by interaction of an anionic polysaccharide with adsorbed proteins, usually exist as gel layers covering the dispersed particles. This gel structure is formed by aggregation of the complexes. Conformational stability of proteins in the complex affects formation and stability of the gel structure and thus its ability to stabilize the dispersed phase [24,28]. The complexation may suppress competition between different proteins for the interfaces and may also cause partial unfolding of protein molecules. The latter can maximize various intermolecular interactions in the adsorbed layer. The net charge of the adsorbed surface layer can also affect the stability of emulsions.

The principal reasons for using protein-polysaccharide complexes as emulsion stabilizers are their high surface activity, their ability to increase the viscosity of the dispersion medium, and their ability to form gel-like charged and thick adsorbed layers. The mechanical strength of the adsorption layer as well as the electrostatic (repulsion of emulsion droplets carrying like charges) and steric (barrier of thick stabilizing layer) effects are the most important factors contributing to the kinetic stability of oil-in-water emulsions [12,14,21,24,28].

Negatively charged, thick gel-like shells around colloid particles may be obtained either by directly using protein-anionic polysaccharide complexes to stabilize emulsions or by adding an anionic polysaccharide to a protein-stabilized emulsion. In the latter case the protein is used as the primary stabilizer during emulsion preparation. The added polysaccharide acts like a protective colloid.

When colloidally dispersed particles are charged, a coulombic repulsion exists between them. However, when the ionic strength is sufficiently large, the electrostatic potential between the colloidal particles (and the electrostatic repulsion between two colloid particles) becomes almost zero. Here, salts (more precisely the oppositely charged ions) act as flocculating agents. By analogy, anionic polysaccharides (especially of high molecular weight and rigidity) can function as efficient flocculating agents of protein suspensions and protein-stabilized emulsions.

Interbiopolymer complexes can be used to encapsulate liquid and solid materials. Generally, for encapsulation, the protein and polysaccharide solutions are mixed at a pH above the protein's IEP. Then the pH is readjusted to a value below the protein's IEP; this causes phase separation and coating of each dispersed particle by a layer of the complex phase formed.

Charged polysaccharide chains control the pH dependence of activities of bound enzymes. For instance, binding of trypsin to anionic polysaccharides shifts the pH optimum of its activity to a higher pH range. This is due to suppression of dissociation of the enzyme's side-chain ionizable groups in the microenvironment of an anionic polysaccharide. The latter creates an abundance of carboxyl groups around the protein molecule in the complex [31].

III. THERMODYNAMIC INCOMPATIBILITY OF PROTEINS AND POLYSACCHARIDES

Biopolymer incompatibility is the rule rather than an exception. Compatibility or miscibility of unlike biopolymers in aqueous solutions has only been exhibited by a few biopolymer pairs. For instance, mixtures containing serum albumin and ovalbumin, 7S- and 11S-globulins of broad beans, and serum albumin or gelatin and pectin have been found to be single-phase mixed solutions over a wide range of pH and protein and salt concentrations [24].

A. Phase Diagrams

The phenomenon of thermodynamic incompatibility of food proteins and polysaccharides is illustrated in Figure 5. It depicts a typical phase diagram for a ternary protein-polysaccharide-water system in the form of rectangular coordinates. The solid line shown in Figure 5 is a binodal curve. The binodal (or the solubility) curve shows the co-solubility profile of biopolymers in a given medium. The binodal curve separates the regions of single- and two-phase systems. The region lying under the binodal curve corresponds to one-phase mixed solutions (system 2, Fig. 1), while the region above the binodal represents two-phase systems (system 4, Fig. 1).

Figure 5 shows that on mixing aqueous solutions of a protein and a polysaccharide, the mixture may either be stable as a single phase (A-B_1) or break down spontaneously into two phases (A-B). For instance, by mixing solutions A and B in the volume proportion BC/AC, a mixture of composition C is obtained. This mixed solution C breaks down into two liquid phases D and E. Points D and E are the binodal points. The binodal branches do not coincide with the biopolymer concentration axes. They are located fairly close to the axes. This means that biopolymers have limited co-solubility in the common solvent, water, and that each biopolymer is mainly concentrated in one of the phases.

The binodal branches are intersected at the critical point F, where the two co-existing phases are of the same composition and volume. At the critical point the com-

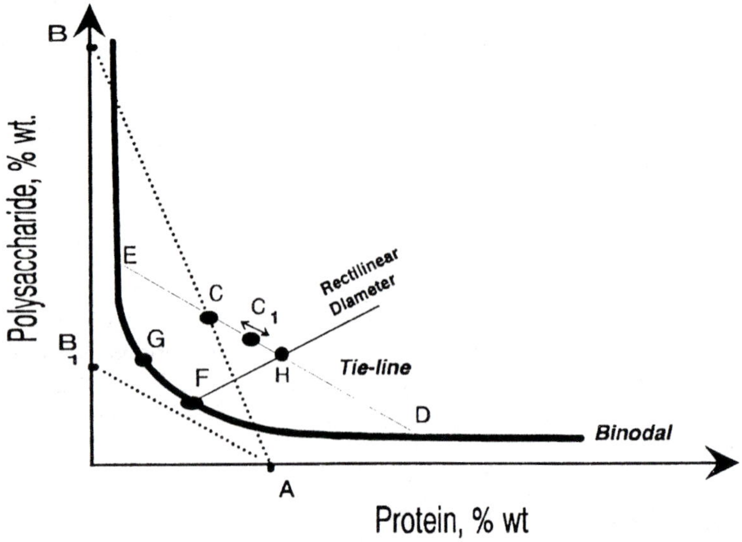

FIGURE 5 Schematic picture of a typical phase diagram of a protein-polysaccharide-water system showing the positions of binodal curve (DFGE), critical point (F), phase-separation threshold G, tie-line (ED), and rectilinear diameter (FH). (From Refs. 25,28.)

position of the phases are equal to that of the initial system. The compositional difference between the coexisting phases usually increases with distance from the critical point F. Point G is the phase separation threshold for a given biopolymer pair, i.e., the minimal total concentration of biopolymers required for phase separation to occur.

The line ED is the tie-line. The tie-line connects the binodal points corresponding to the compositions of the co-existing phases. A point on the tie-line (e.g., C) corresponds to the composition of systems breaking down into phases, each with the same composition as those of D and E. The length ratio of the tie-line segments EC/CD represents the volume ratio of phases D and E. The rectilinear diameter (F-H) is the line connecting the critical point and the midpoint of the tie-line. The mixed solution corresponding to the midpoint of a tie-line breaks down into two phases of the same volume. A phase inversion phenomenon can take place in a water-in-water emulsion when its composition goes from one side of the rectilinear diameter F-H to the other side.

The phase diagram can be regarded as the solubility profile of a given protein in the presence of other biopolymers under given test conditions. Phase diagrams can be used to characterize the effects of nonprotein components of a given food system (carbohydrates, lipids, protein-lipid complexes, etc.), as well as of the effects of variables, such as temperature, salt concentration, and pH, on protein solubility, phase state of the system, and water distribution between the system phases. These data are essential for the prediction of protein behavior in multicomponent food systems [25,26].

B. Construction of Phase Diagrams

To determine a phase diagram, a series of mixed solutions (e.g., A-B, Fig. 5) containing varying concentrations of biopolymers are prepared. Emulsion phases are separated by centrifugation, and the concentration of each polymer in each phase is determined. Figure 5 presents typical results of chemical analysis of co-existing phases obtained under a given phase separation conditions (pH, salt concentration, and temperature) [37]. Chromatography (gel filtration) has been successfully applied to determine the composition of coexisting phases [38]. The phase diagram of the amylopectin-gelatin-water system has recently been determined using Fourier transform infrared spectroscopy [39].

Another technique called the phase-volume ratio method has been used for plotting the phase diagram of protein-protein-water systems [40]. When the point C_1 (Fig. 5) representing the system composition is shifted along the tie-line DE towards the binodal curve, the volume ratio of coexisting phases approaches either zero or unity in the vicinity of the binodal curve. This means that binodal points can be obtained from the system composition–phase-volume ratio relationship. For this the volumes of the two bulk-separated phases are measured for a series of mixed solutions where the concentration fraction of one of the biopolymers is changed from 0 to 1. Then the two binodal points can be found by graphical extrapolation of the experimentally determined dependence of the phase-volume ratio on the concentration of one of the biopolymers. A ratio of system phase volumes of 0.5 corresponds to the system composition at the rectilinear diameter (F-H). The position of the system critical point (F) can be obtained by graphical extrapolation as the point of intersection of the rectilinear diameter with the binodal curve. The position of phase separation threshold G represents the point of contact between the binodal curve and the straight line cutting segments of the same length on the concentration axis. This phase volume ratio method allows determination of the phase diagram without chemical analysis of the phases.

C. Factors Affecting Phase Behavior of Biopolymer Mixtures

The thermodynamics of interaction of biopolymers with each other and with the solvent may be quantitatively described by the values of the second virial coefficient and the cross-second virial coefficient (or the interaction parameter). Both values can be determined using a light scattering technique [24,26]. The interaction parameter quantitatively characterizes the intensity of the thermodynamic interaction between molecules of dissimilar biopolymers in a mixed solution. A positive value of the cross-second virial coefficient is indicative of exclusion of the molecules of one biopolymer from the solution volume occupied by the molecules of the other biopolymer, and a negative value is indicative of mutual miscibility of a given biopolymer pair. The excluded volume effect arises when the molecules of one of the biopolymers cannot have access to the volume occupied by the molecules of another incompatible biopolymer; this causes a reduction in the excess entropy of mixing. Even a small positive value of the cross-second virial coefficient is enough to prevent miscibility.

The light scattering method is also a powerful tool for analyzing the effect of low molecular weight food components on the solubility of biopolymers. For instance, Figure 6 shows that the second virial coefficients for aqueous solutions of legumin (11S broad bean globulin) and dextran increase from a negative value to zero with increasing addition of sucrose. This indicates that in the presence of sucrose the dextran-solvent

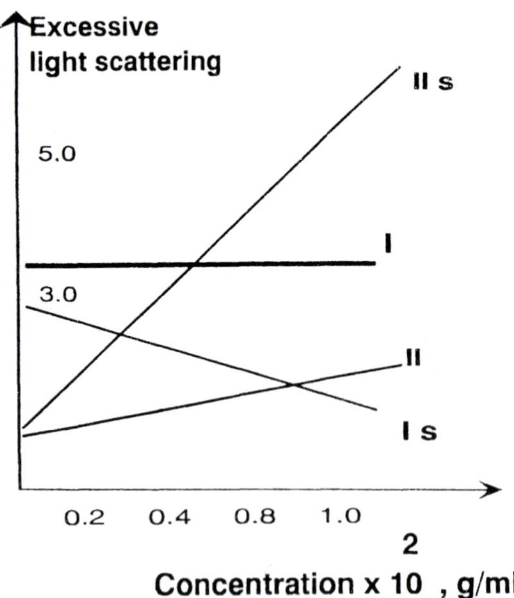

FIGURE 6 The concentration dependence of light scattering intensity in an aqueous solution (0.1 mol/dm^3 phosphate buffer, pH 7.0) of (I) 11S *Vicia faba* globulin and (II) dextran with 50% sucrose (Is and IIs) and without sucrose [28]. The second virial coefficient can be determined as the slope of the concentration dependence of the scattered light intensity of the biopolymer solution.

interaction becomes energetically more favorable than the interaction among dextran molecules themselves. A second virial coefficient equal to zero corresponds to an ideal solution. This means that all three main types of intermolecular interactions, namely, solute-solute, solute-solvent, and solvent-solvent interactions, are energetically equal to each other. Figure 6 shows that sucrose solutions are better solvents than 0.1 M phosphate buffer for both for *Vicia faba* globulin and dextran. These biopolymers, however, have limited compatibility in both water and sucrose solutions.

It should be noted that sucrose can be regarded as a monomer of dextran. This means that going from a simple sucrose to a complex dextran solution as a solvent for the protein (i.e., changing from a monomer to a polymer solution) results in a transition from a good to a very poor solvent or from complete to limited miscibility of solutes [28].

D. Phase Separation Conditions

Figure 7 shows that there are significant differences in the compatibility of various pairs of biopolymers. A smaller area under the binodal curve or lower values of coordinates of the critical point or phase separation threshold would indicate lower compatibility of a given biopolymer pair.

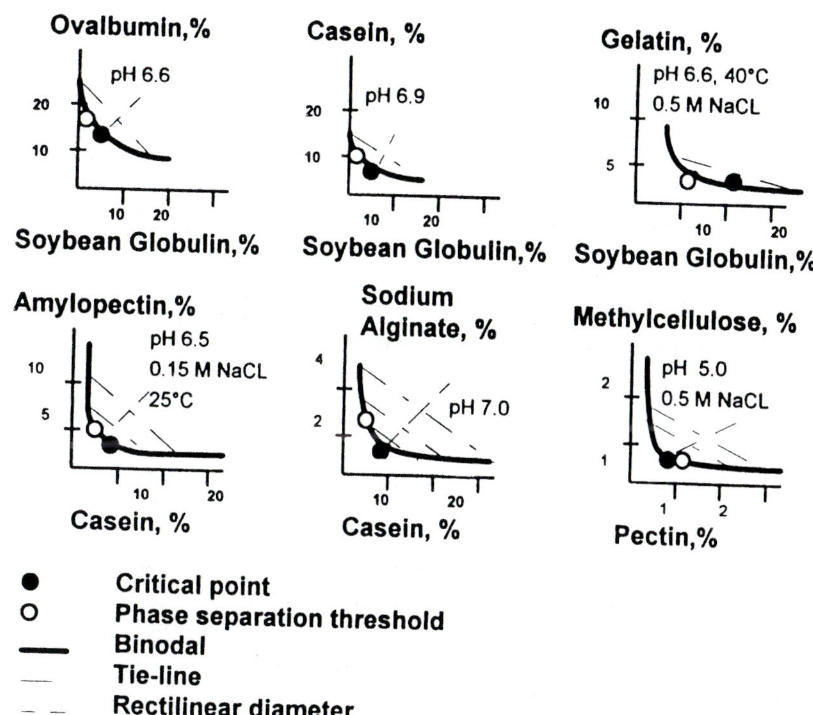

FIGURE 7 Phase diagrams of biopolymer¹-biopolymer²-water systems. (From Refs. 14,25,41.)

The phase separation threshold usually exceeds 15% for mixtures of various globular proteins. It is usually 4% or higher for globular protein-polysaccharide mixtures and less than 3% for casein-polysaccharide or gelatin-polysaccharide mixtures. The phase separation threshold mainly depends on the excluded volume of the biopolymer molecules. It decreases with an increase in the molecular weight and the rigidity of the polysaccharide used. It can also be decreased by denaturation of globular proteins. This is due to an increase in hydrophobicity and in the particle size of aggregated polypeptide chains resulting from thermal denaturation [14,24,25,41].

The differences in excluded volume effects between incompatible biopolymers determine the results of biopolymer competition for the bulk of the mixed solution. The biopolymer with a higher excluded volume expels that with the lower from the solution. This competition for space or the mutual exclusion effect of macromolecules determines the critical conditions of a system phase separation and contributes to the asymmetry of the phase diagrams of protein-polysaccharide systems. The degree of asymmetry of phase diagrams reflects a markedly high hydrophilicity and a large effective volume of polysaccharide (especially linear) macromolecules compared to compact molecules of globular proteins. Accordingly, a less concentrated phase rich in polysaccharide is usually in equilibrium with a concentrated phase rich in protein.

The differences in excluded volume effects between biopolymers are responsible for differences in critical concentrations (up to 10-fold and more) for the transition from their dilute to moderately concentrated solutions. These contribute to the asymmetry of phase diagrams.

The binodal curve usually lies closer to the concentration axis of a biopolymer of low molecular weight. The water content is always higher in the phase of a more hydrophilic biopolymer with a higher excluded volume effect. The asymmetry of phase diagrams may be characterized by (a) the ratio of critical point coordinates, (b) the angle made by tie-lines with the concentration axis of one of the system components, or (c) the length of the binodal segment (FG) between the critical point and the phase separation threshold. Normally, the self-association of macromolecules and formation of ordered supermolecular protein structures in concentrated protein solutions may change the affinity to water and the excluded volume of biopolymer molecules.

E. Some Applications of the Biopolymer Incompatibility Phenomenon: Molecular and Environmental Factors Affecting Protein Functionality

1. Concentration of Protein Solutions

The greater the phase diagram asymmetry, the larger the difference in water content between the phases. Accordingly, phase separation of a mixed solution is accompanied by a nonequal partitioning of water between the phases formed. Because of this, one of the initial biopolymer solutions is concentrated while the other phase is diluted. Usually, highly concentrated protein-rich phase is in equilibrium with a diluted polysaccharide phase. This phenomenon of water transfer between immiscible water solutions of biopolymers can be used as a method for concentrating protein solutions. This method is called "membraneless osmosis" [14,37,41].

Figure 8 shows an example of the phase diagram for a mixture of skimmed milk (A) with pectin (B). The diagram gives the volume proportion (AC/CB or AC_1/C_1B) of pectin solution to be added to skimmed milk to prepare a mixture of composition C (or

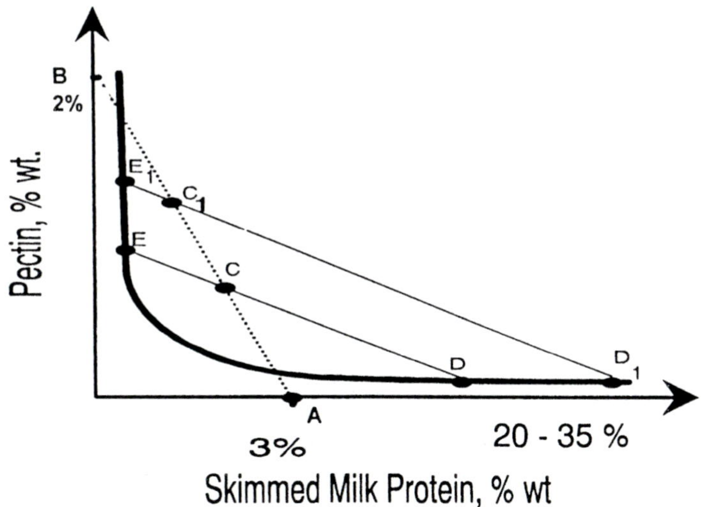

FIGURE 8 Phase diagram of the skimmed milk–pectin (high ester)–water system at pH 6.5 and 25°C. (From Ref. 12,14.)

C_1), which breaks down into phases D and E (or D_1 and E_1). Phase D (or D_1) mainly contains casein, while phase E (or E_1) contains most of the pectin and milk serum proteins. The phase diagram shows that the protein concentration in phase D or D_1 is higher than that of the initial protein solution A. Accordingly, the volume of the protein-rich phase D (or D_1) is smaller (the tie-line segment EC or E_1C_1, respectively) than that of the polysaccharide-rich phase E_1 (the segment CD or C_1D_1).

The protein-rich phase remains liquid up to a high concentration since the use of high-ester pectin results in a transfer of calcium ions from the protein-rich phase into the pectin-rich phase. In the case of neutral polysaccharides, calcium ions and casein are concentrated simultaneously, so that the liquid protein-rich phase becomes gelled [37].

It should be noted that there is a similarity of phase separation conditions of systems containing anionic and neutral polysaccharides. Neutral polysaccharides form soluble complexes with proteins in aqueous media with a low ionic strength and at pH values different from the protein IEP. Dissociation of these complexes with increasing ionic strength or heating leads to the incompatibility of biopolymers and phase separation of their mixed solutions. Phase separation of mixed protein–neutral polysaccharide mixtures usually takes place at salt concentrations exceeding some critical value (normally 0.1–0.2 M). Phase diagrams for these systems usually have a low critical point. Generally, the incompatibility of proteins with neutral polysaccharides is increased with salt concentration, polysaccharide molecular weight, and temperature [12,14,23,24,41].

In the case of incompatibility of an added protein or polysaccharide with the macromolecular food system components, the phase separation is accompanied by redistribution of water and ions between the aqueous phases. Thus, membraneless osmosis determines water partition, rheological, and other physicochemical properties of aqueous phases in many foods [43].

2. Fractionation of Biopolymers

Another important point is biopolymer fractionation. Proteins vary significantly in the boundary conditions for phase separation of their mixed solutions with polysaccharides. This opens the way for protein fractionation and purification by stepwise addition of predicted amounts of a polysaccharide [12]. A diminishing molecular weight usually results in an increase in co-solubility of biopolymers, i.e., an increase in total biopolymer concentration is required for phase separation. Food polysaccharides are usually polydisperse compounds containing macromolecules of different molecular weight, structure, and chemical composition (e.g., esterification, chemical modification, or charge density). Therefore, phase separation of protein-polysaccharide mixed solutions is usually accompanied by fractionation of both proteins and polysaccharides.

3. Functional Properties of Food Macromolecules

When biopolymers are incompatible, the thermodynamic activity of each is increased in a mixed solution. This means that proteins behave as if they were in a highly concentrated solution. This results in specific changes in the functional properties of macromolecular components of food systems [23–26,42,43].

Emulsions Stabilized by Proteins

An increase in thermodynamic activity of a protein due to the addition of an incompatible polysaccharide enhances its adsorption at interfaces. In the case of emulsions, this increases their stability. This also means that the function of polysaccharides as emulsion stabilizers is not only attributable to their ability to increase the viscosity of the continuous phase, but also to their effect on protein adsorption at the oil/water interface. At high concentration levels, polysaccharides can induce phase separation within the bulk of the continuous water phase and promote encapsulation of oil-in-water emulsion droplets by the protein-rich phase. Interactions between proteins and lipids may result in a reduction of both the conformational stability and the hydrophilicity of proteins, which are important factors promoting emulsion stability. Presumably, these might be the reasons for marginal differences in the emulsifying properties of various proteins compared with their foaming properties [25,26,42].

Mixed Gels

Thermodynamic incompatibility strongly affects the structure formation of multicomponent gels. The addition of an incompatible polysaccharide to a globular protein solution has no effect on the equilibrium of native $\rightleftharpoons$ denatured transition, but intensifies the aggregation of denatured protein molecules by increasing the protein's thermodynamic activity. The effect of an added incompatible biopolymer on gel properties is more pronounced in mixed solutions of a biopolymer with an unfolded conformation, such as gelatin or agarose.

As mentioned above (see Secs. I.B and III.D), incompatible biopolymers are in competition for the bulk of their mixed solutions. As a result of mutual exclusion of each of the biopolymers from a volume occupied by molecules of the other, the effective concentration of both biopolymers is increased in the mixed solutions. Incompatible biopolymers behave as if they were in a highly concentrated mixed single-phase solution (Fig. 1, system 2). Accordingly, gelation capacity and mechanical properties of gels of protein-polysaccharide mixtures differ greatly from those of their individual gelling agents. The critical concentration for gelation of each gelling agent is lower in a mixed solution. The elastic module is greater in single-phase biopolymer mixtures than in the gels of the individual gelling agents. For instance, gelation of a gelatin solution is strongly accelerated and gel strength is increased by the addition of a small amount of polysaccharide, such as dextran, maltodextrin, or alginate. When the quantity of the added polysaccharide exceeds the critical concentration required for phase separation, the mixed biopolymer solution breaks down into two liquid phases. Gelation of two-phases systems gives rise to filled or composite gels, e.g., gelatin gels filled by liquid dispersed droplets of dextran-rich phase. Elastic properties of such filled gels decrease with increasing volume fraction of the filler. Filled mixed gels are the most realistic model for many foods. By varying the combination of hydrocolloids and the composition of their mixtures, it is possible to observe synergistic and antagonistic effects typical of mechanical properties of mixed and filled gels and real food systems [12,43–45].

Mixed Dispersed Systems

Real foods are usually heterophase systems that can contain different lipid and aqueous, liquid and solid, crystalline and amorphous (gel-like or glassy state) phases. Low-fat systems can be given as an example of a successful formulated food. They are based either on two-phase aqueous gelatin-maltodextrin-water systems or on small spherical particles of protein gels. Some features of aqueous two-phase biopolymer systems are of great importance for structure formation in foods, particularly low-fat spreads [25,28,42,43]. First there are interfacial layers (or depletion layers; Sec. I.B) between the aqueous phases. The formation of interfacial layers of low biopolymer concentration and low viscosity reflects a trend towards minimizing contact between noncompatible macromolecules. The thickness of the interfacial layers corresponds to the large size of the macromolecules.

An important specific feature of two-phase gelatin-polysaccharide mixtures is the high hydrophobicity of the gelatin gel surface formed by contact with a nonpolar phase such as oil or air. Both phases of the gelatin-polysaccharide-water system contain gelatin and polysaccharide in different proportions. Gelatin is more surface active at the oil-water interface than maltodextrin and can dominate at the interface of both aqueous phases. In other words, the interfaces of both immiscible aqueous phases should be highly hydrophobic. Accordingly, lipids added to a water-in-water emulsion are concentrated at the interface between the two aqueous phases. Thin lipid layers between aqueous phases could form a continuous three-dimensional honeycomblike structure. Thus, the highly hydrophobic interface of a water-in-water emulsion may provide the physical basis for formation of a continuous lipid phase. This honeycomblike lipid structure filled with the immiscible aqueous phases can be fixed by solidification of all three phases of the system by cooling. Both aqueous phases form thermoreversible gels with melting temperatures close to mouth temperature, and both phases can form in the mouth highly viscous melted droplets covered by thin lipid layers. Sensorial texture perception of such a system (thermoreversible gel granules covered by lipid layers) can imitate droplets of mayonnaise. This mechanism of structure formation may also be of importance for many other foods, such as ice cream mixtures and different fat replacers.

The honeycomblike structure of butter replacers can also be arranged using small gel granules. Gelatin gel granules can be used for microscopic modeling of honeycomb-like structure formation. Gelatin gels are not wetted with water but are perfectly wetted with oil. This phenomenon was discovered during the development of the method for producing a caviar analog [43]. Because of this phenomenon oil cannot be separated from the surface of gelatin gel granules formed in air and oil media by washing with water. Gelatin granules covered by thin layers of oil form large aggregates in water. The aggregation of spherical gel granules leads to a honeycomblike structure with a continuous lipid phase. Thus, the factors affecting lipid functionality in protein-containing systems can be related to phase separation and the properties of hydrophobic surfaces of aqueous gel particles [23,28,42].

4. Formation of Food Structure

Figure 9 shows structure formation processes occurring during food processing. This scheme clarifies the spinneretless spinning phenomenon, which is responsible for the formation of granular, fibrous, or lamellar structures of many foods. Liquid-dispersed particles of water-in-water emulsions are deformed by shear forces. Low interfacial tension typical of water-in-water emulsions is an important factor contributing to the easy deformability of liquid-dispersed particles. The threadlike liquid particles formed are unstable. They break down into smaller droplets, which can be deformed and broken down again. Spherical and deformed droplets may also coalesce to form larger droplets and longer liquid filaments. In a flowing emulsion the dynamic equilibrium establishes itself between droplet deformation, break down, and coalescence. The structure of a system, i.e., the shape of the liquid particles, is fixed by gelation of one or both liquid

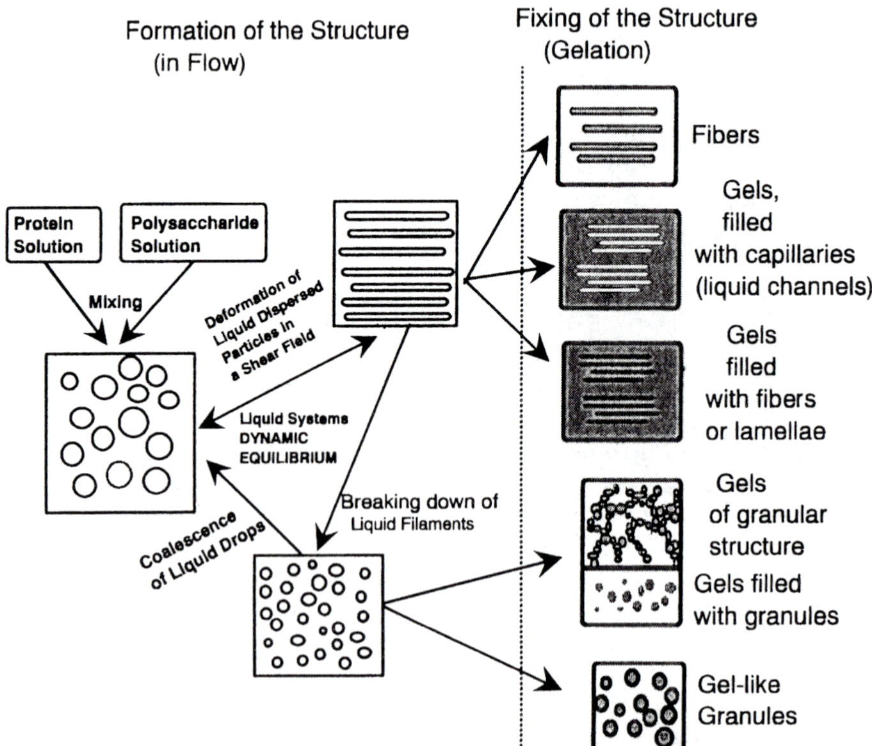

FIGURE 9 General scheme of structure formation in food processing. (From Refs. 12,46,47.)

phases of a system in flow. Controlling the processes of deformation, breakdown, and gelation of dispersed particles makes possible the production of foods with various structures. Figure 9 shows that the processing of heterophase liquid systems provides different types of food structures from fat replacers to textured proteins produced by thermoplastic extrusion [12,22,46–48]. Generally, when two biopolymers are incompatible in dilute solution in a mutual solvent they are also incompatible and exhibit phase separation in the bulk of their mixed melts. Phase separation or mixtures of melted water-plasticized proteins and polysaccharides as well as deformation and orientation of liquid-dispersed particles in flow are the two main elements of the mechanism of structure formation during thermoplastic extrusion [12,42,46,47]. It should be stressed that the general nature of the phenomena of thermodynamic incompatibility of biopolymers and easy deformability of dispersed particles of two-phase liquid systems accounts for their importance in the formation of food structures [12,24,43,47]. Phase-separated food systems can be organized into a large variety of morphologies. Protein or polysaccharide fibers, anisotropic gels filled with oriented liquid or gel-like filaments, isotropic gels filled with granules or composed by granules, as well as small spherical gel-like particles (for fat replacers) and large spherical or deformed granules (for granular caviar or rice analogs) may be produced [42,43,48].

IV. FUTURE DIRECTIONS

We are still near the starting point in this field. Quite a long distance remains before our knowledge of protein interactions in multicomponent water media becomes sufficient to permit control and improvement of the functional properties of proteins in food systems. The conditions of immiscibility and phase equilibrium have been studied in a quite limited number of biopolymer mixtures: (a) gelatin and polysaccharide, (b) globular protein and polysaccharide, (c) protein and another protein, (d) native and thermally denatured proteins, or (e) a native globular protein and its partial hydrolysate.

The phase behaviors of many important model systems remain unstudied. This particularly relates to systems containing (a) more than two biopolymers, such as a protein and a mixture of polysaccharides or a mixture of proteins and a polysaccharide or mixtures of proteins and polysaccharides, (b) mixtures containing denatured proteins, (c) partially hydrolyzed proteins, (d) protein-polysaccharide complexes and conjugates, or (e) enzymes (proteolytic and amylolytic) and their partition coefficients between the phases.

The influences of other food components (especially of low molecular weight, such as lipids, sugars, and polyvalent metal ions) and operating conditions, such as high shear forces, high pressure, and high temperature (e.g., pasteurization and sterilization), on the phase behavior of protein-polysaccharide mixtures remain unstudied.

Further basic studies on the nature of interactions among biopolymers and on the phase behavior of their mixtures will undoubtedly provide a sound scientific basis for improving conventional food technologies and designing new food formulations and manufacturing processes.

It should also be noted that this chapter has been limited to physicochemical interactions between food macromolecules. The main reason for this is that chemical protein-polysaccharide interactions in food systems have not yet been sufficiently studied [17–19]. A new trend in this area is covalent protein-polysaccharide hybrids, which are of great interest as functional additives and food ingredients [49,50].

REFERENCES

1. M. W. Beijerinck, Ueber eine Eigentümlichkeit der löslichen Stärke, *Zentralbl. Bakteriol. Parasitenkd. Infektionskr.* 2:697 (1896).
2. M. W. Beijerinck, Ueber Emulsionsbildung bei der Vermischung wasseriger Lösungen gewisser gelatinierender Kolloide, *Z. Chem. Ind. Kolloide* (*Kolloid Zeitschrift*) 7:16 (1910).
3. F. W. Tiebackx, Gleichzeitige Ausflockung zweier Kolloide, *Z. Chem. Ind. Kolloide* (*Kolloid Zeitschrift*) *8*:198 (1911).
4. W. Ostwald and R. H. Hertel, Kolloidchemische Reaktionen zwischen Solen von Eiweisskörpern und polymeren Kohlehydraten, *Koll. Z. 47*:258, 357 (1929).
5. H. G. Bungenberg de Jong, Crystallisation-coacervation-flocculation, *Colloid Science* (H. R. Kruyt, ed.), Elsevier, Amsterdam, 1949, p. 232.
6. A. Dobry and F. Boyer-Kawenoki, Phase separation in polymer solution, *J. Polym. Sci.* 2:90 (1947).
7. A. Dobry and F. Boyer-Kawenoki, Sur l'incompatibilité des macromolécules en solution aqueuse, *Bull. Soc. Chim. Belg. 57*:280 (1948).
8. P.-A. Albertsson, *Partition of Cell Particles and Macromolecules*, Wiley Interscience, New York, 1972.
9. H. Walter, G. Johansson, and D. E. Brooks, Partitioning in aqueous two-phase: recent results, *Anal. Biochem. 197*:1 (1991).
10. P. J. Flory, *Principles of Polymer Chemistry*, Cornell University Press, Ithaca, NY, 1953.
11. D. R. Paul and S. Newman, eds. *Polymer Blends*, Vols. 1 and 2, Academic, New York, 1978.
12. V. Tolstoguzov, Functional properties of protein-polysaccharide mixtures, *Functional Properties of Food Macromolecules* (J. R. Mitchell and D. A. Ledward, eds.), Elsevier Applied Science Publishers, London, 1986, p. 385.
13. V. B. Tolstoguzov, E. E. Braudo, and E. S. Vajnerman, Physikalisch-chemische Aspekte der Herstellung kunstlicher Nahrungsmittel, *Nahrung 19*:973 (1975).
14. V. Tolstoguzov, V. Grinberg, and A. Gulov, Some physicochemical approaches to the problem of protein texturisation, *J. Agric. Food Chem. 33*:151 (1985).
15. J. E. Hidalgo and P. M. T. Hansen, Selective precipitation of whey proteins with carboxymethyl cellulose, *J. Dairy Sci. 54*:1270 (1971).
16. A. P. Imeson, D. A. Ledward, and J. R. Mitchell, On the nature of the interactions between some anionic polysaccharides and proteins, *J. Sci. Food Agric. 28*:661 (1977).
17. G. Stainsby, Proteinaceous gelling systems and their complexes with polysaccharides, *Food Chem. 6*:3 (1980).
18. D. A. Ledward, Creating textures from mixed biopolymer systems, *Trends Food Sci. Technol. 4*:402 (1993).
19. D. A. Ledward, Protein-polysaccharide interactions, *Protein Functionality in Food Systems* (N. Hettiarachchy and G. Ziegler, eds.), Marcel Dekker, New York, 1994, p. 225.
20. S. K. Samant, R. S. Singhal, P. R. Kulkarni, and D. V. Rege, Protein-polysaccharide interactions: a new approach in food formulations, *Int. J. Food Sci. Technol. 28*:547 (1993).
21. V. B. Tolstoguzov, E. E. Braudo, and A. N. Gurov, Functional properties of food protein and their control, *Nahrung 25*:231, 817 (1981).
22. V. Tolstoguzov, Development of texture in meat products through thermodynamic incompatibility, *Developments in Meat Science—5* (R. A. Lawrie, ed.), Elsevier Applied Science, London, 1991, p. 159.
23. V. Tolstoguzov, Interactions of gelatin with polysaccharides, *Gums and Stabilisers for the Food Industry* (G. O. Phillips, P. A. Williams, and D. J. Wedlock, eds.), IRL Press, Oxford, 1990, p. 157.
24. V. Tolstoguzov, Functional properties of food proteins and role of protein-polysaccharide interaction, *Food Hydrocoll. 4*:429 (1991).

25. V. Tolstoguzov, Thermodynamic aspects of food protein functionality, *Food Hydrocolloids: Structures, Properties and Functions* (K. Nishinari and E. Doi, eds.), Plenum Press, New York, 1994, p. 327.
26. V. Tolstoguzov, The functional properties of food proteins, *Gums and Stabilisers for the Food Industry*, Vol. 6 (G. O. Phillips, P. A. Williams, and D. J. Wedlock, eds.), IRL Press, Oxford, 1992, p. 241.
27. V. Tolstoguzov, Functional properties of food proteins. Role of interactions in protein systems, *Food Proteins, Structure and Functionality* (K. D. Schwenke and R. Mothes, eds.), VCH, Weinheim, 1993, p. 203.
28. V. Tolstoguzov, Some physico-chemical aspects of protein processing into foods, *Gums and Stabilisers for the Food Industry*, Vol. 7 (G. O. Phillips, P. A. Williams, and D. J. Wedlock, eds.), IRL Press, Oxford, 1994, p. 115.
29. A. Y. Sherys, A. N. Gurov, and V. B. Tolstoguzov, Water-insoluble triple complexes: bovine serum albumin-bivalent metal cation-alginate, *Carbohydr. Polym. 10*:87 (1989).
30. A. N. Gurov, E. S. Vajnerman, and V. B. Tolstoguzov, Interaction of proteins with dextransulfate in aqueous medium. 2. Nonequilibrium phenomena, *Staerke 29*:186 (1979).
31. Z. A. Streltsova and V. B. Tolstoguzov, Eigenschaften von Eiweisstoffen in Komplexen mit sauren Polysacchariden und anderen Polyelektrolyten, *Coll. Polym. Sci. 255*:1054 (1977).
32. S. Poole, S. I. West, and C. L. Walters, Protein-protein interactions: their importance in the foaming of heterogeneous protein systems, *J. Sci. Food Agric. 35*:701 (1984).
33. D. C. Clark, A. R. Mackie, L. J. Smith, and D. R. Wilson, The interaction of bovine serum albumin and lysozyme and its effect on foam composition, *Food Hydrocoll. 2*:209 (1984).
34. S. Damodaran and J. E. Kinsella, Role of electrostatic forces in the interaction of soy proteins with lysozyme, *Cereal Chem. 63*:381 (1986).
35. E. S. Tokaev, A. N. Gurov, I. A. Rogov, and V. B. Tolstoguzov, Properties of oil/water emulsions stabilized by casein-acid polysaccharide mixtures, *Nahrung 31*:825 (1987).
36. T. V. Burova, N. V. Grinberg, V. Y. Grinberg, A. L. Leontiev, and V. B. Tolstoguzov, Effects of polysaccharides upon the functional properties of 11S globulin from broad beans, *Carbohydr. Polym. 18*:101 (1992).
37. Y. A. Antonov, V. Y. Grinberg, N. A. Zhuravskaya, and V. B. Tolstoguzov, Concentration of protein skimmed milk by the method of membraneless isobaric osmosis, *Carbohydr. Polym. 2*:81 (1982).
38. A. S. Medin and J. C. Janson, Studies on aqueous polymer two-phase systems containing agarose, *Carbohydr. Polym. 22*:127 (1993).
39. C. M. Durrani, D. A. Prystupa, A. M. Donald, and A. H. Clark, Phase diagram of mixtures of polymers in aqueous solution using Fourier transform infrared stectroscopy, *Macromolecules 26*:981 (1993).
40. V. I. Polyakov, V. Y. Grinberg, and V. B. Tolstoguzov, Application of phase-volume-ratio method for determining the phase diagram of water-casein-soybean globulins systems, *Polym. Bull. 2*:757 (1980).
41. V. Tolstoguzov, Concentration and purification of proteins by means of two-phase systems. Membraneless osmosis process. *Food Hydrocoll. 2*:195 (1988).
42. V. Tolstoguzov, Thermodynamic incompatibility of food macromolecules, *Food Colloids and Polymers: Structure and Dynamics* (P. Walstra and E. Dickinson, eds.), Royal Society of Chemistry, Cambridge, 1993, p. 94.
43. V. Tolstoguzov, Some physico-chemical aspects of protein processing in foods. Multicomponent gels, *Food Hydrocoll. 8*:317 (1995).
44. V. Tolstoguzov and E. Braudo, Fabricated foodstuffs as multicomponent gels, *J. Texture Stud. 14*:183 (1983).
45. E. E. Braudo, A. M. Gotlieb, I. G. Plashina, and V. B. Tolstoguzov, Protein-containing multicomponent gels, *Nahrung 30*:355 (1986).

46. V. Tolstoguzov, Creation of fibrous structures by spinneretless spinning, *Food Structure—Its Creation and Evaluation* (J. M. V. Blanshard and J. R. Mitchell, eds.) Butterworths, London, 1988, p. 181.
47. V. Tolstoguzov, Thermoplastic extrusion—the mechanism of the formation of extrudate structure and properties, *J. Am. Oil Chem. Soc. 70*:417 (1993).
48. V. Tolstoguzov, Some physico-chemical aspects of protein processing into foodstuffs, *Food Hydrocoll. 2*:339 (1988).
49. A. Kato, T. Sato, and K. Kobayashi, Emulsifying properties of protein-polysaccharide complexes and hybrids, *Agric. Biol. Chem. 53*:2147 (1989).
50. E. Dickinson and V. B. Galazka, Emulsion stabilization by ionic and covalent complexes of beta-lactoglobulin with polysaccharides, *Food Hydrocoll. 5*:281 (1991).

7

Structure-Function Relationships of Caseins

Douglas G. Dalgleish
University of Guelph
Guelph, Ontario, Canada

I. INTRODUCTION

Protein structures depend on their molecular composition, whereas functional properties are generally defined on the basis of the bulk behavior of solutions of the protein. To relate these is not simple, because the gap between molecular structure, on the one hand, and, for example, the rheological properties, on the other, is quite large. Nevertheless, it is possible to at least qualitatively explain some of the properties of caseins with reference to their molecular composition and properties. A second problem arises from the assumption that observations of individual proteins in the laboratory are translatable into industrial practice. This is by no means certain, because laboratory experiments are performed on proteins which are generally prepared in high purity, in small batches, by nonstandard methods, and these preparations may not be directly comparable to those that are used in a large scale in the food industry, where large batches are prepared, and there may be significant batch-to-batch variation in the behavior of the protein preparation. Thus, a freeze-dried preparation of laboratory caseinate prepared from milk fresh from the cow may have properties quite distinct from those of sodium caseinate prepared industrially; analogously, sodium caseinate preparations from different sources may differ in composition and properties. Indirect or anecdotal evidence suggests that this is an effect connected with changes in the quaternary structure of the casein, but detailed experimental evidence is lacking.

Industrially, caseins come in a variety of forms, from insoluble materials such as rennet caseins, through particulate materials such as calcium caseinates, to almost completely soluble sodium and potassium caseinates. Although each of these has found a place in the preparation of foods, they have different functional properties, which nevertheless can be at least qualitatively explained by the properties of the individual casein molecules.

This chapter deals with the molecular behavior of the caseins and the general functional properties arising from it. It is not our intention to detail all of the uses that caseins find in the formulation of foods, or even nonfood products, although some cases may be considered in detail. Rather, it is the intention to show how the behavior of caseins in foods relates to their molecular properties, with some idea of how the properties of the proteins may be modified during production. Moreover, if we are to attempt molecular explanations of the properties, it becomes very difficult to discuss semi-solid or solid materials such as processed cheese and cheese analogs. Most of this chapter is concerned with the properties of caseins in solutions, emulsions, or simple gels.

Caseins, as far as is known, have in nature the simple function of providing nourishment to neonatal mammals. As such, they have no enzymatic activity that can be related to the structures of the molecules. Very simply, they have evolved to aggregate, along with inorganic materials such as calcium phosphate, into particles (casein micelles), which remain stable in the maternal milk until consumed by the young animal, after which they clot in the stomach as the result of proteolytic action or acidification, or both. Fragments of the caseins after digestion may possess, perhaps fortuitously, biological activity, either in calcium transport or as opioid agonists or antagonists. However, outside its transient existence in milk as a source of nutrient, casein appears to have no natural function.

It is arguable whether caseins possess defined fixed three-dimensional structures. Caseins are proteins, and the four different types of casein found in bovine milk (α_{s1}, α_{s2}, β, and κ) have molecular masses in the region of 20 kDa [1]. Each of the four proteins has a distinct primary structure, although there are similarities between them; however, there is debate about the extent to which these primary structures translate into rigidly defined secondary and tertiary structures. It is possible that the molecules are very flexible and do not adopt any particular conformation in solution. Certainly, it has not been possible to crystallize caseins, so no experimental three-dimensional structures are available. Equally certainly, it is established that caseins aggregate strongly to give quaternary structures of some size [2], but once again the mechanisms of the aggregation reactions and the precise nature of the products are by no means universally agreed upon.

From this, we can foresee that the functional behavior of caseins in foods will relate less to phenomena associated with tertiary or secondary structure than to properties where the unique primary structures and flexible nature of the proteins are important. These properties mainly relate to the formation of aggregates, such as in the formation of curds during the manufacture of cheese, and also the abilities of the proteins to adsorb strongly to hydrophobic interfaces and to play a large part in stabilizing emulsions and similar preparations. The physical chemistry of these interactions has been discussed in detail in a recent publication [3]. A discussion of these applications will follow, but it is first of all necessary to understand the structures of the proteins as far as they are known.

II. STRUCTURES OF INDIVIDUAL CASEINS

Bovine milk contains four types of caseins; typically, the concentration in milk is about 25 g/liter, and the four caseins are present in the approximate ratio $\alpha_{s1}:\alpha_{s2}:\beta:\kappa$ = 4:1:4:1 [4]. As secreted, they exist in the form of casein micelles, which are particles whose diameters are in the range of 50–250 nm, and consist of caseins complexed with calcium phosphate [5]. Acidification of the micellar complexes to a pH of 4.6 causes the calcium phosphate to dissolve and the proteins to precipitate; the caseins are the only proteins in milk normally precipitated at this pH, so that they can be collected and redissolved by neutralization to give products such as sodium or potassium caseinates [6]. Although all of the different caseins are present in the redissolved caseinate, there are unfortunately no easy ways to separate them, because of their strong tendency to self-associate. Laboratory methods exist, where the tendency to aggregate is offset by incorporation of a dissociating agent, but these do not give products that can be used in the food industry because of the nature of the dissociating agents, such as high concentrations of urea, which have to be used. Therefore, until recently, only whole caseinate was available to the food industry. Recently, food-grade methods have been developed for the isolation of β-casein from caseinate, but so far the protein has not found a specific use.

Therefore, it is safe to say that all casein used in food is a mixture of the four proteins; basically only its quaternary structure is different in different preparations. As an extreme simplification, we may say that the choice for the food processor is between a fully soluble casein (as in neutralized sodium caseinate) or a particulate form (casein micelles, calcium caseinate), with a wide variation of possibilities contained within these two extremes.

The uniqueness of the caseins as food ingredients arises from their compositions (full sequence information is available and has been frequently published: the reader is referred to Ref. 1). First, they are phosphoproteins, in which some of the seryl residues are substituted with a phosphate group, although the numbers differ from one casein to another (Table 1). In the α_s and β caseins, the phosphate residues are not spread randomly along the structure but tend to be grouped in particular regions containing "phosphate centers" containing the sequence -SerP-SerP-SerP-X-SerP-. Not only do these centers possess a considerable charge at pH values around neutral (because of the double negative charge on the phosphate groups), but they also bind calcium ions strongly (as does inorganic phosphate). The binding of Ca^{2+} to the caseins depends strongly on these residues; when the charge is titrated by acidification or the protein is dephosphorylated, the tendency to bind Ca^{2+} decreases. The exception to this behavior is κ-casein, which has only one (rarely two) phosphoseryl residues and binds little Ca^{2+}, even at neutral pH.

The major caseins (the α_{s1} and β-caseins) contain neither cysteine nor cystine; as a consequence they lack the capacity to form inter- or intramolecular disulfide bonds.

Table 1 Some Characteristics of Caseins

Protein	% in total casein	Mol. Wt. (daltons)	SerP per mole	Glyco?	Cysteine per mole	Charge at pH 6.6
α_{s1}-Casein	40–45	23551[a]	8–9	No	0	−20.90
α_{s2}-Casein	10	25238[b]	10–13	No	2	−13.80
β-Casein	35–40	24028[c]	5	No	0	−12.80
κ-Casein	9–15	19038[d]	1–2	Yes	2	−3.00

[a]For α_{s1}-casein C with 8 serine phosphates.
[b]For α_{s2}-casein A with 11 serine phosphates.
[c]For β-casein A^1.
[d]For κ-casein B unglycosylated.

However, the minor caseins (α_{s2} and κ) each contain two cysteinyl residues per molecule, and these are known to give rise to intermolecular disulfide bonds. There is little evidence for the formation of mixed complexes of κ- and α_{s2}-caseins, especially in unheated milk; only homodimers or homopolymers seem to be formed. Aggregation of the α_{s2}-casein in this way is limited to the formation of dimers, but the κ-casein can form extended polymers of perhaps 30 units, linked by intermolecular disulfide bonds [7]. It seems likely, although it has not been conclusively proved, that κ-casein occurs naturally in a polymeric form even within the casein micelle.

The third important factor shown from the primary structures of the caseins is that they are all hydrophobic proteins, that is, they contain many amino acids with nonpolar side chains. Moreover, the distribution of these, especially in β and κ-caseins, is by no means random along the peptide sequence. In β-casein most of the charge of the molecule, and the polar groups is contained in the first 40 amino acids from the N-terminal; the remainder of the molecule is strongly hydrophobic (Fig. 1). This makes β-casein an amphipathic protein, having hydrophilic and hydrophobic ends, at least as shown from the linear sequence of amino acids, although how this translates into tertiary structure remains to be seen. The same is true for κ-casein, although in this case it is the N-terminal region of the protein (residues 1–105) that is mainly hydrophobic and the C-terminal region (residues 106–169) that is hydrophilic. This hydrophilic tendency is enhanced by the presence of four glycosylation sites, which may be substituted by short oligosaccharides. About half of the molecules are substituted in this way, so glycosylation

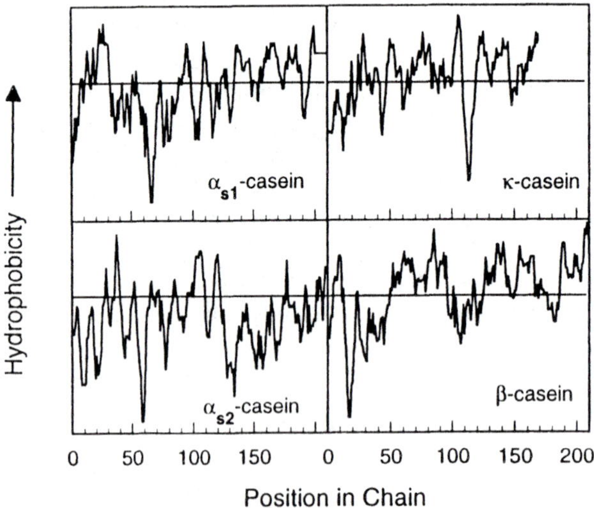

FIGURE 1 Hydrophobicities of the caseins along their peptide chains. The values are calculated by running averages of five residues along the chains. In each protein, the region above the horizontal line is the most hydrophobic portion of the protein. Thus, most of α_{a2}-casein is hydrophilic, while most of κ-casein is hydrophobic, with the exception of the region surrounding the rennet-sensitive bond 105–106. β-Casein also shows much hydrophobic character, with only the N-terminal (where the phosphoseryl residues are located) being hydrophilic.

is not necessary for the "function" of the κ-casein. The distribution of hydrophilic and hydrophobic residues in the α_s-caseins is more uniform than in the other caseins, but they still are capable of showing hydrophobic patches and behavior associated with them (Fig. 1).

A final point of importance in considering the primary structures of the proteins is that they (especially β-casein) contain relatively large amounts of proline; this residue is well established to be difficult to incorporate into the accepted components of the secondary structure of the protein (α-helix and parallel and antiparallel β-pleated sheet). As a result, the caseins in general are expected to possess little secondary structure, and by and large this has been found to be the case when the proteins were examined using circular dichroism or FTIR spectroscopy [8,9]. From these observations has arisen the notion that the caseins are essentially disordered proteins, without specific structure—even that they are flexible and have virtually no fixed structure at all. This implies that in solution a number of configurational states could be accessible to the molecules and that the application of even small forces could cause the molecules to adopt conformations best adapted to the circumstances, e.g., as they adsorb to an interface or form aggregates.

This interpretation has been disputed, inasmuch as calculations have been made using structure-predicting software to produce hypothetical structures for the individual casein molecules, on the assumption that only one conformational state exists for each type of casein. This has been elaborated in a series of papers [10,11], but independent experimental confirmation of the validity, and especially the uniqueness, of these calculated structures is so far lacking. Because the caseins lack biological activity, and because they cannot be crystallized, it is difficult to make an independent estimate of the conformations of the proteins, so that there are few tests that can be applied to the calculated structures to show how they compare to the actual structures of the proteins. One of the arguments against defined structures for the caseins is that they show no heat of denaturation in differential scanning calorimetry (DSC) experiments; there seems to be no energy difference between "native" and "denatured" states [12]. Thus it is entirely possible that many states exist with very closely separated energy levels and that the caseins simply move between these states under normal thermal energies. Such arguments would of course apply less to caseins that were fixed in a specific structure, such as in the casein micelle, or on an oil/water interface, where conformational constraints will apply to a greater extent. However, the conformations of the caseins in these states are completely unknown.

III. AGGREGATION OF THE CASEINS AND QUATERNARY STRUCTURE

Because of the hydrophobic regions that form part of their structure, the caseins will associate with other hydrophobic material, i.e., they will interact with oil/water interfaces, but they will also aggregate with themselves. Solutions containing purified β-casein aggregate to form particles containing each about 30 individual protein molecules, although the number appears to depend on the conditions [2,13]. This seems to be a kind of micellization reaction as is common in surfactant chemistry, where an equilibrium exists between monomeric and polymeric material, and there is a critical micelle concentration below which the protein exists as monomers; this is just what is found. That the interaction in this case is likely to be hydrophobically driven is shown by (a) lowering temperature causes the aggregates to dissociate to monomer [14] and (b) enzymatic removal of the hydrophobic C-terminal of the protein prevents the reaction [15].

Of the other caseins, κ-casein also aggregates extensively, even when its cysteinyl residues are chemically blocked. Thus, although aggregation of this casein can be enhanced by the formation of intermolecular disulfide bonds, these are not necessary for the aggregation to occur. The α_s-caseins in general self-aggregate less, tending to form aggregates with only a few protein molecules [2].

Because the caseins are relatively flexible and form extended chains in solution, they demonstrate an ability to hold considerable amounts of water. Solutions of caseins generally exhibit high viscosity (compared for example with similar solutions of whey proteins) because the effective volume occupied by the molecules is higher than that of an equivalently sized globular protein. However, because caseins do not form gels, efficient utilization of this effect is difficult to achieve.

At neutral pH, the aggregates of caseins are self-limiting, i.e., the caseins do not aggregate sufficiently to precipitate. However, this is changed if the protein is treated with calcium ions. In the presence of sufficient Ca^{2+}, isolated α_s- and β-caseins precipitate. It is possible to explain this reaction as arising simply from a neutralization of the negative charge on the protein as Ca^{2+} binds. When the charge has been sufficiently decreased, there is insufficient interprotein repulsion to keep the molecules apart, and they simply precipitate [16]. This interpretation appears valid for the individual caseins at least, although it needs to be modified when all of the caseins are present and is not relevant to the behavior of κ-casein, which does not precipitate in the presence of calcium.

So far, this discussion has dealt with individual caseins, but it seems clear that the situation is more complex when the four caseins are present together. In the absence of calcium, caseinate can form particles weighing about 600,000 daltons, apparently involving all of the different caseins. It has not been established whether these particles are homogeneous or heterogeneous, but there is some evidence that dissociation of native casein micelles gives rise to particles having at least two distinct compositions [17]. The relationship between the particles produced by aggregation of caseins or dissociation of micelles is not, however, established. Further aggregation of the casein oligomers can be caused by the addition of Ca^{2+}. In this case, however, the casein does not precipitate, but forms stable particles of calcium caseinate (unless large excesses of Ca^{2+} are used). These are often misnamed casein micelles but they differ from these particles; as will be seen, native casein micelles contain inorganic phosphate as well as casein and calcium, and their properties depend on it [18]. The casein-Ca^{2+} complexes can, however, be used in circumstances where there may be a reason to exclude phosphate.

The reason that the casein-Ca^{2+} complexes are limited in size is that the κ-casein acts as a colloidal stabilizer by positioning itself on the surface of the particle, and when the growing particles are completely covered by the κ-casein, growth stops. It follows from this that the final size of the aggregates depends on the relative amounts of κ-casein and the other caseins. It also implies that the κ-casein is free to move so as to take up position on the surface of the particle. The stabilization of these particles may be assumed to be similar to that of the native casein micelles, where the κ-casein on the surface forms a sterically stabilizing "hairy layer," which protrudes into solution by several nanometers [19].

These complexes share some, but by no means all, of their properties with casein micelles. Although the size distributions of the two materials are similar, they show different stabilities to heat and to the addition of ethanol [18]. Evidently, the native micelle possesses some structure not found in the simpler calcium caseinate aggregates.

IV. CASEIN MICELLES

The definition of the structure of the casein micelle has been, and remains, a somewhat contentious subject. Various models have been put forward, none of which is completely satisfactory, although to some extent this need not hinder the understanding of their functional properties. The micelles contain virtually all of the casein in freshly secreted milk, but subsequent storage of the milk at low temperature (4°C) can cause the micelles to partially dissociate by releasing a proportion of their β-casein into the serum, suggesting that this material is held inside the micelle by hydrophobic interactions. Although this phenomena can be reversed by rewarming the milk, it is not certain whether the rewarmed micelle has an altered structure. Combined with the caseins in the micelle is inorganic calcium phosphate, and it is this material that gives casein micelles their characteristic behavior [5]. Casein micelles have not generally been available industrially, since their efficient isolation had to await modern techniques of membrane filtration, to separate them from the whey proteins. These modern techniques do indeed produce a form of "phosphocaseinate," which is related to the original micelles and shares many of their properties. Because it is manufactured using a partial diafiltration of milk, the material tends to be lower in calcium phosphate than are native casein micelles, although much higher than in calcium caseinates. This material has not yet found wide use, and industrially, casein micelles are used in the form of skim milk powders, or of milks, rather than being used as isolated materials.

What is agreed generally about the structures of casein micelles is that they appear to be generally spherical under the electron microscope, but also appear to be composed of smaller domains, so that the overall structure resembles a raspberry [20]. What these domains are, however, is not established. One school of thought maintains that they are true "submicelles," i.e., that casein micelles are composed of similar or identical subunits, each of which in turn consists of several (about 30) casein molecules [21]. These submicelles would then be linked together by the micellar calcium phosphate [22,23]. The submicelles themselves are believed to be held together by hydrophobic interactions between the proteins (Fig. 2A). This type of model is based on the well-known tendency of the caseins to self-associate, so that the aggregates found in solution are thought to simply be aggregate by the incorporation of calcium phosphate. A second model alternatively allows for a more detailed contact between calcium phosphate and casein, in that it is supposed that the serine phosphate groups of the caseins actually form part of the calcium phosphate [5]; on this basis the "submicelles" may either be thought of as being nodal points within an extended gel structure or as calcium phosphate surrounded by caseins; in the latter case, the submicelles will be held together by hydrophobic interactions to form the micelle (Fig. 2B).

Neither of these two models is completely satisfactory in describing the behavior of micelles. For example, the original submicellar model does not account for the integrity of the overall micellar structure when cooled or when the pH is reduced, when it is known that some of the caseins can dissociate from the structure. Nor is it possible to

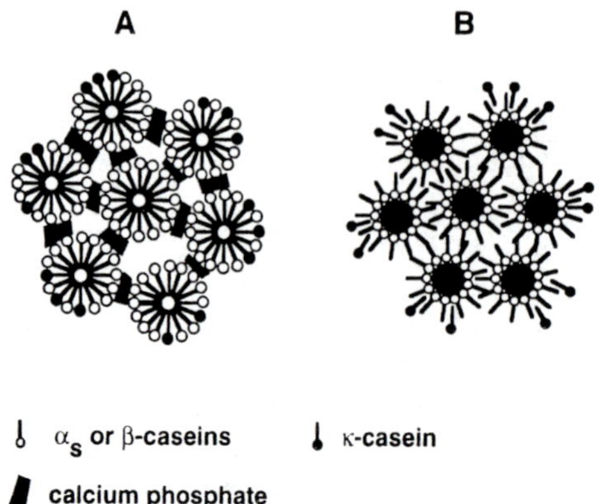

Figure 2 Highly schematic diagrams of the possible structures of casein micelles: (A) based on the models of Schmidt [23] and Walstra [22]; (B) based on the model of Holt [5]. In the first model, caseins are shown hydrophobically interacting to make "submicelles," which then interact with calcium phosphate to form micelles. In the second, the caseins bind to the calcium phosphate via their phosphoserine residues to form the submicelles, which then interact hydrophobically. κ-Casein molecules are also shown interacting hydrophobically with the submicelles. The models shown are very freely adapted from the sources quoted.

explain the behavior of the micelle by supposing it is made up of identical subunits, since in that case much of the κ-casein would have to be buried within the structure and would not be available to stabilize the micellar surface. On the other hand, it is difficult to see how the second model can be assembled from preaggregated caseins. Simply, it is a question of balance between different types of aggregation. If assembly of micelles occurs in the secretory vesicle by casein being present originally and calcium phosphate being formed later, there is a strong case to be made for the linking together of these aggregates by the calcium phosphate as it forms. However, it is also possible that the presence of calcium phosphate could induce conformational changes, and even differences in the aggregation mechanism, of the original casein aggregates. This could in fact lead to a different kind of "submicelle" being formed as aggregation to the micellar state occurs. This process might cause the breakage and reformation of some of the hydrophobic interactions, but since these interactions are not specific, being basically entropic in nature, it is entirely possible that this could occur.

It might be expected that dissociation of the micelles by removal of calcium phosphate would allow the resolution of some of these structural problems; however, it is established that the manner of removal of the calcium phosphate causes different products to be formed. Thus, removal of calcium phosphate by treatment with EDTA [24] or dialysis against buffers [25], or by acidification [26], causes release of caseins in different ways; it is not easy to determine where, if anywhere, true submicelles can be isolated.

There is more general agreement, that, whatever the interior structure of the micelle, the properties of the micellar surface can be described in more detail. Most, if not all, of the κ-casein in the micelles is to be found on the surfaces of the particles, where it acts as a stabilizer [27]. Not all of the surface is κ-casein; the other caseins, mainly the

α_s fraction, are also present, but the β-casein is mainly buried within the interior of the particle. The stabilizing action of the κ-casein arises mainly from the C-terminal third of the protein (the glycomacropeptide), which protrudes from the micelle surface into solution to provide steric stabilization, although it is also charged. It has been established in several experiments that this extended "hairy layer" exists, and that it is apparently 5 nm [28], or even as much as 10 nm [29], thick. When the micelles are treated with chymosin, the κ-casein is split between residues 105 and 106, the stabilizing macropeptide is removed from the micellar surface, and the particles aggregate because their stabilizing surface is lost [30]. Because the glycomacropeptide is charged, its removal also decreases the apparent surface charge (the ζ-potential) of the micelles, although this charge is not in itself sufficient to explain the micellar stability. Because of the presence of this extended layer of macropeptide, casein micelles are highly hydrated, possessing as much as 3–4 g of water per g of protein [31]. This is partly the result of the overall porous nature of the micellar structure (as we have seen, it can be regarded as a microgel) and partly because the macropeptide layer occupies a great deal of volume for a small mass of protein.

This renneting of κ-casein causes one of the most profound changes that can be induced into the casein system. Because none of the other caseins are stable in the presence of calcium or calcium phosphate and the stabilizing ability of the κ-casein has been destroyed, the renneted micelles coagulate to form a curd. This is considered in detail in a later section. In many cheeses, the renneting is accompanied by the action of acid-forming microorganisms, so that the pH of the milk is being decreased during renneting and curd-forming. This has an effect also on the structures of the casein micelles. As pH decreases, the micellar calcium phosphate is progressively dissolved, and since this is the "cement" that holds the micelles together, changes must occur in the micellar structure. However, these are not fully understood. At temperatures below about 30°C, dissociation of some proteins, particularly β-casein, from the acidified micelles occurs [26]. There seems to be no tendency, however, for the micelles to disperse completely into their submicellar units during this process; even though some of the proteins are dissociated, some framework of the particles remains. This is very different from the behavior found when the micellar calcium phosphate is dissociated by chemical means (by chelating the calcium with EDTA) at neutral pH; in this case the micelles are much more completely dissociated [24]. Presumably the two processes differ in the charges found on the caseins; dissociation of the calcium phosphate at neutral pH leaves the caseins in a highly charged state, with doubly negatively charged phosphoserine residues. However, dissociation of the calcium phosphate by pH occurs as the charges on the caseins are being titrated, so that there will be smaller repulsive forces between the protein molecules and the hydrophobic interactions will be sufficient to hold them together.

The structures of the casein micelles are also altered by the application of heat [32]. In particular, heating of milk or milk-based preparations causes denaturation of the milk serum proteins and the binding of them to the κ-casein of the micelles, which is on or near to the surface of the particles [33]. This changes the structure of the heated micelle in two ways. First, under suitable conditions of pH, the κ-casein–serum protein complex can dissociate from the micelle to give a change in composition of the particle [34]. Second, there must be a change in the surface structure of the micelle, because of the additional proteins present. Exactly what this change comprises is not certain and is likely to depend on the extent of heating and the method used. However, it should be noted that the interaction of the whey proteins with the κ-casein occurs via the formation of disulfide bonds, and these must involve the penetration of the whey proteins through the hairy macropeptide layer, because the cysteinyl residues of the κ-casein are found on the buried *para*-κ-casein moiety of the molecule. Thus, we may expect the added whey proteins to bind within the original surface of the micelle.

The most profound effects of this are that the casein micelles become much less susceptible to coagulation by rennet; it appears that this is less an effect upon the action of chymosin itself than on the subsequent coagulation of the micelles to form a curd [33,35]. Therefore, it may be that the presence of whey proteins near the surface creates a repulsive force, which prevents the aggregation reaction from proceeding. An alternative suggestion is that as the micelle is heated, the mineral composition changes as a result of the deposition of calcium phosphate from the milk serum, and it is this that prevents the micelles from aggregating, either because of its presence on the micellar surface or because the concentration of calcium ion in the solution is decreased. Acidification of the heated milk allows rennetability to be restored, but even then the structure of the cheese curd is different, suggesting that permanent changes have occurred in the micellar structure [36]. However, if the milk is heated at slightly decreased pH, then rennetability is maintained.

V. EMULSIFYING PROPERTIES OF CASEINS

Among food proteins, the soluble caseins are distinguished by their excellent emulsifying properties. All of the individual caseins, except perhaps κ-casein, show a strong tendency to adsorb to both air/water and oil/water interfaces, and thus they find an important use in the manufacture of stable emulsions for toppings and coffee whiteners, as well as in products such as cream liqueurs and products for infant nutrition, where long-term emulsion stability is essential. Emulsions are generally formed by the high-pressure homogenization of oil, protein, and in many cases some small molecule surfactants as well. The action of the homogenizer breaks up the oil phase into small droplets, whose interfaces are rapidly covered with surfactant, which prevents recoalescence of the emulsion droplets. What defines the sizes of the droplets is the power of the homogenizer, but also in large part the concentration of the protein and its surfactant properties.

The emulsifying properties of caseinates arise from the structures of the proteins: it was described earlier how the proteins were amphipathic, judged on the basis of their primary structure. Because they possess extensive hydrophobic regions, they would be expected to bind readily to hydrophobic materials, such as oil/water interfaces, and this is indeed found to be the case [37]. The decrease in interfacial tension caused by the adsorption of β-casein is greater than can be achieved by most other proteins, showing the high surface activity of this protein. Experiments on individual caseins [38] and sodium caseinate [39] have shown that they will cover an interface to the extent of 2–3 mg m^{-2}, similarly to other proteins (e.g., whey proteins). This interfacial layer appears to be a monolayer, so that even if the protein is originally present as an aggregate, as is most likely with casein in solution, it seems that dissociation of these aggregates occurs as the protein adsorbs to the oil/water interface. Thus, the emulsion droplets formed between caseinates and oil in most food systems where high homogenization pressures are used are likely to be covered by a monolayer of protein only. Some experiments at higher concentrations of casein do suggest that it is possible to form multilayers as well

[40], but this may also depend on the type of caseinate used; caseinate made in the laboratory may differ from commercial spray-dried caseinates.

In emulsions made incorporating whole caseinates, there seems to be little selection between the caseins, i.e., they are all adsorbed to the oil/water interface in proportion to their composition in solution [41]. However, there seems to be a distinction between whole caseinate and mixtures of purified individual caseins in this respect, since the latter show competitive adsorption between α_{s1}- and β-caseins [38], while the former shows much less. This may suggest that the quaternary structures of the caseins in the two solutions differ, possibly because of the presence of κ- and α_{s2}-caseins altering the structures of the casein aggregates.

The sizes of oil-in-water emulsion droplets formed using caseins depend on the relative amounts of protein and oil and on the homogenization technique used. Unlike some other proteins, casein is capable of extending widely over an oil/water interface, so that it can produce a stable emulsion when the amount of protein on the surface is as little as 1 mg m^{-2} [39]. With more protein, surface coverage rises to a value of 2–3 mg m^{-2} as the proteins change conformation. Such behavior seems to be peculiar to casein, presumably because of its flexibility (Fig. 3). It does, however, allow for a good

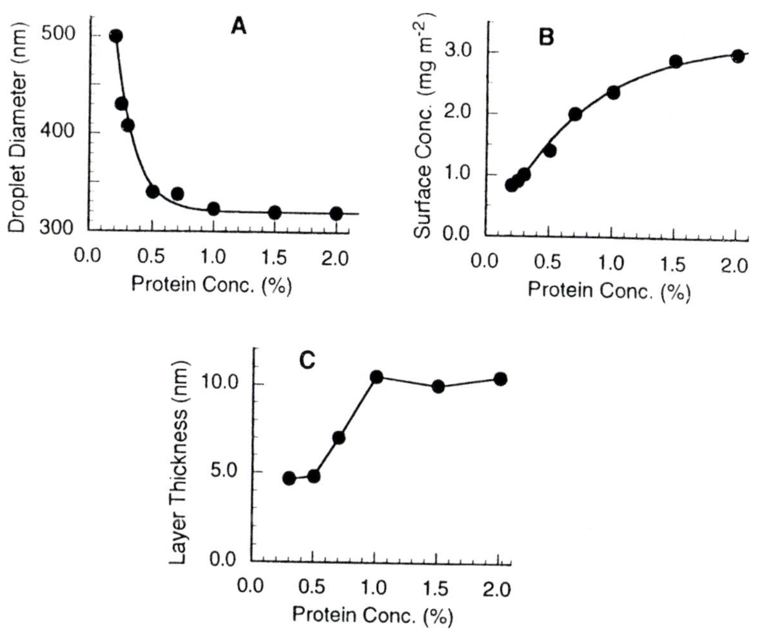

FIGURE 3 Behavior of caseinate emulsions. (A) Diameters of the droplets in an emulsion (20% soy oil) made using different concentrations of caseinate, in a Microfluidics Microfluidizer®, demonstrating that droplet diameter depends on casein only when it is in low concentration. (B) Surface load of casein in the same emulsions, showing strong variation with casein concentration and minimum value for surface coverage of about 0.8 mg.m^{-2} at low casein concentration. (C) Thickness of the adsorbed layer of casein, showing that there are two values where the casein is spread on the surface or is permitted to protrude more into the solution from the position on the interface.

deal of variability in the composition of emulsions using caseinate. It must be remembered, however, that no emulsion can remain stable in the presence of insufficient protein, which for casein is about 1 mg m^{-2}. Below this value, emulsion droplets will coalesce until the area of the interface has decreased sufficiently to allow the spread protein to cover all of it.

Emulsions based on caseinates or individual caseins are very stable, because adsorbed casein molecules are likely to possess the two important attributes for successful stabilization of colloidal material; they are charged (especially because of their phosphoserine content) and they are conformationally flexible. The charge will tend to prevent close approach of the emulsion droplets, because like-charged droplets will repel one another, and the flexibility of the protein allows it to adopt an extended conformation after it is adsorbed. So both the primary and tertiary structures favor stability. The adsorption of the proteins through their hydrophobic regions leaves the charged portions to project into solution. Details of the adsorption of β-casein have been established experimentally; parts of the molecule protrude as much as 10 nm from the interface to which it adsorbs [42]. This allows for steric stabilization of a classic kind. Moreover, the part of the protein that extends into solution carries the five phosphoseryl residues of the protein, so that the projecting layer is highly charged as well. In fact, the high charge may help to hold the protein in its extended conformation, because of intermolecular repulsion, since the adsorbed layer collapses somewhat when the ionic strength is raised, consistent with effects of shielding of interacting ions [43]. The conformation of the protein on the interface appears to be such that its N-terminal protrudes into solution, while the bulk of the protein is held close to the interface [44]. That is, most of the mass of the protein molecule is within 1–2 nm of the interface, and the N-terminal 40 residues forms the protruding layer for the remaining 8 nm [45]. As a result of this extended layer, caseinate emulsions, analogously to casein micelles, are surrounded by a highly hydrated "hairy layer." The collapse of the adsorbed layer is particularly evident when the charge on the protein is reduced by the binding of Ca^{2+}. Not only does the ζ-potential of the particles decrease, but the thickness of the adsorbed layer also decreases. The presence of ions in moderate amounts does not greatly affect the adsorption of casein to the interface.

Of the individual caseins, β-casein has been studied most extensively in respect to its interfacial properties. However, it has been established that all of the other caseins are surface active, although they do not form adsorbed layers that extend quite so far into the solution, presumably because of the extremely amphiphilic nature of β-casein. Whole caseinate gives adsorbed layers similar to those measured for β-casein, even though the other caseins are present in the adsorbed layer [39]. These structural factors explain why caseinate is such a good stabilizer of emulsions: all of the proteins are charged and give a hydrated sterically stabilizing layer and the conformational mobility allows the casein to spread widely over the interface. Even the low level of adsorption, which gives a thinner protruding layer of protein (5 instead of 10 nm) does not seem to produce particularly unstable emulsions. In fact, stability to precipitation by Ca^{2+} may even be increased in emulsions where the adsorbed casein is well spread [46]. Nevertheless, it is possible to stabilize emulsions using less casein than would normally be required for other proteins.

Caseins are also remarkable for their ability to stabilize emulsions in the presence of substantial concentrations of ethanol. The proteins, in the absence of calcium, are quite stable in the presence of ethanol at concentrations up to about 40%. The source of

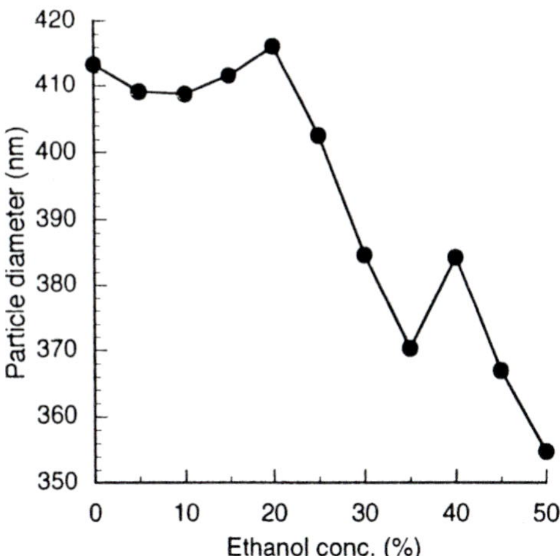

FIGURE 4 Effect of ethanol on the diameter of emulsion droplets (soy oil) stabilized by caseinate. Low concentrations of ethanol have no effect on the particle diameter, but as ethanol concentration increases, the layer of casein on the particle surface becomes thinner as a result of the incompatibility of the protein and the solvent. Because the decrease is much larger than that found when the casein layer is totally destroyed (Fig. 3C), it is suspected that high concentrations of ethanol also begin to dissolve the emulsion droplet.

this stability is not fully understood, although studies on casein micelles have demonstrated that the surface layer of κ-casein is collapsed when ethanol is added to milk [47]. Whether an analogous effect occurs in emulsions is not clearly demonstrated, but some preliminary measurements of the diameters of caseinate-stabilized emulsions in ethanolic media (Fig. 4) do suggest that the extended layer of casein does collapse, even though this does not cause immediate destabilization of the emulsion; at high concentrations of ethanol, in fact, the oil becomes partly dissolved also; it appears that the emulsion is destabilized by a process of Ostwald ripening [65].

The protein that is adsorbed to the oil/water interface shares some of the properties of the original caseins. Specifically, it binds calcium, and the emulsions can be destabilized by the presence of too much calcium (especially if ethanol is also present). Also, the emulsions are acid-unstable, just as are the proteins. Reducing the pH of a casein-stabilized emulsion below 5.5 will generally provoke destabilization, which is difficult to avoid without the use of some additional stabilizer.

VI. EMULSIONS CONTAINING CASEINS AND EMULSIFIERS

In many food emulsions, such as whipped toppings and coffee whiteners, the caseins are not the only emulsifiers present, because small molecule emulsifiers, e.g., lecithins, Tweens, or GMS, are also incorporated in the formulation. This is done for a number of reasons, but two are of great importance. First, the small molecule emulsifiers can cover interfaces that casein does not, resulting in an emulsion with smaller particles, leading to greater stability. However, more important in some products is the effect of the small molecules in destabilizing the emulsion [48]. A topping must destabilize as air is whipped into it to allow partial coalescence of the emulsion droplets around the air bubbles. If emulsions using caseinate alone are used, the stability may be too high for the adsorbed layer of protein to break and allow the adsorption of the fat to the surface of the air bubble; however, if small molecule surfactants are included, they help to form a very weak interfacial film which is easily broken, so that the interaction between air and oil is facilitated. This phenomenon is also important in the formation of ice cream and will be enlarged upon later.

The small surfactant molecules seem to exert their effects in different ways. Some owe at least part of their action to the displacement of casein from the interface; Tweens are particularly notable for this effect. Oil-soluble surfactants such as GMS, however, do not displace the caseins so efficiently, although they reduce the surface coverage somewhat. Lecithins appear to act in a completely different way [49]; if casein is at less than saturation coverage, the lecithin may co-adsorb and allow the caseins to adopt conformations that protrude further into solution (Fig. 5). It does not appear that lecithins displace caseins from the interface, and their mode of action is not clear [50]. However, it is known that the nature of the oil phase is important in the lecithin-casein-oil system, and also that the particular lecithin used is critical. For example, di-oleyl-phosphatidyl choline (DOPC) appears to have a specific interaction with casein, which is not shared with the other phosphatidylcholines. In contrast to lecithins, Tween displaces casein and makes the interfacial layer thinner and may even lead to an uneven emulsion surface where there are portions denuded of protein (Fig. 5); these particles would be particularly relevant to whipped systems, because not only are they mechanically unstable, but they

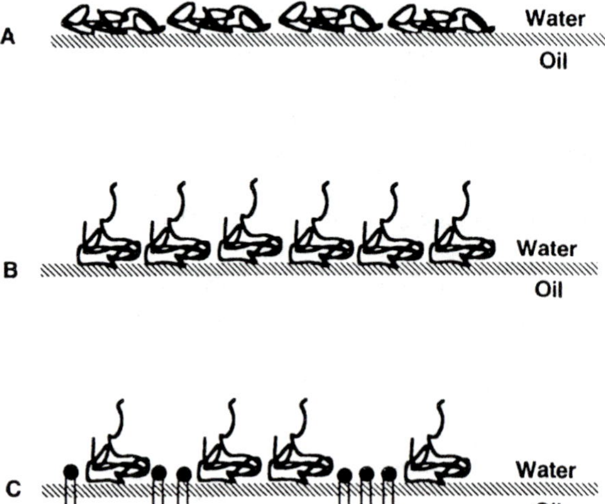

FIGURE 5 Principles of the conformations of casein adsorbed to oil/water interfaces. (A) Casein is in low concentration and must spread widely over the interface. (B) Casein is in high concentration, and each molecule is not now required to spread so widely and can protrude into solution. (C) With low casein concentration in the presence of phospholipid, some adsorption of the phospholipid may occur to allow the casein to spread less widely and to adopt a conformation more typical of high concentration.

also lack any component of steric stabilization. Thus, the presence of surfactants, and the choice of the surfactants, may affect the emulsifying properties of the caseins in different ways.

VII. FOAMING PROPERTIES OF THE CASEINS

Caseins are less effective when they are incorporated into foams. Although they adsorb well to air-water interfaces, and even though they interact well with one another, the caseins do not create the strong lamellar layers between the air bubbles that are characteristic of a stable foam. Foams can be produced with good overrun, but the stability is such that they generally collapse within a short time. These phenomena can be readily explained on the basis of the known properties of the caseins. As has been described for emulsions, the caseins adsorb strongly to an air/water interface with their hydrophilic portions protruding into solution [51]. Because interactions between caseins are mainly hydrophobic in nature, as is the interaction between casein and interface, the act of adsorption will decrease the tendency of the protein molecules to interact with one another. Because the hydrophilic portions in solution do not bind to one another, they will not produce rigidity within the interbubble lamella of the foam. Furthermore, the interfacial viscosity generated by adsorbed caseins is low, because they lack the capacity to form strong intermolecular bonds (e.g., disulfides). These factors give the air/water interface little mechanical strength, so that the interfacial layer is easily broken. The foaming capacity may be contrasted with that of β-lactoglobulin, where much stronger interfacial layers are formed and the interactions between individual molecules are strengthened by the formation of disulfide bonds.

For these reasons, caseins are not widely used on their own as foaming agents. However, we must contrast the use of the proteins alone with their use in combination with other materials, especially fat, in the production of stable foams. In such cases, it is not the casein that provides the rigidity, but the semi-crystalline fat or oil, which is partly emulsified by casein. Such behavior is typical of both caseins and casein micelles, in that they can acquire strength by becoming adsorbed to some more solid material.

VIII. EMULSIONS FORMED USING CASEIN MICELLES

The naturally occurring form of casein is the casein micelle, and these particles may be used in emulsion systems. The most important of these is, of course, homogenized milk, which is an important product in many forms and an ingredient in some other food products such as infant formulas and nutritional drinks. It is at once apparent that casein micelles are not such effective emulsifiers as are caseinates or individual caseins. This is because each particle contains large numbers of casein molecules, which travel through the solution together, so on average a micelle has to travel further through solution before it meets a fat surface than does an individual protein molecule; in effect, the concentration of available casein is reduced. Moreover, while casein aggregates are held together by hydrophobic interactions, which may be and indeed are broken when the protein reaches an interface, casein micelles are held together by calcium phosphate, which is likely to resist attempts to mechanically dissociate the micelle. This results in lesser emulsifying properties of casein micelles relative to molecular caseins.

It is not certain that casein micelles themselves can adsorb to any extent on a hydrophobic interface without first being disrupted. Nor is it certain whether casein micelles can be disrupted by the act of homogenization, i.e., homogenized and unhomogenized skim milks seem to be quite similar in respect to particle sizes, so that mechanical disruption, if it does occur, is mainly reversible [52]. However, when the micelles are homogenized in the presence of fat or oil, it is evident that some disruption of these particles must occur because of the local turbulences, which may tear the particles apart, and also because the protein and fat surfaces are thrown together with some force during the homogenization process. This may in effect smash the micelles open; once the structure of the micelle begins to break up, then hydrophobic regions of the micelle will adsorb to the oil/water interface. The result of moderate homogenization (e.g., in a valve homogenizer) is that the milk fat droplets are surrounded by a layer that contains semi-intact casein micelles and casein aggregates (Fig. 6A), and possibly whey proteins, but this may depend on the method used [53]. In extreme conditions (such as in the use of a Microfluidizer), the casein on the surface may approach a monolayer and there is considerable sharing of the casein among the small fat globules; few intact micelles are present (Fig. 6B). Therefore, a homogenized fat globule may have a rather heterogeneous surface depending on the way it has been homogenized, the fat: protein ratio, and the proportions of micellar casein and other proteins in the mixture.

Since the fat/water interface is coated with fragmented casein micelles, it seems logical to determine which of the caseins holds these micellar fragments onto the interface. This determination is by no means simple, because attempts to remove the caseins not directly adsorbed to the interface by treatment with agents that dissociate casein micelles (EDTA and urea) can lead to the desired dissociation of micellar fragments, but also to interchange between the caseins. Some experiments of this nature have suggested that the κ- and α_{s2}-caseins bind to the interface and that the α_{s1}- and β-caseins play a smaller part. This surprising conclusion requires confirmation, because intuitively the micelles would be expected to adsorb by their β-casein, which is not only the most surface active of the caseins, but is mainly found in the interior of the micelle.

There is a lack of agreement about the ratio of whey protein to caseins in the adsorbed surface layers on the fat globules of homogenized milks. Whey proteins are almost as good as caseins in their surfactant properties. Some studies have found that large amounts of whey protein are present in homogenized fat globules, and in other cases much smaller amounts have been found. There seems to be a correlation with the severity of homogenization; the more severe the homogenization of the milk, the less whey protein and the more casein are adsorbed. Thus, milk severely homogenized using a Microfluidizer seems to contain no adsorbed whey protein, while milk systems produced in valve homogenizers at lower pressures show adsorbed whey proteins. Such results could be explained if the micelles are indeed broken up by very high-pressure homogenization, because this would increase their effective concentration and allow the casein to compete more effectively with the whey proteins.

The result of the change in micellar structure on adsorption is generally that homogenized milk is less stable than whole milk to heating or retorting, and this becomes especially acute when the milk is concentrated [54]. Because the casein micelles are not in their native conformation, it may be presumed that they are less protected by κ-casein and therefore coagulate more readily. Whether for this or another reason, it is known that homogenized milk coagulates more readily than skim milk when it is renneted [55], which is consistent with a reduced ability of the κ-casein to stabilize the particles in the homogenized milk. In whole or skim milk the micelles are coated with their natural surface of κ-casein, but in the homogenized milk there may be insufficient κ-casein to

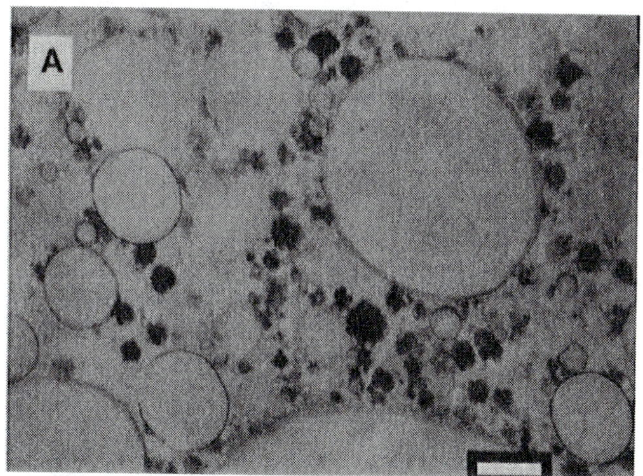

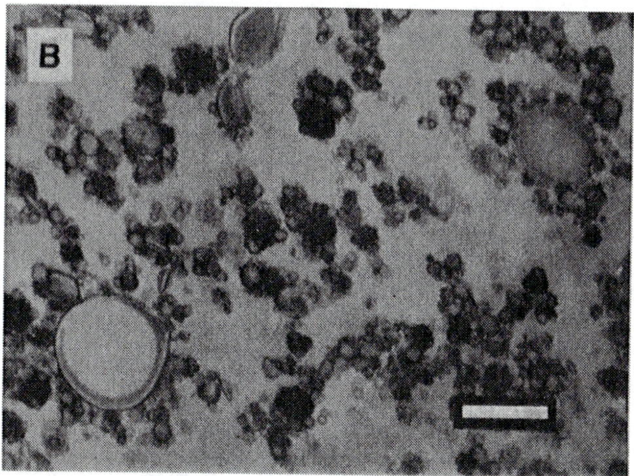

FIGURE 6 Transmission electron micrographs of milks homogenized in different ways. (A) Commercially prepared homogenized milk (4% fat) produced using a valve homogenizer; the working pressures are not known. The bar is 300 nm in length. Note the presence of casein micelles and fragments [dark materials surrounding the fat globules (unstained)]. (B) Milk homogenized using a Microfluidizer® M110S (Microfluidics Inc., Newton, MA), showing much smaller fat globules, no free casein micelles, and extensive sharing of casein micellar fragments between the fat globules. The scale bar is 500 nm long.

cover the particulate surfaces to the same density as in micelles. Thus, breakdown of fewer molecules of κ-casein will be required to bring about aggregation.

The effect of heat on homogenized milk also depends on the disrupted forms of the micelles, which are to be found on the surfaces of the fat globules. As with undisrupted casein micelles, heating causes interaction of the whey proteins with the κ-casein, and possibly the α_{s2}-casein, which is associated with the fat surface. Qualitatively, this reaction is the same with micelles both on and off the interface; however, quantitative studies have shown that there are differences in the interactions of the adsorbed micellar material with the denatured whey proteins, indicating why the order of processing steps (i.e., heating before homogenization rather than vice versa) is important. When the milk is homogenized before heating, there is a greater amount of whey protein bound to the casein adsorbed to the surfaces of the fat globules.

An important function for casein micelles is in the formulation of ice cream mixes [56]. These are mixtures of cream, homogenized so that the cream is stabilized by casein micelles as has been described. This rather concentrated dispersion of fat is of course stable, similarly to homogenized milks. The formulation also contains some small molecule surfactant, such as Tween, which also adsorbs to the fat/water interface. (It is well known that in homogenized creams, if insufficient casein is present to fully cover the fat interface created during homogenization, a semi-stable emulsion is formed, which can be readily destabilized when heated or the pH is changed, e.g., when put into coffee.) As the ice cream mix is cooled, the fat partly crystallizes. While the emulsion is held without agitation, it is stable, but when air is beaten into the mixture, the regions of the fat/water interface covered by small molecules break open, although this does not release the fat from the droplets because it is partly crystalline. The broken fat droplets, however, bind to the air/water interface, and interact one with another by contact between protruding fat crystals (partial coalescence). The result is a structure of air bubbles stabilized by adsorbed fat globules, which in turn are stabilized by micellar casein. This forms the basis of ice cream, although there are of course many other significant processes, such as the formation of glasses and ice crystals, during the manufacture of the product.

IX. COAGULATION OF CASEINS AND CASEIN MICELLES

Caseinates can be induced to coagulate in a number of ways. Some caseinates are already insoluble (e.g., acid casein or rennet casein), but the other forms of caseinate, including casein micellar material, can be caused to coagulate, precipitate, or gel by the action of acid (isoelectric precipitation), ethanol, or proteolytic enzymes. The first and the last of these are the basis for the formation of an extensive range of products, e.g., cheeses of many types, yogurts, and cottage cheese. Caseins do not themselves form strong gels, in contrast to the whey proteins, by simple action of heat, and so although they can be used to modify viscosity because of their extended structures and water-retaining capacity, they cannot be used as gelling agents. Incorporation of other materials, e.g., whey proteins, can be used to heat-coagulate the caseins, as in the heat treatment of milks, but in that case it is likely that the whey proteins act as cross-linking agents between the casein micelles.

A. Coagulation by Ethanol

The response (or lack of it) of casein to ethanol is the basis of the formulation of cream liqueurs. By combining the emulsifying properties of caseins and their partial resistance to precipitation by ethanol, it is possible to produce cream liqueurs containing 20% ethanol, which are highly stable, providing that the concentration of calcium in the product is kept low. This tendency of calcium to induce instability has its source in the casein micelle, where studies of the pH dependence of ethanol stability have shown that the stability decreases as decreasing pH releases calcium ions into the solution [57]. Ethanol acts basically by being a poor solvent for the protein; although at moderate concentrations of ethanol the proteins in the casein micelle do not precipitate, it is established that the hairy layer of κ-casein on the micellar surface is collapsed by ethanol [47]. Steric stabilization of the micelle is therefore diminished, and the particle becomes more susceptible to precipitation. In an analogous way, it seems that ethanol is capable of collapsing the extended layer of casein on the surface of an emulsion, although only at higher concentrations of ethanol than are required for the micelles in milk, where there is calcium present naturally. Precipitation of caseinate emulsions by ethanol is almost completely avoided by chelating the residual calcium (from the cream or from the caseinate) by adding citrate to the product.

B. Rennet Coagulation

As discussed above, casein micelles can be coagulated by rennet to form a gel; when the gel is cut, however, it shrinks, expelling water (the process of syneresis) to give cheese curd [58]. During the original coagulation, the casein micelles seem to remain structurally distinct, although they are linked together. The action of rennet does not break up the casein micelle, it simply degrades the surface layer. The surface of the micelle is covered with "hairs" of glycomacropeptide of κ-casein as a steric stabilizer, and there comes a point in the removal of these hairs at which the micelles become unstable and begin to aggregate.

There have been many studies devoted to the exact mechanism of the destabilization, and it seems that breakdown of only a small amount of the micellar κ-casein is insufficient to cause instability; only when a large proportion of the κ-casein on an individual micelle is broken down does the particle become capable of aggregation [59]. It is possible to explain the reaction simply on structural or steric grounds. In the early stages of the reaction, the removal of one hair may not have a particularly destabilizing effect. However, once the chymosin has destroyed sufficient κ-casein to produce a hole in the surface layer, then aggregation can occur via a similar hole in the surface of a second micelle, and so on (Fig. 7). That being the case, if the attack of chymosin on κ-casein is random, then only towards the end of the enzymatic reaction are there likely to be holes large enough to allow aggregation of two micelles to occur. It is therefore

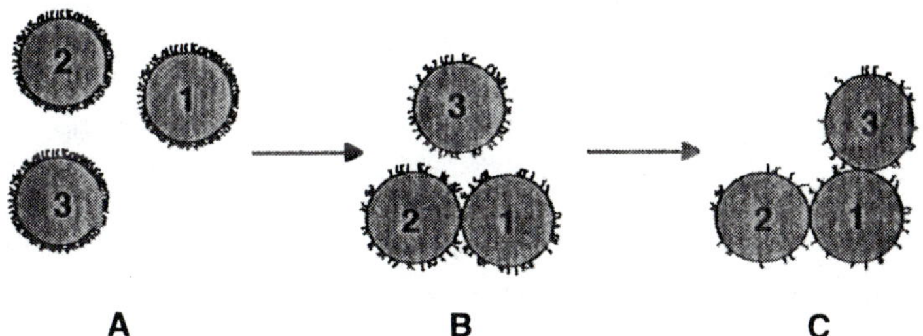

Figure 7 Schematic of micelle aggregation during renneting. (A) Micelles are separate and their surface layers of κ-casein are intact. (B) Renneting has partially destroyed the surface layers and micelles can interact but only through part of the surface (i.e., micelles 1 and 2 can interact through a small part of their surfaces, but micelle 3 lacks a gap on the surface of sufficient size to allow interaction). (C) As renneting proceeds, gaps are apparent on the surfaces of all micelles, and more extensive aggregation can occur. In the limit, all κ-casein is destroyed and micelles become infinitely capable of aggregation.

possible to explain renneting on purely geometric and structural grounds. Alternative models have considered the general overall interaction potential between casein micelles as the renneting proceeds [60]; in effect, this description of the reaction is very similar to the geometric model in its mathematical form.

The overall effect is that casein micelles "suddenly" aggregate after a lag stage following the addition of rennet to the milk. As described above, this phenomenon can be described as a result of a continuous process, and there is no need to invoke any radical change in micellar structure whereby micelles become capable of aggregating. It is interesting to speculate whether any such change in structure might occur, although there is at present no direct evidence for it (and a more complex argument might not commend itself to followers of William of Ockham). Certainly, as the renneting reaction proceeds, the surface of the casein micelle becomes increasingly hydrophobic because of the exposure of *para-κ*-casein, to a point at which the exterior of the micelle may be more hydrophobic than the interior. This situation could give rise to a massive conformational change in the micelle, which in turn could lead to the whole micelle, rather than part of it, becoming capable of aggregation. There is no evidence for or against this hypothesis, but it would offer an alternative to those discussed above. It would of course be determined by the mobility of caseins or casein clusters within the micelle, which has been shown to be possible because of the movement of β-casein as milk is cooled.

Aggregation of renneted micelles appears to depend on hydrophobic interactions between them because of the observation that curd does not form at temperatures below 15°C, which is typical of such interactions. However, the reaction is not completely hydrophobic in nature because micellar aggregation will not occur in the absence of calcium ions (hence the well-known practice of adding calcium chloride to cheese milk). Calcium is not necessary for aggregation of *para-κ*-casein, so it is evident that the coagulation of the micelles must involve interactions other than simply hydrophobic interactions between *para-κ*-casein fragments. Since the binding of calcium ions to *para-κ*-casein is low, presumably other caseins to which Ca^{2+} can bind must be involved, i.e., all of the phosphorylated caseins. Once again, we come back to the need to understand the structure of the casein micelle, both before and after renneting, and what factors cause the structure to change. Among these are the deposition of whey proteins and calcium phosphate during heating of milk before renneting. Although the action of the whey proteins is only partly defined, it may be in part to reduce the hydrophobicity of the renneted surface; however, it may also serve to prevent conformational changes to the casein micelle by holding together adjacent molecules on the surface. Although it is known that rennetability can be restored by the acidification of the milk, this also changes the structure of the curd, so it may be inferred that the micellar structure is also changed.

In the preliminary formation of rennet curd, the micelles more or less maintain their identity. After the curd is cut, however, and syneresis occurs, the packing of the micelles becomes much closer, and eventually the micelles fuse as the curd is treated and the cheese matures. Nevertheless, there seems to be still an influence of the original structures, because the differences in cheese varieties can arise from structural differences in the micelles as well as the way in which the curd is processed. For example, the pH at cutting determines the properties of the finished cheese. Thus, although we cannot always be precise on the details of the mechanism, it seems evident that the micellar structure of the caseins determines the properties of the product.

C. Acidification of Milk

Acidification is the second major way of forming gels or coagula in milk, and during the course of the process the micellar structure is severely disrupted. Rapid acidification of milk by direct addition of acid leads to the formation of a precipitate, which can later be neutralized to give sodium or other caseinates. However, slower acidification, by addition of glucono-δ-lactone (GDL) or more commonly by the action of bacterial cultures, can cause the formation of more structured products. Acidification causes two major processes to occur; first, the calcium and phosphate are dissolved out of the micelle as a result of the protonation of the ionized phosphate groups and the structure of the micelle is profoundly altered. In addition, at pH values around 5.0, the caseins precipitate isoelectrically. Surprisingly, the first of these processes does not necessarily lead to micellar dissociation. As long as the temperature during acidification remains above about 25°C, which is generally the case, the caseins remain within the particles—whether these are to be called casein micelles is a debatable point. Although the micellar structure must be seriously weakened by the loss of the calcium phosphate, it does not fall apart, although rearrangements of the caseins may occur. As acidification proceeds, these casein particles precipitate or under quiescent conditions form rather fragile gels, which undergo extensive syneresis when cut and warmed.

In products such as yogurts, the casein texture is altered by heating the milk before acidification and also running the process at a high temperature, about 40°C. Although acidification does not proceed differently in these systems, the structure of the final gel or precipitate is different from simply acidified milks, at least partly because of the residual micellar structure [61] (although the texture of yogurts is also controlled by the addition of gums and stabilizers).

It has not been determined how the details of the micellar conformation change during acidification, since, as we have seen, the structure of the casein micelle is itself not fully established. The layer of macropeptide becomes thinner, presumably because the κ-casein molecules adopt a more compact conformation; this has been observed directly, but other evidence for conformational change is less direct. There are complex changes in the micellar hydration as pH decreases [62], presumably arising from the removal of the calcium phosphate, and this effect has been detected also in studies of the rheology of the acid gelation of the milk [63,64]. It is especially noticeable that the rheology of gels formed from partly renneted milk shows a distinct change in the region of pH 5–5.5, the precise value depending on the amount of rennet used. As the gel is formed, its apparent viscosity increases, but then is seen to decrease as acidification proceeds, before rising again as the final acid gel is formed. This change in the rheology reflects changes in the gel structure, and in turn this must reflect changes in the interactions between the caseins. Taken in combination with the changes in the hydration of the micelles, this indicates a considerable change in the structure and properties of the caseinate particles.

X. CONCLUDING REMARKS

Perhaps the preceding sections have illustrated the complexity of trying to relate structure to function in caseinate systems, not least because the structures in question (either of individual caseins, casein aggregates, or casein micelles) are far from clearly defined, even if absolute definition is possible. Indeed it may be that the most important attribute that the caseins possess is that of flexibility. This property makes them good emulsifying agents, allowing them to act as surfactants and also to form extended sterically stabilizing layers around the emulsion droplets. It may even be possible that flexibility is inherent in the casein micelle itself leading to breakdown in defined ways during homogenization or to the formation of different gel structures during renneting and/or acidification. Emulsification and gelation are the major functional characteristics of caseins; other aspects such as foaming and solubility are less significant, the first because of the general weakness of casein foams and the second being dependent on manufacture rather than on the properties of the protein per se.

So what of the future? Casein fractionation is gradually becoming commercially feasible, so that individual proteins may become available. This has the potential to allow exploitation of separate aspects of the proteins; for example, β-casein may be the most surface active of the proteins, but could a functional emulsion be made with κ-casein so that its renneting potential could be exploited? Or can nonstandard mixtures of caseins be used to prepare synthetic micelles of altered composition, the aggregation or emulsifying properties of which might be different from intact micelles? Along the same lines, what are the uses of the phosphocaseinate fraction that is now available directly from microfiltration of skim milks? As a last thought, there is the possibility of extracting protein complexes formed from the interaction of caseins (κ and α_{s2}) with whey proteins during heating. Could these materials combine the emulsifying, foaming, and gelation characteristics of the component proteins, but show synergistic behavior?

REFERENCES

1. H. E. Swaisgood, Chemistry of the caseins, *Advanced Dairy Chemistry*, Vol. 1: *Proteins* (P. F. Fox, ed.), Elsevier Applied Science, London, 1992, p. 63.
2. H. S. Rollema, Casein association and micelle formation, *Advanced Dairy Chemistry*, Vol. 1: *Proteins* (P. F. Fox, ed.), Elsevier Applied Science, London, 1992, p. 111.
3. O. Robin, S. T. Turgeon, and P. Paquin, Functional properties of milk proteins, *Dairy Science and Technology Handbook*, Vol. 1 (Y. H. Hui, ed.), VCH Verlag, New York, 1992, p. 277.
4. D. T. Davies and A. J. R. Law, The content and composition of protein in creamery milks in south-west Scotland, *J. Dairy Res. 47*:83 (1980).
5. C. Holt, Structure and stability of bovine casein micelles, *Adv. Prot. Chem. 43*:63 (1992).
6. C. R. Southward, Utilization of milk components: casein, *Modern Dairy Technology*, Vol. 1 (R. K. Robinson, ed.), Chapman and Hall, London, 1994, p. 375.
7. H. J. Vreeman, The association of bovine SH-κ-casein at pH 7.0, *J. Dairy Res. 46*:272 (1979).
8. E. R. B. Graham, N. M. Malcolm, and H. A. McKenzie, On the isolation and conformation of bovine β-casein A, *Int. J. Biol. Macromol. 6*:155 (1984).
9. D. M. Byler and H. Susi, Examination of the secondary structure of proteins by deconvolved FTIR spectra, *Biopolymers 25*:469 (1986).
10. T. F. Kumosinski, E. M. Brown, and H. M. Farrell, Jr., Three-dimensional molecular modeling of bovine caseins: α_{s1} casein, *J. Dairy Sci. 74*:2889 (1991).
11. T. F. Kumosinski, E. M. Brown, and H. M. Farrell, Jr., Three-dimensional molecular modeling of bovine caseins: κ-casein, *J. Dairy Sci. 74*:2879 (1991).
12. M. Paulsson and P. Dejmek, Thermal denaturation of whey proteins in mixtures with caseins studied by differential scanning calorimetry, *J. Dairy Sci. 73*:590 (1990).
13. K. Kajiwara, R. Niki, H. Urakawa, Y. Hiragi, N. Donkai, and M. Nagura, Micellar structure of β-casein observed by small-angle x-ray scattering, *Biochim. Biophys. Acta 955*:128 (1988).

14. A. L. Andrews, D. Atkinson, M. T. A. Evans, E. G. Finer, J. P. Green, M. C. Phillips, and R. N. Robertson, The conformation and aggregation of bovine β-casein A. I: Molecular aspects of thermal aggregation, *Biopolymers 18*:1105 (1979).
15. G. P. Berry and L. K. Creamer, The association of bovine β-casein. The importance of the C-terminal region, *Biochemistry 14*:3542 (1975).
16. D. S. Horne and D. G. Dalgleish, Electrostatic interaction and the kinetics of protein aggregation: α_{s1}-casein, *Int. J. Biol. Macromol. 2*:154 (1980).
17. T. Ono, S. Odagiri, and T. Tagaki, Separation of the submicelles of from micellar casein by high performance gel chromatography on a TSK-GEL G4000SW column, *J. Dairy Res. 50*: 37 (1983).
18. D. G. Schmidt, P. Both, and J. Koops, Properties of artificial casein micelles. 3. Relationship between salt composition, size and stability towards ethanol, dialysis and heat, *Neth. Milk Dairy J. 33*:40 (1979).
19. P. Walstra, V. A. Bloomfield, G. J. Wei, and R. Jenness, Effect of chymosin action on the hydrodynamic diameter of casein micelles, *Biochim. Biophys. Acta 669*:258 (1981).
20. M. Kalab, B. E. Phibbs-Todd, and P. Allan-Wojtas, Milk gel structure XIII. Rotary shadowing of casein micelles for electron microscopy, *Milchwissenschaft 37*:513 (1982).
21. C. W. Slattery and R. Evard, A model for the formation and structure of casein micelles from subunits of variable composition, *Biochim. Biophys. Acta 317*:529 (1973).
22. P. Walstra, On the stability of casein micelles, *J. Dairy Sci. 73*:1965 (1990).
23. D. G. Schmidt, Association of caseins and casein micelle structure, *Developments in Dairy Chemistry—1. Proteins* (P. F. Fox, ed.), Applied Science Publishers, New York, 1982, p. 61.
24. M. C. A. Griffin, R. L. J. Lyster, and J. C. Price, The disaggregation of calcium-depleted casein micelles, *Eur. J. Biochem. 174*:339 (1988).
25. C. Holt, D. T. Davies, and A. J. R. Law, Effects of colloidal calcium phosphate content and free calcium ion concentration in the milk serum on the dissociation of bovine casein micelles, *J. Dairy Res. 53*:557 (1986).
26. D. G. Dalgleish and A. J. R. Law, pH-Induced dissociation of bovine casein micelles. I. Analysis of liberated caseins, *J. Dairy Res. 55*:529 (1988).
27. D. G. Dalgleish, D. S. Horne, and A. J. R. Law, Size-related differences in bovine casein micelles, *Biochim. Biophys. Acta 991*:383 (1989).
28. D. S. Horne, Steric stabilization and casein micelle stability, *J. Colloid Interf. Sci. 111*:250 (1986).
29. C. Holt and D. G. Dalgleish, Electrophoretic and hydrodynamic properties of bovine casein micelles interpreted in terms of particles with an outer hairy layer, *J. Colloid Interf. Sci. 114*: 513 (1986).
30. D. G. Dalgleish, The enzymatic coagulation of milk, *Advanced Dairy Chemistry*, Vol. 1: *Proteins* (P. F. Fox, ed.), Elsevier Applied Science, London, 1992, p. 579.
31. P. Walstra, The voluminosity of bovine casein micelles and some of its implications, *J. Dairy Res. 46*:317 (1979).
32. H. S. Rollema and J. A. Brinkhuis, A H-NMR study of bovine casein micelles; influence of pH, temperature and calcium ions on micellar structure, *J. Dairy Res. 56*:417 (1989).
33. D. G. Dalgleish, Denaturation and aggregation of β-lactoglobulin in heated milk. *J. Agric. Food Chem. 38*:1995 (1990).
34. H. Singh and P. F. Fox, Heat stability of milk: pH-dependent dissociation of micellar κ-casein on heating milk at ultra high temperatures, *J. Dairy Res. 52*:529 (1985).
35. A. C. M. van Hooydonk, P. G. de Koster, and I. J. Boerrigter, The renneting properties of heated milk, *Neth. Milk Dairy J. 41*:3 (1987).
36. H. Singh, S. I. Shalabi, P. F. Fox, A. Flynn, and A. Barry, Rennet coagulation of heated milk: influence of pH adjustment before or after heating, *J. Dairy Res. 55*:205 (1988).
37. E. Dickinson, Protein-stabilized emulsions, *J. Food Eng. 22*:59 (1994).

38. E. Dickinson, S. E. Rolfe, and D. G. Dalgleish, Competitive adsorption of α_{s1}-casein and β-casein in oil-in-water emulsions, *Food Hydrocoll.* 2:397 (1988).
39. Y. Fang and D. G. Dalgleish, Dimensions of the adsorbed layers in oil in water emulsions stabilized by caseins, *J. Colloid Interf. Sci. 156*:329 (1993).
40. M. Britten and H. Giroux, Interfacial properties of milk protein-stabilized emulsions as influenced by protein concentration, *J. Agric. Food Chem. 41*:1187 (1993).
41. J. A. Hunt and D. G. Dalgleish, Adsorption behaviour of whey protein isolate and caseinate in soya oil-in-water emulsions, *Food Hydrocoll. 8*:175 (1994).
42. D. G. Dalgleish, The conformations of proteins on solid/water interfaces—caseins and phosvitin on polystyrene latices, *Colloids Surf. 46*:141 (1990).
43. D. V. Brooksbank, C. M. Davidson, D. S. Horne, and J. Leaver, Influence of electrostatic interactions on β-casein layers adsorbed on polystyrene latices, *J. Chem. Soc. Farad. Trans. 89*:3419 (1993).
44. J. Leaver and D. G. Dalgleish, Topography of β-casein adsorbed at an oil/water interface as determined from the kinetics of trypsin-catalysed hydrolysis, *Biochim. Biophys. Acta 1041*: 217 (1990).
45. E. Dickinson, D. S. Horne, J. Phipps, and R. Richardson, A neutron reflectivity study of the adsorption of β-casein at fluid interfaces, *Langmuir 9*:242 (1993).
46. S. O. Agboola and D. G. Dalgleish, Calcium-induced destabilization of oil-in-water emulsions stabilized by caseinate or by β-lactoglobulin, *J. Food Sci. 60*:399 (1995).
47. D. S. Horne, Steric effects in the coagulation of casein micelles by ethanol, *Biopolymers 23*: 989 (1984).
48. H. D. Goff and W. K. Jordan, Action of emulsifiers in promoting fat destabilization during the manufacture of ice cream, *J. Dairy Sci. 72*:18 (1989).
49. Y. Fang and D. G. Dalgleish, Casein adsorption on the surfaces of oil-in-water emulsions modified by lecithin, *Colloids Surf. B 1*:357 (1993).
50. J.-L. Courthaudon, E. Dickinson, and W. W. Christie, Competitive adsorption of lecithin and β-casein in oil-in-water emulsions, *J. Agric. Food Chem. 39*:1365 (1991).
51. D. E. Graham and M. C. Phillips, Proteins at liquid interfaces II. Adsorption isotherms, *J. Colloid Interf. Sci. 70*:415 (1979).
52. P. Walstra, Effect of homogenization on milk plasma, *Neth. Milk Dairy J. 34*:81 (1980).
53. P. Walstra and H. Oortwijn, The membranes of recombined fat globules. 3. Mode of formation, *Neth. Milk Dairy J. 36*:103 (1982).
54. A. W. M. Sweetsur and D. D. Muir, Effect of homogenization on the heat stability of milk, *J. Dairy Res. 50*:291 (1983).
55. D. G. Dalgleish and E. W. Robson, Coagulation of homogenized milk particles by rennet, *J. Dairy Res. 51*:417 (1984).
56. K. G. Berger, Ice cream, *Food Emulsions*, 2nd ed. (K. Larsson and S. E. Friberg, eds.), Marcel Dekker, New York, 1990, p. 367.
57. D. S. Horne and T. G. Parker, Factors affecting the ethanol stability of bovine milk. II. The origin of the pH transition, *J. Dairy Res. 48*:285 (1981).
58. P. Walstra, The syneresis of curd, *Cheese: Chemistry, Physics and Microbiology*, 2nd ed., Vol. 1 (P. F. Fox, ed.), Chapman and Hall, London, 1993, p. 141.
59. D. G. Dalgleish, Proteolysis and aggregation of casein micelles treated with immobilized or soluble chymosin, *J. Dairy Res. 46*:653 (1979).
60. D. F. Darling and A. C. M. van Hooydonk, Derivation of a mathematical model for the mechanism of casein micelle coagulation by rennet, *J. Dairy Res. 48*:189 (1981).
61. J. Visser, A. Minihan, P. Smits, S. B. Tjan, and I. Heertje, Effects of pH and temperature on the milk salt system, *Neth. Milk Dairy J. 40*:351 (1986).
62. A. C. M. van Hooydonk, H. G. Hagedoorn, and L. J. Boerrigter, pH-induced physico-chemical changes of casein micelles in milk and their effect on renneting. 1. Effect of acidification on physico-chemical properties, *Neth. Milk Dairy J. 40*:281 (1986).

63. D. G. Dalgleish and D. S. Horne, Studies of gelation of acidified and renneted milks using diffusing wave spectroscopy, *Milchwissenschaft 46*:417 (1991).
64. Y. Noel, C. Durier, N. Lehembre, and A. Kobilinsky, [Multifactorial study of combined enzymatic and lactic milk coagulation measured by viscoelasticimetry], *Lait 71*:15 (1991).
65. S. O. Agboola and D. G. Dalgleish, Effects of pH and ethanol on the kinetics of destabilization of oil-in-water emulsions containing milk proteins, *J. Sci. Food Agric.*, in press (1997).

8

Structure-Function Relationships of Whey Proteins

Philippe Cayot and Denis Lorient
Université de Bourgogne, Dijon, France

I. INTRODUCTION

Whey from the cheese industry is increasingly being used either as animal feed or as a source of protein for human nutrition. Two types of whey can be distinguished: (1) soft whey, which comes from rennet coagulation of milk at pH 6.6 (e.g., in cheddar or emmental manufacturing) and (2) the acid whey obtained from fresh soft cheese production (e.g., cream cheese, Camembert, or Petit Suisse), after acid coagulation of milk, and, in the case of cottage cheese, following heating of the curd. Marketed wheys come in liquid and powder forms, with different compositions (Table 1).

Whey protein is a mixture of proteins with numerous and diverse functional properties and therefore may have many potential uses (Fig. 1). The main proteins are β-lactoglobulin and α-lactalbumin. They represent approximately 70% of the total whey proteins and are responsible for the hydration, gelling, and surface-active properties (emulsifying and foaming properties) of the whey protein ingredients. Several industrial applications and uses of whey products in the food industries are listed in Table 2. These

Table 1 Composition of Different Whey Products (% w/w)

Whey concentrates	Water	Lactose	Lipids	Minerals	Nitrogen	β-Lactoglobulin	α-Lactalbumin
Liquid	93.5	4.5	0.3	0.6–0.8	1.0	0.45	0.2
Powder	4.0	70	—	9–12.0	13.0	—	—
Curd	42.0	36.0	—	8.0	9.0	—	—
Concentrate UF	4.0	9.5	6.0	3.0	75.0	32.0	15

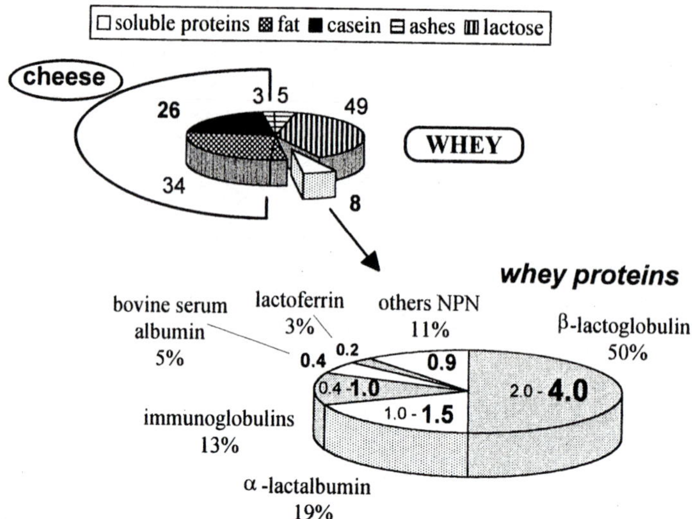

FIGURE 1 Composition of cheese whey. Numbers indicate concentration in g/liter. NPN = products composed of nonprotein nitrogen.

protein ingredients are also used for their nutritional and therapeutic properties in low-calorie diets and in intensive care enteral nutrition.

The aim of this chapter is to describe the structures and main functional properties of the major whey proteins in order to try to establish relationships between protein structure and functionality. Particularly, the effects of technological treatments on protein structure and their effects on functional properties and the behavior of whey protein ingredients (concentrates and isolates) in food products will be discussed.

II. STRUCTURE OF WHEY PROTEINS

A. Primary and Spatial Structures of Native Whey Proteins

β-Lactoglobulin (2–4 g/liter in whey) has a molecular mass of 18.3 kDa. It contains 162 amino acid residues with one thiol group and two disulfide bonds (Table 3). Among seven genetic variants, the A and B variants are abundant. They differ from each other by two amino acid residues (residue 64 is Asp for A and Gly for B; residue 118 is Val for A and Ala for B). The A variant is therefore the most negatively charged at the pH of the milk (6.6) and can be separated from the B variant by electrophoresis or by anion exchange chromatography.

The primary structures of β-lactoglobulin from various ruminant species are known [1–3] (Fig. 2a). The secondary and tertiary structures show a high degree of organization with a great proportion of β sheets (43–50% of the residues) and only 10–15% of α helix and 15–20% of β turns (Fig. 3A) [4]. The monomer form of β-lactoglobulin resembles that of a cone or a calyx with a hydrophobic pocket capable of binding vitamin A and fatty acids. In this respect, the structure of β-lactoglobulin is analogous to that of the retinol-binding protein from blood plasma. Its compact structure is due to a stacking

TABLE 2 Uses of Whey Proteins in Human Foods

Industrial applications	Functional properties expected	Proteins used
Bread making Biscuit manufacturing Breakfast cereals	Waterholding Fat dispersibility Emulsion stabilization Overrun of foam Gelling properties Browning Aroma enhancement	WPC or WPC + caseinates WPI WPI, coprecipitates Whey
Pasta	Binding and texturing effect Browning	Coprecipitates
Confectionary Chocolate confectionary	Emulsion manufacturing Overrun of foam Browning, aroma Antioxidizing effect	WPC + hydrolyzed caseinates WPC Whey Coprecipitates
Ice cream	Emulsion stability Overrun of foams Gelling properties	WPC + caseinates and total milk proteins
Meat products Delicatessen Meat	Emulsion making Waterholding (creamy and smooth texture) Adhesive or binding properties	WPC, WPI alone or in mixture with caseinate
Sauces Soups Ready-to-eat food	Emulsion stability Waterholding	WPC + caseinates + egg yolk WPC + caseinates + whole egg
Milk products (cheese, yogurts, "light" butter) Alcoholic beverages	Emulsion stability Waterholding Gelling properties Cream stabilization Cloudy aspect	Caseinates WPC + caseinates WPI WPC + caseinates WPC or WPI
Nutritional uses	Protein intake Enteral nutrition	Whey, WPC, or WPI WPC hydrolysates
Cosmetics	Skin protection Antimicrobial properties	Lactoferrin, WPC hydrolysates Lactoferrin, lactoperoxidase

Source: Centre Interprofessionnel de Documentation et d'information Laitières, Paris.

up of nine β sheets and to the two disulfide bridges. This compact structure is resistant to complete proteolysis by digestive proteases. At the pH of milk, β-lactoglobulin exists as a dimer constituted of two stacked cones [5].

α-Lactalbumin is a small molecule with 123 amino acid residues and four disulfide bridges (Table 3 and Fig. 2b) and has a molecular mass of 14.2 kDa [6,7]. The α-lactalbumin content of whey is about 1.5 g/liter. It exhibits a high sequence homology with lysozyme. As a part of the enzyme galactosyl transferase, it takes part in lactose biosynthesis [8]. The two genetic A and B variants differ from each other by one residue; the residue at position 10 is Gln for A and Arg for B. This protein has a very low content

TABLE 3 Physicochemical Characteristics of Major Whey Proteins

Characteristic	β-Lactoglobulin	α-Lactalbumin	Bovine serum albumin	Immunoglobulins
Molar mass (g/mol)	18,362	14,174	69	150,000–1,000,000
Cysteyl residue/mol	5	8	35	—
Amino acid residues/mol	162	123	582	>1000
Disulfide bonds/mol	2	4	17	4.x
Thiol function/mol	1	0	1	—
Lysyl residues/mol	15	12	59	—
Arginyl residues/mol	3	1	23	
Histidyl residues/mol	2	3	18	
Glutamyl residues/mol	16	8	59	—
Aspartyl residues/mol	10	9	39	—
pl (isoelectric point)	52	4.5–4.8	4.7–4.9	5.5–8.3
Average hydrophobicity (kJ/residue)	508	468	468	458

of organized secondary structure: 30% α helix and 9% β sheets [9]. As a result, it has great flexibility. However, the presence of one bound Ca^{2+} ion and four disulfide bridges maintain it in a compact ellipsoidal structure (Fig. 3B) with a small hydrophobic box.

The two other major whey proteins are bovine serum albumin (BSA) and the immunoglobulins (Ig). BSA is a large protein with 582 amino acid residues (Table 3) [10]. It functions as a carrier protein for transport of nonpolar molecules in biological fluids. The protein is compact on the C-terminal side and can be reversibly denatured by heat or by adding acid or base at 40–50°C.

The immunoglobulins form a heterogeneous family of glycoproteins of 148–1000 kDa with antibody properties (Table 3) [11]. Though they show denaturation temperatures higher than those of β-lactoglobulin and of α-lactalbumin, they are very heat sensitive in presence of BSA, probably because of interaction with the free thiol group.

The lactoferrin is a metalloprotein with a molecular mass of 80–92 kDa; it contains 703 amino acid residues and 16 disulfide bridges [12,13] and one mole of bound Fe^{3+} cation [14]. Because of its high pI, it forms a complex with BSA or β-lactoglobulin. Because of the bound iron, it exhibits high resistance to heat denaturation (see Sec. IV) and antibacterial properties similar to those of lysozyme and lactoperoxidase.

Some minor proteins also play an important role in the functional properties of whey proteins. The proteose peptone fraction, which consists of proteolytic fragments of β-casein [15,16], contains surface-active peptides. Numerous enzymes, like lactoperoxidase, alkaline phosphatase, catalase, sulfhydryl oxidase, and plasmin, are also found in whey.

B. Comparative Physicochemical Properties of Whey

Table 3 presents the physicochemical properties of whey proteins [11,17]. The protein conformation and physicochemical properties depend on the ionic environment, i.e., salt

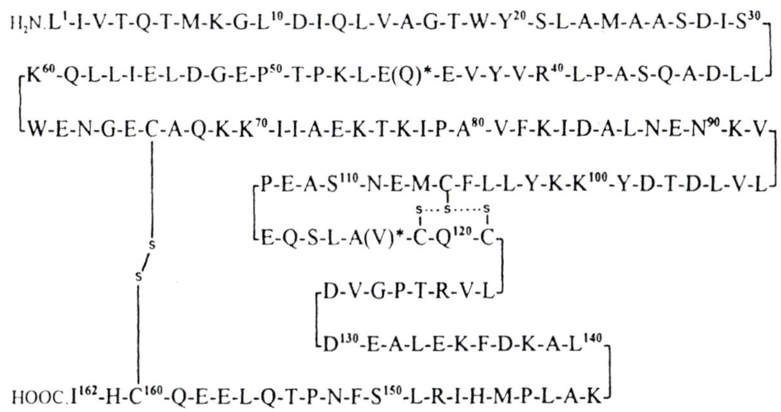

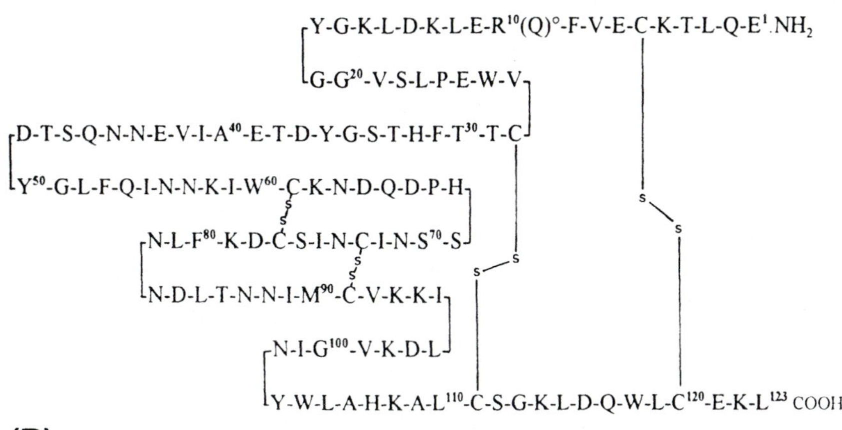

Figure 2 Primary structures of (a) β-lactoglobulin B (* = variant A) and (b) α-lactalbumin B (° = variant A).

concentration and pH. A comparison of the physicochemical properties of two major whey proteins is possible only if the experimental results are obtained under the same conditions [17].

The association properties of β-lactoglobulin depend on the pH [18]. At pH 5–8, β-lactoglobulin exists as a dimer, at pH 3–5 the dimers associate to form octomers, and at extreme pH values (<2 or >8) β-lactoglobulin exists mainly as monomers. At pH > 9, the molecule is irreversibly denatured [18]. Because of this pH-dependent association-dissociation behavior, no significant changes occur in the intrinsic viscosity and in the mobility of side chains (measured by ESR method after labeling the protein with nitroxide radicals) of β-lactoglobulin when pH is changed [19–21]. This very rigid and compact protein is less affected by ionic interactions.

At the oil/water interface, however, β-lactoglobulin undergoes significant structural changes. Rearrangements occur during film formation. The interfacial adsorption of

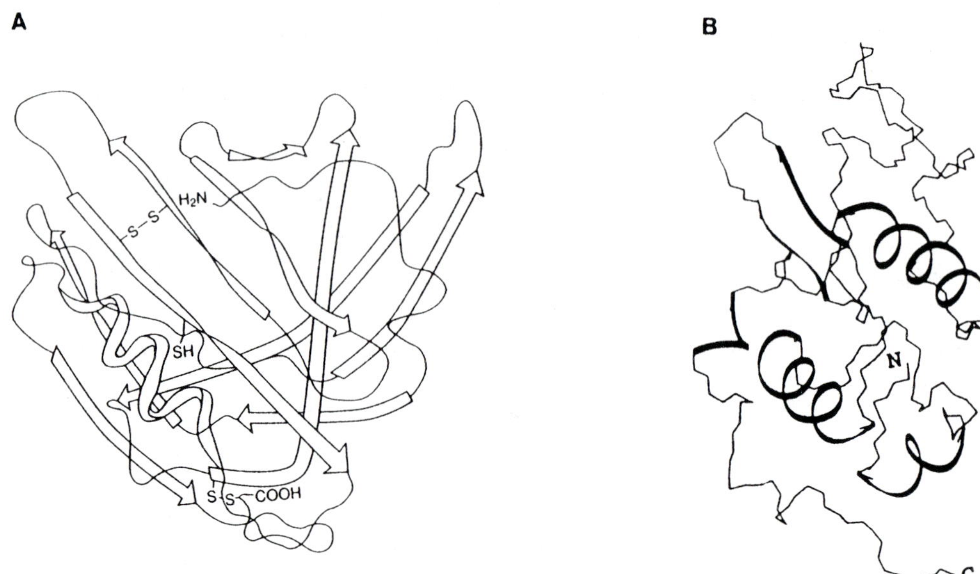

FIGURE 3 Tertiary structures of β-lactoglobulin (A) and apo-α-lactalbumin (B).

β-lactoglobulin can be regarded as a dynamic process involving partial and slow unfolding of one part of the molecule followed by binding of this region to the interface [22–24]. In a quiescent system, a monolayer forms within 6 hours [25]. This adsorption to the interface is pH dependent. At pH 9.0, the adsorption behavior of the denatured β-lactoglobulin resembles that of a flexible protein, such as β-casein.

Although β-lactoglobulin has a high mean residue hydrophobicity, it is very soluble in water [26–30], because most of the nonpolar residues are buried in the protein's interior and a majority of the polar groups are exposed at the surface [26].

In the case of α-lactalbumin, its conformational stability is related to its Ca^{2+}-binding properties. At pH < 4, Ca^{2+} is released and the apo-α-lactalbumin becomes flexible and can be easily proteolysed. At pH near the pI, the rotational diffusivity of the molecule is maximal and is greater than that of β-lactoglobulin [20,21] (Fig. 4). When bound to the air/water and oil/water interfaces, it exerts a maximum surface pressure even at low surface coverage. It is highly soluble even at its pI, and its solubility is not highly influenced by ionic strength and pH variations. Generally, the solubility of proteins increases with the ionic strength at a pH close to the pI, whereas the opposite behavior occurs at acid and basic pH [27,28].

It is often difficult to compare the surface properties of different whey proteins or of whey proteins with other proteins because these properties depend very closely on the denaturation state, the technological treatment used for protein extraction, and, above all, the solution conditions (ionic strength and pH). Nevertheless, β-lactoglobulin, which easily denatures at interfaces [31], is a better surface-active protein than α-lactalbumin [32]. This property is extremely important in the utilization of β-lactoglobulin in fabricated foods, especially as an egg white substitute. The surface-active properties are often closely related to the solubility [33,34]. The loss of solubility does not allow the molecules to diffuse and migrate to the interface. However, although a reasonably good cor-

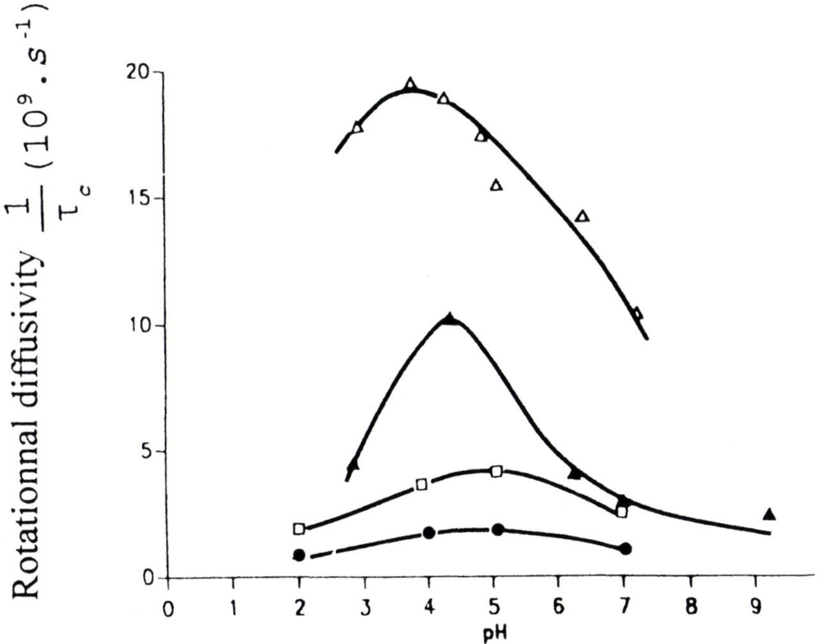

Figure 4 Effect of pH on the rotational diffusivity ($1/\tau_c$) of milk proteins. △: Caseinates (0.5 g/liter); ▲: caseinates (50 g/liter); □: α-lactalbumin (25 g/liter); ●: β-lactoglobulin (25 g/liter).

relation exists between surface properties and surface hydrophobicity, there are several exceptions [33,35].

III. STRUCTURE-FUNCTIONAL RELATIONSHIPS

The structural properties of whey proteins depend on several environmental factors (pH, presence of salts and other proteins) and also on technological treatments applied to milk. Thus, the impact of processing-induced structure modification on the functional properties should also be taken into account. This can be achieved by first studying the functional properties of the native protein and comparing them to those of the treated protein.

A. Solubility and Water Adsorption

The solubility of a protein depends on its water-binding capacity and the physical state. Water can be bound by weak-noncovalent-associations (i.e. physicochemical interactions), such as ion-dipole (ionic hydration), dipole-dipole (hydrogen bonding), and dipole–induced dipole (hydrophobic hydration) interactions, by entropic constraints (which lead to the formation of a caging structure—cryptan—around hydrophobic groups) named hydrophobic hydration, or by capillary forces porous particles. The rates of water adsorption and swelling depend on the number of hydration sites, the conformation, and the properties of the protein surface [26]. Whey protein concentrates are often very soluble even though their water-binding capacities are relatively low.

The water sorption of whey powders depends on water activity (a_w). For $a_w < 0.25$, the monolayer of structural water represents 7 g water/100 g powder for β-lactoglobulin [27]. At higher a_w, multilayer formation occurs and the water content reaches a transition area of 18–25 g water/100 g dry product (vitrous state). At $a_w > 0.75$, the hydration reaches 25–60 g water/100 g dry product. For β-lactoglobulin, the water binding is about 25 g/100 g and 30 g/100 g at $a_w = 0.8$ and $a_w = 0.92$, respectively [27]. Heat denaturation generally increases water binding, because of partial unfolding, which causes exposure of additional hydrophilic groups to water [28].

The water-binding capacity (maximum water quantity retained in the protein powder) depends on the structural state of the protein. In the dimer state, the water-binding capacity of β-lactoglobulin is lower than that of the octomer (40 g instead of 60 g H_2O/100 g protein). α-Lactalbumin, which is a more flexible molecule, is better hydrated than β-lactoglobulin (57 g/100 g protein). Whey powders, depending on the type of whey and the drying process, have water-holding capacities in the range of 70–147 g H_2O/100 g protein.

The solubility can predict, to some extent, the functional properties of protein powders. It depends very much on the environmental conditions. At the pI, the equality between positive and negative electric charges causes very intensive electrostatic attractions, resulting in a partial loss of solubility (total loss if the protein is denaturated). The presence of salts increases the solubility at the pI, whereas the opposite behavior occurs at acid or alkaline pH. The solubilities of the native whey proteins are not greatly influenced by salts [27,28]. Maximum solubility occurs at 0.1–0.15 M NaCl [36]. Kinsella [28] reported that milk powders containing large amounts of proteose-peptones were not very soluble. This could explain the antifoaming properties of these proteins.

Heating usually provokes protein-protein interactions, which cause aggregation and a loss of solubility at pI [27]. However, because they are soluble at acid pH, whey protein concentrates prepared by ultrafiltration can be used as protein-enriched soft drinks (for example some fruit juices).

B. Viscosity

Whey protein solutions are not very viscous even at high protein concentrations. The pH (pH < 8) has little influence on the native protein conformation (structure III), whereas the viscosity decreases when pH reaches the pI [37]. In many hydrated gel-type food products, the absence of syneresis and a homogeneous smooth texture are the main quality criteria that determine their acceptability. Whey proteins can bind water and stabilize food products after cooking (e.g., meat emulsions, sausages, cakes, etc.) when the pH is close to neutrality.

C. Gelation

Protein gelation can be achieved by the addition of salts or enzymes to a concentrated protein solution and also at extreme pH or heat. The heat-induced gelation of whey proteins follows a two-step mechanism: initial unfolding of the molecule followed by aggregation. The unfolding step exposes the buried hydrogen-bonding groups that bind water. When the protein concentration is 6–12%, protein-protein interactions cause formation of a highly hydrated three-dimensional network [38,39]. When protein-protein interactions are very strong, the water is released from the gel (syneresis) and the gel collapses. An equilibrium between attractive (network formation) and repulsive forces (which prevent the collapse of the gel and help retain water by capillarity) is required to form a stable gel network. Thermal gelation of whey proteins is influenced not only by the protein concentration, but also by the pH and salt content (e.g., Ca^{2+}). Strong, viscoelastic, and translucent gels are formed at acid (except at pH $< 3 - 4$ for gel strength) and alkaline pH. Coagulum-type opaque and weak gels are formed at pI.

D. Emulsifying Properties

Whey proteins behave as surfactants; they decrease the interfacial tension at the oil/water interface and form a thin cohesive film around oil droplets. They also stabilize emulsions against flocculation and coalescence.

It has been difficult to assess the emulsifying properties of whey proteins due to the following conditions: (a) different sources of whey protein concentrates (WPC) or whey protein isolates (WPI) and variations in their composition, (b) variations in the analytical methods used, and (c) variations in protein and mineral compositions.

1. Source of the Proteins

Liao and Mangino [40] stated that the solubility of different WPC from acid whey varied from 25 to 82%. The WPC samples exhibited emulsifying capacity in the range of 38–52 ml oil/g protein. These variations could be due to variability in the mineral composition, especially K, P, and Mg contents. In contrast, Kim et al. [41] observed a relatively constant emulsifying capacity of WPC and WPI from sweet whey (~52 ml oil/g protein); they also found a relationship between the emulsifying capacity and the soluble β-lactoglobulin content.

2. Influence and Analytical Methods

Kim et al. [41] also noticed that the measurement of emulsifying capacity in model systems did not quantitatively reflect the performance of WPC in food systems. Using the turbidometric method of Pearce and Kinsella [42], Kato and Nakai [43] observed a significant correlation between surface activity and surface hydrophobicity for the β-lactoglobulin and the BSA. Haque and Kinsella [44] pointed out that several other factors, such as pH, energy input during emulsification, and protein concentration, also influence the surface activity of proteins.

With regard to the measurement of the emulsion stability, it is often worth modifying the formulation of the product instead of inducing destabilization of the commercial product [45]. The stability is often determined by the conductivity method [33,46]. The measurement of the droplet size and size distribution by laser diffraction is very useful in calculating the interfacial area and therefore the interfacial protein concentration. These values give valuable information on the emulsifying capacity and stability of protein-stabilized emulsions.

3. Comparison of the Emulsifying Properties of Different Protein Concentrates

As a general rule, random coil–type proteins such as caseinates make better emulsifiers than globular proteins such as whey proteins [33]. Whey proteins are nevertheless very efficient emulsion stabilizers. However, the emulsifying properties of proteins can be interpreted properly only if the pH and temperature are specified. Morr and Ha [11] observed that at pH 6 and 8 sodium caseinates imparted a better emulsifying stability than did the whey proteins, whereas at pH 7 and at 65°C, the results were inversed. It appears difficult to establish a general rule for the emulsifying behavior of whey protein concentrates. It seems that other factors also contribute to their emulsifying properties. For instance, the solubility also influences the emulsifying properties [47], since it allows a rapid diffusion and adsorption of proteins to interfaces. Thermal denaturation of whey proteins during commercial preparation of WPC powders decreases their solubility and thus impairs their emulsifying properties. In contrast, partial denaturation leads to the improvement of these properties [21,33]. Thus, it is easier to obtain consistent results with purified proteins than with commercial preparations.

4. Emulsifying Properties of Purified Whey Proteins

The rate of adsorption of globular proteins to the oil/water interface is slower than that of small surfactant molecules; this is partly because of the need to undergo conformational change at the interface in order to remain at the interface. In the absorbed film, when the surface load is low, protein-protein interactions between emulsion droplets may lead to flocculation. However, if a surfactant is also used during emulsification, the flocculation of oil droplets can be avoided. The rate of decrease of surface tension can give valuable information on the adsorption rate. According to Kinsella [28], an unfolded protein such as β-casein forms a film more rapidly at the oil/water interface than does β-lactoglobulin or BSA. At the butter oil/water interface, the surface activities of various proteins follow the order β-lactoglobulin < α_s-casein < α-lactalbumin < BSA < casein micelles < β-casein. A good relationship exists between interfacial activity and emulsifying properties or surface hydrophobicity for some pure proteins. Incidently, highly surface-active proteins, such as β-casein, are also the most flexible and possess an optimum distribution of hydrophilic and hydrophobic regions.

During adsorption onto an interface, globular proteins generally retain part of their tertiary and secondary structures [34]. Thus, the adsorbed protein film contains only partly deformed protein molecules. On the other hand, "hard" proteins undergo very little conformational change at the interface [48]. By comparing the denaturation enthalpies of a protein in solution and in emulsion using the DSC method, it can be shown that the α-lactalbumin has nearly the same conformation at the interface and in solution; this is not the case of β-lactoglobulin or BSA.

The interfacial adsorption of the β-lactoglobulin depends on pH. At pH 9, the monomeric structure is destroyed and the adsorbed protein behaves like a flexible protein with loops and trains. At pH 7, the conformation of monomers and dimers is altered in the adsorbed state and the protein forms a rigid interfacial film. At pH 4.65, closer to pI, β-lactoglobulin adsorbs at a faster rate [25]. However, the size of the fat globules formed in the emulsion is very large at pH 4–6, even though the rate of adsorption is faster at this pH range than at any other pH [49]. Shimizu [24] has shown with an immunochemical technique that the conformational change in β-lactoglobulin at the interface involves one part of the molecule; the α-helix region was not denatured [50].

The stabilization of emulsions by proteins depends on many factors, such as the structure and net charge of the protein surfactant, pH and ionic strength, the viscosity of the aqueous phase, droplet density, temperature, and the physical and mechanical properties of the interfacial film (thickness, elasticity, and viscosity). The upper limit for the bulk phase protein concentration necessary to obtain a maximum surface or interfacial concentration is lower for whey proteins than for caseins. Globular whey proteins are adsorbed in the form of a monolayer and form a thin film, while caseins associate and form a multilayer film [51]. On the other hand, even with very small protein concentrations, whey proteins are efficient in creating stable emulsions [52].

pH plays a more important role in casein-stabilized emulsions than in whey protein–stabilized emulsions. The emulsion stabilization by β-lactoglobulin, α-lactalbumin, or whey protein isolate at a pH close to pI is lower than at pH 7 or 2 (Fig. 5). This effect is more significant for β-lactoglobulin than for the other whey proteins [21,53]. This observation can be linked to the fact that the diameter of the fat globules is maximum at pI [54] and that the interparticle electrostatic repulsions are very weak.

5. Emulsifying Properties of Whey Protein Mixtures: Competitive Adsorption

During emulsion formation with a protein mixture, some protein components may adsorb more readily than the others. Furthermore, a protein may sometimes be desorbed from the interface by another protein or by a surfactant because of differences in their affinity to the interface, i.e., competition for adsorption [47].

Competitive Adsorption of Whey Proteins

The ability of a protein to displace another protein from an interface has been clearly demonstrated with α_{s1}- and β-caseins. However, this behavior is not as obvious with whey proteins because the exchange is much slower with these globular proteins [55,56]. β-Lactoglobulin is displaced with difficulty from the oil/water interface by α-lactalbumin, whereas the opposite process is easier. If a mixture of the two major whey proteins is used in an emulsion, it can be observed (by electrophoresis of the two proteins of the cream and serum phases after separation by centrifugation) that β-lactoglobulin is more tightly adsorbed to the interface than α-lactalbumin [53,57]. The polymerization of β-lactoglobulin at the interface via a sulfhydryl-disulfide interchange reaction could be the cause of this irreversible adsorption [55]. The displacement of the β-lactoglobulin

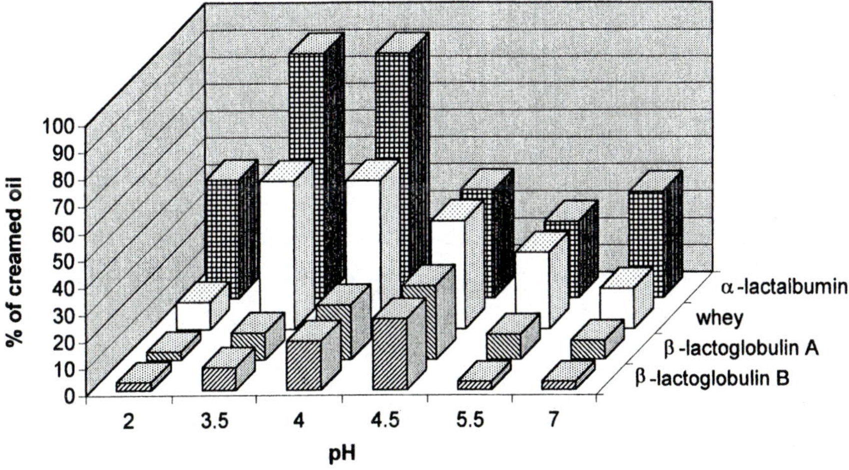

Figure 5 Effect of pH on coalescence of emulsion made with whey proteins after centrifugation at 1000 *g* for 100 minutes.

becomes all the more difficult if the emulsion is aged. Both the intermolecular disulfide bridge formation and the irreversibility of the adsorption are significantly increased with partially denatured β-lactoglobulin [58]. This explains the fact that β-lactoglobulin, when used alone, possesses better emulsifying properties than a 1:1 mixture of β-lactoglobulin and α-lactalbumin or whey protein concentrate [53]. Because of polymerization, the adsorbed β-lactoglobulin forms a thick film at the interface at neutral and alkaline pH, but not at acid pH [59] (Table 4).

Competitive Adsorption Between Whey Proteins and Other Proteins

β-Casein can displace, although slowly and only partially, freshly adsorbed β-lactoglobulin at the dodecane/water interface [56,59]. The exchange is very rapid if the intermolecular disulfide bridges of β-lactoglobulin are chemically stabilized. β-Lactoglobulin can displace α_{s1}-casein but cannot displace either κ-casein or ovalbumin [48].

Competitive Adsorption Between Whey Proteins and Nonprotein Surfactants

Small surfactant molecules (e.g., SDS, fatty acids, sucroester) interfere with the adsorption of whey proteins at interfaces. Some can either partially or totally displace proteins from the oil/water interface and thus influence the rheological properties of the protein film and, consequently, the stability of the emulsion [47].

Co-adsorption Mechanism

A mixture of β-lactoglobulin and α-lactalbumin can form a stable emulsion (Fig. 5). In this case, β-lactoglobulin constitutes the primary layer at the interface and α-lactalbumin the secondary layer. This explains why β-lactoglobulin remains at the interface when a whey protein film is desorbed by a surfactant (e.g., SDS). A similar phenomenon is observed when the addition of κ-casein to an emulsion stabilized by β-lactoglobulin results in adsorption of κ-casein without the loss of β-lactoglobulin [59].

E. Foaming Properties

The molecular factors affecting the ability of a protein to form and stabilize a foam are similar to those for emulsion formation, that is, rapid diffusion of the molecule to the air/water interface, amphiphilic structure, solubility, and flexibility [60,61].

TABLE 4 Relative Concentration of Whey Proteins in Protein Fraction Adsorbed on Fat Globule Surface of Emulsions Made with Milk Proteins at Various pH

Relative proportion adsorbed on interface (%)	pH				Proportion in milk (% m/m)
	3	5	7	9	
β-Lactoglobulin	12.9	16.6	46.1	61.9	10
α-Lactalbumin	48.3	24.4	11.0	9.9	3.8
Bovine serum albumin	1.0	10.6	2.7	1.0	1.3
Casein + immunoglobulins	34.1	40.0	31.3	20.1	83.4
Transferrin + lactotransferrin	3.7	8.4	8.9	7.4	2.2

Source: Adapted from Ref. 36.

κ-Casein rapidly adsorbs to the air/water interface and creates large bubbles, probably because of its highly flexible structure. In contrast, whey proteins adsorb and spread very slowly at the interface and create smaller foam particles [28]. Because the interfacial tension at the oil/water interface (20–40 mN/m) is lower than that of the air/water interface (73 mN/m), globular proteins are generally more denatured at the air/water interface than at the oil/water interface. Thus, the extent of denaturation of a protein at the air/water interface is critical to its foaming properties.

The foaming capacity of whey proteins varies depending on its method of preparation. Elimination of insoluble particles from acid or sweet whey (by centrifugation or filtration) increases the overrun of the whey protein solution. Similarly, microfiltration of whey eliminates foam depressors, such as lipoproteins. At least, the presence of the proteose-peptones explains the variability of the foaming properties of commercial powders of whey proteins [62]. The negative effect of the residual lipids which decrease the foaming ability are also apparent [63].

The ability of protein to stabilize foams is related to the rheological properties of the interfacial film—if the film is thick, it increases the interfacial viscosity and the drainage is slowed. This property is apparently incompatible with the qualities required for foaming (solubility, mobility, and flexibility) but the interfacial film must also be viscoelastic to resist the constraints. A limited surface denaturation is desirable to obtain satisfactory viscosity, rigidity, and cohesion of the interfacial film, but the surface denaturation must not lead to a protein aggregation (excessive denaturation). For example, β-lactoglobulin and α-lactalbumin give the best stabilization of foams at a pH close to pI ([28] for the whipping method; [64] for the bubbling method, Fig. 6) since the solubility is a little lower than the solubility at a pH far from pI; the film cohesion would be improved by the decrease of electrostatic repulsive forces between the protein molecules.

β-Lactoglobulin is a better foam stabilizer than α-lactalbumin and whey protein concentrates (Fig. 6). This is the reason why the β-lactoglobulin content of the protein concentrates is so essential for WPC foam properties [63] and why the purified industrial

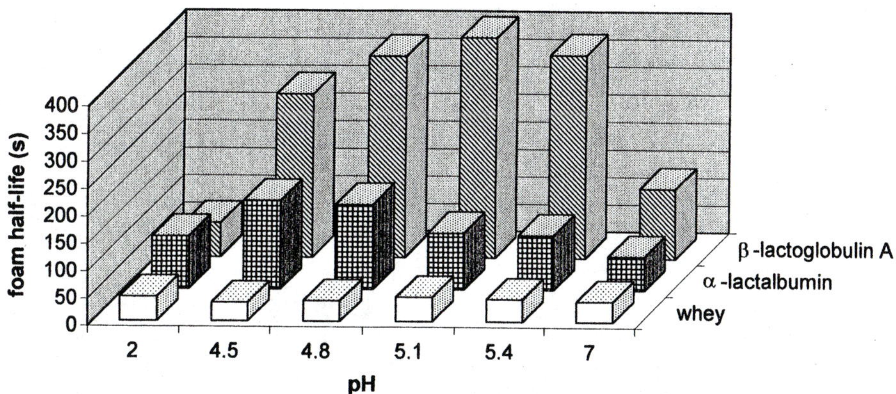

FIGURE 6 Effect of pH on the stability of foams made with 0.1% (w/v) whey protein solutions. Foams were made using a bubbling method, and the half-lives of the foams were determined by the conductimetric method.

β-lactoglobulin is used as a foaming ingredient principally in confection (Process Spherosyl–Rhône-Poulenc–France). The presence of proteose-peptones (hydrophobic molecules) has a negative effect on foam stability. Nevertheless, some insoluble particles occasionally have a beneficial effect on the foam quality [65]. Foam stability is maximum at a pH close to pI if the proteins remain soluble. After a long heating the stability is increased for high pH values (probable role of the deprotonation of SH groups at alkaline pH; nucleophilic group). The role of thiolate function and of the equilibrium between sulfhydryl groups and disulfide bridges during the denaturation is developed in Chapter 4, and shows the beneficial effect of a partial physical or chemical denaturation on the foam stability.

Adding some ingredients to whey protein solutions can improve the foam stability. The addition of sucrose improves the foam stabilization by whey proteins only at pH 6; at pH 8.5, sucrose inhibits interfacial denaturation of whey proteins and thus decreases their foam stability. Some polysaccharides, such as dextran sulfate, improve the stability of β-lactoglobulin foam at neutral pH [66]. The addition of certain basic proteins, such as lysozyme and clupeine, also improves the stability of whey protein foams, especially when sucrose is also included [67]. The positively charged lysozyme and clupeine interact with the negatively charged whey proteins; this strengthens the cohesion of the interfacial film and increases its viscosity. Salts, by neutralizing these electrostatic interactions, inhibit this effect.

IV. EFFECTS OF PROCESSING ON PROTEIN STRUCTURE AND FUNCTIONAL PROPERTIES

As a general rule, proteins, especially milk proteins, are processed prior to their use in foods. For example, industrially prepared milk proteins are generally prepared from pasteurized milk. Pasteurization may cause protein denaturation, and therefore most of the commercial milk proteins are not necessarily native proteins. Hence, before attempting to understand the structure-function relationships of milk proteins, it is first of all necessary to understand structural changes in milk proteins during various technological treatments.

A. Protein Denaturation: New Structures

Milk is processed by different methods to stabilize and preserve it not only in the liquid form, but also in the manufacture of products like cheese, yogurt, caseinate, and whey protein isolate. These treatments can be divided into three categories: thermal, mechanical, and biochemical.

1. Thermal Treatments

Cooling

Refrigeration increases the reactivity of the sulfhydryl group of β-lactoglobulin [68]. It is generally accepted that cooling decreases hydrophobic interactions. Since hydrophobic interactions are responsible for the dimerization of β-lactoglobulin at natural pH of milk [5,18], cooling can induce dissociation of the dimer at neutral pH and thus enhance the reactivity of the SH group [69]. The SH group may follow two possible chemical reactions at low temperatures:

1. Low temperatures could increase the sulfhydryl-disulfide interchange reaction between β-lactoglobulin molecules, resulting in the formation of dimers, trimers, and polymers. However, there is no experimental evidence that disulfide-bonded polymers of β-lactoglobulin occur either in refrigerated milk or in neutral β-lactoglobulin solutions.
2. Because of an increase in oxygen solubility at low temperatures, the SH groups may be oxidized, resulting in formation of disulfide cross-linked β-lactoglobulin dimers.

At $3.7 \leq pH \leq 5.2$, refrigeration (3°C) induces a reversible association of eight β-lactoglobulin units (i.e., octomerization) [18,70–72], and at $pH \geq 8$ refrigeration accelerates irreversible denaturation of the β-lactoglobulin monomer [73,74].

At pH 3 and −4°C β-lactoglobulin is denatured by 3.5 M guanidine hydrochloride [75], whereas only a minor change in its tertiary structure occurs at 33°C. These observations demonstrate that under certain conditions, refrigeration is not inconsequential in terms of structural changes in β-lactoglobulin.

Heating

When a β-lactoglobulin solution is heated between 50 and 80°C at pH 6.85, the reactivity of the thiol group increases; this may be due to dissociation of the binary units [76] besides the SH activation energy increase. If intermolecular hydrophobic interactions increase with temperature (facilitating association of protein units), it is possible that hydrogen bonds have a destabilizing effect when heating and at last provoke β-lactoglobulin unit dissociation. Dissociation at $5.2 \leq pH < 7.4$ seems to occur at a temperature above 30°C [77,78]. At pH = 7.5, unfolding is clearly increased beyond 60°C. Above 70°C, new disulfide bonds appear [79] creating β-lactoglobulin polymerization [80,81]. This phenomenon was also observed with soybean proteins (globular proteins that possess S-S and SH functions) [82]. β-Lactoglobulin polymers seem to aggregate after S-S exchange in complexes [80].

When pH is decreased (especially from 7.5 to 6.5), the thermodynamic stability of β-lactoglobulin is enhanced [83]. Whatever the ionic strength, the denaturation temperature value increases from pH = 9 to an optimum at pH = 3 [83,84]. When pH is below 6.5, the pH of solutions is very far from the 9.35 pK value of the β-lactoglobulin thiol group [75]. In acid medium, the number of deprotonated sulfhydryl groups (nucleophilic) is insignificant. That is why at $pH < 6.5$, even with high temperature, no disulfide bond exchange is possible. In order to demonstrate that polymerization of β-lactoglobulin proceeds from a nucleophilic attack by thiolate function on disulfide bonds, β-lactoglobulin was heated 40 min at 75°C (pH = 6.9) with *N*-ethylmaleimide (SH group protector). No polymerization was observed [79]. To confirm the universality of these results, nearly the same experiment (pH = 8; 90°C during 30 min) was performed with varied globular proteins containing SH and S-S bonds. Once again, no polymer was obtained [85]. Conversely, the total reduction of disulfide bonds prevents polymer formation [85,86]. These results [79,80,82,85,86] lead to the conclusion that the SH group is responsible for polymerization in S-S–containing proteins. At $pH > 8$, the SH reactivity increases so rapidly [75] that the SH–S-S interchange reaction can occur even without heating. As a result of this polymerization, the buried hydrophobic groups may become exposed, which can cause aggregation [75,79,87,88].

In an acid medium, no disulfide bond formation occurs even though aggregation can occur at pH < 6.5 [78]. At pH < 3, heat denaturation of β-lactoglobulin appears to be reversible when the temperature does not exceed 120°C and the treatment is of short duration [74]. At pH > 3.5, the same treatment causes irreversible denaturation. When β-lactoglobulin is heated for 30 minutes at 90°C at pH 2.5, no change in solubility occurs. However, when the pH is raised to 4.5 after this treatment, only 55 or 60% of the protein remains soluble [89,90]. Thus, though no insoluble aggregate appears after heating in an acid medium, some denaturation does occur. Nevertheless, most of the β-lactoglobulin remains in native or quasi-native state. The partial reversibility of unfolding also has been shown to occur at pH 2 and 5.5 [91].

Besides pH, ions also play a role in thermal denaturation of β-lactoglobulin. The aggregation during heating (80°C, 10 min, 6.0 < pH < 7.6) increases with calcium chloride concentration [78]. The addition of calcium salt before or after heating markedly increases flocculation of β-lactoglobulin at pH 6.5 (addition of calcium salt at room temperature does not lead to flocculation) [78]. The addition of calcium chloride actually induces a change in β-lactoglobulin conformation, which does not occur with sodium chloride [92]. This conformational change induced by Ca^{2+} may facilitate aggregation during subsequent heating. Ca^{2+} ions also increase the reactivity of SH groups [92]. Flocculation of β-lactoglobulin by Ca^{2+} addition could also occur due to Ca^{2+} bridging between two carboxyl groups. Although sodium chloride has no major effect on β-lactoglobulin conformation, at pH > 6 it facilitates heat denaturation and aggregation of β-lactoglobulin [78,93–96]. In fact, generally all salts induce aggregation, which indicate that it is ionic strength and not the salt per se that is responsible for aggregation. Salts promote aggregation via hydrophobic interactions owing to a reduction in electrostatic repulsion. At acidic pH values, addition of 0.5 or 1 M NaCl prevents thermal aggregation, while above 2 M NaCl induces aggregation [93].

The addition of citrate or phosphate to β-lactoglobulin solution decreases its aggregation during heat treatment [97]; this is because of chelation of calcium ions by citrate. Moreover, kosmotropic salts, such as citrate, phosphate, sulfate, tartrate, and chloride (in decreasing order of efficiency), also seem to prevent SH–S-S interchange (i.e., polymerization) [75].

Lactoferrin and α-lactalbumin are whey metalloproteins containing no SH group but with 17 [12,13] and 4 disulfide bonds [6,7], respectively; they also contain two Fe^{3+} ions [12,13,98,99] and one Ca^{2+} [100,101], respectively. The apo-lactoferrin (with no Fe^{3+} ion) [102,103] and the apo-α-lactalbumin [104–107] are less thermostable than the native or the corresponding holo-protein forms. This suggests that these ions protect these proteins against thermal denaturation. The denaturation temperature of α-lactalbumin (~65°C) [106,108–113] is lower than that of β-lactoglobulin (71–75°C) [76,108,112]. Nevertheless, most of the heated pure α-lactalbumin recovers its initial structure after heat treatment. The denaturation recorded on DSC corresponds essentially to a transition in the secondary structure (58°C) [110,114]. The α-lactalbumin could be considered to be more heat resistant than β-lactoglobulin owing to reversibility of its denaturation [108], even though DSC results suggest otherwise [83]. Partial polymerization of α-lactalbumin in sodium phosphate buffer at pH 7 occurs when it is heated for 10 minutes at 100°C [110].

Nevertheless, when α-lactalbumin is heated in the presence of β-lactoglobulin at neutral pH, α-lactalbumin is easily and irreversibly denatured. At temperatures above 70°C at 6.5 < pH < 7.5, α-lactalbumin/β-lactoglobulin complexes appear because of a disulfide interchange reaction between the proteins [68,115–118]. In whey protein concentrates, α-lactalbumin is more extensively heat denatured than β-lactoglobulin [119] because of its lower denaturation temperature. However, these results are not totally reliable. The conclusion seems to depend on the analytical technique used. Immunological methods indicate opposite results, i.e., β-lactoglobulin is more denatured than α-lactalbumin [120]. The greater extent of denaturation of β-lactoglobulin might be related to both β-lactoglobulin–α-lactalbumin copolymerization and β-lactoglobulin polymerization, whereas the denaturation of the α-lactalbumin involves only copolymerization.

The thermodenaturation of lactoferrin is increased in the presence of β-lactoglobulin [121] and could follow the same chemical pathway as that of α-lactalbumin. It means that inclusion of proteins possessing free SH groups (such as β-lactoglobulin and bovine serum albumin) in solutions of globular proteins containing no free SH groups but only S-S bonds can cause irreversible thermal denaturation of the latter.

When sweet whey is heated above 70°C, mainly dimers of β-lactoglobulin and copolymers of α-lactalbumin–β-lactoglobulin are formed [117,122,123]. As temperature is increased, not only β-lactoglobulin and α-lactalbumin, but also immunoglobulin and bovine serum albumin disappear [124–126] to form various copolymers [124,127] and/or complexes [128]. An increase of calcium concentration in whey protein concentrate increases the amount of insoluble precipitates produced during heating [119].

At pH < 7.0, depending on the actual pH, aggregation of whey proteins can follow two different mechanisms [129]. At $4.0 \leq \text{pH} \leq 6.5$, the SH–S-S interchange reaction is not possible. However, aggregation of most of the proteins can still occur in the presence of Ca^{2+} [78]. When the pH of whey protein concentrate is lowered from 7.0 to 5.8 before heating, formation of both soluble aggregates and insoluble precipitate decreases [119]. Precipitation occurs even without Ca^{2+}. The precipitate and the soluble aggregates (insoluble with 2% of trichloroacetic acid), formed during heating of the calcium-free whey protein solution at $4.5 \leq \text{pH} \leq 6$, contains predominantly β-lactoglobulin and only a small amount of α-lactalbumin [105,129]. However, extensive precipitation occurs when the pH of the heated whey protein solution is subsequently adjusted to close to the pI of β-lactoglobulin or α-lactalbumin or if Ca^{2+} is added [99,130].

At pH < 4.0, a small amount of α-lactalbumin forms aggregates with β-lactoglobulin [129] and with BSA [130]. These aggregates are soluble in solution adjusted to pH 2.5 or at pH 6 but are insoluble at the pI of these proteins [131]. Aggregation probably occurs via hydrophobic interactions [124]. Heating causes a decrease in secondary structure (α-helix and β-sheet) content of α-lactalbumin and β-lactoglobulin [132] and the exposure of hydrophobic regions. Figure 7 shows denaturation mechanisms at various pH values.

Homogenization

During homogenization [133], a new surface-active film is formed around fat globules for two reasons. The milk fat globule membrane is fragile and is broken easily under turbulent flow during homogenization. The decrease in the diameter of fat globules increases the interfacial area, which necessitates additional surface-active material [134].

When agitation is weak (for example, convection during heating), whey proteins cover the oil/water interface of newly formed milk fat globules [135–139], especially

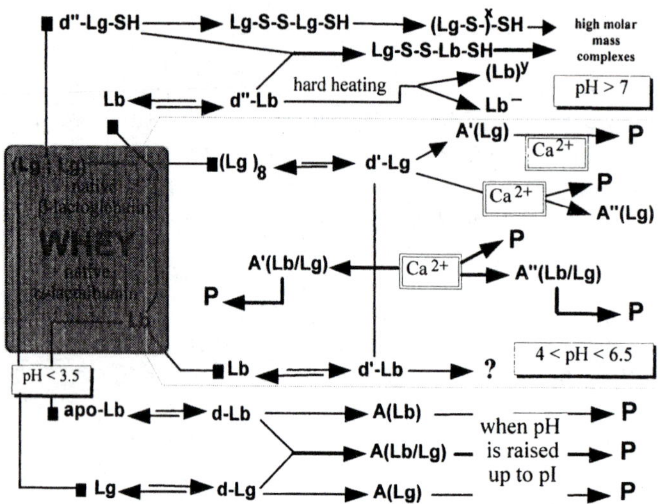

FIGURE 7 Summary of mechanisms involved in whey protein denaturation during heating (⇆ or →). Lg = β-Lactoglobulin; Lb = α-lactalbumin; d = denatured; A = aggregates; P = precipitates; Lb-S-S-Lb-SH = Lb dimers; $(Lg\text{-}S\text{-})^X\text{-}SH$ = Lb polymers; Lg-S-S-Lb-SH = Lg/Lb copolymers.

when the temperature exceeds 72°C [140]. Under turbulent conditions, the interface material essentially consists of caseins micelles [137,141,142]. However, although whey proteins constitute only 5% of the total proteins, they cover about 25% of the newly formed fat globules surface in homogenized milk [134]. Upon adsorption to the fat globule surface, whey proteins undergo irreversible denaturation [143,144]; this also involves the SH–S-S interchange reaction [59,145–147]. When emulsions made with whey proteins or β-lactoglobulin are stored for a few days, this (co-)polymerization on the interface can lead to surface gelation [145,148]. The gelled film cannot be desorbed from the interface by detergents, such as Tween 20 [56,149].

High-Pressure Treatment

When whey proteins are subjected to high pressure, only β-lactoglobulin is denatured, not α-lactalbumin [150]. As discussed earlier, β-lactoglobulin has the potential to form a dimer via oxidation of thiol groups. Since α-lactalbumin has no free SH group, condensation of this protein by oxidation cannot occur. However, in contrast to thermal denaturation at pH > 6.5, high pressure treatment should not induce the SH–S-S interchange reaction [151,152].

3. Biochemical Treatment

pH Modification

When pH is decreased to around 4, the conformation of α-lactalbumin changes precipitously, and it becomes more sensitive to heat- or enzyme-induced denaturation [104]. The Ca^{2+} loss could be responsible for this structural change. The Ca^{2+} ion prevents unfolding when one or two of the four disulfide bonds are reduced and plays an important role in the maintenance of the structure of α-lactalbumin [153,154]. When Ca^{2+} is removed, the apo-α-lactalbumin unfolds more easily at the oil/water interface than the native form [31].

The degree of association of β-lactoglobulin depends on the pH value [18–73]. At pH > 8 [18,72,73], it is reversibly denatured even without heat treatment; this is because of exposure of the SH group, which facilitates the SH–S-S interchange reaction [126].

Dielectric Constant

In ethanolic solutions with a dielectric constant of $\epsilon = 60$, the secondary structure of β-lactoglobulin is altered [155,156]. At $\epsilon = 78$ (in water), β-lactoglobulin is essentially a β-sheet–type protein. At $\epsilon < 60$, the secondary structure is mainly made up of α helix. At $\epsilon < 50$, the retinol-binding capacity of β-lactoglobulin is lost. This property, which reflects a structural peculiarity, is recovered when β-lactoglobulin is returned to an aqueous solution.

Generally, denatured protein substrates are easily hydrolyzed by proteases [157, 158]. This is also true for β-lactoglobulin; the degree of hydrolysis of β-lactoglobulin by pepsin is quite low in water [159,160], however, it increases dramatically in 20–35% (v/v) ethanol, but decreases at higher ethanol concentrations [161].

Enzyme Treatment

The improvement of functionality properties of proteins by enzymic hydrolysis is discussed in Chapter 15. Here, only the relationships between structure-enzymic hydrolysis are analyzed.

β-Lactoglobulin is not significantly hydrolyzed either by pepsin at pH 2 or 2.5 [159,160] or by chymotrypsin at pH 8 [159]. α-Lactalbumin is not significantly hydrolyzed by papain at pH 8 [160] or by trypsin at pH 7.5 [162]. The degree of hydrolysis of β-lactoglobulin by papain [160] or by trypsin [162] is only slightly higher than that of α-lactalbumin, but it is still very low. After a heat treatment above 80°C, the denatured β-lactoglobulin is more easily hydrolyzed by papain, pepsin, and chymotrypsin than the native protein [159,160]. Heat denaturation facilitates increased accessibility of hydrophobic regions, such as phenylalanine, leucine, tyrosine or tryptophan residues, to pepsin and chymotrypsin. On the contrary, heated α-lactalbumin does not show increased susceptibility to pepsin or papain hydrolysis [160]. A combination of heat treatment and enzyme hydrolysis could be used to improve the functional properties of whey proteins.

B. Functional Properties of Processed Whey Proteins

Processing technologies can sometimes improve the functional properties of proteins; however, more often than not they impair the functional properties. The effects of some commonly used processing technologies on functional properties of whey proteins are discussed below.

1. Thermotropic Gelation

Protein Thermodenaturation Before Gelation

Many reports on factors affecting gelation properties of isolated or mixed whey proteins have been published. Publications on the gelation properties of denatured whey proteins are, however, less prevalent. Nevertheless, their functional properties could be affected by prior modifications. When pasteurized whey protein isolates are stored for 2 years or when they are dry-heated at 80°C for a week, the denatured WPI solutions (12%) form weaker gels than those of freshly pasteurized WPI. It seems that the gel strength is significantly correlated to the β-lactoglobulin monomer concentration. Prolonged storage or dry heating increases the concentration of β-lactoglobulin polymers as well as β-lactoglobulin–α-lactalbumin copolymers [163]. Notwithstanding these undesirable effects of prior heat treatment on gelation, it is quite possible that such heat-denatured proteins also may provide some opportunities to develop gels with new qualities or properties. For instance, when a two-step heating method is used, it is possible to obtain a translucent and firm gel using lysozyme (with dithiothreitol) [164], bovine serum albumin [165], and β-lactoglobulin [166]. This would be impossible with a one-step heating method. In this case, the initial heat denaturation with no or very little NaCl ($pH \geq 7$) presumably creates linear polymers, which during the second heating step with high sodium chloride concentration facilitates formation of an organized network.

Reduced Protein Gelation

Although α-lactalbumin does not form a thermally induced gel even at concentrations up to 20% [167], an α-lactalbumin gel can be obtained with a moderate glutathione concentration. Partially reduced α-lactalbumin can influence the gel strength of a β-lactoglobulin–α-lactalbumin–glutathione mixture gel [168].

2. Emulsifying Properties

Heating a Fresh Emulsion

Heating a freshly prepared whey protein–stabilized emulsion increases its stability. The heating induces gelation of the adsorbed film; this increases its mechanical strength [57] and thereby the stability of the emulsion [54].

Moderate heat treatment creates small aggregates at pH 5 and small polymers at pH 7. The heat-denatured β-lactoglobulin prepared under these heating conditions possesses better emulsifying properties than the native protein [169]. It is likely that limited thermal modification of β-lactoglobulin (small "soft" complexes) facilitates interfacial gelation of this protein and thus increases its ability to form a cohesive film.

On the other hand, when β-lactoglobulin is heated for a long time (especially at pH 5, but also at pH 7), its emulsifying properties decrease [21,52]. Similarly, thermal denaturation of whey protein isolates (at >65°C, pH 7) decreases their capacity to stabilize an emulsion [170]. The heat denaturation at high temperature or over a long time creates high molecular weight complexes. These large whey protein complexes cannot rapidly adsorb to the oil/water interface. Also because of the rigidity of these complexes ("hard" complexes), large emulsion particles are formed [54]. In contrast, in the case of α-lactalbumin, heating of the protein at pH 7 for a long time increases its emulsifying properties [171]. Heating of α-lactalbumin at pH 2, 5, or 9 before emulsification also increases its emulsion stability [171]. It should be noted that, as discussed earlier, the structural consequences of heating on α-lactalbumin are different if β-lactoglobulin is present.

The effects of thermal treatment on proteins depend nevertheless on the medium. For example, heating of a whey protein solution (2.5% w/v) at 90°C for 5 minutes at extreme values of pH increases its emulsifying properties [170]. This behavior is different from that of the sample heated at pH 6.5.

The emulsifying properties of undenatured and heat-denatured WPI blends increase proportionally with the amount of denatured WPI in the blend [172]. The reasons for this are not very clear, however, it appears that mixtures of monomeric and polymerized proteins form a cohesive interfacial film.

Reduction of Whey Protein

Reduction and S-carboxymethylation of α-lactalbumin or β-lactoglobulin decrease their emulsifying activity index, but increase the emulsion stability [173]. In contrast, the emulsifying activity index increases when 75% of the disulfide bonds present in these proteins are reduced by sulfitolysis. However, the stability of an emulsion prepared with a 0.1% (w/v) protein solution at pH 7 decreases with the degree of sulfitolysis of β-lactoglobulin (Fig. 8) [174]. At the same time, there is an increase in interfacial protein concentration with the degree of sulfitolysis (Fig. 8), which could be due to an increase in molecular flexibility.

Whey Proteins in Presence of Ethanol

The emulsifying capacity of WPI increases in the presence of 20% (w/w) ethanol [52], probably because of changes in secondary structures and the flexibility of the proteins [155,156].

3. Foaming Properties

Heated α-lactalbumin (1% w/v) at pH 5, 7, or 9 has better foaming capacity than native protein [21,171]. Heating of α-lactalbumin at pH 7 or 9 gives more stable foams than the native protein (Fig. 9) [171]. On the other hand, heating at pH 2 (apo-foam) or at pH 5 (near the pI) decreases the foaming properties of α-lactalbumin, especially when the heat treatment is severe. The thermal denaturation, which is extensive at pH 5 and more so at pH 2, could induce hydrophobic aggregation [9,107], which is detrimental to stability of foams. In the case of β-lactoglobulin, heat denaturation has a negative effect on the overrun at all pH values but has a positive effect on foam stability, especially at pH 2 and 5 [171]. At pH 2 and 5, no SH–S-S interchange reaction during heating is

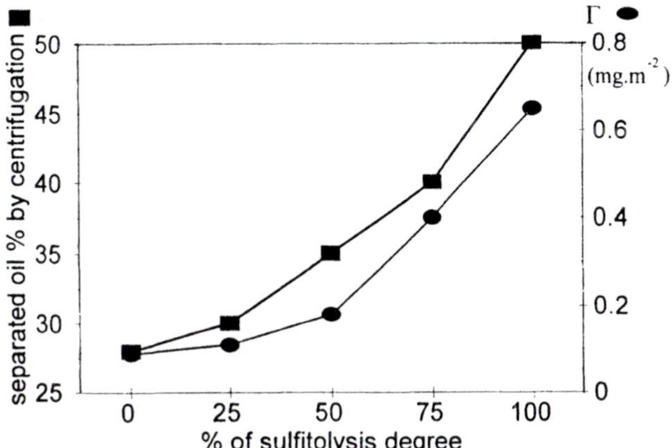

FIGURE 8 Effect of the extent of cleavage of disulfide bonds (sulfitolysis) on the emulsifying properties of β-lactoglobulin. ●, Protein concentration at the O/W interface; ■, emulsion stability.

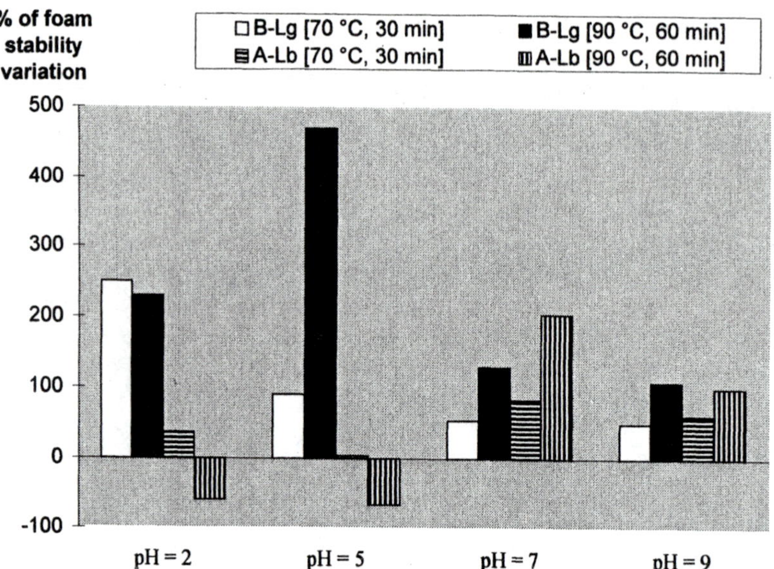

FIGURE 9 Foam stability of heat denatured whey protein solutions (1 g/liter) at pH 7 (half-life of the foam was determined with conductimetric analysis and a bubbling method).

possible; therefore, aggregation of β-lactoglobulin probably occurs via hydrophobic interactions. Thus, the improvement in foam stability might be related to the protein's ability to form a cohesive film.

Moderate heating (50–65°C for 30 min) improves the foaming properties of whey protein concentrates (as determined by a whipping method). The foaming properties are improved more by heating at pH 5 than at pH 7. At pH 5, heating of WPI at 80°C for 10 minutes before whipping improves the stability by 65%; the stability decreases slightly by heating a few minutes at pH 4 or 7 [28]. Above 70°C, the functional properties of whey protein concentrates decrease [28,175]. Heating of whey protein obtained by ultrafiltration at 90°C, pH 6.5 (5 min, 2.5% w/v) decreases the stability of its foam at pH 2, 3, 6, 7, and 8 compared to that of the native protein [170]. Although there is little difference between results with pure whey proteins and mixtures (WPC or WPI), it is not meaningful to compare these results, because the bubbling method, which uses a low protein concentration, does not supply as much energy as the whipping method [52,170]. When whey proteins are heated at extreme pH (pH 3 or 10), the foam stability increases over a wide range of pH (from 4 to 8) [176].

Foaming properties of β-lactoglobulin seem to be highly correlated to the secondary structure (especially β-sheet) [177]. However, it is not clear whether the improvement in the foaming properties (at pH $\geq$ 4) of whey protein heated at pH 3 or 5 is due to an increase in β-sheet content. It seems that heating at pH 2 exposes the tryptophan residues to the solvent [84]. However, since tryptophan residues are not located in the β-sheets regions of β-lactoglobulin [4,178] as well as of α-lactalbumin [9,179], it cannot be concluded that an increase in the exposure of hydrophobic regions is accompanied by an increase in β-sheet content.

It seems that foaming properties of emulsified WPC (obtained by ultrafiltration) are positively correlated to the number of reactive sulfhydryl groups [180]. Storage of the emulsion (which induces polymerization on interface [147]) or prior heating of whey proteins in an alkaline medium severely impairs its foaming properties.

Reduced and S-carboxymethylated β-lactoglobulin and α-lactalbumin have better foaming capacities than active proteins. But these modified whey proteins stabilize somewhat less aqueous foam produced by a bubbling method than native proteins [173]. Upon sulfitolysis, β-lactoglobulin [174] and whey proteins [29] show higher surface pressure at 50% modification. The increase in flexibility caused by breakage of disulfide bonds could facilitate adsorption at the air/water interface and thus improve the foaming capacity. Because sulfonation increases electrostatic repulsion among the protein molecules, which would be unfavorable for foam stability [181,182], only a 50% sulfitolysed β-lactoglobulin [174] (Fig. 10) or 75% sulfitolysed whey protein mixture [29] exhibits better foam stability and capacity. It is important to note that sulfitolysis seems to be favorable for foaming properties (Fig. 10), whereas it is unfavorable for emulsifying properties (Fig. 8). For foam stability, the flexibility of protein surfactant is essential, whereas the SH–S-S interchange seems to be important for the creation of a strong cohesive film around fat globules.

A small quantity of ethanol increases the stability of foams made with β-lactoglobulin solutions. The reason for this seems to be related to the lowering of surface tension [183] without structural modification of protein. At higher ethanol concentration, however, the foam is not very stable, probably because of a modification of the structure.

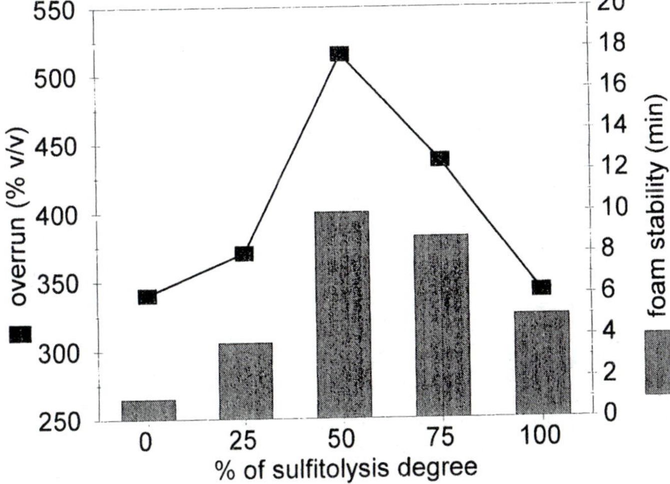

FIGURE 10 Effect of the extent of cleavage of disulfide bonds (sulfitolysis) of β-lactoglobulin on its foaming properties.

V. CONCLUSION

Extensive knowledge of the structure of the major whey proteins is very useful for a better understanding of the main functional properties. The effects of the technological treatments are fairly well understood; however, there is a lack of knowledge about the interactions between these protein molecules and other components (lipids, carbohydrates, salts, surfactants, etc.) in food systems. Until now, protein molecules were considered rigid systems with fixed spatial conformation and not as mobile smooth polymers with rapid motions. Several functional properties, such as interfacial adsorption of proteins, cannot be explained from a tertiary structure viewpoint; the kinetics of adsorption and dynamic changes in the protein structure should also be taken into account. These new approaches will be necessary to understand and predict the effects of new technologies on the functional behaviors of whey protein ingredients in food formulations.

REFERENCES

1. G. Braunitzer, R. Chen, B. Schrank, and A. Stangl, Automatische Sequenzanalyse eines Proteins (β-lactoglobulin AB), *Hoppe-Seyler's Z. Physiol. Chem. 353*:832 (1972).
2. H. A. McKenzie, G. B. Ralston, and D. C. Shaw, Location of sulphydryl and disulfide groups in bovine β-lactoglobulins and effects of urea, *Biochemistry (Easton) 11*:4539 (1972).
3. G. Brignon and B. Ribadeau-Dumas, Localisation de la chaîne peptidique de la β-lactoglobuline bovine de la substitution Glu/Gln différenciant les variants génétiques B et D, *FEBS Lett. 33*:73 (1973).
4. M. Z. Papiz, L. Sawyer, E. E. Eliopoulos, A. C. T. North, J. B. C. Findlay, R. Sivaprasadarao, T. A. Jones, M. E. Newcomer, and P. J. Kraulis, The structure of β-lactoglobulin and its similarity to plasma retinol-binding protein, *Nature 324*:383 (1986).
5. H. L. Monaco, G. Zanotti, P. Spadon, M. Bolognesi, L. Sawyer, and E. E. Eliopoulos, Crystal structure of the trigonal form of bovine beta-lactoglobulin and of its complex with retinol at 2.5 Å resolution, *J. Mol. Biol. 197*:695 (1987).
6. T. C. Vanaman, K. Brew, and R. L. Hill, The disulfide bonds of bovine α-lactalbumin, *J. Biol. Chem. 245*:4583 (1970).
7. K. Brew, F. J. Castellino, T. C. Vanaman, and R. L. Hill, The complete amino acid sequence of bovine α-lactalbumin, *J. Biol. Chem. 245*:4570 (1970).
8. W. J. Browne, A. C. T. North, D. C. Phillips, K. Brew, T. C. Vanaman, and R. L. Hill, A possible three-dimensional structure of bovine α-lactalbumin based on that of hen's egg-white lysozyme, *J. Mol. Biol. 42*:65 (1969).
9. A. T. Alexandrescu, P. A. Evans, M. Pitkeathly, J. Baum, and C. M. Dobson, Structure and dynamics of the acid-denatured molten globule state of α-lactalbumin: a two-dimensional NMR study, *Biochemistry (Easton) 32*:1707 (1993).
10. J. R. Brown, Serum albumin: amino acid sequence, *Albumin Structure, Function and Uses* (V. M. Rosenoer, M. Oratz, and M. A. Rotschild, eds.), Pergamon Press, Oxford, 1977, p. 27.
11. C. V. Moor and E. Y. W. Ha, Whey protein concentrates and isolates: processing and functional properties, *CRC Crit. Rev. Food Sci. Nutr. 33*:431 (1993).
12. A. Pierce, D. Colavizza, M. Benaissa, P. Maes, A. Tartar, J. Montreuil, and G. Spik, Molecular cloning and sequence analysis of bovine lactotransferrin, *Eur. J. Biochem. 196*:177 (1991).
13. B. F. Anderson, H. M. Baker, G. E. Norris, D. W. Rice, and E. H. Backer, Structure of human lactoferrin: crystallographic structure analysis and refinement at 2.6 Å resolution, *J. Mol. Biol. 209*:711 (1989).

14. Y. Nagasako, H. Saito, Y. Tamura, S. Shimamura, and M. Tomita, Iron-binding properties of bovine lactoferrin in iron-rich solution, *J. Dairy Sci. 76*:1876 (1993).
15. A. T. Andrews, The composition, structure and origin of proteose peptone component 5 of bovine milk, *Eur. J. Biochem. 90*:59 (1978).
16. A. T. Andrews, The formation and the structure of proteose peptone components, *J. Dairy Res. 46*:215 (1979).
17. J. C. Cheftel and D. Lorient, Les propriétés fonctionnelles des protéines laitières et leur amélioration, *Lait 62*:435 (1982).
18. H. Pessen, J. M. Purcell, and H. M. Farrell, Jr., Proton relaxation rates of water in dilute solutions of β-lactoglobulin. Determination of cross relaxation and correlation with structural changes by the use of two genetic variants of self associating globular protein, *Biochim. Biophys. Acta 828*:1 (1985).
19. B. Colas, J.L. Courthaudon, M. Le Meste, and D. Simatos, Functional properties of caseinates: role of the flexibility of the protein and of its hydratation level on surface properties, Proceedings of international seminar, Functional properties of food proteins, Budapest, 1989, pp. 186–194.
20. B. Closs, Flexibilité des molécules protéines, *Influence de la Structure sur les Propriétés de Surface des Protéines du Lactosérum*, Thèse de Doctorat de l'Université de Bourgogne, 1990, p. 104.
21. D. Lorient, B. Closs, and J.-L. Courthaudon, Connaissance nouvelles sur les propriétés fonctionnelles des protéines du lait et des dérivés, *Lait 71*:141 (1991).
22. D. G. Dalgleish and J. Leaver, Dimensions and possible structures of proteins at oil-water interfaces, *Food Polymers, Gels and Colloids* (E. Dickinson, ed.), Royal Society of Chemistry, Cambridge, 1991, p. 113.
23. D. G. Dalgleish, S. E. Euston, J. A. Hunt, and E. Dickinson, Competitive adsorption of β-lactoglobulin in mixed protein emulsions, *Food Polymers, Gels and Colloids* (E. Dickinson, ed.), Royal Society of Chemistry, Cambridge, 1991, p. 485.
24. M. Shimizu, Structure of proteins adsorbed at an emulsified oil surface, *Food Macromolecules and Colloids* (E. Dickinson and D. Lorient, eds.), Royal Society of Chemistry, Cambridge, 1995, p. 34.
25. R. D. Waniska and J. E. Kinsella, Surface properties of β-lactoglobulin: adsorption and rearrangement during film formation, *J. Agric. Food Chem. 33*:1143 (1985).
26. M. E. Mangino, Physicochemical aspects of whey proteins functionality, *J. Dairy Sci. 67*: 2711 (1984).
27. O. Robin, S. Turgeon, and P. Paquin, Functional properties of milk proteins, *Dairy Science and Technology Handbook. 1. Principles and Properties* (Y. H. Hui, ed.), VCH Publishers, New York, 1993, p. 277.
28. J. E. Kinsella, Milk proteins: physicochemical and functional properties, *CRC Crit. Rev. Food Sci. Nutri. 21*:197 (1984).
29. N. K. D. Kella, S. T. Yang, and J. E. Kinsella, Effect of disulfide bond cleavage on structural and interfacial properties of whey proteins, *J. Agric. Food Chem. 37*:1203 (1989).
30. R. A. M. Delaney, Composition, properties and uses of whey protein concentrates, *J. Soc. Dairy Technol. 29*:91 (1976).
31. Y. Matsumura, S. Mitsui, E. Dickinson, and T. Mori, Competitive adsorption of α-lactalbumin in the molten globule state, *Food Hydrocoll. 8*:555 (1994).
32. B. Closs, Comportements interfacial des protéines sériques, *Influence de la Structure sur les Propriétés de Surface des Protéines du Lactosérum*, Thése de Doctorat de l'Université de Bourgogne, 1990, p. 123.
33. J. Leman and J. E. Kinsella, Surface activity, film formation, and emulsifying properties of milk proteins, *CRC Crit. Rev. Food Sci. Nutri. 28*:115 (1989).
34. E. Dickinson, Structure and composition of adsorbed protein layers and the relationship to emulsion stability, *J. Chem. Soc. Faraday Trans. 88*:2973 (1992).

35. G. Stainsby, Foaming and emulsification, *Functional Properties of Food Macromolecules* (J. R. Mitchell and D. A. Ledward, eds.), Elsevier Applied Science, London, 1986, p. 315.
36. J. E. Kinsella and D. M. Whitehead, Proteins in whey: chemical, physical and functional properties, *Adv. Food Sci. Nutri. 33*:343 (1989).
37. J. Castle, E. Dickinson, A. Murray, B. J. Murray and G. Stainby, Surface behaviour of adsorbed films of food proteins, *Gums and Stabilizers for Food Industry*, Elsevier Applied Science, London, 1986, p. 409.
38. Q. Tang, O. J. McCarthy, and P. A. Munro, Oscillatory rheological study of the gelation mechanism of whey protein concentrate solutions: effect of physiological variables on gel formation, *J. Dairy Res. 60*:543 (1993).
39. Y. L. Hsieh, J. M. Regenstein, and A. M. Rao, Gel point of whey and egg proteins using dynamic rheological data, *J. Food Sci. 58*:116 (1993).
40. S. Liao and M. E. Mangino, Characterization of the composition, physicochemical and functional properties of acid whey protein concentrates, *J. Food Sci. 52*:1033 (1987).
41. Y. A. Kim, G. W. Chism, and M. E. Mangino, determination of the beta-lactoglobulin, alpha-lactalbumin and bovine serum albumin of whey protein concentrates and their relationship to protein functionality, *J. Food Sci. 52*:124 (1987).
42. N. K. Pearce and J. E. Kinsella, Emulsifying properties of proteins: evaluation of a turbidimetric technique, *J. Agric. Food Chem. 26*:715 (1978).
43. A. Kato and S. Nakai, Hydrophobicity determined by a fluorescence probe method and its correlation with surface properties of proteins, *Biochim. Biophys. Acta 624*:13 (1980).
44. Z. Haque and J. E. Kinsella, Emulsifying properties of food proteins: bovine serum albumin, *J. Food Sci 53*:416 (1988).
45. M. E. Mangino, A collaborative study to develop a standardized food protein solubility procedure, *J. Food Sci. 50*:1715 (1985).
46. P. Suttiprasit, K. Al-Malah, and J. McGuire, On evaluating the emulsifying properties of protein using conductivity measurements, *Food Hydrocoll. 7*:241 (1993).
47. E. Dickinson, Food colloids—an overview, *Colloids Surfaces 42*:191 (1992).
48. D. G. Dalgleish, Structure and properties of adsorbed layers in emulsions containing milk proteins, *Food Macromolecules and Colloids* (E. Dickinson and D. Lorient, eds.), Royal Society of Chemistry, Cambridge, 1995, p. 23.
49. K. R. Langley, D. Millard, and W. E. Evans, Emulsifying capacity of whey proteins produced by ion-exchange chromatography, *J. Dairy Res. 55*:197 (1988).
50. P. Walstra and A. L. de Ross, Protein at air-water and oil-water interfaces: static and dynamic aspects, *Food Rev. Int. 9*:503 (1993).
51. M. Britten and H. J. Giroux, Interfacial properties of milk protein-stabilized emulsions as influenced by protein concentration, *J. Agric. Food Chem. 41*:1187 (1993).
52. J. Foley and C. O'Connell, Comparative emulsifying properties of sodium caseinate and whey protein isolate in 18% oil in aqueous systems, *J. Dairy Res. 57*:377 (1990).
53. B. Closs, Le comportement émulsifiant, *Influence de la Structure sur les Propriétés de Surface des Protéines du Lactosérum*, Thèse de Doctorat de l'Université de Bourgogne, 1990, p. 136.
54. G. Muschiolik, S., Dräger, H. M. Rawel, P. A. Gunning, and D. C. Clark, Investigations of the function of whey protein preparation in oil-in-water emulsions, *Food Macromolecules and Colloids* (E. Dickinson and D. Lorient, eds.), Royal Society of Chemistry, Cambridge, 1995, p. 248.
55. E. Dickinson, S. E. Rolfe and D. G. Dalgleish, Competitive adsorption of α_{s1}- and β-casein in oil-in-water emulsions, *Food Hydrocoll. 3*:193 (1988).
56. E. Dickinson, S. E. Rolfe, and D. G. Dagleish, Competitive adsorption in oil-in-water emulsions containing α-lactalbumin and β-lactoglobulin, *Food Hydrocoll. 3*:193 (1989).
57. J. Leman, Z. Haque, and J. E. Kinsella, Creaming stability of fluid emulsions containing different milk protein preparations, *Milchwissenschaft 43*:286 (1988).

58. E. Dickinson and S.-T. Hong, Interfacial and stability properties of emulsions: influence of protein heat treatment and emulsifiers, *Food Macromolecules and Colloids* (E. Dickinson and D. Lorient, eds.), Royal Society of Chemistry, Cambridge, 1995, p. 269.
59. E. Dickinson, A. Maufret, S. E. Rolfe, and C. M. Woskett, Adsorption at interfaces in dairy systems, *J. Soc. Dairy Technol. 42*:18 (1989).
60. A. Townsend and S. Nakai, Relationship betweenhydrophobicity and foaming characteristics of food proteins, *J. Food Sci. 48*:589 (1983).
61. M. Le Meste, B. Colas, D. Simatos, B. Closs, J.-L. Courthaudon, and D. Lorient, Contribution of protein flexibility to the foaming properties of casein, *J. Food Sci. 55*:1445 (1990).
62. S. E. Hawks, L. G. Phillips, R. R. Rasmussen, D. M. Bardano, and J. E. Kinsella, Effects of processing treatment and cheese-making parameters on foaming properties of whey protein isolates, *J. Dairy Res. 76*:2468 (1993).
63. E. Charbonnel, P. J. Wilde, and D. C. Clark, Preparation and characterization of a model whey protein isolate to determine the contribution of the protein components to functional properties (Posters), Symposium of Royal Society of Chemistry, Food Chemistry Group, Dijon, March 1994.
64. B. Closs, Le comportement moussant, *Influence de la Structure sur les Propriétés de Surface des Protéines du Lactosérum*, Thèse de Doctorat de l'Université de Bourgogne, 1990, p. 151.
65. M. Britten and L. Lavoie, Foaming properties of protein as affected by concentration, *J. Food Sci. 57*:1219 (1992).
66. E. Izgi and E. Dickinson, Determination of protein foam stability in presence of polysaccharide, *Food Macromolecules and Colloids* (E. Dickinson and D. Lorient, eds.), Royal Society of Chemistry, Cambridge, 1995, p. 312.
67. J. E. Kinsella and D. M. Whitehead, Properties of chemically modified proteins, *Advances in Food Emulsions and Foams* (E. Dickinson and G. Stansby, eds.), Elsevier Applied Science, New York, 1988, p. 163.
68. H. A. McKenzie, β-Lactoglobulins, *Milk Proteins: Chemistry and Molecular Biology* (H. A. McKenzie, ed.), Academic Press, New York, 1971, p. 257.
69. S. N. Timasheff and R. Townend, Molecular interactions in β-lactoglobulin. V. The association of the genetic species of β-lactoglobulin below the isoelectric point. VI. The dissociation of the genetic species of β-lactoglobulin at acid pH's, *J. Am. Chem. Soc. 83*:464 (1961).
70. S. N. Timasheff and R. Townend, Structure of the β-lactoglobulin tetramer, *Nature 203*:517 (1964).
71. T. F. Kumosinski and S. N. Timasheff, Molecular interactions in β-lactoglobulin. X. The stoichiometry of the β-lactoglobulin mixed tetramerization, *J. Am. Chem. Soc. 88*:5635 (1966).
72. M. D. Waissbluth and R. A. Grieger, Activation volumes of fast protein reactions: the binding of bromophenol blue to β-lactoglobulin B, *Arch. Biochem. Biophys. 159*:639 (1973).
73. M. D. Waissbluth and R. A. Grieger, Alkaline denaturation of β-lactoglobulins. Activation parameters and effect on dye binding site, *Biochemistry (Easton) 13*:1285 (1974).
74. A. I. Azuaga, M. L. Galisteo, O. L. Mayorga, M. Cortijo, and P. L. Mateo, Heat and cold denaturation of β-lactoglobulin B, *FEBS Lett. 309*:258 (1992).
75. N. K. Kella and J. E. Kinsella, Structural stability of β-lactoglobulin in the presence of kosmotropic salts, *Int. J. Pept. Protein Res. 32*:396 (1988).
76. M. Dupont, Effet des variations de pH sur l'équilibre primaire de la thermodénaturation de la β-lactoglobuline bovine, *Compt. Rend. Acad. Sci. Paris 257*:3495 (1963).
77. M. Dupont, Comparison de la thermodénaturation des β-lactoglobulines A et B à pH = 6,85, *Biochim Biophys. Acta 94*:573 (1965).
78. J. N. de Wit, Structure and functional behaviour of whey proteins, *Neth. Milk Dairy J. 35*: 47 (1981).

79. K. Watanabe and H. Klostermeyer, Heat induced changes in sulfhydryl and disulfide levels of β-lactoglobulin A and formation of polymers, *J. Dairy Res. 43*:411 (1976).
80. M. McSwiney, H. Singh, and O. H. Campanella, Thermal aggregation and gelation of bovine β-lactoglobulin, *Food Hydrocoll. 8*:441 (1994).
81. N. K. Kella and J. E. Kinsella, Enhanced thermodynamic stability of β-lactoglobulin at low pH, *Biochem. J. 255*:113 (1988).
82. S. Utsumi, S. Damodaran, and J. E. Kinsella, Heat induced interaction between soybean proteins, *J. Agric. Food Chem. 32*:1406 (1984).
83. M. Paulsson, P. O. Hegg, and H. B. Castberg, Thermal stability of whey proteins studied by differential scanning calorimetry, *Thermochim. Acta 95*:435 (1985).
84. D. Renard, Caractérisation de la dénaturation et de l'agrégation thermique de la β-lactoglobuline, *Etude de l'Agrégation et de la Gélification des Protéines Globulaires: Application à la β-Lactoglobuline*, Thèse de Doctorat de l'Université de Nantes, 1994, p. 114.
85. C.-H. Wang and S. Damodaran, Thermal gelation of globular proteins: weight-average molecular weight dependence of gel, *J. Agric. Food Chem. 38*:1157 (1990).
86. J. N. de Wit and G. Klarenbeek, A differential scanning calorimetric study of the thermal behaviour of β-lactoglobulin at temperature up to 160°C, *J. Dairy Res. 48*:293 (1981).
87. C. Tanford, Protein denaturation, *Advances in Protein Chemistry*, vol. 23 (C. B. Anfinsen, M. L. Anson, J. T. Edsall, and F. M. Richards, eds.), Academic Press, New York, 1968, p. 121.
88. C. Tanford, Protein denaturation, *Advances in Protein Chemistry*, vol. 24 (C. B. Anfinsen, M. L. Anson, J. T. Edsall, and F. M. Richards, eds.), Academic Press, New York, 1970, p. 1.
89. V. R. Harwalkar, Measurement of thermal denaturation of β-lactoglobulin at pH 2.5, *J. Dairy Sci. 63*:1043 (1980).
90. V. R. Harwalkar, Kinetic study of thermal denaturation of β-lactoglobulin at pH 2.5, *J. Dairy Sci. 63*:1052 (1980).
91. S. Lapanje and N. Polkar, Calorimetric and circular dichroic studies of the thermal denaturation of β-lactoglobulin, *Biophys. Chem. 34*:155 (1989).
92. S. Jeyarajah and J. C. Allen, Calcium binding and salt-induced structural changes of native and preheated β-lactoglobulin, *J. Agric. Food Chem. 42*:80 (1994).
93. V. R. Harwalkar and M. Kalab, Thermal denaturation and aggregation of β-lactoglobulin at pH 2.5. Effect of ionic strength and protein concentration. *Milchwissenschaft 40*:31 (1985).
94. V. R. Harwalkar and C.-Y. Ma, Evaluation of interaction of β-lactoglobulin by differential scanning calorimetry, *Protein Interactions* (H. Visser, ed.), Symposium on Protein Interactions, American Chemical Society, Atlanta, 1991, pp. 359–378.
95. M. Paulsson, and H. Visser, Heat-induced interactions of milk proteins studied by differential scanning calorimetry, *Protein Interactions* (H. Visser, ed.), Symposium on Protein Interactions, American Chemical Society, Atlanta, 1991, pp. 117–134.
96. Y. L. Xiong, K. A. Dawson, and L. Wan, Thermal aggregation of β-lactoglobulin: effect of pH, ionic environment, and thiol reagent, *J. Dairy Res. 76*:70 (1993).
97. O. De Rham and S. Chanton, Role of ionic environment in insolubilization of whey protein during heat treatment of whey products, *J. Dairy Sci. 67*:939 (1984).
98. K. Német and I. Simonovits, The biological role of lactoferrin, *Haematologia 18*:3 (1985).
99 J. Montreuil, J. Mazurier, D. Legrand, and G. Spik, Lactoferrin, *Proteins of Iron Storage and Transport* (G. Spick, J. Montreuil, R. R. Crichton, and J. Mazurier, eds.), Elsevier, Amsterdam, 1985, pp. 25–38.
100. Y. Hiroaka, T. Segawa, K. Kuwajima, S. Sugai, and N. Moroi, α-Lactalbumin: a calcium metalloprotein, *Biochem. Biophys. Res. Commun. 95*:1098 (1980).

101. K. Kuwajima, Y. Harashima, and S. Sugar, Influence of Ca^{2+} binding on the structure and stability of bovine α-lactalbumin studied by circular dichroism and nuclear magnetic resonance spectra, *Int. J. Peptide Prot. Res. 27*:18 (1986).
102. M. A. Paulsson, U. Svensson, A. R. Kishore, and A. S. Naidu, Thermal behaviour of bovine lactoferrin in water and its relation to bacterial interaction and antibacterial activity, *J. Dairy Sci. 76*:3711 (1993).
103. L. Sánchez, J. M. Pieró, N. Castillo, M. D. Pèrez, J. M. Ena, and M. Calvo, Kinetic parameters for denaturation of bovine milk lactoferrin, *J. Food Sci. 57*:873 (1992).
104. G. Miranda, G. Haze, P. Scanff, and J. P. Pélissier, Hydrolysis of α-lactalbumin by chymosin and pepsin, *Lait 69*:451 (1989).
105. V. Bernal and P. Jelen, Effects of pH and calcium on thermal behaviour of isolated whey proteins, *J. Food Sci. 56*:1119 (1991).
106. V. Bernal and P. Jelen, Effect of calcium binding on the thermal denaturation of bovine α-lactalbumin, *J. Dairy Sci. 67*:2452 (1984).
107. D. Xie, V. Bhakuni, and E. Freine, Are the molten globule and the unfolded states of apo-α-lactalbumin enthalpically equivalent? *J. Mol. Biol. 232*: 5 (1993).
108. M. P. Ruegg, V. Morr, and B. Blanc, A calorimetric study of the thermal denaturation of whey proteins in simulated milk ultrafiltrate, *J. Dairy Res. 44*:509 (1977).
109. J. N. de Wit and G. Klarenbeek, Effect of various heat treatments on the structure and stability of whey proteins, *J. Dairy Sci 67*:2701 (1984).
110. V. Bernal and P. Jelen, Thermal stability of whey proteins. A calorimetric study. *J. Dairy Sci. 68*:2847 (1985).
111. L. C. Chaplin and R. L. J. Lyster, Irreversible heat denaturation of bovine α-lactalbumin, *J. Dairy Res. 53*:249 (1986).
112. P. Gough and R. Jenness, Heat denaturation of β-lactoglobulins A and B, *J. Dairy Sci. 45*: 1033 (1962).
113. R. A. K. Owusu, A Three-state heat-denaturation of bovine α-lactalbumin, *Food Chem. 52*: 131 (1995).
114. A. O. Barel, J. P. Prieels, E. Maes, Y. Looze, and J. Léonis, Comparative physicochemical studies of human α-lactalbumin and human lysozyme, *Biochim. Biophys. Acta 257*:288 (1972).
115. A. Baer, M. Oruz, and B. Blanc, Serological studies on heat-induced interactions of α-lactalbumin and milk proteins, *J. Dairy Res. 43*:419 (1976).
116. T. S. Melo and A. F. Hansen, The effect of UHT steam injection on model systems of α-lactalbumin and β-lactoglobulin, *J. Dairy Sci. 61*:710 (1978).
117. A. A. Elfgam and J. W. Wheelock, Interactions of bovine β-lactoglobulin and α-lactalbumin during heating, *J. Dairy Sci. 61*:28 (1978).
118. A. A. Elfgam and J. W. Wheelock, Heat interactions between α-lactalbumin and β-lactoglobulin and casein in bovine milk, *J. Dairy Sci. 61*:159 (1978).
119. C. M. Hollar, N. Parris, A. Hsieh, and K. D. Cockley, Factors affecting the denaturation and aggregation of whey proteins in heated whey protein concentrate mixtures, *J. Dairy Sci. 78*:260 (1995).
120. D. Levieux, Heat denaturation of whey proteins comparative: studies with physical and immunological methods, *Ann. Rech. Vét. 11*:89 (1980).
121. M. Paulsson and U. Elofsson, Thermal gelation of β-lactoglobulin and lactoferrin, *Milchwissenschaft 49*:547 (1994).
122. N. Parris, J. M. Purcell, and S. M. Ptashkin, Thermal denaturation of whey proteins in skim milk, *J. Agric. Food Chem. 39*:2167 (1991).
123. M. E. Hines and E. A. Foegeding, Interactions of α-lactalbumin and bovine serum albumin with β-lactoglobulin in thermally induced gelation, *J. Agric. Food Chem. 41*:341 (1993).
124. K. Shimada and J.-C. Cheftel, Sulfhydryl group/disulfide bond interchange reactions during heat-induced gelation of whey protein isolate, *J. Agric. Food Chem. 37*:161 (1989).

125. N. Parris, S. G. Anema, H. Singh, and L. K. Creamer, Aggregation of whey proteins in heated sweet whey, *J. Agric. Food Chem. 41*:460 (1993).
126. E. Dumay and J.-C. Cheftel, Chauffage d'un concentré protéique de β-lactoglobuline en milieu faiblement alcalin. Effet sur la solubilité et le comportement chromatographique de la β-lactoglobuline et de l'α-lactalbumine, *Sci. Aliments 9*:561 (1989).
127. S. Oh and T. Richardson, Heat-induced interactions of bovine serum albumin and immunoglobulin, *J. Dairy Sci. 74*:1786 (1991).
128. N. Matsudomi, T. Oshita, K. Kobayashi, and J. E. Kinsella, α-Lactalbumin enhances the gelation properties of bovine serum albumin, *J. Agric. Food Chem. 41*:1053 (1993).
129. V. R. Harwalkar, Comparison of physico-chemical properties of different thermally denatured whey proteins, *Milchwissenschaft 34*:419 (1979).
130. V. R. Harwalkar, Kinetic study of thermal denaturation of proteins in whey, *Milchwissenschaft 40*:31 (1986).
131. A. R. Hill, Thermal precipitation of whey proteins, *Milchwissenschaft 43*:565 (1988).
132. M. Nonaka, E. Li-Chan, and S. Nakai, Raman spectroscopic study of thermally induced gelation of whey proteins, *J. Agric. Food Chem. 41*:1176 (1993).
133. A. Gaulin, Appareil et Procédé pour la Stabilisation du Lait. Brevet n° 295596, 1899.
134. P. Walstra, Physical chemistry of fat milk globules, *Developments in Dairy Chemistry—2* (P. F. Fox, ed.), Applied Science Publishers, London, 1983, pp. 119–158.
135. A. V. MacPherson, M. C. Dash, and B. J. Kitchen, Isolation and composition of milk fat globule membrane material. I. From pasteurized milks and creams, *J. Dairy Res. 51*:279 (1984).
136. A. V. MacPherson, M. C. Dash, and B. J. Kitchen, Isolation and composition of milk fat globule membrane material. II. From homogenized and ultra heat treated milks, *J. Dairy Res. 51*:289 (1984).
137. D. G. Dalgleish and J. M. Banks, The formation of complexes between serum proteins and fat globules during heating of whole milk, *Milchwissenschaft 46*:75 (1991).
138. A. V. Houlihan, P. A. Goddard, B. J. Kitchen, and C. J. Masters, Changes in structure of the bovine milk fat globule membrane on heating whole milk, *J. Dairy Res. 59*:321 (1992).
139. A. V. Houlihan, P. A. Goddard, S. M. Nottingham, B. J. Kitchen, and C. J. Masters, Interactions between the bovine milk fat globule membrane and skim milk components on heating whole milk. *J. Dairy Res. 59*:187 (1992).
140. H.-H. Y. Kim and R. Jimenez-Flores, Heat-induced interactions between the proteins of milk fat globule membrane and skim milk, *J. Dairy Sci. 78*:24 (1995).
141. S. K. Sharma and D. G. Dalgleish, Interactions between milk serum proteins and synthetic fat globule membrane during heating of homogenized whole milk. *J. Agric. Food Chem. 41*:1407 (1993).
142. S. K. Sharma and D. G. Dalgleish, Interactions between milk serum proteins and synthetic fat globule membranes during heating of homogenized whole milk, *J. Agric Food Chem. 41*:1407 (1993).
143. P. Walstra and A. L. De Roos, Proteins at air-water and oil-water interfaces: static and dynamic aspects, *Food Rev. Int. 9*:503 (1993).
144. M. Rosenberg and S. L. Lee, Microstructure of whey protein/anhydrous milk fat emulsions, *Food Struct. 12*:267 (1993).
145. E. Dickinson, Emulsion formation, *An Introduction to Food Colloids*, Oxford Science Publishers, Oxford, 1992, pp. 115–119.
146. J. D. MacClements, F. J. Monahan, and J. E. Kinsella, Disulfide bond formation affects stability of whey protein isolate emulsion, *J. Food Sci. 58*:1036 (1993).
147. F. J. Monahan, J. D. MacClements, and J. E. Kinsella, Polymerization of whey proteins in whey protein-stabilized emulsions, *J. Agric. Food Chem. 41*:1826 (1993).
148. E. Dickinson and Y. Matsumura, Time-dependent polymerization of β-lactoglobulin through disulfide bonds at the oil-water interface in emulsions, *Int. J. Biol. Macromol. 13*:26 (1991).

149. J. Chen, J. Evison, and E. Dickinson, Surfactant/Protein competitive adsorption and electrophoretic mobility of oil-in-water emulsions, *Food Macromolecules and Colloids* (E. Dickinson and D. Lorient, eds.), Royal Society of Chemistry, Cambridge, 1995, pp. 256–260.
150. T. Nakamura, H. Sado, and Y. Syukunobe, Production of low antigenic whey protein hydrolysates by enzymatic hydrolysis and denaturation, *Milchwissenschaft 48*:141 (1993).
151. J. C. Cheftel, Applications des hautes pressions en technologie alimentaire, *Ind. Aliment. Agric. 108*:141 (1991).
152. K. Heremans, I. Les hautes pressions. Aspects théoriques des effets des hautes pressions sur les biomacromolécules, *Dossiers Scientifique de l'Institut Français pour la Nutrition 3*: 3 (1993).
153. J. J. Ewbank and T. E. Creighton, Pathway of disulfide-coupled unfolding and refolding of bovine α-lactalbumin, *Biochemistry* (*Easton*) *32*:3677 (1993).
154. J. J. Ewbank and T. E. Creighton, Structural characterization of the disulfide folding intermediates of bovine α-lactalbumin, *Biochemistry* (*Easton*) *32*:3677 (1993).
155. E. Dufour, C. Bertrand-Harb, and T. Haertlé, Reversible effects of medium dielectric constant on structural transformation of β-lactoglobulin and its retinol binding, *Biopolymers 33*: 589 (1993).
156. E. Dufour, P. Robert, D. Bertrand, and T. Haertlé, Conformation changes of β-lactoglobulin: and ATR infrared spectroscopic study of the effect of pH and ethanol, *J. Protein Chem. 13*: 143 (1994).
157. Y. Kang, A. G. Marangoni, and R. Y. Yada, Effect of two polar-aqueous solvent systems on the structure-function relationship of proteases. I. Pepsin, *J. Food Biochem. 17*:353 (1994).
158. Y. Kang, A. G. Marangoni, and R. Y. Yada, Effect of two polar-aqueous solvent systems on the structure-function relationship of proteases. II. Chymosin and *Mucor miehei* proteinase, *J. Food Biochem. 17*:371 (1994).
159. M. I. Reddy, N. K. D. Kella, and J. E. Kinsella, Structural and conformational basis of the resistance of β-lactoglobulin to peptic and chymotryptic digestion, *J. Agric. Food Chem. 36*:737 (1988).
160. D. G. Schmidt and B. W. Van Markwijk, Enzymatic hydrolysis of whey proteins. Influence of heat treatment of α-lactalbumin and β-lactoglobulin on their proteolysis by pepsin and papain. *Neth. Milk Dairy J. 47*:15 (1993).
161. M. Dalgalarrondo, E. Dufour, J.-M. Chobert, C. Bertrand-Harb, and T. Haertlé, Proteolysis of β-lactoglobulin and β-casein by pepsin in ethanolic media, *Int. Dairy J. 5*:1 (1995).
162. B. Closs, Hydrolyse enzymatique des protéines sériques, *Influence de la Structure sur les Propriétés de Surface des Protéines du Lactosérum,* Thèse de Doctorat de l'Université de Bourgogne, 1990, p. 188.
163. D. Rector, N. Matsudomi, and J. E. Kinsella, Changes in gelling behaviour of whey protein isolate and β-lactoglobulin during storage: possible mechanism(s), *J. Food Sci. 56*:782 (1991).
164. F. Tani, M. Murata, T. Higasa, M. Goto, N. Kitabatake, and E. Doi, Heat induced transparent gel from hen egg lysozyme by a two-step heating method, *Biosci. Biotechnol. Biochem. 57*: 209 (1993).
165. M. Michigo, F. Tani, T. Higasa, N. Kitabatake, and E. Doi, Heat induced transparent gel formation of bovine serum albumin, *Biosci. Biotechnol. Biochem. 57*:43 (1993).
166. N. Kitabatake and E. Doi, Improvement of protein gel by physical and enzymatic treatment, *Food Rev. Int. 9*:443 (1993).
167. M. Paulsson, P.-O. Hegg, and H. B. Castberg, Heat induced gelation of individual whey proteins: a dynamic rheological study, *J. Food Sci. 51*:87 (1986).
168. A. M. Legowo, T. Imade, and S. Hayakawa, Heat induced gelation of the mixtures of α-lactalbumin and β-lactoglobulin in the presence of glutathione, *Food Res. Int. 26*:103 (1993).

169. K. P. Das and J. E. Kinsella, Effect of heat denaturation on the adsorption of β-lactoglobulin at the oil/water interface and on coalescence stability of emulsions, *J. Colloid Interface Sci. 139*:551 (1990).
170. B. Lieke and G. Konrad, Thermishe Modifizierung von UF-Molkenprotein. 1. Enifluß auf die physico-funktionellen Eigenschaften, *Milchwissenschaft 48*:567 (1993).
171. B. Closs, Influence des traitement thermique, *Influence de la Structure sur les Propriétés de Surface des Protéines du Lactosérum,* Thèse de Doctorat de l'Université de Bourgogne, 1990, p. 161.
172. M. Britten, H. J. Giroux, Y. Jean, and N. Rodrigue, Composite blends from heat-denatured and undenatured whey proteins: emulsifying properties, *Int. Dairy J. 41*:25 (1994).
173. B. Closs, Réduction des ponts disulfures des protéines sériques, *Influence de la Structure sur les Propriétés de Surface des Protéines du Lactosérum*, Thèse de Doctorat de l'Université de Bourgogne, 1990, p. 179.
174. N. Blondel, Influence de la structure conformationnelle de la β-lactoglobuline sur les propriétés de surface, *Diplôme d' Étude Approfonde*, Université de Bourgogne, 1992.
175. L. G. Phillips, W. Schuman, and J. E. Kinsella, pH and heat treatment effects on foaming of whey protein isolate, *J. Food Sci. 55*:1116 (1990).
176. T. Haggett, *NZ J. Dairy Sci. Technol. 14*:198 (1979).
177. L. G. Phillips, S. E. Hawks, and J. B. German, Structural characteristics and foaming properties of β-lactoglobulin: effects of shear rate and temperature, *J. Agric. Food Chem. 43*: 613 (1995).
178. L. Sawyer, M. Z. Papiz, A. T. C. North, and E. E. Eliopoulos, Structure and function of bovine β-lactoglobulin, *Biochem. Soc. Trans. 13*:265 (1985).
179. P. K. Warne, F. A. Momany, S. V. Rumball, R. W. Tuttle, and H. A. Scheraga, Computation of structures of homologous α-lactalbumin from lysozyme. *Biochemistry (Easton) 13*:768 (1974).
180. R. Peltonen-Shallaby and M. E. Mangino, Composition factors that affect the emulsifying and foaming properties of whey protein concentrates, *J. Food Chem. 51*:91 (1986).
181. P. S. Halling, Protein-stabilized foams and emulsions, *CRC Crit. Rev. Food Sci. Nutr. 2*:155 (1981).
182. M. C. Phillips, Protein conformation at the liquid interfaces and its role in stabilizing emulsions and foams, *Food Technol (Chicago) 35*:50 (1981).
183. M. Ahmed and E. Dickinson, Effect of ethanol on the foaming properties of food macromolecules, *Food Polymers, Gels and Colloids* (E. Dickinson, ed.), Royal Society of Chemistry, Cambridge, 1991, p. 503.

9

Structure-Function Relationships of Soy Proteins

SHIGERU UTSUMI, YASUKI MATSUMURA, AND TOMOHIKO MORI
Research Institute for Food Science
Kyoto University, Kyoto, Japan

I. INTRODUCTION

Seed proteins play an important role in food consumption worldwide. Soy proteins have been utilized for many kinds of traditional foods. The use of soy protein products as functional ingredients is gaining increasing acceptance in food manufacturing from the standpoints of human nutrition and health. The applicability of soy proteins in foods is based on their functionality. Typical functions of soy proteins are gelation, emulsification, foaming, cohesion-adhesion, elasticity, viscosity, solubility, water absorption and binding, fat absorption, and flavor binding. These are influenced by environmental factors and the conditions of protein preparation as they affect the intrinsic physical and chemical properties of the protein. The details of these points have been thoroughly reviewed by Kinsella [1].

In recent decades, considerable information about the structural properties of soy proteins at the primary, secondary, tertiary, and quaternary structural levels has been accumulated. As for function, the basic mechanisms involved in gelation and emulsification properties are becoming well understood. Analysis of gelling properties has been developed in relation to improving the textural properties of foods using soy proteins. This new and basic knowledge in addition to that already understood is important for assessing and extending the applications of soy proteins in foods and for developing new functional ingredients. Therefore, this chapter focuses on recent developments in the understanding of the structure and functionality relationships of soy proteins.

II. STRUCTURE OF SOY PROTEINS

The seed proteins of legumes, including soybeans, are albumins and globulins [2]. Globulins, the dominant storage proteins, account for about 50–90% of seed proteins. Storage globulins are grouped into two types according to their sedimentation coefficients: 7S globulins (vicilin, 7.1–8.7S) and 11S globulins (legumin, 10.1–14S) [2]. The ratio of 11S to 7S globulins varies among cultivars. It is about 0.5–1.7 in soybean [3].

A. 7S Globulins

The 7S globulins of soybean are classified into three major fractions with different physicochemical properties, designated β-conglycinin, γ-conglycinin, and basic 7S globulin [4]. β-Conglycinin is the most prevalent of these three and accounts for 30–50% of the total seed proteins. β-Conglycinin is one of the vicilin-type proteins, which are widely distributed in many legume and nonlegume seeds. Basic 7S globulin and γ-conglycinin account for less than a few percent.

Basic 7S globulin (Bg) is a glycoprotein having a higher isoelectric point (pH 9.05–9.26) than the other globulin species [5]. Bg has a molecular mass of 168 kDa [5] and is composed of four subunits consisting of a high molecular weight polypeptide (26 kDa) and a low molecular weight polypeptide (16 kDa), which are linked by disulfide bridge(s) [6]. The basic 7S globulin gene and cDNA have been cloned, and the amino acid sequence is available [7,8]. Bg is synthesized as a precursor polypeptide [7] similarly to 11S globulin (see below). However, Bg exhibits no sequence homology with 11S globulin [4]. Mature soybean releases a large amount of Bg when immersed in water at 50–60°C [9]. It seems likely that this protein is a kind of heat-shock protein, although the function of Bg is unknown.

γ-Conglycinin is a glycoprotein (carbohydrate content ~5%) and is a trimer with a molecular mass of 170 kDa [10]; it is composed of three identical subunits [11]. The N-terminal amino acid sequence of γ-conglycinin is not like those of the other globulins [4]. The serological properties of γ-conglycinin are also different from other globulins, in that it exhibits no dissociation-association behavior with a change of ionic strength from 0.1 to 0.5 [12].

β-Conglycinin is a trimer with a molecular mass of 150–200 kDa. As a constituent subunit, four subunits are identified: three major subunits α' (72 kDa), α (68 kDa), and β (52 kDa) [13,14] and one minor subunit γ similar in size to that of β subunit [14]. The amino acid sequences of all of the subunits are similar, although the N-terminal of the γ subunit is blocked [4].

Genes and cDNAs that encode the β-conglycinin subunit have been cloned and sequenced (see review in Ref. 15). The amino acid composition of each mature subunit deduced from the nucleotide sequence is shown in Table 1. The contents of cysteine, methionine and tryptophan residues are one, four, and two in α', all one in α, and all zero in β, respectively. Therefore, the order of the nutritional value of the three subunits is $\alpha' > \alpha > \beta$. The cysteine residues of α' and α subunits are present near the N-terminal.

β-Conglycinins is a glycoprotein [16]. Thanh and Shibasaki analyzed the carbohydrate contents of isolated constituent subunits and proposed that the α and α' subunits contain two carbohydrate moieties and the β subunit one [14]. Yamauchi et al. analyzed glycopeptides from the pronase digest of β-conglycinin and observed the sequences of Asn-Gly-Thr and Asn-Ala-Thr [17]. The same group estimated five or six carbohy-

TABLE 1 Amino Acid Compositions and Molecular Masses of Constituent Subunits of β-Conglycinin

Amino acids	α'		α		β	
	Count	Mol%	Count	Mol%	Count	Mol%
Ala	23	3.98	23	4.23	22	5.28
Arg	38	6.58	43	7.91	29	6.97
Asn	37	6.41	37	6.81	33	7.93
Asp	28	4.85	27	7.97	21	5.04
Cys	1	0.17	1	0.18	0	0.00
Gln	52	9.01	45	8.28	33	7.93
Glu	79	13.69	77	14.18	37	8.89
Gly	29	5.02	24	4.41	18	4.32
His	20	3.46	6	1.10	8	1.92
Ile	28	4.85	30	5.52	26	6.25
Leu	41	7.10	45	8.28	42	10.09
Lys	38	6.58	31	5.70	21	5.04
Met	4	0.69	1	0.18	0	0.00
Phe	29	5.02	27	4.97	28	6.73
Pro	33	5.71	38	6.99	21	5.04
Ser	40	6.93	39	7.18	31	7.45
Thr	14	2.42	11	2.02	10	2.40
Trp	2	0.34	1	0.18	0	0.00
Tyr	13	2.25	13	2.39	12	2.88
Val	28	4.85	24	4.41	24	5.76
Total	577	99.91	543	99.89	416	99.92
M.M.	67,237		63,149		48,015	

drate groups to be contained in one mole of β-conglycinin molecule [18]. The structure of Asn-carbohydrate groups is Asn(GlcNAc)$_2$-(Man)$_{7,8,9}$ [19]. The deduced amino acid sequences of the constituent subunits indicate that the α and α' subunits have two consensus N-glycosylation sequences (α: N199G200T201, N455A456T457; α': N215G216T217, N489A490T491) and the β subunit one (N328A329T330). The site of the β subunit corresponds to the latter sites of the α and α' subunits. These facts indicate that all potential sites of β-conglycinin subunits are N-glycosylated.

β-Conglycinin exhibits molecular heterogeneity [20,21]. Thanh and Shibasaki isolated six molecular species and identified their subunit composition as $\alpha'\beta_2$, $\alpha\beta_2$, $\alpha\alpha'\beta$, $\alpha_2\beta$, $\alpha_2\alpha'$, and α_3 [20]. In addition, Yamauchi et al. observed the occurrence of another species, β_3 [21]. Since the β_3 species associates with glycinin in the course of purification, it was not detected by Thanh and Shibasaki. Each subunit has a different amino acid composition (Table 1) and exhibits different behavior on heating [22]. Therefore, each molecular species probably exhibits distinguishable functional properties, although such studies have not yet been carried out.

Thanh and Shibasaki [23] isolated the constituent subunits of β-conglycinin in the presence of urea and renaturated β-conglycinin from the various combinations of the three isolated subunits. All of the molecular species identified from soybean were renaturated, and the association of the subunits into the 7S form did not follow exactly the random assumption [23]. They obtained a 7S form composed of only α', which was not observed in seeds, and did not obtain β_3, since the renaturation condition was not suitable for the formation of β_3 [21,23]. The β subunit has a tendency to form large aggregates at low ionic strength [21].

β-Conglycinin trimers exhibit association-dissociation behavior depending on the pH and ionic strength of the solution [24]. This 7S protein has a trimer structure at neutral pH at high ionic strength ($I > 0.5$) or at acidic pH ($pH < 4.8$) and exists as a hexamer (10S) at low ionic strength ($I < 0.2$) in the pH region 4.8–11.0 [24]. The forces responsible for the association-dissociation (7S $\rightleftharpoons$ 9S) are presumed to be electrostatic in nature since the 9S form dissociates at high ionic strength [25]. Pedrosa and Ferreira examined pressure-induced dissociation of pea vicilin [26]. They suggested the possibility of participation of salt bridges in intersubunit interaction. Based on the effect of pH on the pressure-induced dissociation, it was suggested that lysine residues at subunit interfaces participate in salt bridges, which stabilize the trimer. Since each subunit behaves differently as a function of ionic strength and pH, each β-conglycinin molecular species having different subunit composition might exhibit variations in their association-dissociation behavior.

Seeds of many legumes and nonlegumes contain 7S globulins [2]. The genes and cDNAs encoding these proteins have been cloned and their nucleotide sequences have been determined [15]. Wright aligned the amino acid sequences deduced from the nucleotide sequences to determine the extent of homology among the 7S globulins [3,27]. He observed 53% absolute homology between the α' subunit of β-conglycinin and pea vicilin. If conservative substitutions (e.g., Arg and Lys, Gly and Pro) were considered, the homology increased to 72%. Similarly, phaseolin and pea vicilin shared 46 and 65% homology, respectively. From such comparisons, he suggested that 7S globulins are comprised of a series of alternating conserved (four) and variable (five and N-terminal extension) regions. The alignment indicated that the N-terminal half and the C-terminal half shared similarity with each other. Gibbs et al. confirmed this and suggested that the 7S globulin gene evolved from duplication of one ancestral gene [28].

Tulloch and Blagrove indicated that β-conglycinin appears as a flat disk about 8.5 nm in diameter and about 3.5 nm thick and observed the similar dimensions in the case of phaseolin and pea vicilin, respectively [29]. Plietz et al. determined the structure of phaseolin by small-angle x-ray scattering to be consisting of three Y-shaped subunits with a dimension of 12.5 nm $\times$ 12.5 nm $\times$ 3.75 nm [30]. However, they observed considerable differences between the structures of canavalin, jack bean 7S globulin, and phaseolin. I'Anson et al. reported a dimension of 12.7 $\times$ 12.7 $\times$ 4.06 nm for pea vicilin and a model containing three touching teardrop-shaped subunits comprising two unequal-sized, touching spherical domains from x-ray scattering data [31]. The scanning tunneling microscopy images of pea vicilin did not contradict that derived from x-ray scattering [32].

Crystallization of phaseolin and canavalin was successfully achieved by Suzuki et al. [33] and Sumner and Howell [34], respectively. β-Conglycinin has not yet been crystallized. The difficulty in the crystallization of β-conglycinin may be due to the heterogeneity of the molecular species as described above. It is necessary to isolate molecular species composed of one kind of subunit (e.g., α_3 or β_3) for obtaining crystals. However, there is a possibility that crystals from such preparations cannot diffract x-ray sufficiently, since each subunit exhibits microheterogeneity, especially the β subunit [35].

The determination of the three-dimensional structure of phaseolin was performed at 2.2 Å resolution [36]. Phaseolin polypeptide is composed of two structurally similar units, each made up of a β barrel having a "jellyroll" folding topology and a α-helix domain (Fig. 1A). The phaseolin trimer stereo pair shown in Figure 1B indicate that it has a large intramolecular channel. The dimension of the phaseolin trimer is $9 \times 9 \times 3.5$ nm [37]. The three-dimensional structure of canavalin was determined at 2.6 Å resolution

A

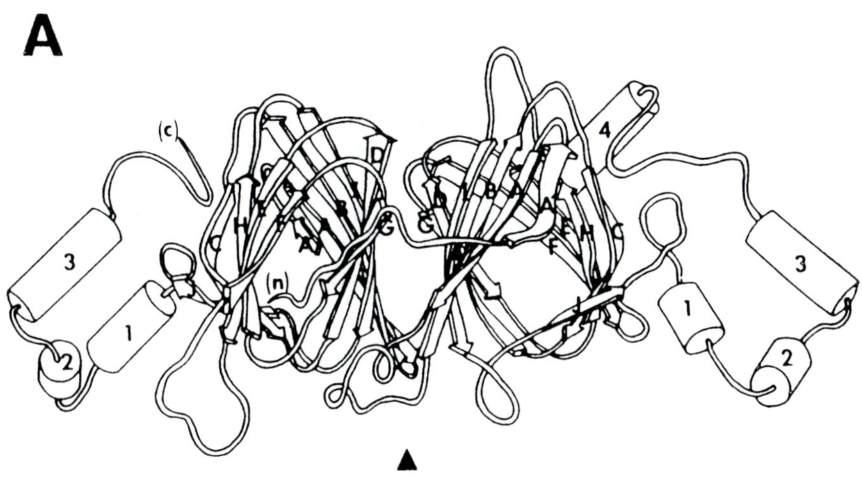

B

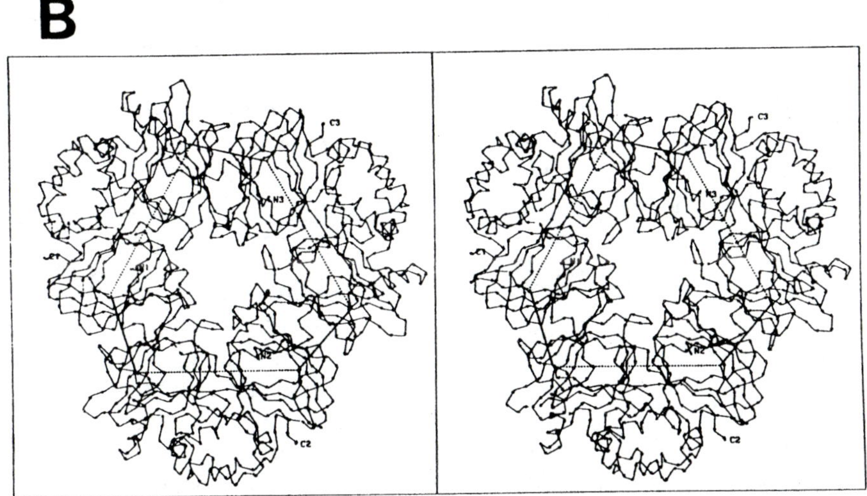

FIGURE 1 (A) Schematic diagram of the phaseolin polypeptide. Plane arrows and cylinders are β-sheets and α-helices, respectively. The N- and C-termini are labeled (n and c). The view is down the molecular threefold axis (indicated as a solid triangle) and from the center of the tetramer outward. (B) Stereo pair of the phaseolin trimer. (From Ref. 37.)

[38] and was shown to be very similar to that of phaseolin. The canavalin trimer is a disk-shaped molecule of about 8.8 nm in diameter with a thickness of about 4.0 nm. The dimensions of phaseolin and canavalin are similar and consistent with those of phaseolin, β-conglycinin, and vicilin predicted by electron microscopy [29]. It is worth noting that the canavalin polypeptide was proteolyzed with trypsin before crystallization [38]. Although the overall shapes of phaseolin and canavalin resemble the model based on x-ray scattering, the dimensions are smaller than the model [30].

The extended α-helix–containing loops account for the association of the subunits into the trimer [36,38]. These subunit interfaces involve the apposition of interdigitating hydrophobic surfaces, and these likely drive trimer formation [38]. This is consistent with the speculation from the association-dissociation behavior that the subunits of the 7S trimer are held primarily by hydrophobic forces [25].

Lawrence et al. observed 14 well-formed salt bridges in the phaseolin trimer [36]. Of these, three are involved in trimer stabilization; Arg53-Glu283, Asp92-Arg346, and Arg179-Glu364. Ko et al. observed six salt bridges in the interfaces between subunits as Arg167-Glu81, Arg168-Glu321, Arg301-Asp107, Arg376-Asp125, Lys414-Asp200, and Arg424-Glu196 [38]. Thus, salt bridges containing arginine residue are dominant in the intersubunit interaction of the phaseolin and the canavalin trimers. The discrepancy between these observations and the participation of lysine residues theorized from the pH effect on pressure dissociation of pea vicilin [26] may be due to the difference of plant species.

B. 11S Globulins

The 11S globulins, legumin-type proteins, are widely distributed in many legume and nonlegume seeds and are generally simple proteins, although there are some exceptions; for example, lupin 11S globulin is a glycosylated protein [39]. The 11S globulin of soybean, glycinin, is a hexamer with a molecular mass of 300–380 kDa. Each subunit is composed of an acidic polypeptide (acidic pI) with a molecular mass of ~35 kDa and a basic polypeptide (basic pI) with a molecular mass of ~20 kDa. The acidic and basic polypeptides are linked together by a disulfide bond [40]. Initially a single polypeptide precursor is synthesized and then processed posttranslationally to form the acidic and the basic polypeptides [15]. The processing site between the acidic and the basic polypeptides of all glycinin subunits is Asn-Gly. Therefore, all subunits have Asn at the C-terminus of the acidic polypeptide and Gly at the N-terminus of the basic polypeptide. Nielsen claimed that some glycinin subunits have a linker peptide between the acidic and the basic polypeptides [41]. However, Momma et al. found no evidence for the existence of a linker sequence [42]. As a constituent subunit, five subunits are identified: $A_{1a}B_{1b}$ (G1), A_2B_{1a} (G2), $A_{1b}B_2$ (G3), $A_5A_4B_3$ (G4), and A_3B_4 (G5) [41]. (Nielsen sometimes employed the labels of $A_{1a}B_2$ and $A_{1b}B_{1b}$ instead of $A_{1a}B_{1b}$ and $A_{1b}B_2$ [43]. However, according to the N-terminal sequence of each polypeptide determined by Nielsen's group [44], the counterpart of A_{1a} must be B_{1b}—therefore the different naming of G1–G5.) Among these subunits, the $A_5A_4B_3$ subunit is synthesized as a single polypeptide precursor similarly to the others, but the acidic polypeptide is cleaved to produce A_5 (97 residues) and A_4 (257 residues) polypeptides [15]. The A_5 polypeptide is derived from the N-terminal region of the acidic polypeptide.

The genes and cDNAs encoding these subunits have been cloned and sequenced (see review in Ref. 15). According to the extent of the homology, the constituent subunits are classified into two groups: group I ($A_{1a}B_{1b}$, A_2B_{1a}, $A_{1b}B_2$) and group II ($A_5A_4B_3$, A_3B_4) [43]. The homology of each subunit is more than 84% in the group and less than 49% between the groups. The amino acid composition and the molecular mass of each mature subunit are shown in Table 2. The cysteine and methionine residue contents are higher in group I subunits than in group II subunits. Another characteristic of these groups is the difference in the molecular masses: group I < group II.

The number and the position of the disulfide bond between the acidic and the basic polypeptides of four ($A_{1a}B_{1b}$, A_2B_{1a}, A_3B_4, $A_5A_4B_3$) of the five subunits were determined by Staswick et al. [40]. Only one disulfide bond was found to be involved in linking the acidic and the basic polypeptides of each subunit, and they were in analogous positions (Fig. 2), i.e., between the cysteine residues at position 7 from the N-terminus of the basic polypeptide and at position 86 from the N-terminus of the acidic polypeptide in the case of A_2B_{1a}. All of the 11S globulin determined so far have cysteine residues in homologous positions [15,36]. These facts strongly suggest that the disulfide bond between the acidic and the basic polypeptides of 11S globulin subunits is located in a homologous position regardless of the origin of the 11S globulins. Staswick et al. suggested that one of the internal disulfide bonds within the acidic polypeptide A_2 is Cys10-Cys51 [40]. However, Wright [3] and Utsumi [15] suggested Cys10-Cys43 in A_2 instead of Cys10-Cys51 from the conservativity of these cysteine residues among 11S globulins of legume and nonlegume seeds. Moreover, they implied that the cysteine residue at around position 10 from the N-terminus and the one at residue 33 downstream from the first could be involved in an internal disulfide bond in all the 11S globulin subunits. Determination of the number of free sulfhydryl residues per mole of glycinin has been attempted by many groups [45–47]. Two to 15 sulfhydryl residues were detected. This means that the constituent subunits of glycinin must contain more disulfide bond(s). As shown in Figure 2, the cysteine residues located near the C-terminus part are conserved in all subunits. If this cysteine residue participates in a disulfide bond, it is more likely that the counterpart is a conserved one, but there is no other conserved cysteine residue. Therefore, it is reasonable to assume that the cysteine residue in each subunit is present as a free sulfhydryl residue. This is supported by the fact that pea and broad bean legumins have cysteine residues at this position but not at other positions, except for the four conserved cysteine residues described above [36]. Therefore, the cysteine residues in the immediate neighborhood of the C-terminus of the acidic polypeptides A_3 and A_4 are naturally free. Each subunit of group I has two cysteine residues in the hypervariable region [48] in the vicinity of the C-terminus of the acidic polypeptides. The hypervariable region has strong hydrophilic nature, indicating that this region is exposed on the surface of the molecule. This implies that the formation of a disulfide bond by these two cysteine residues does not cause the structural difference between groups I and II. These assumptions indicate that the cysteine residues at around position 50 from the N-terminus of the acidic polypeptide A_{1a}, A_{1b}, and A_2 are free. As described above, comparison of the position of the cysteine residues of each subunit suggests that each subunit has two free sulfhydryl residues. Wolf also suggested two sulfhydryl residues per subunit [47]. However, we cannot eliminate the possibility that the cysteine residues that are assumed to be free are forming a disulfide bond in either group I or II subunits.

Glycinin exhibits polymorphism of the subunit composition among the cultivars [49]. Soybean cultivars were classified into five groups according to the subunit composition of glycinins as determined by isoelectric focusing. Moreover, the amino acid sequence of each subunit is different among the cultivars [50,51]. Glycinin also exhibits

Table 2 Amino Acid Compositions and Molecular Masses of Constituent Subunits and Polypeptides of Glycinin

Amino acids	$A_{1a}B_{1b}$				$A_{1b}B_2$				A_2B_{1a}				A_3B_4				$A_5A_4B_3$			
	T[a]		A[a]	B[a]	T		A	B	T		A	B	T		A	B	T		A	B
	C[b]	Mol%	C	C	C	Mol%	C	C	C	Mol%	C	C	C	Mol%	C	C	C	Mol%	C	C
Ala	27	5.67	11	16	28	6.06	12	16	31	6.63	14	17	18	3.65	8	10	22	4.08	9	13
Arg	27	5.67	17	10	29	6.27	17	12	29	6.20	20	9	33	6.70	21	12	36	6.67	25	11
Asn	37	7.77	19	18	36	7.79	17	19	40	8.56	21	19	33	6.70	17	16	33	6.12	18	15
Asp	17	3.57	11	6	16	3.46	10	6	18	3.85	12	6	24	4.87	20	4	30	5.56	27	3
Cys	8	1.68	6	2	8	1.73	6	2	8	1.71	6	2	6	1.21	4	2	6	1.11	4	2
Gln	48	10.08	36	12	49	10.60	35	14	51	10.92	37	14	45	9.14	34	11	48	8.90	35	13
Glu	41	8.61	32	9	38	8.22	30	8	37	7.92	29	8	42	8.53	34	8	55	10.20	44	11
Gly	35	7.35	25	10	31	6.70	21	10	34	7.28	23	11	40	8.13	25	15	37	6.86	24	13
His	8	1.68	5	3	6	1.29	4	2	4	0.85	2	2	15	3.04	12	3	15	2.78	10	5
Ile	26	5.46	15	11	24	5.19	14	10	23	4.92	13	10	17	3.45	11	6	21	3.89	14	7
Leu	33	6.93	16	17	31	6.70	14	17	33	7.06	15	18	34	6.91	18	16	37	6.86	20	17
Lys	24	5.04	18	6	18	3.89	13	5	18	3.85	12	6	18	3.65	12	6	27	5.00	19	8
Met	6	1.26	3	3	5	1.08	3	2	7	1.49	5	2	4	0.81	2	2	2	0.37	2	0
Phe	20	4.20	10	10	26	5.62	15	11	19	4.06	10	9	15	3.04	10	5	14	2.59	7	7
Pro	29	6.09	19	10	29	6.27	19	10	26	5.56	16	10	37	7.52	28	9	37	6.86	28	9
Ser	32	6.72	18	14	32	6.92	17	15	30	6.42	15	15	38	7.72	25	13	43	7.97	30	13
Thr	20	4.20	10	10	18	3.89	10	8	18	3.85	10	8	20	4.06	14	6	20	3.71	14	6
Trp	4	0.84	3	1	3	0.64	2	1	4	0.85	3	1	4	0.81	3	1	6	1.11	4	2
Tyr	11	2.31	7	4	10	2.16	7	3	11	2.35	6	5	15	3.04	7	8	15	2.78	6	9
Val	23	4.83	10	13	25	5.41	11	14	26	5.56	13	13	34	6.91	15	19	35	6.49	14	21
Total	476	99.96	291	185	462	99.89	277	185	467	99.89	282	185	492	99.89	320	172	539	99.91	354	185
M.M.	53,621		33,159	20,480	52,191		31,717	20,492	52,443		32,135	20,326	55,421		36,391	19,049	61,199		10,540 29,953	20,742

[a]T, A, and B represent whole subunit, acidic, and basic polypeptides, respectively.
[b]Count.

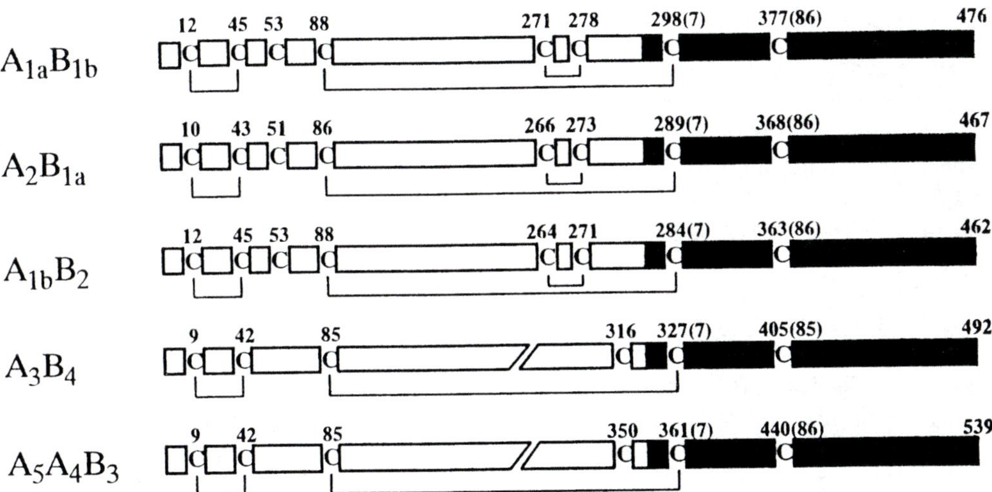

FIGURE 2 Schematic representation of the proposed positions of free cysteine residues and disulfide bonds in glycinin subunits. Cysteine residues are designated by C. The numbers of the residues from the N-terminus are described for the cysteine residues and C-terminus. The numbers in parentheses are the positions from the N-terminus of the basic polypeptides. Open and black areas are the acidic and the basic polypeptides, respectively.

molecular heterogeneity, and hence the occurrence of molecular species with different subunit composition and molecular mass [52]. Each subunit exhibits different behavior on heating [53,54] (see the following section). This means that glycinins having different subunit compositions exhibit distinguishable functional properties. In fact, glycinins from different cultivars differ in their functional properties [45].

The polymorphism of the subunit composition and the molecular heterogeneity of glycinin indicate that the specificity of the subunit interaction for constructing glycinin molecule is not so strict. This together with the homology of the primary structure of glycinin subunits suggests that the exchange of the acidic and the basic polypeptides of each subunit and the formation of glycinin molecule composed of a single subunit are possible. This was proved by the formation of pseudoglycinins having artificial polypeptide or subunit compositions from the isolated acidic and basic polypeptide [53,55] and from the isolated subunits [54].

It is known that glycinin hexamers dissociate to their constituent polypeptides, subunits, and half-molecules (trimer having the size of 7–8S) under various environmental conditions such as pH, ionic strength, and temperature (heating) [56,57]. Wolf and Briggs observed that glycinin partially dissociated to components with the size of 3S (presumably subunits) and the half-molecule under low ionic strength (μ = 0.01) at pH 7.6, and that the half-molecule associated reversibly to the size of 11S [56]. Utsumi et al. demonstrated the occurrence of two molecular species in the glycinin, one being dissociable and the other undissociable at low ionic strength [58]. The conformation of each species was a little different, the dissociable species being more random and unstable than the undissociable species at low ionic strength. The dissociable species contained more $A_5A_4B_3$ and less A_3B_4 than the undissociable species, although the contribution of these subunits was unclear.

The genes and cDNAs encoding 11S globulins of many legumes and nonlegumes have been cloned and their nucleotide sequences determined [15]. The primary structure of the 11S globulins is highly conserved among the plant species [43]. There is 34–67% absolute homology among 11S globulins from various legumes and nonlegumes. The four cysteine residues (C12, C45, C88, C298 in $A_{1a}B_{1b}$) forming inter- and intradisulfide bonds are conserved among all of the 11S globulins known so far [15,36]. These facts strongly suggest that 11S globulins have a structure similar to each other regardless of the origin of the 11S globulins. Wright aligned the amino acid sequences of the 11S globulins to maximize their homology and suggested that they are composed of a series of alternating conserved and variable domains; i.e., four conserved and five variable regions [3,27]. Three (C12, C45, C88) of these four conserved cysteine residues are located in the first conserved region.

Electron microscopy [29,59,60] and small-angle x-ray scattering studies [61,62] of glycinin [59], pea legumin [29], sunflower helianthinin [60,61], pumpkin cucurbitin [29], rape cruciferin [61], and broad bean legumin [62] demonstrated that the dimension of 11S globulin is 10.4 to 12.6 × 10.4 to 12.6 × 7.5 to 9 nm. From these data, some structural models composed of two trimers (half-molecule) have been proposed; e.g., a two-layered parallel hexagonal model [59] and a trigonal antiprism model [60,61]. Miles et al. [63] and I'Anson et al. [64] claimed that there are some discrepancies between the models and the data due to the assumption that each subunit is a sphere with equal-size. The staining agent used for electron microscopy has a tendency to interact only with hydrophilic areas [59,60]. The pattern of penetration of such an agent suggested that the forces for the interaction between the trimers (half-molecule) are electrostatic and hydrogen bonds and that the trimer is held by hydrophobic interaction [59,60].

Crystallization of hemp edestin [65], Brazil nut excelsin [65], and cucurbitin [66] have been achieved. However, comprehensive x-ray analyses have not yet been undertaken, because the primary structures of edestin, excelsin, and cucurbitin have not yet been determined [3]. Another reason is the small size and the disorderliness of these crystals [67]. However, Patel et al. successfully obtained crystals of edestin showing reasonable stability in the x-ray beam (the limit of resolution was not more than about 3.5 Å) [67]. They calculated that edestin has an open ring structure having a diameter of 145 Å and a thickness of 90 Å and suggested that the ring has a large channel through its center.

The 11S globulins from legumes have not yet been crystallized. The difficulty in the crystallization of these proteins may be due to molecular heterogeneity [15]. Success in the high-level expression of glycinin in *Escherichia coli* [68] and *Saccharomyces cerevisiae* [69] should facilitate crystallization, because such an expression system enables the production of a single molecular species. In fact, Utsumi et al. succeeded in crystallizing proglycinin expressed in *E. coli* (the expressed proteins accumulated as proglycinins since *E. coli* does not have the enzyme responsible for the processing of proglycinins to the mature form) [70].

C. Relationship Between 7S and 11S Globulins

From the comparison of the amino acid sequences of the 7S and 11S globulins, Plietz et al. [71], Wright [27], and Gibbs et al. [28] concluded that the genes of both globulins are derived from a common ancestral gene. The extent of homology between both globulins was variable among the authors and was not so strict. Lawrence et al. [36] demonstrated that the entire region of the 11S globulin can be mapped onto the 7S sequences in a way that preserves the structure of the 7S globulins, as shown in Figure 3. Their alignment was based on the canonical 7S elements: Pro67 and Gly81 in the N-terminal half domain and Pro254 and Gly269 in the C-terminal half domain. These residues are totally conserved both between species and between domains in the 7S globulins [36]. They observed 30 residues that are globally conserved or conservatively exchanged

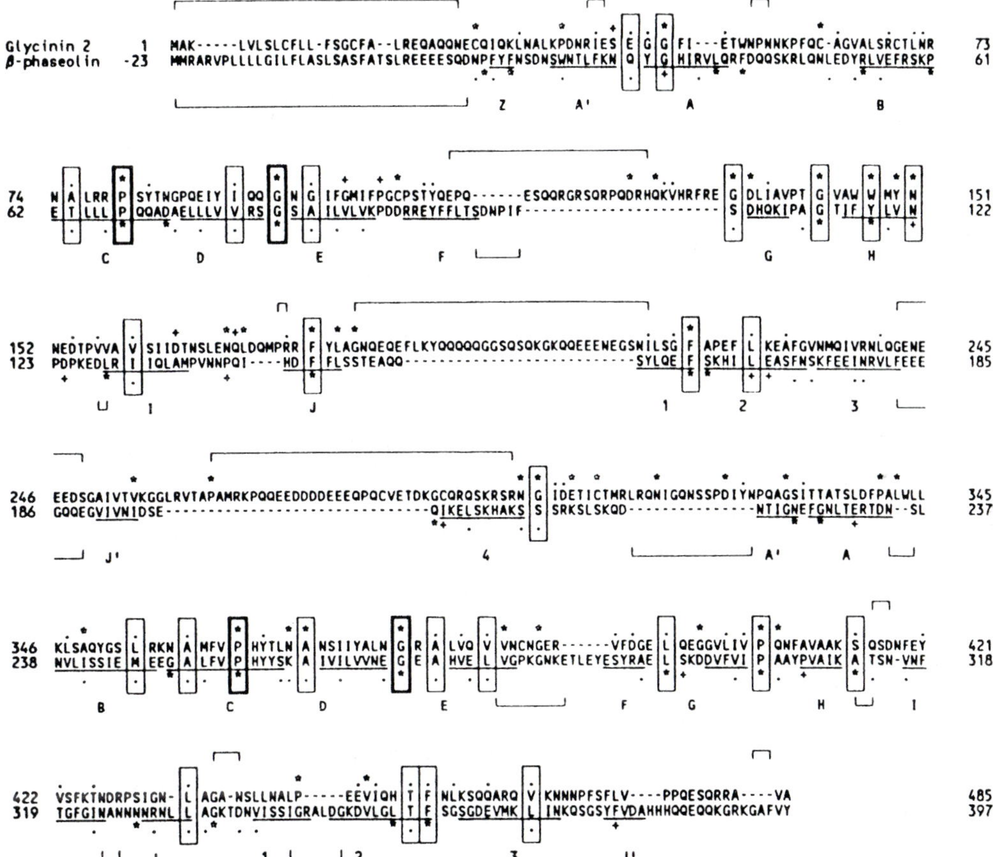

FIGURE 3 Model alignment of the 7S and 11S globulins. The glycinin 2 (A_2B_{1a}) and β-phaseolin sequences were aligned based on the canonical 7S elements. *, Residue strictly conserved within respective 11S/7S family; $^+$, residue strictly conserved in all but one species within respective 11S/7S family; •, residue tolerates conservative replacement within respective 11S/7S family; ⌴, site of length variation within 7S family; ‾, site of length variation within 11S family; □, residue pair globally conserved across 11S and 7S families; **□**, globally conserved Pro/Gly motif. Underscores with alphabet and number indicate residues forming β-sheet and α-helix, respectively. The numbers for residues are from N-terminus of preproglycinin and mature phaseolin. (From Ref. 36.)

across the 7S and 11S globulins. These globally conserved residues correspond predominantly in the 7S structure to residues forming part of the intermonomer packing or to residues in the interstrand loops [36]. Considering the presence of an intradisulfide bond at the N-terminus of the 11S acidic polypeptide and the insertion of the hypervariable region in the 11S acidic polypeptide, the authors suggest that the 11S N- and C-terminal halves are paired oppositely to the 7S modules [36]. The subunits of phaseolin are arranged with 32-point group symmetry [37]. If the 11S globulin sequence can be aligned like the 7S globulins, the 11S globulins likely consist of two 7S-like trimers, indicating that the 11S globulin would also exhibit 32 symmetry [36]. This speculation has been proved by the fact that hemp edestin exhibits 32 symmetry [67].

III. FUNCTIONALITY OF SOY PROTEINS

A. Gelation

The gel-forming ability of soy proteins is one of the most important functional properties used in traditional foods such as tofu and as ingredients in several processed foods. The two globulins glycinin and β-conglycinin have different structures and gel properties. The gelling properties of glycinin and β-conglycinin have been investigated individually, whereas the interaction between two globulins has been demonstrated in mixed systems such as soy protein isolates. Here we will review gel properties of glycinin, β-conglycinin, and the mixed system of two globulins.

1. Glycinin

As described in Chapter 4, in the heat-set gelation system, thermal denaturation of native protein molecule is a prerequisite to the subsequent association of denatured molecules and the formation of the gel network structure [72]. Therefore, the thermal behavior of glycinin has been studied under various conditions using differential scanning calorimetry (DSC). German et al. [73] showed that the thermal transition of glycinin occurred at 72°C. However, thermal behavior of glycinin changes according to the solution conditions, such as protein concentration, pH, ionic strength, etc. Damodaran [74] found that, at high concentrations, NaCl and NaBr progressively increased the values of denaturation temperature of glycinin solution above 100°C, whereas $NaClO_4$ and NaSCN decreased the thermal stability of the proteins. Therefore, when we analyze the gelation behavior of glycinin, we should set the heating temperature very carefully such that it is the same as the thermal denaturation temperature of the protein as measured by DSC under the solution conditions used for gelation. The gelation behavior of glycinin, which is a multisubunit protein, should be more complicated than that of monomeric proteins, because association and dissociation may occur during the gelation process.

Thermal association-dissociation behavior of glycinin at 100°C under high ionic strength (0.5) and at neutral pH (7.6) was investigated by sucrose density gradient centrifugation and gel electrophoresis [57]. Soluble aggregates with a molecular weight of 8×10^6 were formed when 0.5 and 5% protein solutions were heated for 1 minute. At the lower protein concentration, subsequent heating caused the disappearance of the soluble aggregate followed by complete dissociation into acidic and basic polypeptides. As a result, 0.5% glycinin solution became turbid because of the formation of insoluble aggregates or precipitates of separated hydrophobic basic polypeptides.

Based on these results, Mori et al. [57] proposed the overall scheme of heat-induced changes in glycinin, as shown in Figure 4. In step 1, glycinin aggregates (MW 8×10^6) are formed when glycinin solution is heated both at low and high protein concentrations. On subsequent heating, at low protein concentration, the soluble aggregate disaggregates to acidic and basic polypeptides (step 2′), while at high protein concentration it undergoes association resulting in gel formation (step 2). Whether the heated protein solution undergoes disaggregation or gel formation is apparently governed by the concentration of the soluble aggregate, i.e., the more the protein concentration, the greater the probability of gel formation (step 2). The decrease in turbidity of glycinin gel with increasing protein concentration [75] supports the idea of this branching of step 2 and 2′. Although the formation of network structure is completed by step 2, the stabilization of the network structure through further formation of noncovalent bonding and disulfide cross-links by subsequent heating (step 3) proceeds. The changes in physical properties of glycinin gels during step 3 were demonstrated by rheological analyses [76].

The soluble aggregates and gel network were visualized by transmission electron microscopy to understand the mechanism of network formation in the thermal gelation process of glycinin [77]. Based on these results, Nakamura et al. [77] proposed a string-of-beads model in which glycinin molecules associate to form a linear string (Fig. 5). In this model, a "bead" corresponds to a whole glycinin molecule and not to its constituent subunits, i.e., the glycinin molecule is partially unfolded, but still in a globular form without dissociation. It has been discovered that the thickness of the strands comprising soluble aggregates and the gel network is in the range of 10–12 nm [77], similar to the diameter of the native molecule of glycinin (hollow oblate cylinder), whose dimensions were reported by Badley et al. [59] to be $11 \times 11 \times 7.5$ nm. Glycinin molecules associate linearly to form strands I (~48 nm long) and II (~94 nm long). The strand II associates with itself and/or with strand I to form both branched and unbranched strands (strand III, 170–200 nm long). The gel network could then form from these strand III units. Hermansson [78] also proposed a string-of-beads model for glycinin gelation by a process similar to that described above, with the exception of the mechanism of trigger of association. Hermansson suggested that the glycinin molecules dissociate into constituent subunits and reassemble into a glycinin form with the formation of gels.

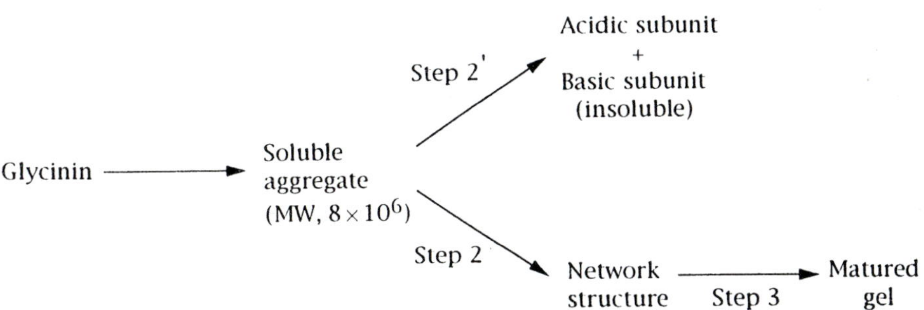

FIGURE 4 Overall scheme of thermal change of glycinin at pH 7.6 and 0.5 ionic strength. (From Ref. 57.)

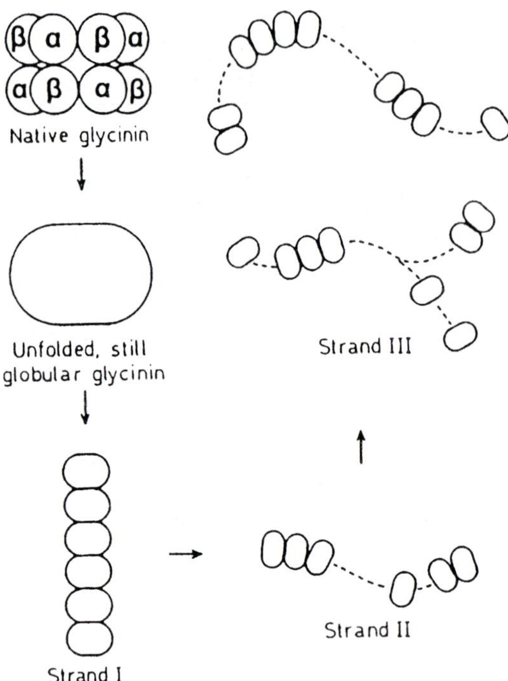

FIGURE 5 Schematic representation of formation of soluble aggregates in the course of gelation of glycinin. The α and β indicate acidic and basic polypeptides, respectively. The description of strands I, II, and III is given in the text. The strands II and III are represented in an abbreviated form with dashed lines. (From Ref. 77.)

It is noteworthy that the gelation model of glycinin shown in Figures 4 and 5 is based on the results under a certain condition, such as high ionic strength (0.5), relatively high protein concentration, high heating temperature (100°C), and neutral pH. A different gelation behavior, particularly the involvement of association-dissociation of subunits, is possible under other heating conditions.

To investigate the molecular forces involved in the gelation process of glycinin, the effects of salts, reduced agents, denaturants, and water-miscible solvents on the heat-induced gelation have been studied. Babajimopoulos et al. [79], from the thermodynamic calculation based on the results of the effects of neutral salts on gelation, concluded that the major forces involved in the gelation of soy proteins were hydrogen bonding and van der Waals interactions; the contribution of hydrophobic and electrostatic interaction is negligible.

Nakamura et al. [77] found that in the presence of N-ethylmaleimide (NEM), a blocking agent of the sulfhydryl group, 5% glycinin solution did not form a gel when heated for 10 minutes and only linear strands were observed, whereas network structure was visible within 4 minutes in the absence of NEM. The results imply that the disulfide bridge formation through intermolecular disulfide exchange reactions may participate in the formation of strands, particularly branched ones, followed by subsequent network formation and gel formation.

The same group [57,77] suggested the contribution of hydrophobic interaction to the formation of soluble aggregates of glycinin, because soluble aggregates were found to be disaggregated in the presence of SDS. SDS is known to destabilize mainly hydrophobic interactions.

The contradiction between the results of Babajimopoulos et al. [79] and Mori et al. [57] or Nakamura et al. [77] in relation to the involvement of molecular forces in glycinin gel formation may be attributed to the different heating temperatures used by both groups. Glycinin solution was heated at 80°C in the study of Babajimopoulos et al. [79], whereas Mori et al. [57] and Nakamura et al. [77] used a temperature of 100°C. Heating at 80°C, a temperature less than denaturation point of glycinin, leads to the formation of reversible gels, which, as the name implies, melt and flow again upon heating. On the other hand, glycinin gel formed at 100°C shows no such thermoreversibility.

Heating at 100°C should cause more unfolding and more exposure of hydrophobic regions on the surface of the glycinin molecule. In this case, therefore, hydrophobic interaction may predominantly contribute to the formation and stabilization of gel network structure. Thermoreversibility would not be expected for such hydrophobic interaction–stabilized gel, because the strength of hydrophobic interaction increases with increasing temperature up to 110°C [80]. For gelation of glycinin at 80°C, the situation is different. Because of the lesser extent of unfolding and lesser exposure of hydrophobic regions on the molecular surface when heated at 80°C, hydrogen bonding seems to be the predominant force in the formed gel. The breakage of hydrogen bonding and resultant imbalance between attractive and repulsive forces with increasing temperature may be responsible for the thermoreversibility of glycinin gel formed by heating at 80°C.

Recently, transglutaminase (TGase)-catalyzed gelation of proteins has attracted fundamental and practical interest. This enzyme can mediate ϵ-(γ-glutamyl)lysyl cross-links between proteins [81]. Caseins are good substrates of the enzyme, because they have flexible structures and their glutamyl and lysyl residues are available to the enzyme. Although globular proteins, such as bovine serum albumin and whey proteins, with native conformation are scarcely attacked by TGase, TGase-catalyzed gelation of 11S globulins has been demonstrated [82,83].

From the fundamental and/or commercial viewpoints, it is interesting to compare physical properties between conventional thermally induced protein gels and TGase-catalyzed gels. It was demonstrated that the minimum concentration required to form a self-supporting gel of 11S globulin catalyzed by TGase was lower than the minimum concentration for thermal gelation [84]. This means that the covalent bonding induced by TGase is more effective in forming a self-supporting network.

Uniaxial compression-decompression tests were performed to compare gel rigidity and elastic properties of TGase-catalyzed and thermally induced gels of glycinin and broad bean legumin [85]. The protein concentration was 12% with both methods of gelation. For the TGase-catalyzed gel, microbial TGase (from *Streptoverticillium*) was added to the protein solution at a ratio of 1:50 (w/w) and the solution was incubated at 37°C for 5 hours. For thermal gelation, the protein solution was heated at 100°C for 1 hour.

Maximum forces generated by deformation of TGase-catalyzed and thermally induced gels to various deformation levels were compared. The maximum forces for TGase-catalyzed gels were higher than those for thermally induced gels at any deformation level, indicating more rigidity of TGase-catalyzed gels. The modulus of deformability for TGase-catalyzed glycinin gel was 4.4 × 10^4 Pa, significantly higher than that (1.5 × 10^4 Pa) of the thermally induced one.

From the compression-decompression curve (Fig. 6a), the ratio of irrecoverable work (hatched area) to the total work (area under the compression curve) was calculated and used as a measure of deviation from ideal elasticity. The results at deformation levels of 20 and 42% are shown in Figure 6b. At the same deformation level, the ratio of irrecoverable work to total work of TGase-catalyzed gels was lower than those of thermally induced gels. This indicated that physical networks of TGase-catalyzed gels were more resistant to large deformation than the networks of thermally induced ones.

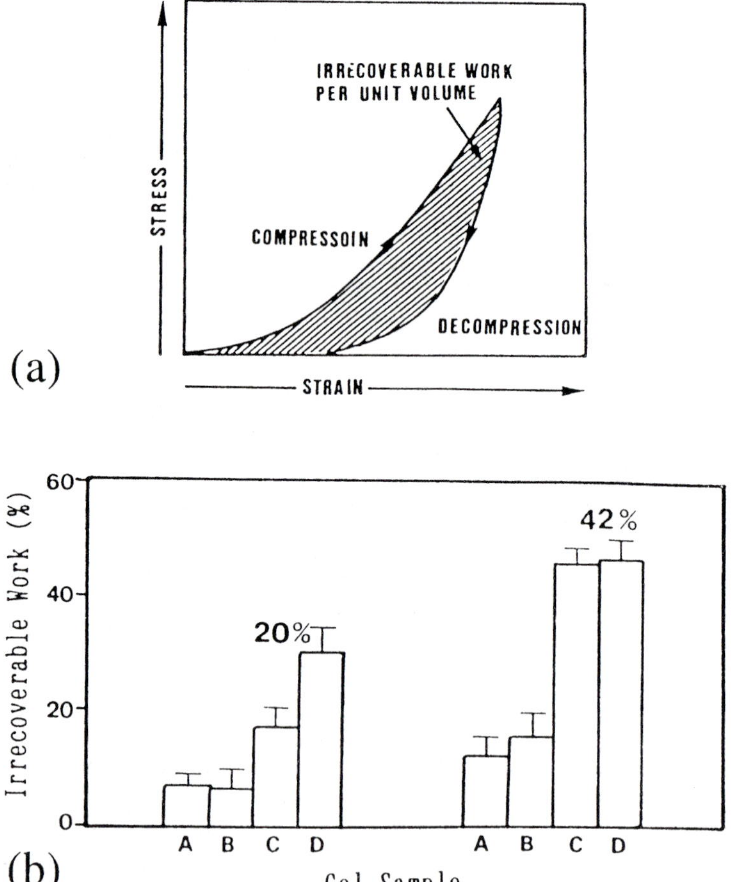

FIGURE 6 (a) Schematic of stress-strain relationships in compression-decompression cycle. The hatched area is the irrecoverable work. The area under the compression curve is the total work. (b) Ratio of irrecoverable work and total work of glycinin and legumin gels subjected to 20 and 42% deformation levels (means with standard deviation bars). (A) Transglutaminase-catalyzed glycinin gel; (B) transglutaminase-catalyzed legumin gel; (C) thermally induced glycinin gel; (D) thermally induced legumin gel. (From Ref. 85.)

The microstructure of the gels visualized by scanning electron microscopy correlated with data on the rheological properties [85]. TGase-catalyzed gels had developed network structures formed by thick strands. This may contribute to the elasticity and stiffness of TGase-catalyzed gels. In thermally induced gels, such developed network structure with thick strands was not observed.

The combination of TGase reaction and heat treatment should be useful to modify gel properties. Kang et al. [86] examined the effect of pre–heat treatment on TGase-catalyzed gelation of glycinin, whereas the heat treatment following TGase reaction was attempted to improve physical properties of soy protein gels [87].

2. β-Conglycinin

β-Conglycinin is the other major component of soy proteins and consists of three subunits α′, α, and β, as described in the previous section. These subunits are not linked to each other via disulfide bonds, unlike the case of glycinin.

The literature is scanty concerning the thermally induced gels of β-conglycinin in comparison with glycinin gels because of the difficulty of obtaining the pure fraction of β-conglycinin. For instance, a β-conglycinin fraction prepared according to the method of Thanh and Shibasaki [88] was found to contain 21% glycinin [74]. The contaminated glycinin probably modifies the gelation behavior of β-conglycinin, since the interaction between subunits of two globulins was induced during heat treatment (this point will be described in the next section). Therefore, we mainly review the studies in which highly purified fraction of β-conglycinin was used.

Nakamura et al. [22] investigated the gelation behavior of β-conglycinin (95% pure) in a protein solution heated at 100°C and at 0.5 ionic strength. The minimum concentration for the formation of self-supporting gel of β-conglycinin was 7.5%, whereas 2.5% glycinin solution formed gel at the same condition.

Figure 7 shows the effects of heating time and protein concentration on the gel hardness of β-conglycinin and glycinin. The hardness values of β-conglycinin and gly-

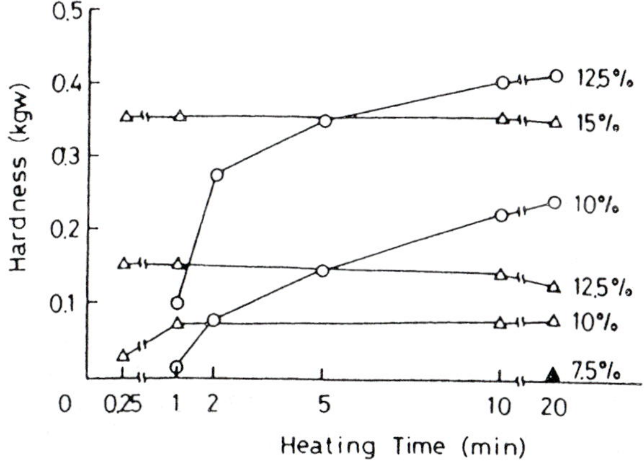

FIGURE 7 Effects of heating time and protein concentration on gel hardness of β-conglycinin and glycinin. Protein solutions were heated at 100°C, pH 7.6, and 0.5 ionic strength. Triangles and circles denote β-conglycinin and glycinin, respectively. (From Ref. 22.)

cinin gels increased with increasing protein concentration. Although values of glycinin gels increased with the heating time and then reached a plateau, the values of β-conglycinin tended to be independent of heating time. If the values of hardness were compared at a plateau level, the values of glycinin gels were much higher than those of β-conglycinin gels at the same protein concentration. This contradicts the previous result [74], showing that β-conglycinin possesses better gelling properties than glycinin gels. The lower gelling properties of glycinin gels reported by Damodaran [74] could be ascribed to the heating temperature used for gelation. The heating temperature, 80°C, might be too low to unfold the glycinin molecule completely and facilitate formation of a well-developed network structure, whereas the complete thermal transition of β-conglycinin could occur at 80°C.

No sulfhydryl/disulfide exchange reaction was found to participate in the gelation of β-conglycinin by SDS-PAGE analysis of soluble aggregates formed prior to gel formation [22]. Electron microscopic observation showed that the network structure of β-conglycinin gel was a randomly aggregated assembly of clusters, while network structure of glycinin gel was built up of regular strands with a thickness of 10–12 nm. These observations indicate that the nature of the β-conglycinin gel network is different from that of the glycinin gel network. The aggregate-type network structure and the absence of disulfide cross-links may be responsible for low rigidity of β-conglycinin gels (Fig. 7).

Recently, Nagano et al. [89] devised a simple method to isolate β-conglycinin (and glycinin) with high purity. Dynamic rheological measurement was used to study the gelation properties of purified β-conglycinin. Figure 8 shows the storage modulus (G′) values as a function of time at different temperatures. The G′ did not rise when heated at temperature <62°C. This temperature was the incipient temperature of thermal denaturation of β-conglycinin detected by DSC measurements. The G′ of β-conglycinin gel heated for 120 minutes increased with increasing heating temperature above 65°C showed a maximum around 70°C, and then decreased. Since the endothermic peak temperature also occurred at 70°C in DSC measurements, it is clear that a rigid gel with a well-developed network structure can be obtained by heating the β-conglycinin solution at around its denaturation temperature. Nagano et al. [90] also demonstrated the hydrophobic interactions and hydrogen bonds are very important for the formation of β-conglycinin gels, as will be described in detail in Section IV. This is in agreement with the suggestion made by Utsumi and Kinsella [91].

Fourier transformed infrared spectroscopy (FTIR) measurements were performed to clarify changes in the secondary structure of β-conglycinin (and glycinin) molecules with heat treatment [92]. Differential spectra obtained by subtracting the spectrum of the solution (15% protein concentration) from that of the gel revealed absorbance change at 1618 cm^{-1} with gel formation. The percentage change in absorbance at 1618 cm^{-1} with gel formation as a function of heating time are shown in Figure 9 for both glycinin and β-conglycinin. The band at 1618 cm^{-1} began to increase at around 80°C for glycinin and at around 65°C for β-conglycinin. A good correlation was observed between the changes in storage modulus G′ and the increase in absorption at 1618 cm^{-1}. Although the band at 1618 cm^{-1} is one of the adsorption bands for the β-sheet structure [93], it is not clear whether the increase in absorbance at 1618 cm^{-1} means an increase in β-sheet structure content during gelation. Wang and Damodaran [94] showed that although there was a net reduction in the β-sheet content of soy proteins after thermal denatur-

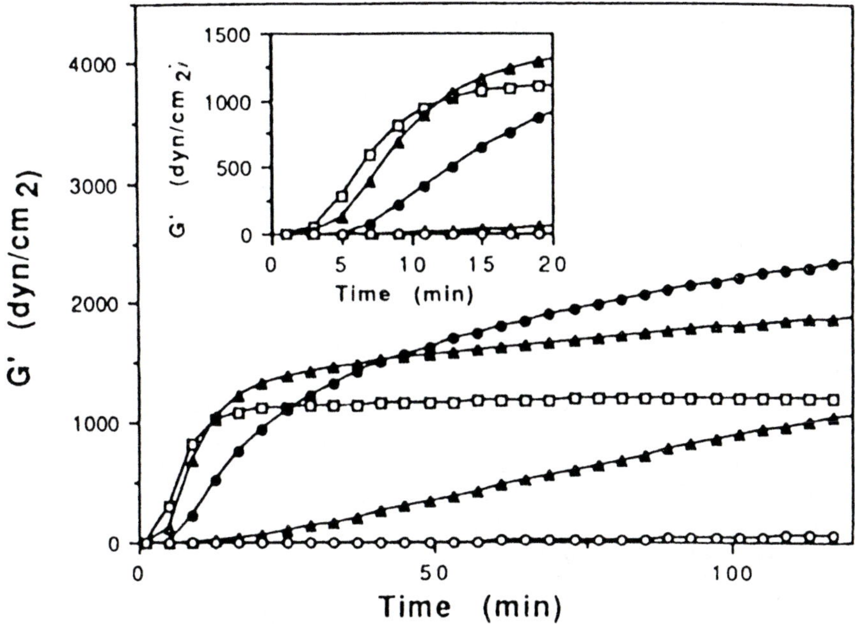

Figure 8 Gelation curves of 10% β-conglycinin solution in 35 mM phosphate buffer, pH 7.5 at different temperatures: (□) 80°C; (△) 75°C; (●) 70°C; (▲) 65°C; (○) 62°C. (From Ref. 89.)

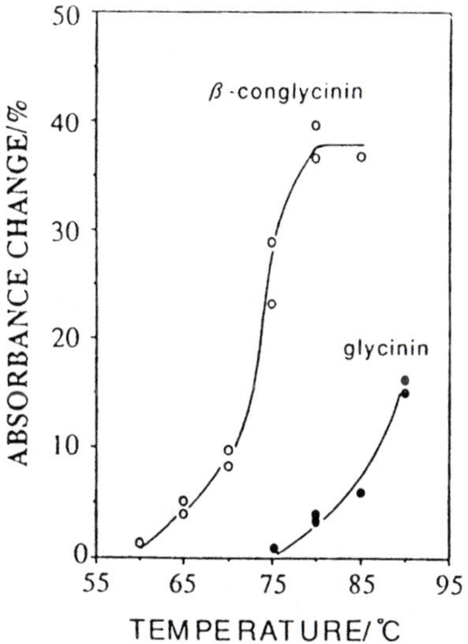

Figure 9 The percentage absorbance change at 1618 cm^{-1} in FTIR spectra as a function of heating temperature. (○), Glycinin; (△), β-conglycinin. (From Ref. 92.)

ation, they retained about 25% β-sheet in the gelled state, as determined by circular dichroic analysis of proteins in fluids expressed from the gel. It was hypothesized that, in globular protein gels, intermolecular hydrogen bonding between segments of β-sheets may act to form junction zones in the gel network [94].

3. Mixed Systems

In a mixed system such as soy protein isolate, the gelation behavior is influenced by the interaction of individual components. Therefore, the interaction among major soy protein components, i.e., glycinin and β-conglycinin, during gelation has been studied.

Babajimopoulos et al. [79] demonstrated that soy protein isolate exhibited better gelling properties at 80°C than either of the constituent protein fractions. This reflects the interaction between the subunits of the constituent glycinin and β-conglycinin during heating. Nakamura et al. [95] also found interaction of both globulins in the mixed system during heat-induced gelation at 100°C under high ionic strength (0.5). However, in their results, the gelation of glycinin was suppressed by β-conglycinin. For instance, although the lowest protein concentrations for the formation of a self-supporting gel for glycinin and β-conglycinin were 2.5 and 7.5%, respectively, as shown before, for the mixed system at a 1:1 ratio of the two globulins, the lowest concentration was 7.5%. Furthermore, the gel hardness of a mixed system was in between that of glycinin and β-conglycinin at most protein concentrations. The difference between these two studies may be attributable to the different heating conditions employed and the presence or absence of reducing agents.

The interaction between glycinin and β-conglycinin was also demonstrated in commercial soy protein isolates. Arrese et al. [96] studied the effect of the degree of denaturation on the solubility in water, gelation capacity, etc., of commercial soy protein isolates. From solubility experiments, it was shown that the amount of insoluble aggregates formed was proportional to the degree of denaturation of soy protein isolates. These insoluble materials were composed of β subunits of β-conglycinin and basic polypeptides of glycinin. The soy protein isolate, which has a higher proportion of β subunit and basic polypeptides in water-soluble fraction, exhibited good gel-forming ability at 80°C. In contrast, the soy protein isolate with a high amount of insoluble β subunit and basic polypeptides showed very low or no gelation capacity. This indicates the importance of the presence of a soluble form of β subunit of β-conglycinin and basic polypeptides of glycinin in soy protein isolates for gelation. As will be described in Section IV, these subunits may interact electrostatically and produce macroaggregates that lead to gel formation of soy protein isolate upon heating [97].

B. Emulsification

Soy milk has become a popular beverage in several countries. In addition, isolated soy protein acts as a macromolecular surfactant to stabilize oil-in-water emulsion systems in food products such as soups, sausages, and coffee whitener. Despite their commercial importance, fundamental knowledge about the emulsifying properties of soy proteins is scanty in comparison with that about their gelling properties.

Tornberg [98] demonstrated that soy protein isolate had a better emulsifying activity than those of whey protein concentrate and sodium caseinate. Kato et al. [99] also showed that the emulsifying activity of β-conglycinin was much better than that of ovalbumin, which is often used as a standard protein, with poor emulsifying activity; they also reported that the emulsifying activity of β-conglycinin was low in comparison with those of κ-casein, β-lactoglobulin, and bovine serum albumin.

The emulsifying properties of glycinin, β-conglycinin, and acid precipitated soy protein (APSP) were systematically investigated by Rivas and Sherman [100–103]. Interfacial tension measurements at the liquid paraffin/water interface [102] showed that β-conglycinin exhibited the best ability to lower the interfacial tension at any pH value; APSP was better than glycinin in relation to the decrease of interfacial tension. This suggests that β-conglycinin has the best emulsifying activity.

The substantial difference between β-conglycinin and glycinin in their ability to reduce interfacial tension can be attributed to the chemical and structural features of these proteins. β-Conglycinin is more hydrophobic [104] and has a lower molecular weight than glycinin. Because of a greater number of hydrophobic patches on its surface, β-conglycinin may adsorb more rapidly to the interface than glycinin. In addition, conformational rearrangement of amino acid residues of glycinin may be slow at the interface because of the presence of intra- and intersubunit disulfide bonds. Repulsion between charged molecules in the vicinity of the interface may also retard adsorption of glycinin to a greater extent because of its high net charge.

Rivas and Sherman [103] also found differences in the interfacial rheological properties of adsorbed films of glycinin, β-conglycinin, and APSP at the corn oil/water interface. The values of instantaneous elastic modulus (E_0) of β-conglycinin films were slightly larger than those of APSP films, but glycinin films showed E_0 values 2–3 times lower than those of β-conglycinin. This indicates that β-conglycinin can form a highly viscoelastic film with ordered structure [105]. The greater hydrophobic interaction among the adsorbed β-conglycinin molecules may contribute to the rigidity of the film.

Both interfacial tension [102] and interfacial rheology [103] data suggest that β-conglycinin has a better emulsion-stabilizing ability as well as emulsifying activity than does glycinin. The emulsifying properties of the β-conglycinin–rich fraction and the glycinin-rich fraction were investigated by Aoki et al. [106]. The emulsifying capacity and emulsion stability of the β-conglycinin–rich fraction generally showed higher values than those of the glycinin-rich fraction.

The ratio of β-conglycinin and glycinin in soybean seeds was found to increase during maturation [107]. Yao et al. [108] investigated the emulsion stability of isolated soy protein from three stages of seed maturity. It was shown that the protein isolate from the mature bean provided better emulsion stability than the immature one. This result suggests that the ratio of β-conglycinin to glycinin is a key factor affecting the emulsifying properties of soy protein isolates.

Emulsifying properties of proteins are closely related to the conformation of and interactions among adsorbed molecules at the oil/water interface. The interfacial denaturation of and the subsequent interaction between oligomeric β-conglycinin and glycinin may be an extremely complicated process in comparison with those of monomeric proteins. No reports exist in the literature on the association-dissociation behavior of soybean protein subunits at the interface and the contribution of chemical bonding to the viscoelasticity of the adsorbed film. Such information is critical to understanding the emulsifying behavior of soy proteins at the molecular level.

IV. STRUCTURE-FUNCTION RELATIONSHIPS

The gelling ability of soy proteins and the physical properties of the gels formed are of importance with respect to their use in foods. In particular, the thermal gelation property is very important because manufactured foods are thermally processed. The gelling properties are influenced by various factors such as the structure and properties of the protein per se and the environmental conditions affecting the physicochemical properties of the protein. This section focuses on the relationships between gelling properties and the physicochemical properties of soy proteins. The relationship of the microstructure of the gels to the gel properties is also discussed.

A. Glycinin

The subunit composition of glycinins isolated from the seeds of various cultivars of soybean vary among the cultivars [49]. The physical properties of gels, i.e., hardness measured with a texturometer (General Foods Corp., GXT-2) and turbidity determined by a chromatoscanner (Shimadzu Co., Ltd, Model CS-910, Japan) [45], of glycinins from different cultivars are different. The hardness of the gels is directly proportional to the percentage of A_3 subunit, which is the largest constituent acidic polypeptide of glycinin. The role of A_3 in increasing the gel hardness will be discussed later in more detail. The turbidity of the gels has a tendency to increase with increasing sulfhydryl group content of glycinins. The source of turbidity is the basic polypeptides of glycinin, which are dissociated from glycinin during heating [57]. The basic polypeptides themselves are intrinsically water insoluble; they are substantially insoluble under the condition at which glycinin is soluble. The dissociation of basic polypeptides is accelerated by 2-mercaptoethanol and depressed by a sulfhydryl-blocking agent, such as N-ethylmaleimide. Differences in reactivity of the cysteine residues depending on their location (surface or internal) in the glycinin molecule, which may partly depend on the soybean cultivar, could be a factor in the extent of turbidity in the gel. Further, protein concentration affects gel turbidity [75]. Glycinin solution at 0.5% protein concentration becomes turbid followed by precipitation of the protein on heating, where glycinins dissociate completely into their constituent polypeptides. A 5% glycinin solution forms a gel with only slight turbidity. The turbidity of the gel decreases with increasing protein concentration; the gel formed at 20% protein concentration is transparent macroscopically. These observations suggest that the extent of dissociation of basic polypeptides depends on the protein concentration, which in turn affects the turbidity of the gels.

The contribution of the various constituent subunits to the physical properties of glycinin gels can be investigated by using pseudoglycinins [54]. Preparation of pseudoglycinins is performed as follows. First, the subunits of glycinin are isolated by a column chromatography. The elution bands, namely, IS-I, IS-II, and IS-III, which contain the acidic polypeptides of A_1, A_2, and A_3, respectively, are collected. Pseudoglycinins are reconstituted from each constituent subunit using a dialysis procedure. The pseudoglycinins thus obtained are composed of one of the constituent subunits only. They are similar to the native glycinin in molecular size, subunit structure, and secondary structure. The gels formed from these pseudoglycinins by heating a 5–10% protein solution exhibit lower turbidity in the cases of IS-I and IS-III pseudoglycinins than does native glycinin. In contrast to this, the gels made from the IS-II pseudoglycinin exhibit much higher turbidity. Thus, the turbidity of glycinin gels might be closely related to the properties of IS-II constituent subunit. The acidic polypeptide A_2, which is the major component of IS-II, is known to have the largest number of cysteine and cystine residues. On the other hand, the gel hardness is higher for the IS-III pseudoglycinin gel and lower for the IS-II pseudoglycinin gel than the native glycinin gel. The hardness of the IS-I pseudoglycinin gel is similar to that of the native glycinin gel. These observations suggest that the IS-III (or A_3) subunit plays an important role in the hardness of glycinin gel.

The factors affecting network structure formation and the physical properties of gels have been reviewed by Clark and Lee-Tuffnell [72] and Hermansson [109]. Insights into the microstructure of gels and the relationship between microstructure and physical properties of gels can improve our understanding of relationship between structure and functionality of gels. The microstructure of gels is analyzed either by transmission (TEM) or scanning (SEM) electron microscope. Observations by TEM of the gels from the pseudoglycinins indicate that the microstructure of gels comprises protein networks with strands as thick as that of the diameter of native glycinin [54]. The network structures of the pseudoglycinin gels are different. The gels of pseudoglycinins composed of IS-I, IS-II, and IS-III show similar, rough and random, and elaborate network structures, respectively, compared to the native gels. Furthermore, the length and extent of branching of the network strands in the pseudoglycinins of IS-I, IS-II, and IS-III are similar, longer and fewer, and shorter and much greater, respectively, than those in the native glycinin. Critical analysis of the hardness of the gels in the light of their microstructure indicate that the gel hardness of glycinin is influenced by the length and extent of branching of the strands of network structure of the gel, i.e., the elaborateness of gel networks. Consequently, one can assume that the contribution of IS-III (or A_3) subunit to the gel hardness is mediated by its influence on the microstructure of gel.

There are several reports on the microstructure of gels in views of understanding the structure-function relationships. The involvement of secondary structure in microstructure formation and mechanical properties of glycinin gel has been reported [94,110]. Both ionic strength and heating temperature influence the microstructure of glycinin gel [78]. Differences in network structures are found in glycinin gels and gels of other seed globulins [111,112]. In β-lactoglobulin gels, it has been demonstrated that the scale of network structure [coarse (particulate gels) or fine (fine-stranded gels)] and the type of network strands [thin (a string of beads) or thick (beads fused together)] influence the fracture and viscoelastic properties of the gels [113]. Thus, the structure of glycinin seem to influence both the formation of the gel structure and the physical properties of the gel. It is hypothesized that in globular protein gels intermolecular hydrogen bonding between segments of β-sheets oriented either in parallel or antiparallel configurations may act as junction zones in the gel network [94]. Additional research is necessary in this area.

Protein structure affects formation of intermolecular bondings in protein gels. The types of intermolecular bondings that occur in the course of gelation in turn influence the physical and rheological properties of gels. In the case of glycinin, disulfide bonding and noncovalent bonds such as hydrophobic interaction and ionic and hydrogen bondings are involved in gel formation. The types and extent of intermolecular bondings formed in the constituent strands within the gel network structure are analyzed by examining the effects of perturbing agents that cleave or weaken specific bondings on the network structure and physical properties of gels. From such analyses, it is shown that junction points in glycinin gels involve disulfide bonding and hydrophobic interactions between constituent strands within the gel network [114]. These intermolecular bondings continue to form within the strands of the network with increase in heating time, resulting in an increase in gel hardness, while no significant change occurs in the gel network structure. This demonstrates the importance of the chemical properties of the protein, in addition to the microstructures of the gel network, to the mechanical properties of protein gels. Creep measurement, a rheological method, is also useful in studying the elastic, viscous, plastic, and viscoelastic components of a gel structure. Creep measurements at different temperatures on glycinin gels indicate that the mechanical strength of the gel formed by heating at 90°C is derived mainly from hydrophobic interactions and hydrogen bondings [115]. In glycinin gels made by heating at 100°C, disulfide bondings contribute to the elastic component, suggesting that the junction zones possibly involve disulfide bonds [76]. These rheological measurements on gels formed by heating at 90 or 100°C indicate that the extent of structural changes in the protein affects the nature of intermolecular bondings formed in the gel network and ultimately the physical properties of the gel itself. Rheological analyses may be useful for understanding gel microstructures, as demonstrated in the case of ice cream [116].

B. β-Conglycinin

The contribution of the constituent subunits of β-conglycinin to the physical properties of β-conglycinin gels is not clear. It has been shown that each subunit exhibits different self-association behaviors in reconstitution systems [23]. This, together with observation of thermal interaction behaviors of β-conglycinin subunits [22,117], suggests that each subunit may contribute differently to the properties of β-conglycinin gel. However, this remains to be investigated.

It is presumed that noncovalent bonds, such as hydrophobic interaction and hydrogen and ionic bondings, participate in the intermolecular interactions in gel networks of β-conglycinin. The gelation of β-conglycinin at 15% protein concentration is inhibited in the presence of 1 M NaSCN, which destabilize hydrophobic interactions. When the gel formed by heating at 80°C and cooling at 20°C are again heated from 20 to 80°C, the storage modulus G′ of the gel decreases and reaches the same level as the storage modulus of β-conglycinin solution heated at 80°C for 30 minutes. These observations [90] indicate that hydrophobic interaction and hydrogen bondings are involved in the increased storage modulus G′ of β-conglycinin gel. The increase and decrease of the storage modulus G′, on cooling and reheating, respectively, show that hydrogen bondings contribute to gel strength at the lower temperature. Thus, hardness of β-conglycinin gels seems to be dependent on the temperature. β-Conglycinin forms a transparent gel at all protein concentrations, while glycinin forms turbid gels and the turbidity increases with decreasing protein concentration [22]. These differences are probably due to differences in the physicochemical properties of the subunits of glycinin and β-conglycinin; β-conglycinin subunits are soluble under conditions at which β-conglycinin is soluble [23].

Spectroscopic methods are used to obtain insight into the molecular conformation of globular proteins in heat-induced gels. Raman and infrared (IR) spectroscopy are useful for investigating secondary structures of proteins in the gel state, since they can be performed at high protein concentrations. FTIR analysis of β-conglycinin in sol and gel states at acidic pH shows that the β-strand is exposed in the gel state and increases with decreasing pH [118]. Rheological analysis of the gels shows that the storage modulus G′ of the gels increases with decreasing pH. The increase of β-strand and storage modulus G′ is also seen in the gels formed at increasing heating temperature [92]. These results indicate that the amount of exposed β-strands is related to the storage modulus G′ of the gels and gel strength. The gel networks are probably formed and/or stabilized by cross-links with intermolecular β-sheet structures [94].

C. Mixed Systems

Soy proteins consist of two major components: β-conglycinin and glycinin. These two globulins have different structures and molecular properties, possess different gelling abilities, and have different gel properties. Investigations into the gel properties of mixed systems reveal information on contributions of the two globulins to the physical properties of the soy protein isolate gels [119]. In the mixed systems prepared by mixing the acid-precipitated proteins and β-conglycinin, hardness and unfracturability of the gels increase remarkably with heating temperature above 93°C. The elasticity of the gels decreases gradually with an increase in the heating temperature (80–100°C). The mixed system with a glycinin:β-conglycinin ratio of 2.41 exhibits higher gel hardness at heating temperature above 93°C than one having a ratio of 0.88. There is no significant difference in gel hardness between the two systems over the temperature range of 80–93°C. Unfracturability of the gels is higher in the mixed system having a higher ratio than in one having a lower ratio over the heating temperature range of 80–100°C. Gel elasticity is higher in the mixed system with a lower glycinin content than that with a higher glycinin content over the temperature range of 80–100°C. In the mixed system, the gel properties are thus changeable depending on the glycinin:β-conglycinin ratio and heating temperature. Although complex interactions occur in the mixed system, some insight is still obtainable with regard to the specific contribution of each globulin fraction to gel properties. Glycinin is apparently related to hardness and unfracturability of the gels. β-Conglycinin largely contributes to the elasticity of the gels. In addition, it has been shown that the basic polypeptides of glycinin preferentially associate with the β-subunits of β-conglycinin via electrostatic interactions, and that glycinin and β-conglycinin interact noncovalently with each other to form composite aggregates during gel formation [1,73,95,120]. These interactions and their extents are likely to be influenced by glycinin:β-conglycinin ratio [120], and they may play a role in the manifestation of gel properties in the mixed system. Environmental conditions such as heating temperature and ionic strength are additional factors to be considered, since they affect conformational changes in soy proteins and also the denaturation temperature of the globulins. The extent of structural changes and subsequent interactions among globulins vary depending on combinations of heating temperature and ionic strength; such variations cause wide variations in gel properties. On the other hand, the glycinin:β-conglycinin ratio affects the turbidity of soy protein gels in the mixed system; the gel turbidity decreases with increasing β-conglycinin content [95]. The gel turbidity is mainly due to basic polypeptides coming from dissociation of glycinin. β-Conglycinin may suppress the dissociation of glycinin into the constituent subunits or prevent the insolubilization of the basic polypeptides through the interactions between glycinin and β-conglycinin in the mixed system [120].

In the mixed system, disulfide bonding and various noncovalent bonds and interactions between the subunits of glycinin and β-conglycinin are involved in determining the properties of the gel. Among these molecular forces, disulfide bonds play an important role in gelation. Evidence for this role comes from the effects of 2-mercaptoethanol, which cleaves disulfide bonds, on the formation of gel network and gel hardness

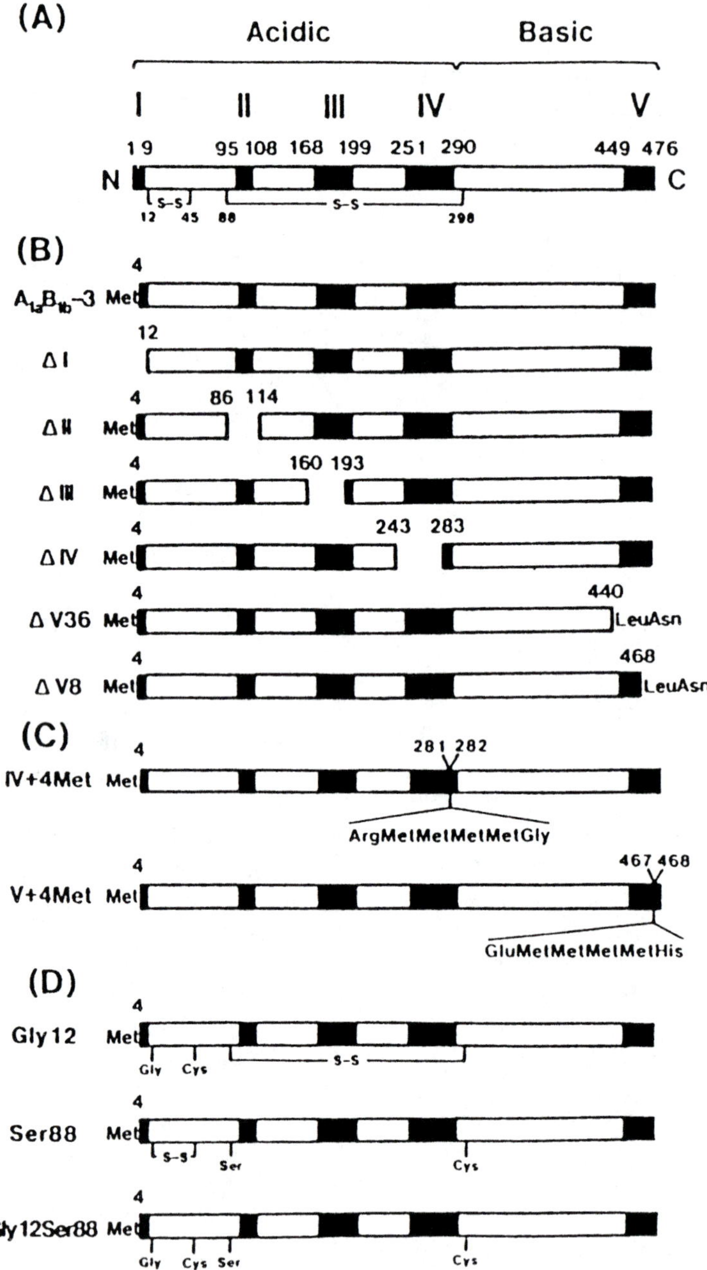

FIGURE 10 (A) The variable and conserved domains of the glycinin $A_{1a}B_{1b}$ subunit aligned by Wright [3]. Black and open areas are variable and conserved regions, respectively. The number of the residues from the N-terminus for the variable regions (I–V) are shown above the alignment. Acidic and basic polypeptide regions are indicated. (B) Construction of the deleted proglycinins and unmodified proglycinin $A_{1a}B_{1b}$-3. $A_{1a}B_{1b}$-3 lacks the N-terminal three amino acids; ΔI, the N-terminal eleven; ΔII, from the 87th to the 113th; ΔIII, from the 161st to the 192nd; ΔIV, from the 244th to the 282nd; ΔV36, from the 441st to the C-terminus; and ΔV8, from the 469th to the C-terminus. N-terminal methionine was retained

[1,57,114]. Also, correlations have been shown between disulfide bond formation and gel firmness from direct determination of sulfhydryl and disulfide bond contents in soy protein isolate gels [121]. The free sulfhydryl groups present in the unheated soy protein isolate play an important role in the formation of a firm gel. The sulfhydryl groups are present in α' and α subunits of β-conglycinin and acidic and basic polypeptides of glycinin; these can either undergo oxidation and/or catalyze the SH–S-S interchange reaction [47]. It is likely that the sulfhydryl group content of soy protein isolate varies widely depending on the procedures used for its preparation; thus, variations in SH content can also cause variations in the properties of soy protein gels.

Protein engineering is a useful technique that can be applied to understanding the relationships between structure and functionality of soy proteins. The site-directed manipulation of protein structure makes it possible to alter the stability or the physicochemical properties of the protein; through precise manipulation of the structure and properties of the protein, a basic understanding of the relationship between protein structure and gelation properties can be developed. In addition, rigorous scientific methodologies for evaluating gel properties need to be developed. Particularly, there is a need to develop techniques useful for evaluating, distinguishing, and characterizing textural properties of the gel [119].

D. Investigation by Protein Engineering

Protein engineering is a technique used to modify the primary structure of proteins by gene manipulation. The primary sequences of a target protein can be modified systematically and consciously. Therefore, protein engineering is a powerful method of elucidating the relationships between the structure and the functional properties of food proteins.

The following relationships at the molecular level in glycinin have been found:

1. Heat instability of the constituent subunits of glycinin is related to the heat-induced gel-forming ability [45].
2. Hydrophobicity is an important factor in the emulsifying properties [122].
3. The surface properties of a protein depend on the conformational stability—the more unstable, the higher the emulsifying properties [123].
4. The topology of free sulfhydryl residues are closely related to the heat-induced gel-forming ability [15].

Assuming that genetically modified proteins should be able to form the correct conformation (extreme modifications are not desired), the following modifications were designed: (a) deletion of a hydrophilic region (a variable region) to increase the relative hydrophobicity (Fig. 10B) [124]; (b) insertion of a hydrophobic oligopeptide (tetra me-

---

in $A_{1a}B_{1b}$-3, ΔII, ΔIII, ΔIV, ΔV36, and ΔV8, and was cleaved in ΔI. ΔV36, and ΔV8 have two extra amino acids, Leu-Asn, derived from universal terminator at their C-terminus. (C) Construction of the tetramethionine–inserted proglycinins. IV+4Met has Arg-Met-Met-Met-Met-Gly between Pro281 and Arg282. V+4Met has Glu-Met-Met-Met-Met-His between Pro467 and Gln468. (D) Construction of the disulfide bond–deleted proglycinins. Cys12 and Cys88 are substituted with Gly and Ser in Gly12 and Ser88, respectively. Both Cys12 and Cys88 are, respectively, substituted with Gly and Ser in Gly12Ser88. (Adapted from Refs. 15, 122, and 124.)

thionine) into a hydrophilic region (a variable region) in order to destabilize the molecule and increase the relative hydrophobicity (Fig. 10C) [124]; (c) deletion of disulfide bond(s) to change the topology of free sulfhydryl residues and destabilize the molecule (Fig. 10D) [125]. Although glycinin expressed in the *E. coli* expression system accumulates as proglycinin (trimer) instead of glycinin because of the absence of the processing enzyme, this approach was nonetheless employed for this investigation because the expressed proglycinins had a secondary structure similar to that of glycinin and exhibited several properties common to soy glycinin, namely, cryoprecipitation and Ca^{2+}-induced precipitation [68].

Expression plasmids for the modified proglycinins shown in Figure 10 were constructed using the expression plasmid for $A_{1a}B_{1b}$-3, of which expression level was ~20% of total *E. coli* proteins [68]. For the modified proglycinins to assume a conformation similar to that of native proglycinin, the following three criteria should be satisfied: (a) high-level expression in *E. coli* (or stable under high ionic strength condition after formation of the proper conformation); (b) solubility comparable to that of globulins; (c) self-assembly into trimers [124–126]. Among the modified proglycinins, ΔI, ΔV8, IV+4Met, V+4Met, Gly12, Ser88, and Gly12Ser88 satisfied the three criteria [124,125], although Gly12Ser88 was unstable under purification conditions (low ionic condition).

Emulsifying activities of the purified modified proglycinins (ΔI, ΔV8, IV+4Met, V+4Met, Gly12, Ser88) are shown in Table 3 [124,125]. All modified proglycinins exhibited higher emulsifying activities than did native glycinin. ΔV8 and V+4Met exhibited twice the value of the native glycinin, and those of ΔI, IV+4Met, Gly12 and Ser88 were similar to that of the unmodified proglycinin $A_{1a}B_{1b}$-3. The residues between positions 10 and 17 from the C-terminus of the $A_{1a}B_{1b}$ subunit are hydrophobic in nature, although this region is surrounded by strong hydrophilic residues. This suggests that the hydrophobicity of the C-terminal region may be closely related to the emulsifying properties of glycinin.

All of the modified proglycinins could form gels when heated at 100°C. Figure 11 shows the hardness of the gels from the modified and unmodified proglycinins and the

TABLE 3 Emulsifying Activities of Native Glycinin and Unmodified and Modified Proglycinins

Samples	Emulsifying activity[a] (%)
Native glycinin	100
Unmodified	123
ΔI	125
ΔV8	203
IV + 4Met	127
V + 4Met	200
Gly12	115
Ser88	133

[a]Emulsifying activity was expressed as relative value (%) compared with the native glycinin.
Source: Refs: 124 and 125.

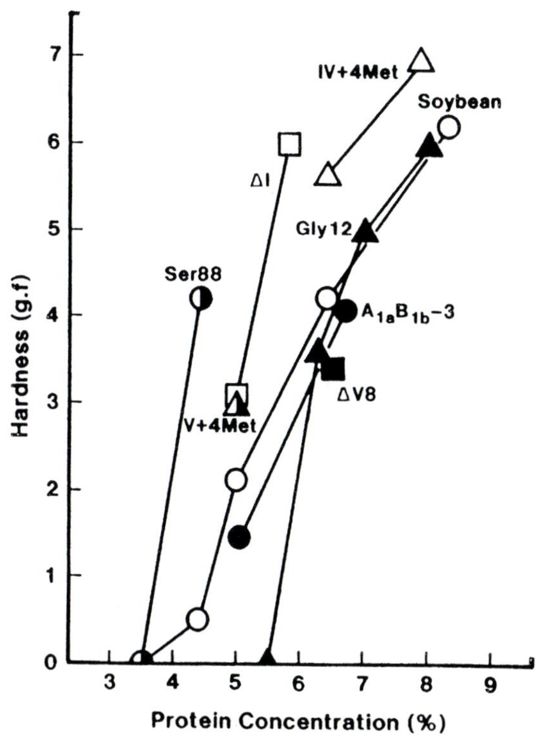

FIGURE 11 Hardness of gels from the native glycinin, the unmodified and modified proglycinins. Protein samples in 3.5 mM K-Pi buffer (pH 7.6) were boiled for 30 minutes. (Adapted from Refs. 124 and 125.)

native glycinin. The hardness of the unmodified proglycinin gels was somewhat lower than that of the native glycinin. The gels of ΔV8 were slightly softer than that of the native glycinin. Gly12 formed gels at protein concentration higher than 6% with gel hardness similar to that of the native glycinin. However, Gly12 could not form a gel at a concentration of 5.6% protein, which is sufficient for the unmodified proglycinin and the native glycinin to form a gel. On the other hand, the gel hardness of ΔI, IV+4Met, V+4Met, and Ser88 was higher than that of the native glycinin; Ser88 could form hard gels at a concentration of 4.4% protein, whereas the native glycinin forms only a very soft gel at this protein concentration. These results indicate that (a) the disulfide bond 12—45 plays an important role in the initiation of the disulfide-exchange reaction for gelation, although it seems that this disulfide bond may not be essential for gelation at higher protein concentration; (b) the topology of disulfide bonds is closely related to the heat-induced gel-forming ability and gel properties; (c) the factors that induce desirable changes in structure for gelation and emulsification properties are quite different.

Utsumi's group has succeeded in obtaining crystals of these modified proglycinins [127]. Future x-ray analysis of these crystals would shed light on the relationships between the structure and the functional properties of glycinin.

REFERENCES

1. J. E. Kinsella, Functional properties of soy proteins, *J. Am. Oil. Chem. Soc. 56*:242 (1979).
2. E. Derbyshire, D. J. Wright, and D. Boulter, Legumin and vicilin, storage proteins of legume seeds, *Phytochemistry 15*:3 (1976).
3. D. J. Wright, The seed globulins, *Developments in Food Proteins 5* (B. J. F. Hudson, ed.), Elsevier, London, 1987, p. 81.
4. H. Hirano, H. Kagawa, Y. Kamata, and F. Yamauchi, Structural homology among the major 7S globulin subunits of soybean seed storage proteins, *Phytochemistry 26*:41 (1987).
5. F. Yamauchi, K. Sato, and T. Yamagishi, Isolation and partial characterization of a salt extractable globulin from soybean seeds, *Agric. Biol. Chem. 48*:645 (1984).
6. W. Sato, T. Yamagishi, Y. Kamata, and F. Yamauchi, Subunit structure and immunological properties of a basic 7S globulin from soybean seeds, *Phytochemistry 26*:903 (1987).
7. H. Kagawa and H. Hirano, Sequence of a cDNA encoding soybean basic 7S globulin, *Nucleic Acids Res. 17*:8868 (1989).
8. Y. Watanabe and H. Hirano, Nucleotide sequence of the basic 7S globulin gene from soybean, *Plant Physiol. 105*:1019 (1994).
9. H. Hirano, H. Kagawa, and K. Okubo, Characterization of proteins released from legume seeds in hot water, *Phytochemistry 31*:731 (1992).
10. W. Sato, Y. Kamata, M. Fukuda, and F. Yamauchi, Improved isolation method and some properties of soybean gamma-conglycinin, *Phytochemistry 23*:1523 (1984).
11. F. Yamauchi, W. Sato, and Y. Kamata, Subunit structure of γ-conglycinin in soybean seeds, *Phytochemistry 24*:1503 (1985).
12. I. Koshiyama and D. Fukushima, Purification and some properties of γ-conglycinin in soybean seeds, *Phytochemistry 15*:161 (1976).
13. N. C. Nielsen, Structure of soy proteins, *New Protein Foods 5: Seed Storage Proteins* (A. M. Altshul and H. L. Wilcke, eds.), Academic Press, Orlando, FL, 1985, p. 27.
14. V. H. Thanh and K. Shibasaki, Beta-conglycinin from soybean proteins, *Biochim. Biophys. Acta 490*:370 (1977).
15. S. Utsumi, Plant food protein engineering, *Advances in Food Nutrition Research 36* (J. E. Kinsella, ed.), Academic Press, San Diego, CA, 1992, p. 89.
16. I. Koshiyama, Carbohydrate component in 7S protein of soybean casein fraction, *Agric. Biol. Chem. 30*:646 (1966).
17. F. Yamauchi, V. H. Thanh, M. Kawase, and K. Shibasaki, Separation of the glycopeptides from soybean 7S protein: their amino acid sequences, *Agric. Biol. Chem. 40*:691 (1979).
18. F. Yamauchi, M. Kawase, M. Kanbe, and K. Shibasaki, Separation of the β-aspartamido-carbohydrate fractions from soybean 7S protein: protein-carbohydrate linkage, *Agric. Biol. Chem. 39*:873 (1975).
19. F. Yamauchi and T. Yamagishi, Carbohydrate sequence of a soybean 7S protein, *Agric. Biol. Chem. 43*:505 (1979).
20. V. H. Thanh and K. Shibasaki, Major proteins of soybean seeds. Subunit structure of β-conglycinin, *J. Agric. Food Chem. 26*:692 (1978).
21. F. Yamauchi, M. Sato, W. Sato, Y. Kamata, and K. Shibasaki, Isolation and purification of a new type of β-conglycinin in soybean globulins, *Agric. Biol. Chem. 45*:2863 (1981).
22. T. Nakamura, S. Utsumi, and T. Mori, Mechanism of heat-induced gelation and gel properties of soybean 7S globulins, *Agric. Biol. Chem. 50*:1287 (1986).
23. V. H. Thanh and K. Shibasaki, Major proteins of soybean seeds. Reconstitution of β-conglycinin from its subunits, *J. Agric. Food Chem. 26*:695 (1978).
24. V. H. Thanh and K. Shibasaki, Major proteins of soybean seeds. Reversible and irreversible dissociation of β-conglycinin, *J. Agric. Food Chem. 27*:805 (1979).
25. J. E. Kinsella, S. Damodaran, and B. German, Physicochemical and functional properties of oilseed proteins with emphasis on soy proteins, *New Protein Foods 5: Seed Storage*

Proteins (A. M. Altschul and H. L. Wilcke, eds.), Academic Press, Orlando, FL, 1985, p. 107.
26. C. Pedrosa and S. T. Ferreira, Deterministic pressure-induced dissociation of vicilin, the 7S storage globulin from pea seeds: effects of pH and cosolvents on oligomer stability, *Biochemistry 33*:4046 (1994).
27. D. J. Wright, The seed globulins, *Developments in Food Proteins 6* (B. J. F. Hudson, ed.), Elsevier, London, 1988, p. 119.
28. P. E. M. Gibbs, K. B. Strongin, and A. McPherson, Evolution of legume seed storage proteins-A domain common to legumins and vicilins is duplicated in vicilins, *Mol. Biol. Evol. 6*:614 (1989).
29. P. Tulloch and R. J. Blagrove, Electron microscopy of seed-storage globulins, *Arch. Biochem. Biophys. 241*:521 (1985).
30. P. Plietz, G. Damaschun, J. J. Müler, and B. Schlesier, Comparison of the structure of the 7S globulins from *Phaseolus vulgaris* in solution with the crystal structure of 7S globulin from *Canavalia ensiformis* by small angle X-ray scattering, *FEBS Lett. 162*:43 (1983).
31. K. J. l'Anson, M. J. Miles, J. R. Bacon, H. J. Carr, N. Lambert, V. J. Morris, and D. J. Wright, Structure of the 7S globulin (vicilin) from pea (*Pisum sativum*), *Int. J. Biol. Macromol. 10*:311 (1988).
32. M. E. Welland, M. J. Miles, N. Lambert, V. J. Morris, J. H. Coombs, and J. B. Pethica, Structure of the globular protein vicilin revealed by scanning tunnelling microscopy, *Int. J. Biol. Macromol. 11*:29 (1989).
33. E. Suzuki, A. Van Donkelaar, J. N. Varghese, G. G. Lilley, R. J. Blagrove, and P. M. Colman, Crystallization of phaseolin from *Phaseolus vulgaris*, *J. Biol. Chem. 258*:2634 (1983).
34. J. B. Sumner and S. F. Howell, The isolation of a fourth crystallizable jack bean globulin through the digestion of canavalin with trypsin, *J. Biol. Chem. 113*:607 (1936).
35. B. F. Ladin, M. L. Tierney, D. W. Meinke, P. Hosángadi, M. Veith, and R. N. Beachy, Developmental regulation of β-conglycinin in soybean axes and cotyledons, *Plant Physiol. 84*:35 (1987).
36. M. C. Lawrence, T. Izard, M. Beauchat, R. J. Blagrove, and P. M. Colman, Structure of phaseolin at 2.2 Å resolution. Implications for a common vicilin/legumin structure and the genetic engineering of seed storage proteins, *J. Mol. Biol. 238*:748 (1994).
37. M. C. Lawrence, E. Suzuki, J. N. Varghese, P. C. Davis, A. Van Donkelaar, P. A. Tulloch, and P. M. Colman, The three-dimensional structure of the seed storage protein phaseolin at 3 Å resolution, *EMBO J. 9*:9 (1990).
38. T.-P. Ko, J. D. Ng, and A. McPherson, The three-dimensional structure of canavalin from jack bean (*Canavalia ensiformis*), *Plant Physiol. 101*:729 (1993).
39. M. Duranti, N. Guerrieri, T. Takahashi, and P. Cerletti, The legumin-like storage protein of *Lupinus albus* seeds, *Phytochemistry 27*:15 (1988).
40. P. E. Staswick, M. A. Hermodson, and N. C. Nielsen, Identification of the cysteines which link the acidic and basic components of the glycinin subunits, *J. Biol. Chem. 259*:13431 (1984).
41. N. C. Nielsen, The structure and complexity of the 11S polypeptides in soybeans, *J. Am. Oil Chem. Soc. 62*:1680 (1985).
42. T. Momma, T. Negoro, K. Udaka, and C. Fukazawa, A complete cDNA coding for the sequence of glycinin A_2B_{1a} subunit precursor, *FEBS Lett. 188*:117 (1985).
43. N. C. Nielsen, C. D. Dickinson, T.-J. Cho, V. H. Thanh, B. J. Scallon, R. L. Fischer, T. L. Sims, G. N. Drews, and R. B. Goldberg, Characterization of the glycinin gene family in soybean, *Plant Cell 1*:313 (1989).
44. N. E. Tumer, J. D. Richter, and N. C. Nielsen, Structural characterization of the glycinin precursors, *J. Biol. Chem. 257*:4016 (1982).
45. T. Nakamura, S. Utsumi, K. Kitamura, K. Harada, and T. Mori, Cultivar differences in gelling characteristics of soybean glycinin, *J. Agric. Food Chem. 32*:647 (1984).

46. M. Draper and N. Catsimpoolas, Disulfide and sulfhydryl groups in glycinin, *Cereal Chem. 55*:16 (1978).
47. W. J. Wolf, Sulfhydryl content of glycinin: effect of reducing agents, *J. Agric. Food Chem. 41*:168 (1993).
48. P. Argos, S. V. L. Narayana, and N. C. Nielsen, Structural similarity between legumin and vicilin storage proteins from legumes, *EMBO J. 4*:1111 (1985).
49. T. Mori, S. Utsumi, H. Inaba, K. Kitamura, and K. Harada, Differences in subunit composition of glycinin among soybean cultivars, *J. Agric. Food Chem. 29*:20 (1981).
50. S. Utsumi, M. Kohno, T. Mori, and M. Kito, An alternate cDNA encoding glycinin $A_{1a}B_x$ subunit, *J. Agric. Food Chem. 35*:210 (1987).
51. S. Utsumi, C.-S. Kim, M. Kohno, and M. Kito, Polymorphism and expression of cDNAs encoding glycinin subunits, *Agric. Biol. Chem. 51*:3267 (1987).
52. S. Utsumi, H. Inaba, and T. Mori, Heterogeneity of soybean glycinin, *Phytochemistry 20*: 585 (1981).
53. T. Mori, T. Nakamura, and S. Utsumi, Formation of pseudoglycinins and their gel hardness, *J. Agric. Food Chem. 30*:828 (1982).
54. T. Nakamura, S. Utsumi, and T. Mori, Formation of pseudoglycinins from intermediary subunits of glycinin and their gel properties and network structure, *Agric. Biol. Chem. 49*: 2733 (1985).
55. S. Utsumi, H. Inaba, and T. Mori, Formation of pseudo- and hybrid-11S globulins from subunits of soybean and broad bean 11S globulins, *Agric. Biol. Chem. 44*:1891 (1980).
56. W. J. Wolf and D. R. Briggs, Studies on the cold-insoluble fraction of the water-extractable soybean proteins. II. Factors influencing conformation changes in the 11S component, *Arch. Biochem. Biophys. 76*:377 (1958).
57. T. Mori, T. Nakamura, and S. Utsumi, Gelation mechanism of soybean 11S globulin: formation of soluble aggregates as transient intermediates, *J. Food Sci. 47*:26 (1982).
58. S. Utsumi, T. Nakamura, K. Harada, and T. Mori, Occurrence of dissociable and undissociable soybean glycinin, *Agric. Biol. Chem. 51*:2139 (1987).
59. R. A. Badley, D. Atkinson, H. Hauser, D. Oldani, J. P. Green, and J. M. Stubbs, The structure, physical and chemical properties of the soybean protein glycinin, *Biochim. Biophys. Acta. 412*:214 (1975).
60. R. Reichelt, K.-D. Schwenke, T. König, W. Pähtz, and G. Wangermann, Electron microscopic studies for estimation of the quaternary structure of the 11S globulin (Helianthinin) from sunflower seed (*Helianthus annuus* L.), *Biochem. Physiol. Pflanzen 175*:653 (1980).
61. P. Plietz, G. Damaschun, J. J. Müler, and K.-D. Schwenke, The structure of 11-S globulins from sunflower and rape seed, *Eur. J. Biochem. 130*:315 (1983).
62. P. Plietz, D. Zirwer, B. Schlesier, K. Gast, and G. Damaschun, Shape, symmetry, hydration and secondary structure of the legumin from *Vicia faba* in solution, *Biochim. Biophys. Acta 784*:140 (1984).
63. M. J. Miles, V. J. Morris, D. J. Wright, and J. R. Bacon, A study of the quaternary structure of glycinin, *Biochim. Biophys. Acta. 827*:119 (1985).
64. K. J. I'Anson, J. R. Bacon, N. Lambert, M. J. Miles, V. J. Morris, D. J. Wright, and C. Nave, Synchrotron radiation wide-angle X-ray scattering studies of glycinin solutions, *Int. J. Biol. Macromol. 9*:368 (1987).
65. A. M. H. Schepman, T. Wichertjes, and E. F. J. Van Bruggen, Visibility of subunits in crystals of oligomeric proteins. Electron microscopy and optical diffraction of edestin and excelsin, *Biochim. Biophys. Acta. 271*:279 (1972).
66. P. M. Colman, E. Suzuki, and A. Van Donkelaar, The structure of cucurbitin: subunit symmetry and organization in situ, *Eur. J. Biochem. 103*:585 (1980).
67. S. Patel, R. Cudney, and A. McPherson, Crystallographic characterization and molecular symmetry of edestin, a legumin from hemp, *J. Mol. Biol. 235*:361 (1994).

68. C.-S. Kim, S. Kamiya, J. Kanamori, S. Utsumi, and M. Kito, High-level expression, purification and functional properties of soybean proglycinin from *Escherichia coli*, *Agric. Biol. Chem. 54*:1543 (1990).
69. S. Utsumi, J. Kanamori, C.-S. Kim, T. Sato, and M. Kito, Properties and distribution of soybean proglycinin expressed in *Saccharomyces cerevisiae*, *J. Agric. Food Chem. 39*:1179 (1991).
70. S. Utsumi, A. B. Gidamis, B. Mikami, and M. Kito, Crystallization and preliminary X-ray crystallographic analysis of the soybean proglycinin expressed in *Escherichia coli*, *J. Mol. Biol. 233*:177 (1993).
71. P. Plietz, B. Drescher, and G. Damaschun, Relationship between the amino acid sequence and the domain structure of the subunits of the 11S seed globulins, *Int. J. Biol. Macromol. 9*:161 (1987).
72. A. H. Clark and C. D. Lee-Tuffnell, Gelation of globular proteins, *Functional Properties of Food Macromolecules* (J. R. Mitchell and D. A. Ledword, eds.), Elsevier Applied Science, London, 1986, p. 203.
73. B. German, S. Damodaran, and J. E. Kinsella, Thermal dissociation and association of soy proteins, *J. Agric. Food Chem. 30*:807 (1982).
74. S. Damodaran, Refolding of thermally unfolded soy proteins during the cooling regime of the gelation process: effect on gelation, *J. Agric. Food Chem. 36*:262 (1988).
75. S. Utsumi, T. Nakamura, and T. Mori, A micro-method for measurements of gel properties of soybean 11S globulin, *Agric. Biol. Chem. 46*:1293 (1982).
76. T. Mori, M. Mohri, N. Artik, and Y. Matsumura, Rheological properties of heat-induced gel of soybean 11S globulin under high ionic strength, *J. Texture Studies 19*:361 (1989).
77. T. Nakamura, S. Utsumi, and T. Mori, Network structure formation in thermally induced gelation of glycinin, *J. Agric. Food Chem. 32*:349 (1984).
78. A-M. Hermansson, Structure of soya glycinin and conglycinin gels, *J. Sci. Food Agric. 36*:822 (1985).
79. M. Babajimopoulos, S. Damodaran, S. S. H. Rizvi, and J. E. Kinsella, Effects of various anions on the rheological and gelling behavior of soy proteins: thermodynamic observations, *J. Agric. Food Chem. 31*:1270 (1983).
80. R. L. Baldwin, Temperature dependence of the hydrophobic interaction in protein folding, *Proc. Natl. Acad. Sci. USA 83*:69 (1986).
81. K. Ikura, T. Kometani, M. Yoshikawa, R. Sasaki, and H. Chiba, Crosslinking of casein components by transglutaminase, *Agric. Biol. Chem. 44*:1567 (1980).
82. N. Nio, M. Motoki, and K. Takinami, Gelation of casein and soybean globulins by transglutaminase, *Agric. Biol. Chem. 49*:2283 (1985).
83. M. Nonaka, H. Tanaka, A. Okiyama, M. Motoki, H. Ando, K. Umeda, and A. Matsuura, Polymerization of several proteins by Ca^{2+}-independent transglutaminase derived from microorganisms, *Agric. Biol. Chem. 53*:2619 (1989).
84. Y. Chanyongvorakul, Y. Matsumura, Y. Sakamoto, M. Motoki, and T. Mori, Gelation of beans 11S globulins by Ca^{2+}-independent transglutaminase, *Biosci. Biotech. Biochem. 58*: 864 (1994).
85. Y. Chanyongvorakul, Y. Matsumura, M. Nonaka, M. Motoki, and T. Mori, Physical properties of soy bean and broad bean 11S globulin gels formed by transglutaminase reaction, *J. Food Sci. 60*:483 (1995).
86. I-J. Kang, Y. Matsumura, K. Ikura, M. Motoki, H. Sakamoto, and T. Mori, Gelation and gel properties of soybean glycinin in a transglutaminase-catalyzed system, *J. Agric. Food Chem. 42*:159 (1994).
87. M. Nonaka, S. Toiguchi, H. Sakamoto, H. Kawajiri, T. Soeda, and M. Motoki, Changes caused by microbial transglutaminase on physical properties of thermally induced soy protein gels, *Food Hydrocoll. 8*:1 (1994).

88. V. H. Thanh and K. Shibasaki, Major proteins of soybean seeds: a straightforward fractionation and their characterization, *J. Agric. Food Chem. 24*:1117 (1976).
89. T. Nagano, H. Hirotsuka, H. Mori, K. Kohyama, and K. Nishinari, Dynamic viscoelastic study on the gelation of 7S globulin from soybeans, *J. Agric. Food Chem. 40*:941 (1992).
90. T. Nagano, H. Mori, and K. Nishinari, Effect of heating and cooling on the gelation kinetics of 7S globulin from soybeans, *J. Agric. Food Chem. 42*:1415 (1994).
91. S. Utsumi and J. E. Kinsella, Forces involved in soy protein gelation: effects of various reagents on the formation, hardness and solubility of heat-induced gels made from 7S, 11S and soy isolate, *J. Food Sci. 50*:1278 (1985).
92. T. Nagano, T. Akasaka, and K. Nishinari, Dynamic viscoelastic properties of glycinin and β-conglycinin gels from soybeans, *Biopolymers 34*:1303 (1994).
93. W. K. Surewicz, H. H. Mantsch, and D. Chapman, Determination of protein secondary structure by fourier transform infrared spectroscopy: a critical assessment, *Biochemistry 32*:389 (1993).
94. C. H. Wang and S. Damodaran, Thermal gelation of globular proteins: influence of protein conformation on gel strength, *J. Agric. Food Chem. 39*:433 (1991).
95. T. Nakamura, S. Utsumi, and T. Mori, Interactions during heat-induced gelation in a mixed system of soybean 7S and 11S globulins, *Agric. Biol. Chem. 50*:2429 (1986).
96. E. L. Arrese, D. A. Sorgentini, J. R. Wagner, and M. C. Avon, Electrophoretic, solubility and functional properties of commercial soy protein isolates, *J. Agric. Food Chem. 39*:1029 (1991).
97. S. Utsumi, S. Damodaran, and J. E. Kinsella, Heat-induced interactions between soybean proteins: preferential association of 11S basic subunits and β subunits of 7S, *J. Agric. Food Chem. 32*:1406 (1984).
98. E. Tornberg, Functional characterization of protein stabilized emulsions: emulsifying behavour of proteins in a valve homogeneizer, *J. Sci. Food Agric. 29*:867 (1978).
99. A. Kato, Y. Osako, N. Matsudomi, and K. Kobayashi, Changes in the emulsifying and foaming properties of proteins during heat denaturation, *Agric. Biol. Chem. 47*:33 (1983).
100. H. J. Rivas and P. Sherman, Soy and meat proteins as food emulsion stabilizers. 1. Viscoelastic properties of corn oil-in-water emulsions incorporating soy or meat protein, *J. Texture Studies 14*:251 (1983).
101. H. J. Rivas and P. Sherman, Soy and meat proteins as food emulsion stabilizers. 2. Influence of emulsification temperature, NaCl and methanol on the viscoelastic properties of corn oil-in-water emulsions incorporating acid precipitated soy protein, *J. Texture Studies 14*:267 (1983).
102. H. J. Rivas and P. Sherman, Soy and meat proteins as food emulsion stabilizers. 3. The influence of soy and meat protein fractions on oil-water interfacial tension, *J. Dispers. Sci. Technol. 5*:143 (1984).
103. H. J. Rivas and P. Sherman, Soy and meat proteins as food emulsion stabilizers. 4. The stability and interfacial rheology of O/W emulsions stabilized by soy and meat protein fractions, *Colloids Surfaces 11*:155 (1984).
104. S. Hayakawa and S. Nakai, Relationships of hydrophobicity and net charge to the solubility of milk and soy proteins, *J. Food Sci. 50*:486 (1985).
105. E. Dickinson, Structure and composition of adsorbed protein layers and the relationship to emulsion stability, *J. Chem. Soc. Faraday Trans. 88*:2973 (1992).
106. H. Aoki, O. Taneyama, and M. Inami, Emulsifying properties of soy protein: characteristics of 7S and 11S proteins, *J. Food Sci. 45*:534 (1980).
107. J. J. Yao, L. S. Wei, and M. P. Steinberg, Effects of maturity on chemical composition and storage stability of soybeans, *J. Am. Oil Chem. Soc. 60*:1245 (1983).
108. J. J. Yao, K. Tanteeratarm, and L. S. Wei, Effects of maturation and storage solubility and gelation properties of isolated soy proteins.

109. A.-M. Hermansson, Gel structure of food biopolymers, *Food Structure—Its Creation and Evaluation* (J. M. V. Blanshard and J. R. Mitchell, eds.), Butterworths, London, 1988 p. 25.
110. Y.-C. Ker, R.-H. Chen, and C.-S. Wu, Relationships of secondary structure, microstructure, and mechanical properties of heat-induced gel of soy 11S globulin, *Biosci. Biotech. Biochem. 57*:536 (1993).
111. B.-A. Zheng, Y. Matsumura, and T. Mori, Thermal gelation mechanism of legumin from broad beans, *J. Food Sci. 56*:722 (1991).
112. N. Yuno-Ohta, H. Maeda, M. Okada, and H. Ohta, Heat-induced gels of rice globulin: comparison of gel properties with soybean and sesame globulins, *J. Food Sci. 59*:366 (1994).
113. M. Stading, M. Langton, and A.-M. Hermansson, Microstructure and rheological behaviour of particulate β-lactoglobulin gels, *Food Hydrocoll. 7*:195 (1993).
114. T. Mori, T. Nakamura, and S. Utsumi, Behavior of intermolecular bond formation in the late stage of heat-induced gelation of glycinin, *J. Agric. Food Chem. 34*:33 (1986).
115. K. Kamada, D. Rector, and J. E. Kinsella, Influence of temperature of measurement on creep phenomena in glycinin gels, *J. Food Sci. 53*:589 (1988).
116. F. Shama and P. Sherman, The texture of ice cream: 2. Rheological properties of frozen ice cream, *J. Food Sci. 31*:699 (1966).
117. S. Damodaran and J. E. Kinsella, Effect of conglycinin on the thermal aggregation of glycinin, *J. Agric. Food Chem. 30*:812 (1982).
118. T. Nagano, H. Mori, and K. Nishinari, Rheological properties and conformational states of β-conglycinin gels at acidic pH, *Biopolymers 34*:293 (1994).
119. I. J. Kang, Y. Matsumura, and T. Mori, Characterization of texture and mechanical properties of heat-induced soy protein gels, *J. Am. Oil Chem. Soc. 68*:339 (1991).
120. S. Damodaran and J. E. Kinsella, Effect of conglycinin on the thermal aggregation of glycinin, *J. Agric. Food Chem. 30*:812 (1982).
121. K. Shimada and J. C. Cheftel, Determination of sulfhydryl groups and disulfide bonds in heat-induced gels of soy protein isolate, *J. Agric. Food Chem. 36*:147 (1988).
122. S. Utsumi and M. Kito, Improvement of food protein functions by chemical, physical, and biological modifications, *Comments Agric. Food Chem. 2*:261 (1991).
123. A. Kato and K. Yutani, Correlation of surface properties with conformational stabilities of wild-type and six mutant tryptophan synthase α-subunits substituted at the same position, *Protein Eng. 2*:153 (1988).
124. C.-S. Kim, S. Kamiya, T. Sato, S. Utsumi, and M. Kito, Improvement of nutritional value and functional properties of soybean glycinin by protein engineering, *Protein Eng. 3*:725 (1990).
125. S. Utsumi, A. B. Gidamis, J. Kanamori, I. J. Kang, and M. Kito, Effects of deletion of disulfide bonds by protein engineering on the conformation and functional properties of soybean proglycinin, *J. Agric. Food Chem. 41*:687 (1993).
126. A. B. Gidamis, P. Wright, Z. U. Haque, T. Katsube, M. Kito, and S. Utsumi, Modification tolerability of soybean proglycinin, *Biosci. Biotech. Biochem. 59*:1593 (1995).
127. A. B. Gidamis, B. Mikami, T. Katsube, S. Utsumi, and M. Kito, Crystallization and preliminary X-ray analysis of soybean proglycinins modified by protein engineering, *Biosci. Biotech. Biochem. 58*:703 (1994).

10

Structure-Function Relationships of Wheat Proteins

F. MacRitchie
CSIRO Plant Industry, North Ryde, New South Wales, Australia

D. Lafiandra
University of Tuscia, Viterbo, Italy

I. INTRODUCTION

Although the processing of wheat for different foods has been practiced for thousands of years, there is currently an increased need to understand the relationships between protein composition and grain/flour functionality. This arises first from the more sophisticated processing brought about by automation, with the resulting demands for compliance with strict limits for properties of ingredients. Second, there is an increasing range of food products, all with special requirements. Research into the basis of functionality can be applied directly to optimize processing in the commercial situation or more indirectly to breed wheat varieties having properties that match specific end-use requirements. For example, hard-grain wheats (durum wheats) giving strong doughs are preferred for pasta making. In contrast, soft-grain wheats having weak, extensible doughs are sought for cookie (biscuit) manufacture. Wheats with properties between these extremes are best suited for various types of bread.

In this chapter, we will examine how different physical properties are determined by variation in the protein composition of the grain. Wheat grain contains an extremely complex mixture of proteins, with at least 100 different species usually being present. We begin by considering how these proteins are grouped into similar classes. Then the methods for analyzing and characterizing different classes will be discussed. The main portion of the chapter is devoted to describing the approaches used to establish relationships between protein composition and structure and functional properties. Based on these results, recent developments in polymer physics are applied to interpret grain/flour properties at macroscopic and molecular levels, culminating with some general conclusions and suggestions for future research directions.

II. CLASSIFICATION OF WHEAT PROTEINS

Several different criteria have been used to classify wheat proteins, e.g., those based on solubility [1] and those based on chemical (and therefore genetic) similarity [2]. From a functionality point of view, it is convenient to divide wheat proteins into two main classes, generally referred to as monomeric and polymeric proteins, depending on whether they consist of single- or multiple-chain polypeptides.

A. Monomeric Proteins

The monomeric (or single-chain) proteins comprise two main groups: the gliadins and the albumins/globulins. Gliadins are storage proteins, whereas albumins and globulins are metabolic and include various enzymes.

1. Gliadins

Gliadins are usually divided into α-, β-, γ-, and ω-gliadins, based on their mobility on 1D Acid-PAGE. Criteria for delineating these four groups have been described by Sapirstein and Bushuk [3]. Molecular weights of gliadins range from 30,000 to 80,000. The ω-gliadins are clearly separated from other wheat polypeptides in SDS-PAGE because their molecular weights (70,000–80,000) do not overlap with others. They are deficient in sulfur, but the other gliadins normally have an even number of cysteine residues, which form intramolecular disulfide bonds.

2. Monomeric Albumins/Globulins

Albumins (soluble in water) and globulins (soluble in dilute salt solution) are a mixture of low molecular weight compounds, many of which are enzymes. They are mostly of lower molecular weight than the gliadins (20,000–30,000). Their amino acid composition is distinctly different from the gluten proteins (gliadins and glutenins), as seen in Table 1 [4,5]. Gluten proteins are characterized by unusually high contents of glutamic acid (mainly in the amidated form as glutamine) and proline, whereas the albumins and globulins have much lower glutamic acid content (mainly in the acid form) but are higher in the essential amino acid lysine.

B. Polymeric Proteins

Three main groups of proteins constitute the multichain or polymeric proteins.

1. Glutenins

Glutenins form the major portion (approximately 85%) of the polymeric proteins. Together with gliadins, they are found in discrete protein bodies in grain or flour. Glutenins are similar to gliadins in chemical composition (see Table 1), indicating similarities in genetic ancestry. They are made up of two main groups of polypeptide chains, termed the high molecular weight (HMW) and low molecular weight (LMW) glutenin subunits. Recent results have indicated that the situation is even more complex, as it has been shown that glutenin polymers can incorporate mutated gliadin components possessing

TABLE 1 Comparison of Amino Acid Analyses for Glutenin, Gliadin, and Monomeric Albumins

Amino acid	Glutenin (mol%)	Gliadin (mol%)	Albumin, monomeric (mol%)
Cys (half)	2.6	3.3	8.1
Met	1.4	1.2	2.6
Asp	3.7	2.8	7.6
Thr	3.4	2.4	2.4
Ser	6.9	6.1	6.4
Glu	28.9	34.6	10.8
Pro	11.9	16.2	7.5
Gly	7.5	3.1	8.3
Ala	4.4	3.3	8.4
Val	4.8	4.8	11.3
Ile	3.7	4.3	1.7
Leu	6.5	6.9	7.6
Tyr	2.5	1.8	3.4
Phe	3.6	4.3	0.1
Lys	2.0	0.6	5.0
His	1.9	1.9	0.02
Arg	3.0	2.0	5.7
Trp	1.3	0.4	3.0

Source: Refs. 4, 5.

an odd number of cysteine residues. The HMW or A subunits fall roughly in the size range 80,000–120,000. The LMW subunits comprise two main subgroups: the B (MW = 40,000–55,000) and the C (MW = 30,000–40,000) subunits. These ranges have been deduced from mobilities on SDS-PAGE gels calibrated with molecular weight standards and are usually slightly higher than those calculated from DNA sequencing. Moreover, it has been found that electrophoretic mobility in SDS-PAGE does not rank HMW subunits exactly according to size. However, a system of nomenclature introduced by Payne and Lawrence [6] in which HMW subunits are numbered in order of increasing mobility has proved valuable for identifying different subunits and allowing comparison between different laboratories. The LMW subunits are not so easily resolved by SDS-PAGE. In a normal pure bread wheat variety, the maximum number of HMW subunits is 5, whereas the number of LMW subunits generally falls in the range of 9–16. The different subunits combine through disulfide bonds to form heterogeneous and polydispersed polymers. These polymers vary in molecular weight from hundreds of thousands up to several million, although accurate determinations are difficult because of their low solubility and very large molecular size.

2. HMW Albumins

The polymeric proteins next in abundance are the HMW albumins. These are mainly β-amylases and do not occur in the protein bodies. It seems probable, therefore, that their subunits form polymers with themselves and not with the glutenins.

3. Triticins

The other group of polymeric proteins are the triticins, which are globulin-type proteins. They also appear to consist of polymers between their own subunits [7]. However, participation of both HMW albumin and triticin subunits in polymers with glutenin subunits has not been ruled out.

III. METHODS FOR ELUCIDATING WHEAT PROTEIN COMPOSITION/STRUCTURE

In order to derive relationships between structure and functionality, it is imperative to have techniques to measure both wheat protein structure as well as functional properties with reliability. Functionality in relation to foods depends primarily on certain physical properties such as grain hardness (important in milling) and dough rheological properties that relate to the processing of different breads. Wheat flour contains starch and protein, these two natural polymeric materials making up over 90% of the flour (75–80% starch, 10–15% protein). However, in most food products, starch occurs in the form of small granules (1–40 μm in diameter) and in systems, such as dough, is a dispersed phase acting as a filler. On the other hand, protein forms a continuous network in a developed dough and is responsible for its viscoelastic properties. The parameter that is most important for determining physical properties such as the viscoelasticity of polymers is the molecular weight distribution (MWD). Therefore, we will pay special attention to measurement of the MWD of the wheat protein complex.

A. Molecular Weight Distribution

Since the application of high-performance liquid chromatography (HPLC) to cereal proteins was pioneered, mainly by Bietz and coworkers [8], this technique has been adopted by many groups and is proving one of the most useful for characterizing wheat proteins. Figure 1 shows a size-exclusion (SE-HPLC) profile for total wheat protein using an SEC 4000 (Beckman) column. Three main peaks are resolved corresponding, in order of increasing elution time, to polymeric proteins, gliadins, and albumins/globulins. The protein was solubilized by sonication of a flour suspension in SDS-buffer solution. This procedure normally solubilizes 95% or more of the total protein. However, two problems need to be recognized. First, a portion of the polymeric protein elutes at the void volume in current SE-HPLC columns, i.e., this protein is not resolved and only a minimum estimate can be made for its molecular weight. Second, the sonication needed for solubilization causes molecular scission of some of the largest molecular-sized proteins, essentially that part of the protein that is not solubilized in the absence of sonication [9]. Fortunately, the effect of sonication is selective in that it only breaks the largest molecules. Furthermore, the probability of scission is greatest at the center of molecules and diminishes rapidly as the distance from the center increases. For a given time and intensity of sonication (or any other process causing shear degradation), only molecules above a critical size are broken down and the degradation products are roughly half the size [9]. By optimizing the time and intensity of sonication, it is possible to maximize solubilization without affecting the distribution of the proteins into their main classes.

The molecular weight ranges, based on calibration with pure protein standards, for the different protein classes calculated from Figure 1 are (a) polymeric proteins > 100,000, (b) gliadins, 30,000–80,000, and (c) albumins/globulins, 20,000–30,000.

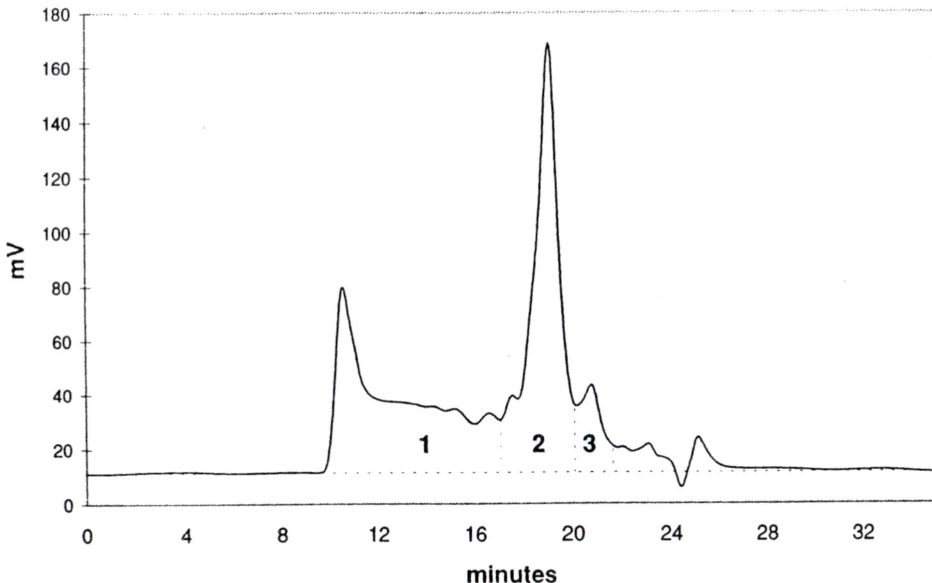

Figure 1 SE-HPLC profile of total wheat protein extracted using sonication of an SDS-buffer suspension of flour. Protein was run on a SEC-4000 column with an eluting solution of acetonitrile/water (1:1) containing 0.5% trifluoroacetic acid. The three main peaks in increasing elution time correspond to polymeric proteins, gliadins, and albumins/globulins.

B. Polymeric Protein Subunit Composition

The polypeptides (subunits) that form the polymeric proteins are determined after splitting the interchain disulfide bonds using a reducing agent such as 2-mercaptoethanol (ME) or dithiothreitol (DTT). Two main methods have then been employed to analyze the subunit composition: SDS-PAGE in conjunction with densitometry and reversed-phase HPLC (RP-HPLC).

1. Glutenin Subunits

Because the LMW (B and C) glutenin subunits are similar in size to the gliadins, methodology is required to separate the monomeric proteins before the glutenin subunit composition can be determined. Using SDS-PAGE, two approaches have been used. The first involves a two-step electrophoresis [10]. The total protein is loaded on the gel. Monomeric proteins run in front of the polymeric proteins, which are too large to significantly penetrate the gel and form a smear close to the starting slot. The part of the gel containing the smear is then separated, after which the protein is dissolved and reduced (i.e., interchain disulfide bonds are broken) and run on a second gel. The other approach has been to separate the monomeric proteins by solubilizing them (using a solvent such as propanol or dimethyl sulfoxide) so as to leave the polymeric proteins in the residue [11,12]. The polymeric proteins are then reduced and either SDS-PAGE or RP-HPLC is used to resolve the subunits and allow quantitation.

Both approaches have some disadvantages. In the two-step electrophoretic method, it is difficult to obtain a clear position of the gel where there is no overlap between the larger monomeric and smaller polymeric proteins. In the one-step electrophoretic procedure, a portion of the smaller polymeric proteins is invariably removed in the solubilization step.

A comparison of the two methods is shown in Figure 2. The HMW (A) glutenin subunits are clearly resolved. The LMW (B and C) subunits form two groups of bands with higher mobilities. Another minor group of subunits, the LMW D subunits, has been identified in some wheat varieties, and these have electrophoretic mobilities between the A and B subunits [13].

2. HMW Albumin Subunits

Fractionation of total seed proteins by diagonal electrophoresis (unreduced × reduced) showed that HMW albumin bands occur in both polymeric and monomeric forms in the native state [14]. Four main bands were identified corresponding to polypeptides of molecular weights 65,000, 63,000, 60,000, and 45,000. Except for the 45,000 polypeptide, all were shown to be β-amylases by immunoblotting.

3. Triticin Subunits

SDS-PAGE of unreduced total protein extracts from hexaploid wheat shows three bands of low mobility in a zone of heavy background streaking. The proteins associated with these bands have been studied by Singh and Shepherd [7]. Their amino acid composition and solubility properties are characteristic of globulins. Using diagonal electrophoresis, they have been shown to be disulfide-linked heterotetramers made up of four subunits of molecular weights 22,000, 23,000, 52,000, and 58,000 in several different combinations. The molecular weights of the native molecules are therefore lower than other polymeric proteins.

C. New Approaches to Molecular Weight Measurement of Polymeric Proteins

As referred to in Section III.A, two problems need to be overcome in order to measure the MWD of polymeric proteins. The first is to find a method for completely solubilizing the protein without altering it. Second, a method is needed for reliable and precise measurements of molecular weights of these large proteins. The factors that limit solubility of glutenins are the deficiency of ionizable groups and the very large molecular size. Surfactants such as SDS apparently increase solubility by complexing with the protein to impart a high electrical charge at the molecular surface. However, although solubility is enhanced, a significant portion of the protein remains insoluble. One possible way of achieving complete solubility would be by chemical modification of protein groups so as to place a very high net electrical charge on the molecules. Such modification would not significantly change the molecular size, provided there was no degradation.

More precise determinations of molecular size range are needed. Because the polymeric protein is a polydispersed system, methods giving a number-average or weight-average MW are not very useful. However, methods such as the newer light scattering techniques using a LASER system appear promising for size-distribution determinations. Another method with potential is field flow fractionation (FFF). The advantage of this technique is that, unlike many methods of molecular weight measurement, its resolution

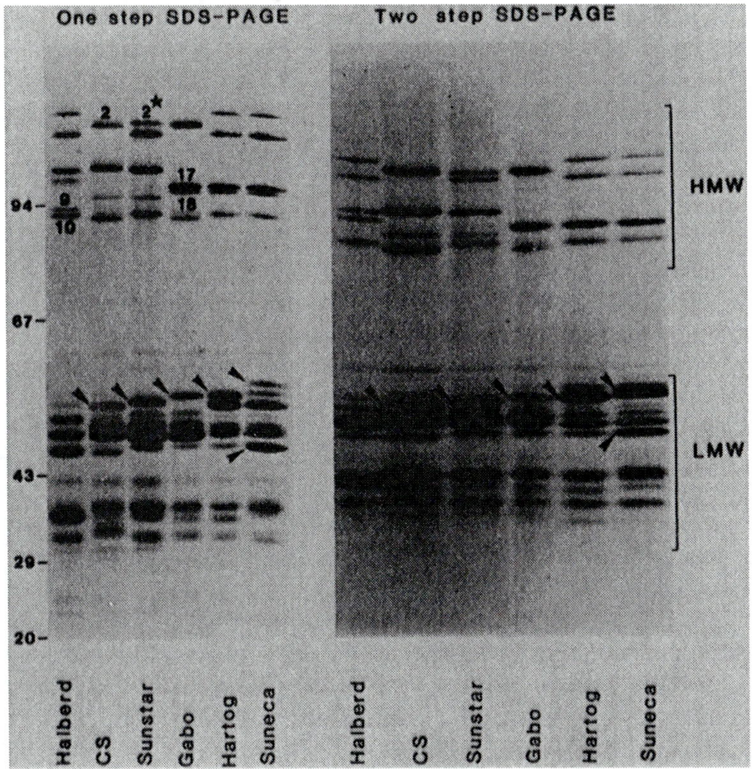

FIGURE 2 One-step and two-step SDS-PAGE patterns of reduced glutenins for six cultivars showing polymorphism in HMW and LMW glutenin subunits. CS, Chinese Spring. (From Ref. 11.)

improves with increasing molecular size. Preliminary measurements of wheat proteins [15] have demonstrated its possibilities, and this should be a valuable method when it is further refined.

IV. NEW METHODOLOGIES

Most of the information about the biochemical and genetical aspects of storage protein components has been achieved through the extensive use of electrophoretic (one- and two-dimensional separations) and chromatographic techniques (RP-HPLC). More recently, the use of molecular biology tools has made it possible to integrate information obtained with classical biochemical techniques. In particular, the polymerase chain reaction (PCR), a technique that allows the specific amplification of a target DNA segment using a pair of flanking oligonucleotides as primers [16] is becoming a widespread research technique, making it possible to characterize different allelic forms of storage protein–encoding genes, to obtain further information on their polymorphism, and to identify quality-related alleles. Identification of cultivars with different qualitative characteristics has been possible using this technique. For instance, durum wheat cultivars with good and poor technological properties were distinguished by selective amplification of γ-gliadin genes or low molecular weight glutenin subunit genes [17,18]. Similarly, in bread wheat, through selection of appropriate primers that could amplify given regions of genes, procedures have been developed that make it possible to discriminate cultivars possessing subunit 5 from those possessing subunit 2 [19] or cultivars possessing subunit 10 from those possessing subunit 12 [20]. In particular, one of the primers used by D'Ovidio and Anderson [19] to discriminate bread wheat cultivars possessing subunit 5 from those with subunit 2 was based on the presence of the codon corresponding to the cysteine residue present at the beginning of the repetitive domain in subunit 5; use of these primers also differentiated bread wheat cultivars that had HMW glutenin subunits with electrophoretic mobility similar to that observed for subunit 5.

Gene cloning and sequencing of HMW glutenin subunit genes have produced more detailed information on corresponding subunits and made it possible to detect discrepancies between the migration of certain subunits in SDS-PAGE and their molecular weight. For instance, the migration of allelic pairs 2/5 and 10/12, encoded at the *Glu-D1* locus, have been shown to be anomalous [21]. In fact, subunit 5 has higher mobility than the smaller allelic subunit 2; similarly, subunit 10 has lower mobility than the larger subunit 12. In addition to exhibiting discrepancies between molecular weight and migration, the molecular weights of HMW glutenin subunits, as determined by their migration on SDS-PAGE, are overestimated. PCR was used to selectively amplify and measure the size of DNA fragments corresponding to allelic subunits present at the *Glu-D1* locus [22]. Molecular weights of corresponding subunits were determined, and results indicated that problems such as overestimation or anomalous migration, normally observed in SDS-PAGE separations, had been eliminated.

Locus-specific primers for both low and high molecular weight glutenin subunit genes have also been identified [23,24] making it possible to amplify only selected fragments present on one of the homologous group 1 chromosomes. Primer sets specific for LMW glutenin subunit genes were also used to identify varieties carrying the 1RS.1BL translocation, recognized by the absence of a PCR amplification product when primers specific for the *Glu-B3* locus were used [23].

Similarly, primers selectively amplifying HMW glutenin subunit genes were recently used to investigate genetic and structural features of a gene encoding a novel HMW glutenin subunit, termed 2.1*, with an unusually high molecular weight [25] (Fig. 3). Using locus-specific primers for homeoallelic genes present at the *Glu-1* loci [24], amplification was only obtained when primers specific for x-type genes at the *Glu-A1* locus were used. In Figure 4, amplification products obtained in different bread wheat genotypes possessing different allelic variants at the *Glu-A1* locus (null, 1, and 2*) are shown along with the line possessing subunit 2.1*. These results confirmed that the structural gene corresponding to the novel subunit was on chromosome 1A. This allowed allelism to be established without any need to perform time-consuming crosses and wait for results on segregating material as is usually necessary. PCR experiments resulted also in the amplification of a silenced gene (null allele) present in some materials. Digestion with proper restriction enzymes of amplified gene fragments revealed that the novel gene differed from prevailing *Glu-A1* alleles 2* and 1 by an extra fragment of about 600 base pairs present in the central repetitive domain. Digestion patterns also indicated that the novel gene has most likely evolved as a result of unequal crossing over, presumably within the gene controlling subunit 2*.

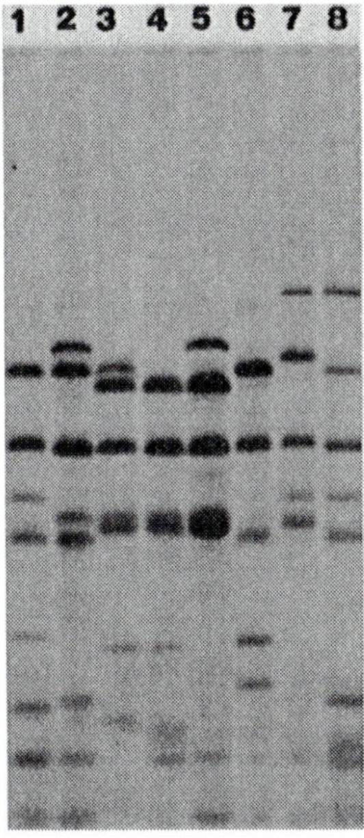

FIGURE 3 One-dimensional SDS-PAGE separation of HMW glutenin subunits from different hexaploid genotypes. 1, Chinese Spring (null); 2, Pandas (1); 3, Cheyenne (2*); 4, PK-16475 (null); 5, PK-16476 (1); 6, PK-16437 (2*); 7, MG-7249 (2*); 8, PK-15684 (2.1*).

V. CORRELATIVE APPROACHES TO ELUCIDATING COMPOSITION/STRUCTURE-FUNCTION RELATIONSHIPS

The functional properties of wheat flours are influenced by both genetic and environmental factors. Systematic variation of each of these factors may be used to give information on the causes of differences in these properties.

A. Variation of Genotype

Many studies have been made in which the functional properties of a range of wheat genotypes have been measured and correlations sought with specific aspects of protein composition. Typically, the presence of certain combinations of electrophoretic bands has been used as the measure of protein composition. This approach has produced some important results. For example, careful observations by Payne and coworkers [26,27] showed that allelic variation in HMW glutenin subunits could be related to dough strength, leading to a greater understanding of the genetic basis of wheat quality. Based on comparisons between the subunit composition of varieties and their quality assessed by the SDS-sedimentation test, scores were assigned to different HMW subunits. These scores ranged from 0 (for a null allele) to 4.

Particular varieties were then given a HMW score by addition of the scores for individual subunits. The HMW score has been used for screening in breeding programs and has been successful in Britain for increasing the strength of varieties.

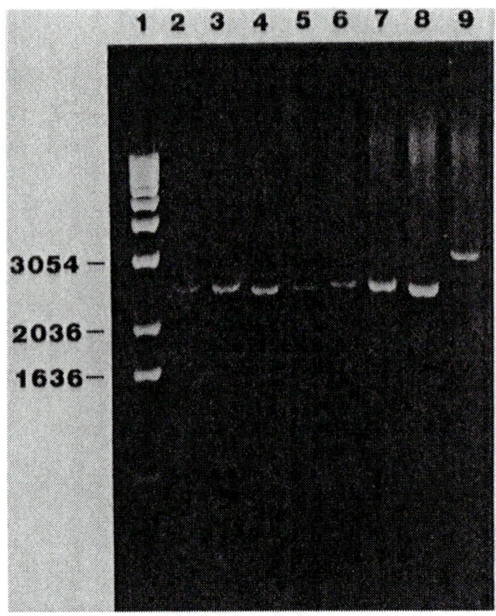

FIGURE 4 Agarose gel separation of PCR amplified fragments coding for HMW glutenin subunits encoded at the *Glu-A1* locus. 1, Molecular size marker; 2, Chinese Spring; 3, Pandas; 4, Cheyenne; 5, PK-16475; 6, PK-16476; 7, PK-16437; 8, MG-7249; 9, PK-15684.

Another line of work [28] drew attention to an apparent link between allelic variation of some γ-gliadins and pasta-making quality. Varieties possessing a γ-gliadin designated as γ-45 had generally good quality, whereas its counterpart γ-42 was associated with poor quality. Both HMW score and γ-gliadins are fairly reliable markers of quality and have been successfully used in early screening. It has now become clear that the genes for γ-gliadins are linked to genes for two different sets of LMW glutenin subunits, which are responsible for the cause-effect relationship [29,30].

Although these examples have greatly stimulated research and led to advances in knowledge, the premise that specific protein bands can be used to fully explain differences in quality is probably a naive concept. There will be well over 100 different polypeptides in a given variety of common wheat. Of these, a maximum of five HMW glutenin subunits are found. The number of LMW glutenin subunits are usually in the range 9–16. The main polypeptides comprising the remainder are HMW albumin and triticin subunits and the monomeric gliadins, albumins, and gobulins. The importance that has been demonstrated for the HMW glutenin subunits arises because they, like the other polymeric protein subunits, have the capacity to vary the molecular weight distribution (MWD) of the protein through their disulfide-bonding capacity.

The parameters according to which wheat specifications are based are mainly physical properties, and these are largely determined by the MWD. The MWD of flour protein can be altered in two main ways, as shown in Figure 5: by varying the ratio of polymeric to monomeric proteins (Fig. 5a) or by varying the size distribution of the polymeric proteins (Fig. 5b). Both of these variables are under genetic control but can also be affected by environmental conditions. Thus, whereas monomeric proteins do not behave

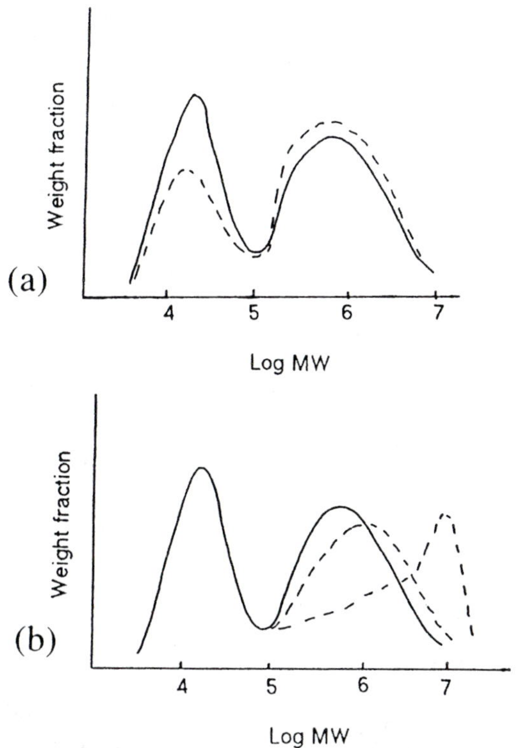

FIGURE 5 Schematic representation of the two ways in which molecular size distribution of flour protein can be altered: (a) by changes in the proportions of monomeric and polymeric proteins or (b) by change in the size distribution of polymeric proteins.

very differently from each other and only exert effects through varying their amounts, polymeric subunits exert their effects by both variation in their amounts and in their capacity to form smaller or larger polymers.

A study of 15 genotypes grown at one site but with six levels of nitrogen fertilizer gave some insight into relationships between protein composition and several measurements of functionality [31]. Regression analysis showed that two compositional variables were particularly related to quality parameters, namely, the percentage of polymeric protein in the total protein and the percentage of polymeric protein in the flour. Correlation coefficients of those two variables with several quality parameters are summarized in Table 2. It is seen that the percentage of polymeric protein in the flour correlated best with farinograph dough development time (FDDT), extensograph extensibility (Ext), and loaf volume in a fixed-mixing-time long-fermentation bake test with 20 ppm of bromate (LV20). On the other hand, the percentage of polymeric protein in the total protein related better to mixograph dough development time (MDDT), extensograph maximum resistance (R_{max}), and loaf volume in an optimized rapid-bake test (LV). These results emphasize that no single measure of composition can explain all aspects of dough and baking properties. As seen from Table 2, the type of dough mixer and the specific baking test influence how mixing and baking properties are affected by different compositional

Table 2 Linear Regression Correlation Coefficients of Quality Parameters with Several Measurements of Protein Composition for n = 84

Quality parameter	Correlation coefficients		
	vs. %PPP	vs. %FPP	vs. %FP
Ext	0.392***	0.831***	0.744***
FDDT	0.083	0.826***	0.809***
LV20 15cv	−0.135	0.684***	0.725***
14cv[a]	−0.007	0.897**	0.877***
R_{max}	0.665***	0.392***	0.241*
MDDT	0.605***	0.127	−0.012
LV	0.616***	0.436***	0.297**

[a]One variety that underperformed because of excessive mixing requirements was omitted from statistical analyis.

%PPP, Percent polymeric protein in the protein; %FPP, percent polymeric protein in the flour; %FP, percent protein in flour; Ext, extensograph extensibility; FDDT, farinograph dough development time; LV20, loaf volume in a long fermentation bake-test; R_{max}, maximum extensograph resistance; MDDT, mixograph dough development time; LV, loaf volume in rapid optimized bake-test; *, **, ***, significantly correlated at 5, 1 and 0.1% probability respectively.

Source: Ref. 31.

variables. Furthermore, different parameters measured by the same instrument (e.g., extensograph) can depend differently on protein composition.

One of the conclusions that can be drawn from Table 2 is that FDDT, Ext, and LV20 relate to flour protein level and are therefore sensitive to environment, whereas MDDT, R_{max}, and LV are influenced less by environment and more by genotype. Of course, all parameters are influenced by both factors. Another feature is that higher correlations are found with the percentage of flour polymeric protein than the percentage of polymeric protein in the protein. An interesting pattern was observed when R_{max} (also MDDT and LV) was plotted against percentage of polymeric protein in the protein (Fig. 6). A clustering of points for different varieties is seen. One conclusion that can be drawn from this is that the percentage of polymeric protein is a varietal characteristic. If we focus on two varieties, Halberd and Israel M68, another characteristic is evident; points for Halberd form a group well above the line of best fit, whereas those for Israel M68 are in a group below the line. Electrophoresis of these two varieties shows that Halberd has a dearth of LMW glutenin subunits whereas Israel M68 is low in HMW glutenin subunits. In fact, the average ratio of HMW to LMW subunits, found by densitometric analysis of SDS-PAGE patterns, was 0.34 for Halberd compared to 0.18 for Israel M68. The greater values of R_{max} for Halberd relative to Israel M68 as a result of increasing the HMW-to-LMW glutenin subunit ratio appears to be due to a shift in the MWD to higher molecular weights (see Figs. 3 and 13).

Subsequent studies have found that R_{max} does not always show a high correlation with the percentage of polymeric protein in the total protein when different sets of cultivars are examined [32]. However, in these cases, it is found that high correlations are usually obtained when R_{max} (or MDDT) is plotted against the percentage of polymeric protein that is unextractable without using sonication (R_{max} and MDDT are considered to be measures of what is termed dough strength). This quantity (% UPP) can be evaluated accurately using SE-HPLC [32]. Because of the inverse relationship between poly-

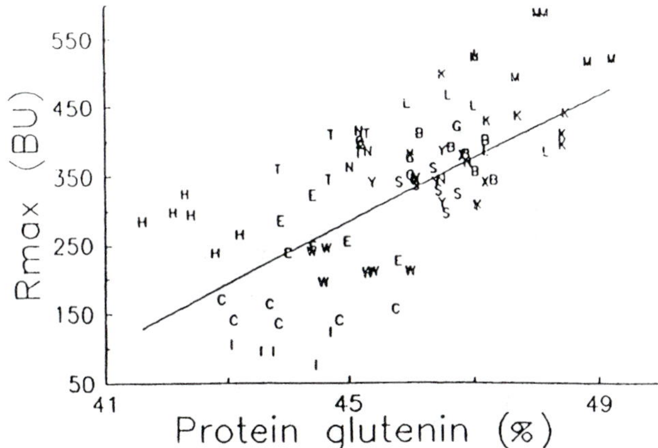

FIGURE 6 Plot of R_{max} versus percentage of polymeric protein for 15 varieties grown at six nitrogen levels (n = 84, six samples being omitted because of insufficient quantity). Varieties are identified by symbols as follows: C, Chile 1B; N, Condor; K, Cook; E, Egret; B, Gabo; G, Gamenya; H, Halberd; I, Israel M68; M, Mexico 8156; L, Olympic; S, Osprey; X, Oxley; T, Timgalen; Y, Wyuna, and W, WW15. The line of best fit is shown. Correlation coefficient for linear regression was 0.665***. (From Ref. 31.)

mer extractability and molecular weight, the percentage of unextractable polymeric protein (% UPP) serves as a relative measure of molecular size distribution. This is illustrated schematically in Figure 7, where the relationships between R_{max} and percent polymeric protein and percent unextractable polymeric protein are compared.

B. Variation of Environment

As we have seen, many properties of wheat varieties are governed by the protein composition, which, in turn, is under genetic control. However, these properties are also affected by environmental factors such as availability of soil nutrients and climatic conditions during growth. We will be concerned here with using variation in some of these factors to obtain information about structure-function relationships. We will consider the results from deliberately changing environmental variables in a systematic way.

1. Variation of Nitrogen Fertilizer

The obvious effect of increasing nitrogen fertilizer level is to increase wheat protein content. However, protein composition is also altered with changing level of grain protein. In general, as protein level increases for a given variety, the proportion of gliadin increases while the proportion of monomeric albumins/globulins decreases [31]. The proportion of polymeric proteins remains fairly constant. Of course, the total amounts of all three groups increase, but with different slopes, as seen in Figure 8. In general, dough strength would be expected to increase with increasing flour protein content, since starch and other nonprotein components simply act as fillers. However, because of the change in balance of the proteins arising from the enhancement of gliadin content, dough strength does not necessarily always increase with increasing level of total protein, and it often decreases.

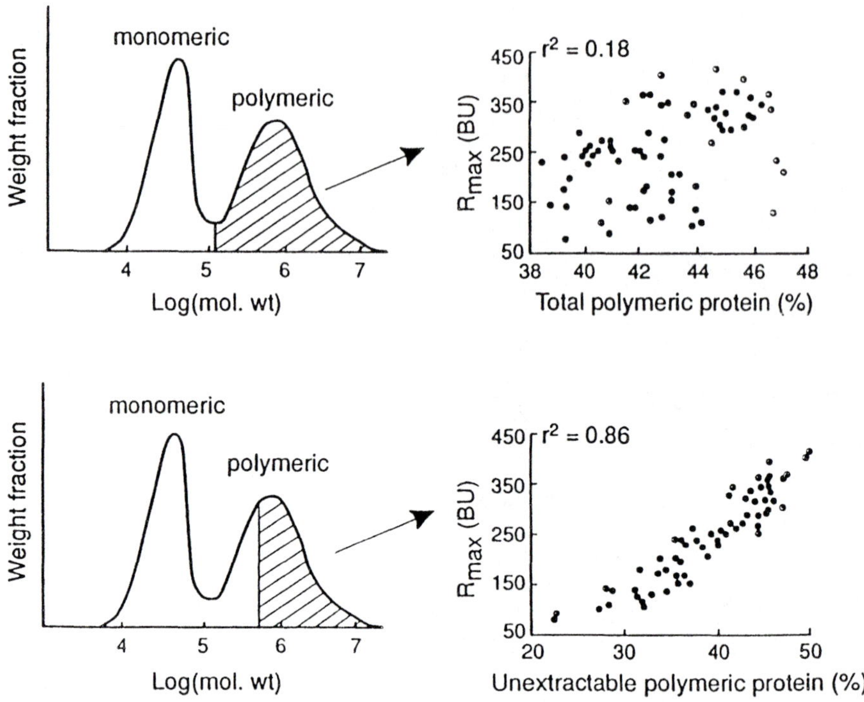

FIGURE 7 Schematic representation of the total and unextractable polymeric protein and plots of R_{max} against these two variables for 74 recombinant inbred lines. (From Ref. 32.)

2. Variation of Sulfur Fertilizer

Variation in the availability of sulfur fertilizer has been shown to produce large changes in dough properties [33]. Sulfur deficiency causes increases in dough strength (as measured by R_{max} and MDDT) and decreased Ext. These effects can be related to changes in the distribution of the different protein groups [34]. The most sulfur-poor proteins of wheat are the ω-gliadins followed by the HMW glutenin subunits. As a result, the amounts of these two groups of polypeptides increase with increasing sulfur deficiency to the detriment of the other relatively sulfur-rich group, including the LMW glutenin subunits. The shift in the ratio of HMW to LMW glutenin subunits appears to be an overriding factor in the change of dough properties caused by variation in sulfur availability. Interaction between nitrogen and sulfur fertilizer is also important. For example, high levels of nitrogen fertilizer can accentuate sulfur deficiency effects that are not so evident at lower nitrogen levels [35].

Apart from the practical effects on wheat quality, the use of varying sulfur levels provides an independent way to alter the protein composition and study the effects on functionality independent of genotype. This approach was used in a study of the single variety Olympic [36], in which nitrogen and sulfur fertilizer levels were both varied systematically. The correlation between several measures of protein composition and flour quality parameters is summarized in Table 3. It should be noted that dough strength (as measured by R_{max}) increases with decreasing percentage of polymeric protein as shown by the high negative correlations between these two variables. The decrease of

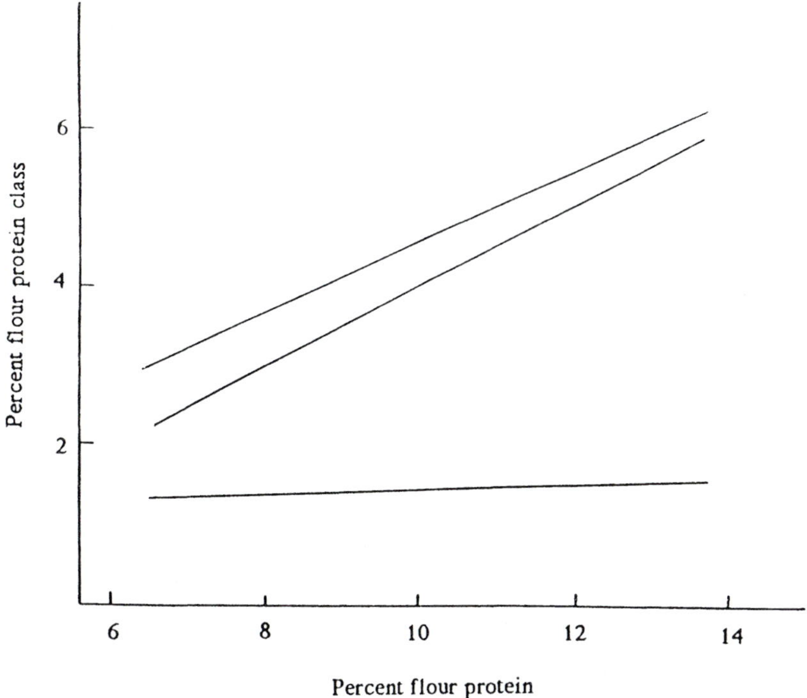

FIGURE 8 Changes in the three main classes of flour protein—polymeric protein (top line), gliadin (middle line), and albumin/globulin (bottom line)—with change of flour protein content. The lines are lines of best fit for the same 84 flour samples described in Figure 6. Equations from linear regressions are % polymeric protein in flour = 0.082 + 0.448 × flour protein; % gliadin in flour = −1.15 + 0.517 × flour protein; % albumin/globulin = 1.07 + 0.0345 × flour protein. (From Ref. 31.)

polymeric protein occurs with decreasing flour sulfur content and arises because the sulfur-rich LMW glutenin subunits are more abundant than the HMW subunits, at least in normal lines and in the absence of sulfur deficiency. However, the accompanying increase in the HMW-to-LMW ratio more than compensates for the decrease in total glutenin by shifting the MWD of the glutenin to higher molecular weights. This is supported by the high correlation between the HMW-to-LMW ratio and % UPP (0.699***).

3. Temperature Effects During Grain Development

Other environmental factors, such as distribution of precipitation, late season frosts, and duration of grain filling, can affect grain quality in addition to the levels of nitrogen and sulfur present in the soil. However, it appears that variation in temperature during grain filling is possibly the most important factor affecting dough viscoelastic properties. In particular, it has been reported that increases up to 30°C in daily mean temperature during grain filling generally increased dough strength and that temperatures above 30°C produced weaker doughs [37]. The identification of heat-shock elements upstream of the coding region of certain gliadin proteins but not glutenin led to the hypothesis that gliadin

Table 3 Linear Correlation Coefficients Between Quality Parameters and Several Measures of Flour Protein Composition for 24 Flour Samples of the Wheat Variety Olympic Grown Under Differing Sulfur/Nitrogen Fertilizer Regimes

Quality parameter	F.P.	%TPP	%UPP	%FPP	HMW/LMW
R_{max}	0.155	−0.721***	0.824***	−0.107	0.795***
MDDT	−0.432*	−0.364	0.785***	−0.643***	0.795***
LV (rapid)	−0.168	−0.532**	0.660***	−0.413*	0.650***
Ext	0.570**	0.336	−0.540**	0.798***	−0.591**
FDDT	0.773***	−0.068	−0.381	0.923***	−0.344
LV (long ferm.)	0.865***	−0.123	−0.205	0.938***	−0.122
FBD	−0.647***	0.875***	−0.559**	−0.394	−0.790***
Range	6.7–12.1	42.2–52.5	44.4–56.4	3.5–5.8	0.31–0.89

The range for each protein measurement is shown. R_{max}, Extensograph maximum resistance; MDDT, mixograph peak dough development time; LV (rapid), loaf volume in an optimized rapid baking test; Ext, extensograph extensibility; FDDT, farinograph dough development time; LV (long ferm.), fixed mixing time, long fermentation baking test; FBD, farinograph breakdown; F.P., flour protein; %TPP, percentage of total polymeric protein; %UPP, percentage of unextractable polymeric protein; %FPP, percentage of flour polymeric protein; HMW/LMW, ratio of HMW/LMW glutenin subunits by RP-HPLC; *, **, ***, significantly correlated at 5, 1 and 0.1% probability, respectively.
Source: Ref. 36.

synthesis, as a result of heat stress, continues at a greater rate than glutenin synthesis; consequently, the mature grain has a higher ratio of gliadin to glutenin and produces weaker dough [38].

Recent results [39] do not support this as a general hypothesis, as it has been reported that some varieties showed no significant increase in the gliadin-to-glutenin ratio in response to heat stress treatment. Similar results were obtained by Bernardin et al. [40], who examined the effects of a gradual daily increase in temperature, reaching a maximum of 40°C, throughout the grain-filling period on the storage proteins deposited in the wheat grain.

Weakening in dough strength was observed in bread wheat grown in different locations in Italy which had experienced heat stress during the grain-filling period [41]. SE-HPLC measurements on these materials showed that high temperatures affected the relative quantity of insoluble polymeric proteins in heat-stressed samples, which was significantly less than that observed in control samples. This suggested that the polymer size distribution was shifted to lower values as a result of the heat stress.

Changes in the polymeric fraction observed in heat-shocked samples may therefore be due to the effects of temperature on the mechanism by which intermolecular disulfide bonds are formed during the deposition of storage proteins. The endoplasmic reticulum is the site of folding and disulfide formation in secretory and vacuolar proteins (including gliadin and glutenin proteins), and these processes may be assisted or catalyzed by molecular chaperones or protein disulfide isomerase, respectively. Heat-shock episodes during grain filling might interfere with these processes with consequent effects on the assembly of individual glutenins into polymers, which can in turn account for the observed changes in flour technological properties.

Differences in the experimental approaches used by different research groups might account for discrepancies observed in the results. According to Stone and Nicolas [39], a gradual temperature rise usually observed under field conditions as opposed to a sudden heat shock induces a different response with respect to protein synthesis and cell viability, and it is likely that the effect observed with a sudden increase in temperature will exaggerate the effect on quality. Other possible sources of discrepancies are timing during grain filling when heat shock is applied and duration of stress.

VI. DIRECT TESTING OF PROTEIN COMPONENTS

A. Protein Fractions

Protein fractions have been prepared and their effects on functionality tested by addition to a base flour or by interchanging corresponding fractions between flours of contrasting properties [42]. Evidence from interchange experiments using crude gliadin and glutenin fractions has been interpreted as showing the glutenin component to be mainly responsible for differences in bread-making potential [43]. The fractions used in these experiments were not well characterized, and their composition varied from one laboratory to another.

Where fractions have been added to a base flour, the effects of varying protein composition has been very clear. Fractions rich in monomeric proteins decrease dough strength (MDDT) and loaf volume in an optimized bake test. Fractions rich in polymeric proteins, by contrast, increase dough strength and enhance test bake loaf volume [44]. Protein fractions imparting strong dough properties have larger proportions of the more unextractable and thus larger molecular-sized proteins as shown by SE-HPLC [45] or SDS-PAGE [46]. Differences in dough strength and baking performance between varieties can be explained by summing the composition of all fractions from each variety. The results of Table 4 illustrate how dough strength (MDDT) and baking performance relate to the percentage of glutenin in the flour and the percentage of HMW glutenin subunits, the latter influencing the molecular size distribution.

B. Purified Polypeptides and Proteins

There are not many examples where the effects of pure proteins have been tested. Strictly, only monomeric (single-chain) proteins can be purified from wheat flour. The effects of

TABLE 4 Quality and Protein Composition for Six Cultivars

Cultivar	%F.P.	LV (ml)	MDDT (min)	%A1	%(A1+A3)
Mexico 8156	12.1	197	7.2	13.2	40.1
Cook	11.8	190	4.4	11.6	38.2
Yecora Rojo	10.9	205	4.7	12.5	34.0
Halberd	10.9	143	3.3	10.0	34.1
Anza	9.9	158	2.4	9.6	30.6
Burgas	15.0	127	1.3	8.7	23.2

%FP, Percent flour protein; LV, loaf volume in rapid optimized baking test using 30.2 g dry flour; MDDT, mixograph dough development time; %A1, percentage of protein in region A1 (HMW glutenin subunits) by densitometry of SDS-PAGE; % (A1+A3), percent of protein in regions A1+A3 (mainly LMW glutenin subunits) by densitometry of SDS-PAGE.
Source: Ref. 46.

different purified gliadins have been tested by addition to a base flour and measurement of the mixing characteristics using a small-scale (2 g flour) micromixer. As expected, all gliadins decreased dough strength, but differences were noted between individual gliadins. The order in the effect of weakening of dough strength was $\omega{-}1 > \omega{-}2 = \alpha = \beta > \gamma$ [47].

When we consider polymeric proteins, certain problems are encountered. First, evidence suggests that the most important of the polymeric proteins, the glutenins, consist of a mixture of proteins heterogeneous in both subunit composition and sequence as well as size. Nevertheless, as we have seen, certain glutenin subunits are associated with greater contributions to dough strength than others. Since glutenin subunits only exert their effects through their participation in large multichain molecules, direct testing of individual subunits cannot be achieved as in the case of monomeric proteins.

One approach to this problem in a dough system has been attempted by Bekes et al. [48]. A protocol was developed during dough mixing in which a reducing agent (potassium iodate) is used to cause a partial reduction of polymeric protein followed by addition of an oxidizing agent to restore the dough properties as closely as possible to the original. Using this procedure, it has been shown that when glutenin subunits are added prior to the oxidation step, they are incorporated into the polymeric structure. They increase dough strength (measured by mixing parameters), whereas with simple addition to the dough without the reduction and oxidation steps, the subunits behave similarly to monomeric proteins and reduce dough strength. Using this procedure, subunits from nonwheat sources or modified subunits from chimeric genes can be tested. If these unusual subunits can be shown to make useful contributions to functionality, there is a possibility that they could be introduced into wheat with beneficial effects using genetic engineering techniques. It should be remembered, however, that only wheat proteins produce doughs with the unique viscoelastic properties required for processes such as bread making. In addition, the scope for tailoring functional properties to a given process by manipulating the protein variability that already exists in wheat has probably not been fully exploited. This should be done before introducing modified components.

VII. USE OF GENETIC VARIANTS

The common wheats used in bread making are hexaploids. They contain 42 chromosomes made up of three genomes, denoted A, B, and D, each having seven pairs of chromosomes (numbered 1–7). Tetraploids (2 genomes, 28 chromosomes) include durum wheats (A and B genomes) favored for pasta processing, while diploids (14 chromosomes), although not in common use commercially, are valuable as sources of novel genes for introduction into hexaploids and tetraploids.

Although some proteins are coded by genes on each of the seven groups of chromosomes, the important proteins that contribute to functionality are coded by genes located on chromosomes 1 and 6, as depicted in Figure 9. Chromosomes consist of long and short arms separated by a centromere. In particular, the HMW glutenin subunits are coded by genes at the complex *Glu-1* loci (*Glu-A1*, *Glu-B1*, *Glu-D1*) on the long arms of group 1 chromosomes. Each locus contains two tightly linked genes, encoding for a subunit of high and low M_r, designated as x- and y-types, respectively. Some high-M_r glutenin genes are silent; usually both x- and y-type subunits are expressed at the *Glu-D1* locus, two or one at the *Glu-B1* locus, and one or none at the *Glu-A1* locus. When one subunit is present at these last two loci, it is invariably an x-type. LMW glutenin

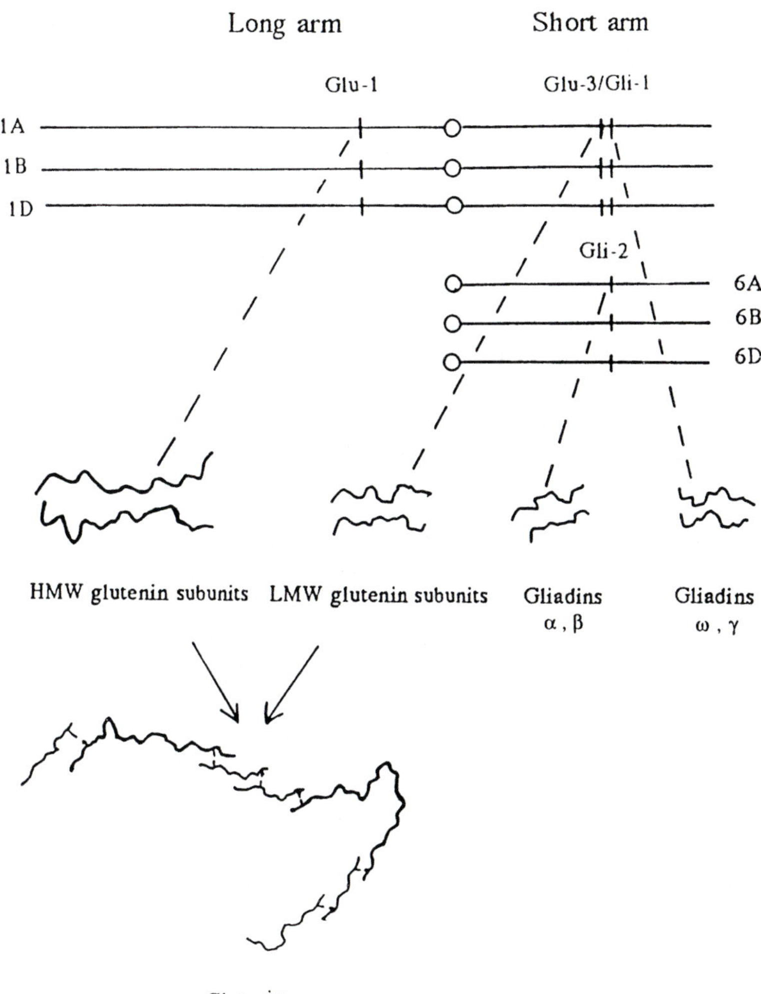

FIGURE 9 Schematic illustration of location of genes coding for the maor proteins that determine flour functionality. HMW and LMW glutenin subunits are controlled by loci on different arms of the group 1 chromosomes, and these subunits polymerize through disulfide cross links during grain development.

subunits are controlled by genes present on the short arms of group 1 chromosomes. However, the genes coding for the LMW glutenin subunits are tightly linked to the genes coding for gliadins, mainly ω and γ. These complex loci are usually denoted *Gli-1/ Glu-3*. Other gliadins, mainly α and β, are coded by genes on the short arms of group 6 chromosomes (*Gli-2* loci).

The location of other loci such as those controlling triticins [49], HMW albumins [14] as well as other gliadins, LMW glutenin subunits, and monomeric albumins and globulins has been elucidated. However, to maintain simplicity, we will concentrate on the nine main loci controlling the major proteins that determine functionality: the three

Glu-1 loci—*Glu-A1*, *Glu-B1*, *Glu-D1*—the three *Glu-3/Gli-1* loci—*Glu-A3/Gli-A1*, *Glu-B3/Gli-B1*, *Glu-D3/Gli-D1*—and the three *Gli-2* loci—*Gli-A2*, *Gli-B2*, and *Gli-D2*. As shown in Figure 9, the glutenin subunits polymerize through disulfide bonds during seed development, whereas the other gene products, the gliadins, form only intrachain disulfide bonds and remain as monomeric proteins.

A. Glutenin Subunits

The manner in which glutenin subunits are incorporated in glutenin polymers is controversial [50], but the limited evidence favors a quasi-random process in which both HMW and LMW subunits participate. As a result, in normal wheat lines, it is not easy to elucidate the effects of either type of subunit singly. Because the loci for HMW and LMW subunits are on different arms of the chromosomes, however, it is possible to separate their effects. In this way, a set of near isogenic lines in which the number of HMW subunits varies from zero to five has been produced by crossing of mutants from two cultivars with null alleles [51]. A set of lines in a common background (variety Gabo) has also been produced in which LMW subunits vary from a maximum to close to zero by substituting one, two, or all three short arms of the group 1 chromosomes of wheat by the short arms of chromosome 1 of rye (single, double, and triple wheat/rye translocation lines). A comparison between the effects of these two sets of lines on R_{max} is shown in Figure 10—i.e., one set in which the number of HMW subunits are varied (LMW subunits constant) and the other in which the number of LMW subunits are varied (HMW subunits constant). On a unit weight basis, R_{max} changes at a greater rate when HMW subunits are varied than when LMW subunits are varied. This is consistent with the conclusions reached in Section V.A that HMW subunits have a greater effect on R_{max}, evidently due to their influence on molecular size distribution.

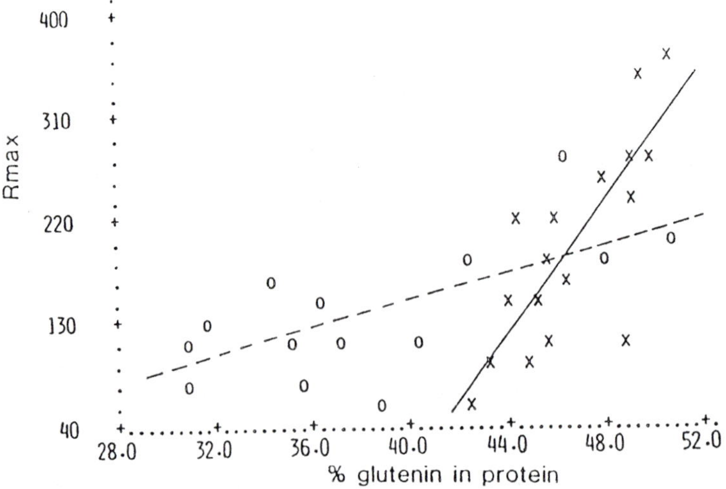

FIGURE 10 Plots of R_{max} against the percent of polymeric protein for two sets of genetic lines, one varying in the number of LMW glutenin subunits (o) and the other varying in the number of HMW glutenin subunits (×). (From Ref. 52.)

Increasing the number of genes actively expressing HMW glutenin subunits has been suggested as an alternative approach to increasing dough strength. One way of achieving this is by introducing HMW glutenin subunit genes from wild wheat progenitors *T. urartu* or *T. dicoccoides*, in which genotypes expressing both x- and y-type subunits at the *Glu-A1* locus are present. In durum wheat this has been attained, and rheological measurements did indeed show an increase in dough strength when both x- and y-type subunits at the *Glu-A1* locus were present [53]. Similar evidence has also been obtained in bread wheat; RP-HPLC analysis of HMW glutenin subunits has identified a Swedish breeding line possessing both x- and y-type subunits at the *Glu-A1* locus (Fig. 11) [54]. Quality data indicate the superior effect of these compared to single-subunit allelic counterparts 2* and 1.

B. Effects of Null Gliadin Loci

1. *Glu-3/Gli-1* Nulls

Genetic variants in which *Gli-1/Glu-3* loci have been deleted are available. These are similar to the rye translocation lines in that LMW glutenin subunits as well as gliadins are absent. However, they differ from the rye translocation lines because secalins from rye are not introduced. Secalins are monomeric proteins that would be expected to have similar effects on functionality to gliadins. Similarly to the rye translocations, single, double, and triple *Glu-3/Gli-1* null lines have been produced, although a thorough study of the functional properties of these lines has not yet been reported.

The effect of deletion of a *Glu-3/Gli-1* locus will depend on the relative amounts of LMW glutenin subunits and gliadins. This ratio may differ from one genome to another and also, for a given genome, from one wheat variety to another. For example, if LMW glutenin subunits are more abundant than gliadins at a given locus, the polymeric-to-monomeric protein ratio may be expected to decrease if this locus is deleted, thus contributing to a decrease in dough strength. In the contrary case (i.e., where gliadins occur in greater quantity than LMW glutenin subunits), a null allele would produce an increase in dough strength. Thus, by choosing the appropriate null alleles, an opportunity is opened to shift the balance towards either a greater proportion of polymeric or monomeric proteins and thus manipulate dough properties in predictable ways. In a study of progeny from a cross between cultivars with contrasting alleles at each of the *Glu-3/Gli-1* loci, 74 recombinant inbred lines homozygous for each of the loci were studied for dough properties [55]. In this set, it was found that, for the alleles analyzed, different *Glu-3* loci could be ranked in order of their effects in increasing R_{max}. The order found was *Glu-B3* > *Glu-A3* > *Glu-D3*. The corresponding order for *Glu-1* alleles (HMW subunits) was *Glu-D1* > *Glu-B1* > *Glu-A1*.

2. *Gli-2* Nulls

The genes coding for α- and β-gliadins on the short arms of the group 6 chromosomes are not linked to genes for LMW glutenin subunits as are those on group 1, although there is evidence that small amounts of LMW subunits may be associated with genes on group 6. The prediction of the effects of *Gli-2* nulls is therefore simpler. In cases where null lines have been studied, deletion of specific types of polypeptides does not lead to a compensation in amounts of those polypeptides. For example, in the set of HMW glutenin subunit null lines described in Section VII.A, an approximately linear correlation was found between the amount and number of subunits present [51]. Since

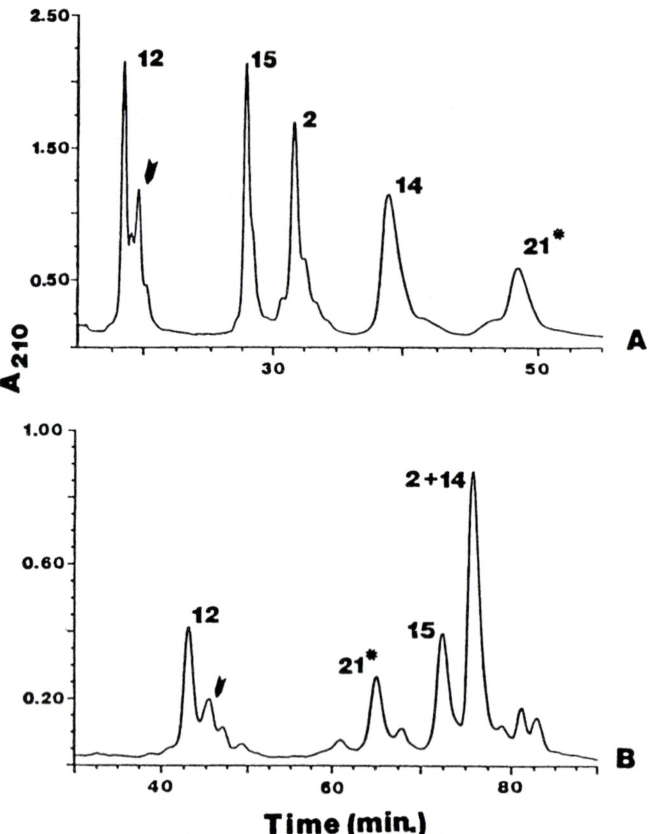

FIGURE 11 RP-HPLC separation of HMW glutenin subunits reduced and alkylated with 4-vinylpyridine (top) and reduced (bottom) from a wheat breeding line (W 29323). Subunits are numbered according to the designation proposed by Payne and Lawrence [6]; arrowheads indicate y-type subunit present at the *Glu-A1* locus.

the total grain protein content was not affected, it would appear that the protein deficit resulting from the deletions must be shared by all other polypeptides. Applying this idea to *Gli-2* nulls, it follows that the resulting decrease in gliadins would shift the polymeric-to-monomeric protein ratio to higher values. For single nulls, the extent of this shift will depend on the number/amount of gliadins associated with the particular genome and with the particular variety. Effects of double and triple nulls may, of course, be expected to be greater than the single ones.

Testing of *Gli-2* nulls and comparison with their parental lines is in the early stages. One potential application is to introduce these null alleles into lines in order to increase dough strength. For example, the weak/sticky dough properties of rye translocation lines appear to be mainly due to the reduction in glutenin arising from the loss of the *Glu-3* locus [56]. By eliminating α- and β-gliadins coded for by the *Gli-2* loci, glutenin levels should tend to be restored without greatly affecting the subunit composition, thus making these agronomically valuable lines suitable for bread making.

VIII. APPLICATION OF POLYMER THEORY TO WHEAT FLOUR FUNCTIONALITY

From the previous discussion, it is clear that dough rheological properties are governed by the protein. When flour and water are mixed in suitable proportions, micrography shows that discrete lumps of protein are stretched out and unite in a continuous network structure throughout the dough, thus imparting viscoelastic properties. The physicochemical changes occurring during dough mixing have been described in other articles [57]. In this section, we will discuss the rheological properties of dough that has optimal characteristics for processing. In a recording dough mixer such as a mixograph, this stage corresponds to the peak in the mixing curve separating the initial development from the subsequent breakdown stage.

A. Fundamental Rheology

Rheology is the study of the stress-strain-time relationships of a material. Two extremes of behavior are represented by the ideal solid and the ideal liquid. An ideal solid, in which stress is proportional to strain, exhibits elasticity. An ideal liquid, in which stress is proportional to the rate of strain, shows viscous behavior. Wheat flour doughs combine the properties of both a solid and a liquid, i.e., they exhibit viscoelasticity. This means that when a stress is applied, part of the strain (change in dimensions) is recoverable (elastic component) and part is nonrecoverable (viscous flow).

Fundamental rheological measurements can be used to predict how dough will behave over a wide range of stress-strain-time relationships. On the other hand, these measurements can be used to deduce a model for a dough system at both macroscopic and molecular levels. Here we are particularly interested in finding general explanations of dough behavior in molecular terms. This assists in understanding problems of processing, but even further gives a basis for manipulating the proteins genetically in plant-breeding situations.

B. Physical Dough Testing

Several standard instruments are used in cereal laboratories to characterize doughs. In particular, the Brabender extensograph will be considered. This instrument measures both dough strength (R_{max}) and extensibility (Ext). It should be recognized that fundamental parameters are difficult to obtain because of the complex geometry of the test sample and the nonconstant strain rate used. However, since specifications are based on instruments such as the extensograph rather than fundamental rheology, our aim will be to seek molecular explanations of dough behavior for these measurements.

The extensograph measurement resembles a tensile stress test. R_{max} corresponds to the ultimate tensile strength and Ext to the elongation at break or draw ratio, the latter term being the ratio of the length of the test piece at break to the initial length. If we consider the parameters R_{max} and Ext, certain key results have emerged in regard to the effects of protein composition as discussed in previous sections:

1. R_{max} is related to the fraction, in the total protein, of polymeric protein above a critical size.
2. Ext is dependent on the percentage of total polymeric protein in the flour.

Thus, although R_{max} and Ext are obtained from the same measurement, the dependence of each parameter on protein composition is obviously different.

1. Extensograph Maximum Resistance (R_{max})

Tensile strength (σ) of polymers has been described by an equation by Flory [58].

$$\sigma = \sigma_0(1 - M_T/M_n) \tag{1}$$

where σ_0 is the limiting tensile strength at high molecular weight, M_T is some threshold molecular weight, and M_n is the number average molecular weight.

In a study by Bersted and Anderson [59], good agreement was found with this equation for monodispersed polymers but not for polydispersed polymers. These authors modified the Flory equation by postulating that only molecules above a critical molecular size contributed to tensile strength. This follows from the general property of all polymers that, above a certain size in response to stress, resistances above that due to normal friction occur. This has been interpreted in terms of molecular entanglements at widely spaced points in the long chains. An alternative description of the effect has been that molecules move by a snakelike displacement of segments, called reptation [60]. For a simple physical picture, the concept of entanglements will be adopted here.

Bersted and Anderson proposed the equation:

$$\sigma = \sigma_0(1 - M_T/M_n^*)\Phi \tag{2}$$

where M_T is the threshold molecular weight for effective entanglements, Φ is the fraction of polymers with $M > M_T$, and M_n^* is the number average molecular weight of this fraction.

Polymer molecules with $M < M_T$ do not participate in effective entanglements, and this portion acts as a diluent of volume fraction $(1 - \Phi)$. Equation 2 was found to predict the tensile strength of polydispersed polymers reasonably well.

Measurements of R_{max} are consistent with the Bersted-Anderson model for polydispersed polymers. The results depicted in Figure 7 show that R_{max} correlates poorly with the percentage of polymeric protein but correlates highly with the percentage of unextractable polymeric protein. This is interpreted to mean that not all polymeric protein contributes to dough strength but only a fraction of the most insoluble and thus highest molecular weight polymeric protein. Expressed in another way, there is a critical molecular weight of polymeric protein, with only that portion of protein of larger size contributing to dough strength.

2. Extensibility

Unlike R_{max}, Ext does not correlate highly with unextractable polymeric protein. It is generally found that this parameter correlates highly with flour protein content and usually slightly higher with flour polymeric protein content. Some insight into this property is given by the theory of Termonia and Smith [61] describing the draw ratio of polymers.

In this theory, two kinetic processes are considered to be important in determining the extension properties of polymers. The theory can be illustrated in a simplified way by the behavior of a single polymer strand in an entangled network, as depicted in Figure 12. The initial state of the strand approximates to a random coil configuration (Fig. 12a). When a tensile stress is applied to the material, the coiled chain is stretched, breaking secondary valence bonds with other chains. This is the first of the kinetic processes and is described by the Eyring activation theory with appropriate values for the energy of

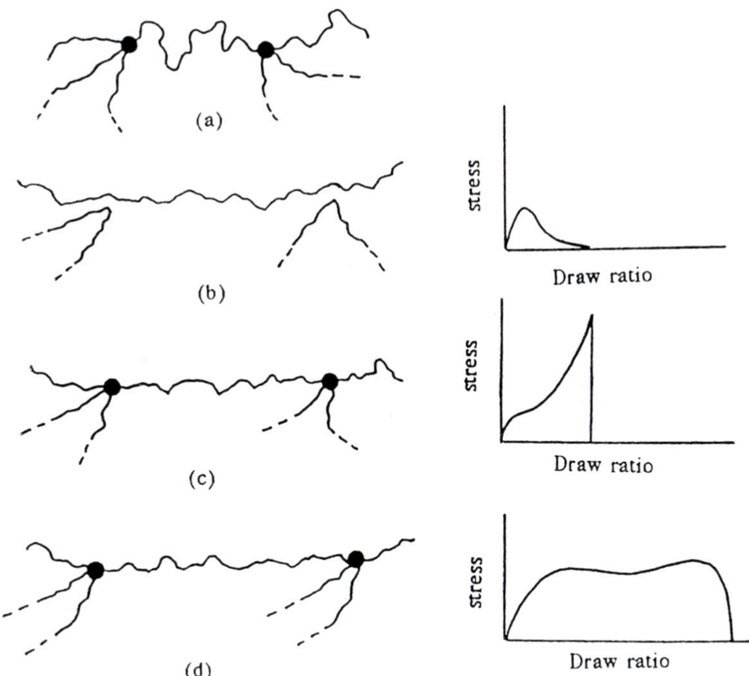

FIGURE 12 Schematic illustration of the effect on the draw ratio (extensibility) of the rate of chain slippage through entanglements relative to the elongation rate. (a) Initial state; (b) slippage rate ≫ elongation rate; (c) elongation rate ≫ slippage rate; (d) optimum slippage rate.

activation for breaking these bonds. When the strands between entanglement nodes are fully stretched, further motion relative to other molecular strands can only occur by slippage through entanglement points. This is the second kinetic process considered with its appropriate activation energy. The behavior in regard to the draw ratio is then determined by the relative rates of the two processes: elongation and slippage of chains through entanglements.

When the rate of chain slippage is much greater than the rate of elongation of the sample, chain slippage occurs easily, thus giving little support to the structure. The stress-strain behavior is shown in Figure 12b. Both strength and draw ratio are low. For the opposite extreme, when the rate of slippage through entanglements is much lower than the rate of elongation, high stress is placed on the strand, giving rise to high tensile strength. However, the draw ratio is low owing to the breakage of strands between entanglement points (Fig. 12c). When the relative rates of chain slippage and elongation are optimum, moderate tensile strength is observed and the draw ratio is maximal (Fig 12d). In the Termonia-Smith theory, the activation step is considered to involve the slippage of not more than one or two chain segments. It has been inferred from measurements of activation energies that many kinetic processes in polymer systems depend on a segment of the molecule rather than the whole molecule. These segments behave to some extent as independent units and move as such in processes like flow, although they are not completely independent since they are connected to other segments through covalent bonds. In the case of proteins, a segment probably consists of some 5–10 amino acid residues.

C. General Conclusions from Extensograph Measurements

The extensograph does not perform a true tensile stress test. First, the sample geometry is rather complex. Second, a constant extension rate is used but not a constant strain rate which would require an exponentially increasing extension rate. In spite of these restrictions, extensograph measurements are amenable to interpretation in molecular terms.

The entanglement network depends on the molecular weight distribution of the polymer, this being the polymeric protein in the case of dough. A physical picture of the effect of changing the MWD of polymeric protein is shown in Figure 13. In the two distributions shown, if we assume that the percentage of protein with $M > M_T$ is the

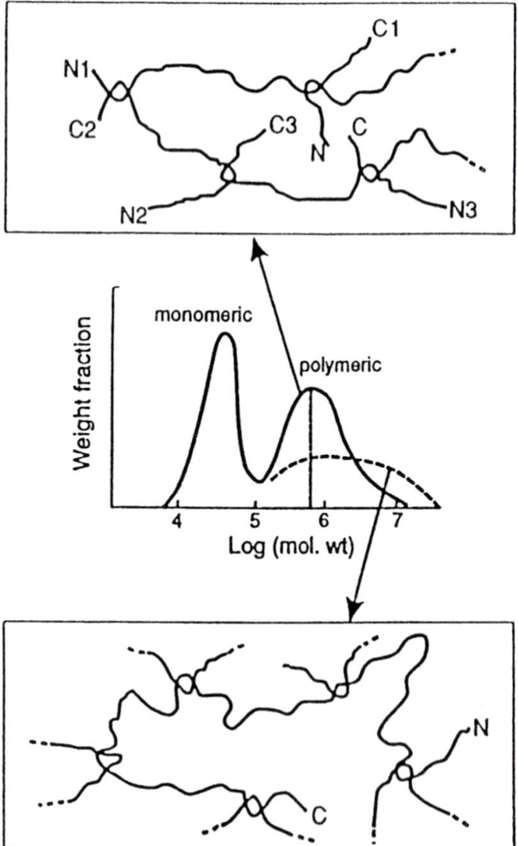

Figure 13 Schematic illustration of how the molecular size distribution changes the nature of the entanglement network. C and N denote the C- and N-terminal residues of the molecules.

same for each, the entanglement density will also be approximately the same. However, one distribution (full line) contains a greater concentration of smaller molecules than the other (dashed line). Therefore, there will be, on average, a smaller number of entanglements per molecule and a greater concentration of chain ends. For a given extension rate, slippage through entanglement points will proceed relatively easily, favoring high extensibility, as in Figure 12d. For the other case, individual molecules on average participate in a greater number of entanglements. There may be a danger, at this rate of extension, that chains will not slip through entanglements sufficiently rapidly, causing chain scission and reduction of extensibility as in Figure 12c.

Some explanation is needed for the different dependencies of R_{max} and Ext on protein content. Suppose we compare doughs from two flours, one with 6% flour protein and one with 12%. The hydrated gluten strands in the dough determine the stress-strain properties. We can approximate the combined gluten strands to a cylinder of gluten protein. The 12% flour will have a cylinder with double the cross-sectional area of the 6% flour, assuming the same length for each. At the point of break in the extension, it seems reasonable to expect that the stress will be about the same; i.e., a critical value for the stress must be reached to cause rupture. This will occur at roughly double the extension for the 12% flour as for the 6% flour. However, the stress near or at the point of break will be similar for the two cases. These considerations help to explain why Ext usually increases linearly with flour protein content (Sec. V.A) whereas R_{max} is not highly correlated with this parameter.

IX. FUTURE DIRECTIONS

A. Solubilization and Molecular Size Measurement

In spite of the advances that have been made in elucidating general relationships between wheat protein structure and functionality, several obstacles to further understanding can be identified. Complete solubilization of wheat protein without alteration of structure remains elusive. This is a goal that needs to be achieved for characterization of the protein, including measurement of the true molecular weight distribution of the polymeric protein. There is a good case for systematic studies of solubility properties based on the determination of the enthalpies and entropies of dissolution. One approach to attaining complete solubilization may be by chemical modification of protein groups, providing this can be attained without significantly altering the molecular weight. In particular, conversion of unionizable to ionizable groups may be a useful approach in view of the paucity of ionizable groups in wheat proteins.

More precise methodology for molecular weight measurement is an area requiring attention, particularly if means for obtaining complete solubilization are devised. SE-HPLC has been useful, but further development of columns capable of resolving the largest glutenins is needed. Field flow fractionation (FFF) is a technique showing promise [62] as, unlike most methods for molecular size determination, resolution improves with increasing molecular size. Laser light scattering and capillary electrophoresis are other methods being developed with potential for these measurements.

B. Quantification of Polymeric Subunit Composition

As we have seen, the structure and properties of polymeric proteins are dependent on the types and amounts of the constituent subunits. Procedures have been developed for quantifying polymeric subunits [9–11], but the shortcomings of these methods have been described (Sec. III.B). Developing methods for achieving a clean separation of polymeric and monomeric proteins in order to accurately quantify polymeric subunit composition is an important area to be pursued.

C. Genetic/Biochemical Basis of Functionality

One of the aims in understanding the molecular basis of functionality is to enable development of strategies to be used in breeding for modifying properties of wheat varieties in predictable ways [63]. The use of special genetic variants has been especially useful in the aspiration of placing this science on a firm footing. In this chapter we have focused on stress-strain properties of doughs to illustrate general principles. However, many other properties (e.g., dough-mixing requirements and mixing stability as well as final quality of the product, be it bread, cookies, noodles, pastries, etc.) are important when considering wheat specifications. Each of these quality parameters may depend in different ways on aspects of protein composition just as R_{max} and Ext have been shown to do. Quality is therefore not simply a question of good versus poor. It is a more complex concept in which a delicate balance of proteins needs to be sought to optimize end-use performance of the grain or flour. Of course, protein is not the only flour component with importance. Nevertheless, dough rheological properties are dominated by the protein. The ultimate goal in a breeding program in regard to processing quality would be to know the genetic make-up required to produce a wheat with specified dough properties. Then, with a data base for different alleles and the types and quantities of polypeptides corresponding to each, the optimum allelic composition could be designed.

In the pursuit of this aim, there are several pertinent questions that would be useful to explore:

1. What is the extent of variation in the relative amounts of LMW glutenin subunits and gliadins at the tightly linked *Gli-1* and *Glu-3* loci? As described previously, this variation will determine the net effect of these loci on dough properties. An excess of LMW glutenin subunits over gliadins will lead to stronger dough properties and vice versa. Careful selection of appropriate loci for introduction into breeding programs can thus be used to alter the protein balance so as to shift the functional properties in predetermined ways.
2. What is the biochemical/genetic basis for the variation in relative quantities of wheat protein classes? This question has been discussed by Singh et al. [64]. Briefly, this work pointed to the result that cultivars with lower proportions of glutenin had additional gliadins and/or fewer glutenins as seen on SDS-PAGE. By contrast, cultivars with higher proportions of glutenins (and therefore exhibiting greater dough strength) had fewer gliadins and additional glutenin subunits. Of course, this simple picture of a general relationship between the amounts of a given protein class and the number of genes coding for the constituent polypeptides can be modified if certain polypeptides are expressed in unusually high amounts. One example where this has been observed is the HMW glutenin subunit 7.
3. What is the effect of deleting specific loci on the accumulation of polypeptides controlled by the remaining loci? In general, it appears that deletion of a small number of loci does not greatly affect the total protein synthesized. Furthermore, there appears to be no compensation for those polypeptides deleted in

terms of more of the same class being synthesized. For example, when five HMW glutenin subunits from a set of near isogenic lines were progressively deleted, the total amount of the HMW subunits decreased linearly [51]. This is an important result since it gives direction to another approach to adjusting the protein balance in a controlled way. Expressed simply, deletion of glutenin subunits will shift the flour properties to lower strength, while deletion of gliadins will have the opposite effect. The use of *Gli-2* nulls for restoring strength in lines having weak dough properties was discussed in Section VII.B.

The nature of glutenin in terms of the structure of the subunits and the manner in which they are combined in the native molecule is an area in which future research is sure to be concentrated. Molecular biology tools are likely to prove valuable in unraveling additional features of both high and low molecular weight glutenin subunits. PCR has proved very useful in this sense, generating information not attainable with classical biochemical techniques; this along with some advantages such as speed, use of nonradioactive material, and the possibility of using small amounts of material, make this technique attractive and useful in wheat quality studies.

REFERENCES

1. T. B. Osborne, *The Proteins of the Wheat Kernel*, Publ. No. 84, Carnegie Institute, Washington, DC, 1907.
2. P. R. Shewry, A. S. Tatham, J. Forde, M. Kreis, and B. J. Miflin, The classification and nomenclature of wheat gluten proteins, *J. Cereal Sci. 4*:97 (1986).
3. H. D. Sapirstein and W. Bushuk, Computer-aided analysis of gliadin electrophoregrams I. Improvement of precision of relative mobility determination by using a three band standardization, *Cereal Chem. 2*:372 (1985).
4. J. A. D. Ewart, Amino acid analysis of glutenins and gliadins, *J. Sci. Food Agric. 18*:111 (1967).
5. J. A. D. Ewart, Isolation and characterization of a wheat albumin, *J. Sci. Food Agric. 20*: 730 (1969).
6. P. I. Payne and G. J. Lawrence, Catalogues of alleles of the complex loci, *Glu-A1*, *Glu-B1* and *Glu-D1* which code for high-molecular-weight subunits of glutenin in hexaploid wheat, *Cereal Res. Commun. 11*:29 (1983).
7. N. K. Singh and K. W. Shepherd, The structure and genetic control of a new class of disulphide-linked proteins, *Theor. Appl. Genet. 71*:79 (1983).
8. J. A. Bietz, High performance liquid chromatography of cereal proteins, *Adv. Cereal Sci. Technol. 8*:105 (1986).
9. N. K. Singh and F. MacRitchie, Controlled degradation as a tool for probing wheat protein structure, *Wheat End-Use Properties: Wheat and Flour Characterization for Specific End-Uses* (H. Salovaara, ed.), University of Helsinki, Helsinki, 1989, pp. 321–326.
10. N. K. Singh and K. W. Shepherd, Linkage mapping of the genes controlling endosperm proteins in wheat. I. Genes on the short arms of group 1 chromosomes, *Theor. Appl. Genet. 75*:628 (1988).
11. R. B. Gupta and F. MacRitchie, A rapid one-step one-dimensional SDS-PAGE procedure for analysis of subunit composition of glutenin in wheat, *J. Cereal Sci. 14*:203 (1991).
12. N. K. Singh, K. W. Shepherd, and G. B. Cornish, A simplified SDS-PAGE procedure for separating LMW subunits of glutenin, *J. Cereal Sci. 14*:203 (1991).
13. S. Masci, D. Lafiandra, E. Porceddu, E. J.-L. Lew, H. P. Tao, and D. D. Kasarda, D-glutenin subunits: N-terminal sequences and evidence for the presence of cystine, *Cereal Chem. 70*: 581 (1993).

14. R. B. Gupta, K. W. Shepherd, and F. MacRitchie, Genetic control and biochemical properties of some high molecular weight albumins in bread wheat, *J. Cereal Sci. 13*:221 (1991).
15. M. Gustavsson, K.-G. Wahlund, L. Wannerberger, T. Nylander, and F. MacRitchie, Size characterization of ultra-high-molecular-weight wheat proteins using asymmetrical flow field-flow fractionation, *Wheat Kernel Proteins*, University of Tuscia, 1994, pp. 237–239.
16. R. K. Saiki, D. H. Gelfand, S. Stoffel, S. J. Scharf, R. Higuchi, G. T. Horn, K. B. Mullis, and H. Erlich, Primer directed enzymatic amplification of DNA with thermostable DNA polymerases, *Science 239*:487 (1988).
17. R. D'Ovidio, O. A. Tanzarella, and E. Porceddu, Rapid and efficient detection of genetic polymorphism in wheat through amplification by polymerase chain reaction, *Plant Mol. Biol. 15*:169 (1990).
18. R. D'Ovidio, Single-seed PCR of LMW glutenin genes to distinguish between durum wheat cultivars with good and poor technological properties, *Plant Mol. Biol. 22*:1173 (1993).
19. R. D'Ovidio and O. A. Anderson, PCR analysis to distinguish between alleles of a member of a multigene family correlated with wheat quality, *Theor. Appl. Genet. 88*:759 (1994).
20. R. L. Smith, M. E. Schweder, and R. D. Barnett, Identification of glutenin alleles in wheat and triticale using PCR-generated DNA markers, *Crop Sci. 34*:1373 (1994).
21. F. C. Greene, O. D. Anderson, R. E. Yip, N. G. Halford, J. M. Malpica Romero, and P. R. Shewry, Analysis of possible quality-related sequence variations in the 1D glutenin high molecular weight subunit genes of wheat, Proceedings of 7th Int. Wheat Genetics Symposium, (T. E. Miller and R. M. D. Koebner, eds), 1988, Institute of Plant Science Research, Cambridge, p. 735.
22. R. D'Ovidio, E. Porceddu, and D. Lafiandra, PCR analysis of genes encoding allelic variants of high-molecular-weight glutenin subunits at the *Glu-D1* locus, *Theor. Appl. Genet. 88*:759 (1995).
23. S. Van Campenhout, J. Vander Stappen, L. Sagi, and G. Volckaert, Locus-specific primers for LMW glutenin genes on each of the group 1 chromosomes of hexaploid wheat, *Theor. Appl. Genet. 91*:313 (1995).
24. R. D'Ovidio, S. Masci, and E. Porceddu, Development of a set of oligonucleotide primers specific for genes at the Glu-1 complex loci of wheat, *Theor. Appl. Genet. 88*:175 (1995).
25. M. Tahir, A. Pavoni, G. F. Tucci, T. Turchetta, and D. Lafiandra, Detection and characterization of a glutenin subunit with unusual high Mr at the *Glu-A1* locus in hexaploid wheat, *Theor. Appl. Genet 92*:654 (1996).
26. P. I. Payne, L. M. Holt, and C. N. Law, Structural and genetic studies on the high-molecular-weight subunits of wheat glutenins. Part 1. Allelic variation in subunits amongst varieties of wheat (*Triticum aestivum*), *Theor. Appl. Genet. 60*:229 (1981).
27. P. I. Payne, L. M. Holt, E. A. Jackson, and C. N. Law, Wheat storage proteins: their genetics and potential for manipulation by plant breeding, *Phils. Trans. R. Soc. London Ser. B 304*: 359 (1984).
28. R. Damidaux, J.-C. Autran, and P. Feillet, Gliadin electrophoregrams and measurements of gluten viscoelasticity in durum wheats, *Cereal Foods World 25*:754 (1980).
29. P. J. Payne, E. A. Jackson, L. M. Holt, and C. N. Law, Genetic linkage between endosperm storage protein genes on each of the short arms of chromosomes 1A and 1B in wheat, *Theor. Appl. Genet. 67*:235 (1984).
30. J.-C. Autran, B. Laignelet, and M. H. Morel, Characterization and quantification of low-molecular-weight glutenins in durum wheats, Proceedings of 3rd Int. Workshop Gluten Proteins (R. Lasztity and F. Bekes, eds.), World Scientific, Singapore, 1987, pp. 266–283.
31. R. B. Gupta, I. L. Batey, and F. MacRitchie, Relationships between protein composition and functional properties of wheat flours, *Cereal Chem. 69*:125 (1992).
32. R. B. Gupta, K. Khan, and F. MacRitchie, Biochemical basis of flour properties in bread wheats. I. Effects of variation in the quantity and size distribution of polymeric protein, *J. Cereal Sci. 18*:23 (1993).

33. H. J. Moss, C. W. Wrigley, F. MacRitchie, and P. J. Randall, Sulfur and nitrogen fertilizer effects on wheat. II. Influence on grain quality, *Aust. J. Agric. Res. 32*:213 (1981).
34. J. G. Fullington, D. M. Miskelly, C. W. Wrigley, and D. D. Kasarda, Quality-related endosperm proteins in sulfur-deficient and normal wheat grain, *J. Cereal Sci. 5*:233 (1987).
35. A. R. Wooding, R. J. Martin, and F. MacRitchie, Effect of sulfur-nitrogen treatments on work input requirements for dough mixing in a second season, Proceedings of 44th Australian Cereal Chemistry Conference, Royal Australian Chemical Institute, Melbourne, 1994, pp. 219–222.
36. F. MacRitchie and R. B. Gupta, Functionality-composition relationships of wheat flour as a result of variation in sulfur availability, *Aust. J. Agric. Res. 44*:1767 (1993).
37. P. J. Randall and H. J. Moss, Some effects of temperature regime during grain filling on wheat quality, *Austr. J. Agric. Res. 41*:603 (1990).
38. C. S. Blumenthal, E. W. R. Barlow, and C. W. Wrigley, Growth environment and wheat quality, *J. Cereal Sci. 18*:3 (1993).
39. P. Stone and M. E. Nicolas, Wheat cultivars vary widely in their responses of grain yield and quality to short periods of post-anthesis heat-stress, *Austr. J. Plant Phisiol. 21*:887 (1994).
40. J. E. Bernardin, S. C. Witt, and J. Milenic, Effect of heat stress on the pattern of protein synthesis in wheat endosperm, Proceedings of the 44th Australian Cereal Chemistry Conference (J. F. Panozzo and P. G. Downie, eds.), Australia, 1994, p. 37.
41. M. Ciaffi, L. Tozzi, E. Cannarella, M. Corbellini, B. Borghi, and D. Lafiandra, Effect of heat stress on gluten proteins and dough technological properties, *Wheat Kernel Proteins—Molecular and Functional Aspects*, S. Martino al Cimino, Viterbo, Italy, 1994, p. 277.
42. F. MacRitchie, Studies of gluten from wheat flours, *Cereal Foods World 25*: 382 (1980).
43. K. Chakraborty and K. Khan, Biochemical and breadmaking properties of wheat protein components. I. Compositional differences revealed through quantitation and polyacrylamide gel electrophoresis of protein fractions from various isolation procedures, *Cereal Chem. 65*: 333 (1988).
44. F. MacRitchie, Evaluation of contributions from wheat protein fractions to dough mixing and breadmaking, *J. Cereal Sci 6*:259 (1987).
45. G. Lundh and F. MacRitchie, SE-HPLC characterization of wheat protein fractions differing in contributions to breadmaking quality, *J. Cereal Sci. 10*:247 (1989).
46. F. MacRitchie, D. D. Kasarda, and D. D. Kuzmicky, Characterization of wheat protein fractions differing in contributions to breadmaking quality, *Cereal Chem. 68*:122 (1991).
47. R. J. Fido, F. Bekes, P. W. Gras, and A. Tatham, Effects of gliadin subunits on the dough-mixing properties of wheat flour, *Wheat Kernel Proteins: Molecular and Functional Aspects*, University of Tuscia, 1994, pp. 305–307.
48. F. Bekes, P. W. Gras, R. B. Gupta, D. R. Hickman, and A. S. Tatham, Effects of a high Mr glutenin subunit (1Bx20), *J. Cereal Sci. 19*:3 (1994).
49. N. K. Singh, K. W. Shepherd, P. Langridge, and L. C. Gruen, Purification and biochemical characterization of triticin, a legume-like protein in wheat endosperm, *J. Cereal Sci. 13*:207 (1991).
50. J. A. D. Ewart, Comments on recent hypotheses for glutenin, *Food Chem. 38*:159 (1990).
51. G. J. Lawrence, F. MacRitchie, and C. W. Wrigley, Dough and baking quality of wheat lines deficient in glutenin subunits controlled by the *Glu-A1*, *Glu-B1* and *Glu-D1* loci, *J. Cereal Sci. 7*:109 (1988).
52. Gupta, R. B., MacRitchie, F., Shepherd, K. W. and Ellison, F., Relative contributions of LMW and HMW subunits to dough strength and dough stickiness of bread wheat, Gluten Proteins 1990 (W. Bushok and R. T. Katchuk, eds.), Am. Assoc. Cereal Chem., St. Paul, 1991, pp. 71–80.

53. M. Ciaffi, D. Lafiandra, T. Turchetta, S. Ravaglia, H. Bariana, R. B. Gupta, and F. MacRitchie, Breadmaking potential of durum wheat lines expressing both x- and y-type subunits at the *Glu-A1* locus, *Cereal Chem. 72*:465 (1995).
54. B. Margiotta, M. Urbano, G. Colaprico, E. Johansson, F. Buonocore, R. D'Ovidio, and D. Lafiandra, Bread wheat lines with both x- and y-type subunits at the *Glu-A1* locus, *J. Cereal Sci 23*:203 (1996).
55. R. B. Gupta, J. G. Paul, G. B. Cornish, G. A. Palmer, F. Bekes, and A. J. Rathjen, Allelic variation at glutenin subunit and gliadin loci, *Glu-1*, *Glu-3* and *Gli-1*, of common wheats. I. Its additive and interaction effects on dough properties, *J. Cereal Sci. 19*:9 (1994).
56. A. S. Dhaliwal and F. MacRitchie, Contributions of protein fractions to dough handling properties of wheat/rye translocation cultivars, *J. Cereal Sci. 12*:113 (1990).
57. F. MacRitchie, Physicochemical processes in mixing, *Chemistry and Physics of Baking* (J. M. V. Blanshard, P. J. Frazier, and T. Galliard, eds.), Royal Society of Chemistry, London, 1986, pp. 132–146.
58. P. J. Flory, Tensile strength in relation to molecular weight of high polymers, *J. Am. Chem. Soc. 67*:2048 (1945).
59. B. H. Bersted and T. G. Anderson, Influence of molecular weight and molecular weight distribution on the tensile properties of amorphous polymers, *J. Appl. Phys. 55*:572 (1971).
60. P. G. DeGennes, Reptation of a polymer chain in the presence of fixed obstacles, *J. Chem. Phys. 55*:572 (1971).
61. Y. Termonia and P. Smith, Kinetic model for tensile deformation of polymers. I. Effect of molecular weight, *Macromolecules 20*:835 (1987).
62. K.-G. Wahlund, M. Gustavsson, F. MacRitchie, T. Nylander, and L. Wannerberger, Size characterization of wheat proteins, particularly glutenins, by asymmetrical flow field-flow fractionation, *J. Cereal Sci 23*:113 (1996).
63. F. MacRitchie, Physicochemical properties of wheat proteins in relation to functionality, *Adv. Food Nutr. Res. 36*:1 (1992).
64. N. K. Singh, G. R. Donovan, and F. MacRitchie, Use of sonication and size-exclusion high performance liquid chromatography in the study of wheat flour proteins. II. Relative quantity of glutenin as a measure of breadmaking quality, *Cereal Chem. 67*:161 (1990).

11

Structure and Functionality of Egg Proteins

ETSUSHIRO DOI[†] AND NAOFUMI KITABATAKE
Kyoto University, Kyoto, Japan

I. INTRODUCTION

Whole egg consists of about 13% protein, which contributes to the functional values of the egg as a food ingredient and foodstuff. The shell, the egg white, and the yolk contain 3, 11, and 17.5% protein, respectively. Many egg proteins have provided model systems for studying the basic concepts of protein chemistry and the structure-function relationship. Some egg proteins have unique biological characteristics and are being used as food supplements. A number of reviews of historical background and detailed information about egg proteins have been published [1–3]. In this chapter the physicochemical properties of the major proteins of egg white and yolk, namely, ovalbumin, ovotransferrin, lysozyme, ovomucoid, ovomucin, and immunoglobulin Y, are discussed and an overview of the structure-functional relationships of these proteins as food ingredients and some biological functions is presented.

II. EGG WHITE PROTEINS

A. Ovalbumin

1. Structure

Ovalbumin (egg albumin) is the predominant protein in egg white (egg albumen), and it constitutes about half of the egg white proteins by weight. It can be easily purified in

[†]Deceased.

large quantities by crystallization from ammonium sulfate solution of egg white [4]. After several recrystallizations, a highly pure ovalbumin with homogeneous molecular weight by sedimentation analysis or SDS-polyacrylamide gel electrophoresis can be obtained; however, some reports have shown heterogeneity in such preparations, and this has been attributed to factors such as differences in phosphorylation and glycosylation, the presence of genetic variants, and contamination of S-ovalbumin.

Ovalbumin is a monomeric phosphoglycoprotein. The complete amino acid sequence of 385 residues is known [5–7]. The N-terminus of the protein is acetylated [8], and it contains four cysteine and one cystine residue. Three types of ovalbumin, namely, A_1, A_2, and A_3, exist. The difference between these three types is the degree of phosphorylation. A_1, A_2, and A_3 ovalbumins have two, one, and zero phosphoryl residue(s) per molecule, respectively. Generally, purified ovalbumin is a mixture of A_1, A_2, and A_3 ovalbumins in an approximate ratio of 85:12.3 [9,10]. The isoelectric points of A_1, A_2, and A_3 are 4.75, 4.89, and 4.94, respectively [10], which can be separated by ion-exchange chromatography [10,11]. The phosphoryl residues are at Ser^{68} and Ser^{344} [6], and these phosphoryl residues can be eliminated by acid and alkaline phosphatase [10,11].

Ovalbumin has a heterogeneous carbohydrate chain, which is covalently bound to the amide nitrogen of Asn292 with an N-glycosyl linkage [6]. Several different ovalbumin glycopeptides with a common core structure consisting of mannose → (1-4)-N-acetyl glucosamine- → (1-4)-N-acetyl glucosamine-Asn292 have been identified. The carbohydrate chain has been eliminated from ovalbumin by endo-β-N-acetylglucosaminidase isolated from a *Flavobacterium* species [12,13]. Genetic variants of ovalbumin are also known; these involve a Gln→Glu substitution at residue 289 [14] and an Asn→Asp substitution at residue 311 [15].

S-Ovalbumin is often found in the ovalbumin samples, which shows greater heat stability than ovalbumin, therefore, it is called S-ovalbumin for abbreviation of stable ovalbumin. S-Ovalbumin is rarely found in the white of newly laid very fresh eggs, while its content increases during the storage of eggs even at cold temperature [16–18]. The formation of S-ovalbumin from ovalbumin proceeds through an intermediate species. Stored eggs contain various amounts of the intermediate and S-ovalbumin [19]. The intermediate and S-ovalbumin are concomitantly purified with ovalbumin by crystallization method, and it is difficult to separate them from ovalbumin sample. To avoid the contamination of S-ovalbumin and intermediate species from the final purified ovalbumin, newly laid eggs should be used as a starting material for publication. S-Ovalbumin can be typically formed by heating ovalbumin at pH 9.9 for 16 hours at 55°C [17]. Identification and presence of S-ovalbumin can be confirmed by the difference in the denaturation temperature. The denaturation temperature of ovalbumin and S-ovalbumin is 84.5 and 92.5°C, respectively, as measured by DSC at pH 9 with a heating rate of 10°C/min [18]. The structural transitions occurring during conversion of ovalbumin to S-ovalbumin have been studied extensively. Analyses such as sedimentation rates, molecular weight, UV absorption spectra, specific optical rotation, electrophoretic and serological properties, sulfhydryl and disulfide groups, crystal form, and solubility show no remarkable difference in conformation between ovalbumin and S-ovalbumin [17]. However, Raman difference spectrum [20], the intrinsic viscosity, and Stokes Radius measurements reveal a slight conformational difference between ovalbumin and S-ovalbumin [20]. These studies indicate that S-ovalbumin has a slightly more compact conformation than ovalbumin [21]. The slight changes in elution pattern in ion-exchange chromatography, isoelectric focusing, and the titration curve have indicated that the conversion of ovalbumin to S-ovalbumin may involve deamidation [21,22]. However, a definite conclusion has not been obtained yet. It is of interest and import to elucidate the mechanism of formation of S-ovalbumin and the reasons for its heat stability, because it might provide an insight into and a tool for construction of heat-stable proteins and enzymes by using chemical and genetic modification methods.

Ovalbumin has four cysteine residues. Each of them has different reactivity to different types of chemical reagents. They do not react with 5,5′-dithiobis(2-nitrobenzoic acid) in the native state, but do react after denaturation by thermal, surface, and cold treatments [23–25]. Determination of cysteine groups with this reagent is one of the most sensitive methods for evaluating the extent of denaturation of ovalbumin. A more convenient method to detect the denaturation of ovalbumin also is isoelectric precipitation at pH 4.7 in the presence of 0.1 M sodium acetate and 0.5 M NaCl [16].

X-ray crystallography study has revealed the three-dimensional structures of ovalbumin and plakalbumin, a proteolytically nicked form of ovalbumin [26,27]. From the amino acid sequence and three-dimensional structure analyses, ovalbumin appears to have a structure similar to those of a group of serine protease inhibitors, serpins [28], while ovalbumin itself has no inhibitory activity.

Plakalbumin is obtained from ovalbumin by cleaving the protein with subtilisin (a bacterial proteinase) at the Ala345/Glu346 or Glu346/Ala347 and Ala352/Ser353 sites [29]. Elastase efficiently cleaves ovalbumin at the Ala352/Ser353 site [29]. These cleavage sites are also susceptible to attack by other proteases except trypsin. Pepsin hydrolyzes native ovalbumin at pH 2, which is the optimal pH of the enzyme; however, it cleaves at only one site (His22/Ala23) of ovalbumin molecule at pH 4 [24]. This indicates that this site is exposed on the surface of the molecule.

2. Gelation and Gel Properties

When eggs are cooked, egg proteins undergo denaturation, coagulation, and gelation. Isolated egg white and yolk are used as food ingredients to enrich the nutritional quality as well as to improve the functional properties such as gel strength and water-holding capacity of foods.

The mechanism by which gels are formed from globular proteins, like ovalbumin, differs from that of fibrous proteins, agar, and gelatin.

Egg white is a transparent viscous liquid, but after heating it changes into a turbid gel. The term "egg white" might be derived from this phenomenon. Usual heating of ovalbumin solution also results in a turbid suspension or gel. However, under certain solution and heating conditions, it forms a transparent solution or transparent gel. The turbidity of the heated ovalbumin solution depends on the pH, ionic strength, and protein concentration. Figure 1 shows the appearance of ovalbumin before and after heating at different pH and NaCl concentrations [31]. The protein concentration in each well was the same. It is apparent that the turbidity depends on both pH and NaCl concentration. The solutions remain transparent after heating if the pH of the solution is far from the isoelectric point and the ionic strength is low. Turbidity appears when the pH is near the pI. At acidic and alkaline pH transparent sols and transparent gel are formed.

In general, coagulum-type gels are characterized by a three-dimensional network structure formed by protein aggregates. The aggregation of ovalbumin during heating results mainly from hydrophobically driven protein-protein interactions. The hydrogen

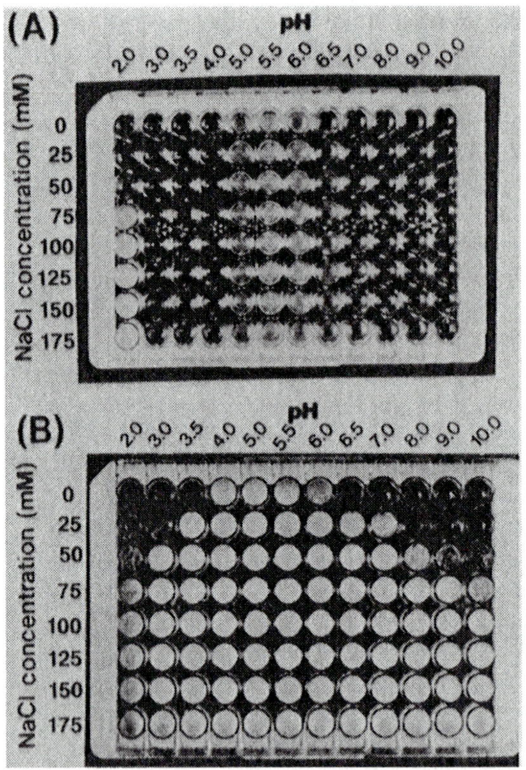

Figure 1 Ovalbumin (A) before and (B) after heating at 80°C for 1 hour. Each well contained 250 μl of ovalbumin solution (70 mg/ml). Ovalbumin was dialyzed against distilled water, and then pH, NaCl concentration, and protein concentration were adjusted. (From Ref. 31.)

bonding and disulfide bridging are apparently not involved in the heat-induced aggregation of ovalbumin [32]. The heated ovalbumin exhibits a greater binding capacity for hydrophobic fluorescent dyes, such as 8-anilino-1-naphthalenesulfonate (ANS) and *cis*-parinaric acid, than the native ovalbumin [33,34]. The far-UV circular dichroism (CD) spectrum of heated ovalbumin reveals no significant loss of secondary structure compared to that of the native ovalbumin, whereas the near-UV CD spectrum indicates a major change in the tertiary structure [35]. A slight increase in the hydrodynamic volume and the Stokes radius suggest only a slight expansion of the molecule. The transmission and scanning electron microscopy of heated ovalbumin reveal formation of linear and random aggregates. Transparent gels or sols contain linear aggregates, whereas turbid gels or suspensions contain random aggregates.

These observations indicate that the conformation of heat-denatured ovalbumin at the secondary structure level is not very different from that of the native molecule, but some of the hydrophobic areas that were buried in the native molecule become exposed after heating. Thus, the heat-denatured ovalbumin seems to be in a so-called "molten-globule state" [36] or "compact denatured state" [37]. The intermolecular interactions between heat-denatured ovalbumin molecules, which are still in a globular shape, are controlled by both the attractive hydrophobic and repulsive electrostatic interactions (Fig. 2). When the electrostatic repulsion is relatively strong and the attractive hydrophobic interaction is restricted, the denatured ovalbumin molecules form soluble linear aggregates; these ordered aggregates look like strings of beads. This type of aggregates are formed when ovalbumin is heated at a pH far from the isoelectric point of ovalbumin and/or at low ionic strength. Under these conditions, ovalbumin molecules are changed, and therefore electrostatic repulsion between them is significant. At high protein concentration, these soluble linear aggregates are cross-linked and form a three-dimensional gel network. As there are no large random aggregates in the network of linear aggregates, the gel appears transparent. When the protein concentration is low, the soluble linear aggregates do not form a gel network, rather a viscous transparent sol. On the other hand, when the electrostatic repulsion is repressed either by adjusting the ph to close too the isoelectric point and/or by increasing the ionic strength, the denatured protein molecules aggregate randomly. This gives a turbid gel or suspension, depending on the protein concentration.

The types of gel structures formed in model systems are shown in Figure 3 [38,39]. At pH values far from the pI and at low ionic strength, linear aggregates are formed (Fig. 3a). With decreasing electrostatic repulsion at low ionic strength or at 7.0 > pH > pI, three-dimensional networks form a transparent gel (Fig. 3b). At high ionic strength or at pH values near the pI, proteins aggregate to form a turbid gel composed of random aggregates (Fig. 3d). At intermediate ionic strength of pH, both linear aggregates and random aggregates are formed. In this case, the linear aggregates form a cross-linked primary gel network and the random aggregates are interdispersed within this network, as shown in Figure 3c. This mixed gel of linear and random aggregates has either a translucent or opaque appearance depending on the relative amounts of the linear and random aggregates. Among these gel types, the transparent and the opaque/translucent gels exhibit higher gel strength and water-holding capacity than the others. This has been shown to be the case for egg white gels. This model for ovalbumin gels can

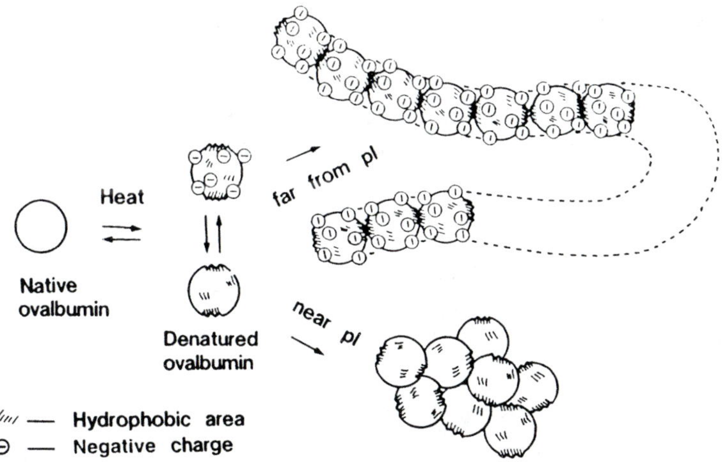

Figure 2 Model for heat denaturation and formation of aggregates of ovalbumin. (From Ref. 38.)

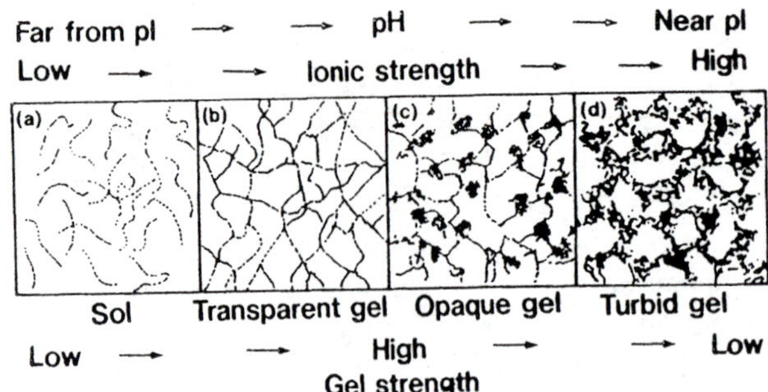

Figure 3 Model for the formation of gel networks by heated ovalbumin. (From Ref. 38.)

also be applied to serum albumin, lysozyme, and milk whey protein gels [40–42]. In these protein gels, change in the net charge or surface hydrophobicity as a function of pH and ionic strength cause alterations in gel properties.

A heat-induced ovalbumin gel melts when high pressure is applied [43]. Partial melting occurs when the gel is subjected to 6000 bars for 20 minutes at room temperature. When the pressure is reduced back to atmospheric pressure, the gel is reformed, indicating that the pressure-induced changes in the gel are reversible. In contrast, however, heat-induced gels of glycinin and cold-set gels of gelatin and agarose do not melt under similar pressure treatment. The gel strength of a transparent ovalbumin gel decreases with an increase of temperature, and the gel melts at higher temperatures. Since, generally, hydrogen bonds are weakened and hydrophobic interactions are strengthened at high temperature, hydrogen bonding may be involved to some extent in the ovalbumin gel network. It should be noted that the gel networks of gelatin or agar are also formed mainly by hydrogen bonding and are weakened (melted) at high temperature; however, unlike ovalbumin gels, no melting occurs at high pressure. This difference between the pressure-induced melting behaviors of ovalbumin and gelatin or agar gels might be attributable to the existence of hydrophobic interactions in the ovalbumin gel.

Heat-set coagulum-type (turbid) protein gels, including an ovalbumin gel, are heat irreversible and, once formed, usually cannot be melted by heating. However, a transparent gel can be melted by heating below 100°C and again reversibly gelled at low temperature. The energy uptake during melting and the energy released during gelling are too small compared to the energy uptake for denaturation of ovalbumin to be detected by DSC. Ovalbumin molecules in some transparent gels seem to bind together weakly, because the hydrophobic attractive and electrostatic repulsive forces are well balanced. Under such conditions, the role of hydrogen bonds that might be formed between denatured molecules become more important. These interactions significantly contribute to network formation; this is evident from the fact that the gel structure changes and the gel melts when heated, because, unlike hydrophobic interactions, hydrogen bonds become weaker at higher temperatures.

3. Surface Denaturation and Foaming Properties

Egg white is an important foaming ingredient for food applications. The foaming properties of proteins are evaluated by their foamability (foaming powers) and foam stabilities. These properties of protein solutions differ with the methods by which the foam is formed [44]. A comparison of the data obtained by different authors can be difficult in some cases. Foamability is related to the rate at which the surface tension of the air/water interface decreases, whereas foam stability is related to the structure of the film of the surface-denatured protein. The former is quantitatively represented by the rate constant of surface tension decay [45,46], and the latter is represented by the surface viscosity. Thus, highly ordered globular protein molecules that do not readily denature at the surface show poor foamability. Foamability of ovalbumin is also relatively low [45], and its foamability is increased by acid treatment, alkali treatment, or heat denaturation [47]. Proteins having a flexible or amphipathic structure (or both), such as casein and gelatin, can readily change their conformation at the air/water interface [45]. Some globular proteins seem to unfold at this interface by orienting themselves with their hydrophobic area toward the air to decrease the surface energy of the system. This surface denaturation of proteins at the air/water interface of foams might cause inactivation of some enzymes. Typically, the surface tension of a globular protein solution continuously decreases for a long time; this is because of slow changes in protein conformation and slow reorientation of the protein groups at the interface. Surface denaturation of ovalbumin has been detected from an increase in the reactivity of the sulfhydryl groups. Figure 4 shows that when an ovalbumin solution is whipped, the ovalbumin molecules are adsorbed at the air/water interface, rearrange themselves and change their conformation to orient their hydrophobic portion in the direction of the gas phase [48]. This change exposes the cysteinyl residues that were buried in the interior of the molecule. The exposed cysteinyl residues undergo oxidation to form disulfide bridges with cysteinyl residues of neighboring ovalbumin molecules at the air/water interface. This reaction proceeds to form aggregates at air/water interface. The sulfhydryl-disulfide interchange reaction between ovalbumin molecules also might take place at the interface. The aggregates formed at the air/water interface seem to produce a surface gel network, and this may be responsible for the stability of ovalbumin foam. However, the aggregates formed by disulfide bridges do not seem to be essential for stabilization of the foam, because the foam prepared with ovalbumin solution in the presence of DTNB was stable for a long time. Thus, the stability of ovalbumin is mainly related to surface denaturation-induced aggregation and surface gelation and, to a lesser extent, to disulfide cross-linking between the denatured aggregated molecules. In other words, conformational changes at the air/water interface strengthen interaction between neighboring molecules via non-

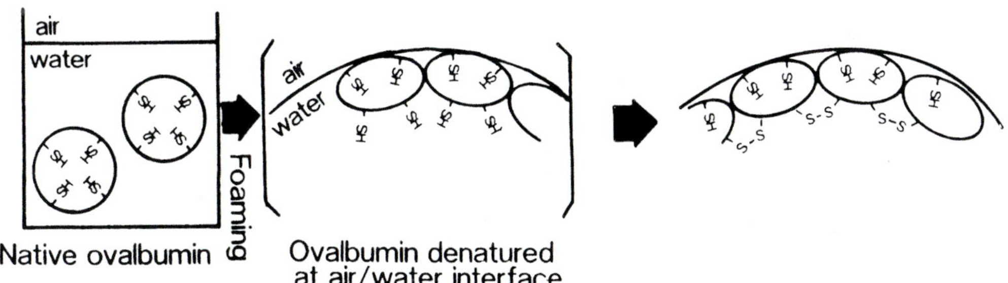

Figure 4 Network formation with ovalbumin molecules surface-denatured at air/water interface by foaming. (From Ref. 48.)

covalent bonding as well as via disulfide bridges. The existence of a surface gel network in foams has been observed in transmission electron microscopy [49,50] and has been confirmed by surface viscosity measurement.

B. Ovotransferrin (Conalbumin)

1. Structure

Ovotransferrin, or conalbumin, is the second most abundant protein in egg white and consists of a single polypeptide chain (78 kDa) [51]. The amino acid sequence of hen egg ovotransferrin, which contains 686 residues, has been determined from its cDNA sequence [52] and from direct amino acid sequencing [53]. Ovotransferrin is obtained from egg white by CM cellulose chromatography [54]. Ovotransferrin is a member of the transferrin family including ovotransferrin, serum transferrin, and lactoferrin. Each transferrin is a monomeric glycoprotein, exhibiting the capacity to bind two Fe^{3+} and two CO_3^{2-} ions per molecule [50–52]. The three transferrins have similar bilobal conformations. Each molecule is divided into two homologous halves, i.e., N-terminal and C-terminal halves. Each lobe is further organized into two domains of about 160 residues each, with the iron site in the cleft between them. Each iron atom is deeply buried, about 10 Å below the protein surface. At pH 7.4, in the presence of bicarbonate, Fe^{3+} binds to either site of ovotransferrin. Fifteen disulfide bridges occur within each lobe, although none pass between them. There are six bridges in the N-lobe and nine in C-lobe [55]. Completely reduced and denatured ovotransferrin in the presence of dithiothreitol and 8 M urea renatures to the native state by a two-step process [56]. The two half-molecules (N- and C-lobes) of ovotransferrin can be obtained separately in the native state by cleaving with trypsin followed by CM sephadex chromatography [57].

2. Denaturation and Gelation

Ovotransferrin is the most easily heat-denaturable egg white protein. The denaturation temperature of ovotransferrin is about 60°C [58]. The functional properties of egg white are most affected by denaturation of ovotransferrin at around 70°C. Binding of metal ions can increase the stability of ovotransferrin. The protein forms an opaque gel on incubation at near-neutral pH and at room temperature with a thiol reagent, such as 2-mercaptoethanol, or glutathione [59,60]. The thiol-induced gelation was dependent on pH, with hard gels formed at pH 7–9 only in the presence of thiol reagents. Under very acidic or alkaline conditions, hard gels were also formed even in the absence of mercaptoethanol. Ovotransferrin is denatured by thiol-dependent cleavage of some disulfide bonds, which is consequently accompanied by an increase in surface hydrophobicity. Thus the denatured molecules may aggregate through intermolecular hydrophobic interactions.

C. Lysozyme

1. Structure

Lysozyme hydrolyzes the β-1,4 linkage between N-acetylglucosamine and N-acetylmuramic acid in the proteoglycan of the bacterial cell [61]. Hen egg white contains the c-type lysozyme, which is the most common type. Lysozyme is a small basic protein (14 kDa, pI = 11), which can be easily purified by crystallization at pH 9.5 in the presence of sodium chloride [62]. The amino acid sequence of hen egg white lysozyme, which contains 129 amino acids has been determined [63,64]. Its three-dimensional structure at 2 Å has been determined.

Lysozyme and α-lactalbumin, which is a mammalian milk protein and a regulatory subunit of lactose synthase, are functionally quite different but structurally highly homologous proteins in terms of their primary and three-dimensional structures. Their gene organizations have been shown to be virtually the same, and their exon structures are identical [65,66]. The catalytic site of the lysozyme is located on exon 2 of gene, coding amino acids 28–82. Exon 3 codes the amino acids 82–108, which gives additional substrate specificity, and exons 1 and 4 code the amino acids that increase the stability of the molecule but are not directly involved in the catalytic function. The lysozyme and α-lactalbumin genes contain three introns at exactly the same positions, although the structures of the introns and of exon 4 differ slightly. These proteins may have diverged from a common ancestor by gene duplication during evolution.

Lysozyme has four disulfide bonds with no free sulfhydryl group. The thermal denaturation temperature of lysozyme is around 70–75°C [58], which depends on pH and other solution conditions. It is about 50 times more heat sensitive in egg white than in phosphate buffer [67]. Some studies have shown that it remains stable over 6 years at room temperature at pH 3.4–9.1 [68]. This remarkable stability may be attributable to the four intramolecular disulfide bonds present in this small protein. X-ray crystallographic studies have revealed that introduction of a Ca^{2+}-binding site, like an EF-hand motif, into human lysozyme by site-directed mutagenesis enhances its structural stability [69,70]. The stability of lysozyme affects its functional properties. Lysozyme has a lower foamability than bovine serum albumin and casein [45]. To produce heat-induced gels of lysozyme, about two of the four disulfide bonds must be reduced [71].

2. Antibacterial Function

The physiological role of lysozyme in egg is to prevent the infection of bacteria, and this antibacterial activity of lysozyme led to its use as a preservative in various foods and as a therapeutic agent [72]. The application of lysozyme as a food preservative and anti-inflammatory drug are currently expanding. Antibacterial action of lysozyme is limited to gram-positive bacteria, but not to gram-negative bacteria. Outer membrane permeability barrier hinders the accessibility of lysozyme to the peptidoglycan layer of the cell wall to perform its action. To increase in the permeability of membranes to lysozyme, introduction of fatty acid by chemical modification technique [73] and hydrophobic pentapeptides by genetic engineering approaches [74] is being attempted.

3. Functional Properties

Lysozyme has a compact and tight conformation due to its intramolecular disulfide bridges. Lysozyme itself does not show superior functional properties in food systems, e.g., foaming, gelling, or emulsification. However, since lysozyme is a very basic protein, it easily interacts with other proteins and other components in food systems, which influences the properties of foods. This effect depends on the pH. For example, the properties of heat-induced ovalbumin gel are affected by the addition of lysozyme [75]. Gel network formation is not affected by mixing ovalbumin and lysozyme at pH 5.5, whereas at pH 7.0 and 8.5 network strength increases. Mixed-protein systems can produce stronger networks than pure proteins. Lysozyme promotes coagulation of egg white at a given pH. Reduction of S-S bridge induces conformational changes and also increases the flexibility of molecule; the gelling properties [41,76] and foaming properties are also significantly improved in a manner similar to that of ovotransferrin.

D. Ovomucoid

Ovomucoid constitutes about 10% of the proteins of egg white and contains about 20–25% carbohydrate [77]. It is a trypsin inhibitor with a molecular weight of 28 kDa [78]. The ovomucoid molecule consists of three tandem domains, each homologous to pancreatic secretory trypsin inhibitor (Kazel). Each domain is cross-linked by three intradomain disulfide bonds. The first and second domains each have two N-linked carbohydrate chains, and there are two types of the third domain, either with or without N-linked carbohydrate chain; the secondary structures of these domains have been studied [79]. Ovomucoid forms a stable association complex with trypsin [78,80]. The chemical structure of ovomucoid carbohydrate chain has already been established [81,82]. Major carbohydrate components are penta-antennary and tetra-antennary complex-type chains consisting of galactose, mannose, and N-acetylglucosamine. These carbohydrate chains are quite different from those of ovalbumin, which are a series of unusual asparagine-linked carbohydrate chains together with a series of high-mannose-type carbohydrate chains [81]. The molecular size of the ovomucoid carbohydrate chain is relatively large, i.e., larger than glucose dodecamer [81]. Ovomucoid remains physically stable, but loses its biological activity when subjected to severe heat treatment, such as 100°C for 60 minutes [83,84]. The reduction and alkylation of the ovomucoid disulfide bonds cause a complete loss of its trypsin inhibitory activity and immunoreactivity with specific antibodies [85–87]. The carbohydrate moiety of ovomucoid contributes to the stability of ovomucoid molecule against tryptic hydrolysis and heat denaturation [88].

E. Ovomucin

Ovomucin is a structurally important sulfated glycoprotein, characterized by a highly viscous and gellike nature. It is not soluble in water but can be solubilized in a dilute salt solution at pH 9 or above. The carbohydrate content of ovomucin is about 30%, consisting of 15–18.6% hexose ((galactose and mannose), 7–12% hexosamine (N-acetylglucosamine + N-acetylgalactosamine) and 2.5–8% sialic acid. Both O- and N-glucosidically linked carbohydrate moieties are present [89–91]. Ovomucin is present in egg white as two different types: one is insoluble ovomucin obtained from the gel fraction or whole thick white, and the other is soluble ovomucin from the liquid fraction of thick white or from thin white [92,93]. Both ovomucins are composed of two kinds of subunit proteins, α- and β-ovomucin, of which the carbohydrates contents are 15 and 50%, respectively [90]. The molecular size of α-ovomucin is about 18 kDa and that of β-ovomucin is about 400 kDa [94]. Relative amounts of α- and β-ovomucin in insoluble ovomucin are 84:20, and those in soluble ovomucin are 40:3. Both soluble and insoluble ovomucin have inhibitory activity against viral hemagglutination and an important role in the foaminess of egg white. The dissociation of β-ovomucin from insoluble ovomucin and its solubilization into the liquid from thick white is considered to be the cause of egg white thinning [95,96].

III. EGG YOLK PROTEINS

Yolk can be separated by centrifugation into sedimented granules and a supernatant, plasma. The compositions of granules and plasma are different. The protein content of granules is much larger than that of plasma. The granules are composed of 70% α- and β-lipoproteins, 16% phosvitin, and 12% low-density lipoprotein [97]. Lipovitellin [98,99], phosvitin [99–102], and vitellogenin are the minor proteins. The granules contain low-density lipoprotein and myelin. On the other hand, the plasma contains low-density lipoproteins and livetins. Besides these two kinds of proteins, there are minor components with some interesting biological properties such as riboflavin-binding protein, biotin-binding protein, and immunoglobulin Y.

Eggs contain the IgM-, IgA-, and IgG-type immunoglobulins in both white and yolk. IgM- and IgA-type immunoglobulins are found in egg white, and their concentration there is about 0.2 and 0.7 mg/ml, respectively. Egg yolk specifically contains IgG-type immunoglobulin, and its concentration is about 10 mg/ml. After hatching, IgG-type immunoglobulin in yolk is transported to the blood of chick and others (IgA- and IgM-type) are moved to intestinal duct. These immunoglobulins protect chicks from bacterial infection. Although the immunoglobulins in yolk correspond to the antibody in IgG class of mammals, some properties are different from those of IgG of mammals. Therefore, immunoglobulin in yolk is classified as a IgY from the viewpoint of the comparative immunology. When a hen is immunized with an antigen, IgY against this antigen is produced and accumulated in the yolk. Compared to the conventional method using rabbit, the production and collection of IgY from egg yolk is convenient and easy. About 40 g of IgY can be prepared from one hen in one year. This amount corresponds to the amount of IgG obtained from the serum of 30 rabbits. Simple methods for purification of IgY from yolk have been reported, and mass production of IgY has been established [103,104]. The IgY is larger than IgG. Cloning of IgY gene has revealed that the constant region of heavy chain consists of four domains (three domains for mammals) and that its structure resembles those of IgM and IgE rather than IgG. The pI of IgY is lower than that of IgG by one unit, and IgY does not activate the complement of mammals. IgY cannot bind to protein A. IgY is expected to be used for medical purposes and/or in food supplements used against gastrointestinal infection and disease and other applications [105–107].

REFERENCES

1. D. T. Osuga and R. E. Feeney, Egg proteins, *Food Proteins*, (J. R. Whitaker and S. R. Tannenbaum, eds.), AVI, Westport, CT, 1977, p. 209.
2. W. D. Powrie and S. Nakai, The chemistry of eggs and egg proteins, *Egg Science and Technology*, 3rd ed. (W. J. Stadelman and O. J. Cotterill, eds.), AVI, Westport, CT, 1986, p. 97.
3. L. C. Eunice and S. Nakai, Biochemical basis for the properties of egg white, *CRC Crit. Rev. Poultry Biol. 2*:21 (1989).
4. S. P. L. Soerensen and M. Hoyrup, *Compt. Rend. Trav. Lab. Carlsberg 12*:12 (1915).
5. L. McReynolds, B. W. O'Malley, A. D. Nisbet, J. E. Fothergill, D. Givol, S. Fields, M. Robertson, and G. G. Brownlee, Sequence of chicken ovalbumin mRNA, *Nature 273*:723 (1978).
6. A. D. Nisbet, R. H. Saundry, A. J. G. Moir, L. A. Fothergill, and J. E. Fothergill, The complete amino acid sequence of hen ovalbumin, *Eur. J. Biochem. 115*:335 (1981).

7. S. L. C. Woo, W. G. Beattie, J. F. Catterall, A. Dugaiczyk, R. Staden, G. G. Brownlee, and B. W. O'Malley, Complete nucleotide sequence of the chicken chromosomal ovalbumin gene and its biological significance, *Biochemistry 20*:6437 (1981).
8. K. Narita and J. Ishii, N-Terminal sequence in ovalbumin, *J. Biochem. 52*:367 (1962).
9. G. E. Perlmann, Enzymatic dephosphorylation of ovalbumin and plakalbumin, *J. Gen. Physiol. 35*:711 (1952).
10. N. Nitabatake, A. Ishida, and E. Doi, Physicochemical and functional properties of hen ovalbumin dephosphorylated by acid phosphatase, *Agric. Biol. Chem. 54*:967 (1988).
11. W. J. Gou and Venkatasubramanian, Metal ion binding properties of hen ovalbumin and S-ovalbumin characterization of the metal ion binding site of ^{31}P NMR and water proton relaxation rate enhancements. *Biochemistry 25*:84 (1986).
12. K. Yamamoto, K. Takegawa, F. Jianqiang, H. Kumagai, and T. Tochikura, *J. Ferment. Technol. 64*:397 (1963).
13. N. Kitabatake, A. Ishida, K. Yamamoto, T. Tochikura, and E. Doi, Deglycosylation of hen ovalbumin in native form by endo- + -N-acetylglucosaminidase, *Agric. Biol. Chem. 52*:2511 (1988).
14. H. Ishihara, N. Takahashi, J. Ito, E. Takeuchi, and S. Tejima, Either high-mannose-type or hybrid-type oligosaccharide is linked to the same asparaine residue in ovalbumin, *Biochim. Biophys. Acta 669*:216 (1981).
15. R. L. Wiseman, J. E. Fothergill, and L. A. Fothergill, Replacement of asparaine by aspartic acid in hen ovalbumin and a difference in immunochemical reactivity, *Biochem. J. 127*:775 (1972).
16. M. B. Smith and J. F. Back, Studies on ovalbumin I. Denaturation by heat, and the heterogeneity of ovalbumin, *Aust. J. Biol. Sci. 17*:261 (1964).
17. M. B. Smith and J. F. Back, Studies on ovalbumin II. The formation and properties of S-ovalbumin, a more stable form of ovalbumin, *Aust. J. Biol. Sci. 18*:365 (1965).
18. M. B. Smith and J. F. Back, Studies on ovalbumin III. Denaturation on ovalbumin and S-ovalbumin, *Aust. J. Biol. Sci. 21*:539 (1968).
19. J. W. Donovan, and C. J. Mapes, A differential scanning calorimetric study of conversion of ovalbumin to S-ovalbumin in eggs, *J. Sci. Food Agric. 27*:197 (1976).
20. S. Kint and Y. Tomimatsu, A raman difference spectroscopic investigation of ovalbumin and S-ovalbumin, *Biopolymers 18*:1073 (1979).
21. R. Nakamura, M. Hirai, and Y. Takemori, Some differences noted between the properties of ovalbumin and S-ovalbumin in native state, *Agric. Biol. Chem. 44*:149 (1980).
22. R. Nakamura and M. Ishimaru, Changes in the shape and surface hydrophobicity of ovalbumin during its transformation to S-ovalbumin, *Agric. Biol. Chem. 45*:2775 (1981).
23. L. A. Fothergill and J. E. Fothergill, Thiol and disulfide contents of hen ovalbumin, *Biochem. J. 116*:555 (1970).
24. N. Kitabatake, K. Indo, and E. Doi, Limited proteolysis of ovalbumin by pepsin, *J. Agric. Food Chem. 36*:417 (1988).
25. T. Watanabe, N. Kitabatake, and E. Doi, Method for the accurate measurement of freezing-induced denaturation of ovalbumin with 5,5′-dithiobis-(2-nitrobenzoic acid), *Biosci. Biotech. Biochem. 58*:359 (1994).
26. P. E. Stein, A. G. W. Leslie, J. T. Finch, and R. W. Carrel, Crystal structure of uncleaved ovalbumin at 1.95 Å resolution, *J. Mol. Biol. 221*:941 (1991).
27. H. T. Wright, H. X. Qian, and R. Huber, Crystal structure of plakalbumin, a proteolytically nicked form of ovalbumin its relationship to the structure of cleaved α-1-protease inhibitor, *J. Mol. Biol. 213*:513 (1990).
28. R. Carrell and J. Travis, α-Antitypsin and the serpins: variation and countervariation, *Trends Biochem. Sci. 10*:20 (1985).
29. K. Linderstrom-Lang and M. Ottesen, Formation of plakalbumin from ovalbumin, *Compt. rend trav. Lab. Carlsberg Ser. Chim. 26*:403 (1949).

30. H. T. Wright, Ovalbumin is an elastase substrate, *J. Biol. Chem. 259*:14335 (1984).
31. N. Kitabatake and Y. Kinekawa, Turbidity measurement of heated egg proteins using a microplate system, *Food Chem. 53*:in press (1995).
32. H. Hatta, N. Kitabatake, and E. Doi, Turbidity and hardness of a heat-induced gel of hen egg ovalbumin, *Agric. Biol. Chem. 50*:2083 (1986).
33. N. Kitabatake, K. Indo, and E. Doi, Changes in interfacial properties of hen egg ovalbumin caused by freeze-drying and spray drying, *J. Agric. Food Chem. 37*:905 (1989).
34. F. Tani, M. Murata, T. Higasa, M. Goto, N. Kitabatake, and E. Doi, The molten globule state of protein molecules in heat-induced transparent food gels, *J. Agric. Food Chem.* in press.
35. T. Koseki, N. Kitabatake, and E. Doi, Irreversible thermal denaturation and formation of linear aggregates of ovalbumin, *Food Hydrocolloids 3*:123 (1989).
36. K. Kuwajima, The molten globule state as a clue for understanding the folding and cooperativity of globular-protein structure, *Proteins 6*:87 (1989).
37. D. O. V. Alonso, K. A. Dill, and D. Stiger, The three states of globular proteins: acid denaturation, *Biopolymer 31*:1631 (1992).
38. E. Doi and N. Kitabatake, Structure of gycinin and ovalbumin gels, *Food Hydrocoll. 3*:327 (1989).
39. E. Doi, Gels and gelling of globular proteins, *Trends Food Sci. Tech. 4*:1 (1993).
40. M. Murata, F. Tani, T. Higasa, N. Kitabatake, and E. Doi, Heat-induced transparent gel formation of bovine serum albumin, *Biosci. Biotech. Biochem. 57*:43 (1993).
41. F. Tani, M. Murata, F. Tani, T. Higasa, M. Goto, N. Kitabatake, and E. Doi, Heat-induced transparent gel from Hen Egg Lysozyme by a Two-step heating method, *Biosci. Biotech. Biochem. 57*:209 (1993).
42. Y. Kinekawa and N. Kitabatake, Turbidity and rheological properties of gels and sols prepared by heating process whey protein, *Biosci. Biotech. Biochem. 57*:43 (1993).
43. E. Doi, A. Shimizu, H. Oe, and N. Kitabatake, Melting of heat-induced ovalbumin gel by pressure, *Food Hydrocolloids 5*:409 (1991).
44. L. G. Phillips, J. B. German, T. E. O'Neil, E. A. Foegeding, V. R. Harwalkar, A. Kilara, B. A. Lewis, M. E. Mangino, C. V. Morr, J. M. Regenstein, D. M. Smith, J. E. Kinsella, Standardized procedure for measuring foaming properties of three proteins, a collaborative study, *J. Food Sci. 55*:1441 (1990).
45. N. Kitabatake and E. Doi, Surface tension and foaming of protein solutions, *J. Food Sci. 47*:1218 (1982).
46. D. E. Graham and M. C. Philips, The conformational of proteins at the air-water interface and their role in stabilizing foams, in *Foams* (R. J. Akers, ed.), Academic Press, London, 1976. p. 237.
47. R. Nakamura, Studies on the foaming property of the chicken egg white. Part VII. On the foaminess of the denatured ovalbumin, *Agric. Biol. Chem. 28*:403 (1964).
48. N. Kitabatake and E. Doi, Conformational change of hen egg ovalbumin during foam formation detected by 5,5′-dithiobis (2-nitrobenzoic acid), *J. Agric. Food Chem. 35*:953 (1987).
49. T. M. Johnson and M. E. Zabik, Egg albumin proteins interactions in an angel food cake system, *J. Food Sci. 46*:1231 (1981).
50. N. Kitabatake, N. Sasaki, and E. Doi, Scanning electron microscopy of freeze-dried protein foams, *Agric. Biol. Chem. 46*:2881 (1982).
51. J. H. Brock, Transferrins, *in Metalloproteins* (P. Harrison, ed.) part 2, Macmillan Press, London, 1985, p. 183.
52. J. M. Jeltsch and P. Chambon, The complete nucleotide sequence of the chicken ovotransferrin mRNA, *Eur. J. Biochem. 122*:291 (1982).
53. J. Williams, T. C. Elleman, I. B. Kingston, A. G. Wilkins, and K. A. Kuhn, The primary structure of hen ovotransferrin, *Eur. J. Biochem. 122*:297 (1982).

54. P. Azari and P. F. Baugh, A simple and rapid procedure for preparation of large quantities of ovotransferrin, *Arch. Biochem. Biophys, 118*:138 (1967).
55. J. Williams, K. Moreton, and D. J. Goodearl, Selective reduction of a disulphide bridge in hen ovotransferrin, *Biochem. J. 228*:661 (1985).
56. M. Hirose, T. Akuta, and N. Takahashi, Renaturation of ovotransferrin under two-step conditions allowing primary folding of the fully reduced form and the subsequent regeneration of the intermolecular disulfides, *J. Biol. Chem. 264*:16867 (1989).
57. H. Oe, E. Doi, and M. Hirose, Amino-terminal and carboxy-terminal half-molecules of ovotransferrin: preparation by a novel procedure and their interactions, *J. Biochem. 103*: 1066 (1988).
58. J. W. Donovan, C. J. Mapes, J. G. Davis, and J. A. Garibaldi, A differential scanning calorimetric study of the stability of egg white to heat denaturation, *J. Sci. Food Agric. 26*: 73 (1975).
59. M. Hirose, H. Oe, and E. Doi, Thiol-dependent gelation of egg white, *Agric. Biol. Chem. 50*:59 (1986).
60. H. Oe, M. Hirose, and E. Doi, Conformation changes and subsequent gelation of conalbumin by a thiol reagent, *Agric. Biol. Chem. 50*:2469 (1986).
61. G. Alderton, W. H. Ward, and H. L. Fevold, Isolation of lysozyme from egg white, *J. Biol. Chem. 157*:43 (1945).
62. G. Alderton and H. L. Fevold, Direct crystallization of lysozyme from egg white and some crystalline salts of lysozyme, *J. Biol. Chem. 164*:1 (1946).
63. R. E. Canfield, The amino acid sequence of egg white lysozyme, *J. Biol. Chem. 238*:2698 (1963).
64. J. Jolles, J. Jauregui-Adell, I. Bernier, and P. Jolles, La structure chimique du lysozyme de blanc d'oeuf de poule: etude detaillee, *Biochim. Biophys. Acta 78*:668 (1963).
65. A. Jung, A. E. Sippel, M. Grez, and G. Schutz, Exons encode functional and structural units of chicken lysozyme, *Proc. Natl. Acad. Sci. USA 77*:5759 (1980).
66. I. Kumagai, S. Takeda, and K. Miura, Functional conversion of the homologous proteins α-lactalbumin and lysozyme by exon exchange, Proc. Natl. Acad. Sci. USA *89*:5887 (1992).
67. F. E. Cunningham and H. Lineweaver, Stabilization of egg-white proteins to pasteurizing temperatures above 60°C, *Food Technol. 19*:1442 (1965).
68. R. E. Feeney, L. R. MacDonnell, and E. D. Ducay, Irreversible inactivation of lysozyme by copper, *Arch. Biochem. Biophys. 61*:72 (1956).
69. R. Kuroki, Y. Taniyama, C. Seko, H. Nakamura, M. Kikuchi, and M. Ikehara, Design and creation of a Ca^{2+} binding site in human lysozyme to enhance structural stability, *Proc. Natl. Acad. Sci. USA 86*:6903 (1989).
70. D. I. Stuart, K. R. Acharya, N. P. C. Walker, S. G. Smith, M. Lewis, and D. C. Phillips, α-Lactalbumin possess a novel calcium binding loop, *Nature 324*:84 (1986).
71. S. Hayakawa and R. Nakamura, Optimization approaches to thermally induced egg white lysome gel, *Agric. Biol. Chem. 51*:771 (1986).
72. V. A. Proctor and P. E. Cunningham, The chemistry of lysozyme and its use as a food preservative and a pharmaceutical, *CRC Crit. Rev. Food Sci. Nutr. 26*:359 (1988).
73. H. R. Ibrahim, A. Kato, and K. Kobayashi, Antimicrobial effects of lysozyme against gram-negative bacteria due to covalent binding of palmitic acid, *J. Agric. Food Chem. 39*:2077 (1991).
74. H. R. Ibrahim, M. Yamada, K. Kobayashi, and A. Kato, Bactericidal action of lysozyme against gram-negative bacteria due to insertion of a hydrophobic pentapeptide into its C-terminus, *Biosci. Biotech. Biochem. 56*:1361 (1992).
75. S. D. Arntfield and A. Bernatsky, Characteristic of heat-induced networks for mixtures of ovalbumin and lysozyme, *J. Agric. Food Chem. 41*:2291 (1993).
76. S. Hayakawa and R. Nakamura, Optimization approaches to thermally induced egg white lysozyme gel, *Agric. Biol. Chem. 43*:23 (1986).

77. R. Montgomery, Glycoproteins, in *The Carbohydrates* (W. Pigman and D. Horton, eds.) Vol. 2B, Academic Press, New York, 1970.
78. I. Kato, J. Shrodo, W. J. Kohr, and Jr. M. Laskowski, Chicken ovomucoid: determination of its amino acid sequence, determination of the trypsin reaction site, and preparation of all three of its domains, *Biochemistry 26*:193 (1987).
79. K. Watanabe, T. Matsuda, Y. Sato, The secondary structure of ovomucoid and its domains as studied by circular dichroism, *Biochim. Biophys. Acta 667*:242 (1981).
80. J. W. Donovan and R. A. Beardslee, Heat stabilization produced by proton-proton association, *J. Biol. Chem. 250*:1966 (1975).
81. K. Yamashita, J. P. Kamerling, and A. Kobata, Structural study of the carbohydrate moiety of hen ovomucoid, *J. Biol. Chem. 257*:12809 (1982).
82. K. Yamashita, J. P. Kamerling, and A. Kobata, Structural studies of the sugar chains of hen ovomucoid. Evidence indicating that they are formed mainly by the alternative biosynthetic pathway of asparagine-linked sugar chains, *J. Biol. Chem. 258*:3099 (1983).
83. F. C. Stevens and R. E. Feeney, Chemical modification of avian ovomucoids, *Biochem. 2*: 1346 (1963).
84. T. Matsuda, K. Watanabe, and R. Nakamura, Immunochemical studies on thermal denaturation of ovomucoid, *Biochim. Biophys. Acta 707*:121 (1982).
85. L. B. Sjoberg and R. E. Feeney, Reduction and reoxidation of turkey ovomucoid—a protein with dual and independent inhibitory activity against trypsin and α-chymotrypsin, *Biochim. Biophys. Acta 168*:79 (1968).
86. T. Matsuda, K. Watanabe, and R. Nakamura, Secondary structure of reduced ovomucoid and renaturation of reduced ovomucoid and its rescued fragments A (1-130) and B (1311-186), *FEBS Lett., 24*:185 (1981).
87. J. I. Morton and H. F. Deutsch, Immunochemical studies of modified ovomucoids, *Arch. Biochem. Biophys. 93*:661 (1961).
88. J. Gu, T. Matsuda, R. Nakamura, H. Ishiguro, I. Ohkubo, M. Sasaki, and N. Takahashi, Chemical deglycosylation of hen ovomucoid: protective effect of carbohydrate moiety on tryptic hydrolysis and heat denaturation, *J. Biochem. 106*:66 (1989).
89. M. B. Smith, T. M. Reynold, C. P. Buckingham, and J. F. Back, Studies on the carbohydrate of egg-white ovomucin, *Aut. J. Biol. Sci. 27*:349 (1974).
90. A. Kato, S. Hirata, H. Sato, and K. Kobayashi, Fractionation and characterization of the sulfated oligosaccharide chains of ovomucin, *Agric. Biol. Chem. 42*:835 (1978).
91. A. Kato, S. Hirata, and K. Kobayashi, Structure of sulfated oligosaccharide chain of ovomucin, *Agric. Biol. Chem. 42*:1025 (1978).
92. A. Kato, R. Nakamura, and Y. Sato, Studies on changes in stored shell eggs. Part VI. Changes in chemical composition of ovomucin during storage, *Agric. Biol. Chem. 34*:1009 (1970).
93. S. Hayakawa and Y. Sato, Physicochemical identity of α-ovomucin or β-ovomucin obtained from the sonicated insoluble and soluble ovomucins, *Agric. Biol. Chem. 41*:1185 (1977).
94. S. Hayakawa and Y. Sato, Subunit structures of sonicated α and β-ovomucin and their molecular weights estimated by sedimentation equilibrium, *Agric. Biol. Chem. 36*:831 (1972).
95. A. Kato and Y. Sato, The release of carbohydrate rich component from ovomucin gel during storage, *Agric. Biol. Chem. 36*:831 (1972).
96. D. S. Robinson and J. B. Monsey, Changes in the composition of ovomucin during liquefaction of thick egg white, *J. Sci. Food Agric. 23*:29 (1972).
97. G. Schmidt, M. J. Bessman, M. D. Hickey, and S. J. Thannhauser, The concentrations of some constituents of egg yolk in its soluble phase, *J. Biol. Chem. 223*:1027 (1956).
98. R. W. Burley and W. H. Cook, *J. Biochem. Physiol. 39*:1295 (1961).
99. J. Kurisaki, K. Yamauchi, H. Isshiki, and S. Ogiwara, Differences between α and β-lipovitellin from hen egg yolk, *Agric. Biol. Chem. 45*:699 (1981).

100. D. K. Mecham and H. S. Olcott, Phosvitin, the principal phosphoprotein of egg yolk, *J. Am. Chem. Soc. 71*:3670 (1949).
101. G. Taborsky and C. C. Mok, Phosvitin homogeneity and molecular weight, *J. Biol. Chem. 242*:1495 (1967).
102. Y. Abe, T. Itoh, and S. Adachi, Fractionation and characterization of hen's egg yolk phosvitin, *J. Food Sci. 47*:1903 (1982).
103. H. Hatta, M. Kim, and T. Yamamoto, A novel isolation method for hen egg yolk antibody, "IgY", *Agric. Biol. Chem. 54*:2531 (1990).
104. E. M. Akita and S. Nakai, Immunoglobulins from egg yolk: isolation and purification, *J. Food Sci. 57*:629 (1992).
105. T. Ebina, K. Tsukada, K. Umezu, M. Nose, K. Tsuda, H. Hatta, M. Kim, and T. Yamamoto, Gastroenteritis in suckling mice caused by human rotavirus can be prevented with egg yolk immunoglobulin (Ig Y) and treated with a protein-bound polysaccharide preparation (PSK), *Microbiol. Immunol. 34*:617 (1990).
106. H. Hatta, K. Tsuda, S. Akachi, M. Kim, T. Yamamoto, and T. Ebina, Oral passive immunization effect of anti-human rotavirus Ig Y and its behavior against proteolytic enzymes, *Biosci. Biotech. Biochem. 576*:1077 (1993).
107. S. Otake, Y. Nishihara, M. Makimura, H. Hatta, M. Kim, T. Yamamoto, and M. Hirasawa, Protection of rats against dental caries by passive immunization with hen-egg-yolk antibody (Ig Y), *J. Dent. Res. 70*:162 (1991).

12

Structure–Function Relationships of Muscle Proteins

YOULING L. XIONG
University of Kentucky, Lexington, Kentucky

I. INTRODUCTION

Proteins, the most important functional components in muscle, confer many of the desirable physicochemical and sensory attributes of muscle foods. As already described in previous chapters, functionality of proteins in food processing refers to any property of the protein that affects palatability of the final product, and it should be distinguished from functionality of proteins in living tissues (e.g., ion transportation through the cell membrane; enzyme catalysis). Muscle proteins comprise 15–22% of the total muscle weight (about 60–88% of mass) and can be divided into three major groups on the basis of solubility characteristics: sarcoplasmic proteins (water-soluble), myofibrillar proteins (salt-soluble), and stromal proteins (insoluble). Of the three groups, myofibrillar proteins play the most critical role during meat processing as they are responsible for the formation of thermally induced cohesive structures and the firm texture of meat products. The functional behavior of myofibrillar proteins is manifested by their ability to produce three-dimensional, viscoelastic gel matrices via protein-protein interactions, to bind water, and to form cohesive and strong membranes on the surface of fat globules in emulsion systems or flexible films around the air/water interface. These functional properties are the major factors contributing to palatability or sensory perception (tenderness, juiciness, mouthfeel, etc.) of processed meat products. A number of sarcoplasmic and stromal proteins also possess desirable functionalities, which are important for producing consumer-acceptable muscle foods. For instance, myoglobin imparts a desirable pinkish-red color to meat, and the specific color is dictated by the chemical state of the heme moiety as well as the structure of globin, the protein moiety. Hydrolyzed collagen derived from the connective tissue has excellent water-binding ability and, therefore, is able to improve the water-holding capacity and tenderness of cooked meat products.

Recent advances in muscle protein research have led to the recognition that many of the functional properties of muscle proteins are related to their structures. In an attempt to establish the structure-function relationships, myofibrillar proteins, particularly myosin, have been extensively studied. At the fibril level, the architecture of the myofibril, i.e., the arrangement of constituting proteins in the myofibrillar assemblage, is a determinant of myofibril swelling, hydration, and protein extraction. At the individual protein level, however, the conformation, shape, and size of the protein molecules have remarkable effects on their functionalities. In meat products, protein functionality is generally induced following major alterations in the native structure of the protein molecules. Typically, this is accomplished by manipulation of heating and cooling procedures, by controlling the total thermal and mechanical energy input, and by selecting proper product ingredients which can directly or indirectly interact with proteins. Slow, progressive unfolding of protein molecules allows for ordered protein-protein interactions and, hence, formation of highly viscoelastic gel networks essential for binding and maintaining the integrity of restructured meats. On the other hand, rapid denaturation of proteins to expose the hydrophobic patches and groups by mechanical actions (e.g., shear) is important for the production of emulsion- and foam-type meat products.

Functional properties of muscle proteins do not bear a simple relationship to their native structures. This is because in meat processing, muscle proteins will undergo a series of structural changes, producing many intermediates when protein denaturation occurs. Structures of partially denatured protein molecules in the "transitional" stage are subject to specific meat-processing conditions, can vary to a large extent, and are difficult to characterize. Yet, these processing-induced structures, not their ultimate precursor (i.e., the native structure), determine the functional performance of proteins. Thus, characterization of structure-function relationships of muscle proteins cannot be accomplished without taking into consideration physicochemical changes in the protein molecules during meat processing. For instance, rigidity and elasticity of comminuted meat products as related to protein gelation and emulsification are often predictable from physicochemical changes in the proteins, which are dependent on the specific ionic environment and heating procedures. Surface polarity, charges, and hydrophobicity, as well as thermal stability, are examples of the physicochemical properties of proteins influenced by meat-processing parameters. A known relationship between protein structure and its functionalities would obviously aid the food processor in designing specific processing protocols for product quality control and new product development.

II. MUSCLE PROTEINS AND THEIR STRUCTURES

A. Sarcoplasmic Proteins

Proteins that are located inside the sarcolemma and are soluble in low salt concentrations (<0.1 M KCl) are referred to as sarcoplasmic proteins. Sarcoplasmic proteins comprise about 30–35% of the total muscle proteins or about 5.5% of the weight of muscle in mature animals [61]. The amount of sarcoplasmic proteins in early embryonic stages may be as high as 70%, but it gradually declines in proportion to the increase in the content of myofibrillar proteins as the animal matures. In the early literature, sarcoplasmic proteins were commonly called myogens, which include most water-soluble proteins. Recent research has resulted in the separation of sarcoplasmic proteins into four different structural components based on their sedimentation velocity in differential centrifugation:

nuclear, mitochondrial, microsomal, and cytoplasmic fractions. They can be obtained as pellets after centrifugation at 1000 *g* (nuclear), 10,000 *g* (mitochondrial), 100,000 *g* (microsomal), and from the supernatant (cytoplasmic). These protein fractions make up more than 100 different sarcoplasmic proteins, including most of the enzymes involved in energy metabolism, such as glycolysis.

Sarcoplasmic proteins occupy about 25% of the cellular space and have a concentration of about 260 mg/ml [61]. Despite their diversity, sarcoplasmic proteins share many common physicochemical properties. For instance, most are of relatively low molecular weight, high isoelectric pH, and have globular or rod-shaped structures. These structural characteristics may be partially responsible for the high solubility of these proteins in water or dilute salt solutions.

Myoglobin is perhaps the single most important sarcoplasmic protein in meat because it is primarily responsible for meat color and, hence, meat quality. The distribution of myoglobin in meat animals varies extensively. It is most abundant in organ tissues, such as chicken gizzard (2–2.6%) and relatively low in skeletal and cardiac muscle (<1%) [42,61]. The specific concentration of myoglobin in tissue is dependent upon animal species, muscle fiber type, degree of exercise, age, sex, and diet. Thus, mammalian muscle (e.g., lamb, beef, pork) contains more myoglobin than poultry muscle, and poultry leg muscle contains more myoglobin than poultry breast muscle. Myoglobin has a molecular weight of around 16,800; the exact molecular weight varies among species. The protein consists of two essential parts: a heme group and a protein moiety called globin (Fig. 1). The heme complex is attached to globin by chelation of a histidine

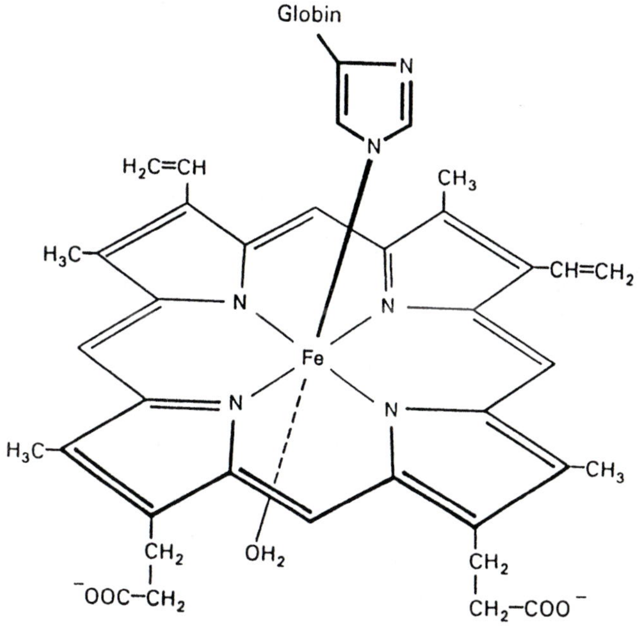

FIGURE 1 Schematic representation of the heme complex of myoglobin.

residue via the central iron atom. Different myoglobins differ in the substitute side chains attached to the porphyrin ring and in the amino acid composition or sequence of the peptide chain. The side chains on the porphyrin ring are either methyl, vinyl, or propyl. The iron located at the center of the porphyrin ring can exist in either ferrous (Fe^{2+}) or ferric (Fe^{3+}) form. The iron has six coordination sites. Four are bound to the four nitrogen atoms of the porphyrin ring, the fifth forms a complex with the nitrogen in the imidazole ring of histidine in the globin molecule, while the sixth is free to bind with different substances—water, oxygen, and other ligands. When myoglobin is in its reducing form, the iron (Fe^{2+}) can either bind with water (deoxymyoglobin) to impart a purplish-red color to meat or complex with molecular oxygen (oxymyoglobin) to produce the desirable cherry-red appearance of fresh meat. In cured meat, the ferrous iron forms a complex with nitric oxide (nitrosylmyoglobin) to yield a pinkish-red color. Under oxidative conditions, the iron is oxidized to its ferric form (Fe^{3+}), which cannot form a complex with oxygen or nitric oxide. The change in the highly conjugated resonance structure of the heme when iron is oxidized results in the development of an undesirable brown color in meat (metmyoglobin).

The conformation of myoglobin varies slightly in different animals or muscles. This is caused by minor shifts in the amino acid sequence in the polypeptide chain, differences in amino acid composition, and variations in length of the polypeptide chain. Mammalian myoglobins lack the single cysteine residue contained in fish myoglobins. There are approximately 153 amino acid residues in myoglobin, 80% of which are located in eight different regions of the α-helix [2,42]. The apolar areas of the apoprotein are in contact with one another in the central region of the molecule stabilized by van der Waals forces. The polar heme group is also located in a nonpolar pocket, but the polar propionic groups of the heme are exposed to the surface. This structural arrangement is responsible for the high solubility of myoglobin in cytosol and in water. The apoprotein globin seems to serve a critical role in protecting the heme against oxidation. The autoxidation rate of oxymyoglobin to metmyoglobin is reportedly 10^8 times slower than the oxidation of heme [97].

B. Myofibrillar Proteins

Myofibrillar proteins, which comprise 55–60% of total protein in muscle, are the structural proteins that make up the myofibrils [61]. Based on their physiological and structural roles in living tissues, myofibrillar proteins can be further divided into three subgroups: (a) the major contractile proteins, including myosin and actin, which are directly responsible for muscle contraction and are the backbone of the myofibril, (b) regulatory proteins, including tropomyosin, the troponin complex, and several other minor proteins, which are involved in the initiation and control of contraction, and (c) cytoskeletal or scaffold proteins, including titin or connectin, nebulin, desmin, and a number of other minor components, which, as their name implies, provide structural support and may function in keeping the myofibril in alignment or register (Table 1). In food processing, where heat is normally applied, myofibrillar proteins play a major structural and functional role, i.e., they tend to interact with one another and with nonprotein ingredients both chemically and physically. These interactions contribute to product consistency as in the case of meat-based soups and gravies, or they can lead to the formation of certain structural components (e.g., gels and emulsions) essential to the stabilization of comminuted and restructured meat products.

Table 1 Some Characteristics of Myofibrillar Proteins from Vertebrate Skeletal Muscle

Protein	Location in myofibril	% of total myofibrillar protein	Molecular weight	Function	α-Helical content (%)	β-Sheet content (%)
Major contractile						
Myosin	A-band (thick filament)	43	500,000	Contraction	57	
Heavy chains (2)			220,000 each			
Light chains (4)			16,000–25,000 each			
Actin (G-form)	I-band (thin filament)	22	42,000	Contraction	<10	~20
Regulatory						
Tropomyosin	Thin filament	8	70,000	Regulates contraction	~100	
α-Chain			34,000			
β-Chain			36,000			
Troponin complex	Thin filament	5			29	20
Troponin-C			18,000	Ca^{2+}-binding		
Troponin-I			21,000	Inhibits myosin-actin interaction		
Troponin-T			35,000	Binds to tropomyosin		
α-Actinin	Z-disk	2	95,000	Cements thin filaments to Z-disk	48–75	
β-Actinin	Free end of thin filament	0.1	130,000	Regulates thin filament length		
γ-Actinin	Thin filament	0.1	35,000	Inhibits G-actin		
Eu-actinin	Z-disk	0.3	42,000	Z-disk density		
C-protein	Thick filament	2	140,000	Attaches to the shaft of thick filament	0	50
M-protein	M-line (center of A-band)	2	165,000	Binds to myosin in the thick filament	13	35
X-protein	Thick filament	0.2	152,000	Binds to myosin		
H-protein	Thick filament	0.18	69,000	Associated with C-protein and myosin	4	
Paratropomyosin	Edges of A-band	0.15	35,000	Possibly involved in postmortem changes		
Creatine kinase M-line		0.1	42,000	Binds to M-protein		
Cytoskeletal						
Titin (connectin)	Throughout sarcomere	8	>1,000,000	Holds thick filaments laterally; links thick filaments to Z-disks	0	0
Nebulin	N-lines	3	600,000	Binds and holds titin along thin filaments		
Desmin	Z-disk	0.18	550,000	Links neighboring Z-disks	45	
Filamin	Z-disk	0.1	230,000	?		
Vimentin	Z-disk	0.1	58,000	Links Z-disks in periphery		
Synemin	Z-disk	0.1	220,000	Association with desmin and vimentin		
Zeugmatin	Z-disk	0.1	550,000	Links thin filaments to Z-disks		

Source: Refs. 37, 56, 61, 124.

1. Myofibrils

Myofibrils of muscle cells are bathed in intracellular fluid, i.e., sarcoplasm, and account for about 80% of the muscle cell volume. They contain the basic structural units responsible for muscle contraction and relaxation in living animals. Contraction of myofibrils that takes place after death is responsible for the development of rigor mortis and meat toughness. Myofibrils are long, thin, and roughly cylindrical rods, 1–2 μm in diameter. Myofibrils extend the entire length of the muscle fiber (cell) which can be several centimeters long. Muscle fibers of meat animals with a diameter of 50 μm could contain as many as 2000 myofibrils [40]. Figure 2 is a diagram showing the ultrastructure of a myofibril from skeletal muscle.

The portion of the myofibril between two adjacent Z-disks (also referred to as Z-lines) is called a sarcomere. The length of the sarcomere varies; in postmortem muscle, it ranges between 1.5 and 2.5, depending on species and muscle types and subjecting to the extent of contraction. The sarcomere is the repeating structural unit of the myofibril, and it includes an A-band and two half I-bands located on both sides of the A-band. Each myofibril is composed of numerous so-called myofilaments. Two types of myofilaments, designated thick and thin filaments, constitute the myofibril. Thick filaments are approximately 14–16 nm in diameter and 1.5 μm in length, and they make up the A-band of the sarcomere. Since the most abundant protein in the thick filament is myosin (about 300 myosin molecules per thick filament), it is also referred to as the myosin filament. The center of the A-band contains a rod portion of myosin molecules without any of the pear-shaped heads. Thin filaments, which are also called actin filaments due to that protein's preponderance, are 6–8 nm in diameter and extend approximately 1.0 μm on either side of the Z-disk. Thin filaments constitute the I-band of the sarcomere and also extend beyond the I-band into the A-band. In muscle at a relaxed state, there is only a small amount of overlap between the thick and thin filaments. However, in severely contracted muscle, such as cold-shortened or thaw-rigor muscle, the two types of myofilaments could overlap by as much as 50%. The length of the A-band is not affected by contraction, but the I-band shortens as the muscle fiber contracts. The degree of A- and I-band overlapping not only affects meat tenderness, it also influences hydration of myofibrils, extraction of myofibrillar proteins, and, ultimately, water-holding capacity and juiciness of meat and meat products. In addition to the major protein constituents—myosin in thick filaments and actin in thin filaments—a number of minor structural and regulatory proteins are present in the myofibril. Among them are the M-line, H-, and C-proteins located in the thick filaments, and tropomyosin and the troponin complex associated with the thin filaments.

The third type of filament located in the myofibril exists as fine lines, which run parallel to both thick and thin filaments. They are referred to as gap filaments because they appear to bridge the gap in the center of the sarcomere and can be observed in highly stretched muscle with the electron microscope [43]. The main protein found in gap filaments is titin (sometimes called connectin), and it is located throughout the sarcomere. The physiological function of titin or gap filaments is not very clear, but it apparently holds the thick filaments laterally in the middle of the A-band. Titin can be broken down to smaller fragments by heating, however, these fragments are large enough to contribute structural strength to the gap filaments [36]. Hence, the gap filaments may be responsible for the "residual" toughness of cooked meat. Recent studies suggest that

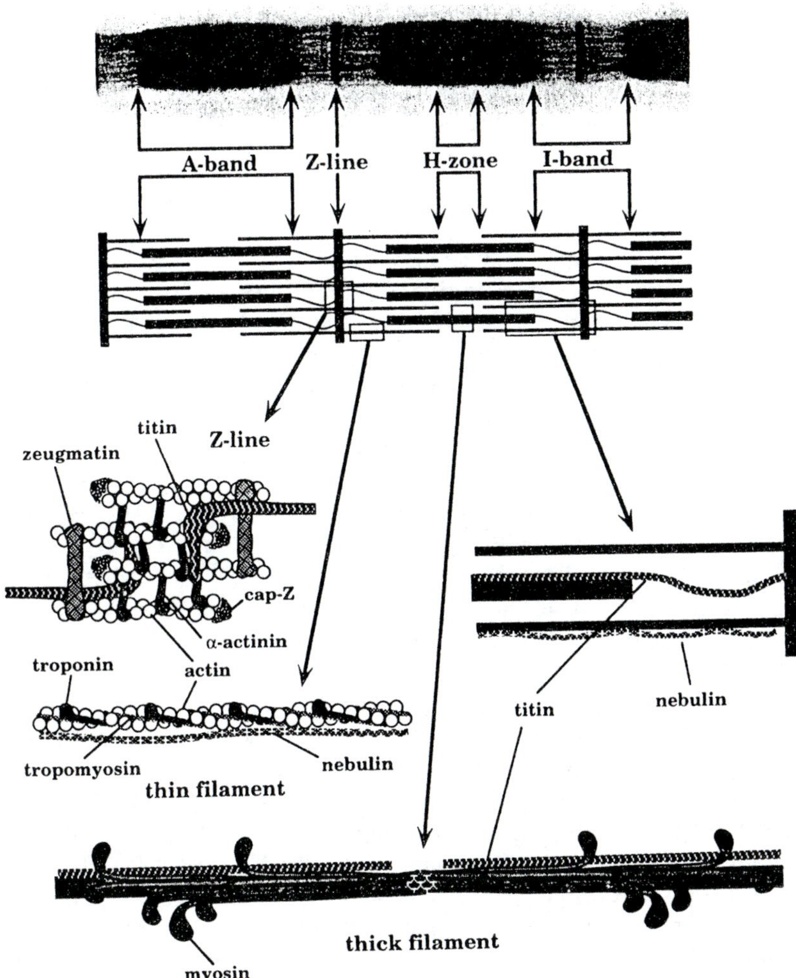

FIGURE 2 Ultrastructure and models of protein arrangement in the myofibril. The drawings illustrate the protein composition and distribution in the different regions of the myofibril. (From Ref. 90.)

gap filaments may have an implication in meat tenderization induced by postmortem aging [39].

The Z-disk possesses quite a complex ultrastructure. It is composed of Z-filaments that serve as bridges between thin filaments of adjacent sarcomeres. Each actin filament on both sides of the Z-disk is attached to four Z-filaments. Z-disks of skeletal muscle have a characteristic zigzag appearance, but the exact structure differs among species and fiber types. Fish muscle myofibrils exhibit only one layer of the zigzag filaments, whereas in *soleus* muscle of mammalian species, four layers of Z-lattice structure are typically seen (61). The width of the Z-disks in red (slow-twitch) and white (fast-twitch) muscle fibers differs considerably. In fast-twitch glycolytic muscle they appear as fine lines. In contrast, in slow-twitch oxidative muscle they appear as thick bands (15). The Z-disks are composed of about 10 proteins, with α-actinin being the predominant constituent. The Z-disks in white muscle fibers are very susceptible to postmortem proteolysis, beginning to lose their structures shortly after death. This postmortem structural changes may lead to improvement in extraction of actomyosin in salted meat [93].

2. Contractile Proteins

Myosin

Myosin is a large, fibrous protein with a molecular weight of about 500,000. It is the most abundant myofibrillar component, constituting approximately 43% of myofibrillar proteins in mammalian and avian muscle tissue [124]. Myosin consists of six polypeptide subunits, two large heavy chains and four light chains arranged into an asymmetrical molecule with two pear-shaped globular heads attached to a long α-helical rodlike tail (Fig. 3). The two globular heads are relatively hydrophobic, are able to bind actin, and exhibit ATPase activity. The rod portion is relatively hydrophilic and is responsible for the assembly of myosin into thick filaments. A myosin molecule is made up of approximately 4500 amino acid residues, of which 40 are cysteine [61]. The native myosin molecule is approximately 150 nm long with a diameter of about 8 nm in the globular region and 1.5–2.0 nm in the rod region. The helical portion of myosin has two hinge points, which make the protein flexible to bind with actin during contraction. Proteolytic treatment by trypsin cleaves myosin at the hinge region yielding two fragments called heavy meromyosin or HMM (head portion) and light meromyosin or LMM (tail portion), respectively. HMM retains all enzyme activity and actin-binding ability. Treatment of HMM with papain results in formation of two additional fragments termed S-1 (globular head) and S-2 (rod portion). The three-dimensional structure of chicken *pectoralis* myosin S-1 subfragment has been characterized using single-crystal x-ray diffraction [64]. The secondary structure of the S-1 subfragment is dominated by α-helices, with approximately 48% of the amino acid residues in this conformation. The isoelectric point of myosin is 5.3, thus, myosin is a negatively charged protein in muscle as well as in meat, whose pH typically ranges from 5.4 to 6.2. Compared with the head portion, which is relatively rich in hydrophobic amino acid residues, the rod portion contains a high proportion of charged side chain groups, such as argininyl, glutamyl, and lysyl residues.

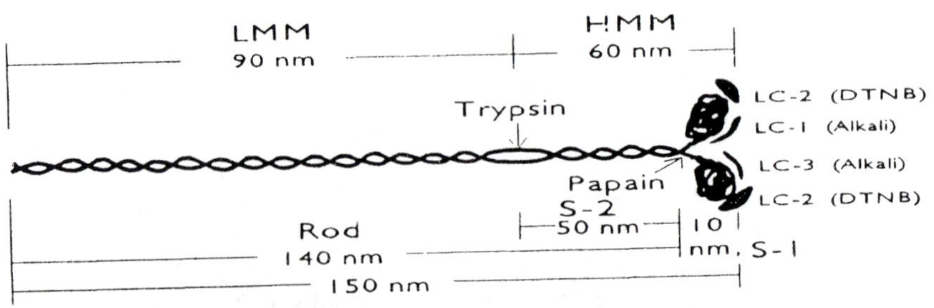

FIGURE 3 Schematic representation of the myosin molecule. Light meromyosin (LMM), heavy meromyosin (HMM), rod, S-1 and S-2 subfragments of HMM, the light chains, and the hinge regions susceptible to trypsin and papain are indicated.

Most (about 27) of the 40 sulfhydryl groups in myosin are located in the head region, and the rest (about 13) are contained in the rod [89]. Neither the head nor the rod portion contains a disulfide bond. In fact, all proteins in the myofibril family are deficient in disulfide bonds.

Myosin contains two identical heavy chains with a size of 220,000 daltons for each and two sets of light chains that range in size from 16,000 to 25,000 daltons, depending on species and fiber type (Table 1). The primary sequence of the myosin heavy chain and the gene encoding the protein have been identified [46,89]. One of light chains in each globular heads is required for ATPase activity, and the other light chain in the head regulates the enzyme action. The four light chains include one alkali 1 light chain, one alkali 2 light chain (so named because they are released from native myosin after treatment with alkali at pH 11) and two identical DTNB light chains [so named because they are released from native myosin after treatment with the sulfhydryl blocking agent 5,5′-dithiobis(2-nitrobenzoic acid)]. The same adult skeletal muscle may contain two or more species of myosin molecules that differ in their light chain composition. For example, one population of myosin may contain two DTNB light chains and two alkali 1 light chains attached to the heads of the heavy chains, while in another population, the two alkali 1 light chains may be substituted for by two alkali 2 light chains [61]. Both alkali 1 and 2 light chains are essential for enzyme function, and the DTNB light chains are regulatory. The two species of alkali chains are closely related but are apparently derived from two different genes. Two highly reactive sulfhydryl groups are present in the head region of myosin heavy chains. Modification of the sulfhydryl groups leads to alterations in ATPase activity, suggesting that they may have a structural or functional role in the enzyme activity of myosin. Each myosin heavy chain may also contain up to four methylated amino acids, including one monomethylysine, two trimethyllysine, and one 3-methyllysine. The exact content of methylated amino acids varies with muscle type and animal species.

Myosin is encoded by genes that apparently vary among species, between red and white muscles, and between skeletal and cardiac muscles, producing various isoforms. In chicken muscle, as many as 31 genes that encode myosin heavy chains have been identified [65]. At least four types of myosin are present in chicken muscle: two embryonic, one fast adult, and one slow adult. The fast and slow myosins differ slightly in amino acid composition. Myosin from white (fast-twitch) fibers contains two residues of 3-methylhistidine, a unique basic amino acid, but myosin from red and cardiac (slow-twitch) muscle is essentially devoid of it. Peptide mapping has revealed different amino acid sequences among the various myosin isoforms. Myosin heavy chains from type I slow red rabbit or bovine fibers have only one type of peptide. In contrast, peptide maps of fast myosin heavy chain demonstrate two different variants, which are substantially different from slow heavy chain and subtly different from each other [125]. The slow and the two fast forms of myosin heavy chain appear to arise from two different genes; however, it is not clear whether the two fast myosin heavy chains are derived from two separate genes or from the same gene. To date, at least nine distinct myosin heavy chain isoforms have been identified in adult mammalian skeletal muscle, including five fast (types IIA, IIB, and IIAB), two slow (type I), and two "developmental" heavy chains [62]. Type IIA heavy chain appears to be slower in electrophoretic mobility than type IIB heavy chain, while type I heavy chain appears to be slightly faster than both type IIA and type IIB heavy chains. Fast myosin heavy chains may also coexist with the heavy slow chain within the same fiber to produce type IIC or type IC fibers.

Adult mammalian and avian skeletal fast muscle contains two distinct alkali light chains (LCIf and LC3f, i.e., alkali 1 and 2 light chains, respectively) and a phosphorylatable light chain (LC2f, analogous to the DTNB light chain), while the slow muscle contains the alkali chain LC1s and a phosphorylatable light chain LC2s and is devoid of light chain 3. The LC1s and LC2s are similar to cardiac myosin light chains LC1v and LC2v. Unlike the heavy chain that is nearly constant in molecular weight among different myosin isoforms, each of the light chains varies considerably in molecular weight depending on fiber types. Electrophoretic analysis shows slower mobility of LC1s and LC2s or LC1v and LC2v than, respectively, LC1f and LC2f, suggesting that the slow light chains have higher molecular weights than the fast light chains [125]. Although a specific type of light chain is usually associated with the same type of heavy chain, coexisting fast and slow light chains have also been observed in fibers containing either fast or slow heavy chain. The existence of various isoforms of myosin heavy chains and light chains and the hexameric structure of myosin make it possible to form a theoretically large number of isomyosins in muscle fibers. Thus, it appears that myosin in both skeletal and cardiac muscles of meat animals is composed of different isozymic forms or isomers, and this may be responsible, to a large extent, for the disparities in physicochemical and functional properties observed between fast and slow and between skeletal and cardiac myofibrillar proteins in meat [107].

Actin

Actin is the second most abundant myofibrillar protein, constituting about 22% of the myofibrillar mass [124]. Actin can exist either as monomers (G-actin) or in a fibrous form (F-actin). In skeletal and cardiac muscle tissues, actin exists as double-helical filaments (F-actin) composed of polymerized globular monomers. Each monomer has a molecular weight of approximately 43,000 daltons. At least six different actin genes are expressed in mammalian and avian species (in skeletal, cardiac, vascular, and enteric tissues and in nonmuscle cells). Despite the existence of multiple actin genes, actin is relatively conserved with a sequence varying little between diverse muscle sources and between muscle fiber types.

Actin is the prevailing protein comprising the thin filaments in myofibrils. Each thin filament contains approximately 400 actin molecules. Actin in rabbit skeletal muscle is comprised of 376 amino acid residues, of which 4.9 and 7.5% are proline and glycine, respectively. The existence of relatively high proportions of both amino acids is probably responsible for the low percentage ($<10\%$) of α-helix structure and for the globular shape of the molecule. In comparison, the rod portion of myosin heavy chain, which is highly helical in structure, contains only 1.5% proline and 3.3% glycine. Each actin molecule contains five sulfhydryl groups and is free of the disulfide bond. In muscle tissue, actin is naturally associated with tropomyosin and the troponin complex. It also contains a myosin-binding site, which allows myosin to form temporary complexes with it during muscle contraction or the permanent myosin-actin complex during rigor mortis development in postmortem meat.

Actomyosin

Actomyosin is not an indigenous protein constituent in the myofibrillar assembly; it is a form of protein complex produced by muscle contraction. As its name implies, actomyosin is the complex of actin and myosin, which is formed after death when high-energy compounds such as ATP are depleted. Thus, rigor mortis or muscle stiffness is a direct result of cross-linking between myosin and actin filaments into the actomyosin complex. In actomyosin, myosin and actin are associated via noncovalent bonds, which can be easily split by high-energy compounds such as ATP or at high ionic strengths. Electrostatic interactions through phosphorous groups play a major role in stabilizing the actin-myosin complex. The formation of actomyosin is prohibited when the myosin-binding site on the actin surface is shielded by tropomyosin. There are two types of actomyosin complexes: one referred to as natural actomyosin, and the other reconstituted actomyosin. Natural actomyosin can be extracted from postrigor muscle and it may contain varying amounts of proteins naturally associated with myosin and actin (e.g., C-protein, tropomyosin, troponin, α-actinin). However, reconstituted actomyosin consists of only myosin and actin and can be prepared by mixing myosin and actin in vitro. The actomyosin complex is biochemically and physicochemically similar to myosin in many aspects. For instance, actomyosin retains most of the myosin ATPase activity and thermally induced gelation properties typically observed on myosin. However, structural changes in the myosin moiety brought about by cross-linking with actin do alter some physical and functional characteristics of myosin. Some functional properties of myosin, e.g., emulsifying capacity, are lost when actomyosin is formed [22]. In general, actomyosin does not exhibit any physicochemical or functional attributes characteristic of F-actin.

3. Regulatory Proteins

The major regulatory proteins are tropomyosin and troponin. They are located on the thin or actin filaments and represent approximately 8 and 5% of the total myofibrillar proteins, respectively (Table 1). In addition, myofibrils contain a number of minor regulatory proteins, including several recently discovered ones, which are believed to be involved in the regulation of filamentous structure of myofibrils. These proteins are distributed in different parts of the myofilaments, e.g., A-bands, I-bands, and Z-disks. Among them are α-, β-, and γ-actinins, C-, M-, H- and X-proteins, creatine kinase, and paratropomyosin (Table 1). While the role of some of these proteins is more or less understood, the precise function of many of them in living tissues is not clear. There is an even greater void in our knowledge about the functionality of these regulatory proteins in meat, particularly their possible contributions to processed muscle foods. Several of these minor regulatory proteins may have a potential involvement in functional behavior of muscle foods. Hence, they are included in the following discussion.

Tropomyosin

As the most abundant regulatory protein in the myofibrillar protein family, tropomyosin is a dimeric molecule consisting of two dissimilar subunits designated α- and β-tropomyosin with molecular weights of about 34,000 and 36,000 daltons, respectively. The α and β subunits can combine to form three possible dimers, i.e., $\alpha\alpha$, $\beta\beta$, and $\alpha\beta$. The distribution of the three dimers varies in different muscles. The $\alpha\alpha$ homodimer prevails in fast white skeletal muscle, the $\alpha\beta$ heteromer has a predominance in slow, red, and mixed muscles, and the $\beta\beta$ homodimer is exclusively found in cardiac muscle [8]. Tropomyosin is a filamentous molecule composed of a coiled coil of two α-helices, each approximately 40 nm in length. The α-helical structure (virtually 100%) throughout the molecule is ostensibly related to the lack of proline in the molecule. Under physiological conditions, it binds to F-actin (at a ratio of 1 tropomyosin:7 G-actin) and troponin-T (1:1) stoichiometrically and regulates myosin ATPase activity. Tropomyosin occupies the peripherical position on the actin double helical-filament, and it shields the myosin-binding sites on actin filament in the relaxed state of muscle. During contraction, which involves calcium, tropomyosin moves to the F-actin groove, thus allowing myosin to bind to the actin filament. At high ionic strengths similar to those in meat processing (e.g., 0.6 M KCl or NaCl), tropomyosin is dissociated from actin. Aqueous solutions of tropomyosin are highly viscous due to its ability to form long end-to-end filaments by ionic interactions. The viscosity is decreased by the addition of salt and increased by the addition of troponin.

Troponin

Troponin is a complex protein comprised of three subunits designated C, I, and T for their ability to bind calcium, to inhibit contraction, and to bind with tropomyosin, respectively. It is the second most abundant regulatory protein in the myofibril and regulates muscle contraction-relaxation based on its interaction with calcium. There is roughly equimolar stoichiometry for these three subunits, but the exact molar ratio varies with species and muscle type.

Similar to tropomyosin, troponin does not directly participate in cross-bridge formation, but it plays an indirect role in the contraction-relaxation cycle. Through cooperative interactions between its three subunits in the presence of calcium, troponin regulates muscle contraction in a very unique manner. Excessive toughness of meat induced by cold-temperature storage or freeze-thaw of prerigor muscle results from an accumulation of calcium ions in cytosol, particularly in red fibers whose sarcoplasmic reticulum is of a porous structure. By binding to high concentrations of calcium, the troponin complex undergoes conformational changes, which ultimately trigger supercontraction of prerigor muscle, resulting in marked increases of meat toughness [13].

Troponin I inhibits the contractile interaction between myosin and actin and myosin ATPase in the presence of tropomyosin. Troponin I is a slightly basic polypeptide (pI 5.5), consisting of 179 amino acid residues with a molecular weight of 20,864 daltons in rabbit skeletal muscle and 21,137 daltons in chicken skeletal muscle [4]. The peptide sequences of rabbit skeletal and cardiac troponin I are nearly identical, but cardiac troponin I has about 20 additional amino acid residues at the N-terminal end [56].

Troponin C is a highly soluble acidic (pI 4.1) protein owing to the large content of aspartic and glutamic acid residues. Troponin C in rabbit skeletal muscle has a molecular weight of 17,800 daltons, and it contains 159 amino acids [61]. A molecule of troponin C contains four calcium-binding sites, two of which have a high affinity for calcium and the other two a low affinity for calcium. Each calcium-binding site is composed of a loop made up about 12 amino acids and surrounded by two α-helical segments. The two high-affinity sites can also bind with magnesium. Thus, the two low-affinity sites, which are exclusively for calcium, appear to control the regulatory function of the protein [56].

Troponin T regulates the contraction process by binding to tropomyosin and other troponin subunits, C and I, on the actin filaments. Troponin T is the largest subunit in the troponin complex. Troponin T in rabbit skeletal muscle consists of 259 amino acids with a molecular weight of 30,503 daltons, and in chicken breast and leg muscle it consists of 287 and 263 amino acids with molecular weights of 33,500 and 30500 daltons, respectively [4]. The N-terminal fragment of troponin T, designated troponin T_1, binds solely and strongly to tropomyosin, while the C-terminal segment, referred to as troponin T_2, interacts with troponin I and troponin C [56]. The event of muscle contraction can be described in terms of cascade reactions. On binding to calcium ions ($>10^{-6}$ M), the conformation of troponin C is changed and the contraction cycle is initiated. The binding results in stronger interactions between the three troponin subunits, C, I, and T, which in turn leads to a stronger binding between troponin T and tropomyosin (which shields the myosin binding site on the actin filament) and disengagement of troponin I from actin. Unlike the vast majority of myofibrillar proteins, troponin T is extremely susceptible to enzymatic breakdown during postmortem aging. Its degradation products have been linked to a 30,000-dalton polypeptide, which emerges in postmortem muscle.

α-Actinin

The major protein found in the Z-disk is α-actinin. Accounting for only 2% of total myofibrillar protein, α-actinin nevertheless plays important regulatory and structural roles in the myofibrils. α-Actinin has a molecular weight of about 95,000 daltons as estimated using SDS-gel electrophoresis. Located in the Z-disk, α-actinin appears to make up the Z-filaments in striated muscle and anchors the thin filaments from opposing sarcomere [61]. In addition, α-actinin presumably regulates growth of the thin filaments, and even modifies the structure of actin. α-Actinin contains a large amount (ca. 75%) of α-helical structure and a relatively high percentage of charged residues. Postmortem aging of meat results in a partial release of α-actinin, especially in fast-twitch fibers, leading to the disruption of the Z-disk and an improvement in meat tenderness [39,93].

C-Protein

The C-protein is associated with myosin in the thick filaments and is an impurity in crude myosin preparations. It is localized into 7–11 bands, each 43 nm apart, which surround the middle of each half of the bipolar thick filaments [61]. There are about 37 C-protein molecules per thick filament; thus, each C-protein band contains three to five molecules. C-protein is relatively high in proline content (7.1%), which explains why the protein contains little or no α-helical structures. The size of the protein varies with species and muscle type and has a molecular weight of 135,000, 145,000, and 150,000 daltons for white, red and cardiac muscles, respectively. C-protein has an elongated ellipsoid shape and is about 35 nm in length. The function of C-protein is not fully understood, although it has been shown to bind to the light meromyosin region of the myosin tail by attaching to the surface of the thick filament shaft [61]. C-protein could provide a mechanism for regulating the interaction and movement of the actomyosin cross-bridges during contraction. There is also speculation that, by forming bands around the myosin filament shaft, C-protein may provide a structural role in the myosin filament. C-protein is largely extracted in salt solution at concentrations above 0.6 M NaCl [55]. However, it is not clear whether its removal contributes to the concomitant extraction of myosin and actin.

M-Line Proteins

Three proteins—myomesin, M-protein, and creatine kinase—are found in the M-line region. Both myomesin and M-protein are integral parts of the M-line with a molecular weight of 185,000 daltons for myomesin and 165,000 daltons for M-protein. Myomesin is tightly bound to the M-line in both skeletal and cardiac myofibrils. It is absent from smooth muscle. M-protein is a single peptide chain with 13% α-helical and 35% β-structure [4]. Myomesin and M-protein may be functional components of myofibrils and may play an important role in providing a scaffold that aids in the assembly and maintenance of the myofibrils as well as the bipolar orientation of the myosin filaments in striated muscle fibers. Transmission electron microscopy has revealed two structural components in the M-line: M-filaments and M-bridges [38]. The M-filaments run parallel to the myofibrils, while the M-bridges are perpendicular to the myofibrils and connect neighboring myosin filaments. Thus, M-bridges may play a role in keeping myosin filaments in register both transversely and longitudinally. Some evidence indicates that creatine kinase is the major constituent of the M-bridges involved in energy regeneration in living tissues.

H- and X-Proteins

These proteins are so named because they correspond to the bands labeled H and X on SDS-gel electrophoresis of crude myosin [86]. Both proteins are associated with myosin at discrete sites on the surface of the thick filaments, and they can overlap with the thin filaments on both ends of the A-band. X-protein is bound to C-protein; thus, it may be a part of the C-protein bands surrounding the myosin filaments. X-protein is present in slow red fibers and absent in fast white fibers. H- and X-proteins account for only a small percentage (0.18 and 0.20%, respectively) of total myofibrillar protein. H-protein is a relatively small protein with a molecular weight of about 69,000 daltons, compared to X-protein which has a molecular weight of 152,000 daltons [86]. Like C-protein, both H- and X-proteins contain a high proportion of polar amino acid residues and, thus, are not hydrophobic. The exact function of H- and X-proteins is not known. However, both proteins are thought to have an enzymic activity, function as regulatory proteins, and participate in the assembly and stabilization of the thick filaments.

4. Cytoskeletal Proteins

By the mid-twentieth century, only myosin, actin, and a number of regulatory proteins had been discovered, some of which were recognized for their important functions in biological systems as well as their influence on meat quality. However, research in the past two decades has led to the identification and characterization of many novel myofibrillar components. Most of the newly discovered proteins function in maintaining the cytoskeletal structure and, thus, are referred to as cytoskeletal proteins. As a group, they appear to provide support and stabilization of the contractile and regulatory proteins either longitudinally or laterally and, therefore, are also called "scaffold proteins." The major proteins in this group include titin (connectin) and nebulin, along with several minor polypeptides such as desmin, filamin, and zeugmatin (Fig. 2; Table 1).

Titin

The thick and thin filaments of the myofibrils are overlapped by a high molecular weight protein that was first isolated and named connectin by Maruyama et al. [47]. This protein was later purified by Wang et al. [98], who named it titin owing to its titantic size. It is now recognized that titin and connectin are the same protein except that titin is free of nebulin and perhaps some other minor contaminants in the connectin preparation. Recent studies have indicated that titin is the major constituent of the fine filaments located in the gap between the actin and myosin filaments, and they can be seen in highly stretched muscle fibers. For this reason, the filaments have been referred to as "gap filaments" [43]. Gap filaments are also called "intermediate filaments" due to their average diameter (100 Å) compared to the actin (60 Å) and myosin (150 Å) filaments. Gap filaments apparently bind to both the Z-disks and the thick filaments. Thus, titin may provide a mechanism to link the thick filament to the Z-disk and interdigitates the thick filaments into the sarcomere. It has also been suggested that titin filaments serve as templates for organization of the thick and thin filaments. The arrangement of titin in the myofibrils indicates that it overrides essentially the entire length of the sarcomere.

Titin is the third most abundant myofibrillar component, accounting for 8–10% of the total myofibrillar protein. It is present in both skeletal and cardiac muscle, but exists in smooth muscle only in small quantities. In SDS-gel electrophoresis, titin appears as a doublet band with an estimated molecular weight of 1 million daltons. However, due to its extremely large size, the exact molecular weight of titin is difficult to determine. It is generally believed that titin is devoid of secondary structures including α-helix. Nevertheless, morphologically, titin resembles elastin and collagen, forming a highly elastic network of very thin filaments. The titin filaments appear to be flexible, and they are assembled together probably by covalent cross-links through lysine derivatives. The physical and chemical characteristics of native titin molecule are poorly understood at the present time, because titin is usually denatured or partially hydrolyzed by proteases during isolation and purification. Titin is susceptible to proteolysis in meat during the aging process. It can be degraded by muscle endogenous enzymes such as calpains and carboxyproteases [39]. The degradation is relatively slow at refrigerator temperatures and rapid during cooking when the temperature reaches 50°C or higher. The contribution of titin to meat toughness has been subjected to much debate, and its exact role ramains obscure.

Nebulin

This protein was initially identified as a contaminant of connectin preparation. Nebulin has been identified histologically as thin, nebulous, continuous transverse arrays, called N_2-lines, near the boundary of the A-I zones and on each side of the Z-disks [61]. In fact, nebulin is the only protein that has been identified to date in the seven N-lines, which include one N_1-line, four N_2-lines, and two N_3-lines. Like titin, nebulin has a high molecular weight estimated to be around 600,000 daltons, and comprises about 5% of the total myofibrillar protein. The physiological function of nebulin is not clear. However, present in the A-I junction, nubulin appears to be involved in steering the actin filaments into a favorable alignment so as to optimize their interactions with the thick filaments [90]. Furthermore, nebulin may be attached to titin in the gap filament, thereby providing structural and regulatory roles in the myofibril assemblage. Nebulin is rapidly degraded in postmortem meat in the initial aging period and is particularly susceptible to calpain. Thus, its possible involvement in meat tenderization during aging has been speculated [93].

Desmin

Like titin and nebulin, desmin is another newly discovered protein with a molecular weight of 55,000 daltons. Although it comprises only 0.18% myofibrillar protein in skeletal muscle, it has received considerable attention due to its ostensible structural role in the orientation and arrangement of myofibrils. Desmin is arranged in a filamentous form around the periphery of the Z-disk, in contrast to α-actinin, which is an integral part the Z-disk. These filaments occupy the space between adjacent myofibrils where they tie the myofibrils into the cytoskeleton of the muscle cell. The content of desmin in smooth muscle (as much as 5%) is much higher than in skeletal muscle, and its role in binding muscle fibers together in the Z-disk region would seem to be even more important. Desmin is highly susceptible to postmortem proteolysis, and its degradation rate is comparable to that of troponin T. Since the disruption of the Z-disks, particularly in fast-twitch fibers, is one of the most noticeable postmortem changes in meat, it seems likely that proteolysis of desmin is involved in meat tenderization during postmortem aging [93].

C. Stromal Proteins

The interstitial space of muscle cells contains three extracellular proteins—collagen, reticulin and elastin—and the supporting ground substance. Collectively, these proteins and proteinaceous substances are called stromal proteins or connective tissue proteins since they make up tissues connecting the muscle cells. The extracellular proteins around the muscle fibers consist mainly of fine reticular and collagenous fibrils, which constitute the endomysium layer, as well as perimysium, which surrounds fiber bundles, and epimysium, which encases the whole muscle. Elastin is mainly associated with blood vessels, capillaries, and nerve systems. Therefore, only small amounts of elastin are generally found in meat. Belonging to the stromal protein group are also those insoluble proteins that are constituents of membranes in many intracellular organelles. However, compared to connective tissue proteins, membranal proteins are of minute quantities, and their contributions to meat quality seem to be negligible.

1. Collagen

A glycoprotein, collagen is the main component of the intercellular tissue of muscle. At least 11 genetic varients of collagen have been identified and characterized [61]. Based on its macromolecular structures, collagen can be divided into three major groups: (a) striated fibrous collagen, which includes types I, II and III collagen, (b) nonfibrous collagen, which contains type IV or basement membrane collagen, and (c) microfibrillar collagen, which encompasses types VI and VII (the matrix micrifibrils), types V, IX, and X (the pericellular collagen), and types VIII and XI, which are yet unclassified. Type I is the most prevalent type and is found as the main collagenous component in skin, tendon, and bone of meat animals. Type II collagen is the major component in cartilage, and type III collagen has a preponderance in vascular tissues, skin, and intestine. The collagens that comprise the intercellular matrix are predominantly types I, II, and III. Types IV and V collagen form fine networks in the basement membranes surrounding the muscle cell.

Each collagen molecule is composed of three polypeptide chains, designated α-chains, which form a triple-helical structure stabilized by H-bonds (Fig. 4). The helices differ from the typical α-helix due to the abundance of proline and hydroxyproline, which interfere with the α-helical structure. The makeup of collagen molecules can be envisioned to be somewhat analogous to triglycerides except that the three fatty acid acyl chains in triglycerides are attached to the glycerol base. The tropocollagen monomers, in turn, constitute the collagen fibrils mainly via the hydrogen bonds. Different α-chains, designated $\alpha 1$, $\alpha 2$, or $\alpha 3$, within the same type of collagen differ in their amino acid composition. The distribution of $\alpha 1$-, $\alpha 2$-, and $\alpha 3$-chains in collagen molecules varies depending on the specific genetic variants. For instance, type I collagen consists of two identical $\alpha 1$(I)- and one $\alpha 2$(I)-chain, type II collagen is composed of three homogeneous

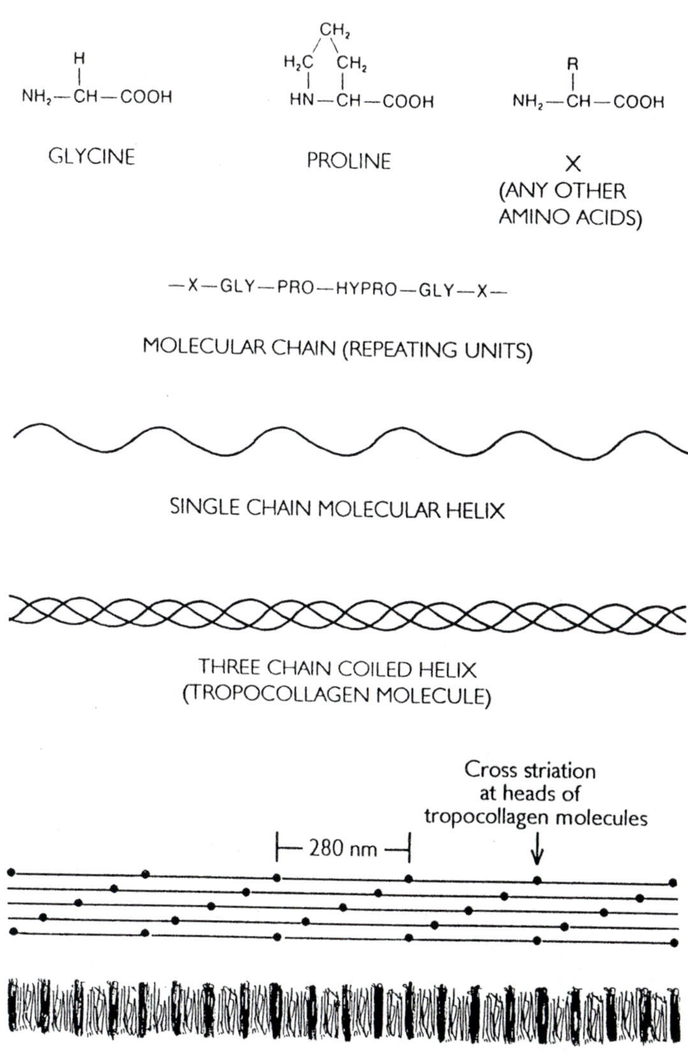

FIGURE 4 Schematic representation of the amino acid sequence and molecular structure of collagen and tropocollagen molecules and of collagen fibril formation.

α1(II)-chains, while type VI collagen is made up of three heterogeneous chains, i.e., one α1(VI)-, one α2(VI)-, and one α3(VI)-chain [61,79].

A newly synthesized procollagen molecule (α-chain) contains peptide extensions at both the N-terminal and the C-terminal with sizes of about 20,000 and 35,000 daltons, respectively. The peptide extensions, which are of a globular structure, are required for the initiation of tropocollagen fibril synthesis, but they are cleaved by limited proteolysis before the helical fibrils are formed in the extracellular space. A procollagen molecule has as many as five distinct domains, which are located in the N- and C-terminal regions and along the long helical portion. Both N- and C-terminal ends may contain oligosaccharides rich in mannose. Collagen is the longest of all protein molecules known to date. Its tropocollagen monomers are approximately 300 nm long and 1.5 nm in diameter. Each of the α-chains is composed of more than 1000 amino acid residues with a molecular weight of about 100,000 daltons, yielding a total molecular weight of about 300,000 daltons for the tropocollagen molecule. The α-chains have a left-handed helix with three amino acid residues per turn, but their trimer (tropocollagen) is a right-handed superhelix. Another unique feature about collagen is that each collagen molecule contains about 33% glycine, 12% proline, and 11% hydroxyproline and is devoid of tryptophan. The central triple-helical region of a collagen molecule is composed of segments of tripeptides with a repeating unit of -Gly-Pro-X- or -Gly-X-Hyp-. The distribution of polar and nonpolar residues in the X position determines the ordered aggregation of molecules into fibrils. The regions composed mainly of ionic amino acids appear as dark bands when viewed under an electron microscope, and they attribute to the striated pattern of the collagen fibrils.

The three layers of connective tissue in muscle—endomysium, perimysium, and epimysium—are made up of different types of collagens. Type I collagen is the major component of the epimysium, both type I and type III are present in the perimysium in large amounts, whereas types II, IV, and V are the primary species in epimysium [79]. These connective tissues collectively impart toughness to meat. Since the intermuscular epimysium tissue is usually trimmed from cooked meat, the contribution of types II, IV, and V collagen to meat toughness would be small compared to type I or type III collagen. Another important meat toughness factor, besides the collagen content, is cross-links formed between α-chains in the tropocollagen molecule and between the tropocollagen monomers that make up the collagen fibrils. When two α-chains are cross-linked, the dimer is referred to as β-component; when three α-chains are cross-linked, the trimer is called the γ-component; and so on. In general, α-chain monomers are salt-soluble, and their covalently linked dimers (β-component) and trimers (γ-component) are acid- or alkali-soluble. The cross-linking involves oxidative deamination of lysine residues, resulting in the formation of acid- and heat-labile dehydrohydroxyl lysinonorlucine and heat-stable hydroxylysino-5-keto-norlucine bonds, and probably some other covalent linkages such as the dehydrohistidino-hydroxymerodesmosine bond, all of which are reducible [40]. Cross-links in collagen increase with an increase in animal age. Furthermore, as the animal age increases, the collagen cross-links are converted from a reducible form to a more stable, nonreducible form. The nature of the nonreducible cross-links is not yet clear. However, the different amounts of nonreducible cross-links may partially explain why meat from young animals, e.g., veal from calves, is more tender than meat from more mature animals, e.g., beef from cows or bulls. The extent of cross-links, as well as the exact composition and structure of collagen, are also dictated by animal species, breed, sex, and nutritional status [40].

2. Elastin and Reticulin

These two stromal proteins are found only in small quantities in muscle relative to collagen. However, since they are also constituents of the connective tissue and may play a role in imparting toughness to meat, they are discussed below.

Elastin

As its name implies, elastin is a protein with rubberlike elastic properties. Elastin is the amorphous component in the elastic fibers and has a molecular weight of about 70,000 daltons. Elastin normally exists in a contracted state, but it can extend to twice its contracted length. Elastin fibers are found in tissues that are involved in continuous deformation such as ligamentum nuchae, in tendon, and in the walls of large arteries. Elastin is also present in inter- and intramuscular connective tissues, particularly in *semitendinosus* muscle. Thus, it may have an implication in meat toughness. Because of their yellow appearance, elastic tissues are referred to as "yellow connective tissue." Morphologically and chemically, elastin fibers are divided into two classes: fibrous and amorphous. The fibrous component, which is rich in polar amino acids and contains a number of disulfides, forms a network, and subsequently, the amorphous component (i.e., elastin), which is mainly composed of nonpolar amino acids and has a molecular weight of about 70,000 daltons, is deposited.

The amino acid composition of elastin is similar to that of collagen. Each elastin molecule contains about 33% glycine, 10–13% proline, and 40% hydrophobic amino acids [61]. However, unlike collagen, which has a regular helical structure, elastin is composed of kinetically free random coil networks. Thus, the molecular structure and conformation of elastic and collagen are distinctly different. Elastin also contains two unusual amino acidxs, desmosine and isodesmosine, which are formed by cross-linking lysine residues. The combination of great hydrophobicity and cross-linking renders elastin highly insoluble and unusually stable. For this reason, it is unlikely that elastin plays any significant functional role in meat or processed meat products. A soluble polypeptide, named tropoelastin, has also been isolated from the elastin fibers of aortas of copper-deficient animals [61]. Compared with the regular, insoluble elastin, tropoelastin is devoid of cross-links and has an elevated lysine content. However, due to its anatomical location in the animal as well as its anomaly, it should not have any significant involvement in meat quality.

Reticulin

Reticulin is a mucoprotein. It contains a considerable amount of lipid rich in myristic acid. Similar to collagen, it forms fine, wavy fibers in the endomysium layer of muscle. In the early days, reticulin was thought to be a separate protein, but today, reticulin is believed to be a special form of collagen. This is due to many factors, including the fact that reticulin and collagen fibers share many physical and chemical characteristics, e.g., ultrastructural periodicity and cross-links between fibrils.

III. PHYSICOCHEMICAL CHANGES IN MUSCLE PROTEINS DURING MEAT PROCESSING

One of the most important criteria used to determine the usefulness of a food protein or a protein isolate is its functional performance in food processing. This is with no exception for muscle proteins. Aside from their high nutritive value, muscle proteins as a whole exhibit excellent functional properties, as manifested by their ability to form viscoelastic networks, to bind water, to entrap flavors, to emulsify fat and oil, and to form deformable, yet stable, foams. The relationships between physicochemical and functional properties of proteins have long been recognized. A muscle protein in its native structure does not normally exhibit any functional characteristic, although its functional behaviors can be related to its native structures, particularly the primary and secondary structures. Instead, "intermediate" structures, i.e., those involved in the transition from the native form to the denatured state during meat processing, seem to be more important and bear a more direct relationships with the protein functionality. Thus, protein functionality is generally induced by external factors (i.e., processing conditions) that can cause considerable structural and conformational changes as well as modifications of other physicochemical attributes of the protein. Examples of external factors that can bring about protein changes are energy input through thermal (heat) and mechanical (shear) means and alterations of the physical and chemical environment (e.g., pH, ionic strength, redox potential). Furthermore, a variety of food ingredients and additives, including lipids, polysaccharides, nonmuscle proteins, antioxidants, minerals, and divalent cations, can interact with muscle proteins, leading to changes in protein structures and conformation, as well as their susceptibility to heat and acid/alkali solutions. These changes, in turn, can result in major alterations in muscle protein functionality.

A. Myofibril Dissociation

In meat processing, where 2–3% salt is typically incorporated in the product formulation, muscle fibers and proteins undergo major structural changes due to electrostatic interactions between proteins and both sodium and chloride ions. Salt-induced structural changes in myofibrils include (a) swelling of myofibrils caused by charge repulsion between myofilaments, which can be seen with high-magnification microscopes, (b) depolymerization of myofilaments particularly the thick filaments, and (c) dissociation of actin from myosin or actomyosin from the myofibril structure, leading to the extraction of myosin, actin, the actomyosin complex, and a number of other myofibrillar constituents [55,60].

Fiber swelling is an important physicochemical process involved in processing of raw meat prior to cooking, and it is responsible for increased hydration and water-holding capacity of processed muscle foods. Myofibrils start to swell at an ionic strength of about 0.4 (~0.4 M NaCl), which can be viewed using phase contrast microscopy. Maximum swelling, accompanied by marked protein extraction and loss of myofibril structure, occurs at 1.0 M NaCl. The disintegration of myofibrils is facilitated by comminution, mixing, tumbling, and massaging of meat and greatly affected by the type and amount of ingredients or additives added to meat. Pyrophosphate, a functional ingredient widely used in meat products, promotes myofibril swelling, disengagement of the thick and thin filaments, and thus, myosin extraction, probably by dissociating the actomyosin cross-bridges. Pyrophosphate also changes the pattern of myofibril extraction. In the absence of pyrophosphate, swelling is accompanied by protein extraction from the middle of the A-band. However, in the presence of pyrophosphate, the A-band is completely extracted, beginning from its end, i.e., the junction of thick and thin filaments [55]. The action of pyrophosphate is somewhat similar to that of ATP, a high-energy phosphonucleotide involved in muscle contraction. Although pyrophosphate accelerates myofibril swelling by reducing the salt concentration required to cause swelling, it does not seem to change the maximal swelling. The dissociation effect by pyrophosphate often requires magnesium, a cofactor for the hydrolytic enzyme pyrophosphatase, which is present in meat. However, some studies have shown that magnesium is not always needed for pyrophosphate to exert its effect [4]. Tripolyphosphate also promotes myofibril swelling and protein extraction in a manner similar to pyrophosphate. However, in meat, tripolyphosphate is thought to be initially hydrolyzed to pyrophosphate before it becomes functionally effective. Furthermore, the contractile state of muscle is of great importance and can influence the extent of myofibril dissociation and protein extraction. Thus, myosin and actin in prerigor, noncontracted muscle can be readily extracted, and this is the basis for myosin and actin preparations for muscle biochemical studies. On the other hand, actomyosin, the major protein species in contracted muscle, is extracted from postrigor muscle or myofibrils.

Muscle fibers contain some structural constraints, which seem to restrict myofibril swelling and protein extraction. As indicated previously, one of the possible constraints is cross-bridges formed between myosin and actin filaments in postrigor meat [55]. Moreover, several other transverse structural components of myofibrils, including M- and Z-lines, are also thought to exert restraining forces to inhibit myofibril swelling and myosin or actomyosin extraction [55]. Xiong and Brekke [113] have noted a marked increase in the total amount of protein extracted from 24-hour postmortem chicken *pectoralis* muscle when compared to muscle at death, coinciding with major structural changes in the Z-lines presumably caused by endogenous proteases. In a more recent study, Hultin et al. [30] noted an increase in myosin and actin extraction from myofibrils after C-protein, X-protein, α-actinin, and three or four other unidentified proteins were removed. Hence, it appears that myofibril swelling and the extraction of contractile proteins are closely influenced by some structural constituents of the myofibril. Because a number of myofibrillar proteins (e.g., myosin, C-protein, and α-actinin) are present in muscle in various isoforms, which differ subtly in structure and amino acid composition, the swelling of myofibrils and extraction of constituent proteins from fast- and slow-twitch myofibrils may be different. Evidence indicates a common existence of the dependence of myofibril swelling and protein extraction on fiber types among different species, including bovine, rabbit [60], and avian (unpublished results) muscles. In the presence of 0.4–0.6 M NaCl, chicken *pectoralis* (white) myofibrils exhibit more extensive structural alteration and greater protein extraction and are more sensitive to pyrophosphate than chicken *gastrocnemius* (red) myofibrils.

B. Thermal Denaturation

Muscle proteins undergo remarkable structural and conformational changes during meat processing, which are necessary for producing desirable functionalities. Processing conditions, particularly heating, promote interactions among protein molecules as well as between proteins and other meat components, including lipids, polysaccharides, minerals, and additives. Myosin, actin, or their complex form actomyosin are destabilized by increasing salt concentrations. The energy absorbed by proteins during thermal denaturation can be measured using deferential scanning calorimetry. Heat flow during the process of protein structure unfolding is manifested by endothermic transitions, each of which may involve dissociation of noncovalent bonds in a particular structural domain or dissociation of myofilaments and protein subunits. Myosin is a multidomain protein; the domains unfold independently of each other [63]. The number of endothermic transitions for myosin varies with species, salt concentration or ionic environment, and buffer conditions. The transitions range from one at 0.1 M KCl, two at an ionic strength of 0.05 M KCl, to three or more at 0.5 M or higher concentrations of KCl or NaCl (Table

2). In processed meats, a 2% (~0.5 M) or higher concentration of salt is typically added to solubilize myosin or actomyosin for functionality. At elevated ionic strengths, myosin is less stable due to disruption of intramolecular electrostatic forces that confer protein conformational stability. Chicken breast myosin suspended in 0.6 M NaCl at pH 6.5 exhibits four cooperative (i.e., independent) endothermic transitions [99]. The maximum transition temperatures (T_m, defined as the temperature at which 50% of the domain is unfolded) are 49, 50, 57, and 67°C. By deconvolution, the transitions can be further separated into 10 independent transition peaks at 40, 44, 46, 48, 51, 54, 57, 62, 63, and 67°C. These transitions almost exactly match the transitions of myosin subfragments, i.e., LMM, rod, the heavy chains, the S-1 and S-2 subunits, HMM, and light chains, and hence, have been ascribed to the denaturation of these subfragments or domains [80]. The addition of salt also lowers the enthalpies and denaturation temperatures in myosin [35]. The multitransition phenomenon associated with myosin and its major subfragment rod is a result of complexity in the protein structure. When the protein structure is less complex, fewer transitions are produced. This is the case for myosin light chains and its S1 subfragment, actin, and myoglobin, which show only one endothermic transition during heat-induced denaturation [6,24,35,118].

Stability and thermal denaturation of sarcoplasmic, stromal, and other myofibrillar proteins under the pH and ionic conditions of meat processing have not been adequately studied. Purified sperm whale myoglobin has a single transition at 78°C [24]. The "exudate" (sarcoplasmic trip) of chicken breast muscle shows four transitions with T_ms ranging between 58 and 72°C [118]. Actin, which exhibits a sharp transition at 78°C at ionic strengths of about 0.15, is destabilized by added salt both in T_m and in apparent enthalpy (ΔH_{app}) of denaturation [35].

C. Aggregation

Aggregation of proteins results from the association of protein molecules, usually in a large scale. Protein aggregation can be measured with a number of analytical tools, ranging from sensitive transmission electron microscopy after either negative staining or rotary shadowing or light scattering spectrophotometry, to less sensitive sedimentation or solubility determinations. Unlike protein oligomers or polymers, protein aggregates are usually much larger in size, i.e., they are of a macroscopic scale and are able to scatter incident light. Aggregation is a dynamic process that may involve both association and dissociation reactions. Aggregation of muscle proteins can occur when protein surface charges approach a zero value or neutrality, e.g., at a pH equaling the isoelectric point of the protein. In this type of aggregate protein molecules are loosely associated with one another, and once the solubility constraints are removed, e.g., by either increasing or decreasing the pH or salt concentration, the aggregates will be dissociated again. In fact, this is the basis on which myofibrillar proteins, which are soluble in salt solutions but not in water, are prepared. Association of myofibrillar proteins probably begins prior to the heating process. Morita et al. [49] reported that in 0.6 M KCl solution at pH 5.4 or 5.7, chicken white myosin monomers cross-linked to produce long filaments. In contrast, red myosin formed short, discrete rods. White myosin filaments were highly intermingled, forming a continuous lattice, while red myosin filaments exhibited a lower degree of cross-linkages.

Protein aggregation occurring during thermal processing of meat is a much more complex physicochemical process. Here, aggregation results from interactions between

TABLE 2 Thermal Denaturation of Selected Muscle Proteins Under Conditions Close to Those in Meat Processing

		Condition			Transition temp. (°C)					
Protein	Species	pH	Γ or M	Heating rate (°C/min)	T_{m1}	T_{m2}	T_{m3}	T_{m4}	ΔH_{cal} (J/g)	Ref.
Myosin	Rabbit	6.0	0.96 M KCl	10	44	51	60		14.9	103
	Chicken breast	6.5	0.60 M NaCl	1	49	50	57	67	19.4	100
	Chicken breast	6.5	0.6 M NaCl	1	48	54	57	63	—	80
	Rabbit	7.0	0.50 M KCl	1.5	43	46	49	54	15.0	6
Myosin heavy chains	Chicken breast	6.5	0.6 M NaCl	1	46	54	64		—	80
LMM	Rabbit	6.0	0.96 M KCl	10	42	61			19.0	103
	Chicken breast	6.5	0.6 M NaCl	1	39	52			—	80
HMM	Rabbit	6.0	0.96 M KCl	10	48	58			15.2	103
	Chicken breast	6.5	0.6 M NaCl	1	48	55	59	63	—	80
Myosin rod	Rabbit	6.0	0.96 M KCl	10	44	61			19.6	103
	Rabbit	6.4	0.50 M KCl	1	42	49	56		17.8	7
	Chicken breast	6.5	0.6 M NaCl	1	45	50	56	63	—	80
	Rabbit	7.0	0.50 M KCl	1	47	57	60		19.8	44
	Rabbit	7.0	0.50 M KCl	1	43	54			20.2	6
S-1	Rabbit	6.0	0.96 M KCl	10	48				9.7	103
	Chicken breast	6.5	0.6 M NaCl	1	47				—	80
	Rabbit	7.0	0.50 M KCl	1.5	46				9.3	6
S-2	Chicken breast	6.5	0.6 M NaCl	1	47	51	54		—	80
	Rabbit	6.6	0.60 M KCl	1	41	51			8.4	91
	Rabbit	7.0	0.50 M KCl	1	45	56			8.6	44
Myosin light chains	Chicken breast	6.5	0.6 M NaCl	1	48	58			—	80
	Rabbit	7.0	0.50 M KCl	1	51				4.6	6
Myoglobin	Sperm whale	?	0 M NaCl	10	78				7.6	24

protein molecules via exposed reactive groups or amino acid side chains, and the aggregates, once formed, generally cannot be dissociated by physical means, i.e., heat-induced protein aggregation is an irreversible process. Protein aggregation is a prerequisite step for producing a number of structural components, including gels and interfacial membranes that stabilize emulsions and foams [34,110]. Protein aggregates are maintained by the same type of forces that stabilize the native structures of proteins. Thus, muscle proteins can associate with one another through hydrophobic, electrostatic, and hydrogen bonds, Van der Waals interactions, and disulfide linkages. However, the relative contributions of each type of bond are different in aggregated proteins than in the native proteins.

The process of protein aggregation and physical properties of the resulting aggregates are dictated by the process of protein denaturation and the environmental factors. In most cases, slow heating permits progressive, sequential unfolding of different protein structural domains and leads to ordered protein-protein interactions essential to the formation of highly viscoelastic protein gels. This explains why meat products manufactured under slow-cooking conditions usually possess finer texture and are able to retain a larger amount of water compared to cooking at a fast rate. The structure of muscle proteins is of particular importance. For a large molecule like myosin, denaturation of the least stable domain, i.e., the S-1 subgroup, enables myosin molecules to associate with one another via head-head interactions possibly through the disulfide bonds. This initial step is believed to be critical to the development of ordered myosin gel matrix because interactions of uncoiled tail (rod) at a higher temperature can follow [78]. Xiong and Brekke [114] observed two transitions in thermal aggregation of chicken *pectoralis* salt-soluble myofibrillar proteins at pH 5.5 and three transitions at pH 6.0 and 6.5. For chicken leg muscle proteins, however, only two transitions were noted at these pH values. Each of the transitions most likely represents molecular association due to unfolding of different domains of myosin with actin and other minor myofibrillar components playing a minor role.

IV. FUNCTIONALITY OF MUSCLE PROTEINS

Protein functionality can be defined as any physicochemical property of protein that allows protein molecules to interact among themselves and with their environment to produce or improve the quality and stability of final products. In meat processing, protein functionality is usually described in terms of hydration, surface properties, binding, and rheological behavior. The ability to bind water, to solubilize or disperse in solution, and to form gels and emulsions are some of the most important functional properties of muscle proteins in meat processing.

A. Water Retention

The ability of meat and meat products to retain moisture before, during, and after processing or cooking plays a crucial role in palatability and consumer acceptance of the product and is usually described in terms of "water-holding capacity." This term is loosely defined in the literature, but it generally refers to the ability of food to retain its natural as well as added water during processing. Physicochemically, the water in meat is present in either the bound or the free state. The bound water is tightly associated with proteins through charged groups and dipolar sites on the protein surface. Hence, its amount in meat is primarily influenced by the amino acid composition of proteins. The free water is held by capillary and surface tension forces and is highly dependent upon the protein structure. Of the three groups of muscle proteins, myofibrillar proteins are responsible for much of the water-holding capacity of meat because of both their structural organization, which is conducive to water retention, and their quantity predominance in muscle.

Myofibrils are composed of 20% protein and 80% water, and they occupy about 80% of cellular space, i.e., the volume of lean meat [40]. Therefore, the majority of water in meat is confined within the myofibrils in the spaces between the thick (myosin) and thin (actin) filaments. There are two major types of forces that contribute to water retention in meat: polarity, including surface charges, and capillary effects. Binding of water to the surface of protein through hydrogen bonds between water molecules and charged and dipolar amino acid residues seems to be insignificant for water retention in meat. The capillary forces, however, are believed to account for most of the water contained in meat. A small part of free water is also present in the extracellular space by capillary forces. The interfilamental spacing, measuring about 320 Å between the thick filaments, is not constant but varies with intrinsic factors, such as fiber type, sarcomere length, and contractile state, and with extrinsic factors, such as pH, ionic strength, osmotic pressure, and the presence or absence of certain divalent cations and polyphosphates. An increase in the electrostatic charges would lead to an increase in the repulsive force between myofilaments and, thus, increases in myofibril swelling and water-holding capacity. Hence, any change in the surroundings of myofibrils that results in increased protein charges or dipoles (e.g., high concentrations of salt and pH away from the protein isoelectric points) would lead to increased water retention in meat.

Water-holding capacity of meat is influenced by the structure of myofibrils. The formation of cross-bridges between myosin and actin in the rigor state is believed to constrain myofibril swelling and, thus, water-holding ability [55]. Certain structural components of the myofibril, such as the Z-disks that connect the actin filaments and the M-lines that link myosin filaments together, probably also provide constraints to swelling. Some other structural myofibrillar constituents, including C-protein, may play a role in regulating water imbibing by myofibrils. Thus, the removal of many of the scaffold proteins, as occurs in postmortem muscle due to proteolysis, can lead to improved water-holding capacity in meat.

B. Solubility

Solubility of proteins is of a primary importance for the manufacture of processed muscle foods, including comminuted, restructured, and formed meats. This is because most functional properties of muscle proteins are related to protein solubility, and, in fact, some are achieved only when the proteins are in a highly soluble state. For instance, protein solubility is a prerequisite step for gelation, emulsification, binding, and some other functional processes of the protein. Solubility of muscle proteins is a function of protein structures, the structure of myofibrils, pH, concentration (ionic strength) of salt added to meat, temperature, time of mixing meat with salt (i.e., time of protein extraction), and many other intrinsic and processing factors. Sarcoplasmic proteins are naturally soluble in muscle. However, solubilization of myofibrillar proteins generally requires relatively high ionic strength ($\Gamma > 0.4$). Thus, it normally occurs as a result of comminution and blending of meat in the presence of salt. Solubility can be defined as the amount of the total muscle protein (%) that goes into solution under specified conditions and is nonsedimentable by a specified centrifugal force. Thus, solubility at saturation represents an equilibrium between the solute (protein) and solvent (water). Unfortunately, there exists no standardized, universally accepted method for determination of protein solubility, and therefore, "solubility" will be an operational definition, subject to variations in protein concentration, centrifugal force, and centrifugation time. Hence, relative differences in protein solubility seen between two muscle samples in one study may not be observed in another study due to differing centrifugation conditions used. The term "solubility" used for protein should also be treated with caution since under certain conditions "solubilized" proteins can exist in a colloidal form and dispersed in solution, rather than as separated molecules in a true solution that does not scatter light. Thus, terms such as "suspendability" and "dispersibility" are sometimes adopted in literature for muscle proteins.

In the meat science literature, the terms "solubility" and "extractability" are frequently used interchangeably, assuming that once a protein is solubilized, it can be readily extracted from muscle fibers or myofibrils with a proper buffer. However, exceptions exist. For instance, proteins that are bound to or confined in cellular components may not be extracted. A number of glycolytic enzymes and many mitochondrial proteins are examples of soluble, yet nonextractable muscle proteins. Most published solubility studies on myofibrillar proteins have been performed in systems where the aqueous phase is much more abundant than it is in meat. Although the results may not be quantitatively applicable to processed meats, they are nevertheless useful for describing the hydration behavior of muscle proteins under the influence of various processing factors.

The structure-solubility relationship of muscle proteins is well known. Most proteins in the "sarcoplasmic protein" family are of a globular structures and relatively small in size (most have molecular weights between 30,000 and 65,000 daltons) [37,108]. They exist as individuals, although many of them are involved in the same metabolic processes. Most sarcoplasmic proteins possess a single cooperative domain, which is stabilized by intramolecular hydrophobic and hydrogen bonds. The surface of protein molecules is made up of charged and noncharged polar groups, and their isoelectric points are close to neutral pH [37]. Such a distribution of amino acids and the tertiary structural organization permit the sarcoplasmic proteins to freely interact with surrounding water, i.e., the proteins are highly hydrophilic and soluble in water or dilute salt solutions. On the other hand, myofibrillar proteins in muscle are associated with one another in a highly organized, integral structural unit, i.e., the myofibril. Thus, although most myofibrillar proteins have a balanced amino acid composition, their isoelectric points are relatively low, with most falling in the pH 5–6 range [37,61]. The low isoelectric pH is related to protein structural arrangement and the tendency to interact between different segments of the polypeptides. Thus, the solubility of myofibrillar proteins under the physiological condition or low ionic strengths is negligible.

Protein solubility is highly dependent on the ionic strength of the extraction buffer. Extraction of myofibrillar proteins begin at an ionic strength close to 0.5, and it reaches a maximum at ionic strength 1.0 [55]. Thus, an increase in salt (NaCl) concentration to above 0.5 M (approximately 2% salt in meat), as widely used in processed meats, would allow solubilization and extraction of myofibrillar proteins necessary for producing desirable product functionalities. Protein extraction is responsible for the dissociation of thick and thin filaments of the myofibril and the ultimate disappearance of the A-band.

It was not very clear until recently that, in fact, most myofibrillar proteins are also soluble in water or extremely dilute salt solutions [83,87]. The solubility is extremely sensitive to salt concentration between ionic strength 0.0003 and 0.001. This is because a minute quantity of ionic compounds is sufficient to increase protein surface charges, thereby augmenting protein-water interactions. For cod muscle myofibrillar proteins, minimal solubility is established between an ionic strength of 0.01 and 0.2. Protein solubility increases rapidly as ionic strength increases from 0.2 to 1.0, and as the ionic strength further increases, the solubility declines presumably due to the "salt-out" effect. The most critical factors controlling the solubility and extractability of myofibrillar proteins appear to be certain physical and chemical constraints that prevent the dissociation of the myofibril structure. This is true for solubilization of proteins in either water or elevated salt concentrations. Examples of possible constraining factors are the cross-bridges formed between myosin and actin in postrigor muscle and the presence of some structural proteins, e.g., M-line proteins, desmin, and N_2-line proteins that are thought to contribute to myofibril integrity by keeping myofibrils in lateral and longitudinal register [55]. Thus, means that can cause disruption and removal of physical constraints in myofibrils would result in increased protein extraction and solubility. In meat processing, this is usually accomplished by the addition of salt and pyrophosphate to meat formulations, increases in pH of the extraction buffer or of the meat ingredient solution, and prolonged mixing and tumbling time. The effect of pyrophosphate and tripolyphosphate on myofobrillar protein solubility is remarkable. At pH 5.5, maximum structural change (i.e., swelling) occurs in 0.8 M NaCl when no sodium pyrophosphate is present and in 0.4 M NaCl when 10 mM pyrophosphate is present [55]. Protein extraction occurs concomitantly with structural changes in myofibrils; however, the pattern of extraction differs with or without pyrophosphate. In the absence of pyrophosphate, myosin extraction begins at the center of the A-band, but in its presence, extraction begins at both ends of the A-band. Because the actomyosin cross-bridges are absent at the center of the A-band but located at the ends of the A-band, it can be surmised that pyrophosphate functions as a lubrication agents, i.e., it behaves like ATP, whose hydrolysis leads to dissociation of the actomyosin complex and, thus, separation of the thick and thin filaments, or myosin and actin (52). Sodium tripolyphosphate has a very similar effect to sodium pyrophosphate, and its is generally thought that tripolyphosphate is hydrolyzed to pyrophosphate before it becomes functionally active.

The amount of extractable myofibrillar protein is generally believed to be greater for prerigor muscle than for postrigor muscle due to the difference in the contraction state. For this reason, myosin samples are usually prepared from prerigor muscle. However, there are exceptions. Xiong and Brekke [116] have found that myofibrillar proteins from postrigor (24 hours postmortem) chicken breast muscle are more soluble and much more readily extracted than from prerigor chicken breast muscle in a buffer containing 0.6 M NaCl at pH 6.0. In contrast, the proteins in chicken leg muscle are more extractable than those in postrigor chicken leg muscle. This discrepancy may be explained by the fact that in chicken breast muscle, which is composed of a preponderance of fast-twitch white fibers, some possible solubility constraints, including proteins in the Z-disks, are removed by muscle endogenous proteases, presumably calpain, during rigor mortis development. In contrast, the Z-line structure in chicken leg muscle changed very little during the first 1 or 2 days of postmortem storage [25]. Structural differences between white and red muscle fibers ostensibly contribute to the different degrees of protein extraction. The extractability of proteins from myofibrils isolated from cold-shortened (by 29%) bovine sternomemdipularis muscle has also been found to be comparable to those from essentially uncontracted muscle (shortened by only 5%) [109], suggesting that the number of actomyosin cross-bridges may not be critical to protein extraction. Therefore, caution should be taken when studying protein solubility because more than one solubility-limiting factor can be involved for a specified experiment condition. The amount of extracted myofibrillar protein also increases with time of extraction, indicating that the disruption of muscle cells and dissociation of myofibrils is a slow process.

Solubility variations between postrigor fast-twitch (white) and slow-twitch (red) muscle proteins are widely observed phenomena. Protein solubility of both white (23.4% solubility at time 0) and red (18.7% solubility at time 0) myofibrils increases, but at different rates, during incubation in 0.6 M NaCl at pH 6.0, indicating that dissolution of myofibrils in different fiber types follows different time-dependent processes [116]. For both muscle types, an almost linear increase in protein solubility is evident within the initial 2 hours of extraction. However, protein extraction during this period is much faster for white (10% solubility increase/hr) than for red (4.3% solubility increase/hr) myofibrillar sample. Maximum solubility is essentially attained within 12 hours for both white (57.7%) and red (37.8%) myofibrillar proteins. Furthermore, pH has a marked influence on protein solubility. Postrigor chicken white myofibrillar protein exhibits a greater solubility than postrigor chicken red myofibrillar protein at pH >5.9, and a lower solubility at pH <5.8. Prerigor white myofibrillar protein is less soluble than postrigor white myofibrillar protein at pH >5.8 and than either prerigor or postrigor red myofibrillar protein within the pH 5.5–6.5 range (Fig. 5). The degree of protein extraction is also affected by divalent cations present in muscle. The effect of calcium and magnesium is quite complex and differs for white and red muscle proteins. The soluble fraction of white myofibrillar protein increases in the presence of 2.5–5 mM $CaCl_2$ or $MgCl_2$, but de-

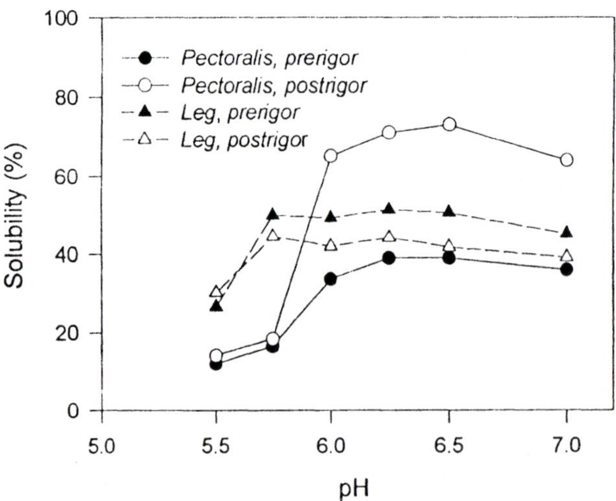

FIGURE 5 Effect of pH of extraction buffer and muscle rigor state on extractability of proteins from chicken breast and leg muscle myofibrils suspended (5 mg/ml) in 0.6 M NaCl, 50 mM sodium phosphate at various pH. (From Ref. 116.)

creases upon further addition of divalent salts. However, the amount of soluble red myofibrillar protein reaches a maximum at 5 mM $CaCl_2$ or $MgCl_2$, with no further increase above 5 mM of either of the divalent salts [117]. The presence of various fast and slow isoforms of calcium- and magnesium-binding protiens (e.g., myosin light chains, troponin) may be part of the reason for the fiber type–dependent solubility differences.

These results indicate that the solubility of myofibrillar protein from white muscle is more sensitive to pH and divalent cations than protein from red muscle. Biochemically, myosin from fast muscles exhibits an alkaline stability, while myosin from slow muscles is relatively stable in acid. Morita et al. [49] have witnessed the formation of longer filaments from chicken fast myosin than from chicken slow myosin in 0.6 M KCl at pH 5.4 and 5.7. This might explain why myofibrillar protein from chicken breast is less soluble or more sedimentable than that from leg at low pH [116]. Three fiber types—slow, fast, and intermediate (type I, type IIA, and type IIB, or β-red, α-red, and α-white, respectively)—and a number of their subgroups are present in chicken thigh muscle, with the slow type being prevalent. In contrast, breast (*pectoralis*) muscle mainly contains fast-twitch white fibers (type IIB) and, to a very minor extent, slow-twitch and intermediate fibers [67].

Parsons and Knight [60] have monitored protein extraction of white and red mammalian myofibrils using phase contrast microscopy. In 0.45 M NaCl plus 10 mM pyrophosphate, myofibrils of rabbit *psoas* muscle (which contains only fast-twitch fibers) are readily extracted. In contrast, myofibrils of rabbit *soleus* and bovine *masseter* muscles (which are composed purely of slow-twitch fibers) are resistant to extraction, requiring higher salt concentrations to depolymerize the myosin filaments. Antibody (antimyosin) binding test on myofibrillar extracts from rabbit *plantaris* muscle (containing a mixture of several fiber types) further confirms that myosin from fast-twitch myofibrils is more readily extracted than myosin from slow-twitch fibers.

The discrepancy in myofibrillar protein solubility between white and red muscles may well be related to two major factors. First, proteins located in the myofibrils must overcome structural barriers and hindrances. Histologically, red myofibrils contain wider Z-bands than white myofibrils, and their major structural protein, α-actinin, has different morphology (isoforms) for red and white myofibrils [77]. Offer and Trinick [55] have shown that weakening and expansion of the Z-bands (from rabbit *psoas* muscle) in salt solutions are accompanied by increased extraction of not only the thin (actin) filaments, but also the thick (myosin) filaments. Furthermore, the existence of various isoforms of several other structural proteins (M-protein, C-protein, H-protein, and X-protein) in different fiber types would also seem to contribute to variations between red and white myofibrils during extraction by affecting the stability of the thick filaments. Evidence suggests that the M-protein provides transverse structural constraints to myosin extraction and is crucial for extraction of total myofibrillar protein. The C-protein and X-protein seem to be less important for myosin and actomyosin extraction. It is thus most likely that the different rates and degrees of myofibrillar protein extraction from red and white fibers are related to the ultrastructures of the Z-bands and the polymorphism of α-actinin and certain subsidiary structural proteins including M-protein. Second, myosin, the most prevalent protein component inside the muscle, exists in various isoforms that are fiber type–specific. Different myosin isoforms are known to differ in physicochemical characteristics, morphology, and solubility. Hence, even if the thick filaments are depolymerized, the different solubilities of white and red myosin are another decisive factor that can influence the total amount of soluble protein in white and red muscles. It is not clear, however, whether myosin from white muscle has a different isoelectric point than myosin from red muscle.

C. Viscosity

Rheological properties as related to flow and deformation are important functional attributes of muscle proteins. The rheological behavior of a protein suspension in muscle foods is often described in terms of viscosity. Viscosity of the aqueous protein phase can influence texture and stability, as well as handling, of meat batters. Proteins are charged polymers capable of binding water and causing fiber swelling by the uptake of water and loosening of the polypeptide matrix. As a consequence of swelling, a protein increases its effective hydrodynamic volume, thereby increasing resistance to shear. The flow behavior of a fluid can be characterized by the stress response of the solution to an imposed shear rate. For Newtonian flow, there is a linear relationship between shear stress and shear rate. The term "dynamic viscosity," defined as the ratio of shear stress to shear rate, is used to describe flow characteristics. However, proteins are large solute molecules, and in solution they generally exhibit non-Newtonian pseudoplastic characteristics. The exact flow behavior can be complex, depending on the intrinsic attributes of the protein, such as molecular size, volume, shape, surface charge and ease of deformation, as well as extrinsic factors, such as pH, temperature, ionic strength, ion type and shear rate. Thus, the term "apparent viscosity" is usually used in the literature to define the flow behavior of muscle protein suspensions.

The unique structure of myosin, i.e., the large length-to-diameter ratio of the rod portion, makes myosin highly viscous in salt solution. The intrinsic viscosity of myosin is 215 cm^3/g [61]. Because of its great viscosity and abundance in muscle, myosin is the major contributor to the viscosity and "thickness" of the aqueous extract in salted meat. Thus, the tacky extract, commonly referred to as "sol," is comprised of salt-soluble myofibrillar proteins as well as sarcoplasmic proteins. However, the viscosity of sarcoplasmic proteins is much less than that of myofibrillar proteins at equal protein concentration, and therefore, their rheological role in the "sol" of salted meat is relatively small compared to myofibrillar proteins.

Studies of rheological properties of chicken myofibrillar proteins have shown a greater viscosity of the white (breast) proteins than of the red (leg) proteins [113]. Viscosity (74 cPa·s) of the white myofibril suspension (2% protein in 0.6 M NaCl) increases rapidly during storage, reaching a maximum of 130 cPa·s within 8 hours. However, viscosity (20 cPa·s) of the red myofibril suspension increases more slowly, reaching a maximum of about 70 cPa·s within 24 hours. The unequal changes in solubility of the two types of myofibrillar proteins during storage are unable to fully account for the disparity since viscosity differences between the two muscle types are also evident for the soluble fraction (i.e., salt-soluble myofibrillar protein or SSP) and viscosity of SSP from both muscle types does not change during storage. Both white and red SSP suspensions in 0.6 M NaCl show pseudoplastic flow, i.e., the rate of the shear stress increase drops as the shear rate increases (Fig. 6). This flow behavior is assumed to result from the dissociation of loosely bound protein aggregates that can form during cold storage and from increased alignment of protein molecules at high shear rates. Shear stress or viscosity of SSP at 0–5°C is time-independent, which means that at a constant shear rate, shear stress of the protein solution does not change with time of shear. However,

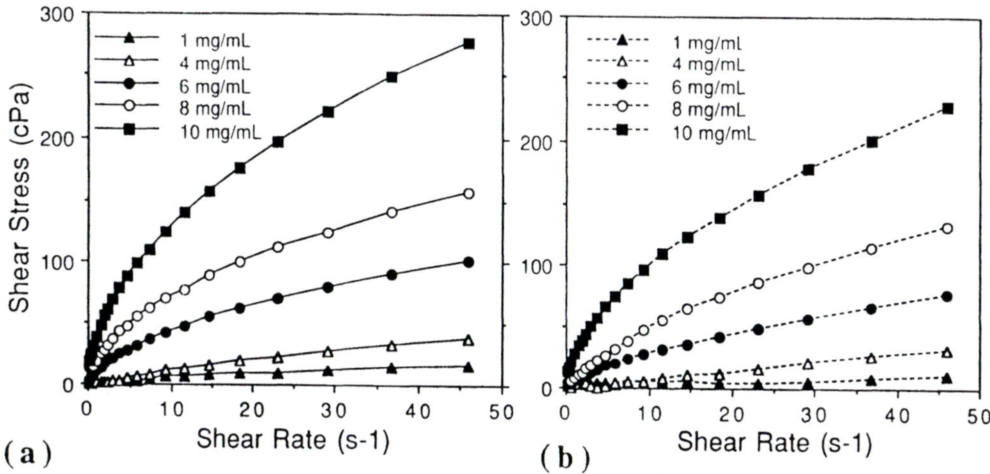

FIGURE 6 Relationships between shear stress and shear rate for chicken (a) *pectoralis* major and (b) thigh salt-soluble protein suspensions (0.6 M NaCl, 50 mM sodium phosphate, pH 6.2) at 5°C. (From Ref. 111.)

time-dependent shear thinning behavior ("thixotropic" flow) has been reported in fish surimi incubated at 35 and 40°C [105], probably due to the disruption of loose networks formed from partially denatured myofibrillar proteins. With equaling shear rates (0.23–46 s^{-1}) imposed, identical protein concentrations (1–10 mg/ml), and pH (5.75–8.0), chicken breast SSP always exhibits greater shear stress and viscosity values than chicken thigh SSP (Fig. 6). The viscous flow of SSP suspensions can be fit with the power law equation as shown below:

$$\tau = k\dot{\gamma}^n$$

where τ (Nm^{-2} or pascal) is shear stress, k (Ns^nm^{-2}) is the consistency index, $\dot{\gamma}$ (s^{-1}) is the shear rate, and n (dimensionless) is the power law index. In the 4-10 mg/ml protein concentration range, the k value is always greater for breast SSP (3.2–53.9 Ns^nm^{-2}) than for thigh SSP (2.6–34.4 Ns^nm^{-2}), indicating that at equal protein concentrations, white muscle proteins are more viscous than red muscle proteins [111]. The SSP suspensions do not relax instantaneously when the imposed shear is stopped. Instead, stress decay takes about 0.3 seconds. Thus, it is assumed that SSP suspensions possess some kind of "structural" feature that can arise due to association among myosin or actomyosin molecules during cold storage. This observation is also consistent with the findings that muscle SSP lacks Newtonian flow. Asghar et al. [3] and Morita et al. [49] noted a greater viscosity of chicken white myosin than red myosin within the pH range 5.4–6.0, suggesting that this discrepancy was probably due to more favored filamentogenesis in white myosin than in red myosin.

Thermal scan analyses show that viscosity differences between white and red myofibrillar proteins depend on the temperature. As the temperature increases from 5 to approximately 15°C, the viscosity of the two types of protein converges, and further heating results in transformation of the viscous sol to a more elastic gel showing a peak stress at 41.7 and 46.1°C for the white and red proteins, respectively [111]. Viscosity differences have also been observed between white and red myofibrils from porcine muscles. The SSP extracted from *longissimus dorsi* muscle (75% white fibers) generally has a greater viscosity (shear stress) than SSP extracted from *serratus ventralis* muscle (43% red fiber) [66], thus further substantiating that intrinsic factors exist between white and red fiber types that produce rheological variations in myofibrillar proteins. The polymorphism phenomenon of myosin and the ability of white myosin to form longer filaments than red myosin may explain, in part, the viscosity differences between the white and red myofibrillar proteins.

D. Gelation

1. General Considerations

Gelation of proteins is a thermodynamic process that occurs widely in food processing. Structurally, a gel is a form of matter intermediate between a solid and a liquid, consisting of strands or chains cross-linked to create a continuous network immersed in a liquid medium [92]. Rheologically, a gel is a substantially diluted system that exhibits no steady flow [17]. Thus, collectively, a gel has been referred to as a continuous network of macroscopic dimensions immersed in a liquid medium and exhibiting no steady-state flow [128]. Examples of protein-based gels include egg custard, gelatin gelly, tofu, and frankfurters. The importance of protein gelation to muscle foods was illustrated in several early studies [21,45] that demonstrated that gels formed from salt-extracted myofibrillar proteins at the junction of meat particles were responsible for the meat binding and texture of cooked sausage products. Numerous later studies have led to the findings that, in addition to binding, protein gels also contribute to emulsion stability, water-holding capacity, and tenderness of many processed meats. For review, see Asghar et al. [4], Ziegler and Foegeding [128], and Xiong [107].

Research on muscle protein gelation has so far been largely confined to the effect of various processing factors on the thermodynamics of gel formation and viscoelastic properties of formed gels. Few studies have attempted to establish a direct relationship between muscle protein structures and physical properties of gels. Difficulties in characterizing structure-gelation relationships exist due to the fact that muscle proteins, many of which are quite complex polymers, undergo major structural changes during meat processing, and these changes usually are highly sensitive to even subtle variations in processing parameters (e.g., pH, heating rate, ionic strength, and ion species), making them difficult to control and predict. However, through exhaustive fundamental studies, it is now recognized that sarcoplasmic proteins, which are relatively simple and of a globular structure, have poor gelling ability and contribute very little to meat binding. On the other hand, myofibrillar proteins, particularly myosin and actomyosin, which are composed of multiple cooperative domains, form highly viscoelastic and rigid gels and are responsible for formation of a "bind" between meat particles in processed meats. Despite the difficulty in precisely predicting gel properties from the protein native structure, many of the gel properties nevertheless can be related to the structure of the "intermediates," i.e., denatured proteins and their polymers or aggregates. This is because protein gelation is a multistage process involving the following steps:

$$\chi P_N \xrightarrow[\text{(denaturation)}]{\text{heating}} \chi P_D \xrightarrow[\text{(aggregation)}]{\text{heating}} (P_D)_{\phi,\psi,\ldots} \xrightarrow[\text{(cross-linking)}]{\text{heating/cooling}} (P_D)_{\phi}\text{-}(P_D)_{\psi}\text{-} \ldots$$

where χ is the total number of protein molecules, ϕ and ψ ($\phi + \psi + \ldots = \chi$) are the number of molecules aggregated at a certain point of the gelation process, P_N is the native protein, and P_D is the denatured protein. The initial denaturation of protein into uncoiled polypeptides is followed by association to form aggregates or strands. When aggregation reaches a certain critical point, a gel with infinite interpeptide cross-linkages and a three-dimensional network structure is created. The final gel state corresponds to aggregates of partly or fully denatured proteins. It should be noted, however, that the above gelation model is applicable to proteins that possess relatively simple structures. For proteins that contain multiple domains of differing thermal stabilities, the actual mechanism of gelation can be much more complicated.

2. Myosin Gelation at Low Ionic Strengths

Myosin can be extracted from prerigor muscle with relatively high–ionic strength ($\Gamma > 0.5$) solutions. The ability of myosin to form a gel and the viscoelastic characteristics of the gel are strongly dependent on the ionic environment. Most studies have found that myosin gels formed at low ionic strengths (0.1–0.3 M KCl, achieved either by dilution or dialysis) are of a higher rigidity (shear modulus) than myosin gels produced at higher ionic strengths (>0.4 M KCl) [16,27,31]. Myosin gels formed at low ionic strengths also possess fine texture and are relatively translucent compared to a coarsely aggregated structure of myosin gels formed at high ionic strengths. The discrepancy is attributed to the fact that low ionic strengths favor the formation of myosin filaments (filamentogenesis) in solution, whereas at high ionic strengths, myosin depolymerize into monomers. In 0.1 M KCl at pH 6.0, myosin gel rigidity is directly proportionate to its filament length formed prior to heating [122]. In this ionic environment, the myosin filaments appear to be heat-stable, namely, they do not dissociate into monomers upon heating. Changes in myosin filaments during heating are confined within the filament architecture and do not involve their length. Without heating, the filaments appear smooth and thin, but when heated to 40 and 60°C, the filaments become rough, larger in diameter, and more entangled, apparently due to structure unfolding in the constituent myosin "residues."

The specific morphology of myosin filaments produced at low ionic strengths depends on a number of factors, including purity of myosin preparation, pH, the presence of ATP and Mg^{2+}, and the speed of lowering the ionic strength. The last factor especially affects the length of the filaments. Short myosin filaments can be obtained by rapid dilution; thus, an array of filaments of various lengths (0.5–3.0μm) can be prepared by controlling the rate of lowering the salt concentration close to the physiological ionic strength [122]. On the other hand, long filaments can be created by slowly lowering the ionic strength, which can be achieved through dialysis. Short myosin filaments formed by simple dilution aggregate randomly and produce weak gels with a coarse, porous structure. In contrast, long myosin filaments formed by dialysis tend to associate side by side to form bundles, and upon heating, they produce more rigid and elastic gels with a fine strandlike network structure. Some minor structural proteins naturally associated with myosin in the thick filaments are difficult to remove during myosin preparation, and these proteins can affect myosin gelation by influencing myosin filamentogenesis. For example, C-protein was found to reduce the diameters of myosin filaments, although it did not affect the filament length. As a result, rigidity of thermally induced myosin filament gels decreased [121]. It is known that C-protein disrupts the regularity of the synthetic myosin filament structure as well as the longitudinal order [48]. G-actin, which does not gel by itself, interferes with myosin gelation in low-salt solutions. Rigidity of a 4.5 mg/ml myosin gel formed in 0.2 M KCl at pH 6.0 decreases fourfold as the myosin-to-actin ratio increases from 100:0 to 67:33 (w/w) [33]. Since the actin effect diminishes in the presence of ATP and it is known that actin binds to sites on the heads of the myosin molecules in low–ionic strength conditions, it appears that gelation of myosin (filaments) results from interfilamental aggregation via head-head interactions of myosin on the surface of the filaments.

The ability of myosin to gel in dilute salt solution varies with muscle fiber types. Most published literature has shown that myosin from fast-twitch glycolytic type IIB (white) fibers forms more rigid and viscoelastic gels than myosin from slow-twitch oxidative type I (red) fibers whether from avian or mammalian species under identical environmental conditions [107]. For both white and red muscle, chicken myosin gels formed at 0.2 M KCl produce a rigidity peak (maximum) at pH close to 6.0 [72], but bovine myosin gels do not exhibit a rigidity peak at this ionic strength [20]. Discrepancies observed between white and red fiber types have been attributed largely to isoforms or myosin. However, conflicting results have also been reported. For instance, chicken leg (red) myosin was found to form more rigid gels at pH 5.0–6.4 than chicken breast (white) myosin [72]. The difference was ascribed to morphological variation in reconstituted myosin filaments, since red myosin formed longer filaments than white myosin in this particular study.

3. Myosin Gelation at High Ionic Strengths

Fundamental studies on gelation of myosin filaments formed in low–ionic strength solutions are useful for elucidation of the relationship between protein structure and morphology and functional behavior of the protein. This is because in muscle systems, where myosin is a prevalent gelling component, protein aggregation in an ordered manner is a prerequisite thermodynamic process for the formation of filamentous, three-dimensional gel networks. Nonetheless, information obtained from this type of investigation is difficult to apply or be extrapolated to actual meat processing, as in situ, much higher concentrations of salt (0.5–1.0 M NaCl) are employed. At an ionic strength of 0.5 or higher, nonheated myosin molecules exist as monomers, rather than filaments, although occasionally myosin aggregates or short filaments could also form when the solution pH is sufficiently low (pH $\leq$ 5.7) [49]. Thus, gelations of myosin at low and high ionic strengths seem to follow different mechanisms.

Studies of myosin gelation in elevated salt solutions ($\geq$0.5 M KCl) have shown that gelation begins upon heating the myosin suspension to about 35–40°C [123]. Increases in gel rigidity over the 35–60°C temperature range coincide with loss in α-helical content [32], indicating that gel networks are produced via protein-protein interactions apparently involving uncoiled tails. The optimum temperature and pH for myosin gelation at 0.5–0.6 M KCl or NaCl are about 65°C and pH 6; the exact values vary slightly depending on intrinsic factors, such as muscle fiber type and purity of myosin preparation. Myosin heavy chains are the main subunit involved in gelation, and the light chains appear to be involved with the effect of pH on gelation [75]. In fact, the remarkably similar gelling characteristics between myosin heavy chains and the whole myosin molecule suggest that the light chains are of little importance in myosin gel formation at high ionic strengths (e.g., 0.6 M KCl). This is in contrast with myosin gelation at low ionic strength. The removal of light chain -2 results in a substantial reduction in rigidity of myosin gel formed in 0.1–0.2 M KCl solutions [50]. Studies of the various fragments of myosin obtained by limited proteolysis have shown that the order of rigidity for gels prepared under identical pH and ionic conditions and on an equal protein concentration basis is myosin > rod > LMM > HMM > S-1 [32,70]. The heat-induced gelation of myosin rod largely resembles that of myosin, both responding to pH and temperature in a similar manner. Mixing the rod and the S-1 fragments in solution before heating does not restore gel rigidity to that of myosin, indicating that the myosin head plays a supporting role in gelation and that the intactness of the myosin heavy chain is important.

The mechanism of myosin gelation under high–ionic strength conditions has been extensively researched. According to Samejima et al. [70], gelation of rabbit myosin involves two sequential reactions: aggregation of the globular head portion with a rigidity transition at 43°C and network formation as a result of thermal unfolding of the helical tail portion, as measured by ORD or CD, with a transition temperature of 55°C. Gelation of myosin is accompanied by loss of the secondary structures. Heat-induced unfolding of α-helices and β-sheets occurs at temperatures as low as 40°C and increases rapidly between 45 and 60°C [32,75,99]. The initial aggregation of myosin probably involves the disulfide bond formation and sulfhydryl-disulfide bond interchanges between peptides, while the network development is mainly through noncovalent interactions, particularly those between nonpolar groups in the hydrophobic regions or patches. Rigidity development in both fish (tilapia) [102] and rabbit [51] myosin gels is closely associated with increases in hydrophobicity due to unraveling of the myosin structure. The sulfhydryl-disulfide interchange reactions appear to be limited to HMM, and do not occur in the LMM portion during myosin gelation [32].

Evidence of structural changes involved in myosin gelation in 0.5–0.6 M KCl at pH 6.0–6.5 has been presented by Yamamoto [120] and Sharp and Offer [78] by electron microscopic analysis. These studies have demonstrated that thermally induced myosin gelation proceeds in at least four steps. Unfolding of the myosin head (S-1 subfragment) occurs at 35°C, leading to formation of dimers and oligomers via head-head interactions. As the temperature increases to 40°C, a globular mass is formed, which consists of a tightly associated head clump with tails radiating outward, resembling the shape of a spider. At 48°C, oligomers coexist with the aggregates formed by the coalescence of two or more oligomers. From 50 to 60°C, these oligomers further aggregate, apparently involving tail-tail cross-linking, to form particles that probably make up the strands of the gel networks. It is important to remember that a gelling point is established only after the aggregation reaction reaches a certain critical point. Under identical thermal conditions, chicken breast myosin (in 0.6 M NaCl, pH 6.5) shows four transitions in the 44–51°C temperature range, which involve unfolding of four independent domains and their aggregation. However, the protein does not gel until 53°C [101].

Molecular interactions involved in myosin gelation in salt solutions are sensitive to the pH of the solution. While myosin is known to be present in a monomeric form at ionic strength greater than 0.5 and pH 6.0 or above, several studies have shown that in 0.6 M KCl, pH 5.4, chicken myosin can form filaments that are somewhat similar in morphology to those formed at low ionic strengths (0.1–0.3 M KCl) [9,10,49]. It is not clear, however, whether myosin filaments formed at both high and low ionic strengths are identical in structure and specific molecular arrangement. The filamentous myosin forms high viscoelastic gels and the gel rigidity increases with the length of the filament. It is possible that gelation of myosin at low pH (5.4) would follow a different mechanism from that in a high pH ($\geq$6.0) condition. It is not clear at present, however, whether and how these myosin filaments would change their structure and morphology during thermal gelation. Also not entirely clear is the mechanism of acid-induced spontaneous gelation of myosin, which does not require heat [19,27].

Gelation properties of myosin are apparently affected by species that contain different myosin isoforms. Fretheim et al. [20] have shown that myosin from bovine *cutaneous trunci* muscle (homogeneously white) is generally, but not always, the better gel former than myosin from *masseter* muscle (homogeneously red). In 0.6 M NaCl, white myosin forms more elastic gel networks than red myosin at pH greater than 5.8, but at pH less than 5.7, the result is reversed. The latter finding at pH less than 5.7 differs from those reported by Asghar et al. [3] on chicken, probably due to different animal species involved. More recently, Culioli et al. [11] reported that bovine *psoas major* myosin (white) gels also exhibit a greater storage modulus than *semimembranosus proprius* (red) gels, thus further substantiating the concept that myosin from white muscle has superior gelling properties to myosin from red muscle (Table 3). It should be noted from Table 3 that the bovine myosin gels were analyzed using a nonfracture small strain shear test and the storage modulus values reflect the elastic characteristic of the gels. On the other hand, the chicken myosin gels, although also analyzed using a fundamental

TABLE 3 Shear Modulus (Rigidity) and Storage Modulus (Elasticity) of Myosin Gels Formed Under Different Gelling Conditions

Species	Gelling condition	Gel viscoelastic values		Ref.
		Red muscle	White muscle	
		Shear modulus (Pa)		
Chicken[a]	4.5 mg/ml protein pH 5.4, 0.2 M KCl	800	685	72
	4.5 mg/ml protein pH 5.4, 0.6 M KCl	300 500	2500 2000	10 49
	4.5 mg/ml protein pH 6.0, 0.2 M KCl	340 1172	1115 684	3 72
	4.5 mg/ml protein pH 6.0, 0.6 M KCl	100 208	110 472	49 3
		Storage modulus (Pa)		
Bovine[b]	10 mg/ml protein pH 6.0, 0.2 M NaCl	400	1600	20
	10 mg/ml protein pH 6.0, 0.6 M NaCl	330	490	20
Bovine[c]	10 mg/ml protein pH 6.0, 0.6 M KCl	241	1200	11

[a]Red muscle = thigh (e.g., *gastrocnemius*) or leg; white muscle = breast (*pectoralis*).
[b]Red muscle = *masseter*; white muscle = *cutaneus trunci*.
[c]Red muscle = *semimembranosus proprius*; white muscle = *psoas major*.

test, were evaluated with a relative large deformation (2 mm), which could disrupt the gel specimen and which does not separate the elastic and viscous components of the deformation response. Therefore, while the trends of the chicken and bovine data may be compared, the exact shear values of the gels between the two species may not be simply correlated with each other.

In spite of the existence of large discrepancies in gelation properties between white and red muscle myofibrillar proteins from meat animals, practically, light (white) and dark (red) meat is often mixed in the same formulation of processed products. Thus, stability and integrity of many meat products are maintained by joint effort of white- and red-type myofibrillar proteins that impart desirable functionalities to the product. The interactions of white and red myosins during gelation have been studied by Fretheim et al. [20] and Choe et al. [10]. Bovine and chicken white and red skeletal myosins showed synergistic effects on gel elasticity or rigidity at pH 5.4, 5.7, and 6.0 in 0.6 M NaCl or KCl. More detailed experiments revealed a stronger synergism between intact red (leg) myosin and white (breast) myosin rod, and between intact white myosin and red myosin rod, than between either myosin and the light meromyosin from the opposite fiber type. The maximum effect varied depending on the specific myosin:rod (or light meromyosin) ratio for different mixtures. No synergism was produced in mixtures of white (or red) myosin and red (or white) heavy meromyosin (which contains the sulfhydryl-dense globular head), suggesting that disulfide bonds may not be involved in heterogeneous myosin complexes or aggregates. These results are interesting because white myosin alone has a different pH optimum for gel rigidity or elasticity than red myosin alone. Gelation of mixed skeletal (rabbit back and leg) and cardiac (porcine heart) myosins and reconstituted actomyosins has been investigated by Samejima et al. [74] to explore the effect of hybridization on the rigidity of heat-induced myosin gels at pH 6.0. Skeletal myosin formed stronger gels than cardiac myosin. Gels of both myosin types were reinforced by actin, with a maximum shear modulus produced at a 1:15 (w/w) myosin:actin ratio. Hybridized actomyosin was formed from the skeletal myosin and cardiac actin; however, the gel rigidity was less than that of skeletal actomyosin. Actomyosin also showed a synergistic effect on myosin gelation, but the myosin:actomyosin ratio for maximum gel rigidity differed for the skeletal and cardiac proteins, i.e., 1:3 for skeletal and 1:5 for cardiac proteins.

4. Gelation of Mixed Myofibrillar Proteins

Unlike gelation of pure myosin, which can occur at both low ($\Gamma < 0.2$) and high ($\Gamma > 0.5$) ionic strengths, gelation of myofibrillar proteins usually requires ionic strengths greater than 0.5, which is typically achieved by adding 2–3% salt in the form of NaCl to the product formulation. The main reason for this is that in order to produce a viscoelastic true gel, myofibrillar proteins must exist in the soluble form. An exception to the high salt requirement was reported by Stone and Stanley [88], who noted gel formation of myofibrillar proteins extracted and suspended in extremely dilute salt solution (0.008–0.012 M NaCl). However, within this ionic strength range, the proteins are also soluble [87,88]. Thus, solubility is a critical factor controlling myofibrillar protein gelation.

As discussed previously in this chapter, solubilization of myofibrillar proteins by mixing meat with salt occurs as a result of disruption of myofibrillar structures. The salt extract thus obtained is comprised of sarcoplasmic proteins and a variety of myofibrillar components, including some intact myofibrils suspended in brine, the actomyosin complex, and dissociated individual proteins such as myosin, actin, tropomyosin, and troponin. The latter portion of the protein extract—a mixture of myofibrillar proteins—is collectively referred to as "salt-soluble protein" (SSP). Because SSP is close in composition to the actual proteins extracted by salt during meat processing, it is believed that the functional behavior of SSP observed in model systems largely reflects what takes place in actual meat processing, i.e., in situ situations.

A SSP suspension can be viewed as a composite system because of its complex nature. While the extraction of actomyosin and the individual myofibrillar components in the presence of salt is usually done in cold, gelation of the extracted proteins generally requires heat. As summarized by Ziegler and Acton [127], natural actomyosin, which contains tropomyosin and some other minor proteins in addition to the actin-myosin complex, undergoes major conformational and structural changes during thermally induced gelation. In the 30–35°C temperature range, native tropomyosin is dissociated from the F-actin backbone, while at 38°C, the F-actin superhelix dissociates into single chains. The myosin light chain subunits dissociate from the heavy chains at approximately 40°C, followed by conformational changes in the head and hinge region of the myosin molecules. The actin-myosin complex is dissociated in the 45–50°C range, followed by a major helix → coil transformation into light meromyosin chains within 50–55°C. These structural changes lead to rapid aggregation and formation of gel networks.

Figure 7 represents a typical rheogram of chicken SSP during gelation. Rheological characteristics of the gel are described by three parameters, the loss modulus (G″), storage modulus (G′), and loss tangent (Tan δ, the ratio of G″/G′), which are obtained from small-amplitude (strain) oscillatory shear or dynamic testing. When the gel is deformed by an imposed strain, part of the energy input is stored and recovered in each shear cycle due to the elastic response of the gel, and part of the energy is dissipated as heat due to internal friction or viscous flow of the sample. Thus, G′ and G″ are measures of

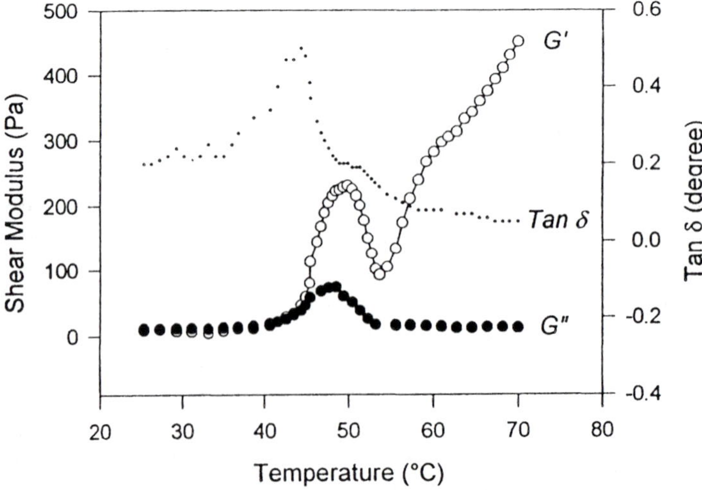

FIGURE 7 Typical rheological curves of salt-soluble myofibrillar protein suspension (20 mg/ml protein in 0.6 M NaCl, pH 6.0) during thermal gelation.

elasticity and viscosity of the gel samples, respectively, and Tan δ reflects the relative contribution of each to the overall rheological characteristics. Although some myofibrillar components (e.g., tropomyosin) may be dissociated from the myofibril at temperatures less than 40°C, no change in shear modulus is detectable below 40°C. As the temperature rises to above 40°C, G″ begins to increase and reaches a maximum at about 48°C. Changes in G′ closely follow changes in G″ and reach a maximum peak value at a slightly higher temperature (50°C). Thus, formation of an elastic gel network (an increase in G′) from 40 to 50°C probably results from aggregation of myosin due to unfolding (hence, an increase in G″) of the head and hinge portions. From 50 to 55°C, G′ decreases drastically. Among various possible reasons, including kinetic variations [104], denaturation of the light meromyosin (myosin tail) seems to be the major contributing factor. It is assumed that the helix → coil transformation can lead to a large increase in fluidity of the semi-gel and may disrupt some of the protein network already formed. Hence, gel elasticity decreases. Pure myosin also exhibits a G′ drop in the 50–55°C temperature range; however, the G′ reduction is relatively minor compared to that of SSP. Therefore, dissociation of the actin-myosin complex, which occurs around 50°C, probably also contributes to the extent of G′ reduction from 50 to 55°C. Indeed, incubation of myofibril or SSP suspension with pyrophosphate (which dissociates the actomyosin complex) before gelation results in less drop in G′ between 50 and 55°C [66,81]. The steady increase in G′ above 55°C involves denaturation of actin and can be ascribed to both an increase in the number of cross-links between protein aggregates or strands and deposition of additional denatured proteins in the existing protein networks so as to reenforce the gel matrix. As the gel approaches the final temperature, Tan δ decreases to almost zero, indicating that a highly elastic myofibrillar protein gel is formed.

Gelation of myofibrillar proteins is affected by a number of intrinsic and extrinsic factors. Gels made from bovine skeletal muscle natural actomyosin have a maximum strength within a relatively wide pH range (5.0–5.5) [1]. The pH for maximum gel strength is higher for bovine cardiac myofibrils (pH 6.0) [73,119]. Tropomyosin and troponin are two highly charged proteins that do not form a gel by themselves in salt solutions and remain soluble even after heating to 65–70°C [71,117]. It appears that neither protein participates in actomyosin gel networks. Thus, substitution of tropomyosin for desensitized actomyosin or myosin on an equal weight basis decreases the rigidity and elastic characteristics of actomyosin or myosin gels [76]. For SSP or myofibril suspensions, gel strength increases with protein concentration and heating up to 65–70°C. There is an apparent correlation between gel strength and surface hydrophobicity (S_0) of chicken SSP, and the development of gel strength also coincides with the increase in turbidity (optical density at 320 nm) of the SSP suspension (Fig. 8). Thus, hydrophobic groups that are exposed during heating are involved in SSP gel networks, and formation of the gel networks obviously results from aggregation of SSP proteins, probably through hydrophobic interactions. Myofibril gel strength is increased by incubation in 0.6 M NaCl, which is partially attributed to increased extraction of actomyosin and myosin [113]. The observed effect of polyphosphate on myofibrillar protein gelation has been controversial. Pyrophosphate was found to increase gel strength of bovine myofibrils due to the combined effects on protein extraction and interaction with the myosin heads [69]. However, in a number of other studies [58,66,95], pyrophosphate treatment resulted in increased fluidity of the protein suspension and decreased myofibrillar protein gel strength. Lipid peroxidation and metal ion–catalyzed oxidation can cause damage to proteins such as deamination and formation of carbonyl and the disul-

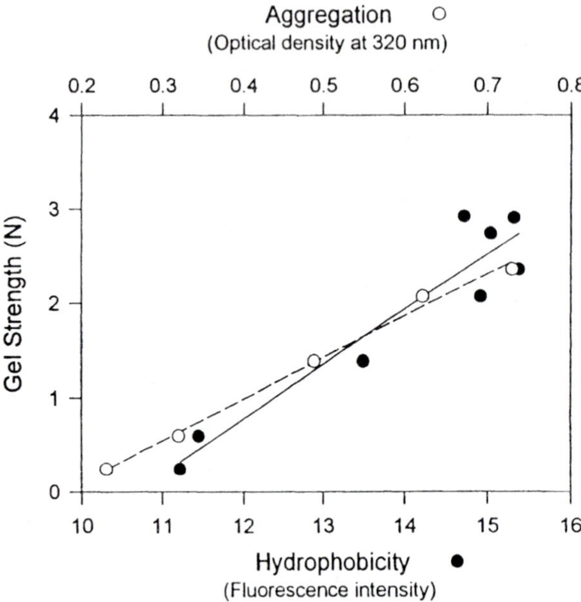

FIGURE 8 Responses of gel strength development to increases in surface hydrophobicity and aggregation of salt-soluble myofibrillar proteins during heating.

fide bonds, thereby leading to structural modification in myofibrillar proteins. Depending on the extent of oxidation, muscle proteins can either form large, insoluble aggregates or small complexes by reacting with oxygen free radicals. Thus, excessive oxidation impairs the gelling ability of SSP, while mild oxidation facilitates gel network formation [14,82]. Heat is not always required for myofibrillar protein gelation. Factors that cause conformational changes conducive to ordered protein-protein association can also effect myofibrillar protein gel formation. Examples are high pressurization (e.g., 2000 kg/cm^2), which has been found to induce gelation of carp actomyosin [57], and weak acids such as glucono-δ-lactone, which induces myofibrillar protein gelation at 4°C by slowly lowering the pH to below 4.5 [54]. It is also possible to form myofibrillar protein gels without a significant alteration in protein structure by using the muscle endogenous enzyme transglutaminase, which catalyzes interpeptide cross-linkages via acyl transfer reactions [68].

There is ample evidence indicating that gelation properties of myofibrils and SSP are specific to fiber types. The higher protein solubility observed in fast-twitch white myofibrils than in slow-twitch red myofibrils partially accounts for the gel property variation. However, even with an equal solubility (supernatant), white SSP still forms more rigid and elastic gels than red SSP as shown in many model systems, suggesting that some isozymic factors probably determine the SSP gelation process and structure of the final gel. Because of the existence of various molecular isoforms associated with many of the myofibrillar proteins including myosin, the myofibril complex may show some fiber type–dependent functionalities which could not be seen in myosin alone. Under identical environmental conditions, white myofibrillar proteins generally form stronger gels than red myofibrillar proteins at the same protein concentrations (20–40 mg/mL), as determined by both destructive extrusion tests and nondestructive, small-strain dynamic measurements [107]. The exact differences appear to depend on the pH of the protein suspension. The optimum gelation pH is 6.0 for white myofibrils and 5.5 for red myofibrils. The solubility difference between the two myofibril types may contribute to the variation in gel strength by altering the volume of the effective gelling components, however, this is not entirely true above pH 6.0. Similarly, leg SSP, the soluble fraction of the myofibrils, always forms weaker and less elastic gels than breast SSP at pH 6.0 under similar ionic conditions (Fig. 9).

Fiber type–dependent myofibrillar protein gelation is extended to other poultry species. In 0.5 M NaCl, SSP from turkey breast muscle produces more rigid gels than SSP from the thigh muscle, in spite of the similar solubility of both SSP preparations [18]. A maximum in gel rigidity is established at pH 6.0 for both white and red SSP, and gels formed at pH 7.0 are weaker than those prepared at pH 5.0, although the protein solubility followed the order of pH 7.0 > pH 6.0 > pH 5.0. These results verify that inherent variations related to protein isoforms exist between different types of myofibrillar protein and that protein solubility is not the sole factor affecting gel structure and gel strength. More recently, Ndi et al. [53] explored gelling properties of duck breast and leg myofibrils, noting a great disparity in gel strength (penetration force) between the two types of myofibrils. In comparison to chicken myofibril gels with an identical pH, temperature, and ionic environment, duck leg gel is very similar to chicken leg gel strength, but duck breast gel is considerably less firm and more fragile than chicken breast gel. In contrast to chicken breast muscle, which is composed almost exclusively of white fibers, duck breast muscle belongs to the histochemically red fiber group, as does leg muscle. Therefore, it is not entirely surprising that duck breast shows less difference from its leg counterpart in gel properties than does chicken breast.

Turbidity measurements show that mixed chicken white and red myofibrillar proteins can co-aggregate, forming possible actomyosin hybrids [106]. However, synergistic

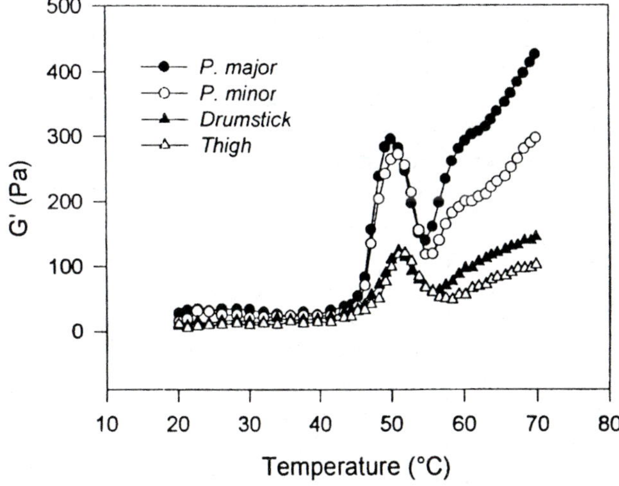

FIGURE 9 Comparison of gelation properties of white and red myofibrillar proteins. White and red myofibrillar proteins are from fast-twitch (chicken *pectoralis* major and minor) and slow-twitch (chicken thigh and drumstick) muscles. (From Ref. 112.)

effects on gel strength do not occur between the two myofibril types, differing from the findings for myosin. This can be attributed to the complex protein composition in myofibrils and the numerous isoforms of myofibrillar components in addition to those found in myosin. Systems containing myosin, actomyosin, or actin have been shown to produce gels with a range of physical properties, depending on the relative proportion of the composite proteins [49,74]. Yamamoto et al. [121] demonstrated that, in the presence of C-protein, rabbit skeletal myosin forms weak gels compared to myosin free of C-protein.

Samejima et al. [73] compared the gelling properties of rabbit skeletal (back and leg) and porcine cardiac myofibrils. At 35–80°C and pH 5–7, the skeletal myofibrils always produced more rigid gels with greater breaking energy than the cardiac myofibrils. Furthermore, the optimum gelation pH was around 5.6 for skeletal myofibrils and 6.0 for cardiac myofibrils. It is also conspicuous that solubility differences between the two types of protein are unable to explain the disparity in gel strength, as often observed in poultry species [18,113,115]. Young et al. [126] examined the rheological characteristics of gels made from myofibrils isolated from bovine fast-twitch (*cutaneous trunci*), slow-twitch (*masseter*), and cardiac muscles, and their data were well in line with the findings of Samejima et al. [73] and Xiong and Blanchard [112]. Specifically, the study showed that the onset gelation temperature was lower for fast myofibrils than for slow or cardiac myofibrils, and the former also produced more rigid gels than the latter. Again, polymorphism of myosin associated with different fiber types was implicated.

E. Emulsification

A classic emulsion is defined as a heterogeneous system consisting of at least two immiscible liquid phases, one of which is dispersed in the other in the form of droplets [12]. Based on the size of the droplets in the dispersed phase, emulsions can be classified into three groups, namely, macroemulsion (0.2–50 μm), miniemulsion (0.1–0.4 μm), and microemulsion (0.01–0.1 μm). Comminuted meat falls into the "macroemulsion" category and resembles an oil-in-water emulsion in certain aspects. However, it is not a true emulsion system in the classic sense. Thus, the term "meat batter" is often used to more appropriately reflect the nature of the multiphase system. Meat batters are stabilized through two mechanisms. The first mechanism is physical entrapment of fat globules within the protein matrix formed largely via protein-protein interactions. In the second mechanism, fat globules are stabilized by an interfacial protein film (membrane) that surrounds them. The interfacial film is "interactive" in the sense that it interacts with the viscoelastic continuous phase to further enhance the emulsion stability. For review, see Jones [34] and Barbut [5].

The ability of proteins to emulsify fat in comminuted meat is of particular importance. Proteins are amphoteric molecules consisting of mixed polar and nonpolar amino acid residues. They are surface-active agents and can concentrate at the fat/water interface. The formation of an interfacial protein layer on the surface of fat globules involves three steps. First, protein molecules must diffuse to the interface. This process obviously is influenced by the solubility and molecular size of the protein, by temperature, and by viscosity of the continuous phase. Second, when a protein molecule arrives at the interface, it must overcome one or more barriers before it can be adsorbed. This means that the protein molecule must compress molecules already adsorbed, i.e., overcoming the interfacial pressure barrier. Adsorption can be influenced by amino acid profile, pH, ionic strength, and temperature. Third, the protein molecule must undergo conformational changes, i.e., unfolding, so that its hydrophobic groups are anchored inside the nonpolar lipid phase while the hydrophilic residues orient toward the polar aqueous phase. By doing this, the protein loses its tertiary and probably some secondary structure, although helical segments may persist. Proteins present at the interface may assume a configuration that contains "spreadable" random coils adsorbed on the surface of the fat globule and loops that extend into the bulk phase. Such structural alterations and arrangements at the oil/water interface are thermodynamically favored because they lead to a reduction of the free energy of the system.

The important role of proteins as emulsifiers in meat emulsion or batter systems is manifested by high-magnification electron microscopic techniques. Scanning electron microscopy reveals the presence of a proteinaceous membrane that coats on the surface of fat globules (Fig. 10). Proteins that comprise the reconstituted fat globule membrane form cross-linkages with the protein matrix in the continuous phase, resulting in a stabilized three-dimensional microstructure of meat batters [94]. The thickness of the interfacial membrane seems to contribute critically to the stability of the emulsion. Thus, increasing the chopping time and temperature of the batter up to about 15°C increases the thickness of the protein film, thereby improving the emulsion stability. The interfacial

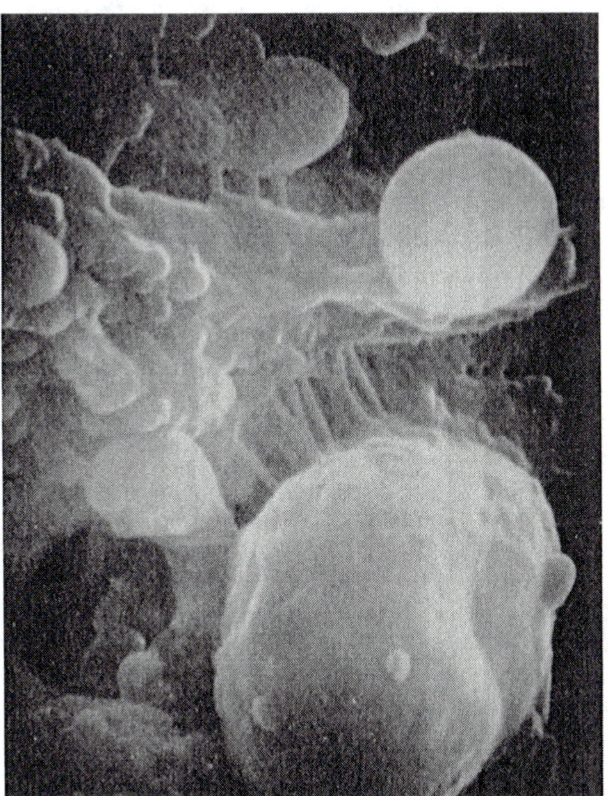

FIGURE 10 Scanning electron microscopic image of a frankfurter, an emulsion-type meat product. Note the cross-linkages between the membranal proteins around the fat globule and proteins in the matrix structures. (From Ref. 94.)

membrane formed during chopping possesses a complex microstructure by itself. Multilayered films have been observed using transmission electron microscopy [5].

In meat emulsions, salt-soluble proteins play the most critical role in forming interfacial films that encapsulate fat particles or oil droplets. The emulsifying capacity of different muscle proteins (5 mg/ml) in 0.3 or 0.6 M NaCl solution was found to follow the order of myosin > actomyosin > sarcoplasmic proteins > actin [22,26]. If ATP or pyrophosphate is added to actomyosin solution, the emulsifying capacity can improve to a level similar to that of myosin. Tropomyosin and troponin are poor emulsifiers and stabilizers compared to myosin or actomyosin. Increases in ionic strength and protein concentration diminish the differences between different proteins. At a protein concentration of 12 mg/ml in 0.5 M KCl, myosin, actin, tropomyosin-troponin, and sarcoplasmic proteins essentially have equal emulsifying capacities [96]. During emulsification, myosin is rapidly and preferentially adsorbed at the fat/water interface, followed by actomyosin and actin. The superior emulsifying properties of myosin to any other muscle proteins at low concentrations are ostensibly derived from two unique structural factors of the myosin molecule. The unbalanced distribution of amino acids in different segments of myosin, i.e., a prevalence of the hydrophobic residues in the head region or HMM S-1 subfragment and a preponderance of polar groups in the tail or LMM portion, makes myosin an ideal emulsifier. Furthermore, myosin has a high length-to-diameter ratio (roughly 40:1), a structure that is conducive to strong protein-protein interaction and molecular flexibility at the interface. The relatively low surface activity exhibited by other proteins may be related to their small size and spherical shape, compared to myosin. Salt-insoluble proteins (in 0.6 M NaCl), which probably contain some cytoskeletal proteins and collagen, also participate in emulsion formation, probably by serving as emulsion stabilizers [23]. Emulsions that are formed by using different protein emulsifiers exhibit different ultrastructures. Thus, myosin-stabilized emulsions consist of small globules of uniform size, while actin-stabilized emulsions are composed of large globules with a greater size range [96].

Despite numerous studies in model systems, the molecular event involved in the interfacial film formation in meat batter is still not very clear. Jones [34] suggests that in raw sausage batters, myosin initially forms a monolayer about 0.16 μm thick on the surface of the fat globule. If the penetration of the globular head portion into the fat globule is taken into consideration, then the thickness of the monolayer would be about 0.13 μm. These myosin molecules are relatively undenatured. Subsequent protein-protein interactions immediately outside the monolayer produce a thick membrane of considerable viscoelasticity and can stabilize the fat globule. A further stabilization of the emulsion system is apparently achieved through thermally induced changes in myosin conformation, i.e., denaturation, and may involve formation of the disulfide bond. Since proteins extracted into the aqueous phase by salt include not only myosin, but also actomyosin, actin, sarcoplasmic proteins, and numerous other minor myofibrillar components, the final interfacial membrane is conceivably a composite, multilayer system composed of a variety of proteins, which are denatured and precipitate on the myosin monolayer. To ensure stability, these multilayers must possess sufficient viscosity, cohesiveness, and elasticity to withstand friction, collision, and other mechanical shock.

Structure-function studies have also led to the finding that emulsifying properties of muscle proteins are well correlated with protein surface hydrophobicity (S_o) [41], especially for high-solubility proteins. For salt-extractable proteins whose solubility is greater than 50%, S_o alone accounts for more than 70 and 80% of the variability in emulsifying activity index and emulsifying capacity, respectively. Solubility parameters are more influential for samples with lower solubility (<50%). Thus, emulsifying properties can be accurately predicted by considering both surface hydrophobicity and solubility. The interaction term $S_o \times$ solubility accounts for more than 71% of the variability in the emulsifying activity index of beef salt extracts. The above relationships hold true for both heated and nonheated samples.

F. Foaming

The foaming of protein solutions is a common occurrence and can be quite a nuisance to protein chemists who are involved in protein isolation and purification both commercially and at a research laboratory level. Difficulties arise when the foam volume expands to the capacity of the container and proteins become denatured as a result of foam formation. However, foaming is a desirable and important functionality of proteins in many prepared foods where proteins play a crucial role in air entrapment. Foam-based foods are primarily nonmeat products, with meringues, souffles, whipped toppings, ice creams, and leavened bakery products being the most popular examples. Although meat-based foam products are relatively scarce, some (e.g., meat mousses) are available in different parts of the world, e.g., in France.

The mechanism of foam formation is believed to be similar to that of emulsion formation except that the discontinuous phase in a foam is air, rather than the fat droplets, and at the interface proteins undergo more extensive structural changes in foams than in emulsions. During foam formation, proteins initially diffuse to the air/liquid (water) interface where they are adsorbed, concentrated, and reorient, resulting in a reduction in surface tension (or an increase in surface pressure). The behavior of proteins at the air/liquid interface is extremely important because the formation of a protein-based flexible, cohesive film around air bubbles is essential for foaming capacity and foam stability. Foaming capacity of proteins is influenced by the nature of the protein and the prevailing environmental conditions. There is a relationship between the molecular flexibility of proteins, film properties, and foam stability. Flexible, disordered proteins are more surface-active than extensively cross-linked, stable, and compact globular proteins. Thus, α-casein can rapidly form films, whereas lysozyme has limited film-forming properties. For detailed discussion of foaming mechanisms, see Chapter 3.

Research on foaming of muscle proteins has been limited, and consequently, the structure-foaming relationship of muscle proteins has not been elucidated. It is presumed that foams formed at sufficiently dilute muscle protein concentrations probably contain a monolayer film in which the protein molecules are extensively, if not completely, unfolded. Because myosin is a highly surface-active protein, it probably plays a crucial role in the formation of the monolayer film. However, at high concentrations, muscle proteins may form concentrated or multilayer films in which protein molecules may retain a substantial portion of their native structure or conformation, particularly the secondary structures. Based on a foaming study in rabbit muscle model system, O'Neill et al. [59] concluded that sarcoplasmic proteins were more surface active than salt-soluble myofibrillar proteins. The rate of surface tension drop was remarkably greater for water-extracted proteins than for salt-extracted proteins during the initial stage of foam formation. This was attributed to the size difference since the sarcoplasmic protein molecules are relatively small and, hence, would migrate more rapidly to the interface than the high molecular weight myofibrillar proteins. The ability to reduce the surface tension by the Weber-Edsall solution extract decreased with extraction time from 8 to 48 hours. This led the authors to suggest that myosin (short-time extraction) was more surface active than actomyosin (long-time extraction). The average equilibrium surface pressure at the air/water interface was 21.5 mNm^{-1} for a water extract, compared to 20.1 mNm^{-1} for a 1 M KCl extract. Although molecular mobility in solution appears to be a limiting factor for protein adsorption at the air/water interface, it is not clear exactly how the different fractions of muscle proteins or components of myofibrils behave at the air/water interface, i.e., changes in conformation and rearrangement in configuration in exchange for a decreased free energy or entropy gain for the foam system.

Protein solubility seems to be required for foam formation. For fish protein concentrate, only the soluble fraction of proteins is responsible for foaming, probably due to its ability to diffuse to and orient at the interface. The insoluble fraction participates in foam by stabilizing the foam product [28]. Although fish protein concentrate lacks most functional properties, its good foaming properties make it a possible substitute for more expensive proteins, such as egg white solids and whey protein isolates. An exception to solubility requirements for muscle protein foams was reported by Huidobro and Tejada [29], who studied foaming properties of fish minces during frozen storage at −18°C for up to a year. The foaming capacity of frozen stored minces was initially inhibited by the presence of the species' own lipids. Thus, foam volume of the minces at the beginning of frozen storage followed the order of blue whiting (low fat species) > horse mackerel (intermediate fat) > and mackerel (high fat). Presumably, lipids act to displace proteins from the air, water interface in a competitive manner by undergoing strong preferential adsorption without forming a resilient foam-stabilizing film. Although protein solubility (in 0.86 M NaCl, pH 7.0) decreased steadily during frozen storage, there was an increase in foaming capacity. On the basis of these observations, the authors concluded that protein solubility in frozen fish mince was not an indispensable requisite for foaming capacity, and a mere dispersion of proteins or their aggregates would be sufficient. This was substantiated by the additional observation that extended frozen storage decreased foaming ability and capacity, apparently resulting from formation of larger, indispersible protein aggregates.

REFERENCES

1. J. C. Acton, M. A. Hanna, and L. D. Satterlee, Heat-induced gelation and protein-protein interaction of actomyosin, *J. Food Biochem.* *5*:101 (1981).
2. E. Antonini, Interrelationship between structure and function in hemoglobin and myoglobin, *Physiol. Rev.* *45*:123 (1965).
3. A. Asghar, J.-I. Morita, K. Samejima, and T. Yasui, Biochemical and functional characteristics of myosin from red and white muscles of chicken as influenced by nutritional stress, *Agric. Biol. Chem.* *48*:2217 (1984).
4. A. Asghar, K. Samejima, and T. Yasui, Functionality of muscle proteins in gelation mechanisms of structured meat products, *CRC Crit. Rev. Food Sci. Nutr.* *22*:27 (1985).
5. S. Barbut, Importance of fat emusification and protein matrix characteristics in meat batter stability, *J. Muscle Foods* *6*:161 (1995).
6. A. Bertazzon and T. Y. Tsong, High-resolution differential scanning calorimetric study of myosin, functional domains, and supramolecular structures, *Biochemistry* *28*:9784 (1989).
7. A. Bertazzon and T. Y. Tsong, Study of effects of pH on the stability of domains in myosin rod by high-resolution differential scanning calorimetry, *Biochemistry* *29*:6453 (1990).

8. D. D. Bronson and F. H. Schachat, Heterogeneity of contractile proteins. Differences in tropomyosin in fast, mixed, and slow skeletal muscles of the rabbit, *J. Biol. Chem. 257*: 3937 (1982).
9. I.-S. Choe, J.-I. Morita, K. Yamamoto, K. Samejima, and T. Yasui, The heat-induced gelation of myosin rods prepared from chicken leg and breast muscles, *Agric. Biol. Chem. 53*:625 (1989).
10. I.-S. Choe, J.-I. Morita, K. Yamamoto, K. Samejima, and T. Yasui, Heat-induced gelation of myosins/subfragments from chicken leg and breast muscles at high ionic strength and low pH, *J. Food Sci. 56*:884 (1991).
11. J. Culioli, C. Boyer, X. Vignon, and A. Ouali, Heat-induced gelation properties of myosin: influence of purification and muscle type, *Sci. Aliments 13*:249 (1993).
12. K. P. Das and J. E. Kinsella, Stability of food emulsions: physicochemical role of protein and nonprotein emulsifiers, *Adv. Food Nutr. Res. 34*:81 (1990).
13. G. L. Davey and K. V. Gilbert, The effect of carcasss posture on cold, heat and thaw shortening in lamb, *J. Food Technol. 8*:445 (1973).
14. E. A. Decker, Y. L. Xiong, J. T. Calvert, A. D. Crum, and S. P. Blanchard, Chemical, physical, and functional properties of oxidized turkey white muscle myofibrillar proteins, *J. Agric. Food Chem. 41*:186 (1993).
15. T. R. Dutson, A. M. Pearson, and R. A. Merkel, Ultrastructural postmortem changes in normal and low quality porcine muscle fibers, *J. Food Sci. 39*:32 (1974).
16. B. Egelandsdal, K. Fretheim, and K. Samejima, Dynamic rheological measurements on heat-induced myosin gels: effect of ionic strength, protein concentration and addition of adenosine triphosphate or pyrophosphate, *J. Sci. Food Agric. 37*:915 (1986).
17. J. D. Ferry, *Viscoelastic Properties of Polymers*, 3rd ed., John Wiley and Sons, New York, 1980.
18. E. A. Foegeding, Functional properties of turkey salt-soluble proteins, *J. Food Sci. 52*:1495 (1987).
19. K. Fretheim, B. Egelandsdal, O. Harbitz, and K. Samejima, Slow lowering of pH induces formation of myosin, *Food Chem. 18*:169 (1985).
20. K. Fretheim, K. Samejima, and B. Egelandsdal, Myosins from red and white bovine muscles: part 1—gelation (elasticity) and water-holding capacity of heat-induced gels, *Food Chem. 22*:107 (1986).
21. T. Fukazawa, Y. Hashimoto, and T. Yasui, Effect of some proteins on the binding quality of an experimental sausage, *J. Food Sci. 26*:541 (1961).
22. S. J. Galluzzo and J. M. Regenstein, Role of chicken breast muscle proteins in meat emulsion formation: myosin, actin and synthetic actomyosin, *J. Food Sci. 43*:1761 (1978).
23. M. T. Gaska and J. M. Regenstein, Timed emulsification studies with chicken breast muscle: soluble and insoluble myofibrillar proteins, *J. Food Sci. 47*:1438 (1982).
24. B. Hagerdal and H. Martens, Influence of water content on the stability of myoglobin to heat treatment, *J. Food Sci. 41*:933 (1976).
25. J. D. Hay, R. W. Currie, and F. H. Wolfe, Effect of postmortem aging on chicken muscle fibrils, *J. Food Sci. 38*:981 (1973).
26. G. B. Hegarty, L. J. Bratzler, and A. M. Pearson, Studies on the emulsifying properties of some intracellular beef muscle proteins, *J. Food Sci. 28*:663 (1963).
27. A.-M. Hermansson, O. Harbitz, and M. Langton, Formation of two types of gels from bovine myosin, *J. Sci. Food Agric. 37*:69 (1986).
28. A.-M. Hermansson, B. Sivik, and C. Skjoldebrand, Functional properties of proteins for foods—factors affecting solubility, foaming and swelling of fish protein concentrate, *Lebensm.-Wiss. Technol. 4*:201 (1971).
29. A. Huidobro and M. Tejada, Foaming capacity of fish minces during frozen storage, *J. Sci. Food Agric. 60*:263 (1992).

30. H. O. Hultin, Y. Feng, and D. W. Stanley, A re-examination of muscle protein solubility, *J. Muscle Foods 6*:91 (1995).
31. M. Ishioroshi, K. Samejima, Y. Arie, and T. Yasui, Heat-induced gelation of myosin: factors of pH and salt concentration, *J. Food Sci. 44*:1280 (1979).
32. M. Ishioroshi, K. Samejima, and T. Yasui, Further studies on the roles of the head and tail regions of the myosin molecule in heat-induced gelation, *J. Food Sci. 47*:114 (1982).
33. M. Ishioroshi, K. Samejima, and T. Yasui, Heat-induced gelation of myosin filaments at a low salt concentration, *Agric. Biol. Chem. 47*:2809 (1983).
34. K. W. Jones, Protein-lipid interaction in processed meats, Proceedings of 37th Annual Reciprocal Meat Conference, Lubbock, Texas, 1984, pp. 52–57.
35. J. M. Kijowski and M. G. Mast, Thermal properties of protein in chicken broiler tissues, *J. Food Sci. 53*:363 (1988).
36. N. L. King. Breakdown of connectin during cooking of meat, *Meat Sci. 11*:27 (1984).
37. N. L. King and J. J. Macfarlane, Muscle proteins, *Advances in Meat Research, Vol. 3—Restructured Meat and Poultry Products* (A. M. Pearson and T. R. Dutson, eds.), Van Nostrand Reinhold Company, New York, 1987, p. 21.
38. G. G. Knappeis and F. Carlson, The structure of the M-line in skeletal muscle, *J. Cell. Biol. 38*:202 (1968).
39. M. Koohmaraie, The role of Ca^{2+}-dependent proteases (calpains) in postmortem proteolysis and meat tenderness, *Biochimie 74*:239 (1992).
40. R. A. Lawrie, *Meat Science*, 5th ed., Pergamon Press, New York, 1991.
41. E. Li-Chan, L. Kwan, and S. Nakai, Physicochemical and functional properties of salt-extractable proteins from chicken breast muscle deboned after different post-mortem holding times, *Can. Inst. Food Sci. Technol. J. 19*:241 (1986).
42. D. J. Livingston and W. D. Brown, The chemistry of myoglobin and its reactions, *Food Technol. 35* (5):244 (1981).
43. R. H. Locker, The non-sliding filaments of the sarcomere, *Meat Sci. 20*:217 (1987).
44. J. L. Lopez-Lacomba, M. Guzman, M. Cortijo, P. L. Mateo, R. Aguirre, S. C. Harvey, and H. C. Cheung, Differential scanning calorimetric study of the thermal unfolding of myosin rod, light meromyosin, and subfragment 2, *Biopolymers 28*:2143 (1989).
45. J. J. Macfarlane, G. R. Schmidt, and R. H. Turner, Binding of meat pieces: a comparison of myosin, actomyosin and sarcoplasmic proteins as binding agents, *J. Food Sci. 42*:1603 (1977).
46. T. Maita, E. Yajima, S. Nagata, T. Miyanishi, S. Nakayama, and G. Matsuda, The primary structure of skeletal muscle myosin heavy chain: IV. Sequence of the rod, and the complete 1,938-residue sequence of the heavy chain. *J. Biochem. 110*:75 (1991).
47. K. Maruyama, K. Nayori, and Y. Nonomura, New elastic protein from muscle, *Nature (London) 262*:58 (1976).
48. C. Moos, G. Offer, R. Starr, and P. Bennett, Interaction of C-protein with myosin, myosin rode and light meromyosin, *J. Mol. Biol. 97*:1 (1975).
49. J.-I. Morita, I.-S. Choe, K. Yamamoto, K. Samejima, and T. Yasui, Heat-induced gelation of myosin from leg and breast muscles of chicken, *Agric. Biol. Chem. 51*:2895 (1987).
50. J.-I. Morita and T. Ogata, Role of light chains in heat-induced gelation of skeletal muscle myosin, *J. Food Sci. 56*:855 (1991).
51. J.-I. Morita and T. Yasui, Involvement of hydrophobic residues in heat-induced gelation of myosin tail subfragments from rabbit skeletal muscle, *Agric. Biol. Chem. 55*:597 (1991).
52. K. M. Nauss, S. Kitagawa, and J. Gergely, Pyrophosphate binding to and adenosine tripolyphosphate activity of myosin and its proteolytic fragments, *J. Biol. Chem. 244*:755 (1969).
53. E. E. Ndi, C. J. Brekke, and G. V. Barbosa-Canovas, Thermal gelation of duck breast and leg muscle proteins, *J. Muscle Foods 5*:27 (1994).
54. T. M. Ngapo, B. H. P. Wilkinson, and R. Chong, 1,5-Glocono-δ-lactone-induced gelation of myofibrillar protein at chilled temperatures, *Meat Sci. 42*:3 (1996).

55. G. Offer and J. Trinick, On the mechanism of water holding in meat: the swelling and shrinking of myofibrils, *Meat Sci. 8*:245 (1983).
56. I. Ohtsuki, K. Maruyama, and S. Ebashi, Regulatory and cytoskeletal proteins of vertebrate skeletal muscle, *Adv. Protein Chem. 38*:1 (1986).
57. M. Okamoto, Y. Kawamura, and R. Hayashi, Application of high pressure to food processing: textural comparison of pressure- and heat-induced gels of food proteins, *Agric. Biol. Chem. 54*:183 (1990).
58. E. O'Neill, P. A. Morrissey, and D. M. Mulvihill, Heat-induced gelation of actomyosin, *Meat Sci. 33*:61 (1993).
59. E. O'Neill, D. M. Mulvihill, and P. A. Morrissey, Surface properties of muscle protein extracts, *Meat Sci. 25*:1 (1989).
60. N. Parsons and P. Knight, Origin of variable extraction of myosin from myofibrils treated with salt and pyrophosphate, *J. Sci. Food Agric. 51*:71 (1990).
61. A. M. Pearson and R. B. Young, *Muscle and Meat Biochemistry*, Academic Press, Inc., San Diego, CA, 1989.
62. D. Pette and R. S. Staron, Cellular and molecular diversities of mammalian skeletal muscle fibers, *Rev. Physiol. Biochem. Pharmacol. 116*:2 (1990).
63. P. L. Privalove and S. J. Gill, Stability of protein structure and hydrophobic interaction, *Adv. Protein Chem. 39*:191 (1988).
64. I. Rayment, W. R. Rypniewski, K. Schmidt-Base, R. Smith, D. R. Tomchick, M. M. Benning, D. A. Winkelmann, G. Wesenberg, and H. M. Holden, Three-dimensional structure of myosin subfragment-1: A molecular motor, *Science 261*:50 (1993).
65. J. Robbins, T. Horan, J. Gulick, and K. Kropp, The chicken myosin heavy chain family, *J. Biol. Chem. 261*:6606 (1986).
66. G. H. Robe and Y. L. Xiong, Dynamic rheological studies of salt-soluble proteins from three porcine muscles, *Food Hydrocolloids 7*:137 (1993).
67. B. W. Rosser and J. C. George, The avian *pectoralis*: histochemical characterization and distribution of muscle fiber types, *Can. J. Zool. 64*:1174 (1986).
68. H. Sakamoto, Y. Kumazawa, S. Toiguchi, K. Seguro, T. Soeda, and M. Motoki, Gel strength enhancement by addition of microbial transglutaminase during onshore surimi manufacture, *J. Food Sci. 60*:300 (1995).
69. K. Samejima, B. Egelandsdal, and K. Fretheim, Heat gelation properties and protein extractability of beef myofibrils, *J. Food Sci. 50*:1540 (1986).
70. K. Samejima, M. Ishioroshi, and T. Yasui, Relative roles of the head and tail portions of the molecule in heat-induced gelation of myosin, *J. Food Sci. 46*:1412 (1981).
71. K. Samejima, M. Ishioroshi, and T. Yasui, Heat induced gelling properties of actomyosin: effect of tropomyosin and troponin, *Agric. Biol. Chem. 46*:535 (1982).
72. K. Samejima, K. Kuwayama, K. Yamamoto, A. Asghar, and T. Yasui, Influence of reconstituted dark and light chicken muscle myosin filaments on the morphology and strength of heat-induced gels, *J. Food Sci. 54*:1158 (1989).
73. K. Samejima, N. H. Lee, M. Ishioroshi, and A. Asghar, Protein extractability and thermal gel formability of myofibrils isolated from skeletal and cardiac muscles at different postmortem periods, *J. Sci. Food Agric. 58*:385 (1992).
74. K. Samejima, Y. Oka, K. Yamamoto, A. Asghar, and T. Yasui, Effects of SH groups, ϵ-NH_2 groups, ATP, and myosin subfragments on heat-induced gelling of cardiac myosin and comparison with skeletal myosin and actomyosin gelling capacity, *Agric. Biol. Chem. 52*:63 (1988).
75. K. Samejima, H. Yamauchi, A. Asghar, and T. Yasui, Role of myosin heavy chains from rabbit skeletal muscle in the heat-induced gelation mechanism, *Agric. Biol. Chem. 48*:2225 (1984).
76. T. Sano, S. F. Noguchi, T. Tsuchiya, and J. J. Matsumoto, Contribution of tropomyosin to fish muscle gel characteristics, *J. Food Sci. 54*:258 (1989).

77. F. H. Schachat, A. C. Canine, M. M. Briggs, and M. C. Reedy, The presence of two skeletal muscle α-actinins correlates with troponin-tropomyosin expression and Z-line width, *J. Cell Biol. 101*:1001 (1985).
78. A. Sharp and G. Offer, The mechanism of formation of gels from myosin molecules, *J. Sci. Food Agric. 58*:63 (1992).
79. T. J. Sims and A. J. Bailey, Connective tissue, *Developments in Meat Science–2* (R. A. Lawrie, ed.), Applied Science Publishers, London, 1981, p. 29.
80. A. B. Smyth, D. M. Smith, V. Vega-Warner, and E. O'Neill, Thermal denaturation and aggregation of chicken breast muscle myosin and subfragments, *J. Agric. Food Chem. 44*: 1005 (1996).
81. S. Srinivasan and Y. L. Xiong, Gelation of beef heart surimi as affected by antioxidants, *J. Food Sci. 61*:707 (1996).
82. S. Srinivasan, Y. L. Xiong, and E. A. Decker, 1996, Inhibition of protein and lipid oxidation in beef heart surimi-like material by antioxidants and combinations of pH, NaCl, and buffer type in the washing media, *J. Agric. Food Chem. 44*:119 (1996).
83. D. W. Stanley, A. P. Stone, and H. O. Hultin, Solubility of beef and chicken myofibrillar proteins in low ionic strength media, *J. Agric. Food Chem. 42*:863 (1994).
84. R. Starr, R. Almond, and G. Offer, Location of C-protein, H-protein and X-protein in rabbit skeletal muscle fiber types, *J. Muscle Res. Cell Motil. 6*:227 (1985).
85. R. Starr and G. Offer, Polypeptide chains of intermediate molecular weight in myosin preparations, *FEBS Lett. 15*:40 (1971).
86. R. Starr and G. Offer, H-protein and X-protein. Two new components of the thick filaments of vertebrate skeletal muscle, *J. Mol. Biol. 170*:675 (1983).
87. G. Stefansson and H. O. Hultin, On the solubility of cod muscle proteins in water, *J. Agric. Food Chem 42*:2656 (1994).
88. A. P. Stone and D. W. Stanley, Muscle protein gelation at low ionic strength, *Food Res. Int. 27*:155 (1994).
89. E. E. Strehler, M.-A. Strehler-Page, J.-C. Perriad, M. Periasamy, and B. Nadal-Ginard, Complete nucleotide and encoded amino acid sequences of a mammalian myosin heavy chain gene. Evidence against intron-dependent evolution of the rod, *J. Mol. Biol. 190*:291 (1986).
90. D. R. Swartz, S.-S. Lim, T. Fassel, and M. L. Greaser, Mechanisms of myofibril assembly, Proceedings of 47th Annual Reciprocal Meat Conference, College Station, Pennsylvania, 1994, pp. 141–161.
91. C. A. Swenson and P. A. Ritchie, Conformational transitions in the subfragment-2 region of myosin, *Biochemistry 19*:5371 (1980).
92. T. Tanaka, Gels, *Sci. Am. 244*:124 (1981).
93. R. G. Taylor, G. H. Geesink, V. F. Thompson, M. Koohmaraie, and D. E. Goll, Is Z-disk degradation responsible for postmortem tenderization? *J. Anim. Sci. 73*:1351 (1995).
94. D. M. Theno and G. R. Schmidt, Microstructural comparison of three commercial frankfurters, *J. Food Sci. 43*:845 (1978).
95. P. J. Torley and O. A. Young, Rheological changes during isothermal holding of salted beef homogenates, *Meat Sci. 39*:23 (1995).
96. R. Tsai, R. G. Cassens, and E. J. Briskey, The emulsifying properties of purified muscle proteins, *J. Food Sci. 37*:286 (1972).
97. J. H. Wang, Hemoglobin and myoglobin, *Oxygenases* (O. Hayaishi, ed.), Academic Press, New York, 1962, p. 469.
98. K. Wang, J. McClure, and A. Tu, Titin: major myofibrillar proteis of striated muscle, *Proc. Antl. Acad. Sci. 76*:3698 (1979).
99. S. F. Wang and D. M. Smith, Dynamic rheological properties and secondary structure of chicken breast myosin as influenced by isothermal heating, *J. Agric. Food Chem. 42*:1434 (1994).

100. S. F. Wang and D. M. Smith, Heat-induced denaturation and rheological properties of chicken breast myosin and F-actin in the presence and absence of pyrophosphate, *J. Agric. Food Chem. 42*:2665 (1994).
101. S. F. Wang and D. M. Smith, Gelation of chicken breast muscle actomyosin as influenced by weight ratio of actin to myosin, *J. Agric. Food Chem. 43*:331 (1995).
102. L. Wicker, T. C. Lanier, J. A. Knopp, and D. D. Hamann, Influence of various salts on heat-induced ANS fluorescence and gel rigidity development of tilapia (*serotherodon aureus*) myosin, *J. Agric. Food Chem. 37*:18 (1989).
103. D. J. Wright and P. Wilding, Differential scanning calorimetric study of muscle and its proteins: myosin and its subfragments, *J. Sci. Food Agric. 35*:357 (1984).
104. J. Q. Wu, D. D. Hamann, and E. A. Foegeding, Myosin gelation kinetic study based on rheological measurements, *J. Agric. Food Chem. 39*:229 (1991).
105. M. M. Wu, T. C. Lanier, and D. D. Hamann, Rigidity and viscosity changes of croaker actomyosin during thermal gelation, *J. Food Sci. 50*:14 (1985).
106. Y. L. Xiong, Thermally induced interactions and gelation of combined myofibrillar protein from white and red broiler muscles, *J. Food Sci. 57*:581 (1992).
107. Y. L. Xiong, Myofibrillar protein from different muscle fiber types: implications of biochemical and functional properties in meat processing, *CRC Crit. Rev. Food Sci. Nutr. 34*: 293 (1994).
108. Y. L. Xiong and A. F. Anglemier, Gel electrophoretic analysis of the protein changes in ground beef stored at 2°C, *J. Food Sci. 52*:287 (1989).
109. Y. L. Xiong and S. P. Blanchard, Functional properties of myofibrillar proteins isolated from cold-shortened and thaw-rigor bovine muscles, *J. Food Sci. 58*:720 (1993).
110. Y. L. Xiong and S. P. Blanchard, Myofibrillar protein gelation: viscoelastic changes related to heating procedures, *J. Food Sci. 59*:734 (1994).
111. Y. L. Xiong and S. P. Blanchard, Rheological properties of salt-soluble protein from white and red skeletal muscles, *J. Agric. Food Chem. 42*:1624 (1994).
112. Y. L. Xiong and S. P. Blanchard, Dynamic gelling properties of myofibrillar protein from skeletal muscles of different chicken parts, *J. Agric. Food Chem. 42*:670 (1994).
113. Y. L. Xiong and C. J. Brekke, Changes in protein solubility and gelation properties of chicken myofibrils during storage, *J. Food Sci. 54*:1141 (1989).
114. Y. L. Xiong and C. J. Brekke, Thermal transitions of salt-soluble proteins from pre- and postrigor chicken muscles, *J. Food Sci. 55*:1540 (1990).
115. Y. L. Xiong and C. J. Brekke, Physicochemical and gelation properties of pre- and postrigor chicken salt-soluble proteins, *J. Food Sci. 55*:1544 (1990).
116. Y. L. Xiong and C. J. Brekke, Protein extractability and thermally induced gelation properties of myofibrils isolated from pre- and postrigor chicken muscles, *J. Food Sci. 56*:210 (1991).
117. Y. L. Xiong and C. J. Brekke, Gelation properties of chicken myofibrils treated with calcium and magnesium chlorides, *J. Muscle Foods 2*:21 (1991).
118. Y. L. Xiong, C. J. Brekke, and H. K. Leung, Thermal denaturation of muscle protein from different species and animal types as studied by differential scanning calorimetry, *Can. Inst. Food Sci. Technol. J. 20*:357 (1987).
119. Y. L. Xiong, E. A. Decker, G. H. Robe, and W. G. Moody, Gelation of crude myofibrillar protein isolated from beef heart under antioxidative conditions, *J. Food Sci. 58*:1241 (1993).
120. K. Yamamoto, Electron microscopy of thermal aggregation of myosin, *J. Biochem. 108*:896 (1990).
121. K. Yamamoto, K. Samejima, and T. Yasui, The structure of myosin filaments and the properties of heat-induced gel in the presence and absence of C-protein, *Agric. Biol. Chem. 51*: 197 (1987).
122. K. Yamamoto, K. Samejima, and T. Yasui, Heat-induced gelation of myosin filaments, *Agric. Biol. Chem. 52*:1803 (1988).

123. T. Yasui, M. Ishioroshi, H. Nakano, and K. Samejima, Changes in shear modulus, ultrastructure and spin-spin relaxation times of water associated with heat-induced gelation of myosin, *J. Food Sci. 44*:1201 (1979).
124. L. D. Yates and M. L. Greaser, Quantitative determination of myosin and actin in rabbit skeletal muscle, *J. Mol. Biol. 168*:123 (1983).
125. O. A. Young and C. L. Davey, Electrophoretic analysis of proteins from single bovine muscle fibers, *Biochem. J. 195*:317 (1981).
126. O. A. Young, P. J. Torley, and D. H. Reid, Thermal scanning rheology of myofibrillar proteins from muscles of defined fiber types, *Meat Sci. 32*:45 (1992).
127. G. R. Ziegler and J. C. Acton, Mechanisms of gel formation by proteins of muscle tissue, *Food Technol. 38* (5):77 (1984).
128. G. R. Ziegler and E. A. Foegeding, The gelation of proteins, *Adv. Food Nutr. Res. 34*:203 (1990).

13

Enzyme and Chemical Modification of Proteins

KLAUS DIETER SCHWENKE
Universität Potsdam, Bergholz-Rehbrücke, Germany

I. INTRODUCTION

Food proteins in their natural state do not necessarily possess optimal nutritional and functional properties. Egg proteins are boiled so as to increase their digestibility. Meat is cooked or roasted so as to make it more palatable and also to improve both its nutritive and its organoleptic properties. The latter is highly affected by the heat-induced generation of flavor compounds. Heat-induced reactions between food proteins and reducing sugars (Maillard reaction) are the chemical bases for producing taste, color, and flavor in baking products. Meat ripening is an example of enzymatic protein modification where some proteases are involved. Proteolytic processes are also responsible for milk clotting and production of cheese products.

In some cases, the modification of food protein caused by food processing leads to undesirable changes in or deterioration of its nutritional and functional qualities. Examples are extreme nonenzymatic browning by strong heating or the formation of lysinoalanine and other products of decomposition of protein amino acids by alkali treatment. It is therefore appropriate to use the term "modification" to connote alteration of the protein structure to achieve desirable changes in nutritive or functional properties.

Recent progress in protein modification has been summarized in some excellent monographies and review articles (e.g., Refs. 1–9). This chapter focuses on relevant methods of food protein modification with regard to functional properties and nutritional implications. Only representative examples are selected here from the enormous number of published papers on chemical and enzymatic modification of food proteins.

Purified proteolytic enzymes are often used to partially hydrolyze food proteins [8]. Both the proteolytic enzymes of the digestive tract and those of plant (e.g., papain) or microbial (e.g., alcalase, neutrase, thermitase) origin have been applied to modify food proteins [8]. This is in contrast to ancient oriental procedures for preparing indigenous fermented foods such as natto or tempeh [10] and also to the use of proteolytic processes in traditional food, e.g., dairy technology.

A. Limited Proteolysis and Functional Properties

A large number of papers describe the effect of proteolytic breakdown of food proteins on their functional properties. They comprise animal proteins such as casein [11–13], soybean proteins [8,14–20], faba bean proteins [21,22], wheat gluten and gluten fractions [23,24], maize proteins [25,26], oat flour [27], cottonseed proteins [28], sunflower [29], and rapeseed proteins [30], among others.

Proteolytic modification has special importance for the improvement of solubility of proteins, e.g., from cereals that are poorly soluble in aqueous media. This effect becomes significant even after very limited hydrolysis. Zein, the maize storage protein that is highly unsoluble at pH 2–5, exhibited good solubility (30–50%) at this pH range when only 1.9% of the peptide bonds were split by treatment with trypsin [25]. Thermitase treatment of wheat gluten up to a degree of hydrolysis (DH) of 9.8% increased the solubility at pH 7 from 7 to 50% [23]. Alcalase- or neutrase-treated oat flours had two to three times greater protein solubility under isoelectric conditions (pH 5). At a constant enzyme-to-substrate ratio, the solubility increased with an increase of DH (3.8–10.4%) [27]. Limited proteolysis of a soybean protein isolate with alcalase at pH 8 and neutrase at pH 7 changed the pH solubility curve from the U-shape to a flat curve, which was relatively independent of pH [15]. Thermitase treatment of a faba bean protein isolate to a DH of 8.3% increased the solubility at the isoelectric pH to more than 40% [21]. Hydrolysis of casein to DH of 2 and 6.7%, respectively, with *Staphylococcus aureus* V8 protease increased the isoelectric solubility to 25 and 50%, respectively [12].

Proteolytic modification improved the moisture sorption and water binding of several proteins [14,19,27]. This is because moisture uptake by proteins is related to the number of ionic groups present, which increases as a result of hydrolysis due to the liberation of amino and carboxyl groups [31]. The moisture uptake of a soybean protein isolate even at 84% relative humidity and room temperature had increased in proportion to the extent of enzyme treatment [14]. A 2- to 2.5-fold increase in water adsorption both for an acid-precipitated soybean protein and the 11S globulin (glycinin) was also observed after limited proteolysis with bromelain [19]. The water-holding capacity of neutrase- or alcalase-treated oat flours increased with an increase in DH, when a particular enzyme/substrate system was compared [27].

Emulsifying properties of proteins are sensitive to proteolytic modification. Limited hydrolysis (DH 2 and 6.7%) of casein decreased the emulsifying activity (EA) at all pH [11,12], whereas the emulsion stability (ES) at DH = 2% was higher than that of native casein [11]. The EA of casein was reported to decrease with increasing net charge and with the decreasing hydrophobicity due to hydrolysis [13]. An increase in EA of soybean protein was found after a short incubation with pepsin, whereas it decreased when the incubation time was increased [16]. A relationship between the peptide chain length in protein hydrolysates and emulsifying properties was discussed [15]. Optimum conditions were found in soybean protein isolates at DH of about 5% [17]. Proteolytic modification can improve or impair the emulsifying properties of peanut proteins depending on the nature of the enzyme and the conditions of incubation [32,33]. Sunflower protein lost its EA and ES after proteolysis irrespective of the method of enzyme inactivation [29].

Studies on partial tryptic digests of 11S globulins from soybean (glycinin) [18] and faba bean [22,34] revealed that the high molecular weight hydrolysis products glycinin-T and legumin-T, respectively, play a key role in emulsification and emulsion stabilization. While EA and ES increased with the formation of legumin-T, they decreased when legumin-T was destroyed by progressive tryptic digestion [22].

The beneficial effect of limited proteolysis of emulsifying activity and emulsion stability may be due to exposure of buried hydrophobic groups, which may improve the hydrophobic-hydrophilic balance for better emulsification [31]. The loss of hydrophilic peptides from the surface of the proteins may directly result in an increase of the surface hydrophobicity, thus favoring surface adsorption [34]. The detrimental effect of excessive digestion is related to the loss of globular structure and optimum size of split peptides, resulting in formation of a thinner protein layer around the oil droplets and a less stable emulsion [31].

Although proteolytic modification generally increases the foaming power, it decreases the foam stability [37]. Alcalase or neutrase treatment of soybean protein isolate up to a DH of 3–6% increased the foaming capacity (FC) severalfold [17]. A great increase in foaming capacity and a marked decrease in foam stability (FS) were reported for faba bean protein isolate that was partially hydrolyzed (DH = 3.2–8.3%) with thermitase [21]. While the FC of wheat gluten was not significantly changed by thermitase treatment (DH = 5.2–9.8%), the FS decreased markedly with increasing DH [23]. Native zein, which produced no foam at pH 1–12, showed significantly improved FC after partial tryptic hydrolysis (DH = 1.4–1.9%) [25]. The foam produced was, however, very unstable. A significant increase in foam expansion was also reported for peanut proteins partially hydrolyzed by papain [32] and for sunflower protein after hydrolysis with pepsin [29]. Obviously, partial cleavage and unfolding of the protein by proteolytic modification favors rapid unfolding and rearrangement of peptides and formation of a film at the water/air interface. The film formed is, however, not thick and viscoelastic enough to provide stability to the foam.

Limited proteolysis generates high molecular weight products via a zipper-type [38] splitting of larger peptides from the surface of the proteins. In the case of 11S plant-storage proteins with molecular masses of about 350 kDa, tryptic breakdown generates products with molecular masses in the range of 200–260 kDa (globulin-T), which still maintained the principal structural organization of the original protein [34,39–42]. This modification causes, however, a structural destabilization of 11S proteins, which may favor dissociation and unfolding at the surface. This behavior might be responsible for the improving effect of globulin-T intermediates on surface functional properties.

B. Progessive Proteolysis

1. Liberation of Biologically Functional Peptides

The discovery of opioid activity of peptides derived from partial digestion of milk proteins [43] and wheat gluten [44] introduced a new criterion for evaluating the "nutritive value" of food proteins [45]. Peptides that are in an inactive state within the amino acid sequence of a protein may be released by digestive processes in vivo and may act as potential physiological modulators of metabolism during the gastrointestinal passage of the diet [45]. In analogy to the endogenous opioid peptides (endorphins), opioid peptides from food proteins have been called "exorphins" [44].

Opioid peptides have been found in the hydrolysates of both casein fractions and whey proteins. Some have an affinity to opiate receptors like enkephalins ("agonists"), others have both affinity for opiate receptors and antiopiate effects like nalaxone ("antagonists"). Some examples of opioid peptides from milk proteins and wheat gluten are given in Table 1.

An excellent review of the state of research on bioactive peptides found in milk proteins has recently been published [48]. It also summarizes peptide sequences other than opiods. These are immunostimulating peptides found both in $\alpha_{S,1}$- and β-casein, angiotension I–converting enzyme–inhibiting peptides (casokinins) found in $\alpha_{S,1}$- and β-casein, anthithrombotic peptides (casoplatelins) found in κ-casein, and mineral-binding peptides characterized by an accumulation of serine phosphate residues isolated from $\alpha_{S,1}$- and β-casein [48]. Some opioid peptides from wheat gluten and gliadin hydrolysates have been isolated and characterized [49,50]. The role of bioactive peptides from food protein hydrolysates as exogenous regulatory substances is still an unresolved problem.

2. Bitter Peptides

Depending on the nature of the protein and the protease used, progressive proteolysis can liberate bitter peptides from proteins, the bitterness of which is a function of amino acid composition and sequence as well as the peptide chain length [51,52]. The bitter taste of peptides derived especially from casein and soybean proteins has been related to the hydrophobicity of the peptides and was first expressed by Ney [53] in the form of the so-called Q-value. For molecular weights up to about 6000 daltons, the Q-value was found to be higher than 1400 kcal/mol for bitter peptides and less than 1300 kcal /mol for nonbitter peptides [53]. Systematic investigation of the threshold value of bitterness for a large number of amino acids, amino acid derivatives, amines, and peptides led Belitz and coworkers [54,55] to derive a quantitative relationship between hydro-

TABLE 1 Opioid Peptides from Milk Proteins and Wheat Gluten

Sequence	Name	Origin and sequence positions in the protein	Ref.
Endogenous peptides:			
Tyr Gly Gly Phe Leu	Leu/enkephalin		
Agonists from milk proteins:			
Arg Tyr Leu Gly Tyr Leu Glu	α-Casein exorphin	Bovine $\alpha_{S,1}$-casein (90–91)	46
Tyr Pro Phe Pro Gly	β-Casomorphin 5	Bovine β-casein (60–65)	43
Tyr Gly Leu Phe-NH_2	α-Lactorphin	Human and bovine α-Lactalbumin (50–53)-NH_2	47
Tyr Leu Leu Phe-NH_2	β-Lactorphin	Bovine β-lactoglobulin (102–105)-NH_2	47
Antagonists from milk proteins:			
Tyr Pro Ser Tyr-OCH_3	Casoxin 4	Bovine $\alpha_{S,1}$-casein (35–38)-OCH_3	47
Exorphins from wheat gluten:			
Gly Tyr Tyr Pro	A4	HMW glutenin	50
Tyr Gly Gly Trp	B4	Not determined	50
Tyr Pro Gln Pro Gln Pro Phe	Gliadorphin	α-Gliadin (43–49)	49

phobicity and bitter taste. They found that the interaction of the amino group with a nucleophilic receptor 3 Å apart from the hydrophobic interaction site gave rise to bitterness. However, this would give a sweet taste when the carboxyl group also interacts. The bitterness was expressed as the logarithm of the threshold value, that is, the molar concentration of the peptide below which the bitterness was not perceived [54].

The problems in understanding the bitterness of protein hydrolysates on the basis of a modified "Q-role" were summarized by Adler-Nissen [52]. The author followed the process of hydrolysis, where the bitterness reaches a maximum when more and more hydrophobic side chains become exposed. Finally, extensive hydrolysis usually results in a decreased overall bitterness, which is underlined by the successful use of exopeptidase for debittering casein hydrolysates. Correspondingly, a qualitative relationship between bitterness and DH was derived [15]. Molecular weight limits between 1000 and 6000 daltons were reported to be necessary for bitterness [15,53,54]. The bitterness of a protein hydrolysate was described on the basis of the hydrophobicity distribution of peptides taking into account the enzyme specificity, the nature of the protein, and the conditions of peptide separation [52].

The practical use of protein hydrolysates for food or medical purposes requires the reduction of bitterness by removal of bitter peptides. This can be realized by different methods including adsorption on hydrophobic supports, azeotropic extraction, use of exopeptidases, masking bitterness, and plastein reaction. These are discussed in detail by Adler-Nissen [8].

3. Plasteinlike Reactions

When protein hydrolysates at high concentration were incubated with a protease, the formation of a precipitate, often gel-like ("plastein"), was observed in many cases [56,57]. The factors that affect the formation of plastein are the size of the peptides in the protein hydrolysate, the substrate concentration, and the pH of the incubation medium. Peptide fractions with an average molecular weight of 1043 and 685 have been shown to produce plasteins much more effectively than lower and higher molecular weight fractions. Optimum substrate concentration was found to be in the range of 20–40% (w/v) [57]. The optimum pH value for plastein synthesis is in the range of pH 4.0–7.0 [57]. It differs conderably for some proteolytic enzymes from the pH optimum of degradation [57].

Condensation and transpeptidation have been discussed as the mechanism of plastein formation [57]. That transpeptidation should be primarily responsible for plastein synthesis has been shown [59,60]. This was deduced from the analysis of α-amino nitrogen [58] and the determination of the molecular weight distribution of peptides, which remained unchanged or showed only slight changes.

Plasteins are more hydrophobic than the original peptide mixture and tend to aggregate at high temperature [57]. This temperature effect is consistent with hydrophobic interactions, which are enhanced at high temperature. In accordance with this, the gel formation of plasteins was suggested as an entropy-driven process [56], where the driving force is the increase in entropy of water after formation of an initial concentration of suitable peptides by either condensation or transpeptidation or both.

An important practical application of the plastein reaction is its use in the debittering of protein hydrolysates, as previously mentioned. Another application is its use for improving the nutritional quality of food proteins by incorporation of essential amino acids, especially methionine and its esters. This has been demonstrated, e.g., for hydrolysates from soybean protein [61,62,64,65], faba bean protein [66], ovalbumin [63], and casein [65], as well as microbial and leaf proteins [62]. Methionine incorporation into soy protein has also been achieved by a one-step process combining protein degradation and plastein synthesis using papain [64]. The initial velocity of papain-catalyzed incorporation of the different amino acid esters into plastein has been shown to be dependent on the amino acid side chain structure [63]. The velocities tended to increase with increasing hydrophobicities of the side chains, except for the cases of β-branched-chain amino acids. Even poor hydrophobic and β-branched-chain amino acids could be effectively incorporated when esterified with longer-chain alcohols [63]. Examples for the enrichment of lysine content in wheat gluten and for the simultaneous incorporation of methionine, lysine, and tryptophan into algal, bacterial, and leaf protein hydrolysates via papain-catalyzed plastein reaction have been reported [62].

Evidence was provided that methionine was incorporated predominantly at the C-terminals of the plastein molecules [57]. Both N-terminal and C-terminal incorporation of methionine into a milk protein hydrolysate was reported, though the latter was three times as high as that at the N-terminal position [67]. The plasteinlike enzymatic peptide modification of casein combined with methionine incorporation significantly reduced the allergenic character of the protein [65].

A modification of the classical plastein reaction into a new process has been reported, which permits the use of a protein itself instead of a protein hydrolysate as a material to be improved [68]. The process consists of a papain-catalyzed aminolysis of a protein by an amino acid ester at pH 8–10 leading to a product to which the amino acid ester was attached covalently. Treatment of $\alpha_{S,1}$-casein, the solubility of which was enhanced by succinylation, in the presence of leucine-dodecylsulfate ester gave a definite amphiphilic product consisting of a 20 kDa macropeptide with the amino acid ester at the C-terminal position [68]. Similarly, gelatin was modified using L-leucine esters with n-alkyl chains ranging from C2 to C12 [68]. The products had a molecular weight distribution in the range of 2–40 kDa with averages of 7.3–8.3 kDa. Approximately 1 mol of leucine dodecyl ester was attached to 7.5 g of protein in the corresponding reaction product. Fish protein concentrate, soy protein isolate, casein, and ovalbumin in a succinylated form were correspondingly modified [68].

The surface functional properties of the products vary depending on their lipophilicity, which is related to the chain length of the n-alkyl moiety attached. Attachment of a C4-C6 n-alkyl ester of L-leucine to gelatin led to a greatly improved whippability as well as to stable foams, while the highest oil-emulsifying activity resulted when a longer-chain n-alkyl ester, especially n-dodecyl ester of L-leucine, was attached. Several of these proteinaceous sufactants were used in food applications. Thus, gelatin-Leu-0C6 (EM6-6) was shown to be a useful ingredient for snow jelly, while gelatin-Leu-0C12 (EM6-12) was especially suitable for ice cream, mayonnaise, and bread [68]. EM6-12 was shown to have antifreeze properties and to be able for acting as cryoprotecting agent [69].

III. GLYCOSYLATION

Glycoproteins are widely distributed in nature and serve a variety of functions in animals, plants, and microorganism [70]. Glycoproteins that are produced chemically, in vitro, by the covalent attachment of mono- or oligosaccharides to proteins have been called "neoglycoproteins" [71]. Introduction of saccharides into proteins makes the latter more hydrophilic. Glycosylation is therefore an interesting tool for improving food protein functionality.

A. Neoglycoproteins

From a practical point of view, the most simple and acceptable way of glycosylation is the attachment of sugar residues to proteins via the Maillard reaction [72]. When ovalbumin was mixed with several disaccharides having glucose at the reducing end and maintained at 50°C and 65% relative humidity for 0–20 days, a progressive blocking of the ϵ-amino groups of lysine and a concomitant change of physicochemical properties of the adducts was observed [73]. Isomaltose and melibiose strongly induced brown coloration and protein polymerization under these conditions, whereas maltose, cellobiose, and lactose did so only weakly. The weaker production of advanced Maillard reaction products in the latter cases indicated that the terminal pyranoside group bonded at the C-4 OH of glucose retarded further degradation to aldehyde components of the Amadori rearrangement products [73].

Similarly, polysaccharides such as dextran and galactomannan could be attached to lysozyme, ovalbumin, and dry egg white proteins by keeping the mixture at 60°C and 79% relative humidity [74–76]. These protein-polysaccharide conjugates have been shown to possess excellent emulsifying properties superior to commercial emulsifiers [75–77].

The heat stability of lysozyme was enhanced by conjugation with galactomannan [76]. Lysozyme-dextran and lysozyme-galactomannan conjugates exhibited antimicrobial activity against both gram-negative and gram-positive bacteria [74–76], while native lysozyme failed to lyse gram-negative bacteria.

The method of reductive alkylation (Fig. 1) was used for attachment of sugar residues to casein [78,79] and pea legumin [80]. The reaction proceeded under moderate alkaline conditions (pH 8.0–8.5) at 37°C, where mixtures of the protein with cyanoborohydride and various amounts of the carbohydrates were kept for different times [78]. The aldoses (glucose, galactose, mannose) were coupled faster than the tested ketose (fructose) to casein, perhaps because of their higher reducing power. The reaction rate of the studied disaccharides (maltose, lactose) was intermediate because of the lower amount of reducing groups per unit mass [79].

$$\text{Glyco–CHO} \xrightleftharpoons{\text{H}_2\text{N-protein}} \text{Glyco–CH=N–protein} \xrightleftharpoons{\text{NaCNBH}_3} \text{Glyco–CH}_2\text{–NH–protein} \quad (1)$$

$$\text{protein–COO}^- + \text{R'–N=C=N–R'} \xrightarrow{\text{pH 4-7}} \text{protein–CO–O–}\underset{\underset{\text{NR'}}{\|}}{\text{C}}\text{–NH–R'}$$

$$\xrightarrow{\text{glucosamine}} \text{protein–CO–NH–glucose} + \text{R'–NH–CO–NH–R'} \quad (2)$$

FIGURE 1 Covalent attachment of carbohydrates to protein: attachment of a reducing sugar to a protein amino group by reductive alkylation (1), attachment of an amino sugar via the activation of a protein carboxyl group by a carbodiimide (2).

β-Lactoglobulin was glycosylated by different methods. Maltose or β-cyclodextrin residues were introduced into the protein by the reaction of carbohydrate cyclic carbonate derivative with the amino groups of the protein [81]. Glucosamine, glucosamine-oxtaose [81], as well as gluconic and melibionic acid [82] were attached to the protein using the carbodiimide method (Fig. 1). Up to 65% of the amino or the carboxyl groups of β-lactoglobulin could be glycosylated using the two methods, respectively. Glycosylation induced a change in the secondary and tertiary structures and the hydrophobicity of the protein, the extent of which depended on the nature of attached sugar residues and on the degree of modification [80,83]. The structural changes were reflected in an increase in the viscosity of protein solutions [79,80,83].

The solubility of all modified caseins was increased in the range of their isoelectric points, unchanged in neutral and alkaline solution, and decreased at low pH values [79]. Because of an increase in the flexibility and unfolded state of glycosylated casein, the glycosylated and galactosylated derivatives adsorbed more rapidly at the interface and also had better foaming capacity and stability than the native casein [84]. Glycosylated β-lactoglobulin retained more water in the foam initially and after drainage and yielded a more stable foam than the native protein, while emulsifying activity and emulsion stability of glycosylated β-lactoglobulin improved slightly [85]. A heat-stabilizing effect of attached gluconic or melibionic acid on β-lactoglobulin was also observed [82]. The foaming capacity as well as the stability of foams and emulsions of legumin were improved by glycosylation. Neutral sugars favored the foaming properties, while the charged carbohydrates improved the emulsifying properties [86].

B. Enzymatic Glycosylation of Protein

Transglutaminase (glutaminyl peptide: amine γ-glutamyltransferase; EC 2.3.2.13) has been shown to catalyze the attachment of glycosyl residues to glutamine residues of protein when the sugar is extended chemically with an alkylamine and could serve as an amine donor for transglutaminase [87]. This method was applied for the glycosylation of β-casein [87] as well as pea legumin and wheat β-gliadins [88]. To prevent ε-(γ-glutamyl)-lysine cross-link formation, the lysine residues of legumin and β-gliadins were first blocked by reductive alkylation. Using 6-aminohexyl 1-thio-β-D-galactopyranoside, 18 and 57 glycosyl units per mole of alkylated β-gliadins and legumin, respectively, could be incorporated [88]. The solubility of neoglycoproteins was markedly increased over that of native proteins in the range of their isoelectric points.

IV. PHOSPHORYLATION

Some important proteins of the regular human diet, e.g., milk casein, egg white albumin, and egg yolk phosvitin, are phosphoproteins. The introduction of phosphoryl residues increases the negative charge and hydration and changes the functional properties of proteins.

A. Chemical Phosphorylation

Different reagents have been proved to be useful to chemically phosphorylate proteins [89]. However, phosphorus oxychloride ($POCl_3$) and sodium trimetaphosphate (STMP) seem to be the main suitable reagents for large-scale phosphorylation of food proteins [89,90].

In an aqueous system, $POCl_3$ is rapidly hydrolyzed in an exothermic reaction via the dichloro and monochloro compounds to the phosphate ion. When $POCl_3$ is directly added to an aqueous protein solution, the quick liberation of protons decreases the pH and the heat produced causes denaturation of the protein. In order to minimize this problem, the phosphorylation with $POCl_3$ is carried out by adding small portions of the reagent dissolved in an organic solvent (e.g., carbon tetrachloride or *n*-hexane) to the aqueous protein solution cooled in an ice bath. The pH is maintained between 8 and 9 by adding sodium hydroxide as required [90]. Using pH stability studies, ^{31}P nuclear magnetic resonance (NMR) spectroscopy, and digestion of proteins followed by analysis of phospho amino acids, evidence for the phosphorylation of serine and threonine, lysine, histidine, and tyrosine residues by $POCl_3$ has been reported (Fig. 2) [90]. The formation of phosphoanhydrides with aspartic and glutamic residues has been observed in phosphorylated lysozyme [90]. Diphosphate and tri (or poly)phosphate residues and intermolecular cross-links in proteins phosphorylated with $POCl_3$ have also been detected [89,91]. Though the nature of these cross-links is not yet elucidated, it is likely that O,O-

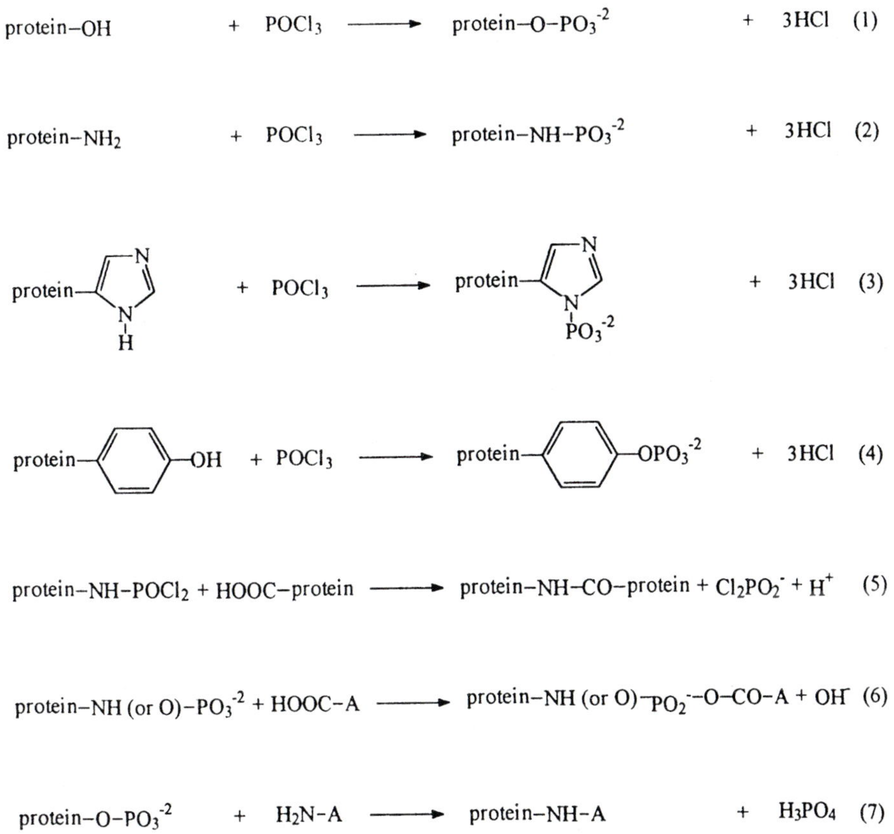

$$\text{protein–OH} + POCl_3 \longrightarrow \text{protein–O–}PO_3^{-2} + 3HCl \quad (1)$$

$$\text{protein–}NH_2 + POCl_3 \longrightarrow \text{protein–NH–}PO_3^{-2} + 3HCl \quad (2)$$

$$\text{protein–(imidazole, N–H)} + POCl_3 \longrightarrow \text{protein–(imidazole, N–}PO_3^{-2}) + 3HCl \quad (3)$$

$$\text{protein–}C_6H_4\text{–OH} + POCl_3 \longrightarrow \text{protein–}C_6H_4\text{–O}PO_3^{-2} + 3HCl \quad (4)$$

$$\text{protein–NH–}POCl_2 + \text{HOOC–protein} \longrightarrow \text{protein–NH–CO–protein} + Cl_2PO_2^{-} + H^{+} \quad (5)$$

$$\text{protein–NH (or O)–}PO_3^{-2} + \text{HOOC–A} \longrightarrow \text{protein–NH (or O)–}PO_2^{-}\text{–O–CO–A} + OH^{-} \quad (6)$$

$$\text{protein–O–}PO_3^{-2} + H_2N\text{–A} \longrightarrow \text{protein–NH–A} + H_3PO_4 \quad (7)$$

FIGURE 2 Phosphorylation of proteins: phosphorylation of serine/threonine hydroxyl groups (1), amino groups (2), imidazol residues of histidine (3) and phenolic hydroxyl groups of tyrosine (4), cross-linking by formation of an isopeptide bond (5), incorporation of an amino acid (A) via the carboxyl (6) or amino function (7). (From Refs. 90, 91.)

phosphodiester and N,N-phosphodiamidate bridges as well as isopeptide bonds (Fig. 2) were present [90]. It is this cross-linking ability of $POCl_3$ that facilitates the incorporation of essential amino acids into proteins. This has been shown with zein [11,90]. Up to 40 moles of phosphorus per mole of zein could be introduced by reaction with $POCl_3$ in the absence of amino acids. When the limiting amino acids tryptophan and lysine were present, up to 12 moles of phosphorus per mole of zein were covalently bound with an additionally attachment of both amino acids [90]. However, threonine, the third limiting amino acid in zein, was not covalently bound under these conditions of reaction. The mechanism of activation leading to covalent incorporation of tryptophan and lysine has been suggested to proceed via a mixed phosphate/carboxyl anhydride or by the amino group of amino acids acting as a nucleophile to displace phosphate from the phosphorylated protein (Fig. 2). The activation of the amino acid by $POCl_3$ can also not be excluded [90].

Nitrogen-bound phosphate is very acid labile. Phosphorylated proteins containing nitrogen-bound phosphate lose part of their phosphate in aqueous solution at 2–37°C, even at neutral pH [92,93]. Since oxygen-bound phosphate is acid stable [89] and the pH of most foods is 3–7, O-phosphorylation can be the modification of choice for food proteins [94].

The nature of phosphate linkage in chemically phosphorylated proteins depends on the nature of the protein [89]. Phosphorylated casein showed only O-phosphate ester bonds [92], while N-phosphate bonds were found in phosphorylated β-lactoglobulin [93]. Both N- and O-attached phosphate residues were detected in phosphorylated lysozyme and other proteins [89,92].

Phosphorylation with $POCl_3$ [91,95] or STMP [96] was shown to be suitable to dissociate nucleoprotein complexes in yeast homogenate so as to obtain yeast protein isolates with a low content in nucleic acids. The phosphorylation with STMP was carried out by incubating proteins, such as soy protein isolate [97] and yeast homogenate [96], under strongly alkaline conditions (pH $\sim$ 11) at 35–38°C. Though an attachment of phosphate residues to serine and lysine residues of soybean protein was reported [97], other authors (92) were unable to detect any covalently bound phosphorus in soy protein and lysozyme using the above-mentioned modification method.

The functional properties of proteins can be improved or impaired by phosphorylation, depending on the nature of the protein used. Increased water solubility at pH 2–9 has been reported for zein phosphorylated with $POCl_3$ [25]. STMP treatment shifted the isoelectric point of a soybean protein isolate to a lower pH, which results in an increase of the solubility at pH > 6 [97]. Similarly the solubility of a yeast protein isolate at pH 7 was increased [96]. In contrast to that, water solubility decreased in those cases where protein cross-linking occurred as in casein and lysozyme [92] and soybean glycinin [98] after treatment with $POCl_3$. Cross-linking could also be one reason for the increase of viscosity of phosphorylated glycinin [98] and casein [92]. Increased water binding was reported for $POCl_3$-treated glycinin [98], casein, and lysozome [92] as well as for STMP-treated glycinin [97].

Both increased and decreased emulsifying activity (EA) was reported for proteins phosphorylated with $POCl_3$ [89]. The EA was decreased in phosphorylated casein [92], but increased in phosphorylated soybean protein isolate (SPI) [99] and zein [11] depending on pH and the level of phosphorylation. The STMP-modified soybean protein [97] and yeast protein isolate [96] showed improved EA. The effect of phosphorylation on the emulsion stability of β-lactoglobulin depended significantly on the concentration of Ca^{2+} present [100].

Foam expansion and foam stability were markedly increased in $POCl_3$-treated soybean glycinin [98], STP-treated soybean [97], and yeast [96] protein isolates. Gel-forming properties of casein [92] and gluten [89] were improved after $POCl_3$ treatment, possibily due to cross-linking of these proteins. Phosphorylated β-lactoglobulin underwent spontaneous gelation at ambient temperature in the presence of 100 mM Ca^{2+} at pH 5 and 7 [100], whereas the unmodified protein required heat treatment to gel.

Earlier works on the digestibility of phosphorylated proteins which show little influence of modification on the in vitro digestibility with pepsin and pancreas proteases have been reviewed by Matheis and Whitaker [89,92]. When casein was phosphorylated with $POCl_3$, the initial rates of both trypsin- and chymotrypsin-catalyzed hydrolysis decreased considerably. The extent of hydrolysis after 24 hours of both trypsin- and α-chymotrypsin–catalyzed digestion was, however, the same for the control casein and phosphorylated casein [92]. The phosphate bonds in phosphorylated proteins could be cleaved by phosphatases [93,99].

No differences were found between in vitro hydrolysis of phosphorylated and nonphosphorylated soybean [97,99] and yeast proteins [91] by digestive enzymes. The in vivo digestibility of phosphorylated proteins was studied in a bioassay with *Tetrahymena thermophili* [11,92]. This microorganism has been shown to be similar to mammals in its requirements for essential amino acids in the presence of similar enzyme systems. There were no differences between casein and phosphorylated casein [92], indicating that phosphorylation did not affect digestion, absorption, or utilization of the amino acids by the organism. An increased growth effect was found for zein, to which tryptophan and lysine were chemically attached by phosphorylation [11].

B. Enzymatic Phosphorylation

1. The Enzymes

Enzymes catalyzing protein phosphorylation are called protein kinases (EC 2.7.1.37). There are two main types of protein kinases with regard to the substrate specificity: histone-type and casein (or phosvitin)-type kinases [94,101,102]. Histone-type kinases catalyze the phosphorylation of histone but are not active toward acidic proteins. They are cAMP (or cGMP) dependent. Casein-type kinases are generally cyclic nucleotide independent and catalyze the phosphorylation of acidic proteins such as casein and phosvitin. cAMP-dependent kinases have the highest specificity toward the amino acid sequences Arg-Arg/Lys-X-Ser/Thr > Arg-X-X-Ser/Thr, where serine or threonine are the phosphoacceptor sites [103]. Phosphorylation of serine is generally preferred over threonine [103]. Casein kinase occurs in two distinct sets of enzymes. The casein kinases in mammary tissue (casein kinase I) normally provides in vivo phosphorylation of nascent proteins. They target sites rich in N-terminal, negatively charged, i.e., acidic or phosphorylated amino acids, as follows: Ser(P)-X_2-Ser/Thr or (Asp/$Glu_{2\text{-}4}$ $X_{2\text{-}0}$)X-Ser/Thr. Casein kinases II are multisubstrate ubiquitous enzymes with the consensus sequence Ser/Thr-(Asp/Glu/Ser(P)$)_{1\text{-}3}$, $X_{2\text{-}0}$) [103].

Both a cAMP-dependent kinase (cAMPdPK) and the two types of casein kinase have been used for the phosphorylation of food proteins. The cAMP-dependent kinase was from bovine cardiac muscle. It exists as an inactive tetrameric holoenzyme with two identical regulatory (R) and two identical catalytic (C) subunits [101]. Two molecules of cAMP are bound to each R subunit resulting in the dissociation of the holoenzyme and activation of the C subunit [104]. The catalytic subunit is commercially available. It has been crystallized [105], and the primary sequence has been determined [106].

2. Food Protein Phosphorylation

The catalytic subunit of cAMPdPK from bovine cardiac muscle was used to phosphorylate soybean proteins [94,107–109]. In accordance with the number of potentially available phosphorylation sites predicted by the amino acid composition, glycinin was phosphorylated to a greater extent than β-conglycinin [94]. About 4.5 moles of phosphorus per mole of protein could be incorporated into glycinin under optimum conditions [98]. This means that no more than 25% of the known phosphorylation sites in glycinin were phosphorylated. Although no phosphorylation of the basic polypeptide chains of glycinin was first observed [107], these polypeptide chains could be phosphorylated after optimization of the method [109]. Phosphorus incorporation into the β and α/α' subunits of β-conglycinin was also found [109]. The greatest level of phosphorylation was reached in all protein preparations studied (glycinin, β-conglycinin, and soybean protein isolate) with denatured proteins [94,108].

A casein kinase-I from mammary glands was used successfully to phosphorylate caseins and, to a smaller extent, β-lactoglobulin and α-lactalbumin [110]. Although this type of kinase ought to favor the phosphorylation of sequences already bearing a phosphoserine residue [103], there was a marked preference of the enzyme for dephosphorylated caseins over native caseins [110].

A casein kinase-II from the yeast *Yarrowia lipolytica* was used for the phosphorylation of glycinin and β-conglycinin [111]. Phosphate was incorporated specifically into the α subunits of β-conglycinin and into the acidic subunits of glycinin. About 0.73 mol P/mol β-conglycinin and 0.20 mol P/mol glycinin were found to be incorporated. The accessibility of the phosphorylation sites of both proteins was improved by succinylation. Although the acidic subunits of glycinin were the major ones phosphorylated after succinylation, the basic subunits became substrates as well [111].

Enzymatically phosphorylated soybean proteins showed improved emulsifying activity and emulsion stability [98,109,112]. Mixed results have been reported for the foaming properties with unchanged foam expansion of phosphorylated glycinin [112] and a phosphorylated protein isolate [109] and an increased foam expansion of phosphorylated glycinin [98] compared with the unmodified proteins. The data on foam stability are also contradictory, pointing both to a decrease for a phosphorylated isolate [109] and to zero effect for the modified glycinin [98,112]. Both the level of phosphorylation and the structural state of proteins induced by the modification must be taken into account to understand these differences. A comparison of the data obtained with $POCl_3$-treated and cAMPdPK-treated glycinin [98] showed a more pronounced effect of chemical modification upon functional properties, which corresponded to the higher level of phosphorylation reached by chemical modification.

V. ACYLATION

The acylation of proteins with acetic, succinic, maleic, or citraconic anhydride is one of the most convenient and most frequently used methods for altering the functional properties of proteins. A large number of papers have been published in this area, which particularly describe the advantages of acetylation and succinylation for improving the functionality of food proteins. Only a limited number of these could be considered in this chapter.

A. Chemical Aspects

The various reagents employed to acylate proteins do not react selectively with one type of functional groups but can react with all nucleophilic groups [1]. These include amino groups (N-terminal α- and lysine ϵ-amino groups), phenolic (tyrosine) and aliphatic (serine and threonine) hydroxyl groups, sulfhydryl groups (cysteine), and imidazole (histidine) groups. However, both the reactions of these groups and the stability of their acyl derivatives differ considerably.

The ϵ-amino group of lysine is the most readily acylated group because of its relatively high reactivity and its steric availability for reaction. Since reactivity increases with increasing nucleophilicity and decreasing protonation, the reaction should be performed under alkaline conditions. However, the acylating reagent, usually a carboxylic anhydride, is quickly hydrolyzed in strong alkaline solution, and this decreases the efficiency of acylation of amino groups. Therefore, in practice, the acylation of amino groups is performed under mild alkaline conditions (pH 7.5–8.5). In the case of acetylation, the basic amino groups are transformed into neutral groups, while succinylation, maleylation, or citraconylation changes the positive charge to the negative charge by the introduced carboxyl group (Fig. 3). In contrast to N-acetyl or N-succinyl bonds, N-maleyl or N-citraconyl bonds are acid lable [1], which permits a reversible blocking of amino groups (Fig. 3). S-Acyl derivatives of cysteine and N-acyl derivatives of histidine residues are also unstable [1]. O-Acetyl tyrosine residues require the use of hydroxylamine for hydrolysis, while O-succinyl tyrosines undergo a spontaneous intramolecularly catalyzed hydrolysis at pH > 5 (Fig. 3) [1]. The reactivity of the anhydrides of homologous carboxylic acids decreases with an increase in the aliphatic chain [113].

The extent of modification of the functional groups of a protein can be varied widely by changing the amount of the acylating agent used. It could be shown that with different proteins, the extent of succinylation or acetylation of hydroxy amino acids is low when unreacted lysine and tyrosine residues are still present. However, O-acylation (esterification) increases sharply when the reaction with the lysyls and tyrosyls is essentially complete. This occurs at a sufficiently high excess of the reagent [114,115]. A considerable amount of S-acylated cysteine residues were detected for example in myofibrillar [116] and whey [117] proteins.

B. Conformational Changes in Globular Proteins

The increase in negative charge of proteins caused by neutralizing amino groups via acetylation or by introducing additional carboxyl groups via succinylation may result in a drastic change in the native protein conformation. Such changes are more pronounced for succinylated proteins than for acetylated proteins [118]. Oligomeric proteins, such as the 11S and 7S storage proteins from oilseeds and grain legumes, tend to dissociate into their subunits after acylation [115,119–122]. The dissociation of the hexameric 11S globulins after succinylation proceeds principally via the trimeric 7S half-molecule [115,121]. Unlike succinylation, acetylation may result in aggregation of the modified protein [119,123]. Using hydrodynamic and spectroscopic methods, it has been shown that there are two principal stages of acylation with regard to conformational changes

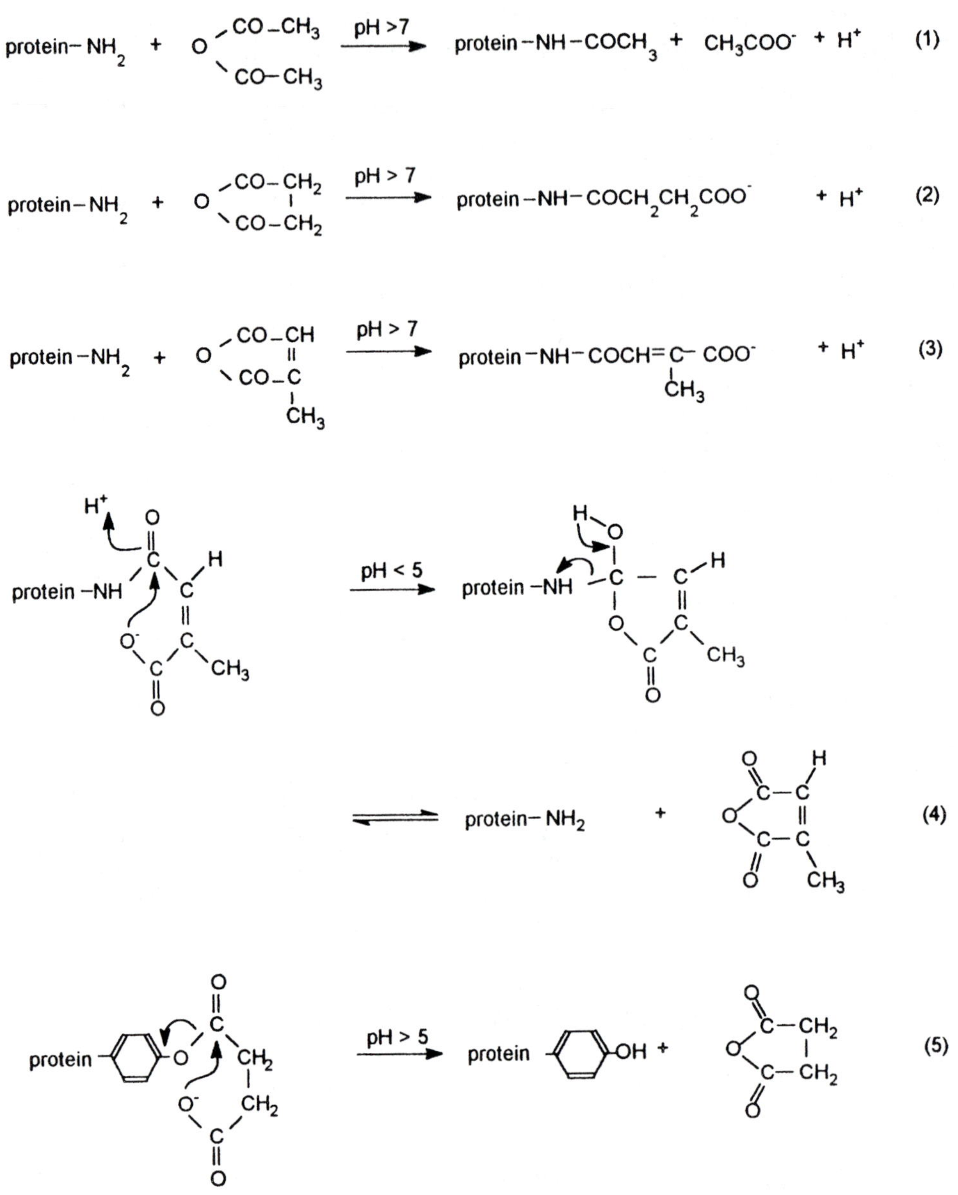

$$\text{protein}-NH_2 + O(CO-CH_3)_2 \xrightarrow{pH>7} \text{protein}-NH-COCH_3 + CH_3COO^- + H^+ \qquad (1)$$

$$\text{protein}-NH_2 + O\begin{matrix}CO-CH_2\\ | \\ CO-CH_2\end{matrix} \xrightarrow{pH>7} \text{protein}-NH-COCH_2CH_2COO^- + H^+ \qquad (2)$$

$$\text{protein}-NH_2 + O\begin{matrix}CO-CH\\ \| \\ CO-C-CH_3\end{matrix} \xrightarrow{pH>7} \text{protein}-NH-COCH{=}C(CH_3)-COO^- + H^+ \qquad (3)$$

protein –NH, H^+, O, C, H, C, O^-, C, CH_3, O $\xrightarrow{pH<5}$ protein –NH, H, O, C, C, H, O, C, CH_3, O

$\rightleftharpoons$ protein$-NH_2$ + O, C, C, H, C, C, CH_3, O (4)

protein, O, C, CH_2, CH_2, O^-, C, O $\xrightarrow{pH>5}$ protein–OH + O, C–CH_2, C–CH_2, O (5)

Figure 3 Acylation and deacylation of functional groups in proteins: acetylation (1) and succinylation (2) of amino groups, citraconylation (3) and decitraconylation (4) of amino groups, and deacylation of O-succinyltyrosyl groups (5).

[115,120,121]. In the first stages, modification up to about 60–75% N-acylation occurred but no large changes in the secondary and tertiary structure, whereas drastic conformational changes occurred in the second stage after passing through a "critical" threshold of acylation. This critical threshold of modification coincides with the appearance of a large amount of O-acylation, which contributes markedly to change in charge and hydrophobicity [115,124]. Corresponding conformational changes also has been documented for a maleylated 7S globulin (vicilin) using measurements of thermal denaturation temperature and enthalpy by differential scanning microcalometry and evaluation of hydrophobicity [125].

The principal course of conformational change found in oligomeric plant proteins after acylation was observed previously in papers on monomeric globular proteins such as bovine serum albumin and lysozyme. These latter proteins showed a sudden change in the rate of sulfitolysis of disulfide bonds, antibody precipitation, and hydrodynamic properties after passing a critical degree of succinylation [126].

C. Changes in Functional Properties

A great number of food proteins have been investigated with regard to the improvement of their functional properties by acylation. These comprised myofibrillar proteins [116,127], egg white [128,129], casein [130,131], wheat gluten [132], proteins from oat [133], peanut [134], cottonseed [135], soybean [136–139], sunflower seed [140,141], field pea [142], faba bean [122,143], and rapeseed [144–147], among other proteins.

Blocking of amino groups by acyl residues changes the isoelectric point of the proteins to lower pH values, which results in a general shift of the solubility profile to the more acid region and to an increase in protein solubility in weak acidic, neutral, and even alkaline solutions [122,130,136]. As shown with β-casein, acyl residues bearing a neutral aliphatic chain induced protein aggregation via hydrophobic interactions, which increased with an increase in chain length [148].

The results for the water-binding capacity of acylated proteins are contradictory. While an improvement in water binding was observed for succinylated or citraconylated wheat gluten [132], acetylated or succinylated pea [142], and peanut proteins [134], among others, a negative effect was reported for sunflower [140] and oat proteins [133]. Obviously, factors such as structural differences between proteins, the isolation procedures used, and the technological treatments applied to proteins prior to modification seem to influence the functional properties. This has also been shown to be the case with surface functionality. Surface functional properties are related to protein solubility, charge, and hydrophobicity, all of which can be changed considerably by acylation [149,150].

An increase in the EC/EA and ES after acetylation or succinylation has been reported for a number of proteins such as isolates or concentrates from soybean [136], faba bean [122,123,151], oat [133], pea [142], rapeseed [144,145], and sunflower [140]. In some cases, the increase was maximal at certain degrees of acylation [135,146]. Studies with purified 11S globulins, which showed the dependence of emulsifying properties on the degree of acylation, were performed, e.g., with soybean glycinin [139], rapeseed cruciferin [147], and faba bean legumin [151]. Surface adsorption measurements led some authors to derive relationships between the modified protein structure and surface functionality [147,152]. The critical association concentration (CAC), corresponding to the critical micelle concentration of low molecular weight detergents, decreased in acetylated faba bean legumin with increased modification [151]. This was also reflected in the increased hydrophobicity of the protein. This change in CAC correlated well with the increase in emulsifying activity [151]. The emulsification parameters (EA, ES, EC) are generally not linearly related to the number of acyl residues introduced.

They should be high, if the hydrophile-lipophile balance (HLB) index of the protein is close to the optimal HLB for the oil. However, it is rather difficult to define the term HLB for proteins.

The high emulsion-stabilizing effect of acetylated faba bean protein isolates has been shown to be connected with the formation of a "mechanical barrier" preventing centrifugal creaming [153]. The surface tension of acylated glycinin dropped with the increase of aliphatic chain length introduced (capronic > caprinic > myristic) [154].

As with emulsifying properties, foaming properties depend on the solubility, charge, and hydrophobicity of proteins [149]. Moreover, they are considerably influenced by the viscosity of the protein solution [149]. Acylation, which alters all these parameters, has a positive effect on foam expansion (foam capacity = FC) of a number of food proteins, e.g., of soybean [136], sunflower [140,141], pea [142], faba bean [122], cottonseed [135], and oat [133]. Egg white, the classical foam-forming protein, showed a drop in FC after succinylation [129]. Succinylation, however, reduced the heat-induced damage to the foaming properties of egg white [128]. The contribution of viscosity to foam formation has been demonstrated with succinylated faba bean protein isolates, both by increasing the protein concentration and by inducing conformational changes by heating or succinylation [155]. Maximum EC values were obtained for the native and moderately succinylated proteins after heat induced denaturation in a sufficiently concentrated (10%) solution [155]. However, when the degree of modification was high enough to cause protein unfolding and viscosity increase [156], maximum FC was obtained even with unheated solutions [155].

The increase in net charge density as a result of acylation, especially succinylation, tends to decrease foam stability (FS) since it prevents optimum protein-protein interactions, which are required for a continuous film around air bubbles. Therefore, a number of succinylated proteins showed a drop of FS with increasing degree of modification [133,142,155]. With some proteins, stable foams were obtained although the extent of succinylation was high [136,140,141]. Obviously, other intrinsic (structural) and extrinsic factors override the charge repulsion in these cases. If certain unique structural factors of a protein promote the formation of a continuous viscoelastic film, then succinylation may even improve the stability of the foam despite an increase in charge repulsion. This was the case in the foaming behavior of succinylated rapeseed cruciferin, where the negative charges were suggested as being responsible for preventing the coalescence of air bubbles [147]. Acylation induces a considerable change in the rheological properties of protein dispersions. This was demonstrated with protein isolates from rapeseed [157,158], faba bean [156,159], soybean [137], and others. This is of interest especially in relation to gel-forming properties and gel stability. Thermally induced faba bean and rapeseed protein gels had maximum gel strength at medium degrees of acetylation [143] and succinylation [158], respectively, where obviously the attractive protein-protein interactions was sufficiently strong to overcome the repulsive electrostatic interactions. While unmodified rapeseed protein isolates required strong alkaline conditions (pH > 9.5) for formation of gels, which were opaque, the succinylated proteins formed transparent gels in the range of pH 5–11 [158].

D. Nutritional Aspects

Succinylation increased the initial rate of proteolysis [160] but reduced the utilization of the modified amino acid residues [117], the protein efficiency ratio (PER), and net protein utilization (NPU) [161,162]. Acylated proteins were absorbed by the intestine but not utilized by the organism [117,162]. Digestion of acetylated or succinylated casein with pepsin/pancreatin gave longer peptides than did native casein [163]. Amino- and carboxypeptidases released N-acetyl- and N-succinyllysine, respectively, from the peptic digest [163]. Considerable differences between the PER values of acetylated (1.6) and succinylated (0.3) casein (2.5 for unmodified casein) were found [130]. These results corresponded to earlier findings [164] that rats were able to deacylate acetylated proteins but not proteins that were modified with other acyl residues.

VI. DEAMIDATION

The deamidation of glutamine and asparagine residues in proteins results in the liberation of carboxyl groups and thereby to an increase in negative charge and hydration. This is especially important in the case of plant storage proteins, which are rich in amidated glutamic and aspartic acid residues.

A. Chemical Deamidation

Glutamine and aspargine amide groups are usually hydrolyzed by acid or base catalysis. However, this reaction is not specific, because it also causes splitting of peptide bonds. Acidic hydrolysis of asparagine generates aspartic acid via a one-step mechanism, whereas alkaline hydrolysis produces a cyclic intramolecular imide as intermediate product. This imide can be hydrolyzed to an aspartic acid residue, or it can react with an amino group and form an isopeptide bond. The latter is less desirable from a nutritional point of view. The rate of deamidation in acidic or alkaline milieu depends on the environment (amino acid sequence in the neighborhood) of the amide group and on the conformation of the protein. The deamidation of glutamine takes place less quickly than that of asparagine [94].

To minimize peptide bond hydrolysis, the deamidation of vegetable proteins has been performed using long-chain alkylsulfate and alkylsulfonate anions, arylsulfonate anions in the form of cation exchange resins, and common anions such as phosphate or bicarbonate as catalysts [94]. Thus, up to 40% of amide groups could be split, but only 1–4% of peptides were hydrolyzed. The deamidated proteins showed substantially improved solubility, water-binding capacity, foam expansion, emulsion capacity, and emulsion viscosity as compared with their unmodified counterparts [94].

Thermal deamidation of proteins in a restricted water environment has been accomplished for soy protein, egg lysozyme, casein, and wheat gliadin [165]. The deamidation percentage varied from about 3 to 18% for all four proteins after heating at 115°C for 2 hours and with a water content of approximately 50% for soy protein and lysozyme, 60% for casein and 6% for gliadin.

Deamidation by acid treatment has been applied successfully for improving the functional properties of wheat gluten [166–168]. Deamidated gluten can be dispersed at low pH and added to fruit-based beverages, with the protein dispersibility primarily dependent upon the degree of deamidation [166]. Treatment of gluten with 0.02 N hydrochloric acid for 30 minutes or with 0.05 N hydrochloric acid for 15 minutes at 121°C gave soluble preparations with significantly improved emulsifying properties. Solubilization was a result of not only deamidation but also rupture of few peptide linkages [167]. Treatment of gluten with increasing concentrations of hydrochloric acid ranging from 0.02 to 0.1 N HCl induced an increase in hydrophobicity, while the bread loaf volume and dough extensibility were decreased [168]. Rheological investigations showed that, compared to controls, the consistency of deamidated glutens was softer. The deviations in relative percent penetrations increased strongly initially with increasing degree of deamidation, while later it tended towards a constant value [169]. When the degree of deamidation was increased to 40%, the helical content of gluten decreased from 30 to 10%. The surface hydrophobicity increased with the degree of deamidation and is negatively correlated with the surface tension and positively correlated with the emulsifying activity and emulsion stability [170]. The cleavage of peptide bonds can be minimized when deamidation is performed under moderate conditions (0.1 N HCl, 70°C). This allows enhancement of the solubility of gluten proteins (total gluten, gliadin, glutenin) under neutral and basic conditions. A 10–20% deamidation level was sufficient for increasing the emulsifying and foaming capacity. The resulting foams and emulsions were as stable as those of bovine serum albumin, soybean, and pea globulins [171]. Deamidation improved the solubility and emulsifying capacity of zein in the pH ranges of 6–11 and 7–11, respectively, whereas it decreased the foam stability [25].

B. Enzymatic Deamidation

1. Action of Proteases

Peptide bond hydrolysis can be avoided by using deamidases, such as transglutaminases and peptidoglutaminases [94]. Moreover, proteases also cause deamidation of proteins [172,173]. About 20% of the amide groups in ovalbumin and lysozyme were deamidated using papain, pronase E, or chymotrypsin at pH 10 and 20°C. Peptide bond hydrolysis ranged from 0 to 8%, with soy protein and gluten receiving the greatest degree of hydrolysis [172,173]. Using chymotrypsin immobilized on controlled pore glass, selected plant and animal proteins could be deamidated to 5–10% at the same conditions of pH and temperature [174]. This procedure increased the emulsifying and foaming properties of food proteins. Gluten solubility at all pH values in the range of 2–12 increased after deamidation [173,174]. Further studies raised serious questions about the use of proteases to deamidate food proteins. The above described reaction conditions (pH 10, 20°C) and enzymes (papain, chymotrypsin, and pronase E) did not prove to be suitable for deamidation of soy protein. Instead, substantial proteolysis and generation of ammonia resulting from nonenzymatic deamidation of free glutamine and deamidation of certain free amino acid in the hydrolysates were found [94,175]. It was suggested that the nonenzymatic deamidation of free glutamine was most likely catalyzed by phosphate and bicarbonate anions in the buffer. Thus, the possibility of application of proteases as deamidating agents for the improvement of protein functional properties remains an unsolved problem and requires further research.

2. Action of Peptidoglutaminases

Peptidoglutaminases (L-glutamine amido hydrolases EC 3.5.1.2.) were first isolated from the soil bacterium *Bacillus circulans* [176]. They are capable of deamidating peptide-bound glutamine. The isolated peptidoglutaminase (PG) contained two distinct enzyme activities, PG I and II. PG I catalyzed the deamidation of C-terminal glutamine in small peptides, whereas PG II catalyzed the deamidation of glutamine within the peptide chain. However, the activity of PG II towards casein and whey was limited [177]. Mixtures of PG I and II were shown to have a very limited deamidating activity toward intact animal and plant proteins but had a high deamidating activity towards hydrolysates of soybean protein, wheat and corn gluten, egg albumin, lactalbumin, and casein [94]. Heat denaturation prior to PG treatment and combined heat treatment/hydrolysis was shown to be effective in enhancing the PG activity for deamidation of food proteins [178]. The modified protein preparations had increased solubility at pH 4–7 as well as at alkaline conditions and increased emulsifying activity, emulsion stability, and foaming power but had no apparent effect on foam stability [179]. In order to prepare deamidated food proteins with improved functionality on a large scale, a batch membrane bioreactor method was developed in which hydrolyzed protein substrate can be circulated through the containment vessel to achieve complete deamidation [94].

A new type of deamidase has been isolated from germinating wheat grains [180], which catalyzes the deamidation of seed storage proteins and also acts on egg lysozyme, horse hemoglobin, reduced RNase, glutamine, and the peptide Gly-L-Glu-Tyr. In contrast to PG I and II, this deamidase is very active towards native proteins.

3. Action of Transglutaminase

Transglutaminase (TG) catalyzes the acyl transfer reaction between the γ-carboxyl groups of glutaminyl residues in proteins and primary amines [94]. When the primary amine is part of the same or another protein, intra- or intermolecular cross-linking occurs (see Sec. VIII and Fig. 4). In the absence of available amine groups, or when the primary amine is present but blocked, water can act as an acyl acceptor and deamidation occurs with the release of ammonia (Fig. 4).

A Ca^{2+}-activated TG from guinea pig liver was mainly used for deamidation and cross-linking studies [94,181–183]. To avoid cross-linking, proteins can be acylated, deaminated, or guanidinated prior to TG catalysis [94]. A reversible blockage of lysine ϵ-amino groups was performed using citraconic anhydride for the deamidation of α_{s1}-casein [182] and pea legumin [183] and maleic anhydride for the deamidation of wheat γ-gliadin [184]. Almost complete citraconylation was achieved, and about 80% of the glutamine was deamidated in α_{s1}-casein [182]. The deamidated protein had improved solubility after the removal of the blocking groups. Using wheat gliadin fractions, it has been shown that the deamidation activity of guinea pig TG [185] and bovine plasma (TG factor XIII) [184] was maximal under acidic conditions. Using a purified γ-gliadin fraction (γ-46-gliadin), it was shown that under optimal conditions (pH 5.5, presence of 2% dioxan) 15% of the glutaminyl residues present in γ-gliadin after 24 hours of enzymatic treatment were deamidated. Though a slight degree of cross-linking took place, the solubility of the deamidated gliadin at pH > 5 was markedly increased. A 25–27% deamidation level has been shown to provide maximum emulsion activity index. The

$$\text{protein–Glu-}(\gamma)\text{–CONH}_2 + H_2O \xrightarrow{\text{TGase}} \text{protein–Glu-}(\gamma)\text{–COOH} + NH_3 \quad (1)$$

$$\text{protein–Glu-}(\gamma)\text{–CONH}_2 + H_2N(\varepsilon)\text{–Lys–protein} \xrightarrow{\text{TGase}} \text{protein–Glu-}(\gamma)\text{–CO–HN}(\varepsilon)\text{–Lys–protein} \quad (2)$$

Figure 4 Transglutaminase (TGase)-catalyzed reactions: deamidation (1), cross-linking (2).

emulsion made with TG-treated gliadin showed much higher stability against coalescence than the chemically deamidated gliadin at a comparable degree of deamidation [185].

Pea legumin, characterized by a closely packed globular structure, was shown to be a poor substrate for TG in spite of its high contents of glutaminyl and lysyl residues. Dissociation and unfolding induced by citraconylation markedly increased the level of TG-catalyzed deamidation [183].

VII. ESTERIFICATION OF CARBOXYL GROUPS

The carboxyl groups of proteins can be esterified with methanol containing small amounts of hydrochloric acid (Fig. 5). In order to minimize conformational changes, the reaction is performed under moderate conditions (0.25°C and 0.02–0.1 M HCl). This treatment can lead to two types of secondary reactions: methanolysis of amide groups in glutamine and asparagine residues and N → O rearrangement (Fig. 5). Between pH 7 and 10, the methylesters are slowly hydrolyzed, but at pH >10 the hydrolysis takes place quickly [1]. The esterification reduces the number of anionic groups in the protein and increases the isoelectric point. Esterification of casein and its incorporation into micellar casein at a level of 2–5% increased the thermal stability of the casein micelles [186]. When bovine β-lactoglobulin was esterified with methanol, ethanol, or *n*-butanol, the surface activity of the protein, determined from surface and interfacial tension measurements at an air/water and oil/water interface, were enhanced [187]. Methylester exhibited the greatest increase in surface hydrophobicity, which was measured by the hydrophobic fluorescence probe 1,8-anilinonaphtalene sulfonate [187]. Esterification with methanol or ethanol changed the rheological properties of wheat gluten [169,188]. With increasing degrees of esterification, the rheological properties of gluten deteriorate and the relaxation time shorten significantly [169]. In the first stage of modification, which seems to correspond to the esterification of the free carboxyl groups, no significant change was observed in the rheological properties. Obviously, free carboxyl groups did not participate significantly in the formation of secondary bonds that stabilized the protein. At higher levels of esterification, the amide groups also might be involved in the modification. The decrease of intrinsic viscosity indicated an increase in compactness of the protein with increasing degree of modification.

$$\text{protein–COOH} + CH_3OH \xrightarrow{HCl} \text{protein–COOCH}_3 + H_2O \quad (1)$$

$$\text{protein–CONH}_2 + CH_3OH \xrightarrow{HCl} \text{protein–COOCH}_3 + NH_3 \quad (2)$$

$$\text{protein–}\underset{}{\overset{\text{OH}}{\text{SER}}}\text{–NH–CO–R'} \xrightarrow{H^+} \text{protein–}\overset{\text{O–COR'}}{\text{SER}}\text{–}NH_3^+ \quad (3)$$

FIGURE 5 Esterification of carboxyl groups (1), amide groups (2), and N → O rearrangement (3) at a serine residue catalyzed by hydrochloric acid.

VIII. PROTEIN CROSS-LINKING

The introduction of artificial cross-links into food proteins [189] may be useful for improving the functional properties. Transglutaminase (TG), peroxidase (POD; EC 1.11.1.7), and polyphenol oxidase (PPO; EC 1.14.18.1) are suitable for enzymatic cross-linking of proteins [190].

TG, which is specific to glutamyl residues of proteins, catalyzes the reaction of a nucleophile, such as the ϵ-amino group of a lysyl residue, with the amide group of the glutaminyl residue (Fig. 4). TG has been applied for the cross-linking of β-lactoglobulin [191], casein [192], soybean glycinin [193], wheat gliadin [184,185], as well as cross-linking between different food proteins, e.g., between myosin, soybean protein, casein, or gluten [194]. Since the extent of cross-linking depends on the number of lysine and glutamine residues on the surface of the protein, heat denaturation prior to the enzyme reaction may increase the number of cross-links as seen for soybean glycinin [194]. Cross-linking can have a positive [194] or negative [192] effect on the gel-forming properties, depending on the nature of the protein. The gelation properties of TG–cross-linked α_{S1}-casein have been successfully applied for preparing films [195]. TG-catalyzed cross-linking changes the pH-dependent solubility of proteins [185,196]. It improves hydration properties [196], while it can impair [196] or improve [185] the emulsion stabilization of different proteins. Films of cross-linked α_{S1}-casein can be digested by chymotrypsin to the same extent, but with a slower speed than the unmodified protein [196].

Peroxidase (POD)/H_2O_2 treatment of proteins oxidizes tyrosine residues into di-tyrosine and trityrosine in different proteins [197]. Addition of POD/H_2O improves the dough-forming properties and the baking performance of wheat flours [198]. The possible mechanism of POD-catalyzed cross-linking, where a number of components such as phenols or their oxidation products (chinons) and protein amino groups are involved, is discussed in some papers (see Ref. 197).

Polyphenol oxidase (PPO) causes strengthening of wheat gluten dough. This occurs due to an oxidative modification of sulfhydryl residues [199]. PPO generates o-chinons either from tyrosine residues or phenolic compounds, which are present in plant food-stuffs and can react with cysteine, lysine, histidine, and tryptophan residues [197]. Thus, the nutritive value of a protein can be impaired both by destruction of essential amino acids and by cross-linking reactions, which reduce the susceptibility of the protein to enzyme hydrolysis [197].

Several studies have demonstrated that lipoxygenase (LO; EC 1.13.11.12) causes cross-linking of proteins. LO improves the dough-forming properties and baking performance of wheat flour. The application of LO in food processing, possible mechanism of LO action on proteins, and its nutritional implications have been discussed by Matheis and Whitaker [190].

There are many bifunctional reagents facilitating chemical cross-linking of proteins [1]. Protein cross-linking also occurs as an undesired side reaction of food processing. This has been reviewed by Feeney and Whitaker [200].

IX. CONCLUSIONS

Protein modification using enzymatic or chemical modification is an important tool for tailoring food proteins into products with very different functional properties. The use of enzymatic modification as a "natural" way of changing protein functionality should have priority over chemical modification in cases where enzymes are available and modification of the protein is extensive enough to give desired functional properties with acceptable processing costs. This is the case with proteolytic modification using microbial proteases, already produced in large quantities. Soybean protein hydrolysates are an example of the application of proteolytic modification for producing highly soluble functional food ingredients [8]. Moreover, plasteinlike reactions open new perspectives for functionalizing food proteins, as well as for improving the nutritional value of food proteins by introducing essential amino acids. Controlled proteolysis can also serve to produce bioactive peptides such as opioid peptides or phosphopeptides from milk proteins. For example, Japanese researchers have produced a casein phosphopeptide that has been claimed to increase the absorption of calcium, iron, and other minerals and prevent calcium loss from the bone [201].

Chemical modification is generally more efficient than enzymatic modification with regard to structural and functional changes. However, extensive blocking of essential amino acids and the occurrence of undesired side reactions may prove to be obstacles to the introduction of chemically modified proteins into food products. A special example of chemical modification, only marginally mentioned in this chapter, is the covalent attachment of essential amino acids to proteins using methods of peptide synthesis [202]. Though an automatization of the procedure seems to be realistic, it remains to be determined whether these methods are acceptable with regard to economics and to safety for human and animal consumption. Chemical modification, especially acylation, phosphorylation, glycosylation, and cross-linking reactions, are, however, interesting as efficient methods for transforming proteins, especially those of vegetable origin, into emulsifying, foaming, or gel-forming materials for nonfood use.

ACKNOWLEDGMENT

The author would like to thank Steffi Dudek for typing the formulas in the figures.

REFERENCES

1. G. E. Means and R. E. Feeney, *Chemical Modification of Proteins.* Holden Day, San Francisco, 1971.
2. R. E. Feeney, *Chemical Modification of Food Proteins. Food Proteins. Improvement Through Chemical and Enzymatic Modification* American Chemical Society, Washington, DC, 1977, p. 3.
3. R. E. Feeney, Tailoring proteins for food and medical use: state of art and interrelationships, *Protein Tailoring for Food and Medical Uses* (R. E. Feeney and J. R. Whitaker, eds.), Marcel Dekker, New York, 1986, p. 1.
4. R. E. Feeney, Chemical modification of proteins: comments and perspectives, *Int. J. Peptide Protein Res. 29*:145 (1987).
5. T. Richardson, Functionality changes in proteins following action of enzymes, *Food Proteins. Improvement Through Chemical and Enzymatic Modification* (R. E. Feeney and J. R. Whitaker, eds.), American Chemical Society, Washington, DC, 1977, p. 185.
6. J. R. Whitaker and A. J. Puigserver, Fundamentals and applications of enzymatic modifications of proteins: an overview, *Modification of Proteins. Food, Nutritional, and Pharmacological Aspects* (R. E. Feeney and J. R. Whitaker, eds.), American Chemical Society, Washington, DC, 1982, p. 57.

7. J. W. Lee and A. Lopez, Modification of plant proteins by immobilized proteases, *CRC Crit. Rev. Food Sci. Nutr. 21*:289 (1984).
8. J. Adler-Nissen, *Enzymic Hydrolysis of Food Proteins*, Elsevier, London, 1986, p. 427.
9. F. F. Shi, Modification of food proteins by non-enzymatic methods, *Biochemistry of Food Proteins* (B. J. F. Hudson, ed.), Elsevier, Amsterdam, 1992, p. 235.
10. N. R. Reddy, M. D. Pierson, S. K. Sathe, and D. K. Salunke, Legume-based fermented foods: their preparation and nutritional quality, *CRC Crit. Rev. Food Sci. Nutr. 17*:335 (1982).
11. J.-M. Chobert, M. Zitohy and J. R. Whitaker, Specific limited hydrolysis and phosphorylation of food proteins for improvement of functional and nutritional properties, *J. Am. Oil Chem. Soc. 64*:1704 (1987).
12. J.-M. Chobert, M. Z. Zitohy and J. R. Whitaker, Solubility and emulsifying properties of caseins modified enzymatically by *Staphylococcus aureus* V8 protease, *J. Agric. Food Chem. 36*:220 (1988).
13. M. I. Mahmoud, W. T. Malone and C. T. Cordle, Enzymatic hydrolysis of casein: effect of degree of hydrolysis on antigenicity and physical properties, *J. Food Sci. 57*:1223 (1992).
14. G. Puski, Modification of functional properties of soy proteins by proteolytic enzyme treatment, *Cereal Chem. 52*:655 (1975).
15. J. Adler-Nissen and H. S. Olsen, The influence of peptide chain length on taste and functional properties of enzymatically modified soy protein, *Functionality and Protein Structure* (A. Pour-El, ed.), American Chemical Society, Washington, DC, 1979, p. 125.
16. F. Zakaria and R. F. McFeeters, Improvement of the emulsifying properties of soy protein by limited pepsin hydrolysis, *Lebensm.-Wiss. Technol. 11*:42 (1978).
17. H. S. Olsen and J. Adler-Nissen, Die Anwendung technischer Enzyme zur Herstellung von Produkten aus Sojaprotein, *Zeitschr. Lebensm.-Technol. Verfahrenstechnol. 31*:359 (1980); *32*:55 (1981).
18. K. Ochia, Y. Kamata, and K. Shibasaki, Effect of tryptic digestion on emulsifying properties of soy protein, *Agric. Biol. Chem. 46*:91 (1982).
19. M. Mohri and S. Matsushita, Improvement of water absorption of soybean protein by treatment with bromelain, *J. Agric. Food Chem. 32*:486 (1984).
20. S. Y. Kim, P. S. W. Park, and K. C. Rhee, Functional properties of proteolytic enzyme modified soy protein isolate, *J. Agric. Food Chem. 38*:651 (1990).
21. U. Behnke, M. Schultz, H. Ruttloff, and H. Schmandke, Veränderungen der funktionellen Eigenschaften von Vicia-faba-Proteinisolaten durch partielle enzymatische Hydrolyse, *Nahrung 26*:313 (1982).
22. J.-P. Krause and K. D. Schwenke, Changes in interfacial properties of legumin from faba beans (Vicia faba L.) by tryptic hydrolysis, *Nahrung 39*:396 (1995).
23. J. I. Tschimirov, K. D. Schwenke, D. Augustat, and V. B. Tolstoguzov, Functional properties of plant proteins. Part V. Influence of partial enzymatic hydrolysis on selected functional properties of wheat gluten, *Nahrung 27*:659 (1983).
24. Y. Popineau and F. Pineau, Emulsifying properties of wheat gliadins and gliadin peptides, *Food Proteins. Structure and Functionality* (K. D. Schwenke and R. Mothes, eds.), VCH, Weinheim, 1993, p. 290.
25. M. A. Casella and J. R. Whitaker, Enzymatically and chemically modified zein for improvement of functional properties, *J. Food Biochem. 14*:453 (1990).
26. A. Mannheim and M. Cheryan, Enzyme-modified proteins from corn gluten meal: preparation and functional properties, *J. Am. Oil Chem. Soc. 69*:1163 (1992).
27. R. Ponnampalam, G. Goulet, J. Amiot, and G. J. Brisson, Some functional and nutritional properties of oat flours as affected by proteolysis, *J. Agric. Food Chem. 35*:279 (1987).
28. A. Arzu, H. Mayorga, J. Gonzalez, and C. Rolz, Enzymatic hydrolysis of cotton seed protein, *J. Agric. Food Chem. 20*:805 (1972).
29. M. Kabirullah and R. B. H. Wills, Functional properties of sunflower protein following partial hydrolysis with proteases, *Lebensm.-Wiss. Technol. 14*:232 (1981).

30. A. M. Hermansson, D. Olsson, and B. Holmberg, Functional properties of proteins for foods—modification studies on rapeseed protein concentrate, *Lebensm.-Wiss. Technol. 7*:176 (1974).
31. R. D. Phillips and L. R. Beuchat, Enzyme modification of proteins, *Protein Functionality in Foods* (J. P. Cherry, ed.), American Chemical Society, Washington, DC, 1981, p. 275.
32. A. A. Sekul, C. H. Vinnett, and R. L. Ory, Some functional properties of peanut proteins partially hydrolyzed with papain, *J. Agric. Food Chem. 26*:855 (1978).
33. L. R. Beuchat, Functional property modification of defatted peanut flour as a result of proteolysis, *Lebensm.-Wiss. Technol. 10*:78 (1977).
34. K. D. Schwenke, A. Staatz, St. Dudek, J.-P. Krause, and J. Noack, Legumin-T from faba beans: isolation, partial characterization and surface functional properties, *Nahrung 39*:193 (1995).
35. Y. Mine, K. Chiba, and M. Tada, Effects of limited proteolysis of ovalbumin on interfacial adsorptivity studied by 31P nuclear magnetic resonance, *J. Agric. Food Chem. 40*:22 (1992).
36. M. Shimizu, T. Takahashi, S. Kaminogawa, and K. Yamauchi, Adsorption onto an oil surface and emulsifying properties of bovine $\alpha_{S,1}$-casein in relation to its molecular structure, *J. Agric. Food Chem. 31*:1214 (1983).
37. P. J. Halling, Protein-stabilized foams and emulsions, *CRC Crit. Rev. Food Sci. Nutr. 15*: 155 (1981).
38. A. Rupley, Susceptibility to attack by proteolytic enzymes, *Meth. Enzymol. 11*:905 (1967).
39. G. W. Plumb, H. J. Carr, V. K. Newby, and N. Lambert, A study of the trypsinolysis of pea 11 S globulin, *Biochim. Biophys. Acta 999*:281 (1989).
40. Y. Kamata and K. Shibasaki, Formation of digestion intermediate of glycinin, *Agric. Biol. Chem. 42*:2323 (1978).
41. A. N. Danilenko, V. Y. Vetrov, A. P. Dmitrochenko, A. L. Leontiev, E. E. Braudo, and V. B. Tolstoguzov, Restricted proteolysis of legumin of broad beans: effect of thermodynamic properties of aqueous solutions and interaction with ficoll, *Nahrung 36*:105 (1992).
42. A. D. Shutov, J. P. Pineda, V. I. Senyuk, V. A. Reva, and I. A. Vaintraub, Action of trypsin on glycinin. Mixed-type proteolysis and its kinetics; molecular mass of glycinin-T, *Eur. J. Biochem. 199*:539 (1991).
43. A. Henschen, F. Lottspeich, V. Brantl, and H. Teschemacher, Novel opioid peptides derived from casein (β-casomorphins). II. Structure of active components from bovine casein peptone, *Hoppe-Seyler's Z. Physiol. Chem. 360*:1217 (1979).
44. C. Zioudrou, R. A. Streaty, and N. A. Klee, Opioid peptides derived from food proteins, The exorphin, *J. Biol. Chem. 254*:2446 (1979).
45. E. Schlimme, H. Meisel, and H. Frister, Bioactive sequences in milk proteins, *Milk Proteins. Nutritional, Clinical, Functional and Technological Aspects* (C. A. Barth and E. Schlimme, eds.), Steinkopf-Verlag, Darmstadt/Springer-Verlag, New York, 1989, p. 143.
46. S. Loukas, D. Varoucha, C. Zioudrou, R. A. Streaty, and W. A. Klee, Opioid activities and structures of α-casein derived exorphins, *Biochemistry 22*:4567 (1983).
47. H. Chiba and M. Yoshikawa, Biologically functional peptides from food proteins: new opioid peptides from milk proteins, *Protein Tailoring for Food and Medical Uses* (R. E. Feeney and J. R. Whitaker, eds.), Marcel Dekker, New York, 1986, p. 123.
48. E. Schlimme and H. Meisel, Bioactive peptides derived from milk proteins. Structural, physiological and analytical aspects, *Nahrung 39*:1 (1995).
49. L. Graf, K. Horvath, E. Walcz, I. Berzetei, and J. Burnier, Effect of two synthetic gliadin peptides on lymphocytes in coeliac disease: identification of a novel class of opioid receptors, *Neuropeptides 9*:113 (1987).
50. S. Fukudome and M. Yoshikawa, Opioid peptides derived from wheat gluten: their isolation and characterization, *FEBS Lett. 296*:107 (1992).
51. Y. Guigoz and J. Solms, Bitter peptides, occurrence and structure, *Chem. Senses Flavor 2*: 71 (1976).

52. J. Adler-Nissen, Relationship of structure to taste of peptides and peptide mixtures, *Protein Tailoring for Food and Medical Uses* (R. E. Feeney and J. R. Whitaker, eds.), Marcel Dekker, New York, 1986, p. 97.
53. K. H. Ney, Bitterness of peptides: amino acid composition and chain length, *Food Taste Chemistry* (J. C. Boudreau, ed.), American Chemical Society, Washington, DC, 1979, p. 149.
54. H. Wieser and H.-D. Belitz, Zusammenhänge zwischen Struktur und Bittergeschmack bei Aminosäuren und Peptiden, II. Peptide und Peptidderivate, *Z. Lebensm.-Unters.-Forsch. 160*: 383 (1976).
55. H.-D. Belitz, W. Chen, H. Jugel, R. Treleano, H. Wieser, J. Gasteiger, and M. Marsili, Sweet and bitter compounds: structure and taste relationship, *Food Taste Chemistry* (J. C. Boudreau, ed.), American Chemical Society, Washington, DC, 1979, p. 93.
56. S. Eriksen and I. S. Fagerson, The plastein reaction and its applications: a review, *J. Food Sci. 41*:490 (1976).
57. M. Fujimaki, S. Arai, and M. Yamashita, Enzymatic protein degradation and resynthesis for protein improvement, *Food Proteins. Improvement Through Chemical and Enzymatic Modification* (R. E. Feeney and J. R. Whitaker, eds.), American Chemical Society, Washington, DC, 1977, p. 156.
58. J. Horowitz and F. Haurowitz, Mechanism of plastein formation, *Biochim. Biophys. Acta 33*:231 (1959).
59. B. Hofsten and G. Lalasidis, Protease-catalyzed formation of plastein products and some of their properties, *J. Agric. Food Chem. 24*:460 (1976).
60. M. Y. Gololobov, V. M. Belikov, S. V. Vitt, E. A. Paskonova, and E. F. Titova, Transpeptidation in concentrated solution of peptic hydrolyzate of ovalbumin, *Nahrung 25*:961 (1981).
61. M. Yamashita, S. Arai, S.-J. Tsai, and M. Fujimaki, Plastein reaction as a method for enhancing the sulfur-containing amino acid level of soybean protein, *J. Agric. Food Chem. 19*:1151 (1971).
62. M. Yamashita, S. Arai, and M. Fujimaki, Plastein reaction for food protein improvement, *J. Agric. Food Chem. 24*:1100 (1976).
63. K. Aso, M. Yamashita, S. Arai, I. Suzuki, and M. Fujimaki, Specificity for incorporation of α-amino acid esters during the plastein reaction by papain, *J. Agric. Food Chem. 25*:1138 (1977).
64. M. Yamashita, S. Arai, Y. Imaizumi, Y. Amano, and M. Fujimaki, A one-step process for incorporation of L-methionine into soy protein by treatment with papain, *J. Agric. Food Chem. 27*:52 (1979).
65. Gy. Hajós, S. Hussein, and E. Gelencsér, Enzymatic peptide modification of food proteins, *Food Proteins: Structure and Functionality* (K. D. Schwenke and R. Mothes, eds.), VCH, Weinheim, 1993, p. 82.
66. H. Winkler, H. Nötzold, and E. Ludwig, Untersuchungen zum Methionineinbau in ein peptidisches Partialhydrolysat aus Ackerbohnenproteinisolat durch Plasteinreaktion mit Thermitase, *Nahrung 28*:1029 (1984).
67. Gy. Hajos, I. E. Elias, and A. Halasz, Methionine enrichment of milk protein by enzymatic peptide modification, *J. Food Sci. 53*:739 (1988).
68. M. Watanabe and S. Arai, Proteinaceous surfactants prepared by covalent attachment of L-leucine n-alkyl esters to food proteins by modification with papain, *Modification of Proteins Food, Nutritional, and Pharmacological Aspects* (R. E. Feeney and J. R. Whitaker, eds.), American Chemical Society, Washington, DC, 1982, p. 199.
69. S. Arai, M. Watanabe, and N. Hirao, Modification to change physical and functional properties of food proteins, *Protein Tailoring for Food and Medical Uses* (R. E. Feeney and J. R. Whitaker, eds.), Marcel Dekker, New York, 1986, p. 75.

70. A. Gottschalk (ed.), *Glycoproteins. Their Composition, Structure and Function*, 2nd ed., Elsevier, Amsterdam, 1972.
71. C. P. Stowell and Y. C. Lee, Neoglycoproteins: the preparation and application of synthetic glycoproteins. *Adv. Carbohydrate Chem. Biochem. 37*:225 (1980).
72. Y. Kato, K. Watanabe, and Y. Sato, Effect of Maillard reaction on some physical properties of ovalbumin, *J. Food Sci. 46*:1835 (1981).
73. Y. Kato, T. Matsuda, N. Kato, and R. Nakamura, Maillard reaction of disaccharides with protein: Suppressive effect of non reducing end pyranoside groups on browning and protein polymerization, *J. Agric. Food Chem. 37*:1077 (1989).
74. S. Nakamura, A. Kato, and K. Kobayashi, New antimicrobial characteristics of lysozyme-dextran conjugate, *J. Agric. Food Chem. 39*:647 (1991).
75. S. Nakamura, A. Kato, and K. Kobayashi, Bifunctional lysozyme-galactomannan conjugate having excellent emulsifying properties and bactericidal effect, *J. Agric. Food Chem. 40*: 735 (1992).
76. S. Nakamura, A. Kato, and K. Kobayashi, Novel macromolecular emulsifier having improved antimicrobial action elaborated by covalent attachment of polysaccharides to lysozyme, *Food Proteins. Structure and Functionality* (K. D. Schwenke and R. Mothes, ed.), VCH, Weinheim, 1993, p. 29.
77. A. Kato, K. Minahi, and K. Kobayashi, Improvement of emulsifying properties of egg white proteins by the attachment of polysaccharide through Maillard reaction in a dry state, *J. Agric. Food Chem. 41*:540 (1993).
78. H. S. Lee, L. C. Sen, A. J. Clifford, J. R. Whitaker, and R. Feeney, Preparation and nutritional properties of caseins covalently modified with sugars, reductive alkylation of lysines with glucose, fructose or lactose, *J. Agric. Food Chem. 27*:1094 (1979).
79. J.-L. Courthaudon, B. Colas, and D. Lorient, Covalent binding of glycosyl residues to bovine casein: effects on solubility and viscosity, *J. Agric. Food Chem. 37*:32 (1989).
80. D. Caer, A. Baniel, M. Subirade, J. Gueguen, and B. Colas, Preparation and physico-chemical properties of glycosylated derivation of pea legumin, *J. Agric. Food Chem. 38*: 1700 (1990).
81. R. D. Waniska and J. E. Kinsella, Preparation of maltosyl, beta-cyclodextrinyl, glucosaminyl and glucosamine octaosyl derivatives of beta-lactoglobulin, *Int. J. Peptide Protein Res. 23*: 573 (1984).
82. N. Kitabatabe, J. L. Cuq, and J. C. Cheftel, Covalent binding of glycosyl residues to β-lactoglobulin: effects of solubility and heat stability, *J. Agric. Food Chem. 33*:125 (1985).
83. R. D. Waniska and J. E. Kinsella, Physico-chemical properties of maltosyl and glucosaminyl derivatives of beta-lactoglobulin, *Int. J. Peptide Protein Res. 23*:467 (1984).
84. B. Closs, J.-L. Couthaudon, and D. Lorient, Effect of chemical glycosylation on the surface properties of the soluble fraction of casein, *J. Food Sci. 55*:437 (1990).
85. R. D. Waniska and J. E. Kinsella, Foaming and emulsifying properties of glycosylated beta-lactoglobulin, *Food Hydrocolloids 2*:439 (1988).
86. A. Baniel, D. Caer, B. Colas, and J. Gueguen, Functional properties of glycosylated derivatives of the 11 S storage protein from pea (Pisum sativum L.), *J. Agric. Food Chem. 40*: 200 (1992).
87. S. C. B. Yan and F. Wold, Neoglycoproteins: in vitro introduction of glycosyl units at glutamines in β-casein using transglutaminase, *Biochemistry 23*:3759 (1984).
88. B. Colas, D. Caer, and E. Fournier, Transglutaminase-catalyzed glycosylation of vegetable proteins: effect of solubility of pea legumin and wheat gliadins, *J. Agric. Food Chem. 41*: 1811 (1993).
89. G. Matheis and J. R. Whitaker, Chemical phosphorylation of food proteins: an overview and a prospectus, *J. Agric. Food Chem. 32*:699 (1984).
90. G. Matheis, Phosphorylation of food proteins with phosphorus oxychloride—improvement of functional and nutritional properties: a review, *Food Chem. 39*:13 (1991).

91. Y.-T. Huang and J. E. Kinsella, Phosphorylation of yeast protein: reduction of ribonucleic acid and isolation of yeast protein concentrate, *Biotechnol. Bioeng. 28*:1690 (1986).
92. G. Matheis, M. H. Penner, R. E. Feeney, and J. R. Whitaker, Phosphorylation of casein and lysozyme by phosphorus oxychloride, *J. Agric. Food Chem. 31*:379 (1983).
93. S. L. Woo, L. K. Creamer, and T. Richardson, Chemical phosphorylation of bovine β-lactoglobulin, *J. Agric. Food Chem. 30*:65 (1982).
94. F. F. Shi, J. S. Hamada, and W. E. Marshall, Deamidation and phosphorylation to improve protein functionality in foods, *Molecular Approaches to Improving Food Quality and Safety* (D. Chatnagar and T. E. Cleveland, eds.), Van Nostrand Reinhold, New York, 1992, p. 37.
95. S. Damodaran and J. E. Kinsella, Dissociation of yeast nucleoprotein complexes by chemical phosphorylation, *J. Agric. Food Chem. 32*:1030 (1984).
96. A. Giez, B. Stasinska, and J. Skupin, A protein isolate for food by phosphorylation of yeast homogenate, *Food Chem. 31*:279 (1989).
97. H.-Y. Sung, H.-J. Chen, T.-Y. Liu, and J.-C. Su, Improvement of the functionalition of soy protein isolate through chemical phosphorylation, *J. Food Sci. 48*:716 (1983).
98. F. F. Shi, Chemical and enzymatic phosphorylation of soy glycinin and their effects on selected functional properties of the protein, *Food Proteins: Structure and Functionality* (K. D. Schwenke and R. Mothes, eds.), VCH, Weinheim, 1993, p. 180.
99. M. Hirotsuka, H. Taniguchi, H. Narita, and M. Kito, Functionality and digestibility of highly phosphorylated soybean protein, *Agric. Biol. Chem. 48*:93 (1984).
100. S. L. Woo and T. Richardson, Functional properties of phosphorylated β-lactoglobulin, *J. Dairy Sci. 66*:984 (1983).
101. T. A. Langan, Protein kinases and protein kinase substrate, *Adv. Cycl. Nucl. Res. 3*:99 (1973).
102. S. K. Hanks, A. M. Quinn, and T. Hunter, The protein kinase family: conserved features and deduced phylogeny of the catalytic domains, *Science 241*:42 (1988).
103. P. J. Kennelly and E. G. Krebs, Consensus sequences as substrate specific determinants for protein kinases and protein phosphatases, *J. Biol. Chem. 266*:15555 (1991).
104. S. B. Smith, J. B. White, J. B. Siegel, and E. G. Krebs, Cyclic AMP-dependent protein kinase. Primary steps of allosteric regulation, *Protein Phosphorylation* (O. R. Rosen and E. G. Krebs, eds.), Cold Spring Harbor Laboratory, Cold Spring Harbor, NY, 1981, p. 55.
105. S. Okuno and H. Fugisawra, Stabilization purification and crystallization of catalytic subunit of cAMP-dependent protein kinase from bovine heart, *Biochim. Biophys. Acta 1038*:204 (1990).
106. S. Shoj, L. H. Ericsson, K. A. Walsh, E. H. Fischer, and K. Tetani, Amino acid sequence of the catalytic subunit of bovine type II adenosine cyclic 3′,5′-phosphate-dependent protein kinase, *Biochemistry 22*:3702 (1983).
107. K. Seguro and M. Motoki, Enzymatic phosphorylation of soybean proteins by protein kinase, *Agric. Biol. Chem. 53*:3263 (1989).
108. L. F. Ross and D. Bhatnagar, Enzymatic phosphorylation of soybean proteins, *J. Agric. Food Chem. 37*:841 (1989).
109. N. F. Campbell, F. F. Shi, and W. E. Marshall, Enzymatic phosphorylation of soy protein isolate for improved functional properties, *J. Agric. Food Chem. 40*:403 (1992).
110. E. W. Bingham and H. M. Farrel, Casein kinase from the golgi apparatus of lactating mammary gland, *J. Biol. Chem. 249*:36 (1974).
111. M.-C. Ralet, D. Fouques, T. Chardot, and J.-C. Meunier, Enzymatic phosphorylation by a native casein kinase II of native and succinylated soy storage protein glycinin and β-conglycinin, *J. Agric. Food Chem. 44*:69 (1996).
112. K. Seguro and M. Motoki, Functional properties of enzymatically phosphorylated soybean protein, *Agric. Biol. Chem. 54*:1271 (1990).
113. G. H. Dixon and H. Neurath, Acylation of the enzymatic site of α-chymotrypsin by esters, acid anhydrides and acid chlorides, *J. Biol. Chem. 225*:1049 (1957).

114. A. D. Gounaris and G. E. Perlman, Succinylation of pepsinogen, *J. Biol. Chem. 242*:2739 (1967).
115. K. D. Schwenke, D. Zirwer, K. Gast, E. Görnitz, K.-J. Linow and J. Gueguen, Changes of the oligomeric structure of legumin from pea (Pisum sativum L.) after succinylation, *Eur. J. Biochem. 194*:621 (1990).
116. T. A. Eisele and C. J. Brekke, Chemical modification and functional properties of acylated beef heart myofibrillar proteins, *J. Food Sci. 46*:1095 (1981).
117. M. Siu and U. Thompson, In vitro and in vivo digestibility of succinylated cheese whey protein concentrate, *J. Agric. Food Chem. 30*:743 (1982).
118. A. S. F. A. Habeeb, H. G. Cassidy, and S. J. Singer, Molecular structural effects produced in proteins by reaction with succinic anhydride, *Biochim. Biophys. Acta 29*:587 (1958).
119. T. Yamauchi, H. Ono, Y. Kamata, and K. Shibasaki, Acetylation of amino groups and its effect on the structure of soybean glycinin, *Agric. Biol. Chem. 43*:1309 (1979).
120. K. Y. Shetty and M. S. N. Rao, Effect on succinylation on the oligomeric structure of arachin, *Int. J. Peptide Protein Res. 11*:305 (1978).
121. K. D. Schwenke, E. J. Rauschal, D. Zirwer, and K.-J. Linow, Structural changes of the 11 S globulin from sunflower seed (Helianthus annus L.) after succinylation, *Int. J. Peptide Protein Res. 25*:347 (1985).
122. E. J. Rauschal, K.-J. Linow, W. Pähtz, and K. D. Schwenke, Chemische Modifizierung von Proteinen. 8. Mitt. Beeinflussung physikochemischer und funktioneller Eigenschaften von Proteinen aus Ackerbohnen durch Succinylierung, *Nahrung 25*:241 (1981).
123. A. Seifert and K. D. Schwenke, Improved approach for characterizing the coalescence stability of legumin stabilized O/W emulsions by analytical ultracentrifugation, *Prog. Colloid Polymer Sci. 99*:3 (1995).
124. K. D. Schwenke, R. Mothes, D. Zirwer, J. Gueguen, and M. Subirade, Modification of the structure of 11 S globulins from plant seeds by succinylation, *Food Proteins: Structure and Functionality* (K. D. Schwenke and R. Mothes, eds.), VCH, Weinheim, 1993, p. 143.
125. M. A. I. Ismond, E. D. Murray, and S. D. Arntfield, Stability of vivilin, a legume storage protein, with stepwise electrostatic modification, *Int. J. Peptide Protein Res. 26*:584 (1985).
126. A. F. S. A. Habeeb, Quantitation of conformational changes on chemical modification of proteins: use of succinylated proteins as a model, *Arch. Biochem. Biophys. 121*:652 (1967).
127. H. S. Groniger, Jr., Preparation and properties of succinylated fish myofibrillar protein, *J. Agric. Food Chem. 21*:978 (1973).
128. Y. Sato and R. Nakamura, Functional properties of acetylated and succinylated egg white, *Agric. Biol. Chem. 41*:2163 (1977).
129. C. Y. Ma, L. M. Poste, and J. Holme, Effects of chemical modification on the physicochemical and cake-baking properties of egg white, *Can. Inst. Food Sci. Technol. J. 19*:17 (1986).
130. L. K. Creamer, J. Roeper, and E. H. Lohrey, Preparation and evaluation of some acid soluble casein derivatives, *NZ J. Dairy Sci. Technol. 6*:107 (1971).
131. Z. Haque and M. Kito, Lipophilization of $\alpha_{S,1}$-casein. 2. Conformational and functional effects, *J. Agric. Food Chem. 31*:1231 (1983).
132. K. J. Barber and J. J. Warthesen, Some functional properties of acylated wheat gluten, *J. Agric. Food Chem. 30*:930 (1982).
133. C.-Y. Ma, Functional properties of acylated oat protein, *J. Food Sci. 49*:1128 (1984).
134. L. R. Beuchat, Functional and electrophoretic characteristics of succinylated peanut flour protein, *J. Agric. Food Chem. 25*:258 (1977).
135. Y. R. Choi, E. W. Lusas and K. C. Rhee, Succinylation of cotton seed flour: effect on the functional properties of protein isolates prepared from modified flour, *J. Food Sci. 46*:954 (1981).
136. K. L. Franzen and J. E. Kinsella, Functional properties of succinylated and acetylated soy protein, *J. Agric. Food Chem. 24*:788 (1976).

137. J. Umeya, N. Mitsuichi, F. Yamauchi, and K. Shibasaki, Effect of acetylation on hardening protein-water suspending systems, *Agric. Biol. Chem. 45*:1577 (1981).
138. H. Aoki, O. Taneyama, N. Orimo, and I. Kitagawa, Effect of liphophilization of soy protein on its emulsion stabilizing properties, *J. Food Sci. 46*:1192 (1981).
139. K. S. Kim and J. S. Rhee, Effect of acetylation on emulsifying properties of glycinin, *J. Agric. Food Chem. 38*:669 (1990).
140. M. Canella, G. Castriotta, and A. Bernardi, Functional and physicochemical properties of succinylated and acetylated sunflower protein, *Lebensm. Wiss. Technol. 12*:95 (1979).
141. M. Kabirullah and R. B. H. Wills, Functional properties of acetylated and succinylated sunflower protein isolate, *J. Food Technol. 17*:235 (1982).
142. E. A. Johnson and C. J. Brekke, Functional properties of acylated pea protein isolates, *J. Food Sci. 48*:722 (1983).
143. H. Schmandke, R. Maune, S. Schuhmann, and M. Schultz, Contribution to the characterization of acetylated protein fractions of vicia faba in view of their functional properties, *Nahrung 25*:99 (1981).
144. L. U. Thompson and Y. S. Cho, Chemical composition and functional properties of acylated low phytate rapeseed protein isolate, *J. Food Sci. 49*:1584 (1984).
145. R. Ponnampalam, J. Deslisle, Y. Gagne, and J. Amiot, Functional and nutritional properties of acylated rapeseed proteins, *J. Am. Oil Chem. Soc. 67*:531 (1990).
146. A. T. Paulson and M. A. Tung, Emulsification properties of succinylated canola protein isolate, *J. Food Sci. 52*:1557 (1987).
147. J. Gueguen, S. Bollecker, K. D. Schwenke, and B. Raab, Effect of succinylation on some physicochemical and functional properties of the 12 S storage protein from rapeseed (Brassica napus L.), *J. Agric. Food Chem. 38*:61 (1990).
148. P. D. Hoagland, Acylated β-caseins. Effect of alkyl group size on calcium sensitivity and on aggregation, *Biochemistry 7*:2542 (1968).
149. S. Nakai, E. Li-Chan, and S. Hayakawa, Contribution of protein hydrophobicity to its functionality, *Nahrung 30*:327 (1986).
150. A. T. Paulson and M. A. Tung, Solubility, hydrophobicity and net charge of succinylated canola protein isolate, *J. Food Sci. 52*:1557 (1987).
151. P. Krause, R. Mothes, and K. D. Schwenke, Some physicochemical and interfacial properties of native and acetylated legumin from faba beans (Vicia faba L.), *J. Agric. Food Chem. 44*: 429 (1996).
152. M. Subirade, J. Gueguen, and K. D. Schwenke, Effect of dissociation and conformational changes on the surface behavior of pea legumin, *J. Colloid Interface Sci. 152*:442 (1992).
153. A. Seifert, M. Schultz, K. Strenge, G. Muschiolik, and H. Schmandke, Mechanical barrier preventing centrifuge creaming of O/W food emulsions stabilized by proteins, *Nahrung 34*: 293 (1990).
154. Z. Haque and M. Kito, Effect of acylation on the emulsifying properties of soybean proteins, *Agric. Biol. Chem. 48*:1099 (1984).
155. K. D. Schwenke, E. J. Rauschal, and K. D. Robowsky, Functional properties of plant proteins. Part. 4. Foaming properties of modified proteins from faba beans, *Nahrung 27*:335 (1983).
156. L. Prahl and K. D. Schwenke, Functional properties of plant proteins. Part. 7. Rheological properties of succinylated protein isolates from faba beans (Vicia faba L.), *Nahrung 30*:311 (1986).
157. A. T. Paulson and M. A. Tung, Rheology and microstructure of succinylated canola protein isolate, *J. Food Sci. 53*:821 (1988).
158. A. T. Paulson and M. A. Tung, Thermally induced gelation of succinylated canola protein isolate, *J. Agric. Food Chem. 37*:319 (1989).
159. G. Schmidt, H. Schmandke, and R. Schöttel, Viscosity behavior of Vicia faba protein isolates and their acetylated derivatives, *Acta Aliment. 15*:175 (1986).

160. E. S. Alford, V. Piriyapan, C. W. Dill, C. R. Young, R. L. Richter, and W. A. Landmann, Effect of succinylation on the proteolysis of food proteins, *J. Food Sci. 49*:614 (1984).
161. T. A. Eisele, C. J. Brekke, and S. M. McCurdy, Nutritional properties and metabolic studies of acylated beef heart myofibrillar proteins, *J. Food Sci. 47*:43 (1982).
162. G. Goulet, R. Ponnampalam, J. Amiot, A. Roy, and G. J. Brisson, Nutritional value of acylated oat protein concentrates, *J. Agric. Food Chem. 35*:589 (1987).
163. T. Matoba, E. Doi, and D. Yonezawa, Digestibility of acetylated and succinylated proteins by pepsin-pancreatin and some intracellular peptidases, *Agric. Biol. Chem. 44*:2323 (1980).
164. J. Bjarnason and K. J. Carpenter, Mechanisms of heat damage in proteins. 1. Models with acylated lysine units, *Br. J. Nutr. 23*:859 (1969).
165. J. Zhang, T. C. Lee and C.-T. Ho, Thermal deamidation of proteins in a restricted water environment, *J. Agric. Food Chem. 41*:1840 (1993).
166. J. W. Finley, Deamidated gluten: a potential fortifier for fruit juices, *J. Food Sci. 40*:1283 (1975).
167. C. W. Wu, S. Nakai, and W. D. Powrie, Preparation and properties of acid solubilized gluten, *J. Agric Food Chem. 24*:504 (1976).
168. C.-Y. Ma, B. D. Oomah, and J. Holme, Effect of deamidation and succinylation on some physico-chemical and baking properties of gluten, *J. Food Sci. 51*:99 (1986).
169. R. Lasztity, Investigation of the rheological properties of gluten II. Visco-elastic properties of chemically modified gluten, *Acta Chim. Acad. Scient. Hung. 62*:75 (1969).
170. N. Matsudomi, T. Sasaki, A. Kato, and K. Kobayashi, Conformation and surface properties of deamidated gluten, *Agric. Biol. Chem. 46*:1583 (1982).
171. Y. Popineau, S. Bollecker, and J. Y. Thebaudin, Characterization biochimique et fonctionnelle des protéines de gluten désamidées artiellement en conditions ménagées, *Sci. Aliments 8*:411 (1988).
172. A. Kato, A. Tanaka, N. Matsudomi, and K. Kobayashi, Deamidation with food proteins by protease in alkaline pH, *J. Agric. Food Chem. 35*:224 (1987).
173. A. Kato, A. Tanaka, A. Y. Lee, N. Matsudomi, and K. Kobayashi, Effects of deamidation with chymotrypsin at pH 10 on the functional properties of proteins, *J. Agric. Food Chem. 35*:285 (1987).
174. A. Kato, Y. Lee, and K. Kobayashi, Deamidation and functional properties of food proteins by the treatment with immobilized chymotrypsin at alkaline pH, *J. Food Sci. 54*:1345 (1989).
175. F. F. Shi, Deamidation during treatment of soy protein with protease, *J. Food Sci. 55*:127 (1990).
176. M. Kikuchi, H. Hayashida, E. Nakano, and K. Sakaguchi, Peptidoglutaminase. Enzymes for selective deamidation of γ-amide of peptide-bound glutamine, *Biochemistry 10*:1222 (1971).
177. B. P. Gill, A. J. O'Shaughnessey, P. Henderson, and D. R. Headon, An assessment of potential of peptidoglutaminase I and II in modifying the charge characteristics of casein and whey proteins, *Irish J. Food Sci. Technol. 9*:33 (1985).
178. J. S. Hamada, Effects of heat and proteolysis on deamidation of food proteins using peptidoglutaminase, *J. Agric. Food Chem. 40*:719 (1992).
179. J. S. Hamada and W. E. Marshall, Preparation and functional properties of enzymatically deamidated soy proteins, *J. Food Sci. 54*:598 (1989).
180. I. A. Vaintraub, L. V. Kotova, and R. Shaha, Protein deamidase from germinating wheat grains, *FEBS Lett. 302*:169 (1992).
181. M. J. Mycek and H. Waelsch, The enzymatic deamidation of proteins, *J. Biol. Chem. 235*:3513 (1960).
182. M. Motoki, K. Seguro, N. Nio, and K. Takinami, Glutamine-specific deamidation of $a_{S,1}$-casein by transglutaminase, *Agric. Biol. Chem. 50*:3025 (1986).
183. C. Larré, Z. M. Kedzior, M. G. Chenu, G. Viroben, and J. Gueguen, Action of transglutaminase on an 11 S seed protein (pea legumin): influence of the substrate conformation, *J. Agric. Food Chem. 40*:1121 (1992).

184. M.-C. Alexandre, Y. Popineau, G. Viroben, M. Chiarello, A. Lelion, and J. Gueguen, Wheat γ-gliadin as substrate for bovine plasma factor XIII, *J. Agric. Food Chem. 41*:2208 (1993).
185. C. Larré, M. Chiarello, J. Y. Blanloeil, M. Chenu, and J. Gueguen, Gliadin modifications catalyzed by guinea pig liver transglutaminase, *J. Food Biochem. 17*:267 (1993).
186. J. Dziuba, The effect of carboxyl groups esterification on certain physical and chemical properties of casein, *Acta Aliment, Polon, 3*:137 (1977).
187. M. I. Halpin and T. Richardson, Selected functionality of β-lactoglobulin upon esterification of side-chain carboxyl groups, *J. Dairy Sci. 68*:3189 (1985).
188. T. Mita and H. Matsumoto, Flow properties of aqueous gluten methylester dispersions, *Cereal Chem. 58*:57 (1981).
189. R. Uy and F. Wold, Introduction of artificial cross-links into proteins, *Protein-Cross-Linking. Biochemical and Molecular Aspects* (M. Friedman, ed.), Plenum Press, New York, 1977, p. 169.
190. G. Matheis and J. R. Whitaker, A review: enzymatic cross-linking of proteins applicable to foods, *J. Food Biochem. 11*:309 (1987).
191. S.-Y. Tanimoto and J. E. Kinsella, Enzymatic modification of proteins: Effect of transglutaminase cross-linking on some physical properties of β-lactoglobulin, *J. Agric. Food Chem. 36*:281 (1988).
192. K. Ikura, T. Kometani, M. Yoshikawa, R. Sasaki, and H. Chiba, Cross-linking of casein components by transglutaminase, *Agric. Biol. Chem. 44*:1567 (1980).
193. I. J. Kang, Y. Matsumura, K. Ikura, M. Motoki, H. Sakamoto, and T. Mori, Gelation and gel properties of soybean in a transglutaminase-catalyzed system, *J. Agric. Food Chem. 42*: 159 (1994).
194. L. Kurth and P. J. Rogers, Transglutaminase catalyzed cross-linking of myosin to soya protein, casein and gluten, *J. Food Sci. 49*:573 (1984).
195. M. Motoki, H. Aso, K. Seguro, and N. Nio, $\alpha_{S,1}$-Casein film prepared using transglutaminase, *Agric. Biol. Chem. 51*:993 (1987).
196. M. Motoki, N. Nio, and K. Takinami, Functional properties of food proteins polymerized by transglutaminase, *Agric. Biol. Chem. 48*:1257 (1984).
197. G. Matheis and J. R. Whitaker, Modification of proteins by polyphenol oxidase and peroxidase and their products, *J. Food Biochem. 8*:137 (1984).
198. R. Kieffer, G. Matheis, H. W. Hofmann, and H.-D. Belitz, Verbesserung der Backeigenschaften von Weizenmehlen durch Zusätze von Peroxidase aus Meerettich, H_2O_2 und Phenolen, *Z. Lebensm-Unters. Forsch. 173*:376 (1981).
199. T. Kuninori, J. Nishiyama, and H. Matsumoto, Effect of mushroom extract on the physical properties of dough, *Cereal Chem. 53*:420 (1976).
200. R. E. Feeney and J. R. Whitaker, Importance of cross-linking reactions in proteins, *Advances in Cereal Science and Technology, Vol. IX* (Y. Pomeranz, ed.), American Association of Cereal Chemists, St. Paul, MN, 1988, p. 21.
201. D. Potter, Functional foods—a major opportunity for the dairy industry? *Dairy Ind. Int. 55*: 16 (1990).
202. A. J. Puigserver, H. Gaertner, L. C. Sen, R. E. Feeney, and J. R. Whitaker, Covalent attachment of essential amino acids to proteins by chemical methods: nutritional and functional significance, modification of proteins. *Food, Nutritional and Pharmacological Aspects* (R. E. Feeney and J. R. Whitaker, eds.), American Chemical Society, Washington, DC, 1982, p. 149.

14

Genetic Engineering of Food Proteins

CARL A. BATT
Cornell University, Ithaca, New York

I. INTRODUCTION

The attributes of any biological system, whether it be as simple as a bacterium or as complex as a human being, are dictated in part by its principal genetic material, deoxyribonucleic acid (DNA). Since the discovery that DNA is the factor that mediates the character of a biological system, great strides have been made in understanding the process. Perhaps one significant outcome of this accumulated knowledge base is the ability to redesign proteins to improve their functional properties in a process termed "protein engineering" (Fig. 1). It can be argued that Mother Nature is the ultimate and most efficient engineer of protein structure and hence should be relied upon for the functional improvement of proteins. However, these functional improvements manifest themselves by conferring, to the host, a selective advantage over its previous version (or at least not confer a selective disadvantage). Natural variation in the population exists, and advantageous traits can be selected by propagation of the desired offspring. This approach has been successful, as documented by the improvement in domestic plant and animal species, but requires an enormous amount of time and patience. Alternatively it has been argued that improvements can be introduced into the population by genetic engineering [1].

The nucleotide sequence of a gene is almost always faithfully translated into a specific amino acid sequence, with the result being a protein. The protein has a distinctive structure, a three-dimensional conformation that is a function of the amino acid sequence and the environment in which it resides. It is feasible to change the structure and therefore the function of a protein by affecting a change in the nucleotide sequence. Recombinant DNA techniques have been developed to selectively and precisely replace nucleotides in a targeted gene sequence. The requirements for establishing a protein engineering system include knowledge of the complete nucleotide sequence (and by deduc-

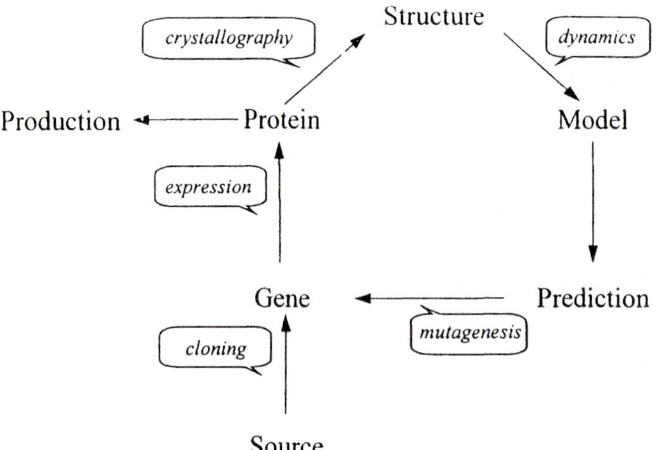

FIGURE 1 The iterative cycle of protein engineering. Major components in the cycle are shown in bold; major technologies are shown in balloons.

tion the amino acid sequence) in addition to some information about the catalytic/functional properties of specific regions within the protein. Furthermore, intelligent design strategies require some structural knowledge to avoid the need for serendipitous events that "gee-whiz" mutagenesis usually requires [2].

The requirements for carrying out protein engineering will be described, with a focus on those proteins that might have utility as foods, food ingredients, or processing aids. Primary examples will be drawn for the milk protein β-lactoglobulin. β-Lactoglobulin serves as an excellent model system because the various requirements to carry out protein engineering have been established, including a high-resolution structure and a system for expression of the recombinant protein. In addition, some examples with enzymes including lysozyme and subtilisin will be offered.

A. Why Engineer?

The variety observed in the organisms found in nature and the products that they produce is due to continual mutation and selection. Through a variety of environmental pressures, they have developed and adapted into a wide array of genera, only a fraction of which have been documented. Depending upon their given niche, these organisms have evolved to succeed in their environment and to survive the stresses they may encounter. The evolutionary process that created the current biodiversity can be considered a form of genetic engineering, and it clearly has been successful for some, as witnessed by their survival and propagation.

The search for new and improved proteins that might have valuable applications in food formulations does not necessarily benefit from evolutionary processes. Nevertheless, there has been and continues to be a school of thought that contends that, given the vast array of environments on the earth, a microorganism, say, can be found to carry out virtually any bioconversion or a protein might be isolated with a desired functional performance. If one subscribes to the existence of extraterrestrial life, an even greater number of environments could be predicted. This approach, which could be described as "ecological screening," has in fact resulted in the identification of a number of very important bacteria and enzymes (or, more generally, proteins). An obscure example is the discovery of a class of proteins that have intense sweetness [3]. Two of these proteins, monellin and thaumatin, have enjoyed a certain degree of commercial success. The rationale behind their evolution includes arguments usually ascribed to the appearance of color and fragrance in flowers, namely, that these properties attract animal vectors that would assist the distribution of their pollen or seeds. Similarly, a sweet protein might attract an animal that would enjoy ingesting a sweet treat and unwittingly then serve as a means for dispersion.

The physical properties of a protein that might make it a good food ingredient do not usually translate into attributes that would confer evolutionary fitness. Producing a protein with a superior ability to gel, for example, would not be an advantage for a dairy cow. Thus, unfortunately, it is not always possible to identify an environment that would select for a protein with improved physical characteristics.

The process by which organisms have evolved requires a mutational event followed by some selective process favoring a given phenotype. Mutational frequencies for a given biological system vary; for example, a mutation occurs in the bacteriophage T2 once in every 1×10^8 gene replications, while a mutation occurs once in every 8×10^8 asexual spores of a common fungus. The mutation rate depends upon the integrity of the mechanisms that repair the damage inflicted on the DNA by environmental chemicals or ionizing irradiation or on the accuracy of DNA replication. The mutational rate can be enhanced by treatment with a variety of mutagens or by the selection of mutants that have some defect in the biological systems mentioned above. Exposure to a mutagen can increase the mutation frequency at least 10,000 times. Selection of the desired phenotype still requires that it be identifiable within the context of the number of candidates that can be feasibly screened. This number depends upon the complexity of the assay procedure and the number of man-years that can be devoted to its identification.

B. Mutation and Selection

The classical approaches toward strain improvement through mutagenesis and selection have resulted in the development of a great number of extremely valuable organisms. The best cases can be made for bacteria where enhanced production of amino acids and other metabolic products has been realized by strain mutagenesis and selection. In plants, breeding of improved varieties is the method of choice, and constant movement toward higher-producing varieties as well as those with disease resistance has been documented for a diverse number of species. In animals the breeding process is much slower due in part to the slower generation time and the much lower number of progeny that can be obtained from a single mating event. In dairy cattle the advent of artificial insemination has provided for wider distribution for animals with improved genetic potential over traditional animal breeding. Unfortunately, breeding and selection cannot deliver progeny without the latent genetic potential to code for the desired improvement. Given enough time, effort, and the appropriate biological selection, an organism may evolve with the desired traits. It is, however, often difficult to create the selectable environment necessary to effect the desired change. Molecular genetics have, however, improved the process by providing tools that allow an objective analysis of genotype well in advance and through a much simpler process than assessing phenotype.

Mutations accumulate on a continuous basis, only some of which are retained and fixed in the population. The functional attributes that are modified by these mutations should have some positive effect, giving the individual with such attributes some selective advantage over the rest of the population, which does not possess the modification. Selective external pressure can be simulated to help enhance the likelihood that a desired mutation will be obtained. For example, clever strategies for the selection of thermostable enzymes have been developed. These strategies depend upon establishing conditions where thermostable function is demanded under a specific set of growth conditions. Thus, for example, by introducing a thermolabile antibiotic resistance marker (in this case an enzyme that modifies the antibiotic kanamycin) into a thermophilic bacterium, mutations that give rise to thermostable drug resistance can be selective by growth at high temperatures under antibiotic selection. Significant increases in the thermostability of kanamycin nucleotidyltransferase were realized by selection of *Bacillus stearothermophilus* strains that carried the normally thermolabile kanamycin resistance gene. Although elegant in concept, few examples can be cited where such simple conditions can demand improvement in the functional properties that would be beneficial for food proteins.

An even more problematic situation exists for milk proteins, perhaps most notably for the whey protein β-lactoglobulin, that is, no functional role has been readily defined. The vast array of β-lactoglobulin variants found in mammals suggests that the structure is sufficiently robust to accommodate these variations yet still retain function. Alternatively, β-lactoglobulin may not serve any immediate function and may be the ancestral remnant of a previously required protein.

II. A PROTEIN-ENGINEERING PRIMER

A directed mutational strategy can in theory yield improvements in a targeted product. It is, however, rarely simple due to the complicated interactions on both a molecular and an organismic level. The normal physiological function of a protein that might also be used as a food ingredient is usually not known, and therefore modification of its structure-function through directed mutagenesis may give unexpected consequences. Conversely, since it may not carry out an essential physiological function, it may tolerate a large degree of structural alteration without deleterious consequences to the host. As described later, they whey protein β-lactoglobulin can be expressed in the milk of transgenic mice. Although mice do not normally express β-lactoglobulin, neither transgenic mice nor their suckling offspring that do express this foreign protein shows any adverse consequences. The following outlines the individual steps required for protein engineering.

A. Structural Determination

What is required for a directed attempt to improve the functional property of a protein? First, a complete knowledge of the encoding nucleotide sequence is needed. In the past this was a formidable task; now with the advent of the polymerase chain reaction (PCR) and automated DNA sequencers, it has been reduced to a relatively trivial task. In addition, structural information that can be used as a guide for altering its function is valuable. Without this latter information, intelligent strategies cannot be developed with predictable conclusions. Given the myriad of potential alterations with both direct as well as secondary structural and hence functional consequences, it is difficult to imagine achieving any predictable improvement.

Protein engineering should be guided by a finite knowledge of the protein's structure. The complexity of interactions that mediate a protein's native structure are too complex to assume that any individual change made at random will have a defined and limited consequence. At best, in the absence of a system to resolve a protein's structure, the ramifications of a given mutation might be evaluated by activity. A single mutation may alter a protein's functionality through either direct or indirect effects. In the case of the latter, assumptions may prove erroneous and the true nature of these changes in fact may be more complex. Protein engineering carried out in the absence of a refined structure may be superficially successful, but it does not advance the learning process.

The structure and hence a correlation with the function of a protein can be determined in a number of ways. The most widely used method is determining the x-ray crystal structure of the protein. Proteins under certain conditions can form crystals, which represent a collection of upward of 10^{10} molecules arranged in an orderly lattice. Crystallization of proteins is not an exact science and the conditions vary for each protein; a researcher can toil for years without success. The first protein to be crystallized, jack bean urease, did so almost spontaneously. Curiously, the structure for any urease was not resolved until recently. Despite the availability of crystals, the structure was left to be solved by P. A. Karplus at Cornell some 50 years after its initial crystallization [4]. Once a crystal is obtained, its structure can be determined by the diffraction pattern obtained using x-ray beams. The diffraction pattern is a reflection of the relative spacing between atoms within the protein and can be resolved to less than 1 Å. Given sufficient data, the position of amino acid side chains and potential intramolecular bonds can be mapped. Once a structure is determined for a given protein, subsequent structures for site-directed mutants can be easily resolved. Resolving the crystallographic data is reduced to simply subtracting the vast number of reflections that are similar between the wild-type and mutant proteins. Then the remaining electron densities can be assigned based upon the knowledge of the type of amino acid substitution.

The data obtained from crystallographic analysis is not an absolute measure of the structure of the protein. Given the crystallographic coordinates that indicate the positions of the atoms in the protein, the structure can be refined by bringing to bear the laws of thermodynamics. The energy of the structure can be minimized to yield the most favorable conformation.

B. Molecular Modeling

The static picture generated by crystallographic studies is complemented by molecular dynamic simulations that explore the range of potential fluctuations or movements a protein may exhibit. Theoretical studies of the possible effects of mutations can also be explored to gain insight into how these changes might give rise to the desired performance properties of the protein.

Molecular modeling is accomplished via computer analysis and high-resolution graphic terminals. The process is iterative, and the predictions must be tested against the original x-ray diffraction data. Energy calculations that account for the interactions (attraction, repulsion, bond angles, etc.) between the various atoms can be made within certain distance constraints. Minimized structures that depict more favorable conformations based upon minimum energies can be predicted. An estimation of the fluctuations in these structures can also be determined by integrating standard estimates of potential motion allowable within the constraints of the polypeptide.

In the end, any structural theory must be tested biologically since the laws of nature mediate the final structure. Analyzing the structure facilitates the prediction of changes that might improve the protein's function. The predicted changes are then made by altering the nucleotide sequence of the gene through site-directed mutagenesis, producing the protein and finally analyzing the effect. It is usual to then determine the structure of the modified protein as described above. This process is cyclical, and several rounds of structure determination, mutagenesis, and functional analysis are routine.

Attempts to derive a protein structure de novo are in their infancy, and it is not currently possible to successfully predict protein conformation from molecular mechanic calculations without extensive information from other sources, for example, NMR. Efforts reported for milk proteins suffer from a naive conceptual appreciation of the problems involved.

C. Mutagenesis

There are a number of constraints that limit either the natural or the engineered evolution of a protein. For example, expression of a protein may be toxic to its host, thus preventing a desired change from being realized. At a minimum, however, any nucleotide sequence change can be made in a directed fashion through the application of oligonucleotide-directed site-specific mutagenesis. Since a prerequisite for knowledge of a nucleotide sequence has already been established, this sequence can then be used to alter the template in a precise manner. Typically these mutations are introduced by the design and synthesis of a short oligonucleotide that contains the desired nucleotide(s) alteration as well as approximately 10–20 nucleotides of flanking sequence. These flanking sequences help to stabilize a hybrid between the target sequence that is to be mutagenized and the mutagenic oligonucleotide. Although chemical synthesis of oligonucleotides has been vastly improved over the past 15 years, it still is not feasible to routinely synthesize a complete gene. Even if an attempt is made to initially synthesize a gene coding for a given protein, subsequent rounds of mutagenesis would still be carried out using short oligonucleotides coupled to enzymatic synthesis of the balance of the gene.

In theory, with an oligonucleotide as a primer and a DNA polymerase to synthesize the balance of the gene, an equimolar pool of "wild-type" and mutant genes should be generated. In practice, however, the population of mutant genes usually is only $<0.01\%$. Screening for these desired mutants is feasible by hybridization, although distinct differences between mutant and wild-type genes is sometimes difficult. Over the past 10 years a number of selection schemes have been developed to enrich for mutant genes [5]. These methods capitalize on the biochemical differences between the wild-type and mutant gene and employ either an enzymatic or a biological selection against the wild-type gene.

D. PCR

The manipulation and analysis of nucleic acids has realized a quantum leap with the advent of DNA amplification using the thermostable DNA polymerase isolated from *Thermus aquaticus* (*Taq* polymerase). Not only has mutagenesis been made simple, but virtually all molecular biology techniques have realized a tremendous benefit. The intellectual genesis of PCR began in California in the middle 1980s and was brought to practical reality by a group at Cetus [6]. PCR involves the selected amplification of a region of DNA as delineated by a set of oligonucleotide primers (Fig. 2). By successively cycling at different temperatures for different periods of time, a series of annealing, extension, and dissociation steps can be carried out with the net result of exponentially amplifying the sequences flanked by these primers. In approximately 30 cycles, with a total time on the order of 2–4 hours, a millionfold amplification of the targeted DNA sequence can be realized. The development of PCR-based methods has largely replaced most of the mutagenic selection methods. PCR is capable of generating a million copies of a mutant gene from a single wild-type template. Therefore, the dramatic increase in the ratio of mutated gene to the original wild type simplifies the subsequent selection process.

E. Expression of Recombinant Proteins

Recombinant proteins can be produced in a variety of heterologous or homologous hosts. Several factors including yields, purification costs, and eventual usage are considered when selecting the particular expression host. Clearly the state of genetic manipulation available for a particular host system is a critical issue in selecting an expression host. For milk proteins, for example, the ideal would be to consider the dairy cow for production of engineered milk proteins, however, the technical complexity of genetically modifying this system is daunting. In most cases either a prokaryote or a lower eukaryote

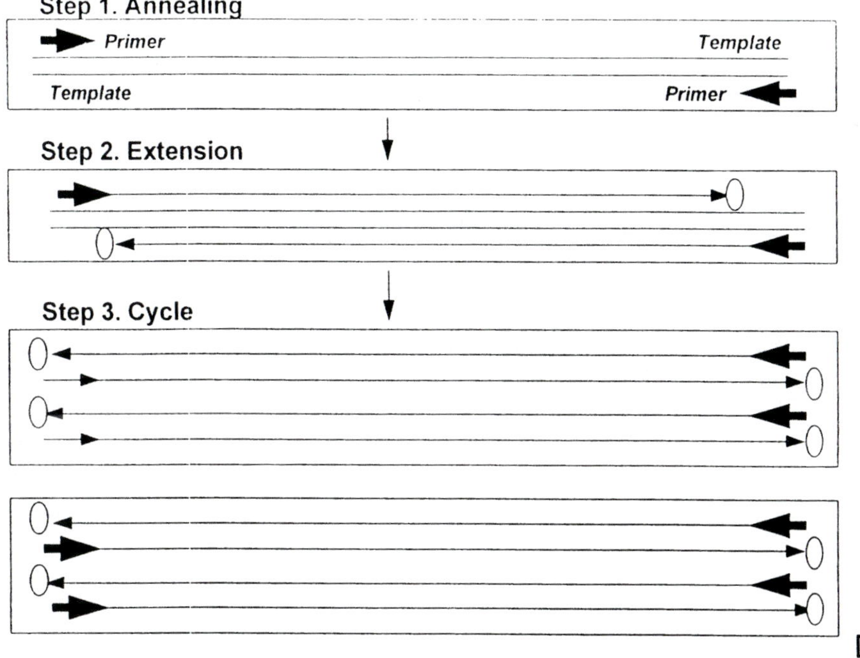

FIGURE 2 Outline of the steps in the polymerase chain reaction. Template is shown as thin line, primers as thick arrows, extension products as thin arrows, and *Taq* DNA polymerase as circle.

is the system of choice for expression of recombinant proteins. The selection of a particular host may have as much to do with the resident expertise of the laboratory as it does with any other particular feature. In addition, decisions to pursue intracellular versus extracellular production of a given protein must be made. Since a myriad of hosts, vectors, and associated issues influence the choice of expression system, the following discussion will be limited to our own experiences with the whey protein β-lactoglobulin.

1. Bacteria

By virtue of the availability of highly developed molecular tools, the gram-negative bacterium *Escherichia coli* has been the initial organism of choice for expressing heterologous proteins. A plethora of plasmids and other expression systems exist that allow the regulated and high-level expression of heterologous proteins. Plasmids that have differing copy numbers and promoters capable of various transcriptional levels can be readily obtained. Further, the transcriptional and translational mechanism is well understood, allowing the design of genetic constructs that have a reasonable chance of success. Finally, secretion (or more properly, release into the periplasmic space) has been reported for a number of heterologous proteins when the appropriate secretion signal is fused to the protein. Release of these proteins from the periplasmic space is usually achieved by osmotic shock with the advantage that cytoplasmic proteins do not contaminate the preparation.

As an example of this process, in our own laboratory the bovine β-lactoglobulin has been expressed in *E. coli* [7]. A variety of highly transcribed promoter systems are available, most of which are based upon either a *trp/lac* hybrid (i.e., *tac*) or the bacteriophage T7. A regulated promoter is usually desirable, especially where the recombinant product is toxic to the cell. Using a regulated promoter, cell mass can be increased under conditions where the promoter is not expressed, usually by withholding the appropriate inducer. When sufficient cell mass is obtained (typically 10^9 cells/ml), the inducer can be added or the conditions adjusted to allow expression. For the *tac* promoter, which has been used to express the β-lactoglobulin gene, a synthetic analog of lactose (isopropyl-thio-β-D-galactoside; IPTG) is added, which switches the *lac* operator to allow transcription to proceed. Expression levels were on the order of 15–20% of the total cellular protein, but unfortunately β-lactoglobulin was produced as inclusion bodies. This is not an uncommon observation for heterologous proteins expressed in *E. coli*. The next challenge was to develop methods for denaturation and renaturation of the recombinant β-lactoglobulin. In fact, once efficient renaturation conditions are established, expression systems that yield inclusions bodies have the advantage of providing a facile first purification step. Inclusion bodies, because of their relatively unique density, can be harvested from other cellular debris by simple centrifugation. Unfortunately, although denaturation is simple, renaturation and establishing the conditions under which the denatured protein can refold is usually complex.

2. Yeast

Yeasts, most often *Saccharomyces*, have also been used to express recombinant proteins, including β-lactoglobulin. Yeasts have the potential to secrete heterologous proteins if the appropriate signal sequences are fused on the targeted gene.

Recently our attention has been focused on expressing β-lactoglobulin in the methanol-utilizing yeast *Pichia pastoris*. This yeast has a remarkable record of expressing heterologous proteins to very high levels. Almost all of the reports of expression in

P. pastoris involve the use of the *aox* promoter system. *aox* encodes the alcohol oxidase, and its promoter is not only tightly regulated, being induced by methanol, but when fully induced it can also direct high level of transcription. Plasmid constructs that are commercially available include the *aox* promoter as well as the αF secretion signal sequence. Selection is typically made using complementation of *his*$^-$ *P. pastoris* auxotrophs by a His marker on the plasmid. Recombinant sequences are forced to integrate into *P. pastoris* chromosome by linearization of the vector. The vector can be linearized by restriction at a site with the His or the Aox gene, directing it toward its respective chromosomal target. Integration into the *his*$^-$ results in Mut$^+$, His$^+$, while integration into the Aox yields *mut*^{s}, His$^+$ transformants. Finally, vectors (i.e., pPIC9K) are available that contain a marker, which can be used to select for *P. pastoris* transformants with multiple copies of the insert. The kanamycin resistance, which in eukaryotes can be selected based upon resistance to the aminoglycoside G418, has been reported to be successfully employed for *P. pastoris*. Selection of transformants through sequentially higher concentrations of G418 yields transformants with multiple copies of the kanamycin resistance gene and hence multiple copies of the *aox*–gene of interest fusion.

We have fused the bovine β-lactoglobulin to the αF secretion signal. The sequence mature β-lactoglobulin protein was inserted into pPIC9 immediately after the αF signal sequence cleavage site. Transformants using integration into both the Mut and *his* loci were obtained. β-Lactoglobulin was observed to be secreted into the medium, and preliminary characterization of the protein using both native and denaturing polyacrylamide electrophoresis reveal it to be indistinguishable from the protein purified from cow's milk. Yields in shake flask are approximately 100 mg/liter. Expectations are that the yield should increase at least 10-fold upon scale-up into the appropriate batch-fed fermenter (M. Meagher, personal communication). The *P. pastoris* system will allow us to extend our studies of β-lactoglobulin to include structural resolution by x-ray crystallography and nuclear magnetic resonance. The latter will be facilitated by the robust growth of *P. pastoris* and expression of *aox* promoter-driven heterologous proteins in a minimal medium. The minimal medium will permit the use of ^{15}N-labeled nitrogen sources that will enhance NMR resolution.

3. Transgenic Tricks

The improvements made to any milk protein including β-lactoglobulin have little practical significance and remain parlor tricks in the absence of a means to produce them economically. Although recombinant fermentations have drastically reduced the costs of producing proteins, these improvements are usually justified by citing high-value proteins with pharmacological effects. In contrast, modified milk proteins, as highlighted in this chapter, would have to compete with the more traditional dairy-derived proteins and more importantly any other food proteins (i.e., those from soy or other plants). A means to the production of modified milk proteins is clear. Why not simply utilize dairy cattle, which for centuries have been the production vehicle for these proteins?

The scope of genetic manipulations extends to the animal kingdom as well. Animals cells grown in tissue culture can be transformed with exogenous DNA and the effect of the transforming DNA on the cell observed. It is not currently feasible to regenerate animals from cells in culture as can be achieved (in certain cases) with plants. In 1982, Ralph Brinster of the University of Pennsylvania and Richard Palmiter of the University of Washington produced transgenic mice by inserting a gene into the male pronucleus of a fertilized egg [8]. The microinjected egg was then implanted into a foster mother and allowed to come to term. In a fraction of the progeny (the system is not 100% efficient) the injected DNA is integrated into the chromosome, resulting in a transgenic animal. To date a number of transgenic animals have been produced, including rabbits, sheep, pigs, fish, and cattle. The major obstacles are in part technical—how to isolate, inject, and reimplant fertilized eggs efficiently and in the end produce transgenic animals at a high frequency. One of the more interesting potential uses of transgenic animals is to produce high-value pharmaceuticals in their milk. The approach is to genetically fuse the gene coding for the desired protein (i.e., TPA) to the expression sequences and secretion signal of a milk protein. Paul Simons and John Clark of the AFRC in Edinburgh in initial reports utilized mice that were made transgenic by the introduction of the ovine β-lactoglobulin gene. A total of 6 kb of genomic DNA was microinjected into the male pronucleus of fertilized eggs, which were then reinserted into foster mothers. Female transgenic offspring were then screened to reveal some that produced β-lactoglobulin in their milk. As expected, β-lactoglobulin production levels varied in part to due to the number of copies of the transgene and probably positional effects. This latter phenomenon is not understood but probably reflects the influence of sequences flanking the inserted transgene on the transcriptional level of the transgene. This group next produced transgenic sheep carrying the gene coding for factor IX (a blood clotting protein) to the gene coding for ovine β-lactoglobulin. The result was a small amount of factor IX secreted in the milk of these animals. This approach takes advantage of the normal biosynthetic machinery in the animal and produces the product in a relatively simple medium such as milk, which has only a small number of well-characterized proteins facilitating purification of the product. Subsequent improvements in the genetic constructs coupled to the selection of higher-producing founder lines has made the production of these pharmaceutical proteins economically viable.

In a similar fashion, we have isolated the appropriate genetic signals to direct expression of bovine β-lactoglobulin and have constructed several minigenes [9]. These minigenes coupled with the technology to generate transgenic cattle will yield the next generation of dairy cows capable of producing whey proteins with improved functional properties.

III. EXAMPLES OF PROTEIN ENGINEERING

Overall, most exercises in protein engineering attempt to do one of the following:

1. Increase thermostability or optimum temperature for activity
2. Alter the optimum pH for activity
3. Modify or eliminate cofactor and/or metal requirements
4. Alter substrate specificity

For food proteins, good examples of the first two are readily available, and in general, examples of all of these goals in protein engineering can be found in the literature. The following sections give examples of successes in engineering improvements in food proteins.

A. Thermostability: T4 Lysozyme

Thermostability is perhaps one of the most important properties for proteins used for industrial applications. Thermostability can be distinguished from the optimum temperature for activity, since the former is the maximum temperature at which a protein does not irreversibly lose either its native structure or activity. The latter is the maximum temperature at which the enzyme can display activity. For industrial applications, thermostability is important because it allows higher operating temperatures at which reactions typically run faster and the chances of contamination are less. As mentioned previously, thermostable enzymes can often be found in thermophilic microorganisms. Within the past 20 years a number of hot spots, including thermal springs and undersea thermal vents, have been explored, within which a number of hyperthermophilic microorganisms have been discovered. Perhaps the most rewarding was *Thermus acquaticus*, which was discovered by Thomas Brock in a hot spring in Yellowstone National Park in Wyoming. This organism has been the source of a number of thermostable enzymes including *Taq* DNA polymerase. *Taq* DNA polymerase is the key component in the aforementioned polymerase chain reaction.

What makes a protein thermostable? A number of features probably contribute to thermostability, but no definitive blueprint for engineering thermostability has been developed. Comparisons of closely related groups of enzymes that include thermolabile and thermostable versions reveal a number of differences, some of which may directly contribute to thermostability. Among these differences, disulfide linkages appear to stabilize a protein, although it is not an absolute or even relative measure of thermostability. It is, however, one route by which thermostable variants of proteins have been engineered.

Lysozyme is a hydrolytic enzyme that cleaves the N-acetyl-neuramic bonds that are found in the cell walls of bacteria. Lysozyme can be isolated from a number of animal sources as well as from bacteriophages. It probably serves as a protective mechanism in the former and to assist in cell lysis in the latter. The thermostability of lysozyme varies depending upon the source.

One of the first demonstrations of protein engineering was achieved by Ron Wetzel's group at Genentech in the early 1980s [10,11]. It remains one of the most intriguing examples of how structural information and biochemical intuition can be used to formulate a protein engineering strategy. T4 lysozyme, isolated from the bacteriophage T4, is relatively thermolabile compared to lysozyme purified from hen egg white. The structure for T4 lysozyme has been resolved providing a road map for designing in disulfide linkages. Through a variety of standard biochemical measurements, hen egg white lysozyme was found to contain four disulfide linkages, whereas T4 had no disulfide linkages. The available T4 structure was used to design a disulfide linkage that might help to stabilize the protein.

The first generation of engineered T4 lysozymes had a single disulfide linkage between a mutated Ile3Cys residue and an existing Cys97. The result was an increase in thermostability with a half-life at 67°C of 28 minutes. The Ile3Cys mutant did, however, exhibit a biphasic thermal denaturation profile suggesting a mixed population of lysozymes. This observation was subsequently resolved by the realization that not only Ile3Cys-Cys97 disulfides but also Ile3Cys-Cys54 disulfide linkages had formed. Removal of the Cys54 yielded a homogeneous population of T4 lysozyme with a thermal inactivation rate of only 15% of the wild-type. This mutant T4 lysozyme was as thermostable as hen egg white lysozyme.

If one disulfide linkage conferred a finite degree of thermostability, would multiple disulfide linkages give an additive effect? This was explored using the same T4 lysozyme system, and two more disulfide linkages were introduced. With three disulfide linkages the thermostability of T4 lysozyme was improved more than 23°C [12]. A knowledge of the structure was, however, essential as one of the engineered proteins included a disulfide bond that spanned the opening to the active site. Thus, although this lysozyme was extremely stable, it had no activity!

B. Thermostability: β-Lactoglobulin

The whey protein β-lactoglobulin is found in the milk of many, but not all, ruminants, in addition to numerous other mammals [13]. Its classification as a whey protein is by virtue of its ability to remain in solution at a relatively low pH (~3.0). Considerable interest exists in its use as a food ingredient. Various preparations of whey proteins have found applications in food formulations for either their nutritional or functional properties. For example, the texture of yogurt is highly dependent upon the incorporation of β-lactoglobulin into the curd. To achieve this the milk substrate is heated to approximately 85°C to aggregate the β-lactoglobulin on the casein micelle. In contrast, the use of whey proteins in beverage formulations is limited especially where thermal processing is required and a clear, nonturbid solution is desirable. The applications are limited due to its tendency to form aggregates upon heating. It is clear from a number of studies, mostly employing various reagents that either reduce disulfide linkages or alter the chemical reactivity of thiol groups, that the thermal-induced reactions leading to the formation of macromolecular complexes are initiated by the free thiol group. β-Lactoglobulin cannot form effective gels when reducing conditions are employed. The redox status of the free thiols is not, however, the exclusive determinant of gel formation, as the ionic strength and pH also affect the process. Clearly, a number of amino acid residues interact directly and through their contribution to various structural domains of the protein affect gel formation.

Bovine β-lactoglobulin undergoes irreversible denaturation when heated at temperatures of >85°C. Although the events involved in this process are not clear, several lines of evidence suggest that the Cys121 plays a critical role in initiating this process. The end result is the formation of macromolecular complexes that under most conditions precipitate from solution. In certain cases where various conditions are optimal, the process can result in the formation of a thermoset gel, which is desirable. In other situations where an aggregate is formed, the insoluble mass is considered undesirable and limits the utility of whey proteins for a number of applications.

The events leading to the macromolecular associations observed with β-lactoglobulin appear to be mediated by disulfide interactions. A large number of indirect experimental results using a variety of reagents to block or otherwise modify free thiols effectively reduce aggregation and/or gelation of the protein. Blocking the free thiol at Cys-121 can prevent gelation and therefore stabilize β-lactoglobulin against thermal induced alterations.

The initial solution proposed to reduce thermoinduced aggregation was to remove the Cys121 and replace it with the structurally similar yet chemically inert alanine [14]. Although a chemically inert alanine residue could be inserted into this position and verified by nucleotide sequence analysis, the recombinant protein could not be purified from inclusion bodies. The alternative for enhancing thermostability was to engage the free thiol in a disulfide bond, thus introducing a third disulfide linkage into the protein. It was envisioned that not only would the free thiol be chemically inactivated but that the additional disulfide might lend greater stability to the protein.

Two potential amino acid candidates were selected for substitution: Ala132 and Leu104 (Fig. 3). Each was chosen for replacement with a cysteine due to the close proximity of their side chains to the free thiol at Cys121. Molecular modeling studies indicated that a cysteine could be substituted at either position and oxidized to form a disulfide linkage with the Cys121. To further characterize the potential stability of these different mutations, molecular dynamics were carried out.

The results of these molecular dynamics studies indicated that the effects of these two possible mutations could be quite different. Thus, variants of β-lactoglobulin were engineered to create the A132C and L140C proteins. In contrast to the C121A, both of these proteins refolded properly, and assignment of the third disulfide linkage could be made by peptide mapping and quantitative measurement of the free thiol content. The conformational stability of the L104C and A132C mutant proteins against thermal denaturation was substantially increased (8–10°C) as compared with wild-type β-lactoglobulin. The midpoint of the denaturation profile of A132C β-lactoglobulin is at a guanidine hydrochloride concentration of 4.5 M as compared to 3.6 and 3.4 M for L104C and

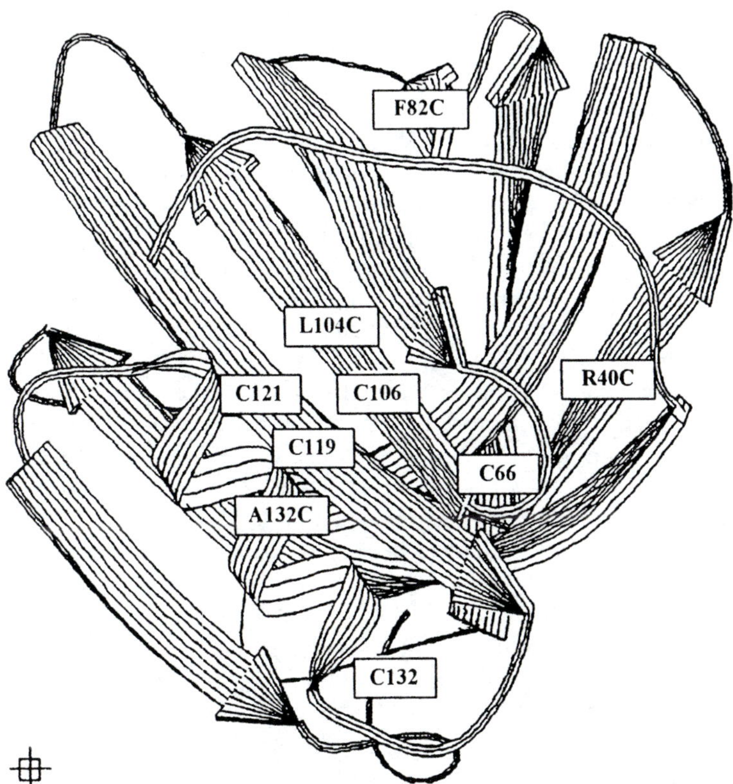

FIGURE 3 Ribbon diagram of β-lactoglobulin depicting the location of the cysteine residues (C121, C106, C119, C66, C160) and amino acid residues mutated to cysteine (A132, L104, and F82, and R40).

wild-type β-lactoglobulins, respectively. More importantly, neither the A132C nor the L104C would polymerize upon heating at 65°C, in contrast to the wild-type protein. Thus, a valuable performance attribute was engineered into the protein by modification of its thermal stability.

C. Increasing Gelation of β-Lactoglobulin

The curd formed during the manufacture of cheese and some other fermented dairy products is the result of a partial denaturation and aggregation of the proteins present in milk. Although the curd is primarily composed of caseins, some of the whey proteins may become entrapped depending upon the process and most notably the temperature to which the milk is heated prior to fermentation. Yogurt, for example, has a significant amount of β-lactoglobulin in its curd due to the high temperatures used in the process, which are sufficient to denature this protein. The incorporation of β-lactoglobulin into the curd is desirable both for the textural and perhaps for the flavor properties of the product.

The design of thermostable variants of β-lactoglobulin coupled with the work of Richardson, who demonstrated that chemical thiolation could increase the gel strength, suggested a strategy for engineering proteins with enhanced gelation ability [15,16]. Rather than decrease the number of free thiols, gelation might be enhanced by increasing the free thiol content. Increasing the gel strength would be desirable for food formulations such as puddings where a firm texture was desired. Additional cysteine residues would need to be positioned far enough away from the Cys121 as to not allow them to form disulfide bonds. Two mutations were designed, F82C and R40C, with both the single and double substitutions introduced by site-directed mutagenesis. One important analytical tool that we needed to develop was an instrument for measuring the gel strength of micro-scale samples. This was accomplished by building a penetrometer, which measured the weight necessary for a flat-tipped needle to puncture the gel inside a capillary tube. Wild-type β-lactoglobulin when heated at 90°C for 15 minutes formed gels with strengths of 14–19 g over a concentration range of 9.4% to 10%, but it did not form a gel below 9%. In contrast, the F82C β-lactoglobulin formed a gel at concentrations down to 8% with a gel strength of 23.7 g. Gels of the R40C/F82C β-lactoglobulin were formed at concentrations as low as 6.8% and had a gel strength of 16.5 g. The F82C and R40C/F82C β-lactoglobulins formed transparent gels, however, the R40C β-lactoglobulin formed a coagulum gel and therefore its gel strength could not be accurately measured.

The key issue with a number of the modifications that have been made to β-lactoglobulin is their performance as food ingredients, especially when added to complex formulations [17]. As a model system, yogurt is simple, yet it could be improved, for example, by reducing whey syneresis. Since the double mutant R40C/F28C formed stronger gels at lower temperatures as compared to the wild-type protein, its functionality as a food additive for yogurt was explored. Very small amounts of R40C/F82C β-lactoglobulin reduced the amount of whey formed in yogurt that was processed at 70°C, a temperature approximately 15°C lower than the standard regimen.

D. Engineering Oxidation Stability in Subtilisin

One of the more elegant examples of engineering a protein to improve its performance is the work of David Estell and coworkers at Genentech [18]. Subtilisin is a protease isolated from the bacterium *Bacillus amyloliquifaciens.* It has served as a model target for a number of protein engineering studies. A robust expression system was available, and the requisite crystal structure is also available. Furthermore, it has easily assayed activity, and a number of chromogenic substrates are available that make careful kinetic measurements facile. As a protease, subtilisin could be potentially used in laundry detergents, but it is sensitive to oxidation, a problem because the addition of oxidizers to detergents is desirable. Through a variety of biochemical and structural studies, a methionine group at position 222 was implicated as the residue that, when oxidized, inactivated the enzyme. The crystal structure of subtilisin suggested that this methionine residue was in the active site, supporting the hypothesis that it was the target for oxidation. Through site-directed mutagenesis the methionine residue was replaced with a variety of other amino acids, some of which actually improved the catalytic activity of the enzyme. Typically for essential active-site amino acids, natural selection has optimized the enzyme and most mutations are deleterious. In fact, a number of amino acid substitutions for Met-222 severely reduced the catalytic activity of the enzyme. As expected, those amino acid substitutions which were structurally or chemically similar to methionine such as cysteine were less deleterious than others, for example, lysine. The highest activity was achieved when cysteine replaced methionine, although this enzyme was still oxidation sensitive. Of these replacements, serine and alanine substitutions resulted in an enzyme that was much more resistant to chemical oxidation. Met222Ser and Met222Ala were able to withstand more than 16 minutes of exposure to 1.0 M hydrogen peroxide without any appreciable loss in activity. In contrast, the wild-type enzyme lost more than 50% of its activity within one minute of exposure to the same concentration of hydrogen peroxide.

Success with subtilisin was achieved because a fundamental base of knowledge was established on the structure of the enzyme and the key amino acid residue, which was adversely affected by oxidation. Recently this modified enzyme was the first engineered protein to be patented in the United States.

E. Altering the pH Optimum of Subtilisin

The pH optimum of an enzyme is determined by the nature of its catalytic mechanism. The pH of the local environment surrounding the catalytic center is a function of the amino acids packed into the active site. In addition, local effects of solvents and ions that might diffuse into the active site as well as distal effects of charged amino acids on the surface of the protein play a role in the pH of the enzyme. Fersht and colleagues capitalized on earlier observations of the effect of chemical modification on the pH optimum of proteases [19]. Again using subtilisin as a model system, mutations were introduced that resulted in charge substitutions either inside of the active site or on the surface. In these experiments, alterations directly involving the catalytic triad of this serine protease were avoided as they would have an obvious deleterious effect on activity. A mutation in the active site, Glu156Lys, increased the pKa of the enzyme approximately 0.6 units. Similar effects were observed when a surface residue substitution of Asp99Lys was introduced. The double mutant resulted in an increase of almost one unit in the pKa.

The effects of charge substitution on the pH optimum of the enzyme are, however, not robust. In most cases the pH shift effects are masked by increasing the ionic strength of the medium. Thus, although substantial improvement in an enzyme's activity at a desired pH might be realized, these improvements are observed only in very dilute ionic strength solutions. Unfortunately, in most food and industrial applications, the ionic strength of the medium is quite high. This work still serves as a valuable example of the power of protein engineering to alter an important performance attribute of an enzyme.

IV. CONCLUSIONS

In the past 20 years, great technical progress has been made, establishing a robust and sophisticated set of tools to dissect the structure-function of proteins. Efforts in this area involve multidisciplinary teams of scientists with distinct skills in molecular biology, biochemistry, and physical chemistry. Universities are excellent establishments for carrying out such activities because they have not only an existing intellectual base but the facilities necessary for practicing the diverse array of methodologies. There are no obvious impediments to the application of these tools to proteins used as ingredients in or processing aids for foods. At a minimum these tools will help to resolve specific issues with regard to the behavior of proteins under certain conditions or the reaction mechanisms that are employed for the conversion of a substrate into a product. We have, for example, confirmed that the free thiol group in β-lactoglobulin is the catalyst in the formation of aggregates at high temperatures. Other more general concepts concerning the influence of the ionization state of active site residues on the pKa of a catalytic residue have also been illuminated. Despite these scientific successes, several challenges still exist that limit the application of these techniques to proteins destined for foods. A primary limitation is the cost of producing these modified proteins. The selection of proteins for use as major ingredients in foods is based upon cost. Plants including soybeans and animal products including dairy products are relatively inexpensive in comparison to recombinant proteins produced even under optimized conditions. As, however, we move these modified proteins back into traditional production vehicles, they will become cost competitive. The current major application of proteins modified by recombinant DNA technology is as processing aids. Although it is difficult to track due to the proprietary nature of these commercial processes, it is reasonable to assume that a number of "engineered" proteins have found their way into the detergent market. As a more applicable example, certain process adjuncts are likely to be improved via a directed mutagenic approach. Investment in this technology will likely yield further improvements in food proteins, and hopefully this path will not be totally forsaken as cost reduction becomes the dominant yet short-sighted goal of the food industry.

REFERENCES

1. Y. Kang and T. Richardson, Genetic engineering of caseins, *Food Technol. 39*:89 (1985).
2. J. R. Knowles, Tinkering with enzymes: What are we learning? *Science 236*:1252 (1987).
3. H. Van Der Wel and K. Loeve, Isolation and characterization of thaumatin I and II, the sweet-tasting proteins from *Thaumatococcus daniellii* Benth, *Eur. J. Biochem. 31*:221 (1972).
4. E. Jabri, M. B. Carr, R. P. Hausinger, and P. A. Karplus, The crystal structure of urease from *Klebsiella aerogenes*, *Science 268*:998 (1995).
5. M. A. Vandeyar, M. P. Weiner, C. J. Hutton, and C. A. Batt, A simple and rapid method for the selection of oligodeoxynucleotide-directed mutants, *Gene 65*:129 (1988).
6. R. Saiki, S. Scharf, F. Faloona, K. Mullis, G. Horn, H. Erlich, and N. Arnheim, Enzymatic amplification of B-globin genomic sequences and restriction site analysis for diagnosis of sickle cell anemia, *Science 230*:1350 (1985).

7. C. A. Batt, L. D. Rabson, D. W. S. Wong, and J. E. Kinsella, Expression of recombinant β-lactoglobulin in *Escherichia coli*, Agr. Biol. Chem. 54:949 (1990).
8. R. Palmiter, R. Brinster, R. Hammer, M. Trumbauer, M. Rosenfeld, N. Birnberg, and R. Evans, Dramatic growth of mice that develop from eggs microinjected with metallothionein-growth hormone fusion genes, *Nature 300*:611 (1982).
9. M. Silva, D. W. S. Wong, and C. A. Batt, Cloning and sequencing of the genomic bovine β-lactoglobulin gene. *Nucleic Acids Res. 18*:3051 (1990).
10. L. J. Perry and R. Wetzel, Disulfide bond engineered into T4 lysozyme: stabilization of the protein toward thermal inactivation, *Science 226*:555 (1984).
11. L. J. Perry and R. Wetzel, Unpaired cysteine-54 interferes with the ability of an engineered disulfide to stabilize T4 lysozyme, *Biochemistry 25*:733 (1986).
12. M. Matsumura, G. Signor, and B. W. Matthews, Substantial increase of protein stability by multiple disulphide bonds, *Nature 342*:291 (1989).
13. S. G. Hambling, A. S. McAlpine, and L. Sawyer, β-Lactoglobulin, *Advanced Dairy Chemistry*, Vol. 1, *Proteins* (P. F. Fox, ed.), Elsevier, London, 1992, pp. 141–190.
14. Y. Cho, W. Gu, S. Watkins, S. P. Lee, J. W. Brady, and C. A. Batt, Thermostable variants of bovine β-lactoglobulin, *Prot. Eng. 7*:263 (1994).
15. S. C. Kim, N. F. Olson, and T. Richardson, Polymerization and gelation of thiolated β-lactoglobulin at ambient temperature induced by oxidation by potassium iodate, *Milchwissenschaft 45*:627 (1990).
16. S. P. Lee, Y. Cho, and C. A. Batt, Enhancing the gelation of β-lactoglobulin. *J. Agric. Food Chem. 41*:1343 (1993).
17. S. P. Lee, D. S. Kim, S. Watkins, and C. A. Batt, Reducing whey syneresis in yogurt by the addition of a thermolabile variant of β-lactoglobulin, *Biosci. Biotech. Biochem. 58*:309 (1994).
18. D. A. Estell, T. P. Graycar, and J. A. Wells, Engineering an enzyme by site-directed mutagenesis to be resistant to chemical oxidation, *J. Biol. Chem. 260*:6518 (1985).
19. A. Russell and A. Fersht, Rational modification of enzyme catalysis by engineering surface charge, *Nature 328*:496 (1987).

15

Functionality of Protein Hydrolysates

Per Munk Nielsen
Novo Nordisk A/S, Bagsvaerd, Denmark

I. INTRODUCTION

The desire to improve the quality of industrially processed food products has provoked great interest recently in the functional properties of protein ingredients. It is well known that proteins from different sources have different properties. The functionality of proteins can be changed by subjecting them to physical and chemical treatments, such as pH, ionic strength, heat, mechanical shear, etc., as well as enzymatic treatments, such as proteolysis or polymerization using transglutaminase. Numerous possibilities exist for improving the functional properties of food proteins by partial enzymatic hydrolysis. This chapter will focus on the factors affecting enzymatic hydrolysis of proteins and the functional, nutritive, and immunological properties of protein hydrolysates.

II. ENZYMATIC HYDROLYSIS

A. The Hydrolysis Reaction

Figure 1 shows a simple reaction involving cleavage of a peptide bond by a protease. In this reaction, one mole of water is added for every peptide bond cleaved. This can affect the mass balance of the amount of protein hydrolyzed and the final amount of hydrolysate

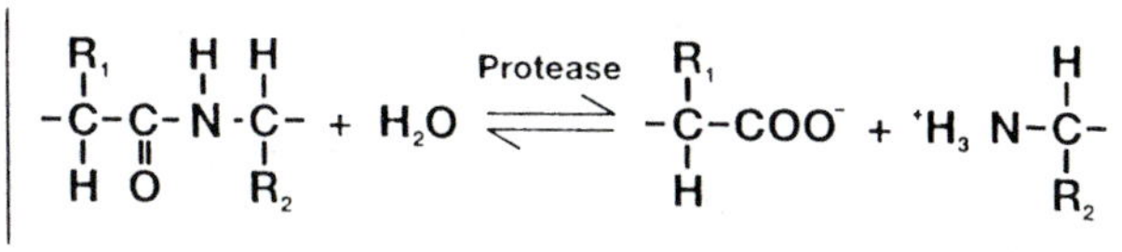

$$-\overset{R_1}{\underset{H}{C}}-\underset{O}{\overset{\|}{C}}-\overset{H}{N}-\overset{H}{\underset{R_2}{C}}- + H_2O \xrightleftharpoons{\text{Protease}} -\overset{R_1}{\underset{H}{C}}-COO^- + {}^+H_3N-\overset{H}{\underset{R_2}{C}}-$$

Figure 1 Hydrolysis of a peptide bond by protease.

obtained. For example, in a hydrolysate produced with a 75% degree of hydrolysis (DH) (i.e., 75% of the original peptide bonds are cleaved), for every four moles of amino acid residues cleaved in the substrate, three moles of water are added. If we assume that the average molecular weight of amino acid residues in proteins is 115 g/mol, then 1 g of protein will yield 1.117 g of protein hydrolysate. This "chemical gain" will increase with an increase in DH. However, this poses a practical problem—a protein hydrolysate cannot be considered a protein isolate, because, according to the definition, an isolate must contain 90% or more protein nitrogen in the dry matter. The Kjeldahl nitrogen factor is often used to calculate the protein content. Protein hydrolysates with very high degree of hydrolysis normally contain less than 90% protein nitrogen. For instance, in a hydrolysate with a DH of 75% from a raw material with 92% protein in dry matter, the end product will contain only 82.3% (92 × 100/11.7) protein on dry weight basis, provided no other alterations in dry matter occur.

B. Control of Degree of Hydrolysis

One of the basic parameters that needs to be controlled in protein hydrolysates is the degree of hydrolysis. This is essential because the properties of protein hydrolysates are closely related to the DH. The pH-stat method, which is generally used to follow biochemical reactions [1], is generally used to monitor the DH during the enzymatic hydrolysis of proteins [2]. The DH is calculated as:

$$\%DH = \frac{100h}{h_{tot}} \qquad (1)$$

where h is the number of peptide bonds cleaved per gram of protein and h_{tot} is the total number of peptide bonds per gram of protein. h_{tot} can be estimated from the amino acid composition of proteins.

When a peptide bond is cleaved, the freed carboxyl and amino groups will be ionized depending on the pH of the reaction solution. Except in the pH range of 5–6, the hydrolysis will release or take up a H^+ ion, which will change the pH of the solution. The number of H^+ ions released (or consumed) will be equivalent to the number of peptide bonds cleaved when the hydrolysis is carried out outside the range of the pK values of the amino and carboxyl groups. The principle involved in the pH-stat method is that, as the pH changes as a result of hydrolysis, the H^+ ions are titrated by adding NaOH and the pH is maintained at the initial pH value. When the hydrolysis is carried out at neutral-to-basic pH, the cleavage of a peptide bond will produce one equivalent acid, which can be titrated by one equivalent base. The amount of base will correspond to the number of bonds cleaved. In a similar way, hydrolysis under acidic conditions will produce base, and titration is then carried out with an acid. The pH-stat method is a very important tool in hydrolysis experiments, as it is very accurate and the DH is read continuously during the reaction. The ionization of carboxyl groups and amino groups together with the limitations in the pH range in which the pH-stat method can be used are illustrated in Figure 2.

The pH-stat method has some drawbacks when used in production scale. These include limitations of the pH range in which it works and technical difficulties associated with the addition of base (or acid) during the reaction. There are several alternative methods available for the measurement of DH (Table 1).

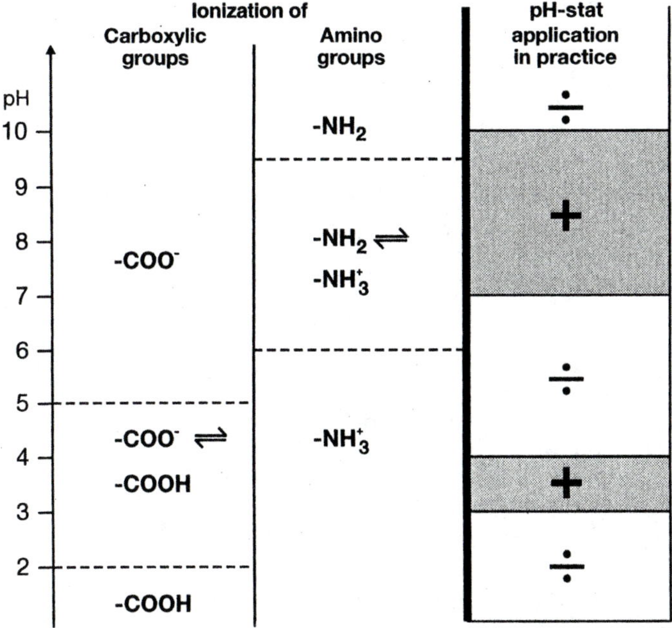

FIGURE 2 pH-dependent ionization of carboxylic and amino groups and the practical limitations in the use of the pH-stat method to monitor the degree of hydrolysis.

When selecting a method for the determination of the DH of a protein hydrolysate, it is very important to know (a) if it can be used in situ during the hydrolysis reaction, (b) if it can be used on a hydrolysate already produced and if the hydrolysis process has been terminated, and (c) if the method is accurate. If process control is desired, then the pH-stat, OPA, osmometry, formole titration, pH change, viscosity, or titration back to the original pH method can be used. These are rapid methods, but accuracy may vary. With the last method, special care should be taken so that the titration does not take the pH from a value of low protease activity to a high activity level. A fast heat treatment/ inactivation step can be considered before titration.

Of the methods listed in Table 1, only the pH-stat and OPA methods are directly correlated to the formation of free amino groups as a result of hydrolysis and thus provide a direct measure of the degree of hydrolysis. Some care should be taken in the case of the OPA method, since OPA does not react with the amino group of proline or hydroxyproline and reacts only partly with cysteine. The osmometry method also correlates very well with the DH [2].

Figure 3 shows the time course of hydrolysis as measured by changes in osmolality and pH during the reaction. Soy protein concentrate was hydrolyzed using Alcalase and Neutrase at an enzyme-to-substrate ratio of 1.5% and 1%, respectively. The protein content was 8%, and the initial pH was adjusted to 8.0 before adding the enzymes. The pH of the solution continuously decreased during hydrolysis. The shape of the curve is characteristic of a system showing the highest rate of reaction in the beginning of the process. About half of the increase in osmolality occurs within 10 minutes of hydrolysis and the pH drops rapidly to a low value.

Table 1 Different Methods of Controlling the Degree of Hydrolysis of Protein Hydrolysates.

Method	Principle	Ref.
pH stat	Keeping pH constant during hydrolysis; amount of titrant is equivalent to DH	2
OPA	O-Phthaldialdehyde reaction with primary amino groups to form a colored detectable compound	3,4
TNBS	2,4,6-Trinitrobenzenesulphonic acid reaction with amino groups to form a colored detectable compound	2,5
Osmometry	Measurement of freezing point depression, which correlates to DH	2
Ninhydrin	Ninhydrin reaction with amino groups to form a colored detectable compound	6
Formole titration	Titration of amino groups with formaldehyde	
Soluble nitrogen	Amount of soluble nitrogen during hydrolysis	7
Brix	Refractive index correlating with soluble dry matter	2
TCA index	Amount of peptides soluble in trichloroacetic acid (above a certain molecular weight the peptides precipitate)	2
Peptide chain length	HPLC method based on gel permeation chromatography	
Change in pH	Follow pH during hydrolysis	8
Viscosity	Follow change in viscosity during hydrolysis	2
Titration to basic pH	Titrate acid formed during hydrolysis (at pH > 5.5) up to pH 8.0	

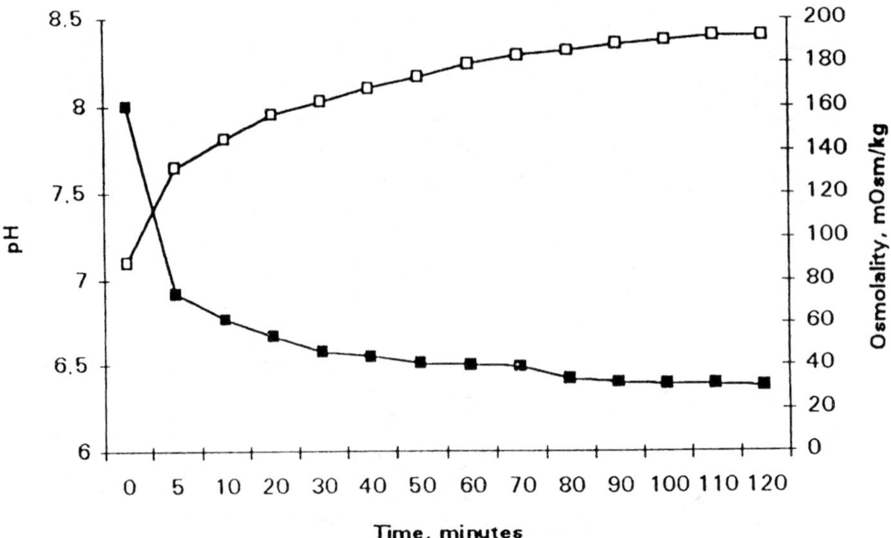

Figure 3 Changes in osmolality and pH during Alcalase hydrolysis of soy protein.

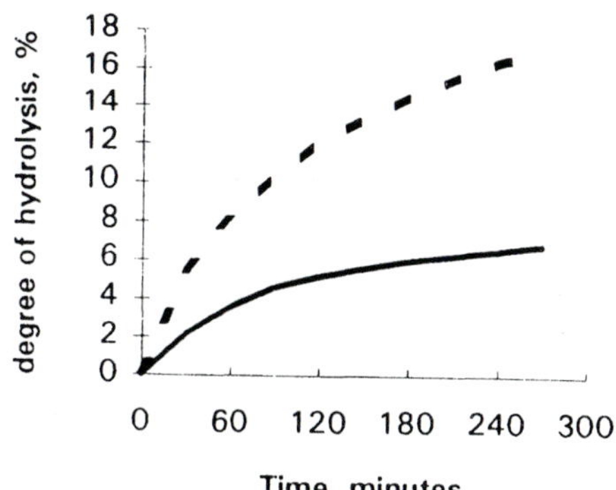

Figure 4 Rates of hydrolysis of soy protein isolates prepared by acid precipitation (----) and ultrafiltration (——) methods. Hydrolysis conditions were: 8% protein concentration, 50°C, pH 8.0 (pH stat) and enzyme (Alcalase)-to-substrate ratio 1:50. (From Ref. 9.)

C. Substrate Preparation

The important of pretreatment of raw material is illustrated in Figure 4, which shows the rates of hydrolysis of two different soy protein isolate preparations. One product is a commercial soy protein isolate produced by isoelectric precipitation and the other is a soy protein isolate produced by ultrafiltration. The isoelectric precipitation induces conformational changes in the protein, and hence it becomes more susceptible to proteolysis. On the other hand, the ultrafiltered soy protein isolate is mostly in the native state, and hence a majority of the peptide bonds are not readily available to the protease as they are buried in the interior of the protein structure [9].

The influence of heat treatment on the hydrolysis of α-lactalbumin and β-lactoglobulin by pepsin and papain, respectively, has been investigated [10]. Unlike β-lactoglobulin, α-lactalbumin was almost unaffected by the heat treatment. Native β-lactoglobulin was not hydrolyzed by pepsin, but it was hydrolyzed after heat treatment at 82°C for 8 minutes. The hydrolysis curve with papain changed from sigmoidal to hyperbolic after the heat treatment.

The effect of heat treatment on the susceptibility of β-lactoglobulin to proteolysis was explained in terms of changes in its structure. In the native state, cleavage of the first peptide bond is the rate-limiting step. This appears to be the reason for the sigmoidal shape of the hydrolysis curve. Once this is accomplished, the protein undergoes unfolding, which exposes additional peptide bonds for further hydrolysis by the enzyme. In the heat-denaturated state, since several peptide bonds are readily available for the enzyme, the hydrolysis curve follows a hyperbolic profile. It should be pointed out that many native protein preparations from plant sources contain protease inhibitors, and these would decrease the efficiency of hydrolysis unless they are inactivated by heat.

III. CHANGES IN FUNCTIONAL PROPERTIES

Hydrolysis of peptide bonds causes several changes in proteins:

1. The NH_3^+ and COO^- content of the protein increases, which increases its solubility.

2. The molecular weight of the protein/polypeptide decreases.
3. The globular structure of the protein is destroyed and the buried hydrophobic groups become exposed.

These changes drastically affect several functional properties of the protein.

A. Solubility

Solubility is an important prerequisite for a protein to perform as a functional ingredient in foods. The solubility of food proteins differs greatly depending on both the source of the protein and the processing treatments applied during isolation.

1. Effect of Degree of Hydrolysis

The solubility of proteins increases with an increase in DH. This is mainly due to a reduction in the molecular weight and an increase in the number of polar groups. Adler-Nissen and Olsen [11] observed a significant increase in the solubility of soy protein isolate after hydrolysis. Unhydrolyzed soy protein isolate had a nitrogen solubility of only about 5% at pH 4.2, whereas after hydrolysis with Alcalase to a DH value of 7.7%, the solubility increased to 75% in the pH range 2–8. When soy protein was hydrolyzed using Neutrase to a DH value of 5.8%, the solubility increased to about 60–70% in the pH range 2–8 (Fig. 5), which was almost the same as that achieved with Alcalase. A slight decrease in solubility occurred at a DH of 1%; this probably was due to the heat treatment given after hydrolysis in order to inactivate the enzymes and to terminate the reaction.

Although the solubility of soy protein generally increases with increasing DH, it does not reach 100% solubility even at very high DH [2]. Figure 6 shows the relationship between the protein solubility index (PSI) and DH at varying pH for soy grits hydrolyzed with the endoprotease Alcalase. The PSI is defined as the percentage of protein soluble at the isoelectric pH. The data in Figure 6 show that the solubility increases with pH and DH but does not reach 100% even at DH 16% and pH 9.0. It is evident that the pH of hydrolysis impacts the solubility of the protein at any given DH. The magnitude of the pH effect on solubility, however, is very much dependent on the enzyme used. For instance, soy proteins hydrolyzed at pH 7 and 8 with Alcalase (from *Bacillus*

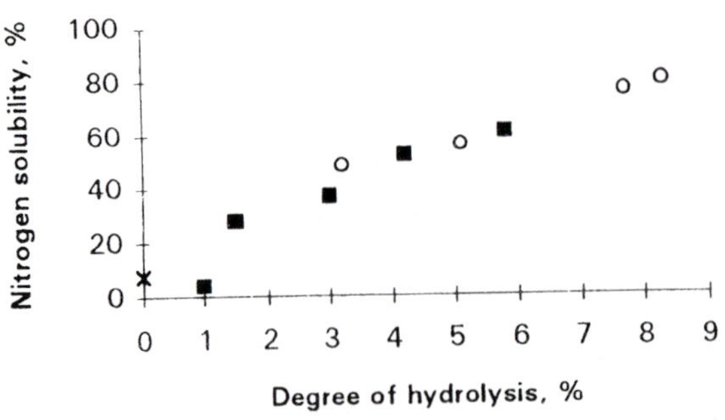

Figure 5 Nitrogen solubility at pH 4.2 as a function of degree of hydrolysis of soy protein by Alcalase (■) and Neutrase (○). The datum point X is for unhydrolyzed soy protein. (From Ref. 11.)

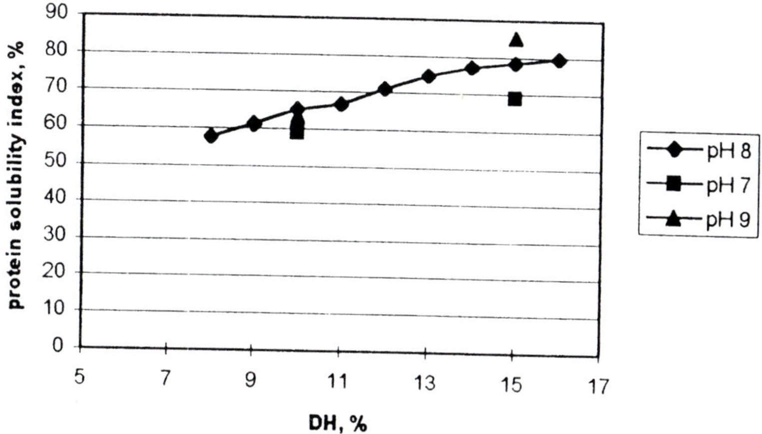

FIGURE 6 Protein solubility index of soy grits as a function of degree of hydrolysis with Alcalase at different pH. (From Ref. 2.)

licheniformis) show a higher solubility than those prepared with Sumizyme (from *Aspergillus niger*) [2].

The effect of proteolysis on protein solubility depends on the protein structure. For instance, compared to the solubility behavior of soy protein as discussed above, the solubility of wheat gluten hydrolysate at its isoelectric pH exceeds 90% at DH > 10% [12], which indicates that wheat gluten is more easily solubilized by enzymatic hydrolysis than soy protein.

Investigations on Alcalase hydrolysis of lean meat showed that 85% solubilization occurred at DH 10–15% [13]. It was found that the solubilization was dependent on the protein content in reaction mixture. Meat could be solubilized to the same extent as meat diluted with water, but it needed a higher DH to achieve the same degree of solubility. Adler-Nissen et al. (14) showed that hydrolysis of potato protein concentrate (having an initial solubility of about 70%) to a DH of 3% did not significantly increase its solubility. This is in contrast to similar experiments with soy protein isolate and fababean protein isolate. This difference was attributed to competition between large insoluble proteins and small soluble proteins and peptide substrates for the enzyme. As a consequence, a large fraction of the enzyme activity was "wasted" on small protein and peptide substrates, which are already soluble.

2. Effect of pH

The pH solubility profile of a hydrolyzed protein is affected by the DH [15]. For instance, the solubility of casein hydrolyzed with *Staphylococcus aureus* V-8 protease increased with the DH at the isoelectric pH (Fig. 7). The solubility increased from 0% at the isoelectric pH for intact casein to 24% at 2% DH and to 53% at 6.7% DH. However, it should be noted that the pH solubility profiles of the 2% and 6.7% DH samples were not similar. The pH of minimum solubility was 4.5 for the intact casein and the 2% DH samples, whereas it shifted to 5.5 for the 6.7% DH sample. This might be attributable to differences in the pIs of the peptides formed at 2% and 6.7% DH. This could also be related to the specificity of the V-8 protease (specific for glutamic acid and aspartic acid).

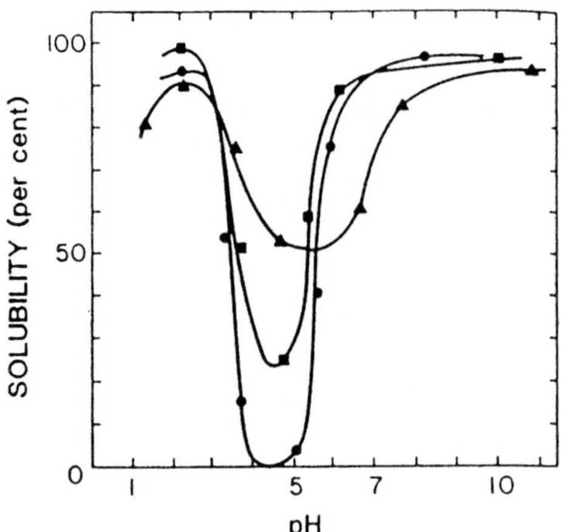

Figure 7 pH-solubility profiles of native casein and of *Staphylococcus aureus* V-8 protease-modified casein. The solubility is expressed as percent of total protein (0.1%) in solution. ●, Native casein; ■, 2% DH; ▲, 6.7% DH. (From Ref. 5.)

3. Effect of Enzyme

The solubility of a protein hydrolysate also depends on the enzyme used [2]. Figure 8 shows the isoelectric pH solubility of casein and soy protein hydrolysates prepared with Alcalase, trypsin, and Sumyzyme RP (an acid protease from *Rhizopus* sp.).

Mullally et al. [16] studied the solubility at pH 6.6 of α-lactalbumin hydrolysates prepared using several commercial proteases. The pH is relevant because it represents

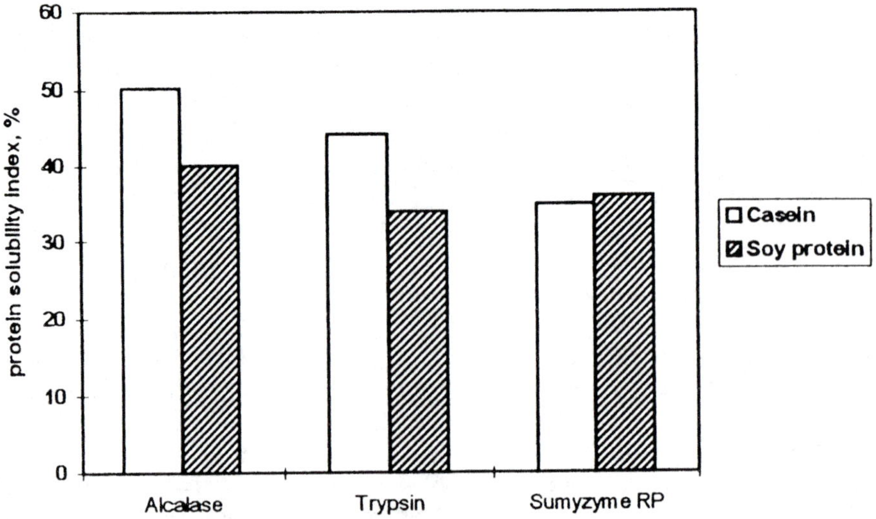

Figure 8 Protein solubility index of casein and soy proteins hydrolyzed to 5% DH by Alcalase, Trypsin, and Sumyzyme RP. (From Ref. 2.)

the typical pH of a range of food products where hydrolysates can be used, e.g., low allergenic human milk substitutes. The enzymes investigated belonged to two groups: (a) corolase PP and pancreatin, which contains trypsin, chymotrypsin, elastase, and aminopeptidase, and (b) PTN 3.0S, PEM 2500S, PEM 800S, and PEM 2700S, which have mainly trypsin and chymotrypsin activities. Hydrolysates of α-lactalbumin prepared by treating with these enzymes for 240 minutes had solubility at pH 6.6 in the range of 78–100% at a DH range of 3.9–15.8%. There was no correlation between the final DH and the solubility, suggesting that the increase in solubility was very dependent on the type of enzyme used.

In industrial situations where commercial grade-proteins, which are often denatured and insolubilized during manufacture, are used, a different approach might be necessary. Monti and Jost [17] investigated the ability of trypsin, papain, and a neutral protease from *Bacillus subtilis* to break down whey protein aggregates formed as a result of heat treatment during production. Trypsin hydrolysis resulted in more than 90% soluble nitrogen at pH 6.0 and 65% at pH 4.5. The hydrolysates made with the other two enzymes had much lower maximum solubility. The hydrolysate made with papain exhibited a maximum solubility of 80% at pH 3.0.

The solubility of protein hydrolysates is particularly important in liquid protein supplements [18,19], in which appearance of an insoluble sediment is undesirable. It should be noted that no unhydrolyzed commercially produced food protein ingredient has 100% solubility at the isoelectric pH.

In industrial-scale protein production, enzymes can be used to solubilize proteins in order to increase the yield. This is particularly useful where protein raw materials are extracted for use as flavors, e.g., meat hydrolysates/soup stock and hydrolyzed yeast.

B. Emulsifying Properties

Proteins are often used as surfactants in emulsion-type processed foods. Examples of this include the use of soy protein and casein in emulsified meat products, such as sausages, egg white in toppings, and whey protein concentrate in dressings.

Partial hydrolysis of proteins generally increases the number of polar groups and hydrophilicity, decreases the molecular weight, alters the globular structure of proteins, and exposes previously buried hydrophobic regions. These changes will affect their emulsifying properties.

1. Effect of DH

Several studies have shown that the emulsifying properties of protein hydrolysates are affected by the DH. Adler-Nissen and Olsen [11] studied emulsifying properties of soy protein isolate hydrolyzed up to 8.3% DH using Alcalase and Neutrase. The emulsifying capacity was determined according to the method of Swift et al. [20]. The results of these studies are shown in Figure 9. The lower emulsifying capacity at a DH of 1% compared to the control sample (Purina 500E) was attributed to the acid treatment used for inactivation of the enzymes. The study showed that the emulsifying capacity could be significantly increased by hydrolyzing soy protein with Alcalase to a DH of about 5%. Neutrase seemed to be less effective, probably because of differences in the specificity of these two proteases. The study also showed that the maximum DH for emulsifying capacity did not correlate with the solubility.

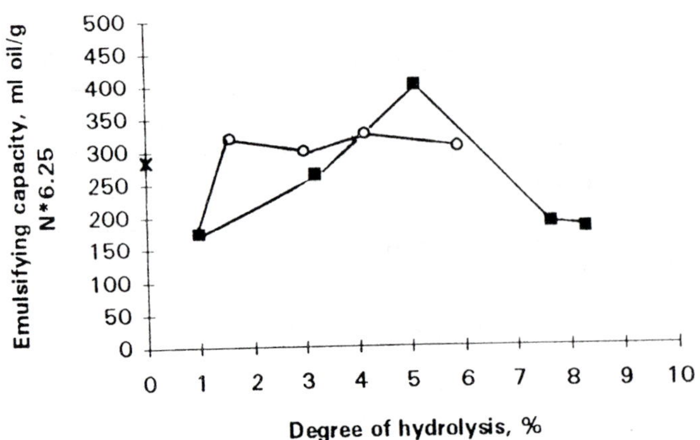

FIGURE 9 Emulsifying capacity at pH 7.0 versus degree of hydrolysis for soy protein hydrolysates prepared by using Alcalase (■) and Neutrase (○). (From Ref. 11.)

Studies by Mietsch et al. [21] confirmed the improvement of emulsifying properties by limited proteolysis. They investigated the effects of Alcalase and Neutrase hydrolysis of soy protein isolate and caseinate. The emulsifying activity of the hydrolysates was significantly higher than the controls, and the results seemed to confirm the existence of an optimum DH for maximum emulsifying activity. However, the stability of the emulsions prepared with hydrolysates did not differ significantly compared to that of the control (unhydrolyzed) samples. It was also shown that the effect of hydrolysis on the improvement of soy protein was much greater than that on caseinate.

Hydrolysis of whey protein concentrate up to DH < 10% with Neutrase also improved its emulsifying capacity [22]. However, at DH > 10% with Alcalase, a decrease in emulsifying capacity occurred.

Mahmoud et al. [23] reported that the emulsifying activity of casein decreased with an increase of DH in the range of 25–67% with porcine pancreatin. However, they did not study the effect at DH between 0 and 25%. Other studies have shown that the emulsifying activity of casein decreased at low DH with the site-specific V-8 protease from *S. aureus* [24]. At 2% and 6.7% DH, a marked reduction in emulsifying activity occurred in a very broad range of pH. Since the V-8 protease is highly specific for glutamate residues, and since the glutamate residues are uniformly distributed in the primary structure of caseins, a poor emulsifying activity is not expected at these DH values. This could be tentatively attributed to some unique properties of the peptides, which had glutamate residues at the C-terminus. In a similar experiment using trypsin for hydrolysis, it was reported that limited hydrolysis of casein (DH = 4.3, 8.9, and 9.9%) improved its emulsifying activity, indicating the importance of the specificity of the enzymes used. However, emulsion stability of these hydrolysates was lower than that of the control. One of the reasons for this could be that the peptides produced were not amphiphilic enough to impart a high stability to the emulsion [25].

2. Effect of pH

It has been shown that the pH during hydrolysis (e.g., pH 8 and 9, with Esperase, trypsin, and V-8 protease) did not have a significant effect on the emulsifying properties of proteins [26]. Generally, the pH of protein solutions during emulsification affects their emulsifying properties via charge effects. The emulsifying capacity of protein hydrolysates is usually low at isoelectric pH, which is relatively low for peptides. However, addition of salt improves the emulsifying properties at the isoelectric pH [27]. Several studies have shown no correlation between emulsifying capacity and high solubility [24,27].

3. Effect of Enzyme

The type of enzyme used for protein hydrolysis seems to influence the emulsifying properties of the hydrolysate. Jost and Monti hydrolyzed whey proteins using different enzymes, such as trypsin, chymotrypsin, papain, pepsin, thermolysin, and a range of nonspecific *Bacillus* proteases. The peptides isolated from these hydrolysates using ultrafiltration had average chain lengths of three to seven amino acid residues. Oligopeptides produced by trypsin and chymotrypsin showed the strongest reduction in interfacial tension. Also, a significant correlation was found between the extent of reduction of interfacial tension and the average chain length. With the exception of the tryptic preparation, all other peptide preparation exhibited very poor emulsifying activity. The emulsion made with the tryptic preparation was stable against creaming and coalescence at $>1\%$ peptide concentrations. The emulsifying properties were better at pH $\geq$ 7. The excellent emulsifying properties of the tryptic preparation may be attributable to the conformation of the peptides, which have lysine or arginine residues at the C-terminus.

Hydrolysis of casein up to 5% DH using immobilized trypsin, chymotrypsin, and Rhozyme-41 (a mixture of various proteolytic enzymes) markedly improved its emulsifying properties [29]. The highest emulsifying activity (3.2-fold increase) was obtained with Rhozyme. On the other hand, the trypsin hydrolyzate of casein exhibited the highest increase in oil-holding capacity (six times higher than the control).

The emulsifying properties of whey protein isolate hydrolyzed with trypsin, chymotrypsin, and pepsin were studied [30]. The emulsions made with the hydrolysates contained smaller oil droplets and more uniform droplet size distribution than that of the unhydrolyzed control. Electron micrographs of the emulsions showed that the fat droplets of emulsions made with chymotrypsin hydrolysates were surrounded by a more uniform layer of protein of moderate thickness compared to the emulsions of hydrolysates made with trypsin and pepsin. These results indicate that chymotrypsin appears to be a better enzyme than either trypsin or pepsin for improving the emulsifying properties of whey protein isolate. Unfortunately, no data on the emulsion stability of these preparations have been reported.

It is evident that the emulsifying properties of protein hydrolysates are influenced by the degree of hydrolysis. There is no doubt that the emulsifying properties of some proteins can be improved by selecting the right enzyme (or combination of enzymes) and by optimizing the DH; the optimum appears to be at relatively low DH (i.e., $<10\%$). This means that the amount of enzyme needed to obtain this relatively low DH should be rather small. Since many food products are emulsion-type products, protein hydrolysates can be used as emulsifying agents in a number of applications, such as salad dressings, spreads, ice cream, coffee whitener, and emulsified meat products like sausages or luncheon meat.

C. Foaming Properties

1. Effect of DH

Figure 10 shows the effect of DH on the foaming properties of soy protein hydrolysates. The foam expansion for the two soy protein preparations is significantly improved by the hydrolysis and maximum foam expansion occurs at a DH of 3.5%. It should be noted that the foam expansion does not decrease in the DH range 3.5–6.0 for the hydrolysate made from ultrafiltered soy protein, which is presumably in the native state, whereas it decreases for the hydrolysate made with acid-precipitated soy protein, which is presumably in a partially denatured state. The data suggest that a soy protein–based whipping agent with properties similar to egg white can be produced using a proper combination of ultrafiltration and DH [11].

The stability of foams made with hydrolysates that have been ultrafiltered to remove small peptides and free amino acids showed a correlation with the residual amount of free amino acids in the hydrolysate. This suggests that small peptides and free amino acids are the major factors adversely affecting the foaming properties of protein hydrolysates [11]. This also means that if an enzyme can produce protein hydrolysates with a narrow molecular weight distribution, it would be the ideal enzyme for this application. The excellent foaming properties of a soy protein hydrolysate made using a microbial acid aspartic protease from *Mucor miehei* (used in cheese production) might be due to the narrow molecular weight distribution of peptides produced by this enzyme [31]. This enzyme is very specific and can hydrolyze very few of the peptide bonds in soy protein. It has been claimed that at 0.5% DH, the soy protein hydrolysate made with enzyme behaves like an egg white substitute with excellent foaming properties [31].

Studies with wheat gluten hydrolysates have also shown that maximum foam expansion occurs at a relatively low DH. An overrun of 600% at DH 2–3% has been reported [12].

2. Effect of pH

A combination of heat and pH treatments has been shown to affect the foaming properties of whey protein hydrolysates [32]. pH had a much more pronounced effect than DH on the foaming properties during heating of the hydrolysate. Heating at pH 8.0 significantly improved the foaming properties. Also, an optimum in overrun and foam firmness was found at 3.1% DH, although the improvement was small.

Figure 11 shows the effect of pH on foam expansion for soy protein hydrolysate produced from ultrafiltrated soy protein isolate [33]. Maximum foam expansion occurred at pH 4–5.

The importance of pH during the preparation of foams from hydrolyzed whey proteins was investigated by Kuehler and Stine [34]. It was found that the greater the net charge on the proteins/peptides, the greater was the tendency to foam. High foam volumes were obtained at pH 2 and 9, and low foam volumes near the isoelectric pH range of whey proteins. The foam stability was low at acid pH, whereas at pH > 6, the stability increased with increasing pH.

3. Effect of Enzyme

The type of protease used for preparing the hydrolysate has been shown to influence its foaming properties [12]. For example, foam expansion for Alcalase-hydrolyzed soy protein isolate was 12 times that of the unhydrolyzed protein, whereas the Neutrase-

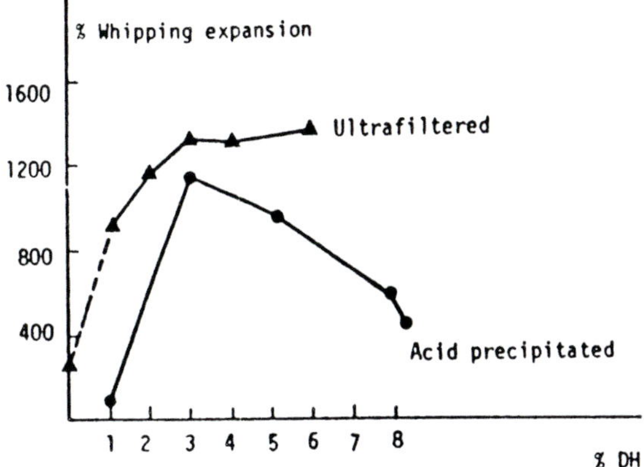

FIGURE 10 Whipping expansion versus degree of hydrolysis for soy protein hydrolysates produced from two types of soy protein isolates: acid precipitated and ultrafiltered SPI. (From Ref. 11.)

hydrolyzed soy protein isolate had only about fourfold higher foam volume than that of the control [12]. However, in both cases, maximum foam volume occurred at 3–4% DH. In another study [22], Alcalase-hydrolyzed whey protein concentrate (DH = 10–25%) exhibited better foaming properties than that hydrolyzed with Neutrase (DH = 5–13%). In this particular case, it is not clear whether the difference was due to differences in DH or the enzymes used.

These results are not in accord with previous reports [35], which showed that the foaming capacity of soy protein hydrolysate made with a *B. subtilis* protease was maximal at about 10% DH. Soy protein hydrolysates made with an enzyme preparation from *Aspergillus oryzae* also showed higher foaming capacity with increasing DH up to 20%.

For whey protein hydrolysates, a correlation between foaming capacity and stability versus DH has been reported [36]. The foam stability decreased and the foaming capacity

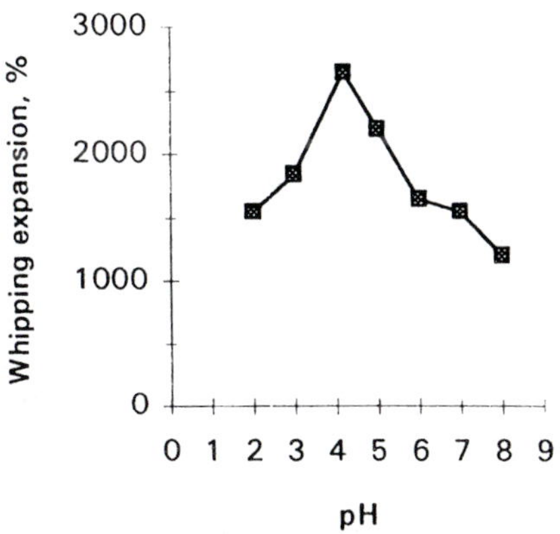

FIGURE 11 Effect of pH on whipping expansion of a 6% DH hydrolysate of soy protein isolate prepared by the ultrafiltration method. (From Ref. 33.)

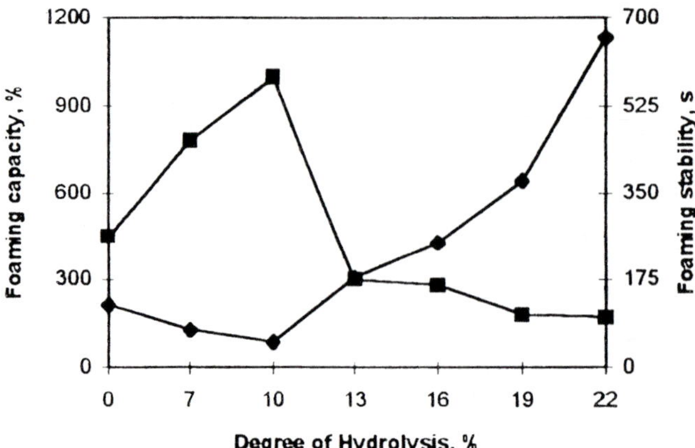

FIGURE 12 Effect of the degree of hydrolysis on foaming capacity (■) and foam stability (♦) of whey proteins. (From Ref. 36.)

increased with increasing DH (Fig. 12). This suggests that the molecular properties affecting the foaming capacity and stability of protein hydrolysates are different. This study also showed that maximum foam expansion occurred at a relatively high DH (10%).

Hydrolysis of zein with trypsin improved its foaming properties [37]. The native zein was not able to form a foam, whereas the hydrolyzed zein (DH 1.42, 1.7, and 1.87%) product did. The stability of the foam was dependent on DH; only the hydrolysate with 1.87% DH formed a stable foam.

Experimental evidence suggests that for every protein, an optimum DH exists to achieve maximal foaming capacity. However, if the stability of the foam also is taken into consideration, the results are a little more complicated. It seems that the molecular weight distribution is important, because low molecular weight peptides and amino acids tend to destabilize the foam. This calls for hydrolyzing proteins with highly specific enzymes or removal of unwanted small peptides in order to maximize the foaming properties. It is also very important to keep in mind that bitter peptides are released at high DH values.

D. Viscosity

Enzymatic hydrolysis can be used as a method for reducing the viscosity of certain proteinaceous products [38]. Studies on viscosity changes during hydrolysis with Alcalase and Neutrase showed considerable variations among proteins [11]. For gelatin, the viscosity gradually decreased with increasing DH up to 7%. The changes in viscosity of soy protein were quite different, as the viscosity decreased dramatically even at 3% DH and remained at the same level at higher DH. The type of enzyme used also had an effect. Alcalase was more efficient in lowering the viscosity than Neutrase.

In contrast to gelatin and soy protein, the viscosity of corn gluten increased steeply during hydrolysis with Neutrase. The differences in the behaviors of these three proteins were ascribed to conformational differences and also to differences in the heterogeneity of protein composition [11]. Gelatin is a homogeneous protein, which shows a good

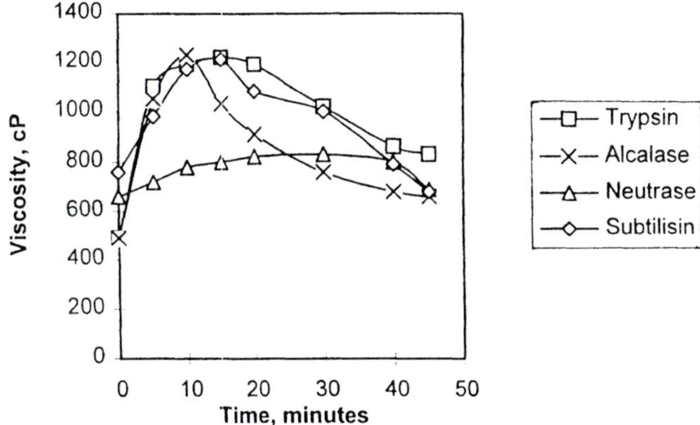

FIGURE 13 Viscosity changes during hydrolysis of mechanically deboned meat using different enzymes. (From Ref. 39.)

correlation between chain length and viscosity, soy protein isolate is a heterogeneous mixture of several soluble proteins, and corn gluten is a highly heterogeneous mixture of slightly soluble protein particles.

Figure 13 shows changes in the viscosity of a 10% mechanically deboned meat (MDM) dispersion during hydrolysis with four different proteases at 35°C [39]. The viscosity increased initially and then decreased with reaction time for all enzymes except Neutrase. The increase in viscosity is attributable to an increase in the solubility of MDM and a consequent increase in its water-binding capacity. Extensive hydrolysis at longer time is due to breakdown of the protein to smaller peptides, which has a detrimental effect on viscosity and water-binding capacity. The results shown in Figure 13 suggest that the viscosity/thickening properties of MDM can be improved by partial hydrolysis. However, one technical problem with this approach is controlling the hydrolysis reaction. The hydrolysis reaction should be terminated when the viscosity is at the maximum level. The only practical way to do this is to terminate the reaction by heat treatment, which will denature the proteins. This could lead to gelation and thus more water binding, both of which are desirable.

Reduction of viscosity in stickwater in a fish meal plant is an example of utilization of proteases as a processing aid [40]. During evaporation of the stickwater in the processing plant, the viscosity limits the maximum amount of dry matter that can be obtained. Proteolysis can decrease the viscosity and thereby increase the capacity of the evaporator.

If partial hydrolysis of a protein solution can increase its viscosity, then this can be used to improve the rheological properties of protein ingredients. Higher viscosity produces an advantage, for instance, in the preparation of emulsified meat products, where a high viscosity of the aqueous solution facilitates emulsification. Dressings and gravies are other products where high viscosity is desirable.

E. Gelation

A classic example of the application of enzymes in the preparation of food gels is the production of cheese. In this process, the milk-coagulating enzyme chymosin hydrolyzes a specific bond (Phe_{105}-Met_{106}) of κ-casein in casein micelles, resulting in destabilization of the micelle structure followed by aggregation and formation of an insoluble coagulum. Several studies have indicated that under certain conditions, hydrolysis of globular proteins, such as whey proteins and soy proteins, also causes gelation.

1. Effect of Enzyme

Enzymatic gelation of partially heat-denatured whey proteins by trypsin, papain, pronase, pepsin, and a protease preparation from *Streptomyces griseus* was studied [41]. Among these, only the pepsin hydrolysate did not form a gel. The enzymes induced gelation of whey proteins. The DH at which gelation occurred were 27.1% for trypsin, 23.1% for papain, 28.2% for pronase, and 15.6% for *S. griseus* protease. To et al. [42] also showed that limited pepsin hydrolysis improved the gel strength of whey protein concentrate. The gel strength depended on the enzyme used. It increased with increasing DH even at DH as high as 61.3% [41]. This is surprising, because at 61.3% DH one would expect breakdown of proteins to dipeptide and amino acid levels; it is not clear how such low molecular weight peptides could form a gel. In fact, Venter et al. [22] showed that whey proteins hydrolyzed to 7.9–24.3% DH could not form a gel.

The possibility of utilizing commercial proteases for coagulation of soy protein has been described [43–45]. Among a range of enzymes investigated, some acidic, neutral, and alkaline proteases were able to coagulate soy proteins. The coagulum retained 80–85% of the protein in the raw material and had a water content after centrifugation of 85–90% and a fat content of 6%. The properties of the coagulum produced by enzymes from *B. amyloliquefaciens*, *B. subtilis*, and *B. polymyxa* were described as having a smooth texture and "umami" taste. On the other hand, several enzymes from *Rhizopus* sp. and *Mucor miehei*, bromelin, and papain developed bitterness [43,44]. Emulsions made using these coagulums exhibited high stability within a broad range of pH [45].

2. Plastein

The formation of gels or protein aggregates is characteristic of the so-called plastein reaction, which can occur when a protein hydrolysate is incubated with a protease. Generally, plasteins have a gel-like structure, which is thixotropic, and are destroyed by mixing or compressing. The most important parameters for this reaction are type of substrate, concentration of the substrate, protease, DH, enzyme for plastein synthesis, and pH during the plastein reaction. Table 2 shows the conditions of plastein reaction by various proteases on various protein substrates.

The three most important parameters for the formation of plasteins are type and size of peptide, concentration of the peptide, and pH of the reaction. It is important that the hydrolysate used in the plastein reaction have a high DH. Table 2 shows the reaction time rather than the DH due to lack of specific information in the literature about this point. However, it can be assumed that the DH was high because of the long hydrolysis time (24 hr) used. The yield of plasteins measured as the amount of sediment after centrifugation has been used to determine the optimum peptide chain length for the formation of plasteins. It was found among seven fractions from gel filtration of pepsin hydrolysate of casein that the fraction with a molecular weight of 380–800 daltons was the best substrate for the plastein reaction [51]. This is in accordance with previous reports [49], which showed that the middle fraction with a molecular weight range of 451–1450 daltons was better than either the peptide fractions with molecular weights lower than 451 daltons or greater than 1450 daltons. The yields of the peptides were 2.1, 36.3, and 1.2%, respectively, measured by the TCA precipitation method.

Table 2 Presentation of References Describing Experiments with Different Conditions for Plastein Reaction

Substrate	Hydrolysis		Plastein			Ref.
	Enzyme	Hours	Enzyme	pH	Protein (%)	
Na caseinate	Trypsin	24	Trypsin	5.0	25.5	46
Soy protein isolate	Pepsin	14	Pepsin	6.5	conc.	47
Na caseinate	Pepsin	48	Pepsin	4.5	30–35	48
Na caseinate	Chymotrypsin	48	Chymotrypsin	4.5	30–35	48
Na caseinate	Papain	48	Papain	4.5	30–35	48
Whey protein concentrate	Pepsin	12	Chymotrypsin	7.0	35	49
Bean protein isolate	Pepsin	0.5–24	Thermitase	7.0	35	50
Na caseinate	Pepsin	24	Pepsin	4.5	40	51
Ovalbumin	Pepsin	168	Chymotrypsin	7.0	40	52
Ovalbumin	Pepsin	4–24	Chymotrypsin	5.0	40	53
Ovalbumin	Pepsin	4–24	Pepsin	5.0	40	53
Ovalbumin	Pepsin	4–24	Papain	5.0	40	53
Whey protein concentrate	Esperase	—	Esperase	6.0	30	54
Whey protein concentrate	Esperase	—	Alcalase	6.0	30	54
Whey protein concentrate	Pepsin	24	Papain	6.0	35	55
Casein	Pepsin	24	Papain	6.0	35	55
Ovalbumin	Pepsin	24	Papain	6.0	35	55
Soy globulin	Pepsin	24	Pepsin	4.0	50	56
Casein	Pepsin	24	Chymotrypsin	7.0	conc.	57
Soy protein	Pepsin	24	Chymotrypsin	7.0	conc.	57
Gluten	Pepsin	24	Chymotrypsin	7.0	conc.	57

Source: Ref. 26.

Table 2 also shows that the concentration of protein required for the plastein reaction is rather high (27.5–40%). Based on the TCA precipitation method, other researchers [58] have reported that a peptide concentration of 30–40% is optimal for the plastein formation of a pepsin hydrolysate of soy protein. This was in accordance with other reports, which reported a value of 25–40% using the TCA method [50] and 20–35% based on a centrifugation method [51]. However, it was pointed out that the formation of aggregates at high substrate concentration is not necessarily due to the plastein reaction [46]; aggregates with similar properties also occur during hydrolysis of casein by trypsin at 5% concentration.

Plastein formation generally takes place at a pH value different from the optimum pH for hydrolysis. This was shown by Adler-Nissen [5], who found that, of the eight different enzymes tested, all but one had a pH optimum for plastein formation 2–3 pH units away from the pH optimum of hydrolysis. Only papain had the same optimum for plastein formation and protein hydrolysis.

Hydrolysis of whey protein concentrate with a glutamic acid–specific protease from *Bacillus lichiniformis* at pH 8.0 and 8% protein concentration has been shown to produce plastein aggregates [60]. The viscosity of the solution increased dramatically during hydrolysis and reached a maximum at 6% DH (Fig. 14).

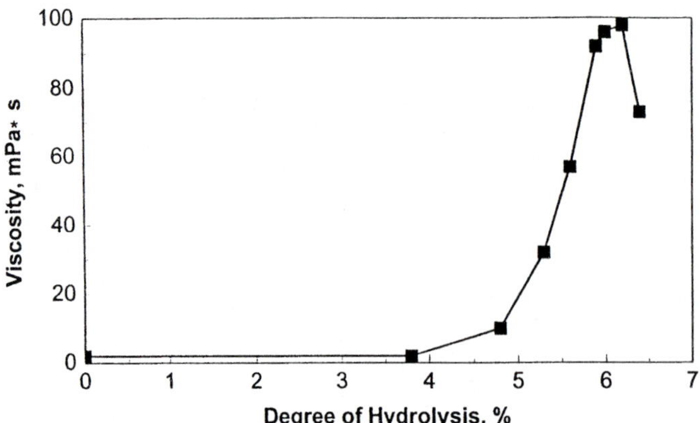

FIGURE 14 Viscosity changes in a whey protein solution during hydrolysis with a protease from *Bacillus licheniformis*. (From Ref. 60.)

The studies described above suggest that the plastein reaction may be used to improve certain functional properties of food proteins. It can also be used to reduce the bitterness of protein hydrolysates [61,62], to covalently incorporate essential amino acids into peptides in order to improve the nutritive value of a protein hydrolysate [46,63,64]. Covalent attachment of essential amino acids to proteins/peptides stabilizes them against degradation during processing and storage and also improves the efficiency of intestinal absorption. The plastein reaction also can be used to reduce the allergenicity of protein hydrolysates [65].

IV. ALLERGENICITY

One well-established application of proteases is to reduce the risk of allergenicity when utilizing milk protein as a substitute for human milk. It has been known for a long time that feeding of cow's milk to newborn babies can provoke an allergic response. The allergenicity of milk proteins can be reduced by hydrolyzing the proteins. This can benefit two groups of consumers: (a) consumers (babies) who are allergic to milk or at a very high risk of developing allergy, and (b) consumers with a higher than average risk of developing allergy. Human milk substitutes based on hydrolyzed proteins have been on the market for over 50 years [60].

Consumers who are allergic to cow's milk may be protected by eliminating it from their diet and using products based on other proteins, e.g., soy protein. However, it has been found that the same group of consumers also have a very high risk of developing allergic reactions to other intact proteins in the diet [67]. For this group of consumers, protein hydrolysates have been shown to be the most acceptable protein supplement.

A. Allergenicity of Milk Proteins

Human and cow's milks differ a great deal in their protein composition. For instance, β-lactoglobulin is found in cow's milk in relatively high amounts (9.8% of the total protein) [68] but is absent from human milk. Among the population allergic to milk,

60–80% are allergic to β-lactoglobulin, 60% to caseins, 50% to α-lactalbumin, and 50% to serum albumin [69].

B. Casein Hydrolysate

The ability of a protein to cause an allergic reaction is related to the size of the protein, the sequence of the amino acids, and the secondary and tertiary structures. One processing step often used together with enzymatic hydrolysis is heat treatment. Heat treatment has been documented to be efficient in reducing the allergenicity of globular proteins, such as whey proteins [70]. Caseins are not affected by heat treatment [71] because they are heat stable.

Mahmoud et al. [23] studied the antigenicity of casein hydrolysate prepared using pancreatic enzymes. The antigenicity of the hydrolysates was analyzed by an ELISA inhibition assay. In the unhydrolyzed casein, the amount of immunologically active casein was 10^6 μg/g protein. At DH = 24% it was reduced to $10^{3.7}$ μg/g protein and at DH = 55% it was $10^{2.7}$ μg/g protein. Further hydrolysis only resulted in a minor reduction in antigenicity (Fig. 15). This reduction in antigenicity was not as large as the reduction discussed by Knights [66], who obtained a reduction of 10^6 compared to intact casein. Analysis of the molecular weight of peptides in the hydrolysate showed that no peptide with a molecular weight greater than 1200 daltons was present, 67% were smaller than 500 daltons, and 32.5% were in the range of 500–1000 daltons. Another casein

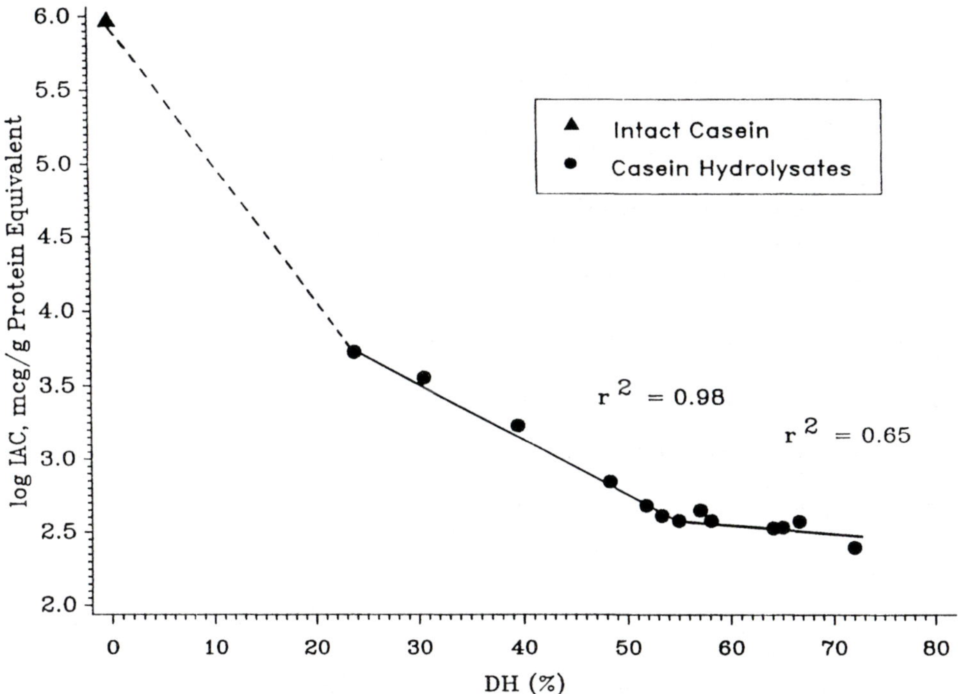

Figure 15 Relationsihp between casein antigenicity loss and degree of hydrolysis. (From Ref. 23.)

hydrolysate with a slightly different molecular weight distribution (in which all peptides were smaller than 5000 daltons and only 0.3% were in the range of 3500–5000 daltons and 0.5% in the range of 2500–3500 daltons) caused anaphylaxis in guinea pigs. This indicated that even a very low amount of antigenically active peptides could cause a reaction in guinea pigs. Other studies [72] have shown that peptides with a molecular weight of less than 1000 daltons produced by chymotrypsin hydrolysis of casein were able to raise antibody in rabbits. Furthermore, certain fragments of β-casein sensitized rabbits.

C. Whey Protein Hydrolysates

Because of their relatively high nutritive value and the relatively low bitterness of their hydrolysates, whey proteins are considered to be ideal substitutes for human milk. To determine the percentage of native β-lactoglobulin remaining in hydrolysates, Pahud et al. [73] analyzed two whey protein hydrolysates with DH 12.9% and 16.1% using a radioimmunoassay. The hydrolysates were ultrafiltered in order to remove undissolved materials and large aggregates. The radioimmunoassay showed that less than 0.04% of the β-lactoglobulin remained in the 16.1% DH hydrolysate and 0.25% in the 12.9% DH hydrolysate. The two hydrolysates were neither able to sensitize guinea pigs not able to cause anaphylactic reaction in animals sensitized with skim milk and intact whey proteins.

Asselin et al. [65] compared the allergenicity of whey protein hydrolysates produced by pepsin, chymotrypsin, trypsin, pancreatin, and combinations of these. Their results showed that DH was not the only important parameter that influenced the allergenicity. A hydrolysate with 20.5% DH produced by pancreatin had a higher residual content of both α-lactalbumin and β-lactoglobulin compared to a hydrolysate with 11.2% DH made using chymotrypsin. It was also noticed that the immunoreactive component increased with prolonged hydrolysis when pepsinolysis was followed by hydrolysis with different pancreatic enzymes. This might be due to aggregates formed by plastein reactions, which had a higher antigenicity compared to nonaggregated peptides. This suggests that the antigenicity of whey proteins can be markedly reduced but not completely eliminated by combined hydrolysis and heat treatments. However, ultrafiltration could be used to reduce and/or eliminate the residual antigens in whey protein hydrolysates [65].

Otani [74] reported the amino acid sequences of some of the antigenically active peptides from α_{s1}-casein, β-casein, and β-lactoglobulin hydrolysates. The three peptides from β-lactoglobulin had the following sequences:

1. -^{41}Val-Tyr-Val-Glu-Glu-Leu-Lys-Pro-Thr-Pro-Glu-Gly-Asp-Leu-Glu-Ile-Leu-Leu-Gln-Lys^{60}-
2. -^{56}Glu-Ile-Leu-Leu-Gln-Lys-Trp-Glu-Asn-Asp-Glu-^{66}
3. -^{125}Thr-Pro-Glu-Val-Asp-Asp-Glu-Ala-Leu-Glu-Lys-Phe-Asp-Lys-Ala-Leu-Lys-Ala-Leu-Pro-Met-^{145}

It was noted that the antigenically active peptides often, but not always, are positioned in or very close to highly hydrophilic regions. This means that the hydrophilic amino acids such as lysine, arginine, glutamate, and asparate residues play important roles in allergen peptides.

The efficiency of the reduction in allergenicity of whey proteins by various proteases can be expressed using the antigenicity reduction index (ARI), which is defined as the ratio of reduction in antigenicity divided by DH [75]. In this case, the antigenicity is measured by indirect competitive inhibition ELISA assay using antibody against whey protein raised in rabbits. The ARI of whey protein hydrolysates prepared with several proteases are shown in Table 3. It should be noted that there is no correlation between the number of cleavage sites in the epitopes of β-lactoglobulin and the efficiency of reduction in the antigenicity of whey protein concentrate. There could be several reasons for this. First, the quality of the antibodies used in the ELISA assay is very important. Second, the specificity of the enzymes cannot be taken literally. For example, the enzyme from *B. licheniformis* is 1000 times more specific to the peptide bonds of glutamate residues than those of aspartate residues [76]. This reduces the real number of cleavage points in β-lactoglobulin epitopes from 15 to 10. An even more illustrative comparison can be made between trypsin and the *Fusarium* protease. These have the same number of cleavage sites in the epitopes of β-lactoglobulin, as they are both specific for lysine and arginine. However, the *Fusarium* protease prefers arginine to lysine, which is opposite to trypsin. As a result, the efficiency of reduction of antigenicity differs markedly. Finally, the proteases seem to have different abilities to "open" a protein structure. For example, unlike other proteases, trypsin can efficiently hydrolyze (and solubilize) heat-denatured whey proteins.

There is no doubt that with increased knowledge and availability of specific enzymes, milk protein hydrolysates with reduced allergenicity can be prepared using a combination of enzymes targeted at hydrolysis of epitopes. In fact, Nakamura et al. [77] showed that using a combination of two nonspecific endoproteases and a DH of up to 25%, the antigenicity of whey proteins can be reduced by 1000 times. It should be pointed out that the antigenicity can be lowered further by including proteases specific for glutamate, aspartate, lysine, and arginine residues. A relatively high degree of hydrolysis is, however, still needed in order to ensure low allergenicity, because peptides as small as four amino acids in length can cause an allergic reaction in consumers allergic to milk, even though these small peptides will not be able to sensitize a person [78].

TABLE 3 Antigenicity Reduction Index (ARI) = Reduction Factor (RF) in Antigenicity of Whey Protein Divided by Degree of Hydrolysis (DH) from Hydrolysis with Different Proteases

Enzyme	Specificity	No. of cleavage sites in β-lactoglobulin epitopes	DH (%)	ARI = RF/DH
Trypsin	lys, arg	6	13.8	362
Chymotrypsin	phe, tyr, trp	3	4.7	19
Subtilisin A	broad	broad	8.2	12
Alcalase	broad	broad	19.1	12
Papain	arg, lys, phe, gly	8	11.7	14
A. saitoi protease, acidic	leu	9	22.9	218
Fusarium protease	arg, (lys)	6	9.1	2
Bacillus licheniformis	glu, asp	15	3.5	143

Source: Ref. 75.

Human milk substitutes with a minimum risk of sensitization for babies and/or a low level of allergic reactions have been on the market for a long time. These products are mainly protein hydrolysates with more than 50% free amino acid content and are known to have only rare treatment failures in infants allergic to cow's milk. However, these products have two major drawbacks: the sensory properties of extensively hydrolyzed proteins are not very good, and the high content of free amino acids results in a high osmolality of the protein solution, which puts restrictions on formulation of a human milk substitute. Also, the emulsifying properties of these hydrolysates, which contain low molecular weight peptides, are very poor. Development of methods to eliminate and/or drastically reduce the levels of low molecular weight allergen peptides is critically needed, because the number of infants allergic to milk has been steadily increasing. Nonetheless, development of protein hydrolysates as a human milk substitute will be a slow process, since the final documentation of safety of the products requires clinical trials. Immunological methods that can be used to predict hypoallergenic performance of these products have been reviewed [79].

V. ORGANOLEPTIC PROPERTIES

Enzymatic hydrolysis very often changes the organoleptic properties of protein products. To what extent the flavor will change is a question of the DH, the protein source, and the enzymes used.

A distinction must be made between flavor changes caused by the formation of peptides and amino acids from the protein and the release of nonprotein flavor components from the raw material. Every raw material has an inherited "built-in" flavor profile, which very much can be ascribed to nonprotein components. Soy protein is a well-described example, and several components influencing the flavor of soy products have been identified [8]. Hydrolysis will change the structure of a protein, causing a decrease in its ability to bind aroma compounds. The flavor-binding characteristics of proteins are briefly covered in Chapter 5. Therefore, only the flavor characteristics of the proteins/peptides and amino acids in relation to hydrolysis will be discussed in this chapter.

Food proteins with a molecular weight of >6000 daltons are as a general rule tasteless [81]. Several smaller peptides have been found to be bitter. About 206 bitter peptides were detected by one group of researchers [82], with the majority being between 2 and 15 amino acid residues and two having more than 20 amino acid residues.

A. Role of Hydrophobicity

Small peptides have been shown to be bitter if they mainly contain predominantly hydrophobic amino acid residues [81]. A relationship between bitterness and the average hydrophobicity of peptides has been proposed [83]. The average hydrophobicity of peptides—the Q-value—is defined as the sum of the free energies of transfer of the amino acid side chains from ethanol to water divided by the number of amino acid residues in the peptide, i.e.:

$$Q = \Sigma \frac{\Delta g}{n}$$

where Δg is the transfer free energy and n is the number of amino acid residues. The hydrophobicity values of amino acid side chains have been reported in the literature [84].

Sensory analyses of peptides isolated from protein hydrolysates have shown that peptides with a Q-value of less than 1300 kcal/mol are not bitter and that those with a Q-value of greater than 1400 kcal/mol are definitely bitter. If the Q-value is between 1300 and 1400 kcal/mol, it is not possible to predict if the peptide would be bitter or not. In keeping with this finding, the hydrolysates of casein, zein, and soy proteins (which are hydrophobic proteins) are bitter, whereas the hydrolysates of less hydrophobic proteins, such as muscle proteins and gelatin, are not [85]. An excellent review of these findings together with later revisions in the hydrophobicity data for amino acid side chains were reported by Adler-Nissen [2]. The revised hydrophobicity data [86,87] show that Ser, Gly, Asp, Asn, Glu, and Gln have significantly lower hydrophobicity than the other amino acids. This would mean that enzymes with specificity for these amino acid residues will produce less bitter peptides, because peptides with a hydrophobic residue at the C-terminus are more bitter than others (88–90). To cause bitterness, the side chain of the amino acid should at least have a three-carbon chain [88]; amino acids with γ-branched side chains (e.g., Leu) are more bitter than those with β-branched chains (e.g., Ile).

Free amino acids taste either sweet, bitter, neutral, or sour [90,91]. Although only the L-amino acids are found in natural proteins, it should be noted that the taste of several amino acids is stereospecific. The amino acids with a hydrophobic side chain have bitter taste [81,92–94], and the bitterness is correlated to the specific volume of the amino acid [95,96].

The level of glutamic acid in the hydrolysate particularly affects its flavor profile [90]. The umami taste of glutamic acid is characterized as a flavor-enhancing property. Monosodium glutamate is often used in food products either in the free amino acid form or as a protein hydrolysate produced either by chemical hydrolysis (acid hydrolysis) or enzymatic hydrolysis. The umami taste is not restricted to glutamic acid, as some dipeptides also have been reported to have umami taste [97]. In these dipeptides, Gly appears at the C-terminus.

B. Improvement of Organoleptic Properties

Bitterness is the major problem affecting the acceptability of protein hydrolysates. Several attempts have been made to limit the formation of bitter peptides and to remove the bitterness-causing peptides or to mask bitterness [98–102]. The development of bitterness can be reduced by controlling the degree of hydrolysis. When bitterness is plotted as a function of DH, the curve passes through a maximum [11]. The DH where the maximum bitterness occurs depends very much on the enzyme used and on the substrate. Nonspecific endoproteases seem to develop a relatively high level of bitterness compared to site-specific enzymes that cleave at hydrophilic amino acid residues, such as trypsin. Casein tends to be very bitter at low DH, whereas gelatin at any DH rarely develops bitterness.

The specificity of the enzymes used must be taken into consideration. Most of the endoproteases with a broad specificity have a tendency of hydrolyzing at hydrophobic amino acid residues, leaving a nonpolar amino acid residue at the C-terminus of the peptides formed. This will lead to relatively high bitterness [103]. However, there is a good possibility of debittering these peptides by hydrolysis with exopeptidases, which often seem to prefer hydrolyzing end-positioned hydrophobic amino acid residues in the substrate. Both aminopeptidases [104] and carboxypeptidases [105] have been shown to debitter bitter peptides. The plastein reaction between bitter peptides also tends to reduce their bitterness [61,62].

Selective separation of bitter peptides has been applied with success using different methods. Because highly hydrophobic peptides usually have very low solubility at around isoelectric pH, they can be removed by isoelectric precipitation [98]. Another method is selective adsorption of hydrophobic peptides using adsorbents such as activated charcoal. Activated charcoal adsorbs mainly hydrophobic compounds. Protein loss can be high in this treatment; however, decolorization of the protein solution during this treatment, a beneficial aspect of this process, may offset the economics of protein loss [106]. Extraction of bitter peptides with aqueous ethanol has been reported to be an efficient method for removal of bitterness in large-scale production [107].

Several methods for masking bitter taste, principally by adding various compounds have been reported. Addition of polyphosphates [108] and gelatin or glycine [109] during hydrolysis has been shown to decrease bitterness in protein hydrolysates. Also, cyclodextrins can mask bitterness by covering the hydrophobic parts of peptides inside their ring structure [110]. Several other ingredients, such as gelatinized starch, proteins such as whey protein concentrate [110], or organic acids [111], also have been shown to mask bitterness. In emulsion-type products, the methods used to prepare the oil-in-water emulsion may influence the bitterness perception. By mixing the hydrolysate with the oil at high shear rate before adding water and other ingredients, it is possible to efficiently cover the hydrophobic parts of the peptides in the oil phase and thereby prevent them from reaching the tastebuds when the product is consumed (P. M. Nielsen, unpublished).

The availability of numerous exo- and endoproteases with a wide range of specificities makes it possible to produce protein hydrolysates with a high degree of hydrolysis [101,112]. At high DH, the organoleptic properties of hydrolysates change. Bitterness can be avoided and the flavor of amino acids, short-chain peptides, as well as all the other components present in the raw material contribute to the flavor. Extensively hydrolyzed proteins produced by chemical hydrolysis are well known and well established in the food industry. The biological alternative that enzymatic hydrolysis offers creates a pool of free amino acids and short-chain peptides, which offer the possibilities for producing a range of differently flavored products. These products may be used in industrial food products like emulsified meat products, bouillon, instant noodles, soups, gravies, snack foods, etc.

VI. CONCLUSION

Technically, enzymatic hydrolysis offers numerous possibilities to alter the functional properties of proteins. Several concepts are already being practiced, and proteases are being increasingly used in new applications. The work with enzymes in general and proteases in particular has shown that the timing for the application of any new concept is very important. It is very clear that the prerequisites for initiating an enzymatic process on an industrial scale depends on several important factors:

Demand for the functionality provided by the process
Availability of raw materials suitable for hydrolysis
Avalability of enzymes with the right properties
Cost/benefit ratio of the process
Production equipment

Knowledge about enzymes

There is no doubt that the functionality of the protein-containing raw materials used in the industry today is critical, as these are often the reason for utilizing proteins in formulated foods, besides their nutritive value. It has also been demonstrated that the technical feasibility of improving the functionality is not limited by the raw materials, as the functionality of many proteins can be improved by hydrolysis. It is, however, important to be aware of changes in raw materials (cost and properties), as they can provide new opportunities in an enzymatic process.

Several food-grade enzymes with different specificities are currently available. The selection of the enzyme is dictated by its cost, but the cost of the enzyme used often accounts for only a very small percentage of the protein hydrolysate production cost.

Enzymatic processes will often need additional unit operations and additional quality control measures. Therefore, capital investment and knowledge about enzymes are the two major impediments that limit implementation of enzyme technologies in food-processing operations.

ACKNOWLEDGMENTS

I would like to thank my colleagues H. C. Holm, H. S. Olsen, B. R. Petersen, and G. Jensen for helpful discussions during the preparation of the manuscript.

REFERENCES

1. C. F. Jacobsen, J. Léonis, K. Linderstrom-Lang, and M. Ottesen, The pH-stat and its use in biochemistry, *Meth. Biochem. Anal. 4*:171 (1957).
2. J. Adler-Nissen, *Enzymatic Hydrolysis of Food Proteins*, Elsevier Applied Science Publishers, London, 1986.
3. V. J. Svedas, I. J. Galaev, I. L. Borisov, and I. V. Berezin, The interaction of amino acids with O-phthaidialdehyde: a kinetic study and spectrophotometric assay of the reaction product, *Anal. Biochm. 101*:188 (1980).
4. F. C. Church, H. E. Swaisgood, D. H. Porter, and G. L. Catignani, Spectrophotometric assay using O-phthaldialdehyde for determination of proteolysis in milk and isolated milk proteins, *J. Dairy Sci. 66*:1219 (1983).
5. J. Adler-Nissen, Determination of the degree of hydrolysis of food protein hydrolysates by trinitrobenzenesulfonic acid, *J. Agric. Food Chem. 27*:1256 (1979).
6. S. Moore and W. H. Stein, Photometric ninhydrin method for use in the chromatography of amino acids, *J. Biol. Chem. 176*:367 (1948).
7. A. Margot, E. Flaschel, and A. Renken, Continuous monitoring of enzymatic whey protein hydrolysis. Correlation of base consumption with soluble nitrogen content. *Proc. Biochem. 29*:257 (1994).
8. S. M. Mozersky and R. A. Panettieri, Is pH drop a valid measure of extent of protein hydrolysis?, *J. Agric. Food Chem. 31*:1313 (1983).
9. H. S. Olsen and J. Adler-Nissen, Application of ultra- and hyperfiltration during production of enzymatically modified proteins, *Synthetic Membranes*: Volume 11. *Hyper- and Ultrafiltration Uses*. (A. F. Turbak, ed.), American Chemical Society, 1981, p. 133.
10. D. G. Schmidt and B. W. van Markwijk, Enzymatic hydrolysis of whey proteins. Influence of heat treatment of (α-lactalbumin and β-lactoglobulin on their proteolysis by pepsin and papain, *Neth. Milk Diary J. 47*:15 (1993).

11. J. Adler-Nissen and H. S. Olsen, The influence of peptide chain length of taste and functional properties of enzymatically modified soy protein, *Functionality and Protein Structure* (A. Pour-El, ed.), American Chemical Society, 1979, p. 125.
12. H. S. Olden, Enzymes in food processing, *Enzymes, Biomass, Food and Feed* (G. Reed and T. W. Nagodawithana, eds.), 1995, p. 663.
13. G. M. O'Meara and P. A. Munro, Effects of reaction variables on the hydrolysis of lean beef tissue by alcalase, *Meat Sci. 11*:227 (1984).
14. J. Adler-Nissen, S. Eriksen, and H. S. Olden, Improvement of the functionality of vegetable proteins by controlled enzymatic hydrolysis, *Qual. Plant Foods Hum. Nutr. 32*:411 (1983).
15. J.-M. Chobert, M. Sitohy, and J. R. Whitaker, Specific limited hydrolysis and phosphorylation of food proteins for improvement of functional and nutritional properties, *J. Am. Oil Chem. Soc. 64*(12):1704 (1987).
16. M. M. Mullally, D. M. O'Callaghan, R. J. FitzGerald, W. J. Donnelly, and J. P. Dalton, Proteolytic and peptidolytic activities in commercial pancreatic protease preparations and their relationship to some whey protein hydrolysate characteristics, *J. Agric. Food Chem. 42*:2973 (1994).
17. J. C. Monti and R. Jost, Enzymatic solubilization of heat-denatured cheese whey protein, *J. Dairy Sci. 61*:1233 (1978).
18. S. Frokjaer, Use of hydrolysates for protein supplementation, *Food Technol 48*(10):86 (1994).
19. M. K. Schmidl, S. L. Taylor, and J. A. Nordlee, Use of hydrolysate-based products in special medical diets, *Food Technol 10*:7 (1994).
20. C. E. Swift, C. Lockett, and A. J. Fryar, Comminuted meat emulsions—the capacity of meats for emulsifying fat, *Food Technol. 15*:408 (1961).
21. F. Mietsch, J. Feher, and A. Halasz, Investigation of functional properties of partially hydrolyzed proteins, *Nahrung 33*:(1)9 (1989).
22. B. G. Venter, A. E. J. McGill, and S. H. Lombard, Changes in the functional properties of whey proteins during non-specific enzymatic hydrolysis. *S. Afr. J. Dairy Sci. 21*(4):75 (1989).
23. M. I. Mahmoud, W. T. Malone, and C. T. Cordle, Enzymatic hydrolysis of casein: effect of degree of hydrolysis on antigenicity and physical properties, *J. Food Sci. 57*(5):1223 (1992).
24. J.-M. Chobert, M. Sitohy, and J. R. Whitaker, Solubility and emulsifying properties of caseins modified enzymatically by *Staphylococcus aureus* V8 protease, *J. Agric. Food Chem. 36*:220 (1988).
25. J.-M. Chobert, C. Bertrand-Harp, and M.-G. Nicolas, Solubility and emulsifying properties of casein and whey proteins modified enzymatically by trypsin. *J. Agric. Food Chem. 36*: 883 (1988).
26. T. Flytkjaer-Hansen and H. Villadsen, Enzymatic hydrolysis of milk proteins (in Danish), Master's thesis, Veterinary University, Copenhagen, 1993.
27. S. L. Turgeon, S. F. Gautier, and P. Paquin, Emulsifying property of whey peptide fractions as a function of pH and ionic strength, *J. Food Sci. 57*(3):601, 634 (1992).
28. R. Jost and J. C. Monti, Emulgateurs peptidiques obtenus par l'hydrolyse enzymatique partielle de la proteine serique du lait, *Le Lait 62*:521 (1982).
29. Z. U. Haque and Z. Mozaffar, Casein hydrolysate. II. Functional properties of peptides, *Food Hydrocolloids 5*(6):559 (1992).
30. J. Lakkis and R. Villota, A study on the foaming and emulsifying properties of whey protein hydrolysates, *AIChE. Symp. Ser. 86*:87 (1990).
31. C. O. L. Boyce, R. P. Lanzilotta, and T. M. Wong, Enzyme modified soy protein for use as an egg white substitute, U.S. Patent 4,632,903 (1986).
32. M. Britten and V. Gaudin, Heat-treated whey protein hydrolysates: emulsifying and foaming properties, *Milchwissenschaft 49*(12):688 (1994).

33. H. S. Olden, Method of producing an egg white substitute material, U.S. Patent 4,431,629 (1984).
34. C. A. Kuehler and C. M. Stine, Effect of enzymatic hydrolysis on some functional properties of whey protein, *J. Food Sci. 39*:379 (1974).
35. L. S. B. Don, A. M. R. Pilosof, and G. B. Bartholomai, Enzymatic modification of soy protein concentrates by fungal and bacterial proteases, JAOCS 68(2):102 (1991).
36. A. Perea, U. Ugaide, I. Rodriguez, and J. S. Serra, Preparation and characterization of whey protein hydrolysates: applications in industrial whey bioconversion processes, *Enzyme Microb. Technol. 15*:418 (1993).
37. M. L. A. Caselia and J. R. Whitaker, Enzymatically and chemically modified zein for improvement of functional properties, *J. Food Biochem. 14*:453 (1990).
38. T. Richardson, Functionality changes in proteins following action of enzymes, *Adv. Chem. Ser. 160*:185 (1977).
39. P. Rasmussen and A. Rancke-Madsen, Method for upgrading of MRM (mechanically recovered meat), Int. Patent Application WO 90/05462 (1990).
40. F. M. Christensen, Review. Enzyme technology versus engineering technology in the food industry, *Biotechnol. Appl. Biochem. 11*:249 (1989).
41. K. Sato, M. Nakamura, T. Nishiya, M. Kawanari, and I. Nakajima, Preparation of a gel of partially heat-denatured whey protein by proteolytic digestion. *Milchwissenschaft 50*(7):389 (1995).
42. B. To, N. B. Heibig, S. Nakai, and C. Y. Ma, Modification of whey protein concentrate to simulate shippability and gelation of egg white, *Can. Inst. Food Sci. Technol. J. 18*(2):150 (1985).
43. K. Murata, I. Kusakabe, H. Kobayashi, M. Akaike, Y. W. Park, and K. Murakami, Studies on the coagulation of soymilk-protein by commercial proteinases, *Agric. Biol. Chem. 51*(2): 385 (1987).
44. K. Murata, I. Kusakabe, H. Kobayashi, H. Kiuchi, and K. Murakami, Selection of commercial enzymes suitable for making soymilk-curd, *Agric. Biol. Chem. 51*(11):2929 (1987).
45. K. Murata, I. Kusakabe, H. Kobayashi, H. Kiuchi, and K. Murakami, Functional properties of three soymilk curds prepared with an enzyme, calcium salt and acid, *Agric. Biol. Chem. 52*(5):1135 (1988).
46. P. C. Lorenzen and E. Schlimme, The plastein reaction: properties in comparison with simple proteolysis, *Milchwissenschaft 47*:499 (1992).
47. E. K. Harnett and L. D. Satterlee, The formation of heat and enzyme induced (plastein) gels from pepsin-hydrolysed soy protein isolate, *J. Food Biochem. 14*:113 (1990).
48. A. T. Andrews and E. Alichanidis, The plastein reaction revisited: evidence for a purely aggregation reaction mechanism, *Food Chem. 35*:243 (1990).
49. V. Sciancalepore and V. Longone, Plastein synthesis by the action of α-chymotrypsin on a peptic hydrolysate of whey protein, *J. Dairy Res. 55*:547 (1988).
50. H. Winkler, H. Nbtzold, and E. Ludwig, Plasteinbildung, Anreicherung hydrophober Aminosauren and Methionin-Einbau in peptische Partialhydrolysate des Ackerbuhnonproteinisolats bei der durch thermitase katalysierten plasteinreaktion, *Nahrung 32*:135 (1988).
51. G. Sukan and A. T. Andrews, Application of the plastein reaction to caseins and to skimmilk powder. 1. Protein hydrolysis and plastein formation, *J. Dairy Res. 49*:265 (1982).
52. M. Y. U. Golobov, M. V. Belikov, S. V. Vitt, E. A. Paskonova, and E. F. Titova, Transpeptidation in concentrated solution of peptic hydrolyzate of ovalbumin, *Nahrung 25*:961 (1981).
53. J. H. Edwards and W. F. Shipe, Characterization of plastein reaction products formed by pepsin, α-chymotrypsin, and papain treatment of egg albumin hydrolysates, *J. Food Sci. 43*:1215 (1978).
54. B. V. Hofsten and G. Lalasidis, Protease-catalyzed formation of plastein products and some of their properties, *J. Agric. Food Chem. 24*(3):460 (1976).

55. S. Arai, M. Yamashita, K. Aso, and M. Fujimaki, A. parameter related to the plastein formation, *J. Food Sci. 40*:342 (1975).
56. K. Aso, M. Yamashita, S. Arai, and M. Fujimaki, Hydrophobic forces as a main factor contributing to plastein chain assembly, *J. Biochem. 76*:341 (1974).
57. M. Fujimaki, M. Yamashita, S. Arai, and H. Kato, Enzymatic modification of proteins in foodstuffs. Part I. Enzymatic proteolysis and plastein synthesis. Application for preparing bland protein-like substances, *Agric. Biol. Chem. 34*:1325 (1970).
58. S.-J. Tsai, M. Yamashita, S. Arai, and M. Fujimaki, Effect of substrate concentration on plastein productivity and some rheological properties of the products, *Agric. Biol. Chem. 38*:1045 (1972).
59. M. Yamashita, S.-J. Tsai, S. Arai, S. Kato, and M. Fujimaki, Enzymatic modification of proteins in foodstuffs. Part V. Plastein yields and their pH-dependence, *Agric. Biol. Chem. 35*:86 (1971).
60. P. Budtz and P. M. Nielsen, Protein preparations, Int. Patent Application W092/13964 (1992).
61. M. Fujimaki, M. Yamashita, S. Arai, and H. Kato, Plastein reaction—its application to debittering of protein hydrolysates, *Agric. Biol. Chem. 34*:483 (1970).
62. G. Lalasdis and L.-B. Sjbberg, Two new methods of debittering protein hydrolysates and a fraction of hydrolysates with a exceptionally high content of essential amino acids, *J. Agric. Food Chem. 26*:742 (1978).
63. M. Yamashita, S. Arai, S.-J. Tsai, and M. Fujimaki, Plastein reaction as a method for enhancing the sulfur-containing amino acid level of soybean proteins, *J. Agric. Food Chem. 19*:1151 (1971).
64. G. Y. Hajós, T. Szarvas, and L. Vámos-Vigyázó, Radioactive methionine incorporation into peptide chains by enzymatic modification, *J. Food Biochem. 14*:381 (1990).
65. J. Asselin, J. Hérbert, and J. Amiot, Effects of in vitro proteolysis on the allergenicity of major whey proteins, *J. Food Sci. 54*:1037 (1989).
66. R. J. Knights, Processing and evaluation of the antigenicity of protein hydrolysates. *Nutrition for Special Needs in Infancy. Protein Hydrolysates*, (F. Lifshitz, ed.), Marcel Dekker, New York, 1985, p. 105.
67. S. L. Bahna, Control of milk allergy: a challenge for physicians, mothers and industry, *Ann. Allergy 41*:1 (1978).
68. P. Waistra and R. Jenness, *Dairy Chemistry and Physics*, John Wiley & Sons, Inc., New York, 1984.
69. K. Aas, The biochemistry of food allergens: What is essential for future research?, *Food Allergy*, (E. Schmidt and D. Reinhardt, eds.), Raven Press, Ltd., New York 1988, p. 1.
70. L. M. J. Heppell, A. J. Cant, and P. J. Kilshaw, Reduction in the antigenicity of whey proteins by heat treatment: a possible strategy for producing a hypoallergenic infant milk formula, *Br. J. Nutr. 51*:29 (1984).
71. R. Jost, Physiochemical treatment of food allergens: applications to cow's milk proteins, *Food Allergy* (E. Schmidt and D. Reinhardt, eds.), Raven Press, Ltd., New York 1988, p. 187.
72. H. Otani, X. Y. Dong, and A. Hosono, Antigen specificity of antibodies raised in rabbits injected with a chymotryptic casein-digest with molecular weight less than 1,000, *Jpn. J. Dairy Food Sci. 39*:A31 (1990).
73. J.-J. Pahud, J. C. Monti, and R. Jost, Allergenicity of whey proteins: its modification by tryptic in vitro hydrolysis of the protein, *J. Pediatr. Gastrointeral. Nutr. 4*:408 (1985).
74. H. Otani, Antigenically reactive regions of bovine milk proteins, *Jpn. Agric. Res. Quarterly 21*(1):135 (1987).
75. M. Flansmose, Enzymatic hydrolysis of β-lactoglobulin and the effect of antigenicity (in Danish), Master's Thesis, Denmark Technical University, Copenhagen, 1993.

76. K. Breddam and M. Meldal, Substrate preferences of glutamic acid specific endopeptidases assessed by synthetic peptide substrates based on intramolecular flourescence quenching, *Eur. J. Biochem. 206*:103 (1992).
77. T. Nakamura, H. Sado, Y. Syukunobe, and T. Hirota, Antigenicity of whey protein hydrolysates prepared by combination of two proteases, *Milchwissenschaft 48*(12):667 (1993).
78. J. G. Bindels and J. Verwimp, Allergenic aspects of infant feeding, *Nutr. Res. Commun.* 2 (1990).
79. C. T. Cordle, Control of food allergies using protein hydrolysates, *Food Technol.* (10):72 (1994).
80. G. MacLeod and J. Ames, Soy flavour and its improvement, *CRC Crit. Rev. Food Sci. Nutr. 27*(4):219 (1988).
81. K. H. Ney, Bitterness of peptides: amino acid composition and chain length, *Food Taste Chemistry* (J. C. Boudreau, ed.), American Chemical Society, Washington, DC, 1979, p. 149.
82. Y. Guigoz and J. Solms, Bitter peptides, occurrence and structure, *Chem. Senses Flav. 2*:71 (1976).
83. K. H. Ney, Voraussage der Bitterkeit von Peptiden aus deren Aminosaurenzusammensetzung, *Z. Lebensm.-Untersuch. Forsch. 147*:64 (1971).
84. C. Tanford, Contribution of hydrophobic interactions to the stability of the globular conformation of proteins, *J. Am. Chem. Soc. 84*:4240 (1962).
85. K. H. Ney, Aminosaure-Zusammentsetzung von Proteinen und die Bitterkeit ihrer Peptide, *Z Lebensm.-Untersuch. Forsch. 149*:321 (1972).
86. Y. Nozaki and C. Tanford, The solubility of amino acids and two glycine peptides in aqueous ethanol and dioxane solutions. Establishment of a hydrophobicity scale, *J. Biol. Chem. 246*:2211 (1971).
87. C. C. Bigelow and M. Channon, Hydrophobicities of amino acids and proteins, *Handbook of Biochemistry and Molecular Biology*, 3rd ed., Vol. 1 (G. D. Fasman, ed.), CRC Press, Cleveland, 1976, p. 209.
88. N. Ishibashi, Y. Arita, H. Kanehisa, K. Kouge, H. Okai, and S. Fukui. Bitterness of leucine-containing peptides, *Agric. Biol. Chem. 52*(1):2389 (1987).
89. T. Nishimura and H. Kato, Taste of free amino acids and peptides, *Food Rev. Int. 4*(2):175 (1988).
90. R. S. Shallenberger, *Taste Chemistry*, Blackie Academic & Professional, Glasgow, 1993.
91. S. Eriksen and I. S. Fagerson, Flavours of amino acids and peptides, *Int. Flav. (Jan/Feb)*: 13 (1976).
92. T. Matoba and T. Hata, Relationship between bitterness of peptides and their chemical structures, *Agric. Biol. Chem. 36*:1423 (1972).
93. P. A. Lehmann, The correlation of sweetness and bitterness of enantiomeric amino acids, *Life Sci. 22*:1631 (1978).
94. N. Ishibashi, I. Ono, K. Kato, T. Shigenaga, I. Shinoda, H. Okai, and S. Fukui, Role of the hydrophobic amino acid residue in the bitterness of peptides, *Agric. Biol. Chem. 52*(1):91 (1988).
95. G. G. Birch and S. E. Kemp, Apparent specific volumes and tastes of amino acids, *Chem. Senses 14*(2):249 (1989).
96. S. Shamil, G. G. Birch, M. Dinovi, and R. Rafka, Structural functions of taste in 5-membered ring structures, *Food Chem. 32*:171 (1989).
97. M. Tada, I. Shinoda, and H. Okai, L-Ornithyltaurine, a new salty peptide, *J. Agric. Food Chem. 32*:992 (1984).
98. J. Adler-Nissen, Control of the proteolytic reaction and of the level of bitterness in protein hydrolysis processes, *J. Chem. Tech. Biotechnol. 34B*:215 (1984).
99. U. Cogan, M. Moshe, and S. Mokady, Debittering and nutritional upgrading of enzymic casein hydrolysates, *J. Sci. Food Agric. 32*:459 (1981).

100. B. Pedersen, Removing bitterness from protein hydrolysates, *Food Technol. 48*(10):96 (1994).
101. P. M. Nielsen, Enzyme technology for production of protein based flavours, Conference Proceedings, Food Ingredients Europe '94, pp. 106–110.
102. T. Godfrey, Enzyme modification for baked goods and flavour proteins *Eur. Food Drink Rev.* (autumn):43–44, 46, 48 (1990).
103. J. Adler-Nissen, Relationship of structure to taste of peptides and peptide mixtures, *Protein Tailoring for Food and Medical Uses* (R. E. Feeney and J. R. Whitaker, eds.), Marcel Dekker, New York, 1986, p. 97.
104. K. M. Clegg, Dietary enzymic hydrolysates of protein, *Biochemical Aspects of New Protein Foods*, Pergamon Press, Oxford, 1978, p. 109.
105. H. Umetsu, H. Matsuoka, and E. Ichishima, Debittering mechanism of bitter peptides from milk casein by wheat carboxy peptidases, *J. Agric. Food Chem. 31*:50 (1983).
106. H. S. Olsen and J. Adler-Nissen, Industrial productions and applications of a soluble enzymatic hydrolysate of soya protein, *Proc. Biochem. 14*(7):6 (1979).
107. R. Chakrabarti, A. method for debittering fish protein hydrolysate, *J. Food Sci. Technol. 20*:154 (1983).
108. F. Tokita, Enzymatische und nicht enzymatische Ausschaltung des Bittergesmacks bei enzymatischen Eieweisshydrolysaten, *Z. Lebensm.-Untersuch.-Forsch. 138*:351 (1969).
109. D. Stanly, Non-bitter protein hydrolysates, *Can. Inst. Food Sci. Technol. J. 14*(1):49 (1981).
110. M. Tamura, N. Mori, T. Miyoshi, S. Koyama, H. Kohri, and H. Okai, Practical debittering using model peptides and related peptides. *Agric. Biol. Chem. 54*:41 (1990).
111. J. Adler-Nissen and H. S. Olsen, Taste and taste evaluation of soy protein hydrolysates, *Chemistry of Foods and Beverages—Recent Developments* (G. Charalambous and G. Inglett, eds.), Academic Press, New York, 1982, p. 149.
112. D. Taylor, D. Pawlett, and A. Brett, Natural meat flavours, FIE Conference Proceedings, 1991, pp. 207–210.

16

High-Pressure Effects on Proteins

K. Heremans
University of Leuven, Leuven, Belgium

J. Van Camp and A. Huyghebaert
University of Ghent, Ghent, Belgium

I. INTRODUCTION

At the beginning of this century, Bridgman [1] observed that the white of an egg coagulated after a pressure treatment for 30 minutes at 700 MPa (100 MPa = 1 kbar). The appearance of the pressure-induced coagulum was quite different from coagulum induced by temperature. In addition, the original publication stated that it "seems to be such that the ease of (the pressure-induced) coagulation increases at low temperatures, contrary to what one might expect." Recently Hayashi et al. [2] analyzed the effect of pressure on egg yolk. It is well known from the preparation of a hard-boiled egg that the yolk becomes solid at a slightly higher temperature than the white. The reverse is true for the effect of pressure: When an egg is subjected to pressure, the yolk becomes solid at a lower pressure than the white.

It is now clear that these characteristics are the consequence of the unique behavior of proteins [3–6]. The phase diagram for the conditions under which the native and the denatured conformations occur reflects the differences between the native and the denatured states of the protein.

The study of the effects of pressure on proteins has received considerable attention in recent years [7–10]. In general, low pressures induce reversible changes such as the dissociation of protein-protein complexes, the binding of ligands, and conformational changes. Pressures higher than about 500 MPa induce denaturation, which is in most cases, irreversible. However, reports on a few proteins indicate that such high pressures may also cause reversible changes.

There is now an increasing interest in the food industry, particularly in Japan, in the high-pressure treatment of food materials as a possible alternative to temperature treatment. It is known that high pressure denatures proteins, solidifies lipids, destabilizes biomembranes, and inactivates microorganisms. The observation that pressure treatment does not cause changes in the taste or the flavor of food materials is of special interest to the food industry. Several symposia and review articles indicate that there is a growing interest in other countries as well [7,11,12].

The present chapter concentrates on a comparison of pressure and temperature effects on proteins. Thermodynamic as well as kinetic aspects will be considered. Phase diagrams of water, lipids, and proteins will be discussed in order to provide a better understanding of the differences between temperature and pressure treatment. Also emphasis is given to two experimental methods that can provide molecular details on the pressure behavior of proteins: infrared spectroscopy for observations in situ on pressurized samples and rheological techniques that probe the macroscopic behavior of gels after compression, i.e., ex situ.

II. PRESSURE VERSUS TEMPERATURE EFFECTS

The physicochemical advantage of using extreme conditions is to explore the effect of temperature and pressure on the conformation, dynamics, and reactions of biomolecules. The unique properties of biomolecules are determined by the delicate balance between internal interactions, which compete with interactions with the solvent. The primary source of the dynamic behavior of biomolecules is the free volume of the system. In the absence of strong mechanical constraints, this free volume may be expected to decrease with increasing pressure. As temperature effects act via increased kinetic energy as well as free volume, it follows that the study of the combined effect of temperature and pressure is a prerequisite for a full understanding of the dynamic behavior of biomolecules. By intuition, pressure effects should be easier to interpret than temperature effects. This follows from the Le Chatelier–Braun principle. Pressure affects primarily the volume of a system, while temperature changes both the volume and the energy of a system.

A. Physical Transformations

The effect of compression of molecules can best be illustrated by considering the effects on liquids. Two types of liquids are of particular interest, i.e., water and hexane, since they represent typical examples of polar and nonpolar materials, respectively. The compressibility of water is considerably smaller than that of most organic liquids. The volume of water is reduced by 10% at 300 MPa and by 15% at 600 MPa. The volume of hexane is reduced by 20% at 300 MPa and by 25% at 600 MPa.

If one performs the compression under adiabatic conditions, then the following relation gives the temperature increase for a given pressure increase:

$$\left(\frac{\delta T}{\delta P}\right)_S = \left(\frac{T}{C_p}\right)\left(\frac{\delta V}{\delta T}\right)_P = \frac{\alpha T}{\rho\, C_p} \tag{1}$$

where α is the thermal expansion, ρ the density, and C_p the heat capacity of the system. The equation predicts for water a temperature increase of 2 K/100 MPa at 25°C and no temperature increase at 4°C. For hexane, the temperature increases 40 K/100 MPa at 18°C. It is clear that a pressure drop results in a temperature decrease of the same order of magnitude.

Pressure has a minor effect on the viscosity of pure water up to about 600 MPa. For the same pressure range, the viscosity of hexane increases 10-fold.

The effect of pressure on the melting temperature (T_m) of compounds is given by the Clausius-Clapeyron equation:

$$\frac{dT_m}{dP} = \frac{T_m \, \Delta V}{\Delta H} \tag{2}$$

Since the volume (ΔV) and enthalpy change (ΔH) on melting are in general positive, one predicts an increase in melting temperature with increasing pressure. For many organic compounds dT_m/dP is approximately 15 K/100 MPa. An increase of the pressure by 100 MPa would then correspond to a decrease in temperature of 15 K. While this statement is true from a thermodynamic point of view, it is certainly not true on the molecular level. A notable exception to this general rule is water. At room temperature, a pressure of about 1 GPa is needed to obtain the phase of ice (VI), which differs from normal ice by its higher density.

A simple illustration of the Clausius-Clapeyron equation may be found in the effect of pressure on the temperatures of transitions in phospholipids. Pressure favors the crystalline state as a result of the Le Chatelier–Braun principle. Whereas the transition temperature of lipids depends on the length of the hydrocarbon chain, the rate at which the temperature changes with pressure is almost independent on the length ($dT/dP = 20$K/100 MPa) [13]. It is of interest to note that a higher degree of unsaturation of the hydrocarbon chain lowers the dT/dP ($dT/dP = 14$ K/100 MPa) values of the lipids [14]. The dT/dp values depend very little on the pressure, except for the formation of lipid interdigitated phases [14].

In biomembranes it has been found that the physical state of the lipids that surround the membrane proteins plays a crucial role in the activity of membrane-bound enzymes. In addition, the integrity of the membrane of living organisms, such as bacteria, is very sensitive to pressure. This explains the sterilization effect of pressure. It has been observed that the activity of membrane-bound enzymes such as the Na^+-K^+-ATPase and the Ca^{2+}-ATPases changes in a nonlinear fashion with temperature and pressure [15].

B. Chemical Transformations

For a chemical reaction A $\rightleftarrows$ B, where A may represent an enzyme in its active conformation and B one in its inactive conformation, the equilibrium constant is given by:

$$K = \frac{[B]}{[A]} \tag{3}$$

The relation with thermodynamic quantities is given by:

$$\Delta G = -RT \ln K \tag{4}$$

ΔG is the change in Gibbs free energy of the reaction. The effect of temperature and pressure dependence is given by:

$$\Delta G = \Delta E + P\Delta V - T\,\Delta S \tag{5}$$

Since the changes in free energy for a substantial shift from A to B are of the order of a few kcal, it is clear that at ambient pressure the term $P\Delta V$ will have a very small effect on chemical reactions. For a reaction volume, ΔV, of the order of 82 ml, which is a large value for a chemical reaction, $P\Delta V = 2$ kcal. Therefore pressures of the order of several

100 MPa are needed to affect chemical reactions in the liquid phase. The relation between ΔV and the change in equilibrium constant with pressure was first given by Planck:

$$\Delta V = \frac{-RT \; d \ln K}{dP} \tag{6}$$

This expression should be compared with the vant Hoff equation for the effect of temperature on the equilibrium constant:

$$\Delta H = \frac{RT^2 \; d \ln K}{dT} \tag{7}$$

It has to be noted that a positive ΔV implies a shift towards the reactants at higher pressure. A higher temperature favors the reaction products for an endothermic reaction. Similar expressions may be derived to describe the effect of pressure on the reaction rate:

$$\Delta V^{\#} = \frac{-RT \; d \ln k}{dP} \tag{8}$$

where $\Delta V^{\#}$ is the activation volume. The equation should be compared with the Arrhenius equation.

An important point with relation to pressure effects on reaction velocities is that the reaction, depending on the mechanism, may be either accelerated or retarded by high pressure. It is well known that an increase in temperature invariably results in an increased reaction rate. Two examples may illustrate this point: (a) the Maillard reaction is strongly inhibited by high pressure, and (b) the oxidation of nonsaturated lipids is accelerated by high pressure [16].

C. Molecular Interpretation of Thermodynamic and Kinetic Parameters

It has been found useful to interpret the observed reaction and activation volumes in terms of intrinsic and solvent effects:

$$\Delta V \text{ (observed)} = \Delta V \text{ (intrinsic)} + \Delta V \text{ (solvent)} \tag{9}$$

Intrinsic effects arise from packing effects, i.e., the increase in molecular ordering as a consequence of the volume decrease at high pressure and the changes in volume due to the formation or rupture of covalent bonds. The role of the solvent becomes apparent in the noncovalent interactions: electrostatic interactions, hydrogen bonding, and hydrophobic interactions.

The formation of an ion in solution is accompanied by an electrostriction of the solvent, i.e., the ordering of the solvent around the electric charge of the ion. This electrostriction explains the changes in volume for the auto-ionization of water:

$$2\,H_2O \rightleftharpoons H_3O^+ + OH^- \tag{10}$$

where $\Delta V = -22$ ml. Whereas water is a special case, this reaction shows that the formation of a monovalent ion is accompanied by a negative volume change of approximately -10 ml. Electrostriction is also the basis for the explanation of the effect of pressure on the pH of aqueous solutions. It may be seen from Table 1 that the shift in

TABLE 1 Effect of Pressure and Temperature on Changes in pH of Basis and Acids

Reaction[a]	ΔV (ml)	ΔpH/100 MPa	ΔH (kcal)	ΔpH/°C
$R\text{-}NH_3^+ \rightleftharpoons R\text{-}NH_2 + H^+$	+ 1	0.0	13	−0.028
$CH_3COOH \rightleftharpoons CH_3COO^- + H^+$	−12	−0.2	0	0.000
$H_2PO_4^- \rightleftharpoons HPO_4^{2-} + H^+$	−25	−0.4	0	−0.003

[a]The first reaction is applicable to Tris, Bis-Tris, and Bis-Tris-propane buffers. The second reaction is valid for the second ionization of citric acid.
Source: Refs. 17, 18.

pH per 100 MPa (ΔpH/100 MPa) is quite considerable in the case of acetic and phosphoric acid. For fruit juices, which are in general quite acid, a treatment at 500 MPa would cause a pH shift of about one unit to the acid side. In practice, the effect may be less pronounced since the volume change of ionization becomes smaller at high pressure. It can also be noted that buffers that are pressure dependent show a very small temperature dependence and vice versa. From the reaction volume of the ionization of water, it follows that when the pH remains constant as a function of pressure, the pOH changes substantially.

An interesting example of the effect of pressure on electrostatic interactions in enzymes is the pressure-induced reversible inactivation of chymotrypsin due to the dissociation of the salt bridge close to the active site [19].

Studies on various model systems show that hydrogen bonds are stabilized by high pressures. This results from the smaller interatomic distances in the hydrogen-bonded atoms. The stabilizing effect of pressure on hydrogen bonding may be seen from a comparison of the effect on the intermolecular interactions in hydrogen-bonded versus non–hydrogen-bonded liquid amides [20]. The fact that high pressure stabilizes hydrogen bonds has important consequences for the secondary structures in proteins such as α-helices and β-structures. The stabilization of the hydrogen bonds at high pressure is also the basis of the extreme stability of nucleic acids under pressure. Hydrophobic interactions, which play a substantial role in the stabilization of the tertiary structure and in protein-protein interactions, are destabilized under high pressure. On the other hand, stacking interactions between aromatic rings show negative volume changes and are stabilized by pressure [13].

III. PHASE DIAGRAMS OF PROTEINS

The biologically active structure of a protein is only stable within restricted conditions of temperature, pressure, and solvent composition. Outside this range, unfolding or denaturation takes place. When the protein concentration is sufficiently high, aggregation may take place followed by gel formation.

A. Thermodynamics of Denaturation

After the initial discoveries of Bridgman [1] about egg white, detailed investigations of other proteins by Suzuki [3], Hawley [4], and Zipp and Kauzmann [5] showed that these observations may be put together in a phase diagram for the denaturation of proteins. As shown in Figure 1, at high temperature, pressure stabilizes the protein against temperature-induced denaturation (zone III). At low temperature (zone I), increasing temperature stabilizes the protein against pressure denaturation. The fact that one can "cook" an egg with pressure is the result of the unique phase diagram of proteins. Besides its observation for proteins, there is also evidence that it occurs in polysaccharides, such as starch [21,22], and in lipids [14]. Similar diagrams have been observed for bacteriophages [23], and microorganisms [24] and for the inactivation of enzymes [6].

The information about the difference between the native and the denatured state of a protein may be obtained from the temperature and pressure dependence of the Gibbs free energy difference of the following process:

$$\text{Native (Folded)} \rightleftharpoons \text{Denatured (Unfolded)} \tag{11}$$

The change in the free energy difference is given by:

$$d(\Delta G) = -\Delta S dT + \Delta V dP \tag{12}$$

Integration of this equation leads to the following result:

$$\begin{aligned} \Delta G = {} & \Delta G_0 + \Delta V_0 (P - P_0) - \Delta S_0 (T - T_0) \\ & + (\Delta\beta/2)(P - P_0)^2 + (\Delta Cp/2T_0)(T - T_0)^2 \\ & + \Delta\alpha (P - P_0)(T - T_0) \end{aligned} \tag{13}$$

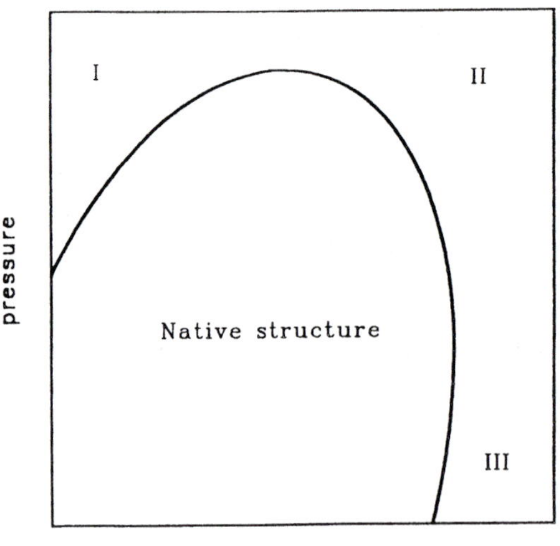

FIGURE 1 A schematic representation of the pressure-temperature phase diagram for the denaturation of proteins or inactivation of enzymes. Pressure denaturation at constant temperature takes place in Zone I. Here ΔH (or $\Delta H^{\#}$) and ΔV (or $\Delta V^{\#}$) are negative. In Zone III (temperature denaturation) ΔH (or $\Delta H^{\#}$) and ΔV (or $\Delta V^{\#}$) are positive. In Zone II (temperature denaturation at high pressure) ΔH (or $\Delta H^{\#}$) are positive and ΔV (or $\Delta V^{\#}$) are negative. (Adapted from Refs. 3–6.)

in which isothermal expansion factor $\Delta\alpha = (\delta\Delta V/\delta T)_P$, isothermal compressibility factor $\Delta\beta = (\delta\Delta V/\delta P)T$, and heat capacity at constant pressure $\Delta C_p = T(\delta\Delta S/\delta T)_P$. The interpretation of these thermodynamic quantities into molecular terms is one of the main tasks of physical biochemists. Several experimental approaches are available. In this chapter, primarily infrared spectroscopy and rheological techniques are used. Other approaches have been reviewed by Silva and Weber [8] and Jonas and Jonas [25].

The Clausius-Clapeyron equation [Eq. (2)] assumes that ΔV and ΔS of the process are pressure and temperature independent. The occurrence of the reentrant phase in the phase diagram of lipids, proteins, and polysaccharides shows that this does not always have to be the case. Removal of these restrictions leads to a more complex equation, the discussion of which is outside the scope of the present chapter.

B. Kinetics of Denaturation

When one obtains thermodynamic or kinetic data as a function of both temperature and pressure, then the activation parameters may be pressure as well as temperature dependent. This was first observed by Suzuki [3] for the denaturation of ovalbumin and carbonylhemoglobin. Hawley [4] has observed changes in enthalpy and volume changes for the denaturation of proteins. In a number of instances it has been found that the activation energy changes sign under hydrostatic pressure and becomes negative [26]. Another consequence is that the reaction rate may become pressure independent at a given temperature. These observations may be understood from the following thermodynamic expression, which may be derived from Maxwell's relations:

$$\left(\frac{\delta\Delta H}{\delta P}\right)_T = \Delta V - T\left(\frac{\delta\Delta V}{\delta T}\right)_P \tag{14}$$

All of these observations are related to the phase diagram for the stability of proteins. Although there is no detailed molecular interpretation of these observations, a change in conformation of the protein is often assumed. For the case of protein denaturation, it has been established, as will be shown below, that temperature and pressure induce different conformational changes.

C. Solvent Effects

In view of the role of solvent composition (water activity) in the denaturation process of proteins, it is important to keep in mind that the structure of a protein is the result of a delicate balance between the intramolecular interactions in the polypeptide chain that compete with the solvent interactions. At the extreme of the spectrum, reports indicate that dry proteins are found to be extremely resistant to temperature and pressure denaturation. In our laboratory, it has been observed that the pressure effect on dry proteins may easily be studied in the diamond anvil cell (unpublished observations). It is clear that the influence of water activity on these processes is crucial. The consequence for the biotechnology application of the high-pressure technique is also evident.

Timasheff [27] has recently reviewed the stabilizing effect of osmolytes (amino acids, sugars, and polyols) on the temperature denaturation of proteins. These compounds are preferentially excluded from the protein surface and thereby affect the binding of water to proteins. Their presence shifts the conformational equilibrium in macromolecules towards a state with the least amount of bound water. In a number of cases the effects are quite nonspecific, as expected for a pure osmotic effect.

In some cases, organic cosolvents such as glycerol and sugars stabilize the protein against pressure-induced denaturation [28]. Oliveira et al. [29] have observed that Arc repressor will not denature in the absence of water. Glycerol has been found to stabilize ribonuclease A against pressure denaturation at 0.7 GPa [30]. The solvation of proteins may also be influenced in reversed micelles. When α-chymotrypsin is solubilized in reversed micelles, high hydrostatic pressures up to 150 MPa stabilize the enzyme against thermal inactivation [31].

What is the highest pressure that a protein can resist in solution? This question is not only of fundamental importance but also has practical implications. There are reports in the literature that some proteins and enzymes may resist pressures up to 1000 MPa while still retaining their biological activity. Stabilization effects of lipids and detergents on the conformation of hydrophobic peptides, such as gramicidin, incorporated in lipids have been reported [32].

For a more fundamental understanding of the effect of pressure and temperature on food components, a detailed study of the differences in structure for pressure and temperature-induced denaturation is needed. Closely related is the question of whether similar phase diagrams apply for the formation of protein gels. Some of the methods that have been employed to study the difference between pressure- and temperature-induced denaturation and gel formation of proteins are discussed in the next section.

IV. METHODOLOGY TO STUDY PROTEIN DENATURATION, AGGREGATION, AND GEL FORMATION

A. Observation Methods During the Pressure and Temperature Treatment

Protein denaturation and gel formation induced by temperature and solvent conditions may be studied by several experimental approaches. The effect of pressure may be studied in a similar way, except that it is not so easy to apply the classical spectroscopic techniques for in situ studies up to about 1000 MPa. This is mainly related to technical difficulties. In the case of circular dichroism spectroscopy, there are also intrinsic reasons: the corrections that would be needed to account for the pressure-induced birefringence in the windows will be much larger than the expected changes in the protein.

In the last few years a high-pressure technique, the diamond anvil cell (DAC), has been developed which allows the observation of biomacromolecules in situ during the compression and decompression phase. With this technique pressures of 1 GPa and more can easily be obtained. The technique can easily be used with Fourier transform infrared (FTIR) spectroscopy and has distinct advantages for the observation of pressure-induced changes in biomolecules as well as in living tissues [33]. Infrared spectroscopy is an absorption technique. Fluorescence impurities present in many samples of biological origin do not interfere, but the strong absorption of water in the region of interest for the protein studies necessitates its replacement by heavy water. Proteins are dissolved in D_2O or any desired buffer solution and mounted in a stainless steel gasket of the diamond cell. Two gem-cut diamonds with polished-off culets are compressed against each other with a metal gasket (0.1 mm thick) leaving a hole (0.5 mm diameter) between them.

The sample volume is of the order of 20 nl, and the total amount of protein needed per run is about 1 mg. The mini-cell from Diacell Products, UK, which has a rated maximum of about 5 GPa, is quite convenient. Any research infrared spectrometer with a liquid nitrogen–cooled MCT detector can be used. The DAC allows the observation of infrared spectral changes induced by pressure and/or temperature. On the same sample, gel formation can be followed in situ from turbidity changes in a UV/VIS spectrophotometer. A crude estimate of changes in viscosity of the sample under pressure can also be made by visual inspection of the cell content under the microscope. This makes it possible to follow the pressure-induced gel formation in proteins and polysaccharides such as starch in situ in the diamond cell. The results that have been obtained in our laboratory for starch are closely correlated with those obtained by Ezaki and Hayashi [34].

Infrared spectroscopy probes the vibration of chemical bonds within the molecules. The bands in the spectrum are characteristic for functional groups, and the band position depends on local molecular interactions. For the CO bond of the peptide group of proteins, hydrogen bonding and coupling of CO vibrations (dipole-dipole coupling) are the predominant factors that determine the features of the amide I′ band of proteins. The frequency of the CO vibration is dependent on the secondary structures of the proteins [35].

A more detailed analysis of the amide I′ band of the protein allows a closer look to the molecular events. This is done with a combination of self-deconvolution and band-fitting, which enables one, in favorable cases, to follow the secondary structures as a function of pressure and temperature. Figure 2 gives the infrared spectrum in the amide I′ region for whey protein concentrate, the principal component of which is β-lactoglobulin. The figure also shows the spectrum after resolution enhancement by Fourier self-deconvolution. Fitting of this spectrum to a sum of Gaussian bands indicates that the spectrum is composed of five main bands, which can be assigned to protein secondary structures. For β-lactoglobulin, these are mainly intramolecular β-sheet and α-helix and unordered structures. In Section V, the effect of temperature and pressure on the spectrum of proteins will be discussed and important differences will be highlighted.

B. Characterization Methods for Pressure- and Temperature-Treated Food Proteins

1. Protein Denaturation and Aggregation

Protein denaturation and aggregation of food proteins as induced at low protein concentrations by high hydrostatic pressure has primarily been studied by spectroscopy, thermal analysis, chromatography, enzymatic digestion, immunology, electrophoresis, microscopy, and solubility.

The amount of free sulfhydryl groups in proteins can been measured by adding 5,5′-dithiobis (2-nitrobenzoic acid) (DTNB), which reacts with SH groups upon release of the colored nitrothiophenolate anion. After precipitation of the proteins with ammonium sulfate, the nitrothiophenolate remaining in solution is determined by measuring the absorbance of the yellow filtrate at 412 nm [36]. The change in surface hydrophobicity as a result of pressure-induced protein denaturation and aggregation can been deduced by hydrophobic fluorescence probes (e.g., anilinonaphtalenesulfonate, ANS) [37,38] and by hydrophobic interaction chromatography [28].

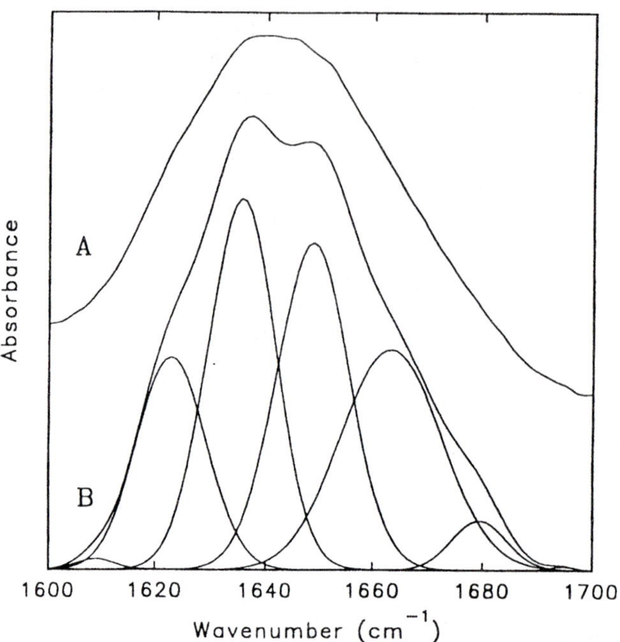

FIGURE 2 Infrared spectrum in the amide I region of whey protein concentrate at 15% in Bis-Tris buffer pH 7. (A) Experimental data. (B) Fourier self-deconvoluted spectrum shows the main components of β-lactoglobulin: intramolecular β-sheet at 1636 cm^{-1} and unordered/α-helix at 1649 cm^{-1}. The fit of the deconvoluted spectrum to a sum of Gaussian components shows the presence of three other secondary structures.

Protein solubility can be taken as a fast detection method for protein denaturation and aggregation. Measurements are performed by determination of the protein (nitrogen) concentration in solution after removal of aggregated protein by direct centrifugation [36], by precipitation with ammonium sulfate [28], or by precipitation near the isoelectric point (IEP) of the proteins [39].

Gel filtration chromatography is also useful to detect denaturation and aggregation of individual proteins [28] but may result in overlapping of native high molecular weight and aggregated low molecular weight proteins in more complex protein mixtures [40]. Immunological detection techniques are more specific, but may suffer from interferences [40,41].

2. Gel Formation of Food Proteins

In the majority of literature references discussing the high-pressure–induced gel formation of food proteins, rheological techniques are used to characterize the gel networks formed. In solid foods rheology, the relationship between the deformation of a material and the forces that act on this material is determined as a function of time during which the force or deformation is applied. Forces are usually transformed into a stress (σ, in Pa), while the deformation is expressed in relative terms as a strain (γ, dimensionless).

The deformations are usually imposed by shear, i.e., parallel to the sample surface, or by uniaxial compression, i.e., perpendicular to the sample surface.

Nondestructive rheological measurements performed at small sample deformations are used to obtain basic information on the type of interactions stabilizing the sample material. On the contrary, rheological measurements performed at large sample deformations are often destructive and allow a general impression of the textural properties of the material (e.g., mouthfeel). In the following the four different types of rheological measurements—compression-decompression tests, relaxation, creep, and oscillation—are discussed. Additional information on the basics of solid foods rheology can be found in Peleg [42] and in Whorlow [43].

Up to now, most rheological tests performed on high-pressure–induced food protein gels generated force-deformation curves by uniaxial compression. During these experiments, the protein gel is compressed at a constant deformation rate, during which the resulting force is monitored as a function of deformation. Various rheological parameters have been derived from these tests, e.g., gel rigidity, gel hardness, or gel strength corresponding to the height at a given level of deformation or the area under the force-deformation curve, and taken as a measure for the total force or the total amount of work, or (when the force is divided by the cross-section area of the gel) the stress needed to deform the sample specimen [2,36,44–48], gel breakability or gel breaking strength as the total force (amount of work) needed to break the gel [2,46], and cohesiveness as the proportion between the work needed for a second deformation and the work needed for the first deformation [44,45].

During a relaxation experiment, the gel is usually deformed by uniaxial compression, after which the force (stress) decay at constant sample deformation is followed as a function of time (Fig. 3a). This force (stress) decay is the result of relaxation phenomena within the gel network whereby highly stressed interactions are replaced by new interactions, which are less hampered by the deformation level imposed. The more weaker interactions present in the gel network, the more pronounced the relaxation process will be. After a sufficient relaxation process (e.g., 10 min or more), the elasticity index can be calculated from the relaxation curve of the gels [47] or from the ratio of the force (stress) remaining in the gel after the relaxation process and the force (stress) measured immediately after compression [48].

Creep tests are static rheological measurements in which a constant stress is imposed on the sample, and the sample deformation (usually expressed as compliance, i.e., the strain, γ, measured relatively to the stress, σ, imposed) is followed as a function of time. Ideal elastic materials deform instantaneously, producing a constant compliance until the stress is released; ideal viscous materials on the contrary deform at a constant rate. In the case of viscoelastic materials, a combination of both models is seen in which an instantaneous compliance is followed by an additional but more slower sample deformation as a function of time (Fig. 3b). The instantaneous compliance in the creep curve may be related to the degree of intermolecular entanglement within the gel network, i.e., higher values refer to a higher degree of mobility for the polypeptide strands to rearrange between the cross-links. The ratio between stress and instantaneous strain has also been used as an elastic modulus [45]. The further increase in compliance as a function of time may reflect the extent of relaxation within the network during creep, i.e., solidlike materials tend to reach a constant compliance, while viscous materials show a further increase in compliance as a function of time (Fig. 3b).

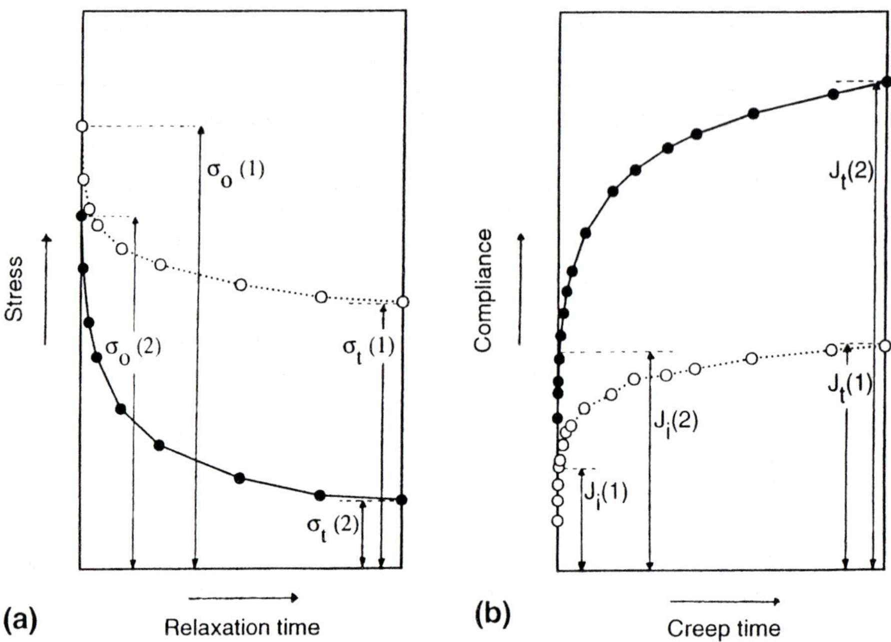

FIGURE 3 Schematic representation of (a) a relaxation experiment (a) and (b) a creep experiment using a viscoelastic sample with a more solidlike nature (○, curve 1) and with a more viscous flow behavior (●, curve 2). For relaxation, the stress is measured after a relaxation time t (σ_t), which can be expressed relatively to the stress obtained immediately after compression (σ_o). During creep, the instantaneous compliance (J_i), and the compliance change in function of time (J_t) are registered.

Oscillation experiments are usually performed by imposing a sinusoidal deformation $\gamma(t)$ with a maximum amplitude γ_o and a frequency $\omega/2\pi$ on the sample specimen:

$$\gamma(t) = \gamma_o \sin \omega t \tag{15}$$

During the experiment, the stress, σ (which also changes sinusoidally as a function of time), needed to maintain this deformation is followed as a function of time:

$$\sigma(t) = \sigma_o \sin (\omega t + \delta) = \sigma_o [\sin (\omega t) \cos \delta + \cos (\omega t) \sin \delta] \tag{16}$$

For an ideal elastic material, no phase difference occurs between stress and strain (phase angle $\delta = 0°$), while in the case of a Newtonian fluid σ differs $(\pi/2)$ radians from γ ($\delta = 90°$). For viscoelastic materials, which are characterized by both an elastic and a viscous behavior, δ (also referred to as loss angle) is higher than 0° but lower than 90°. When σ_o is linearly related to γ_o, it follows that:

$$\gamma(t) = \gamma_o [(\sigma_o/\gamma_o) \cos \delta \sin (\omega t) + (\sigma_o/\gamma_o) \sin \delta \cos (\omega t)] \tag{17}$$

The elastic part contains a storage modulus G′:

$$G'(\omega) = (\sigma_o/\gamma_o) \cos \delta \tag{18}$$

G′ can be taken as a measure of the total number of network interactions that are stable during the course of the rheological measurement, i.e., which are not disrupted and which store the energy applied to the system (e.g., possibly covalent type of interactions like peptide and SS-bonds). The viscous flow behavior is characterized by a loss modulus G″:

$$G''(\omega) = (\sigma_o/\gamma_o) \sin \delta \tag{19}$$

G″ represents the total number of network interactions that are not stable during the rheological measurement, i.e., for which the energy supplied is lost as a result of viscous friction and rupture of network interactions (e.g., noncovalent interactions like ionic, hydrophobic, and hydrogen bonds). Next to G′ and G″, tg δ also serves as a measure to describe the elastic and viscous properties of viscoelastic materials:

$$\text{tg } \delta (\omega) = \frac{G'}{G''} \tag{20}$$

The microstructure of high-pressure–induced food protein gels has also been characterized by scanning electron microscopy [49]. The main disadvantage of the technique is the necessity to work in the absence of water. As a result, the sample preparation step (e.g., fixation in glutaraldehyde; dehydration in graded ethanol series of water/alcohol mixtures; critical point drying through carbon dioxide; and coating of dried samples with gold) may have an influence on the result obtained [47,48]. An alternative technique is photon microscopy, in which the samples are prepared by fast freezing in liquid nitrogen [48].

V. IN SITU INFRARED SPECTROSCOPY STUDIES

It is helpful to consider the possible effect of temperature and pressure on the various organizational levels of proteins. Formally, four levels may be considered. The first level is the sequence of amino acids in the polypeptide chain. Whereas there are a number of reports on the possible effect of temperature on chemical reactions of the polypeptide and the side chains, there are no reports on the effect of pressure on the covalent bonds. The only exception may be the effect of conformational changes on the disulfide exchange reactions. The second level is formed by the hydrogen bonds within and between the peptide chains. In general, one expects stabilization of these structures by pressure, but a higher stability is noted for α-helices, at least in some model systems. This may be due to a larger void between the chains in the β-sheets. The tertiary level is formed by the specific packing of the secondary structures into a more or less globular shape. This level is stabilized by noncovalent interactions. Pressure is expected to affect these interactions. Several compact structures may assemble to form the quarternary structure, which is also stabilized by noncovalent interactions. The interactions between the protein subunits are quite sensitive to pressure [8,9,25]. Here the changes in the secondary structure, which can be detected with infrared spectroscopy in situ, are highlighted.

Systematic studies on the difference between the temperature and pressure denaturation of proteins have, up to quite recently, not be performed. However, a number of papers have reported on temperature denaturation, whereby the infrared spectrum shows new bands around 1620 and 1680 cm^{-1}. These bands are typical for the intermolecular hydrogen bonds that are formed due to the aggregation that follows the denaturation step. Clark and coworkers [50] have observed these bands in bovine serum albumen, insulin, lyzozyme, and chymotrypsin but not in ribonuclease. All of these proteins form gels, and the correlation between the appearance of the high and low frequency bands in the spectrum may therefore be used as a probe for the type of intermolecular interactions in the gel. The case of ribonuclease is a notable exception, which may be related to the degree of unfolding of the protein in the denatured state.

The typical infrared spectrum for temperature denatured proteins has also been observed in β-lactoglobulin [51] and in whey proteins [52]. These bands have also been observed in the spectrum of whey protein concentrate, as shown in Figure 4. Here it is also demonstrated that the main structural components of the spectrum of the native protein disappeared, suggesting a considerable unfolding of the protein. The appearance of the bands characteristic for the intermolecular hydrogen bonding goes together with the development of the turbidity of the solution, suggesting the relation between the gel formation and the presence of the bands.

Pressure-induced denaturation and gel formation has been observed in a number of cases (see next section). The question arises whether the gel formation is also accompanied by the appearance of bands indicative of intermolecular hydrogen bonds as is the case for temperature denaturation. The Raman spectrum of lysozyme [53] and the infrared spectra of chymotrypsinogen [54] did not show these bands. As can be seen from Figure 4, the spectrum of the pressure-induced denaturation of whey proteins at 1000

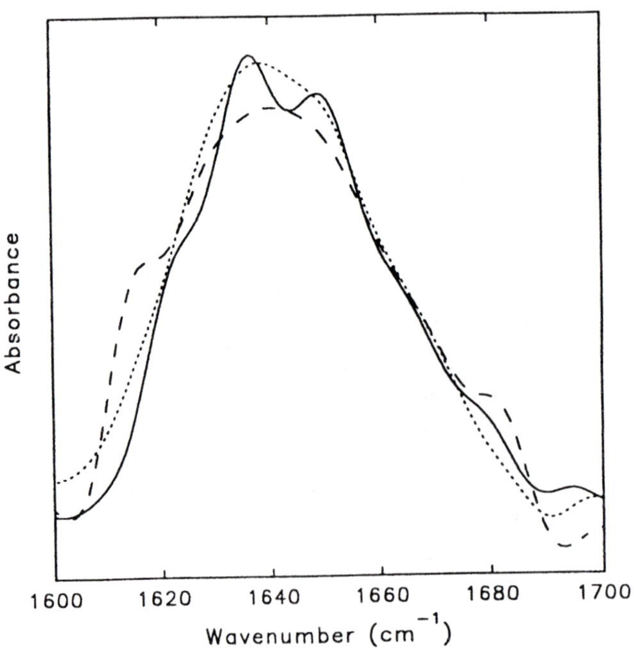

FIGURE 4 Infrared spectrum in the amide I region of whey protein concentrate. Experimental conditions are as in Figure 2. Solid line: Deconvoluted spectrum at ambient conditions. Dotted line: Deconvoluted spectrum at 1 GPa. Broken line: Deconvoluted spectrum at 85°C. Note the presence of bands that may be assigned to intermolecular hydrogen bonding in the spectrum of the temperature-denatured proteins.

MPa shows a similar absence of the bands. A pressure-induced transition is observed at about 200 MPa. This is an unusually low pressure for protein denaturation. It can be noted that the spectrum of the pressure-denatured protein has more of the features of the spectrum of the native protein than the temperature-denatured protein. This suggests that the unfolding for the pressure denaturation is less extensive as for the temperature denaturation. Just as in the case of the temperature-induced gel, the formation of the pressure-induced gel produces a change in turbidity, which can be followed, in situ, in the diamond anvil cell on the same sample used for recording the infrared spectrum.

A final question is what would happen if a temperature-denatured protein were subjected to high pressure. The answer is shown in Figure 5. The bands that are typical for the temperature-denatured protein disappear only at very high pressure, i.e., when the protein is severely distorted. The spectrum at 2 GPa still has the features of the temperature-denatured protein. Upon lowering of the pressure, the original temperature-denatured spectrum reappears. When the protein is first pressurized to form a gel and then heated to high temperatures, intermolecular hydrogen bands appear. These experiments suggest that the unfolding of the protein is less substantial during pressure denaturation. This may also explain the observed partial reversibility of the pressure induced denaturation in contrast to the temperature denaturation. In the latter case the protein is highly, although probably not completely, unfolded and the unfolded conformation is strongly trapped in that conformation by the strong intermolecular hydrogen

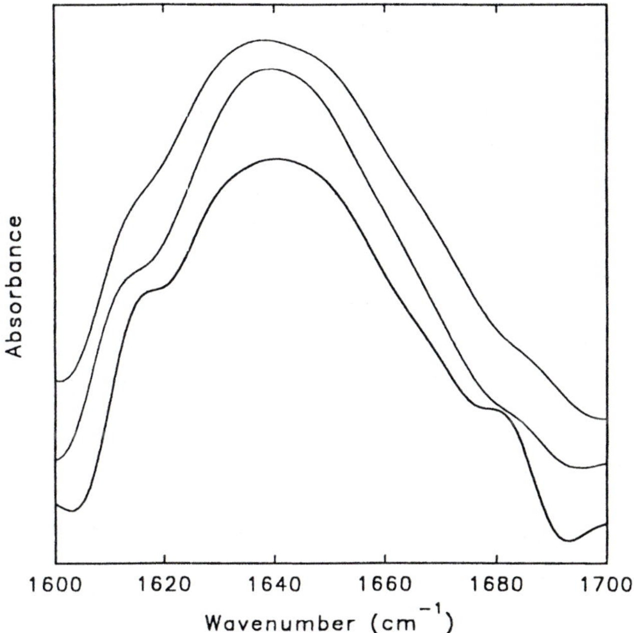

FIGURE 5 Effect of pressure on the infrared spectrum of temperature-denatured whey protein concentrate. Lower spectrum: Deconvoluted spectrum of the temperature-denatured proteins at 85°C. Middle and top spectra: The effect of increasing the hydrostatic pressure on the bands assigned to intermolecular hydrogen bonding at 1 and 2 GPa, respectively. When the pressure is released, the lower spectrum reappears.

bonding. This is not the case for the pressure denatured protein. The gel which forms under these conditions is substantially different. The rheological studies confirm and amplify these results as will be shown in the next section.

VI. EX SITU STIDUES OF PRESSURE-TREATED FOOD PROTEINS

Tables 2 and 3 give an overview of recent publications in the area of pressure-induced denaturation and aggregation (Table 2) and gel formation (Table 3) of food proteins. Details are given concerning the type of food proteins examined, the process conditions applied during pressurization (operating pressure, duration time of the pressure treatment, and initial operating temperature), and the analysis techniques used to characterize the pressurized samples.

The information given in the articles listed in Tables 2 and 3 serves as a basis to discuss recent findings in the field of high-pressure–induced denaturation and aggregation of food proteins, formation of protein gel networks under high pressure, and heat- and acid-induced gelation of food proteins previously treated by high pressure. Also, specific information is given on the pH-lowering effect under pressure in relation to protein functionality and on the time-dependent effects of pressure-induced denaturation, aggregation, and gel formation of proteins as a function of storage time after pressurization.

A. Comparison Between Heat and High-Pressure Treatment of Food Proteins

1. Protein Denaturation and Aggregation

Various studies have demonstrated that significant differences in protein denaturation and aggregation induced by heat as compared to high pressure may occur in a number of food proteins. Although blood plasma and egg white proteins are sensitive to heat and readily form gel networks at moderate operating temperatures (80°C/30 min), no gelation occurs if these proteins are pressurized for 30 minutes to a pressure of 400 MPa. Solutions analyzed 1–3 days after pressurization show alterations in the number of free sulfhydryl groups and display less solubility than unpressurized control solutions. As a consequence, denaturation and aggregation may have occurred, although not sufficient for the formation of a gel network structure [36]. Hayakawa et al. [37] have demonstrated by ANS spectrofluorometry that the conformation of the main protein component of blood plasma and egg white (bovine serum albumin (BSA) and ovalbumin, respectively) remains fairly stable when pressurized at 400 MPa at low protein concentration. This high-pressure stability of ovalbumin and BSA may be positively correlated with the high amount of disulfide bonds stabilizing the three-dimensional structure of both proteins (17 in the case of BSA and 4 in the case of ovalbumin).

β-Lactoglobulin (β-Lg), the main component of the whey proteins, seems far more sensitive towards pressure than ovalbumin and BSA. The work of Dumay et al. [28] has shown that the solubility of pressurized β-Lg in the presence of ammonium sulfate changes significantly (Fig. 6). While the control solution remains soluble up to 2.75 M ammonium sulfate, the solubility of pressurized β-Lg starts to decrease at 1.27 M ammonium sulfate—and this to a greater extent at 5% protein than at 2.5% protein. Ad-

Table 2 Overview of Recent Studies Concerning High-Pressure–Induced Denaturation and Aggregation of Food Proteins, Analyzed After Pressure Treatment

Food Protein	Pressure Treatment	Analysis	Ref.
Egg white and blood plasma proteins	400 MPa/30 min/20°C	Solubility, amount of free sulfhydryl groups (DTNB)	36
Bovine serum albumin, ovalbumin	100–500 MPa/9 min/25°C	Fluorimetry, specific rotation (α-helix), electrophoresis, DSC	37
β-Lactoglobulin	450 MPa/15 min/25°C	DSC, solubility, hydrophobic interaction and gel permeation chromatography	28
β-Lactoglobulin	450 MPa/15 min/25°C	Solubility, SDS-electrophoresis with and without β-mercaptoethanol	39
Whey proteins	200–600 MPa/0–10 min/28°C	Gel filtration HPLC, ELISA	40
α-Lactalbumin, β-lactoglobulin, casein, alcohol dehydrogenase, soy protein, hemoglobin, myoglobin	200 MPa/3 h/30°C	Electrophoresis, digestibility by thermolysin	55
Casein micelles	100–1000 MPa/5 min/20°C	Viscosity, light transmittance (570 nm), SEM, particle size analysis	56
Skim milk proteins	200–600 MPa/0–2h/20°C	Solubility, hydrophobicity (ANS)	38
Bovine colostrum immunoglobulins G, M, and A	100–200 MPa/2–15 h/20–63°C	Radial Immunodiffusion	41
Myosin	70–210 MPa/0–30 min/25°C	Turbidity, sedimentation velocity, TEM	57
Metmyoglobin	750 MPa/20 min/23°C	Electrophoresis, DSC, gel permeation chromatography, UV/visible spectrophotometry	58,59

Table 3 Overview of Recent Studies Concerning High-Pressure–Induced Gel Formation of Food Proteins, Analyzed After Pressure Treatment

Functionality	Food Protein	Pressure Treatment	Analysis	Ref.
Gelation	Egg white, yolk	500–1000 MPa/30 min/25°C	Hardness, breakability, protease digestibility	2
	Egg white, yolk, carp crude actomyosin, rabbit meat, soy protein	100–700 MPa/30 min/25°C	Creep, hardness, cohesiveness, adhesiveness, gumminess	45
	Defatted soy flour, extracted soy protein, concentrated soy protein, isolated soy protein	300–600 MPa/10 min/25°C	Compression stress, shear stress, cohesiveness	44
	Whey proteins, hemoglobin	200–400 MPa/0–60 min/20°C	Gel strength, nonincorporated liquid (NIL)	36
	Whey proteins	400 MPa/30 min/20°C	Oscillation, creep, relaxation, gel strength, SEM, NIL	47
	β-Lactoglobulin + xanthan	450 MPa/30 min/25°C	Gel rigidity, elasticity index, relaxation time, NIL, photon microscopy, SEM	48
	Muscles from carp, Pacific mackerel and chicken breast	500 MPa/10 min/20°C	Hardness, creep/recovery, SEM, TEM	60
	Pacific whiting and Alaska pollock surimi	100–240 MPa/1 h/28–50°C	Shear stress and shear strain at failure	61
Acid-set gelation after pressurization	Skim milk proteins	200–600 MPa/0–120 min/20°C	Gel rigidity, gel breaking strength, water holding capacity, syneresis, protein hydration index, light microscopy	46
Heat-set gelation after pressurization	Myosin	70–210 MPa/0–30 min/25°C	Gel rigidity, TEM and SEM	57

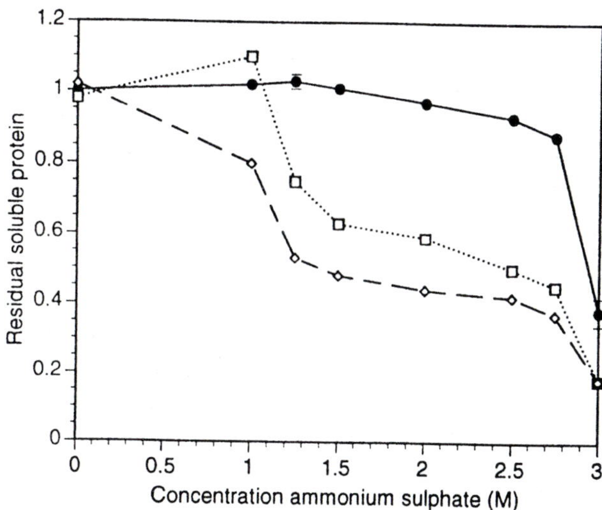

FIGURE 6 Pressure-induced aggregation of β-lactoglobulin isolate (β-Lg) as influenced by protein concentration: residual soluble protein (measured by absorption at 280 nm) after precipitation by ammonium sulfate of β-Lg solutions (pH 7.0); control (unpressurized) β-Lg solution (●); β-Lg solutions containing (w/w) 2.5% (□) or 5% (◇) protein, processed at 450 MPa and 25°C for 15 minutes and then stored unopened at 4°C for 20–27 hours between pressure processing and precipitation. Values are the means of three independent experiments ± standard deviation. Protein content of solutions on precipitation was 0.12% (w/w). (From Ref. 28.)

ditional proof of this pressure-induced unfolding and aggregation of β-Lg was obtained by differential scanning calorimetry, hydrophobic interaction, and gel permeation chromatography [28]. Two studies [40,55] demonstrated that β-Lg is more sensitive towards pressure than α-lactalbumin (α-La). The enzymatic digestion of β-Lg by thermolysin at 200 MPa was markedly increased in contrast to that of α-La, which showed no significant change compared to digestion at atmospheric pressure [55]. The gel permeation HPLC profile of a whey protein concentrate (WPC), pressurized at 200–600 MPa, gave a significant reduction in the peak height of β-Lg, while the amount of α-La showed no significant change after pressurization [40]. These differences in pressure sensitivity are probably attributable to the type of interactions stabilizing the structure of both proteins; while β-Lg contains only two disulfide bonds, α-La is stabilized by four disulfide bonds. Also differences in the relative amount of noncovalent interactions (electrostatic interactions, hydrophobic interactions, hydrogen bonds) may have had an influence.

Significant destabilization of casein micelles in skim milk has been achieved after pressure processing for 5 minutes at 400 MPa and 5°C. The size distribution of the spherical casein micelles changed from 100 to 300 nm prior to processing into chains or clusters of casein submicelles, with a rather wide size distribution ranging from 50 to 500 nm. By heating the skim milk at 30°C, the original size distribution was reinstalled [56]. Johnston et al. [38] applied pressures ranging from 200 to 600 MPa to skim milk and found a significant increase in the exposure of hydrophobic groups after pressure release as measured by ANS fluorescence. An increase in the translucence of milk coupled with a reduction in nonsedimentable protein led the authors to conclude that high pressure resulted in the formation of larger casein fragments possibly containing denatured whey proteins. In a study by Tonello et al. [41], the activity of immunoglobulin G, M, and A in pressure-treated bovine colostrum was measured by radial immunodiffusion. After pressurizing at 200 MPa and room temperature, the activity of the different immunoglobulins decreased by 12% with no significant effect from pressurization time (4–15 hours), prior acidification or the removal of lipids. Increase of the operating temperature to 63°C decreased the activity of immunoglobulin to 40–50% that of the control solution.

Fundamental studies on the denaturation and aggregation of meat proteins have been undertaken by Yamamoto et al. [57] for myosin and by Defaye and Ledward [58,59] for metmyoglobin. The turbidity of a 1% myosin solution containing 20 mM phosphate buffer pH 6.0 and 0.5 M KCl increased after pressurization at 140 MPa for holding times between 10 and 30 minutes. A larger increase was obtained at a pressure of 210 MPa and duration times between 5 and 30 minutes. Transmission electron microscopy demonstrated that increasing pressure and duration time gave higher amounts of one-headed myosin monomers and that aggregation took place by head-to-head interaction between individual myosin molecules [57]. Pressurizing the metmyoglobin molecule at 750 MPa for 20 minutes at 23°C produced dimers stabilized by SDS-labile bonds, as demonstrated by gel filtration chromatography and electrophoresis. The process is highly dependent on pH: dimerization occurs at alkaline pH values (6–10), with a maximum near the isoelectric point (pH 6.9). At acid pH, no dimers are formed [59].

2. Protein Gelation: Rheological Studies

The ability to form gel network structures by high hydrostatic pressure was demonstrated first by Bridgman [1], who coagulated liquid egg white at 600 MPa without additional supply of heat to the pressure vessel. Okamoto et al. [45] compared the force-deformation profiles obtained by uniaxial compression of both heat (10 min at 100°C)– and high-pressure (30 min at 500–700 MPa)–induced egg white gels. Partial coagulation of liquid egg white occurred at 500 MPa, while strong self-supporting gels were formed starting at 600 MPa. The hardness of the gels as well as the elastic modulus increased with pressure but was significantly lower than the heat-induced gels. Hayashi et al. [2] also reported that liquid egg white stored for 5 days prior to pressurization at 600 MPa produced softer gels, which broke easily upon compression as compared to fresh egg white. In the case of soy proteins, Matsumoto et al. [44] and Okamoto et al. [45] found that a minimum pressure of 300 MPa maintained for 10–30 minutes was necessary to induce high-pressure–set gels. Compared to heat treatment (10 min at 100°C), high pressure produced softer gels with a significantly lower elastic modulus [45].

Recently, a thorough comparison between the rheological properties of both heat (80°C/30 min)– and high-pressure (400 MPa/30 min)–induced whey protein gels has been made by Van Camp and Huyghebaert [36,47]. In Figure 7, the resulting gel strength (a), the residual force during relaxation expressed relatively to the initial force obtained immediately after compression (b), the storage (G′) and loss (G″) modulus derived from oscillation (c), and the compliance value during creep (d) are given as a function of the protein concentration for both heat-induced and high-pressure–induced whey protein gels.

Gel strengths were found to increase as a function of the protein concentration, with significantly higher values at a given protein concentration for heat-set as compared to high-pressure–set gels (Fig. 7a). The result may be related to the total number of

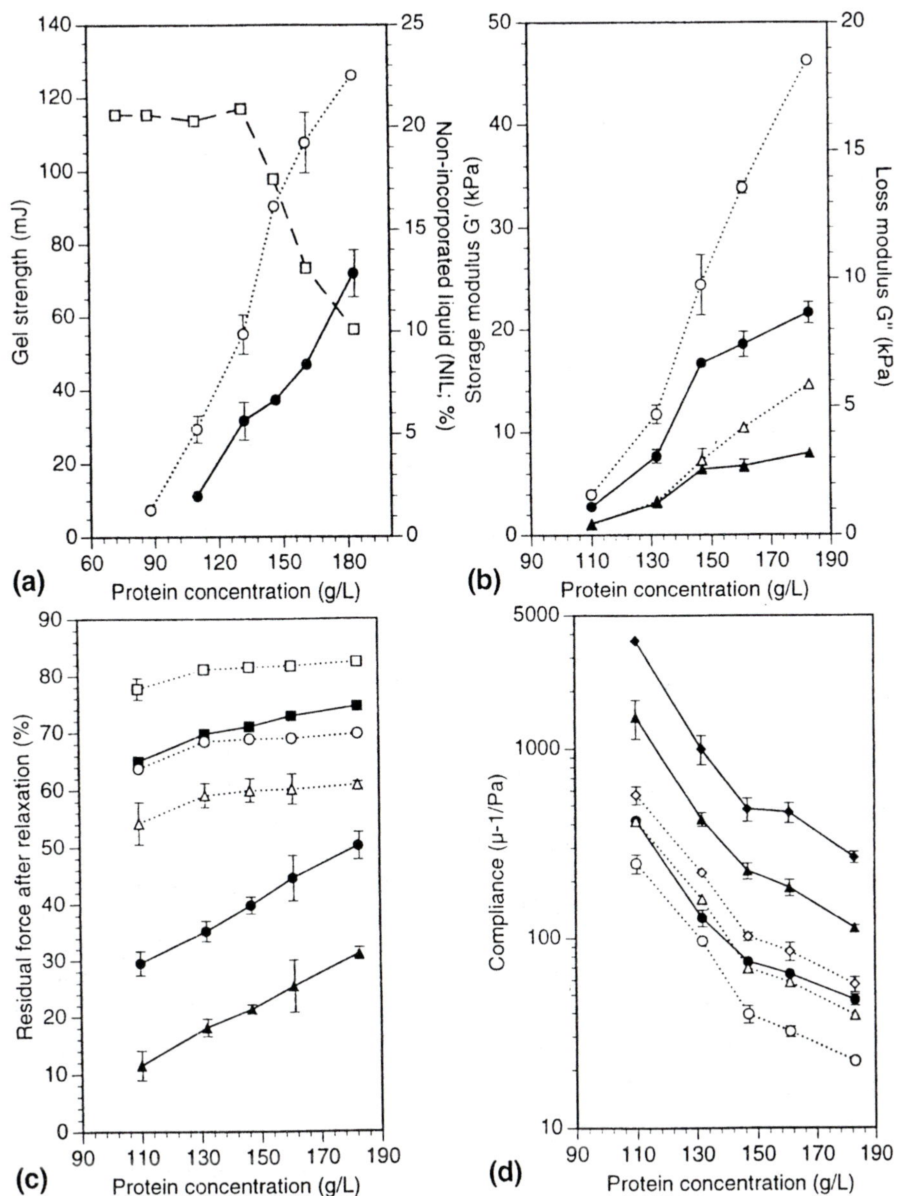

FIGURE 7 Comparison between heat (open symbols)- and high-pressure (filled symbols)-induced whey protein concentrate (WPC) gels as a function of the protein concentration. Gels were formed in a 50 mM phosphate buffer (pH 7) by heating for 30 minutes at 80°C or by pressurization for 30 minutes at 400 MPa. Analysis was performed after 24 hours of storage at 4°C. (a) Gel strength (○ or ●) and NIL (□) for high-pressure-induced WPC gels; (b) storage (○ or ●) and loss (▲ or △) modulus as derived from oscillation; (c) relaxation curves at 0.25 (■ or □), −2.50 (○ or ●), and 10.0 (▲ or △) minute relaxation time; and (d) creep/compliance curves at 0.5 (○ or ●), −50 (▲ or △), and 500 (◆ or ◇) second creep time. (Part (a) from Ref. 36; parts (b)–(d) from Ref. 47.)

interactions formed within the gel network structure: the chance of forming more intermolecular interactions is more pronounced at higher protein concentrations and is more promoted when a higher degree of protein unfolding is achieved (i.e., heat vs. high pressure). The oscillation experiment was performed to have a further indication of the type of interactions stabilizing the gel network. The results in Figure 7b demonstrate that both stable and unstable network interactions increase with increasing protein concentration and are stronger for a given protein concentration in the case of heat-set gels as compared to high-pressure–set gels (i.e., conform to the results of the gel strength measurements). Also, storage moduli are always higher than loss moduli (Fig. 7b), while the relative proportion of loss to storage modulus ($G'':G'$) is higher in the case of high pressure as compared to heat (results not shown). Heat-induced gels are thus composed of more stable network interactions, while the corresponding high-pressure–set gels presumably contain a larger amount of unstable intra- and/or intermolecular interactions.

Figure 7c shows that stress relaxation in the high-pressure–set gels is significantly higher compared to heat-set gels and—in contrast to heat-set gels—shows a pronounced influence from the protein concentration. Rather similar conclusions can be drawn from the creep tests (Fig. 7d). High-pressure–set gels are characterized by a more pronounced increase in compliance as a function of creep time and show higher compliance values for a given protein concentration as compared to heat-set gels. At higher protein concentrations, a reduction in compliance occurs (Fig. 7d).

In contrast to heat treatment, high pressure produces whey protein gels that contain an amount of liquid not incorporated into the gel: the NIL (nonincorporated liquid). This value was maintained at 20% of the initial volume at protein concentrations below 14% (w/w). Higher protein concentrations tend to reduce the amount of NIL, which is concomitant with the increase in gel strength (Fig. 7a). Zasypkin et al. [48] have shown that high-pressure–induced β-lactoglobulin gels can be stabilized against syneresis by the addition of xanthan. It can be postulated that the protein gel network is produced under pressure and that part of the liquid present in this gel network is expelled as a result of volume expansion during pressure release. In the presence of xanthan, expulsion of liquid may be counteracted by the interaction between water and xanthan and by the formation of a more elastic gel stabilized by a higher number of network interactions. The latter has been confirmed by measurements of the elasticity index and by a study of the gel microstructure by photon microscopy and scanning electron microscopy [48].

3. Heat-Set and Acid-Set Gelation After Pressurization

A number of studies report on the acid-set or heat-set gelation of food proteins previously treated by high pressure. Johnston et al. [46] determined several properties of acid-set gels derived from high-pressure–treated skim milk (see Table 3). It was found that the gel rigidity and the gel-breaking strength increased significantly with increasing pressure and duration time of the high-pressure pretreatment. In addition, whey drainage decreased with increasing pressure, reflecting improved resistance of the gel networks towards syneresis. In view of the conformational changes induced in the milk proteins by the high-pressure process [38], suggestions were made by the authors that a more profound disintegration and reaggregation during acid set gelation was stimulated, inducing a significant structural improvement of the resulting protein gel networks. Yamamoto et al. [57] found that a 1% myosin solution containing 20 mM phosphate buffer pH 6.0 and 0.5 M KCl did not form a gel network after pressure treatments up to 210 MPa for 30 minutes. By heating the pressurized solution, a gel was formed that showed no difference in gel rigidity or microstructure from the heat-set gels that received no additional pressure pretreatment. Since intermolecular interaction of myosin tails as induced by heat may be related to helix-coil transitions, the pressure treatments applied probably did not affect the original helical structure of the tail in the myosin monomers [57].

B. Influence of pH Under Pressure: Use of Pressure-Resistant and Pressure-Sensitive Buffers

Due to the electrostriction effect, pressure-sensitive buffers are characterized by a pressure-dependent ionization constant, causing the pH of the aqueous buffer solutions to decrease under pressure. As indicated earlier (see Table 1), the pH of citrate and phosphate buffers becomes more acid at high pressure, while Tris, Bis-Tris, and Bis-Tris-propane buffers keep the pH constant. Also, water is dissociated under pressure, causing acidification in pressurized unbuffered aqueous solutions. It should be pointed out that at high protein concentration, the protein itself may act as a buffer substance.

The effects of various types of buffers on the high-pressure–induced aggregation of a β-Lg isolate have been investigated by Funtenberger et al. [39]. Samples were solubilized in water, a pressure-sensitive phosphate buffer and two pressure-resistant buffers (Bis-Tris and Bis-Tris-propane) to a final protein concentration of 2.5 and 5% (w/w). Pressure experiments were performed at an initial pH of 7.0, using a pressure of 450 MPa for 15 minutes. Protein aggregation was studied after 24 hours of storage at 4°C by determination of the solubility at both pH 4.7 and 7.0. At pH 7.0, the solubility of the pressurized β-Lg decreased significantly only in the presence of Bis-Tris-propane. At pH 4.7, the solubility was higher in the case of water and phosphate as compared to the more pressure-resistant Tris buffers. These results were attributed to a reduction of pH under pressure for water and phosphate solutions, interfering with the formation of intermolecular S-S bonds, which are stabilized at more alkaline pH values. The latter result has been additionally confirmed by electrophoresis, where aggregate bands disappeared in the presence of β-mercaptoethanol.

Information on the gel formation of pressurized food proteins in the presence of pressure-sensitive and pressure-resistant buffers has been obtained by Van Camp and Huyghebaert [36]. In the presence of citrate and Tris buffer, no significant difference in gel strength occurred as compared to distilled water (Fig. 8). In the presence of phosphate buffer, significantly lower gel strength values were registered at both pH 6 and 7 as compared to distilled water, a result that might be due to the increased dissociation of phosphate buffer under pressure. The resulting drop in pH lowers the repulsive forces acting between the protein molecules during the denaturation step, which in turn accelerates the aggregation step within the gelation process and reduces the gel strength. In the presence of Bis-Tris, a pressure-insensitive buffer, stronger gel networks were found, confirming the acidification theory stated above.

C. Influence of Time Scale: Reversibility of High-Pressure Effects on Proteins

High-pressure–induced denaturation, aggregation, and gel formation of food proteins is characterized by a time-dependent behavior in which pressure effects are found to be to some extent reversible. This was demonstrated by Dumay et al. [28], who studied the high-pressure–induced unfolding and aggregation of β-Lg as a function of storage time after compression by measuring the protein solubility in the presence of ammonium

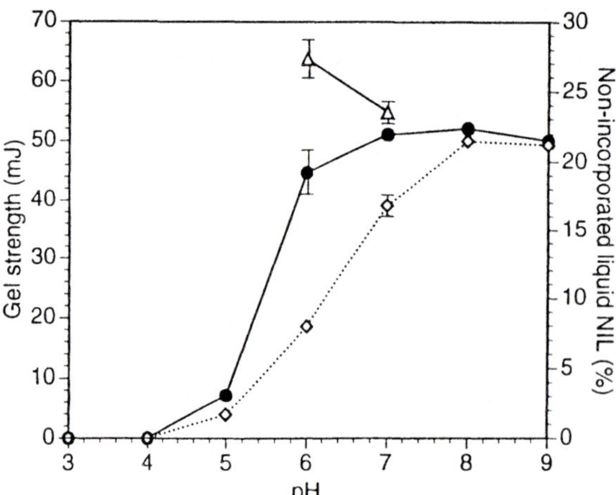

FIGURE 8 Gel strength of high-pressure–induced WPC gels as a function of the pH and in the absence (●) and presence (◇) of 50 mM buffer. For the latter, citrate was used at pH 3, 4, and 5, phosphate at pH 6 and 7, and Tris at pH 8 and 9. The gel strengths in the presence of Bis-Tris are indicated separately (△). The protein concentration was set to 14% (w/w), and the pressure experiment was performed for 30 minutes at 400 MPa operating pressure without additional supply of heat to the pressure vessel. (From Ref. 36.)

sulfate (Fig. 9). Compared to the control solution, a reduction in solubility occurred between 1 and 3 M ammonium sulfate, indicating the formation of aggregates as a result of pressure processing. The process was found to be partially reversible as a function of storage time: compared with the solubility obtained 20 minutes after pressurization, a significant increase in the residual amount of soluble protein was found while additional storage for 5–7 days after pressurization gave only a minor further increase in protein solubility. The same time-dependent changes in protein unfolding and aggregation were found by the use of DSC, hydrophobic interaction chromatography, and gel permeation chromatography [28]. Johnston et al. [38] pressurized skim milk at 600 MPa for 30 minutes and followed the ANS fluorescence binding of the milk proteins up to 8 days of storage at 5°C after pressurization. A significant increase in fluorescence intensity was measured relative to the control solution, which indicates whey protein denaturation and/or destabilization of casein micelles by exposure of hydrophobic side chains. The effect was persistent, as proven by an insignificant change in fluorescence intensity as a function of storage time.

At higher protein concentrations, it was found by Van Camp and Huyghebaert [47] that the storage modulus (G′) of high-pressure–induced whey protein gels increased significantly during the first 3 days of storage at 4°C. Longer storage times (up to 14 days after pressurization) gave only a minor further increase. The loss modulus (G″) remained constant, which suggests that additional stable long-term intermolecular interactions (possibly S-S bonds) were formed as a function of storage time. At large sample deformations, no significant changes in rheological properties could be detected: the gel strength gave no significant difference up to 6 days of storage at 4°C. In addition, the amount of nonincorporated liquid (NIL) did not change significantly as a function of storage time (Table 4). Also, in the study made by Johnston et al. [46] it was found that

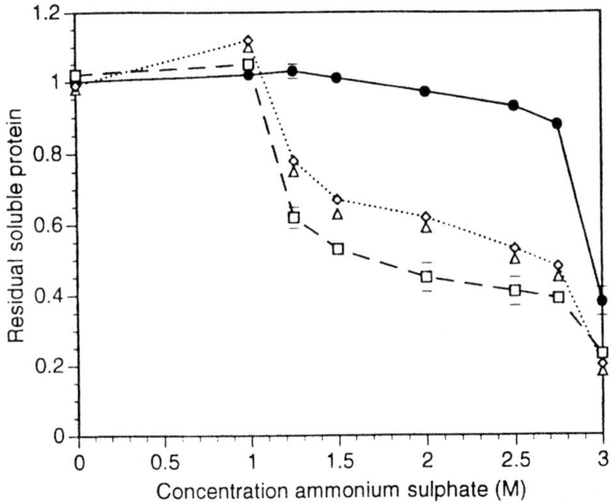

FIGURE 9 Pressure-induced aggregation of β-lactoglobulin isolate (β-Lg) as influenced by storage time after pressure processing: residual soluble protein (measured by absorbance at 280 nm) after precipitation by ammonium sulfate of β-Lg solutions (2.5% protein, w/w, pH 7.0); control (unpressurized) β-Lg solution (●); β-Lg solutions processed at 450 MPa and 25°C for 15 minutes and then precipitated within 20 minutes (□) or stored unopened at 4°C for 20–27 hours (△) or 5–7 days (◇) before precipitation. Means of three independent experiments ± standard deviation. Protein content of solutions on precipitation was 0.12% (w/w). (From Ref. 28.)

the gel rigidity as well as the gel-breaking strength of acid-set gels prepared from skim milk treated for 15 minutes at 600 MPa did not change significantly up to 8 days of storage at 5°C of the pressure treated milk prior to acid-set gelation.

VII. CORRELATION BETWEEN SPECTROSCOPIC AND RHEOLOGICAL STUDIES

The process of gel formation is the macroscopic consequence of denaturation at the molecular level of proteins or other biomacromolecules such as polysaccharides. The native structure (in the case of an enzyme the active conformation) is transformed by denaturation. This denatured state forms a gel or a precipitate according to the chemical and physical circumstances. In many cases, the process is much more complex. Research in this field may therefore be very rewarding in view of possible applications.

Studies with infrared spectroscopy show that there is a correlation between the development of the low- and high-frequency bands in temperature denaturation and the turbidity changes. This suggests a strong correlation between gel formation and intermolecular hydrogen bonding. Detailed studies are going on to probe the difference in temperature/pressure denaturation of proteins and their ability to form gels.

A comparison of the results obtained with infrared spectroscopy with those obtained from rheological techniques reveals some interesting correlations. Table 5 gives an overview. The storage modulus G′ and the loss modulus G″ are higher for the

TABLE 4 Influence of Storage Time on Different Characteristics (G′, G″, GS, NIL) of High-Pressure–Induced (4 kbar; 30 min) WPC Gels[a]

Characteristic	5 h	24 h	72 h	144 h	192 h	312 h
G′ (kPa)	6.92 (0.28)	7.67 (0.32)	9.93 (0.31)	9.37 (0.18)	9.06 (0.30)	9.19 (0.34)
G″ (kPa)	1.19 (0.04)	1.25 (0.07)	1.40 (0.05)	1.28 (0.02)	1.28 (0.03)	1.27 (0.05)
G″/G′	0.17	0.16	0.14	0.13	0.14	0.14
GS (mJ)	32.5 (0.8)	36.1 (1.4)	N.D.	34.7 (1.0)	N.D.	N.D.
NIL (%)	17.2 (1.9)	16.6 (3.2)	N.D.	17.6 (3.2)	N.D.	N.D.

[a]132 g protein/liter in 0.05 mol/liter phosphate buffer at pH 7.0. Oscillation experiments were performed in sixfold; values of the gel strength and NIL are the mean of 12 and 3 repeated determinations, respectively. The standard deviation for each mean is given between brackets. N.D. = not determined.
Source: Ref. 47.

temperature-induced gels than for the pressure-induced gels. Scanning electron microscopy shows that the temperature-induced gels have a higher number of links. Creep and stress relaxation, on the other hand, are higher for the pressure-induced gels. The porosity of these gels is higher. Taken together, these results suggest stronger interactions in the temperature-induced gels in contrast to the weaker interactions of the pressure-induced gels. This correlates rather well with the presence of intermolecular hydrogen bonding in the temperature-induced gels as is observed in the infrared spectra. No such interactions have been observed in pressure-induced gels. A detailed analysis of the changes in conformation suggests that this may be due to the difference in the degree of unfolding.

VIII. CONCLUSIONS

High pressure denatures proteins, solidifies lipids, and destabilizes biomembranes, with the consequent inactivation of microorganisms. On the other hand, the observation that

TABLE 5 Comparison of Rheological and Infrared Data of Pressure- and Temperature-Induced Whey Protein Gels

Characteristic	Pressure-Induced Gel	Temperature-Induced Gel
Storage modulus G′		Higher
Loss modulus G″		Higher
Creep: instantaneous compliance	Higher	
Creep: sample deformation	Higher	
Stress relaxation: force decay	Higher	
SEM: cross-links		Higher
SEM: porosity	Higher	
Strong, long-term interactions		Higher
Weak, short-term interactions	Higher	
FTIR: intermolecular H-bonding	No	Yes
FTIR: degree of unfolding	Small	Large
FTIR: reversibility	Partial	No

pressure treatment does not cause changes in the taste or flavor of food materials is of special interest in view of possible applications in the food industry. Pressure treatment of certain food materials may thus be a possible alternative to temperature treatment. Besides the possibility of forming new textures, the conservation of natural flavor and other ingredients adds to the potential of the technique. A number of books and reviews have been published and should be consulted for more details [7,10,12,16].

The application of high-pressure technology to the processing of food has a long history. In the United States, Hite and coworkers [62] have made extensive investigations into the preservation of fruits and vegetables by inactivating microorganisms with a high-pressure treatment of about 500 Mpa for 30 minutes. The experiments by Bridgman [1] on egg white showed that pressure treatment resulted in coagulation of the white with an appearance like that of a boiled egg. This suggests that microorganisms are inactivated by the action of pressure on the proteins. The observation that protein denaturation profiles may be correlated with the survival of bacteria strongly suggests that this is the primary mechanism [23].

The use of pressure as an alternative to temperature treatment has brought about the need for fundamental studies on the pressure-temperature behavior of macromolecular food constituents. The mechanisms of protein gelation and the sol-gel behavior of polysaccharides are far from being understood. The correlation between the results of spectroscopic techniques with those from rheology is a fruitful area of research.

Pressure-induced effects in food components can be studied in the laboratory with a number of experimental approaches. In this chapter, emphasis was on infrared spectroscopy in situ and rheological studies ex situ. The use of a combined approach was demonstrated to give a deeper level of correlation and interpretation for the phenomena observed.

Future developments will decide whether "Bridgmanization," as the process may be called in honor of the man who showed that pressure can yield products with a new texture, will be a contribution as vital to biotechnology as the process of pasteurization is now.

ACKNOWLEDGMENT

Both laboratories have been supported by the Commission of the European Union, contract No. AIR1-CT92-0296 (project title: High Hydrostatic Pressure Treatment: Its Impact on Spoilage Organisms, Biopolymer Activity, Functionality, and Nutrient Composition of Food Systems).

REFERENCES

1. P. W. Bridgman, The coagulation of albumin by pressure. *J. Biol. Chem. 19*:511 (1914).
2. R. Hayashi, Y. Kawamura, T. Nakasa, and O. Okinaka, Application of high pressure to food processing: pressurization of egg white and yolk, and properties of gels formed, *Agric. Biol. Chem. 53*:2935 (1989).
3. K. Suzuki, Studies on the kinetics of protein denaturation under high pressure, *Rev. Phys. Chem. Japan 29*:91 (1960).
4. S. A. Hawley, Reversible pressure-temperature denaturation of chymotrypsinogen, *Biochemistry 10*:2436 (1971).
5. A. Zipp and W. Kauzmann, Pressure denaturation of metmyoglobin, *Biochemistry 12*:4217 (1973).

6. O. Heinisch, E. Kowalski, K. Goossens, J. Frank, K. Heremans, H. Ludwig, and B. Tauscher, Pressure effects on the stability of lipoxygenase: FTIR and enzyme activity studies, *Z. Lebensm. Untersuch. Forsch. 201*:562 (1995).
7. C. Balny, R. Hayashi, K. Heremans, and P. Masson (eds.), *High Pressure and Biotechnology*, John Libbey Eurotext Ltd, Montrouge, 1992.
8. J. L. Silva and G. Weber, Pressure stability of proteins, *Annu. Rev. Phys. Chem. 44*:89 (1993).
9. M. Gross and R. Jaenicke, Proteins under pressure. The influence of high hydrostatic pressure on structure, function and assembly of proteins and protein complexes, *Eur. J. Biochem. 221*: 617 (1994).
10. V. V. Mozahev, K. Heremans, J. Frank, P. Masson, and C. Balny, Exploiting the effects of high hydrostatic pressure in biotechnological applications, *Trends Biotechnol. 12*:493 (1994).
11. D. G. Hoover, C. Metrick, A. M. Papineau, D. F. Farkas, and D. Knorr, Biological effects of high hydrostatic pressure on food micro-organisms, *Food Technol. 43*:99 (1989).
12. D. A. Ledward, D. E. Johnston, R. G. Earnshaw, and A. P. M. Hasting (eds.), *High Pressure Processing of Foods*, Nottingham University Press, Nottingham, 1995.
13. K. Heremans, High pressure effects on proteins and other biomolecules, *Ann. Rev. Biophys. Bioeng. 11*:1 (1982).
14. R. Winter, A. Landwehr, Th. Brauns, J. Erbes, C. Czeslik, and O. Reis, High pressure effects on the structure and phase behavior of model membrane systems, *High Pressure Effects in Molecular Biophysics and Enzymology* (J. L. Markley, C. Royer, and D. Northrup, eds.) Oxford University Press, New York, 1996, pp. 274–297.
15. K. Heremans and F. Wuytack, Pressure effect on the Arrhenius discontinuity in Ca^{2+}-ATPase from sarcoplasmic reticulum. Evidence for lipid involvement, *FEBS Lett. 117*:161 (1980).
16. B. Tauscher, Pasteurization of food by hydrostatic pressure: chemical aspects, *Z. Lebensm. Untersuch. -Forsch. 200*:3 (1995).
17. R. C. Neuman, Jr., W. Kauzmann, and A. Zipp, Pressure dependence of weak acid ionization in aqueous buffers, *J. Phys. Chem. 77*:2687 (1973).
18. Y. Kitamura and T. Itoh, Reaction volume of protonic ionization for buffering agents. Prediction of pressure dependence of pH and pOH, *J. Sol. Chem. 16*:715 (1987).
19. K. Heremans, The behaviour of proteins under pressure, *High Pressure Chemistry, Biochemistry and Material Science* (R. Winter and J. Jonas, eds), Kluwer Academic, Dordrecht, 1993, pp. 443–469.
20. K. Goossens, L. Smeller, and K. Heremans, Pressure tuning spectroscopy of the low-frequency Raman spectrum of liquid amides, *J. Chem. Phys. 99*:5736 (1993).
21. J. Thevelein, J. A. Van Assche, K. Heremans, and S. Y. Gerlsma, Gelatinisation temperature of starch as influenced by high pressure, *Carbohydrate Res. 93*:304 (1981).
22. A. H. Muhr, R. E. Wetton, and J. M. V. Blanshard, Effect of hydrostatic pressure on starch gelatinisation, as determined by DTA, *Carbohydr. Polym. 2*:91 (1982).
23. H. Ludwig, W. Scigalla, and B. Sojka, Pressure and temperature induced inactivation of microorganisms, *High Pressure Effects in Molecular Biophysics and Enzymology* (J. L. Markley, C. Royer, and D. Northrup, eds.), Oxford University Press, New York, 1996, pp. 346–363.
24. K. Sonoike, T. Setoyama, Y. Kuma, and S. Kobayashi, The effect of pressure and temperature on the death rates of *Lactobacillus casei* and *Escherichia coli*, *High Pressure and Biotechnology* (C. Balny, R. Hayashi, K. Heremans, and P. Masson, eds.), John Libbey Eurotext Ltd, Montrouge, 1992, pp. 297–302.
25. J. Jonas and A. Jonas, High pressure NMR spectroscopy of proteins and membranes, *Annu. Rev. Biophys. Biomol. Struct. 23*:287 (1994).
26. J.-L. Saldana and C. Balny, Device for optical studies of fast reactions in solution as a function of pressure and temperature, *High Pressure and Biotechnology* (C. Balny, R. Hayashi, K. Heremans, and P. Masson, eds.), John Libbey Eurotext Ltd, Montrouge, 1992, pp. 529–531.

27. S. N. Timasheff, The control of protein stability and association by weak interactions with water: How do solvents affect these processes?, *Annu. Rev. Biophys. Biomol. Struct. 22*:67 (1993).
28. E. M. Dumay, M. T. Kalichevsky, and J. C. Cheftel, High pressure unfolding and aggregation of β-Lactoglobulin and the baroprotective effects of sucrose, *J. Agric. Food Chem. 42*:1861 (1994).
29. A. C. Oliveira, L. P. Gaspar, A. T. Da Poian, and J. L. Silva, Arc repressor will not denature under pressure in the absence of water, *J. Mol. Biol. 240*:184 (1994).
30. K. Heremans, K. Goossens, and L. Smeller, Pressure tuning spectroscopy of proteins: FTIR studies in the diamond anvil cell, *High Pressure Effects in Molecular Biophysics and Enzymology* (J. L. Markley, C. Royer, and D. Northrup, eds.), Oxford University Press, New York, 1996, pp. 44–61.
31. R. V. Rariy, N. Bec, J.-L. Saldana, S. N. Nametkin, V. V. Mozhaev, N. L. Klyachko, A. V. Levashov, and C. Balny, High pressure stabilization of α-chymotrypsin entrapped in reversed micelles of Aerosol OT in octane against thermal inactivation, *FEBS Lett. 364*:98 (1995).
32. L. Smeller, K. Goossens, and K. Heremans, The determination of the secondary structure of proteins at high pressure, *Vibrational Spectrosc. 8*:199 (1995).
33. P. T. T. Wong, S. Lacelle, and H. M. Yadzi, Normal and malignant human colonic tissues investigated by pressure-tuning FT-IR spectroscopy, *Appl. Spectrosc. 47*:1830 (1993).
34. S. Ezaki and R. Hayashi, High pressure effects on starch: structural change and retrogradation, *High Pressure and Biotechnology* (C. Balny, R. Hayashi, K. Heremans, and P. Masson, eds.), John Libbey Eurotext Ltd, Montrouge, 1992, pp. 163–165.
35. M. Jackson and H. H. Mantsch, The use and misuse of FTIR spectroscopy in the determination of protein structure, *Crit. Rev. Biochem. Mol. Biol. 30*:95 (1995).
36. J. Van Camp and A. Huyghebaert, High pressure-induced gel formation of a whey protein and haemoglobin protein concentrate, *Lebensm. Wiss. Technol. 28*:111 (1995).
37. I. Hayakawa, J. Kajihara, K. Morikawa, M. Oda, and Y. Fujio, Denaturation of bovine serum albumin (BSA) and ovalbumin by high pressure, heat and chemicals, *J. Food Sci. 57*:288 (1992).
38. D. E. Johnston, A. Austin, and R. J. Murphy, Effects of high hydrostatic pressure on milk, *Milchwissenschaft 47*:760 (1992).
39. S. Funtenberger, E. Dumay, and J. C. Cheftel, Pressure-induced aggregation of a β-lactoglobulin isolate in different pH 7.0 buffers, *Lebensm. Wiss. u Technol. 28*:410 (1995).
40. T. Nakamura, H. Sado, and Y. Syukunobe, Production of low antigenic whey protein hydrolysates by enzymatic hydrolysis and denaturation with high pressure, *Milchwissenschaft 48*:141 (1993).
41. C. Tonello, A. Largeteau, F. Jolibert, A. Deschamps, and G. Demazeau. Pressure effect on microorganisms and immunoglobulins of bovine colostrum, *High Pressure and Biotechnology* (C. Balny, R. Hayashi, K. Heremans, and P. Masson, eds.), John Libbey Eurotext Ltd, Montrouge, 1992, pp. 249–253.
42. M. Peleg, The basics of solid foods rheology, *Food Texture* (P. Moskowitz, ed.), Marcel Dekker, New York, 1987, pp. 3–33.
43. R. W. Whorlow, *Rheological Techniques*, Ellis Horwood Limited, Chichester, 1992.
44. T. Matsumoto and R. Hayashi, Properties of pressure-induced gels of various soy protein products, *Nipp. Nögei. Kaishi. 64*:1455 (1990).
45. M. Okamoto, Y. Kawamura, and R. Hayashi, Application of high pressure to food processing: textural comparison of pressure- and heat-induced gels of food proteins, *Agric. Biol. Chem. 54*:183 (1990).
46. D. E. Johnston, B. A. Austin, and R. J. Murphy, Properties of acid set gels prepared from high pressure treated skim milk, *Milchwissenschaft 48*:206 (1993).
47. J. Van Camp, and A. Huyghebaert. A comparative rheological study between heat and high pressure-induced whey protein gels, *Food Chem 54*:357 (1995).

48. D. V. Zasypkin, E. Dumay, and J. C. Cheftel, Pressure- and heat-induced gelation of mixed β-lactoglobulin/xanthan solutions, *Food Hydrocolloids 10*:203 (1996).
49. A. M. Hermansson and M. Langton, Electron microscopy, *Physical Techniques for the Study of Food Biopolymers* (Ross-Murphy, ed.), Blackie Academic & Professional, Glasgow, 1994, pp. 277–342.
50. A. H. Clark, D. H. P. Saunderson, and A. Suggett, Infrared and laser-Raman spectroscopic studies of thermally-induced globular protein gels, *Int. J. Peptide Protein Res. 17*:353 (1981).
51. H. L. Casal, U. Köhler, and H. H. Mantsch, Structural and conformational changes og β-lactoglobulin B: and infrared spectroscopic stuy of the effect of pH and temperature, *Biochim. Biophys. Acta 957*:11 (1988).
52. N. Parris, J. M. Purcell, and S. M. Ptashkin, Thermal denaturation of whey proteins in skim milk, *J. Agric. Food Chem 39*:2167 (1991).
53. K. Heremans and P. T. T. Wong, Pressure effects on the Raman spectra of proteins: pressure-induced changes in the conformation of lysozyme, *Chem. Phys. Lett. 118*:101 (1985).
54. P. T. T. Wong and K. Heremans, Pressure effect on protein secondary structure and deuterium exchange in chymotrypsinogen: a Fourier transform infrared spectroscopic study, *Biochem. Biophys. Acta 956*:1 (1988).
55. R. Hayashi, Y. Kawamura, and S. Kunugi, Introduction of high pressure to food processing: preferential proteolysis of β-Lactoglobulin in milk whey, *J. Food Sci. 52*:1107 (1987).
56. Y. Shibauchi, H. Yamamoto, and Y. Sagara, Conformational change of casein micelles by high pressure treatment, *High Pressure and Biotechnology* (C. Balny, R. Hayashi, K. Heremans, and P. Masson, eds.), John Libbey Eurotext Ltd, Montrouge, 1992, pp. 239–242.
57. K. Yamamoto, S. Hayashi, and T. Yasui, Hydrostatic pressure-induced aggregation of myosin molecules in 0.5 M KCl at pH 6.0, *High Pressure and Biotechnology* (C. Balny, R. Hayashi, K. Heremans, and P. Masson, eds.), John Libbey Eurotext Ltd, Montrouge, 1992, pp. 229–233.
58. A. B. Defaye and D. A. Ledward, Pressure-induced dimerization of metmyoglobin, *J. Food Sci. 60*:262 (1995).
59. A. B. Defaye, D. A. Ledward, D. B. MacDougall, and R. F. Tester, Renaturation of metmyoglobin subjected to high isostatic pressure, *Food Chem. 52*:19 (1995).
60. K. Yoshioka, Y. Kage, and H. Omura, Effect of high pressure on texture and ultrastructure of fish and chicken muscles and their gels, *High Pressure and Biotechnology* (C. Balny, R. Hayashi, K. Heremans, and P. Masson, eds.), John Libbey Eurotext Ltd, Montrouge, 1992, pp. 325–327.
61. Y. C. Chung, A. Gebrehiwot, D. F. Farkas, and M. T. Morrissey, Gelation of surimi by high hydrostatic pressure, *J. Food Sci. 59*:523 (1994).
62. B. H. Hite, N. J. Giddings, and C. E. Weakly, The effect of pressure on certain microorganisms encountered in preserving fruits and vegetables, *West. Va. Univ. Agr. Expt. Sta. Bull. 146*:1 (1914).

17

Protein and Protein-Polysaccharide Microparticles

Christian Sanchez
ENSRIA-INPL, Vandoeuvre-les-Nancy, France
Paul Paquin
Université Laval, Sainte-Foy, Quebec, Canada

I. INTRODUCTION

People in industrialized countries are increasingly concerned with the health risks associated with obesity. Although obesity is often considered to be a result of high fat intake, it is also due to a marked decline in physical activity of people living in urban settings. Numerous studies have shown that a reduced-fat diet decreases the incidence of artherosclerosis and other cardiovascular diseases. However, despite such evidence, preference for high-fat foods remains high.

The question of why people show this preference for fat is complex. Fats indeed provide a concentrated source of energy and supply essential fatty acids and fat-soluble vitamins to the diet. However, more importantly, the preference for fatty foods is mainly attributable to the sensorial properties that fats impart to foods [1]. Fat acts as a carrier for organic molecules that give food products their characteristic flavor and aroma and markedly contribute to their texture, e.g.; the plasticity and smoothness of solid and semi-solid fats, the creaminess and oily mouthfeel of food emulsions, the softness and freshness of baked goods, the tenderness of meat, etc. The preference for fatty foods is rooted in a physiological process. Indeed, consumption of foods rich in fat is influenced by the endogenous opioid system [2]. Consumption of fat causes an increase in endorphin levels in the brain, and enhanced binding of endorphin to opiate receptors results in the pleasure associated with eating of fatty food (hedonic response).

The demand by consumers for low-fat foods with the sensorial properties of full-fat products has led to the development of a broad range of natural (mainly based on polysaccharides or proteins) and synthetic (based on chemically modified molecules) fat substitutes; these are used as partial or full fat replacers in a number of food products [3]. A fat substitute (also known as fat replacer, fat mimic, fat mimetic, or fat mimicker) is a safe compound with organoleptic properties closely resembling those of fat [4,5]. Fat in foods often takes the form of emulsified fat stabilized by an adsorbed protein film. Therefore, protein- or protein-polysaccharide–based fat substitutes, which are in the form of discrete particles, are particularly interesting. These fat substitutes are more commonly called protein or protein-polysaccharide microparticles. Functional properties of food proteins and polysaccharides have been the subject of numerous studies, and these biopolymers are used as functional ingredients in processed foods. However, physically and chemically modified proteins and/or polysaccharides, such as protein-protein or protein-polysaccharide microprecipitated or microparticulated complexes, are only recently being used as food ingredients. These complexes are highly structured entities formed from polymer dispersions by a combination of treatments such as acidification, heating, shearing, etc. Much remains to be understood about the impact of different treatments and the interactions between different polymer species on the structure and functional properties of microprecipitated or microparticulated complexes. Perhaps the more important parameters affecting sensory properties of microparticles containing food products (e.g., smoothness, powdery, chalky) are the size distribution and shape of microparticles and the viscosity of the microparticle dispersions.

The objective of this chapter is to provide an overview of developments in protein- and protein-polysaccharide–based microparticle technology. Specifically, the physical and chemical environments that transform protein or protein-polysaccharide dispersions into microparticles and the manufacturing processes available for making them are described. A discussion of the functional properties of microparticles and the factors limiting their use in several conventional food products is also presented.

II. STARTING MATERIAL

A. Food Proteins

The most important step in the preparation of protein-based fat replacers is the aggregation of proteins under controlled denaturing conditions. Thus, any denaturable or precipitable food protein is suitable for microparticulation. This includes animal proteins (e.g., milk proteins, egg albumen proteins, fish proteins, and other seafood proteins) and vegetable proteins (e.g., soya proteins, cottonseed proteins, sunflower proteins, peanut proteins). The most commonly used proteins for microprecipitation or microparticulation are milk proteins, egg albumen, and soya proteins. The physicochemical properties of milk, egg, and soya proteins are described in Chapters 7, 8, 9, and 11.

B. Complexing Agents

Complexing agents are those biopolymers (e.g., food proteins and polysaccharides) or low molecular weight molecules (e.g., mono- and diglycerides, emulsifiers) that have an effect, directly or indirectly, on aggregation and microparticulation of the starting protein material. These molecules are used to control the size, the final composition, and the functional properties of microparticles.

1. Biopolymer-Biopolymer Interactions

Biopolymer-biopolymer interactions are controlled by the balance between attractive and repulsive forces and formation of covalent disulfide linkages. Attractive forces include noncovalent electrostatic, van der Waals, and hydrophobic forces and covalent disulfide linkages [6]. Repulsive molecular forces include electrostatic, van der Waals, or London forces as well as hydration and steric forces [6]. The net attraction or repulsion between two or more biopolymers in solution dictates their ability to form complexes or their thermodynamic incompatibility [7]. However, by manipulating factors such as pH and ionic strength, the physicochemical properties of the macromolecules in the mixture (e.g., pI, molecular weight, conformational state), their weight ratio, and thermal (e.g., heating and cooling) and mechanical energy (e.g., high-shear and high-pressure treatments) input, composite protein-protein and protein-polysaccharide complexes can be produced.

Protein-Protein Complexes

Molecular interactions between heated egg albumen proteins, whey proteins, and casein micelles or between heat-denatured whey proteins and calcium caseinate at pH 6.2–6.6 have been exploited to produce protein-based microparticles [8,9]. Sulfhydryl-disulfide interchange reactions and hydrophobic forces are involved in the formation of these composite aggregates [10]. Sulfhydryl-disulfide interchange reaction between β-lactoglobulin (β-lg) and κ-casein naturally occurs in heated milk. This leads to blocking of reactive sites of casein micelles by β-lg, which improves its heat stability. This also blocks the reactive sulfhydryl group of β-lg, which prevents polymerization reactions among whey protein.

Formation of complexes among other proteins is conceivable, provided that they are thermodynamically compatible. For instance, gelatin forms a complex with sodium caseinate [11], but not with egg white proteins at high salt concentration [12]. Whey proteins also interact with rapeseed, pea flour, and peanut proteins. However, it is not known if these complexes can be tailored to form microparticles and used as fat replacers.

Protein-Polysaccharide Complexes

The functional properties of polysaccharides in some food products (Table 1) emphasize the key role played by protein-polysaccharide interactions in the manifestation of textural

TABLE 1 Functional Properties of Polysaccharides in Some Food Products

Functional properties	Food products
Crystallization inhibitor	Ice cream
Emulsifier	Salad dressings
Foam stabilizer	Whipped toppings
Gelling agent	Puddings, desserts, mousses
Stabilizer	Mayonnaise
Syneresis inhibitor	Cheese, frozen foods
Thickening agent	Sauces, gravies

Source: Adapted from Ref. 13.

properties of food. Polysaccharides are hydrocolloids, more commonly referred to as *gums*, which are long-chain, high molecular weight biopolymers with a limited number of repeating monomers. Polysaccharides are highly hydrophilic and highly flexible [13,14]. They markedly enhance the hydrogen-bonded structure of water. Some polysaccharides, such as the acacia, arabic, and tragacanth gums, also possess surface-active properties [14,15].

A mixture of protein and polysaccharide solutions may exist in any of the following three states [16]: (a) a two-phase liquid system in which one of the phases is enriched with protein and the other with polysaccharide (due to limited thermodynamic compatibility of proteins and polysaccharide in aqueous media), (b) a homogeneous stable solution in which there is no interaction between the macromolecular components, or (c) a two-phase system in which both macromolecular components are in the same phase. The latter phenomenon is called *complex coacervation* and is attributed to formation of an insoluble protein-polysaccharide complex via electrostatic interactions. Most fat replacers are complex coarcervates.

The formation of protein-polysaccharide complexes is mainly driven by hydrogen bonding and electrostatic interactions. The conditions that alter the surface charge of macromolecules, by changing either their number or their accessibility (e.g., pH, presence of salts and charged co-biopolymers) greatly affect the extent of interaction between proteins and polysaccharides. Usually, maximal interaction occurs at minimal net charge [17]. In addition to charge, the molecular weights and relative concentrations of the protein and polysaccharide also influence the stability and the functional properties of protein-polysaccharide complexes [18,19].

The phenomenon of protein-polysaccharide complex formation is currently being exploited in protein recovery by precipitation and in food texturization [11,20–23]. Milk proteins, egg albumen proteins, soya proteins, and all the aforementioned proteins, alone or in mixtures, can form complexes with polysaccharides. Among polysaccharides, anionic (xanthan, carrageenans, algins and alginates, pectins, carboxymethyl cellulose, microcrystalline cellulose) and cationic (chitosan, modified starch) polysaccharides are the most suitable. Obviously, all combinations are not possible for thermodynamic compatibility reasons. Analysis of electrokinetic characteristics of protein-polysaccharide complexes, i.e., electrophoretic mobility at various pH, would provide a better understanding of the interactions taking place between various protein-polysaccharide pairs. This would facilitate proper selection of the starting material for fabrication of protein-polysaccharide microparticles [22].

2. Aggregate Extenders

Aggregate extenders play a very important function during microprecipitation or microparticulation. They minimize interactions between biopolymers and control the size of the preformed aggregates. This could occur via two possible mechanisms. An aggregate extender may interact directly with macromolecules and eliminate or shield reactive sites (blocking agent) and lower the interfacial tension, or it may enhance the hydrogen-bonded structure of water, and thus control flow of particles and decrease the probability of their collision.

Anionic polysaccharides (xanthan, carrageenan and locust bean gums, pectin, alginate, etc.), maltodextrins (obtained by either enzymatic or acid hydrolysis of starch) and low molecular weight fatty emulsifiers with hydrophilic-lipophilic balance (HLB) values lower than 8 (lecithin, mono- and diglycerides, calcium steroyl lactylate, sorbitan esters, egg yolk, datem esters) function well as aggregate extenders [5,21,24]. Fatty emulsifiers ameliorate the oily mouthfeel and lubricity of products and have excellent surface properties. On the other hand, xanthan gum gives smaller particles. For example, 95% (mass basis) of coagulated protein particles prepared by extrusion cooking of a whey protein isolate-calcium caseinate mixture at 90°C are smaller than 385 μm [25]. In contrast, under similar experimental conditions, the particle size decreases to <80 μm when 0.5% xanthan gum is included in the initial blend. Combinations of different extenders, e.g., polysaccharide/fatty emulsifier, may provide better efficiency in controlling particle size.

III. PROCESSING CONDITIONS

An accurate description of the conditions employed in various manufacturing processes for the production of protein-based microparticles is difficult, because much of the information is either proprietary or reported only in patents [25]. However, the physicochemical factors affecting denaturation and aggregation of food proteins are known: pH, ionic strength and ionic species, presence of low molecular weight substances, time and temperature of heating, and shear. A combination of these factors is used in the preparation of protein microparticles.

An examination of patents and literature indicates that two processes, namely, protein and protein-polysaccharide *microprecipitation* [11,12,22,24,26,27] and protein and protein-polysaccharide *microparticulation* [5,8,9,21,23,28–32] are generally being used in the production of fat replacers. A flow diagram summarizing the main steps of these processes is given in Figure 1. Briefly, the starting material (food proteins, complexing macromolecules and aggregate extenders) is first pretreated (e.g., concentration, washing, fat extraction, filtration, pasteurization, deaeration), followed by the formation of microparticles (pH adjustment, thermomechanical treatment), and finally posttreatment of the microparticulated material (e.g., concentration, pasteurization, spray-drying, cooling).

Other technologies for production of protein microparticles include concentration of "native" casein micelles [33], atomization with simultaneous formation of protein-polysaccharide complexes [21], or superfine jet mill grinding of protein powders [34]. These processes will not be discussed in this chapter.

A. Pretreatment of the Starting Material

The proteins in the starting material must be in the native state. It should have more than 90–95% solubility and less than 5–10% denaturation, because denatured proteins promote formation of large aggregates, which impart chalky or gritty mouthfeel to the final product. Protein solutions or dispersions that have not undergone any heat treatment are preferable to protein dispersions reconstituted from powders [29]. Moreover, dispersion of protein powder in water requires stirring, which causes foaming (except under vacuum) and interfacial denaturation of the protein.

The protein concentration of the initial mixture is adjusted to the final concentration of the microparticulated product to obtain the desired consistency. If the initial protein dispersion is liquid, such as cheese whey, a preconcentration step involving a conventional process such as ultrafiltration or microfiltration may be necessary. Ultrafiltration (UF) and possibly diafiltration is commonly used for this purpose. In the cases of milk and egg white proteins, the temperature of UF is usually set at 40–50°C to maintain

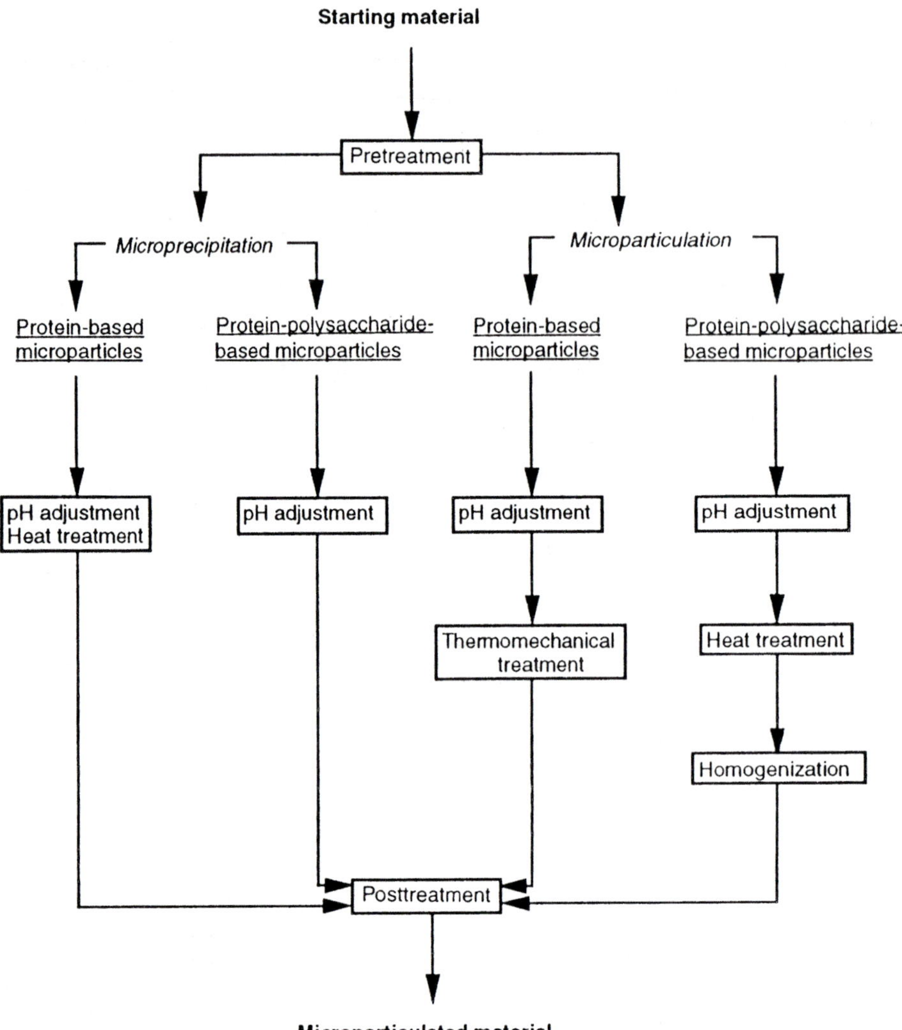

FIGURE 1 Flow diagram of typical protein or protein-polysaccharide microprecipitation or microparticulation processes.

maximum permeation flux rates (by lowering the retentate viscosity) while avoiding heat denaturation of the proteins. Membranes with molecular weight cut-off of about 10,000 for egg white proteins and around 20,000–30,000 for milk and whey proteins are used. Depending on the thermomechanical process applied, e.g., heat-denaturation in scraped-surface heat exchanger, cooking extrusion or heat-denaturation followed by high-pressure homogenization, the total solid content of the starting material is preferably in the range 25–50 wt%, with 15–25 wt% total protein [5,8,25,30–32]. Although these values are tentative, it must be realized that under 15 wt% protein a great deal of water has to be removed to attain the optimal consistency, and beyond 25 wt% protein content processing becomes difficult (extremely high shear rates are required) and the final product is very firm and lacks spreadability.

Milk proteins, and especially whey proteins, show high flavor-binding capacities and act both as carriers of volatile flavors in microparticulated material and as modifiers of perceived flavors in supplemented food [4]. Water washing of microparticles made from whey proteins may be necessary to avoid such detrimental off-flavor effects. Cholesterol and fat may also impart off-tastes, and in this context these molecules are removed from milk proteins by solvent extraction (at 40°C for 4 hours with a mixture containing 94.55 ethanol, 5% water, and 0.05% citric acid) prior to microparticulation [5]. Furthermore, residual lipids interact with milk proteins, which has a significant effect on their aggregation properties.

Food proteins, depending on their source and the method of preparation, show high variability in mineral content (both ion type and concentration). At low concentration, salts may bind to biopolymers, favoring their hydration by decreasing electrostatic interactions between neighboring charged groups (salting-in). Conversely, at high concentration, salts compete with biopolymers for water, suppressing the electrical double layer surrounding the macromolecules and changing their conformation. This promotes biopolymer dehydration and biopolymer-biopolymer interactions and, in the extreme, leads to precipitation (salting-out). These properties greatly affect protein-protein aggregation and protein-polysaccharide complexation [7,35–39]. In this context, the mineral content of the starting material must be adjusted to optimize formation and texture of protein and protein-polysaccharide microparticles.

Although pasteurization of the initial blend is optional, it would prevent undesirable microbial spoilage. Typical pasteurization treatments are 60°C for 30 minutes, 72°C for 15 seconds, or 120°C for 3–4 seconds. However, high-temperature/short–residence time treatments are preferred, since these conditions are amenable to continuous processes and have less deleterious effects on the flavor of the final product [5]. Antimicrobial agents, such as potassium sorbate (≈0.1%), also may be used.

As previously indicated, foaming of a protein dispersion is unavoidable in the preliminary steps of the processes. Deaeration has thus been suggested to improve heat transfer and uniform heating through the dispersion during the thermomechanical step. However, excessive bubbling during deaeration leads to flocculation of proteins denatured at the air/water interface of the foam. Antifoaming agents such as silicone defoamer may be added to the initial dispersion to minimize foaming [26].

B. Microprecipitation and Microparticulation of Proteins and Protein-Polysaccharide Complexes

1. Microprecipitation

Microprecipitation is a process in which denaturation and aggregation of food proteins or complexation of food proteins and polysaccharides is carried out by controlling or manipulating environmental conditions, such as pH, ionic strength and ionic species, temperature, stirring rate, and concentrations of protein, and polysaccharide and aggregate extenders, to obtain the desired particle size and texture without application of strong mechanical treatment.

Extensive intermolecular aggregation reactions during microprecipitation may be prevented by controlling the concentration of protein and protein-polysaccharide

[11,21,25,26]. Thus, for protein microprecipitation, the initial protein concentration must be below the critical gel concentration, usually less than 5–7 wt% for most isoelectric or heat-precipitable food proteins (e.g., whey, egg white, plasma, and soya proteins). For protein-polysaccharide complexes, the optimal concentration is highly variable depending on the nature of macromolecules used and their phase-separation threshold. In addition, when microprecipitation is carried out in a food product, the presence of other ingredients also affects the optimum concentration. Usually, this concentration is in the range of 1–8 wt% [11,21].

Protein Microprecipitation

Microprecipitation of food proteins can be carried out by adjusting the pH of protein solutions close to their isoelectric point (pI) [26]. At pI, proteins become electrically neutral and protein-protein interactions are favored over protein-solvent interactions, resulting in formation of microparticles (or microprecipitates in this case). In the cases of milk, egg white, or soya proteins, the pH adjustment is done with stirring using food-grade acids (hydrochloric, phosphoric, sulfuric, acetic, citric, lactic, tartaric, malic, ascorbic, carbonic, or adipic acids). Stirring during acidification minimizes self-aggregation of proteins and formation of a fibrous material that does not function as a fat replacer [26]. An important requirement is that acidification must be done at a slow rate, because local high concentration of acid may promote excessive protein aggregation. The use of glucono-δ-lactone (GDL) is preferable because its slow aqueous hydrolysis to gluconic acid may prevent this problem. Heat treatment below the protein's denaturation temperature (e.g., lower than 70°C for whey proteins or 55°C for egg white proteins) during stirring would increase protein-protein hydrophobic interactions and would also control the rate of hydrolysis of GDL.

Protein microparticles can also be produced by heating dilute protein solutions under low shear rates ($<200\ s^{-1}$) at natural pH ($\approx$6–7) [24]. Heat-induced aggregation of globular proteins under quiescent conditions proceeds in three distinct stages: (a) a denaturation stage, where native proteins partially unfold and expose their nonpolar side chains and reactive side chains (e.g., sulfhydryl groups), which were initially buried in the interior, (b) an initiation stage where two denatured macromolecules interact through noncovalent (e.g., hydrogen bonds, van der Waals, and hydrophobic interactions) and covalent interactions (e.g., disulfide cross-links) to form an aggregate, and (c) a propagation stage where newly denatured proteins interact with preformed aggregates to produce polymeric structures [40]. In dilute solutions, the propagation stage is stopped because of the insufficient amount of protein, and in concentrated solutions above a critical concentration, gelation occurs. In the above example, the heat-induced aggregation of proteins under stirring is quite different from that occurring under quiescent conditions. This will be described in some detail in the section on protein microparticulation.

Microprecipitation of other food proteins, such as prolamins (from corn, wheat, barley, rice, or sorghum), which are extremely hydrophobic and soluble only in 70–80% ethanol solutions, follow a different mechanism [27]. In such cases, dilution of the alcoholic protein solution with water leads to microprecipitation through strong hydrophobic interactions [25]. The presence of aggregate extenders is particularly essential in preventing excessive aggregation of the proteins.

Thermodynamic incompatibility between gelatin and egg white proteins at high salt concentration (0.6 M) also promotes aqueous phase partitioning and leads to formation of a two-phase system with a continuous phase of gelatin in water and a phase of egg white protein microparticles dispersed in water [12]. Microparticles of egg white proteins are produced by mixing gelatin and egg white proteins at about 1:1 ratio with stirring at 40°C and pH 6.0 to dissolve gelatin, and then heating without stirring at 80°C for one hour to coagulate egg white proteins [12].

Protein-Polysaccharide Microprecipitation

Protein-polysaccharide microprecipitation is simply a particular case of complex coacervation, which results in formation of a dense insoluble and globular or fibrous precipitate. The basic approach to produce complex coacervates involves gradual adjustment of pH to the pI of the complex, followed by its stabilization.

An interesting practical application of this method of insolubilization is that coacervates of spherical or slightly elliptical shape can be obtained by complexing gelatin or whey proteins with acacia gum and/or pectin [11]. In this case, the protein and polysaccharide(s) at ratios 1:5 to 5:1 are dissolved in hot water (45–80°C) and the pH is adjusted (under gentle stirring) to the pI of the coacervates (3.8–4.5). The particles formed are then solidified by cooling at 10–20°C and stabilized by storage at 5°C. The low temperature favors phase separation through electrostatic interactions and hydrogen bonding. An advantage of these globular coacervates obtained by isoelectric precipitation is that they can be formed in situ in a food product. However, they are very sensitive to pH and heat. Furthermore, the impact of other food ingredients on the formation and functional properties of these complexes is not known.

2. Protein and Protein-Polysaccharide Microparticulation

Two processes, namely, isoelectric-mechanical and thermomechanical processes, are used in microparticulation of protein or protein-polysaccharide systems. In these processes, conditions that denature (i.e., heat) or induce complex formation (i.e., isoelectric precipitation) are combined with a strong mechanical treatment, e.g., high shear rates or turbulence and cavitation to produce nonaggregated protein- or protein-polysaccharide–based microparticles of desired size distribution and texture.

Protein Microparticulation

The general steps involved in protein microparticulation are as follows. After addition of aggregate extenders to a concentrated protein dispersion (15–25 wt% protein), the pH is adjusted to values below the pI [30–32,39], around the pI [28], or above the pI [8,25,29], depending on the nature of the protein and the pH of the supplemented food. Heating of the protein dispersion at a pH lower than the pI generally minimizes formation of large aggregates. This is because of electrostatic repulsion between the positively charged protein molecules, which favors protein-solvent interactions over protein-protein interactions [36,41,42]. An example of the pH dependence of the size distribution of microparticles is presented in the Figure 2 for whey proteins coagulated by extrusion cooking. The data indicate high thermal sensitivity of whey proteins at pH greater than 3.8–3.9 [42,43]. At pH = pI, protein-protein interactions are maximal, and therefore acidification must be done under vigorous mixing conditions [28]. Lastly, at pH > pI, where protein molecules carry a net negative charge, electrostatic repulsions do occur. However, despite mechanical treatment and addition of aggregate extenders, sulfhydryl-disulfide interchange reactions at high pH promote formation of large particles [25]. One way to overcome this problem is to create protein-protein complexes during heating,

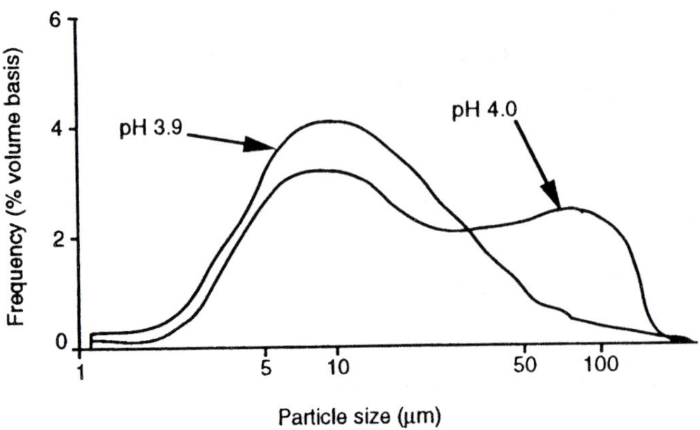

FIGURE 2 Size distribution of a whey protein isolate coagulated by extrusion cooking at pH 3.9 or 4.0, as determined with a Malvern laser diffractometer. (From Ref. 30.)

such as casein micelles–egg white proteins [8], or whey protein–calcium caseinate [25,32], in which egg white proteins and calcium caseinate function like blocking agents.

Heat and mechanical treatment may be simultaneously applied to the protein dispersion. The scraped-surface heat exchanger (SSHE) or the SSHE specifically designed for highly viscous fluids [44] and classical cooking extruders [30] (Fig. 3) are used to heat and apply shear to the protein dispersion. The SSHE or "microcooker" shown in Figure 3 has been specifically designed for the microparticulation of proteins. Its uniqueness lies in a toroidal cavity with bent blades located within a bowl-shaped thermostatted vessel. Efficient mixing is provided by a toroidal flow in the fluid, and control of particle size is ensured by high shear rates (10,000–40,000 s^{-1} at blade rotation rates of 1250–5200 rpm), turbulence, and cavitation (particularly in front of the leading edges of the blades). Although no experimental evidence exists for protein denaturation by shearing in a cooking extruder at 25°C [30], it has been shown that shear rates higher than 5000 s^{-1} are sufficient to unfold proteins [45–47]. Therefore, depending on the time of heating and shearing, protein denaturation in SSHE probably occurs by both heat and shear.

How are food proteins microparticulated? To date, the mechanism of heat aggregation of food proteins under shearing is not well understood. However, it seems that hydrodynamic shear plays only a minor role and fragmentation and erosion of aggregates due to particle-particle and particle-surface collision play a major role in the control of particle size [48]. In fact, the final size of aggregates primarily depends on the balance between growth and the breakage controlled by shear or fragmentation/erosion. This explains why high temperature and prolonged heating produce large aggregates, and strong shearing produce small aggregates [30–47]. In practice, high temperature/short residence time (90–120°C/5 min to 2–4 sec) and intense shearing (≈5,000–40,000 s^{-1} shear rates) are used. These conditions create microparticles with homogeneous size distribution of the desired range. The homogeneity of size distribution may be further improved by preheating the dispersion at temperatures slightly lower than the denaturation temperatures of proteins. This ensures very fast denaturation during microparticulation [29].

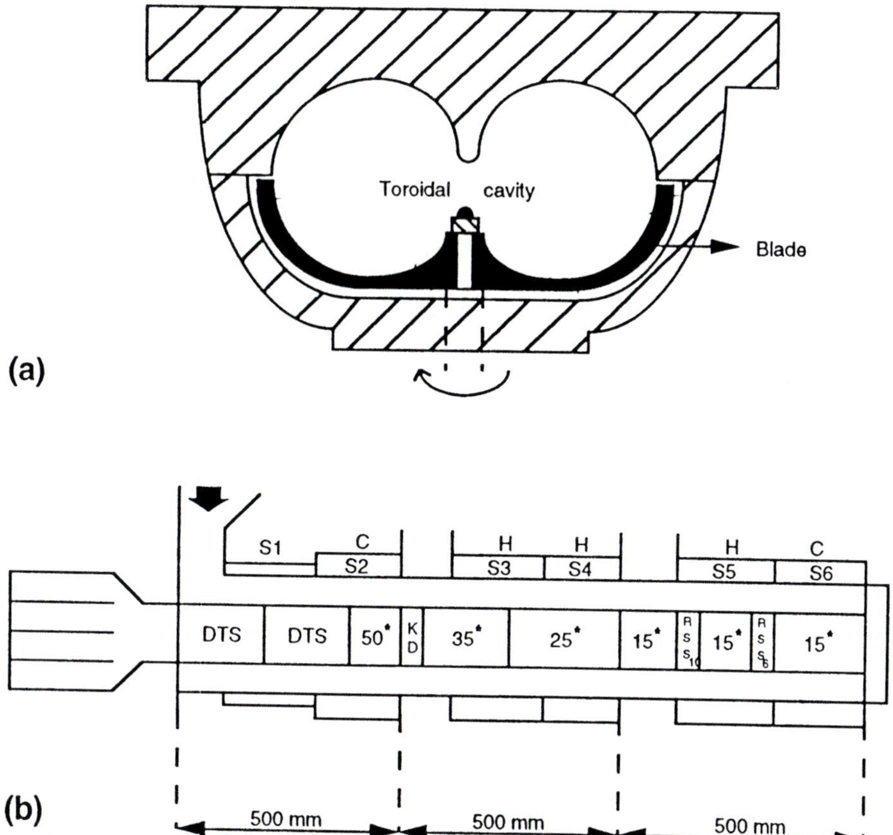

FIGURE 3 Equipment for microparticulation of proteins. (a) Scraped-surface heat exchanger specifically designed for highly viscous fluids or "microcooker." (Adapted from Ref. 44.) (b) Screw profile and heating arrangement of a twin-screw cooking extruder (Clextral BC45). S1 to S6, barrel sections 1–6; C, cooling; H, heating; DT, double thread segment; KD, kneading disks, RS6 or RS10, reverse screw segments (slits of 6 or 10 mm, respectively). 50* to 15* refer to the pitch of screw segments (in mm). Outlet plate (no die). (From Ref. 30.)

Turbulence, cavitation, and shear are also the main phenomena encountered in high-pressure homogenizers such as microfluidizers. The microfluidization technology is a unique type of high-pressure homogenization. The reaction chamber (Fig. 4) is built in such a way that the liquid is divided into two jet streams in which initial high-shear stress occurs, followed by a liquid/liquid contact area at an angle of 180° where the final high-pressure effect takes place. A process has been devised where microfluidization is carried out immediately after heat aggregation of proteins under quiescent conditions [31]. Pressures up to 750 bars (≈11,000 psi) at 25°C and 2–4 passes is normally sufficient to produce microparticles with a desirable size distribution. However, although the initial temperature of the fluid is 25°C, it rises by about 10–15°C after 4 passes at this high pressure. An excessive increase in temperature of microfluidization may favor appearance of microparticles larger than expected.

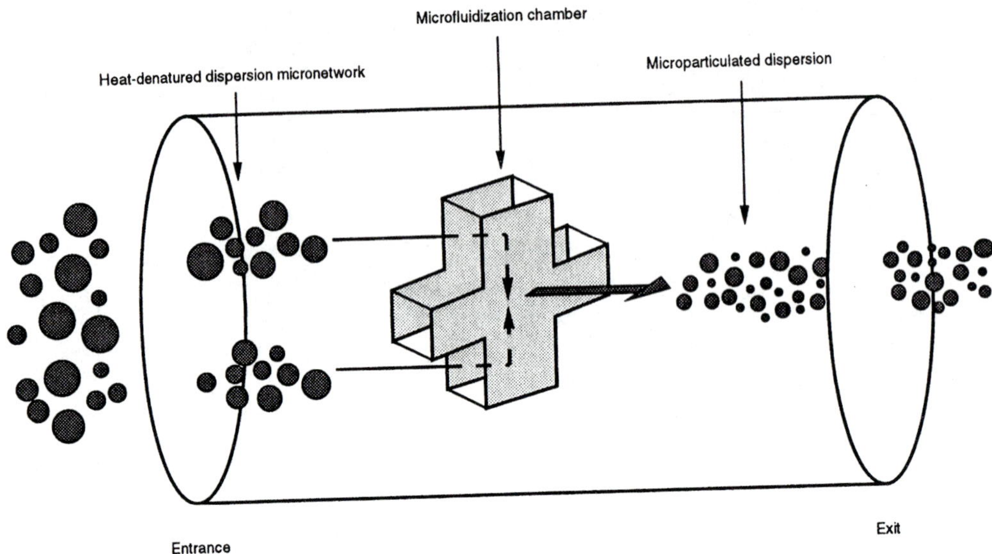

FIGURE 4 Reaction chamber of a high-pressure microfluidizer.

Protein-Polysaccharide Microparticulation

Protein-polysaccharide complexes usually are not spherical particles. They are mostly fibrous materials [21,22]. These fibrous materials, in the "native" state, are unable to mimic fat, and microparticulation (or microfragmentation in this particular case) is required to reduce their size and change their mouthfeel properties. Fibers are obtained by adjusting the pH of a diluted blend (3–7 wt% total solids) containing, for instance, egg white, milk, or soya proteins and anionic or cationic gums (protein:polysaccharide ratio ~2–15) to the pI of the complex. The pI is typically 3.7–4.2 for anionic and 5.5–6.3 for cationic gums [21]. The formation of fibrous coacervates usually begins from the start of pH adjustment, depending on the protein:polysaccharide ratio, and is a spontaneous process.

Usually fibrous coacervates are stable against further changes in pH and ionic strength, but they tend to redissolve when the pH is greater than one pH unit of the isoelectric point of the protein component [21]. Stabilization of the complex is achieved through extensive heat denaturation of the proteins in heat exchanger or in boiling water (90–100°C, 3–5 min) to cause more than 90% denaturation. In the complexed state, proteins are often more heat sensitive than they are in the native state [7,17]. Unfolding during heat treatment increases the number of accessible charged sites and maximizes interactions between proteins and polysaccharides via configurational adjustments [19]. Protein-protein interactions also play a very important role in the stabilization of coacervates, as evidenced by the fact that extensive heat denaturation produces highly rigid complexes. The texture of these complexes may be modified by other means, e.g., by varying the protein:polysaccharide ratio, by separately and carefully heating proteins and polysaccharides before mixing and pH adjustment (which produces softer and finer fibers), or by adding 20–40 w/w% nonionic polysaccharides (starch, agar, guar, or carob gum). The latter entangles with the complex and increases its water-binding properties.

After all operations have been completed, anisotropic particles approximately 2–10 μm in length are produced by repeated passes at pressures between 900–1200 bars (≈13,000–18,000 psi) through high-shear devices such as microfluidizers [21].

The sequence of treatments described above is not a rigid one. For instance, heating of the complex may be carried out after simultaneous pH adjustment and homogenization, or homogenization may follow simultaneous pH adjustment and heating [23]. For instance, whey protein concentrate (≈20 w/w% final protein concentration of the dispersion) and xanthan gum (0.8–2.0 w% gum) may be vigorously mixed with GDL to a final pH of 5.6–5.8, followed by heating with mixing at 93°C for 5 minutes. During heating, the hydrolysis of GDL accelerates and the final pH drops to about 4.1. After cooling and pH adjustment to 6.6 with food-grade alkali, the mixture is microfluidized at 700 bars (1–5 passes at 25°C). Interestingly, this process produces a very homogeneous dispersion of not linear but globular whey protein–xanthan gum microparticles, which could certainly be used as a fat replacer or texturing agent. Acidification by GDL during heating has been used to make mixed dairy gels [49]. GDL certainly modifies the kinetics of heat denaturation of the protein, since the structure of protein is a function of the pH. Proteins probably unfold at a faster rate and excessive aggregation is prevented through whey protein-xanthan gum interactions. The relationships between structure and functional properties of protein-polysaccharide microparticles are not very well understood, and additional research is needed in this area.

C. Posttreatment of Microparticles

The creaminess and fatty mouthfeel of microparticles depend on consistency, and therefore after microprecipitation the dilute dispersion is concentrated. The usual techniques used to concentrate microparticles are thin-film evaporation, ultrafiltration/diafiltration, microfiltration, ion-exchange chromatography, decanting, and centrifugation. Centrifugation (e.g., 5,000–20,000*g* for 20 min at 10–25°C) is a simple method often used in the recovery of microparticles. The efficiency of recovery is improved by acidification of the aqueous dispersion to pH 4–5 [21,24,26,50]. The suitability of these techniques depends on the ingredients present in the dispersion, the desired degree of concentration, and the effectiveness of removing undenatured or uncomplexed molecules [24].

Microparticle-based materials may be pasteurized, cooled, and/or spray-dried. Pasteurization must be carried out at high-temperature/short-time conditions because of the heat sensitivity of the aggregated proteins. Dynamic cooling of microparticles at temperatures lower than 10–20°C is probably one of the most important parameters affecting the textural properties of the final product. Indeed it has been shown that rapid cooling within a few minutes in a heat or SSHE improves hydration of protein-based aggregates and imparts high viscosity to the product [51,52]. On the other hand, static cooling favors inter- and intraprotein-protein interactions and leads to the development of a gel-like structure not suitable for fat replacement. Dynamic cooling must be performed at shear rates greater than 2000 s^{-1} to prevent aggregation of microparticles [28]. However, at such constant high-shear rates, very high-shear stresses develop in the viscous fluid and breakdown of large aggregates cannot be excluded. Whether this disruption markedly affects the properties of crude microparticulated material remains to be proven.

The shelf life of concentrated microparticle dispersions is usually low, typically 3–4 weeks at 5°C, but may be considerably improved, up to a year or more, by spray-drying if preservatives are added. As observed in our laboratory, spray-drying of a highly viscous dispersion containing 20–25% protein is very difficult, and dilution may become necessary. An atomizer inlet temperature of 200°C and an outlet temperature of 85–90°C are convenient for spray-drying. Good stability and dispersibility of the powder may be managed by addition to the initial dispersion of disaccharides such as lactose or saccharose [24]. A protein:disaccharide ratio of around 1:1 is adequate, but higher sugar levels may impair the spray-drying process (sticky powder), and the Maillard reaction between sugars and amines may also impair the quality of the final product. Drying this kind of particle is a very difficult task, and it is a critical step in the process because of the likelihood of aggregation. It is therefore important to check whether dried and rehydrated microparticles perform as effectively as the initial undried microparticulate preparation [25]. Despite this uncertainty, almost all commercially available protein-based fat replacers are sold in the dried form.

IV. FUNCTIONAL PROPERTIES OF MICROPARTICLES

At the present time, the main functionality expected of microparticles is that they emulate the structural and textural properties of fat. It is therefore of paramount importance to relate the structural and physicochemical properties of microparticles to their textural properties so as to optimize their use in a given food product. Moreover, in a number of cases, foods supplemented with microparticles undergo several treatments, such as heating, emulsification, foaming, and shearing. Consequently, it is of practical interest to evaluate the impact of these treatments on the functional properties of microparticulated material.

A. Structural Properties

It must be remembered first that protein and protein-polysaccharide microparticles are only one of the structural components in the microparticulated material. Other molecules such as proteins, lipids, polysaccharides, and emulsifiers are present as well. Fat replacers are composite materials, and their physical properties can be best understood by analyzing their structural properties at the ultra- and microstructure levels, the interactions that occur between the various components, and their tri-dimensional organization.

In general, microparticles adopt different shapes such as spheroidal, slightly elliptical, linear, or irregular, depending on the type of macromolecules used and processing conditions employed. Figure 5 shows the structure of some simple protein microparticles prepared from whey proteins. It is evident that the ultrastructure of the particles is made up of loosely connected globular aggregates. The porous structure of these particles allows water immobilization, which enhances their consistency. Water-binding properties of microparticles are essential to provide efficient replacement of emulsified fat in food products. Microparticles reduce the amount of water in the continuous phase of reduced-fat food products compared to that found in the full-fat products [53]. The microparticles are not always very homogeneous, as shown in Figure 5. In some cases, particles with a dense outer coat of protein and a less dense interior with regions devoid of protein are formed [12]. The ultrastructure of composite complexes, i.e., protein-protein and protein-polysaccharide microparticles, is much less well known, and the molecular arrangement of the components within a particle remains to be characterized. However, immunolocalization methods have provided interesting results as far as identifying, for instance, a layer of heat-denatured egg white protein wrapped around casein micelles [54]. Also,

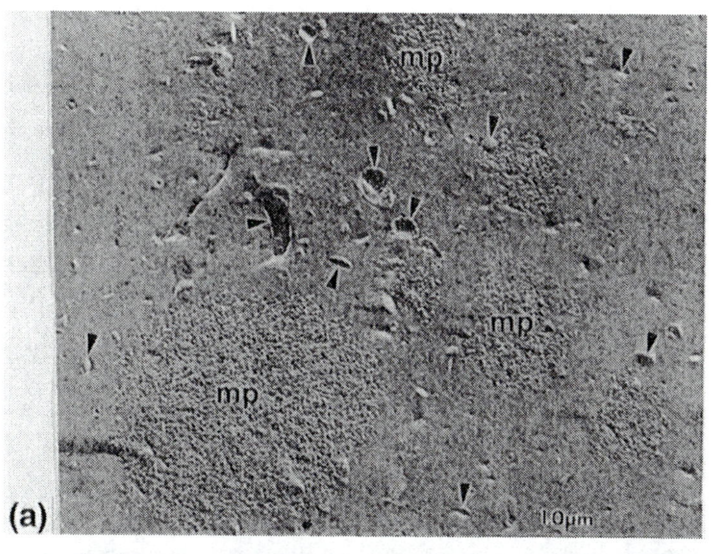

FIGURE 5 Microstructure of protein microparticles. (a) Freeze-fracture electron micrograph of microparticles concentrate (Simplesse 100, Nutrasweet Co.). Note the numerous small lipid particles/vesicles (arrows). (b) Freeze-etch micrograph of isolated protein microparticles (Simplesse 100, Nutrasweet Co.). (From Ref. 71.)

transmission electron microscopy has revealed that whey protein–egg white protein–xanthan microfragmented complexes contain tangled masses of strands of globular protein aggregates lined up along the backbone of xanthan molecules [21,53].

The size distribution of microparticles lies in the range of 0.1–75 μm (Table 2), but generally 85–95% of the particles are 1–3 μm in diameter or length. Weak interactions, such as hydrophobic, electrostatic, and hydrogen bonding, stabilize the micro-macroscopic structures of microparticulated materials. This structure may be conveniently estimated both by microscopic and rheological methods. Figure 6 shows the

TABLE 2 Size Distribution of Protein-Based and Protein-Polysaccharide Complex–Based Microparticles

Microparticulated material	pH of manufacturing	Size distribution of microparticles (μm)	Mean size (d_{VS} = μm)	Ref.
Simple microparticles				
WPC	3.7–4.2	0.1–3	n.r.	5
WPI	4.3–5.0	0.5–50	2	50
SPC	5.3	0.1–20	11.8	28
WPC	4.0		4.5	
WPI	4.8	0.1–10	3.6	24
WPI	3.9	1–75	11.5	30
	4.0	1–175	16.3	
WPC	5.8–6.9	0.1–2	n.r.	29
WPC	4.0	0.5–75	4.8	31
Composite microparticles				
Egg white proteins				
Gelatin	6.0	0.2–2	1–2	12
Casein micelles	6.2–6.6	0.1–3	n.r.	8
Dairy and egg white proteins/xanthane	3.7–4.2 5.5–6.3	2–10	5	21
Gelatin–acacia gum–pectin	3.8–4.5	2–40	8	11
WPI–calcium caseinate	6.7	1–200	8–10	25
	5.0–5.9	1–20	n.r.	32

WPI, whey protein isolate; WPC, whey protein concentrate; SPC, soya protein concentrate; n.r., not revealed; d_{VS} = volume-surface diameter.

structures of two microparticulated materials prepared from whey protein-xanthan and whey protein–κ carrageenan complexes [23]. The former reveals a homogeneous dispersion of spherical particles (~2–4 μm in diameter) interacting in a weak network, which emphasizes the dispersing properties of the xanthan gum. The latter presents a more disordered structure with loose and relatively large aggregates (~5–7 μm in diameter) composed of smaller subunits. Although both materials have a similarly unctuous texture, the structural differences greatly influence their syneresis. The whey protein–xanthan complex only expels 1% liquid after centrifugation at 1000*g* for 15 minutes, whereas about 24% liquid is expelled from the whey protein–κ carrageenan complex under the same conditions.

Most protein and protein-polysaccharide fat replacers are concentrated dispersions. It has been shown that these dispersions have viscoelastic properties characterized by a balance between the elastic and the viscous components. This balance usually prevents formation of large aggregates or a gel-like structure during storage [25]. Microparticle dispersions also exhibit a shear-thinning flow behavior [21,25], that is, their viscosity decreases with increase in shear rate. This property ensures good dispersibility of microparticles. It would also be interesting to verify if the material has time-dependent flow properties, especially thixotropic behavior. This would enable prediction of restructuring characteristics of microparticles in food products.

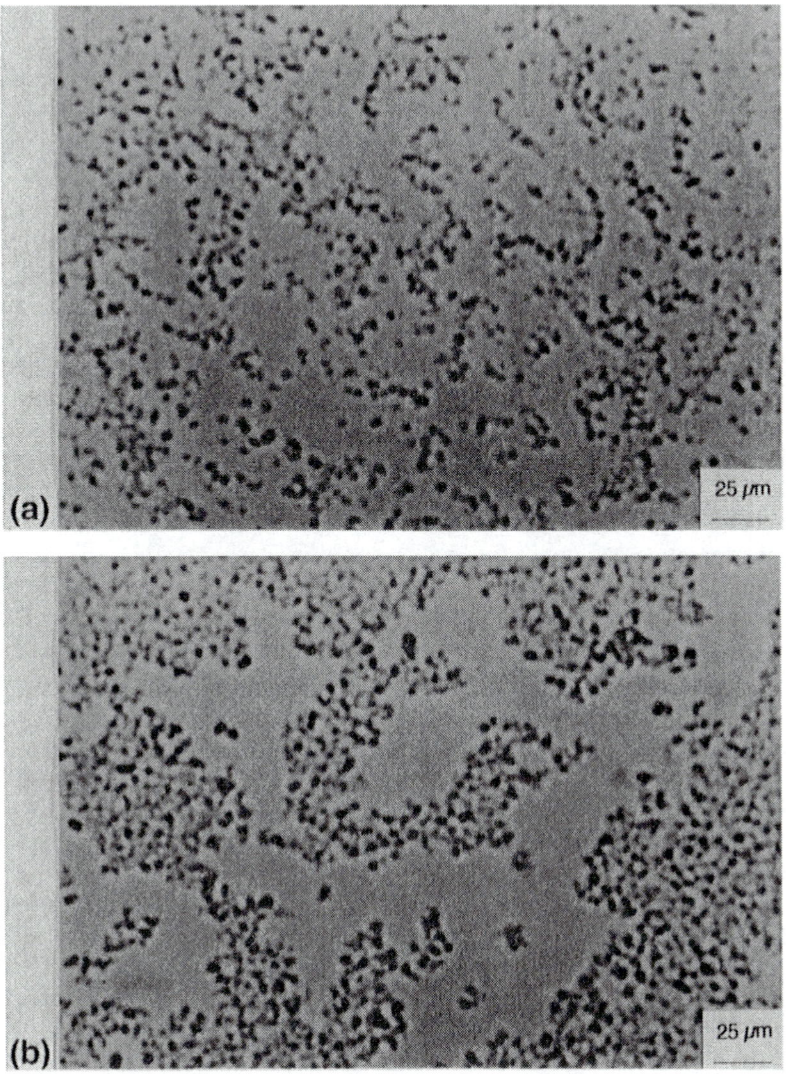

FIGURE 6 Microstructure of protein-polysaccharide microparticles. (a) Optical micrograph of whey protein–xanthan microparticles. (b) Optical micrograph of whey protein–carrageenan microparticles.

B. Sensorial Properties

Sensorial properties of food products include their appearance, flavor, and texture. Food texture is probably the attribute most important to consumers [55]. Textural properties include mechanical (related to stress and strain), geometrical (related to the size, shape, and orientation of particles), and moisture properties (related to the perception of moisture or fat) [56].

Like emulsified fat, microparticulated material gives a creamy and smooth texture to a number of food products (e.g., dairy products). High viscosity and absence of powderiness are the two important factors for creaminess, as demonstrated in soups by Wood [57]. The small size, concentration, hydration properties of microparticles, and presence of water-binding agents (protein, polysaccharide) adequately satisfy these two requirements. The viscous solution controls the flow of microparticles (imparting lubricity) and the time required for perception of creaminess [25,57]. The size of microparticles should preferably be close to that of fat globules for viscous liquids or pastry products, usually in the range of 0.3–3.0 μm [54], although a small proportion (~2–10%) of larger particles (3–175 μm) is acceptable (see Table 2), especially for texturized products, such as cheeses. The threshold for detectability of particles in food products is thought to be about 30–60 μm diameter [53,58]. However, this threshold depends on the properties of particles, on the interactions between particles and other components of products, and also on the viscosity of the continuous phase [53]. For instance, particle diameter larger than 40 μm may be undetectable in a viscous solution, since they would feel chalky or gritty in most dairy products [53]. Table 3 gives an estimate of the relationships between the size distribution and textural attributes of concentrated dispersions of microparticles. Shape and deformability of microparticles could also play a role in determining sensory properties of fat-reduced food products [53].

The mechanical properties of animal fat, such as milk fat, are greatly dependent on the temperature. Milk fat globules are soft at 40°C, semi-solid at 10–15°C, and solid at <5°C, depending on the melting temperature of triglycerides. These melting properties of milk fat markedly contribute to the palatability of dairy foods. Unfortunately, although protein microparticles are deformable, they do not melt at high temperatures. Moreover, it is possible that high temperatures may cause aggregation and dehydration of microparticles, which may increase their "rigidity". Protein microparticles compare better with milk fat globules at temperatures lower than 15°C. This is the reason for the widespread use of protein microparticles in low-fat dairy products and frozen desserts (see Sec. V). On the other hand, some protein–polysaccharide complexes have melting properties that may be manipulated to the convenience of food manufacturers. The melting point of gelatin–acacia gum complexes, for instance, is about 32°C, whereas that of gelatin-pectin complexes is about 70°C. These higher melting temperatures considerably widen the complexes' potential use compared to that of protein-based microparticles.

Generally, foods containing no fat develop different aroma profiles than the same foods with fat. The contribution of microparticles to the aroma profile of foods is poor. Protein-based microparticles seemingly have some flavor-binding properties, but much

TABLE 3 Relationships Between the Size Distribution and Sensory Attributes of Concentrated Dispersions of Microparticles

Size distribution of microparticles (μm)	Sensory attributes
0.01–0.1	Viscous, empty texture
1–2	Creamy
3–5	Powdery
8–10	Gritty

Source: Adapted from Ref. 54.

fewer than fat [59]. Therefore, a minimum fat content in food products is necessary to improve the flavor profile, and changes in flavor formulation of foods containing microparticles seems presently unavoidable.

C. Aggregation and Gelation Properties

It is a well-established fact that microparticulated proteins break down during thermal processing, forming larger aggregates, which impart grittiness. It is known that heat-aggregated proteins possess increased surface hydrophobicity [39,60], that heat favors protein-protein hydrophobic interactions, and that protein aggregates (both soluble and insoluble) probably act as nucleating agents for aggregation of undenatured proteins [48]. Based on these facts, it seems that formation of large aggregates of microparticles during thermal processing might be caused by the denatured and aggregated proteins present in the microparticulated material. However, neither the mechanism nor the precise role of soluble or insoluble protein aggregates in the heat sensitivity of protein microparticles has been elucidated.

Although protein microparticles aggregate upon heating, they do not form a gel. For example, Hung and Smith [61] studied the dynamic viscoelastic properties of a whey protein microparticle dispersion (16 wt% protein containing 73% insoluble aggregates) heated at 90°C for 15 minutes. They found that the storage (G′) and loss (G″) moduli, which represent the solidlike (elastic) and viscous components of a viscoelastic material, respectively, did not change during heating, indicating a lack of network formation. As suggested by Beuschel et al. [62], insoluble whey protein aggregates do not unfold to form a gel matrix on heating, but might act as rigid fillers to increase the rigidity of other food gels.

The thermal stability of microfragmented xanthan–whey protein and xanthan–whey protein-egg white protein complexes has been investigated by Weng-Sherng et al. [21]. It appears that the apparent viscosity of these dispersions (7–15 wt% total solid) increases with heating, indicating aggregation of the complexes. An interesting observation concerns the relationship that exists between the initial viscosity of dispersions and the increase in viscosity after heating. An initial high viscosity may be indicative of subsequent heat gelation of microfragmented material. The prediction of thermal stability of such systems from preliminary viscosity measurements is promising.

As discussed previously, microparticulated proteins and protein-polysaccharide complexes develop viscoelastic properties. In some cases, the balance between the elastic and viscous components is disturbed during storage at low temperature, which leads to an increase in firmness of the product, caused probably by cold "gelation" [25]. This structuring is quite usual for a number of milk protein–based systems, such as acid-type curds or infant formula, and is mediated by protein-protein interactions involving electrostatic interactions and hydrogen bonding.

D. Surface Properties

Food proteins are amphoteric molecules composed of charged groups at their surface and hydrophobic groups buried in the interior. When they approach a newly created oil/water interface (like at the air/water interface), they unfold to expose their hydrophobic groups to the lipid phase and decrease the interfacial tension of the system. The extent of unfolding that occurs at the interface depends on environmental conditions (pH, ionic strength), the rigidity of the three-dimensional protein structure, and the number and location of hydrophobic groups in the molecule [63]. A flexible, non–cross-linked protein (e.g., caseins) unfolds and adsorbs more easily than does a highly structured cross-linked one (e.g., β-lactoglobulin, bovine serum albumin). The former facilitates creation of a large fat surface area, but the latter forms less deformable, stronger films, which prevent coalescence of fat globules and thereby stabilize emulsions.

These properties have led a great number of researchers to try to improve the surface properties of proteins through physical treatments such as partial heat denaturation [64–67]. Because of an increase in the hydrophobic character of the protein surface, partially denatured proteins readily adsorb to interfaces [64]. However, the lack of flexibility of a protein-based aggregate leads to the need for higher protein concentration at the interfaces to adequately decrease interfacial tension. In general, protein aggregates exhibit poorer emulsifying (and foaming) properties than the native proteins, but produce more stable emulsions through extensive protein-protein interactions and development of a network-type structure at interfaces [65,66,68]. In fact, an optimum ratio of native protein to soluble protein aggregate seems to be essential for enhancing both emulsifying capacities of proteins and the stability of the emulsions [67,69]. The soluble aggregate–to–insoluble aggregate ratio is also an important parameter affecting the formation and stabilization of emulsions [70].

Microparticulated proteins are a mixture of native proteins and soluble and insoluble aggregates. Each of these protein fractions may probably possess some interesting surface properties. The effect of nonproteinaceous components on the surface properties of microparticulated material also needs to be studied. In particular, fatty emulsifiers used as aggregate extenders could certainly compete or interact with protein aggregates during interfacial adsorption [65]. Similarly, residual lipids like those organized in "pseudomembranes" (Fig. 5) may have an effect on the surface properties of microparticulated materials [71]. Information on the interfacial properties of protein-polysaccharide complexes is not well known. However, some evidences indicate that the emulsifying capacity and stability increase when a protein forms a complex with a polysaccharide [7]. This association probably reduces the conformational stability of proteins, resulting in a large number of contacts between protein and polysaccharide molecules. This may facilitate exposure of a greater number of hydrophobic groups to the interface and also formation of a thick interfacial film. The higher stability of these emulsions than those of protein-stabilized emulsions is also attributable to the increased number of charges in the adsorption layer arising from ionizable groups of polysaccharide molecules [7]. Because of the scarcity of data, a clear understanding of molecular factors that affect the surface properties of protein-polysaccharide complexes is not possible, and more investigations in this area are needed.

V. FOOD APPLICATIONS OF MICROPARTICLES

Table 4 gives some suggested food applications for microparticles. Protein microparticles presently are used in frozen and, to a limited extent, in thermally processed food products. The main reasons for their limited use in other products are their susceptibility to heat-induced aggregation and their inability to exactly simulate the melting and flavor properties of fat at temperatures above 15–20°C. Protein-polysaccharide microparticles are more stable (especially those from microfragmentation processes) and impart better lubricity to food products than protein-based microparticles. Because of their meltability,

TABLE 4 Food Applications of Protein- or Protein-Polysaccharide–based Microparticle Fat Replacers

Fat replacer	Food product category	Examples
Protein microparticles	Dairy products	Ice cream, yogurt, cream, whippable cream, spread, Cheddar, Gouda, cream cheese and cream cheese spread, fresh cheese
	Oil-based products	Salad dressing, French dressing, mayonnaise
Protein-polysaccharide microparticles	Dairy products	Ice cream, yogurt, cream, whippable cream, spread, Cheddar, Gouda, cream cheese and cream cheese spread, fresh cheese
	Oil-based products	Salad dressing, French dressing, mayonnaise
	Confectionary	Fudge, caramel, marshmallow, nougat, starch jellies, sweet chocolate product
	Baked goods	Sweet bread, sweet roll, coffee cake, donut, pastry, pie shell, icing, topping, filling
	Comminuted meat, meat analog products	Sausage, hot dog, meat loaves

they can be used in a wide range of food products. The sensory properties of reduced-fat and no-fat food products containing microparticles have been described as texturally close to the real ones. However, this is more true for reduced-fat than for no-fat food products; in the latter, the differences are very pronounced. When fat is partially or fully replaced with microparticles, the composition of a food product is quite different from that of its full-fat counterpart. Thus, despite the claims that such food products are texturally close to traditional ones, it would be more appropriate to refer to them as "new food products," distinguished by new composition, functional properties, and stability.

The stability of a food product, that is, the occurrence or lack of structural and/or textural changes during storage, largely determines its acceptability to consumers. For example, in foamed/whipped products like whipped cream or ice cream, the fat phase, because of its stabilizing effect on air cells, contributes decisively to the structure, texture, and stability (shelf life) [71]. Microparticles cannot exhibit these functional properties because of their hydrophilic nature. Moreover, they seem to modify the extent of air incorporation in frozen desserts as well as their draining characteristics [72], two important parameters affecting their quality. These examples are symptomatic of the difficulty of using microparticles as fat replacers in food products in which fat plays a critical function other than imparting creaminess. Clearly, microparticles may adequately mimic fat in viscous liquids (e.g., pourable dressing, liquid cream) and pastes (e.g., spread, cheese spread), in which the structural order of dispersed and continuous phases is not highly complex and creaminess is the only main criterion for acceptance. This, however, may not be the case in gellike materials (e.g., semi-hard and high-fat soft cheeses), where fat actively contributes to their structure and stability.

The structure of cheddar, cottage, and cream cheeses consists mainly of a three-dimensional network of milk protein–milk fat globule aggregates of various sizes and shapes, in which the serum phase is immobilized. Emulsified fat acts as a "pseudoprotein" because of the proteinaceous film around it and participates in the milk coagulation process via copolymerization in the early stages of cheesemaking. It also functions as an "active filler" in the gel matrix of cheese. In semi-hard cheeses (e.g., Cheddar, Gouda), these structural characteristics greatly influence their fracture properties, which in turn affect their sensorial perception. Paquin et al. [31] have shown that protein microparticles are retained in the Cheddar cheese matrix because they fit into the milk protein network (Fig. 7). Microparticles function as "inert fillers," and, when used in cheeses, they increase the moisture content as a result of poor drainage of the curd. Because of this, the fracturability of such cheeses is different from those of regular cheeses [73]. In cream cheese, the network is weaker than in semi-hard cheese, and this weak structural organization improves fracturability and decreases shear stress resistance (allowing for good spreadability). The "yield stress" of milk protein–milk fat globule aggregates depends on the level of interaction between the proteinaceous film of fat globules and aggregated proteins. It is doubtful that microparticles can take part in such interactions. This yield stress is particularly important for cream cheese curd undergoing dynamic cooling after high-pressure homogenization [74]. During homogenization, fat globules cluster because of an insufficient amount of surface-active material available to stabilize the oil/water interface. During dynamic cooling in a scraped-surface heat exchanger, the curd is submitted to high-shear stresses, which break the fat globule aggregates and reduce the degree of aggregation in the curd. It is this controlled disintegration

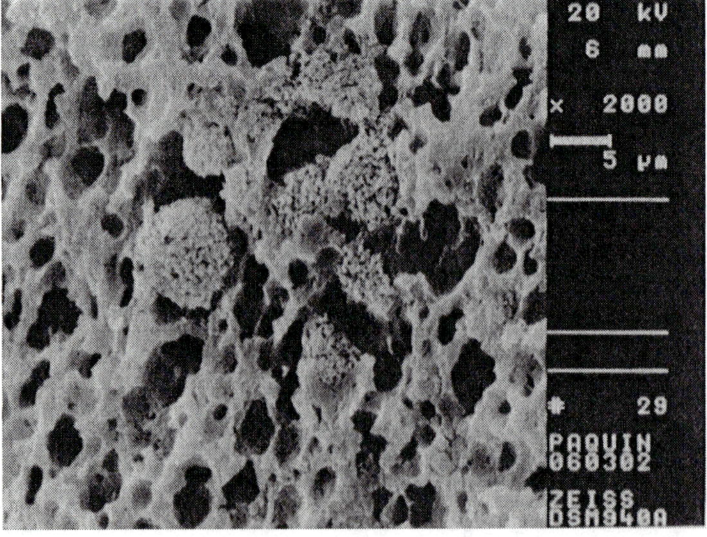

FIGURE 7 Electron micrograph (SEM) of whey protein microparticles inserted into a Cheddar cheese matrix.

of aggregates that ensures optimum textural properties of cheeses. However, if protein microparticles adsorb to the fat globule clusters formed during homogenization, they may weaken the shear stress resistance during dynamic cooling, which may result in a highly unstable product [75]. Moreover, in a normal process, fat globules are greatly destabilized and often disintegrated, which means that the structure and texture of regular cheeses are also determined by fat crystallization in the bulk. The same is true for most dairy products (e.g., cream, ice cream) that are subject to severe mechanical treatments. Microparticles definitely cannot match these properties.

VI. CONCLUSIONS AND FUTURE DEVELOPMENTS

A number of protein- and protein-polysaccharide–based fat substitutes are presently available and are being used in a wide range of low-fat or no-fat products. The textural properties of these microparticle-containing foods compare quite satisfactorily with their full-fat counterparts at low temperatures ($T \leq$ 15–20°C), but not at high temperatures ($T \geq$ 15–20°C), mainly because of their poor melting properties. Moreover, they are prone to aggregation when heated, which limits their use in ready-to-eat foods. Another relevant observation is that fats often play a greater role in the structure development and stabilization of highly structured, high-fat foods. Microparticles cannot mimic fat in such foods. The inability of microparticles to impart to food products the taste of fats explains the low popularity of low-fat or no-fat products in the marketplace.

Despite these drawbacks, microparticles have demonstrated an ability to mimic fat without drastically altering the texture or taste of food products. Microparticles (especially those based on protein-polysaccharide complexes) may find wide use in the future as fat replacers in foods that are traditionally low in fat, provided their aroma profile is reformulated. At present, however, microparticles will likely be used to a greater extent as texturing agents to modify existing or create entirely new and innovative food structures and textures. New technologies must be developed to produce a new generation of microparticles that closely mimic the texture and taste of fats. Recent preliminary investigations have shown that fat can be immobilized in the porous structure of microparticulated heat-denatured proteins. These protein-fat composite complexes could certainly be used as fat substitutes.

REFERENCES

1. A. Drewnoski, Fats and food texture, *Food Texture* (H. R. Moskowitz, ed.), Marcel Dekker Inc., New York, 1987, p. 251.
2. B. F. Hauffmann, Dietary fats still spark controversy, *Inform* 5:346 (1994).
3. B. L. Armbruster and N. Desai, Identification of proteins and complex carbohydrates in some commercial low-fat products by means of immunolocalization techniques, *Food Structure 12*: 289 (1993).
4. J. E. Vanderveen and W. H. Glinsman, Fat substitutes: A regulatory perspective, *Ann. Rev. Nutr. 12*:473 (1992).
5. N. S. Singer, S. Yamamoto, and J. Latella, Protein product base, European Patent Application 0,250,623, John Labatt Ltd. (1988).
6. N. K. Howell, Protein-protein interaction, *Developments in Food Proteins—7* (B. J. F. Hudson, ed.), Applied Science Publishers, London, 1991, p. 291.
7. V. B. Tolstoguzov, Functional properties of food proteins and role of protein-polysaccharide interaction, *Food Hydrocolloids 4*:429 (1991).

8. N. S. Singer, R. Wilcox, J. S. Podolski, H. H. Chang, S. Pookote, J. M. Dunn, and L. Hatchwell, Cream substitute ingredient and food products, *World Patent Applications WO89/05587*, 1989a, (Nutrasweet Co.).
9. C. Queguiner, E. Dumay, and J. C. Cheftel, Compositions protéiques à texture onctueuse, French Patent Application 91/05757, Université Montpellier II (1991).
10. K. Shimada and J. C. Cheftel, Texture characteristics, protein solubility, and sulfhydryl group/disulfide bond contents of heat-induced gels of whey protein isolate, *J. Agric. Food Chem. 36*:1018 (1988).
11. M. A. Bakker, M. M. Koning, and J. Visser, Fatty ingredient, World Patent Application WO94/14334, Unilever PLC, (1993).
12. G. R. Ziegler, Microstructure of mixed gelatin-egg white gels: Impact on rheology and applications to microparticulation, *Biotech. Prog. 7*:283 (1991).
13. M. Glicksman, Gelling hydrocolloids, *Food Product Applications* (J. M. V. Blanshard and J. R. Mitchell, eds.), Butterworths, London, 1979, p. 185.
14. N. Garti and D. Reichman, Hydrocolloids as food emulsifiers and stabilizers, *Food Struct. 12*:411 (1993).
15. E. Dickinson, The role of hydrocolloids in stabilising particulate dispersions and emulsions, *Gums and Stabilisers for the Food Industry—4* (G. O. Phillips, D. J. Wedlock, and P. A. Williams, eds.), IRL Press, Oxford, 1988, p. 249.
16. S. K. Samant, R. S. Singhal, P. R. Kulkarni, and D. V. Rege, Protein-polysaccharide interactions: a new approach in food formulations, *Int. J. Food Sci. Technol. 28*:547 (1993).
17. D. A. Ledward, Protein-polysaccharide interactions, *Food Product Applications* (J. M. V. Blanshard and J. R. Mitchell, eds.), Butterworths, London, 1979, p. 205.
18. A. J. Ganz, How cellulose gum reacts with proteins, *Food Eng. 46*:67 (1974).
19. A. P. Imeson, D. A. Ledward, and J. R. Mitchell, On the nature of interactions between some anionic polysaccharides and proteins, *J. Sci. Food Agric. 28*:661 (1977).
20. J. M. Rispoli, J. P. Sabhlok, A. S. Ho, B. G. Schever, and C. Giuliano, Oil replacement composition, U.S. Patent 4,308,294, General Foods Co. (1981).
21. C. Weng-Sherng, G. A. Henry, S. M. Gaud, M. S. Miller, J. M. Kaiser, E. A. Balmadeca, R. G. Morgan, C. C. Baer, R. P. Borwanker, L. C. Helgeth, J. J. Strandholm, G. L. Hasenheuttl, P. J. Kerwin, C. C. Chen, J. F. Kratochvil, and W. L. Lloyd, Microfragmented ionic polysaccharide/protein complexes dispersions, European Patent Application 0,340,035, Kraft Inc. (1989).
22. M. Paquot, F. Godin, and Q. Dumont de Chassart, Effects of electrokinetic characteristics on the formation of a protein-microcrystalline cellulose complex, *Lebenm- Wiss.u- Technol. 27*: 11 (1994).
23. J. Lefebvre and Urineroité Lora, Study of the different parameters for the production of polysaccharide/protein complexes, Ménidie de maitise (1995).
24. M. J. Hakaart, A. Kunst, and E. Leclercq, Edible compositions of denatured whey proteins, European Patent Application 0,412,590, Unilever NV (1991).
25. J. C. Cheftel and E. Dumay, Microcoagulation of proteins for development of "creaminess," *Food Rev. Int. 9*:473 (1993).
26. J. Mai, D. Breitbart, and D. Fisher, Proteinaceous material, European Patent Application 0,400,714, Unilever NV (1990).
27. L. E. Stark and A. T. Gross, Hydrophobic protein microparticles and preparation thereof., World Patent Application WO90/03123, Enzytech Inc. (1990).
28. C. R. T. Brown, I. T. Norton, and P. Wilding, Protein product, European Patent Application 0,352,144, Unilever NV (1990).
29. C. S. Fang and R. Snook, Proteinaceous fat substitute, World Patent Application WO90/17665, Nutrasweet Co. (1991).
30. C. Queguiner, E. Dumay, C. Salou-Cavalier, and J. C. Cheftel, Microcoagulation of a whey protein isolate by extrusion cooking at acid pH, *J. Food Sci. 57*:610 (1992).

31. P. Paquin, Y. Lebeuf, J. P. Richard, and M. Kalab, Microparticulation of milk proteins by high pressure homogenization to produce a fat substitute, *Protein and Fat Globule Modifications by Heat Treatment, Homogenization and Other Technological Means for High Quality Dairy Products*, Special Issue no. 9303, International Dairy Federation, Brussels, 1993, pp. 389–396.
32. A. J. McCarthy and J. W. Maegli, Protein fat replacer and method of manufacture thereof, U.S. Patent 5,350,590, Beatreme Foods Inc. (1994).
33. M. Habib and J. S. Podolski, Concentrated, substantially non-aggregated casein micelles as a fat/cream substitute, European Patent Application 0,345,226, Nutrasweet Co. (1989).
34. I. Hayakawa, Y. Yamada, and Y. Fujio, Microparticulation by jet mill grinding of protein powders and effects on hydrophobicity, *J. Food Sci. 58*:1026 (1993).
35. A. M. Hermansson, Aggregation and denaturation involved in gel formation, *Functionality and Protein Structure* (A. Pour-El, ed.), American Chemical Society, Washington, DC, 1979, p. 81.
36. A. Kilara and T. Y. Sharkasi, Effects of temperature on food proteins and its implications on functional properties, CRC Crit. Rev. Food Sci. Nutr. 23:323 (1986).
37. P. R. Khun and E. A. Foegeding, Mineral salt effects on whey protein gelation, *J. Agric. Food Chem. 39*:1013 (1991).
38. Y. L. Xiong, Influence of pH and ionic environment on thermal aggregation of whey proteins, *J. Agric. Food Chem. 40*:380 (1992).
39. N. Kitabake and E. Doi, Improvement of protein and gel by physical and enzymatic treatment, *Food Rev. Int. 9*:445 (1993).
40. A. J. Steventon, L. F. Gladden, and P. J. Fryer, A percolation analysis of the concentration dependence of the gelation of whey protein concentrates, *J. Texture Stud. 22*:201 (1991).
41. A. M. Hermansson, Physico-chemical aspects of soy proteins structure formation, *J. Texture Stud. 9*:33 (1978).
42. P. Jelen and W. Buchheim, Stability of whey protein upon heating in acidic conditions, *Milchwissenschaft 39*:215 (1984).
43. J. Patocka, M. Drathen, and P. Jelen, Heat stability of isolated whey protein fractions in highly acidic conditions, *Milchwissenschaft 42*:700 (1987).
44. N. S. Singer, J. Latella, and S. Yamamoto, Protein product, European Patent Application 0,323,529, John Labatt Ltd. (1989).
45. P. Pradipasena and C. Rha, Pseudoplastic and rheopectic properties of a globular protein (β-lactoglobulin) solution, *J. Texture Stud. 8*:311.
46. Y. C. Ker and R. T. Toledo, Influence of shear treatments on consistency and gelling properties of whey protein isolate suspensions, *J. Food Sci. 57*:82 (1992).
47. A. J. Steventon, Thermal aggregation of whey proteins, Ph.D. thesis, University of Cambridge, Cambridge, 1993.
48. S. M. Taylor and P. J. Fryer, The effect of temperature/shear history on the thermal gelation of whey protein concentrates, *Food Hydrocolloids 8*:45 (1994).
49. J. M. Aguilera, Y. L. Xiong, and J. E. Kinsella, Viscoelastic properties of mixed dairy gels, *Food Res. Int. 26*:11 (1993).
50. J. Visser and M. A. Bakker, Edible plastic compositions, European Patent Application 0,347,237, Unilever NV (1989).
51. T. Hori, Effects of freezing and thawing green curds on the rheological properties of cream cheese, *J. Food Sci. 47*:1811 (1982).
52. C. Sanchez, J. L. Beauregard, M. H. Chassagne, J. J. Bimbenet, and J. Hardy, Rheological and textural behaviour of Double cream cheese. Part 2: Effect of curd cooling rate, *J. Food Eng. 23*:595 (1994).
53. M. S. Miller, Proteins as fat substitutes, *Protein Functionality in Food Systems* (N. S. Hettiarachchy and G. R. Ziegler, eds.), Marcel Dekker Inc., New York, 1994, p. 435.

54. N. S. Singer and J. M. Dunn, Protein microparticulation process: The principle and the process, *J. Am. Coll. Nutr.* 9:388 (1990).
55. A. Drewnoski, Sensory properties of fats and fat replacements, *Nutr. Rev.* *50*:17 (1992).
56. G. V. Civille, The sensory properties of products made with microparticulated protein, *J. Am. Coll. Nutr.* 9:427 (1990).
57. F. W. Wood, An approach to understanding creaminess, *Die Stärke* *26*:127 (1974).
58. S. A. Matz, Unctuous foods, *Food Texture*, Avi, Westport, CT, 1962, p. 126.
59. J. P. Schirle-Keller, H. H. Chang, and G. A. Reineccius, Interaction of flavor compounds with microparticulated proteins, *J. Food Sci.* *57*:1448 (1992).
60. S. Nakai, Importance of protein functionality in improving food quality: Roles of hydrophobic interaction, *Comments Agric. Food Chem.* 6:339 (1992).
61. T. Y. Hung and D. M. Smith, Dynamic rheological properties and microstructure of partially insolubilized whey protein concentrate gels, *J. Food Sci.* *58*:1047 (1993).
62. B. C. Beuschel, J. A. Partridge, and D. M. Smith, Insolubilized whey protein concentrate and/or chicken salt-soluble protein gel properties, *J. Food Sci.* *57*:852 (1992a).
63. M. E. Mangino, Properties of whey protein concentrates, *Whey and Lactose Processing* (J. G. Zadow, ed.), Elsevier, London, 1992, p. 232.
64. K. P. Das and J. E. Kinsella, Effect of heat denaturation on the adsorption of β-lactoglobulin at the oil/water interface and on the coalescence stability of emulsions, *J. Colloid Interface Sci.* *139*:551 (1990).
65. M. Britten and H. J. Giroux, Emulsifying properties of whey protein and casein composite blends, *J. Dairy Sci.* *72*:3318 (1991).
66. E. Dickinson and S. T. Hong, Surface coverage of β-lactolobulin at the oil/water interface: Influence of protein heat treatment and various emulsifiers, *J. Agric. Food Chem.* *42*:1602 (1994).
67. H. Zhu and S. Damodaran, Heat-induced conformational changes in whey protein isolate and its relation to foaming properties, *J. Agric. Food Chem.* *42*:846 (1994).
68. H. Singh, P. F. Fox, and M. Cuddigan, Emulsifying properties of protein fractions prepared from heated milk, *Food Chem.* *47*:1 (1993).
69. M. Britten, H. J. Giroux, Y. Jean, and N. Rodrigue, Composite blends from heat-denatured and undenatured whey protein: Emulsifying properties, *Int Dairy J.* *4*:25 (1994).
70. B. C. Beuschel, J. D. Culbertson, J. A. Partridge, and D. M. Smith, Gelation and emulsification properties of partially insolubilized whey protein concentrates, *J. Food Sci.* *57*:605 (1992b).
71. W. Buchheim and W. Hoffman, Microparticulated proteins as fat substitutes, *Europ. Dairy Mag.* 6:30 (1994).
72. K. Schmidt, A. Lundy, J. Reynolds, and L. N. Yee, Carbohydrate or protein based fat mimicker effects on ice milk properties, *J. Food Sci.* *58*.761 (1993).
73. R. C. Lawrence, Incorporation of whey proteins in cheese, *Factors Affecting the Yield of Cheese*, Special Issue no. 9301, International Dairy Federation, Brussels, 1993, pp. 79–81.
74. C. Sanchez, J. L. Beauregard, M. H. Chassagne, A. Duquenoy, and J. Hardy, Effect of processing on rheology and stability of double cream cheese, *Food Res. Int.* (in press).
75. C. Sanchez, J. L. Beauregard, J. J. Bimbenet, and J. Hardy, Rheological properties of double cream cheese containing whey protein concentrate, *J. Food Sci.*, (in press).

18

Edible Protein Films and Coatings

JOHN M. KROCHTA
University of California, Davis, California

I. INTRODUCTION

Edible films and coatings based on proteins, polysaccharides, and/or lipids have much potential for increasing food quality and reducing food-packaging requirements [1–10]. Edible films formed as coatings or placed (preformed) between food components provide possibilities for improving the quality of heterogeneous foods by limiting the migration of moisture, lipids, flavors/aromas, and colors between food components. Edible coatings also have the potential for maintaining the quality of foods after the packaging is opened by protecting against moisture change, oxygen uptake, and aroma loss. In addition, edible films formed as coatings on foods could have an impact on overall packaging requirements. To the degree that an edible film coating acts as an efficient moisture, oxygen, or aroma barrier, the amount and/or complexity of packaging can be reduced. Thus, edible films may be able to help with reducing packaging waste through source reduction. If the barrier characteristics of an edible film coating allow conversion from a multilayer, multicomponent plastic package to a single-component package, package recycling could be improved. Edible coatings also have the potential for carrying food ingredients (e.g., antioxidants, antimicrobials, flavor) and improving the mechanical integrity or handling characteristics of the food. Finally, edible films with adequate mechanical properties could conceivably also serve as edible packaging for select foods [6,9]. The sanitary condition of the edible package would necessarily have to be maintained during storage, transportation, and marketing. The end result would again be source reduction and/or improved recyclability of the remaining elements of the packaging system.

The film-forming ability of collagen and gelatin have been known for some time, although little published property data exist. Globular proteins such as corn zein, wheat gluten, and soy protein were also recognized for their film-forming ability years ago [11–14], but little data on the film properties were available until recently. Similarly, the film-forming ability and film properties of casein and whey protein had not been investigated intensively until the last decade. Recent reviews [9,15] have compiled scattered literature on edible protein-based films and coatings and provided important information on protein nature and recovery, film-production methods, permeability-measurement techniques, and food applications.

The objectives of this chapter are to (a) review research on protein film-formation fundamentals and film properties, (b) relate properties of protein-based films to molecular and environmental factors, (c) compare barrier and mechanical properties of protein-based films to each other, polysaccharide-based films, and synthetic films, (d) summarize applications of protein films in food products, and (e) make conclusions as to the status of protein films and future directions. The emphasis is on information published in the scientific literature, but relevant patent literature (where most of the application concepts have been explored) is included directly or through reference to recent excellent reviews.

Film water vapor permeability (WVP) and oxygen permeability (OP) are the barrier properties most commonly investigated to assess the ability of edible films and coatings to protect foods from the environment and (in the case of WVP) from adjacent food components with different water activity. Tensile strength (TS) and elongation (E) are the properties most commonly investigated to assess ability of edible films and coatings to protect foods against mechanical abuse. Comparison of barrier and mechanical properties among different edible films must take into account the fact that a variety of film compositions and test conditions are used by researchers. Film formulation, temperature, and relative humidity are among the differences that affect the properties of edible films and make comparisons difficult. Nonetheless, such comparisons are important and will be made when possible.

II. EFFECT OF PROTEIN COMPOSITION AND STRUCTURE ON FILM PROPERTIES

In their native states, proteins generally exist as either fibrous proteins, which are water insoluble and serve as the main structural materials of animal tissues, or globular proteins, which are soluble in water or aqueous solutions of acids, bases, or salts and function widely in living systems [16,17]. The fibrous proteins are fully extended and associated closely with each other in parallel structures, generally through hydrogen bonding, to form fibers. The globular proteins fold into complicated spheroidal structures held together by a combination of hydrogen, ionic, hydrophobic, and covalent (disulfide) bonds [18]. The chemical and physical properties of these proteins depend on the relative amounts of the component amino acid residues and their placement along the protein polymer chain.

Of the fibrous proteins, collagen has received the most attention in the production of edible films. Several globular proteins, including wheat gluten, corn zein, soy protein, and whey protein, have been investigated for their film properties. Protein films are generally formed from solutions or dispersions of the protein as the solvent/carrier evaporates. The solvent/carrier is generally limited to water, ethanol, or ethanol-water mixtures [1].

Generally, globular proteins must be denatured by heat, acid, base, and/or solvent in order to form the more extended structures that are required for film formation. Once extended, protein chains can associate through hydrogen, ionic, hydrophobic, and covalent bonding. The chain-to-chain interaction that produces a cohesive film is affected by the degree of chain extension (protein structure) and the nature and sequence of amino acid residues. Uniform distribution of polar, hydrophobic, and/or thiol groups along the polymer chain increases the likelihood of the respective interactions.

Increased polymer chain-to-chain interaction results in films that are stronger but less flexible and less permeable to gases, vapors, and liquids [1,19]. Polymers containing groups that can associate through hydrogen or ionic bonding result in films that are excellent oxygen barriers but that are susceptible to moisture [20]. Thus, protein films are expected to be good oxygen barriers at low relative humidities. Polymers containing a preponderance of hydrophobic groups are poor oxygen barriers but excellent moisture barriers [20]. The more hydrophobic, water-insoluble proteins are thus expected to be better moisture barriers than water-soluble proteins. However, the fact that they are not totally hydrophobic and contain predominantly hydrophilic amino acid residues limits their moisture-barrier properties. Creation of protein-based edible films with low WVP requires addition of lipid components. This is analogous to the situation with synthetic polymers where moisture-sensitive oxygen-barrier polymers must be either copolymerized with a hydrophobic polymer or sandwiched between hydrophobic polymer layers to limit the ability of water to reduce barrier properties.

In addition to the considerations above, increased polymer crystallinity, density, orientation, molecular weight, and cross-linking decrease film permeability [2,20–22]. The complicated, irregular composition and structures of proteins provide unique challenges in forming films with desirable mechanical and permeability properties.

Often, low molecular weight plasticizers must be added to protein films in order to improve film flexibility (decrease brittleness) by reducing chain-to-chain interaction. Unfortunately, this generally increases film permeability as well. Thus, finding plasticizers compatible with the protein and solvent system that optimize film mechanical properties with minimum deterioration of film barrier properties is an important goal. Plasticizers that have been used for edible films include mono-, di-, and oligosaccharides, polyols, and lipids [2].

III. PROTEIN FILM AND COATING SYSTEMS

A. Collagen Films

1. Composition and Structure

Collagen is generally isolated from hides, tendon, cartilage, bone, and connective tissue [23]. The generally accepted basic macromolecule of collagen is tropocollagen (TC). TC has a thin rodlike structure composed of three polypeptide chains of equal length, each chain in the form of a left-handed helix, with the three chains coiled in a slight right-handed superhelix [23]. Collagen is unique in that every third amino acid residue throughout most of the structure is glycine; there exists a characteristic sequence glycine-proline-X, where X is most commonly hydroxyproline [23]. The TC aggregates to give fibrils, with arrangement stabilized by intermolecular cross-linking.

2. Film Formation

In vivo, collagen exists as fibers embedded in mucopolysaccharides and other proteins [23]. Production of films from animal hides can be accomplished using a dry or wet process with some similarity, including (a) alkaline treatment to dehair and remove collagen from carbohydrates and other proteins, (b) acid swelling and homogenization to form a ~4.5% moisture gel (wet process) or ~10% moisture gel dough (dry process), (c) extrusion into a tube, and (d) neutralization of the extruded tube, washing the tube of salts, treating the tube with plasticizer and cross-linkers, and drying to 12–14% moisture, with the order depending on whether the wet or dry process is used [24]. While either process disrupts the larger fiber structure of the collagen, the triple-helical structure and small fibrillar integrity necessary for regeneration (by extrusion and neutralization), good strength, and water-repellant properties are maintained, with the dry process maintaining relatively longer and bigger fibers [24].

3. Film Properties

No data on the WVP of collagen film has been reported in the scientific literature. Leiberman and Gilbert [25] studied the effects of moisture, plasticizer content, denaturing, and cross-linking on the permeability of collagen to gases. Dry collagen films were nearly impervious to dry oxygen, nitrogen, and carbon dioxide, making them much superior to low-density polyethylene (LDPE) and comparable to the best synthetic oxygen-barrier films—ethylene–vinyl alcohol copolymer (EVOH), and polyvinylidene chloride (PVDC)—at 0% relative humidity (RH) [20] (Table 1). Increasing relative humidity had an exponential effect on collagen film gas permeability, similar to cellophane [20,26] (Table 1). Similar to the plasticizing effect of water, aliphatic diol plasticizers increased gas permeability. Interestingly, addition of sorbitol lowered permeability, an effect attributed to interaction with collagen chain–active sites in such a way as to restrict segmental mobility. Heat-denatured collagen also exhibited lower gas permeability, an effect attributed to decreased availability of free volume in the more random structure of the denatured collagen. Finally, Lieberman and Gilbert [25] found that cross-linking collagen with formaldehyde or chromium ions increased gas permeability. This somewhat surprising result was attributed to cross-linked collagen providing more opportunity for moisture uptake, presumably due to greater spacing between collagen chains.

The TS of collagen casings depends on whether longitudinal or transverse sections are tested [24]. The range of TS values from both the wet and dry processes is similar to that for LDPE [27] (Table 2). The E is relatively low, ranging between that of cellophane and oriented polypropylene (OPP) [27] (Table 2). The combination of TS and E are adequate to allow collagen casings and films to function well with meat products.

4. Applications

Collagen is used to make the most commercially successful edible protein films. Collagen casings have largely replaced natural gut casings for sausages [24]. Except for large sausages, where thick casings are required, collagen casings are normally eaten with the sausage. Flat collagen films are used to wrap smoked meats such as hams in order to prevent elastic nets from becoming imbedded during cooking [28]. The collagen film is eaten with the meat product after removal of the netting. Besides providing mechanical integrity to meat products, collagen film is generally seen as reducing oxygen and moisture transport [29].

More recently, additional applications are being explored. Farouk et al. [30] studied the effect of collagen film overwrap on exudation and lipid oxidation in both refrigerated and frozen/thawed beef round steak. After collagen film overwrapping, the samples were either vacuum packaged in high-barrier bags or packaged in polystyrene foam trays with polyvinylchloride (PVC) film. Compared to packaged standards with no collagen film

Table 1 Oxygen Permeabilities of Protein-Based Edible Films Compared to Packaging Films

Film	Film solution	Film test conditions[a]	Permeability ($cm^3 \cdot \mu m/m^2 \cdot d \cdot kPa$)	Ref.
Collagen	Cyanoacetic acid (pH 2.5)	RT, 0%RH	<0.04–0.5*	25
Collagen	Cyanoacetic acid (pH 2.5)	RT, 63%RH	23.3	25
Collagen	Cyanoacetic acid (pH 2.5)	RT, 93%RH	890	25
Zein:Gly (5:1)	75°C, 95% EtOH	25°C, 0%RH	11.8	53
Zein:Gly (4.9:1)	70°C, 95% EtOH	30°C, 0%RH	13.0–44.9	50
Cozeen (Zein:Gly=3.9:1)	EtOH	38°C, 0%RH	67.4	51
Zein:PEG+Gly (2.6:1)	85°C—15 min, 95% EtOH	25°C, 0%RH	38.7–90.3	52
WG:Gly (2.5:1)	75°C—10 min, 52% alk-EtOH	23°C, 0%RH	3.9	78, 79
WG:Gly (3.1:1)	70°C, 52% alk-EtOH	30°C, 0%RH	9.6–24.2	50
WG:Gly (2.5:1)	100°C, 52% alk-EtOH	38°C, 0%RH	6.7	51
WG:Gly (2.5:1)	75°C, 52% alk-EtOH	25°C, 0%RH	6.1	53
SPI:Gly (1.7:1)	60°C—10 min, pH=6[b]	25°C, 0%RH	4.5	88
SPI:Gly (1.7:1)	60°C—10 min, pH=12[b]	25°C, 0%RH	1.6	88
SPI:Gly (2.4:1)	100°C—10 min DW	25°C, 0%RH	6.1	90
WPI:Gly (5.7:1)	90°C—30 min DW	23°C, 50%RH	18.5	108
WPI:Gly (2.3:1)	90°C—30 min DW	23°C, 50%RH	76.1	108
WPI:Sor (2.3:1)	90°C—30 min DW	23°C, 50%RH	4.3	108
LDPE		23°C, 50%RH	1870	20
Cellophane		23°C, 0%RH	0.7	20
Cellophane		23°C, 50%RH	16	26
Cellophane		23°C, 95%RH	252	26
EVOH (70% VOH)		23°C, 0%RH	0.1	20
EVOH (70% VOH)		23°C, 95%RH	12	20
PVDC-based films		23°C, 50%RH	0.4–5.1	20
Polyester		23°C, 50%RH	15.6	54

[a]Based on values for PVDC-based films [20].

[b]Aqueous solution adjusted to pH indicated.

WG = wheat gluten, SPI = soy protein isolate, WPI = whey protein isolate, PEG = polyethylene glycol, Gly = glycerin, Sor = sorbitol, LDPE = low-density polyethylene, EVOH = ethylene-vinyl alcohol copolymer, VOH = vinyl alcohol, PVDC = polyvinylidene chloride, RT = Room temperature, DW = Distilled water.

Table 2 Mechanical Properties of Protein-Based Edible Films Compared to Packaging Films

Film	Film solution/treatment	Film test condition	Tensile strength (MPa)	Elongation (%)	Ref.
Coll:Cell:Gly (3.4:0.8:1)[a]	pH=8/x-link	Not specified	3.3–10.8	25–50	24
Coll:Cell:Sor+Gly (5.3:0.4:1)[b]	pH=4/x-link	Not specified	6.2–9.4	38–57	24
Zein:PEG+Gly (5.9:1)	95% EtOH	23°C, 50%RH	22.3–28.4	6–9	52
Zein:PEG+Gly (3.6:1)	95% EtOH	23°C, 50%RH	6.7–15.7	43–198	52
Zein:PEG+Gly (2.6:1)	95% EtOH	23°C, 50%RH	2.7–6.8	173–213	52
WG:Gly (2.7:1)	52% EtOH, pH=2–4[c], 75°C—10 min	25°C, 50%RH	0.5–0.9	157–229	77
WG:Gly (2.7:1)	52% EtOH, pH=9–13[c], 75°C—10 min	25°C, 50%RH	1.9–4.4	170–208	77
WG:Gly (2.5:1)	75°C—10 min, 52% alk-EtOH	23°C, 50%RH	2.6	238–276	78, 79
WG: Gly (2.5:1)	75°C—10 min, 52% alk-EtOH/lactic	23°C, 50%RH	1.4	417	78
WG:Gly (2.5:1)	75°C—10 min, 52% alk-EtOH/$CaCl_2$	23°C, 50%RH	3.8	162	78
SPI:Gly (1.7:1)	60°C—10 min, pH=6[c]	25°C, 50%RH	3.1	67	88
SPI:Gly (1.7:1)	60°C—10 min, pH=12[c]	25°C, 50%RH	4.5	86	88
SPI:Gly (1.7:1)	pH=11[c], 70°C—30 min	25°C, 50%RH	3.3	100	91
SPI:Gly (1.7:1)	pH=11[c], 70°C—30 min/80°C—24 hr	25°C, 50%RH	10.0	30	91
SPI:Gly (1.7:1)	pH=11[c], 70°C—30 min/95°C—24 hr	25°C, 50%RH	14.0	30	91
SPI:Gly (4.0:1)	85°C DW	23°C, 50%RH	12.8	17	89
WPI:Gly (5.7:1)	90°C—30 min DW	23°C, 50%RH	29.1	4.1	108
WPI:Gly (2.3:1)	90°C—30 min DW	23°C, 50%RH	13.9	30.8	108
WPI:Sor (2.3:1)	90°C—30 min DW	23°C, 50%RH	14.0	1.6	108
WPI:Sor (1.5:1)	90°C—30 min DW	23°C, 50%RH	18.2	5.0	108
LDPE			9–17	500	27
PVC			45–55	120	27
OPP			165	50–75	27
PET			175	70–100	27
Cellophane			48–110	15–25	27

[a]Wet process.
[b]Dry process.
[c]Solution adjusted to pH indicated.
Coll = Collagen, Cell = cellulose, WG = wheat gluten, SPI = soy protein isolate, WPI = whey protein isolate, PEG = polyethylene glycol, Gly = glycerin, Sor = sorbitol, LDPE = low-density polyethylene, PVC = polyvinyl chloride, OPP = oriented polypropylene, PET = polyethylene terephthalate, DW = Distilled water.

overwrap, use of collagen film overwrap reduced exudation with no significant effect on color or lipid oxidation.

A study conducted to assess the potential for elimination of plastic wrappings for meat found that collagen film performed equally well in maintaining quality of frozen beef cubes [31,32]. The advantage is that, as the meat thaws and cooks, the collagen film melts away, eliminating the need for film waste handling.

B. Gelatin Films and Coatings

1. Composition and Structure

Gelatin is formed by hydrolysis of collagen [33]. Thermally reversible gels formed by heating aqueous solutions of gelatin are cross-linked between amino and carboxyl components of amino acid residue side groups [34].

2. Film Formation and Properties

Gelatin gels dry to produce films that are poor moisture barriers. Guilbert [2] found that cross-linking/denaturing gelatin films with calcium ions had no effect on water-barrier properties. Treatment with lactic or tannic acid improved water-barrier properties, although at the cost of less flexible and less transparent films. Guilbert [2] obtained similar results with casein, serum albumin, and ovalbumin films.

3. Applications

Gelatin is used to encapsulate low-moisture or oil-phase food ingredients and pharmaceuticals. Such encapsulation provides protection against oxygen and light, as well as defining ingredient amount or drug dosage [15,35,36]. Laminate coatings that included gelatin protected dried fruits and vegetables from moisture absorption and oxidation [37]. Tannic-acid–treated gelatin films were effective at retaining sorbic acid on the surface of a model food [38]. In addition, gelatin films have been formed as coatings on meats to reduce oxygen, moisture, and oil transport [15,29]. For example, gelatin coating formed on cut poultry before freezing reduced the amount of rancidity developed during storage [39]. The effect was enhanced by adding an antioxidant to the coating.

C. Corn Zein Films and Coatings

1. Composition and Structure

Corn gluten is separated from corn germ, fiber, and starch in the corn wet-milling process [40,41]. Zein is the prolamin (soluble in 70% ethanol) fraction of corn gluten, making up approximately 70% of the corn gluten. Zein resides mainly in the endosperm of the corn kernel. The solubility of zein in aqueous ethanol and insolubility in water are due to high content of the nonpolar amino acids leucine, alanine, and proline. In addition, the high level of glutamine in zein is believed to promote hydrogen bonding between zein chains and contribute to its insolubility in water [40,42].

Native zein is a mixture of proteins that differ in molecular size, solubility, and charge [43]. Approximately 80% of native zein is soluble in 95% ethanol, with the remainder soluble in 60% ethanol. These two fractions are typically designated α-zein and β-zein, respectively. β-Zein has been shown to consist of disulfide–cross-linked α-zein [42]. The reducing environment of the steeping process in corn wet milling cleaves disulfide bonds, leaving only α-zein in commercial corn gluten and zein [44].

Zein is produced commercially by extracting corn gluten with 80–90% aqueous isopropyl alcohol containing 0.25% sodium hydroxide at 60–70°C [40,42]. Centrifugally clarified extract is chilled to precipitate the zein. Additional extractions and precipitations increase zein purity. Recently, aqueous ultrapure zein lattices have been developed that have many potential uses in food systems [45,46].

2. Film Formation

Edible films can be formed by drying aqueous ethanol solutions of zein [3]. Formation of films is believed to involve development of hydrophobic, hydrogen, and limited disulfide bonds between zein chains in the film matrix [15]. The resulting films are brittle and therefore require plasticizer addition for increased flexibility [42]. However, addition of plasticizer increases permeability to water vapor and oxygen [47]. Edible films can also be formed from aqueous lattices of ultrapure zein [45,46]

3. Film Properties

Zein films are relatively good water vapor barriers compared to other edible films [2], but much poorer than LDPE [48] and EVOH [49]. Films made from zein have WVPs lower than or similar to those of films made from cellophane [26], methyl cellulose, ethyl cellulose, ethyl cellulose–polyvinylpyrrolidone, hydroxypropylmethyl cellulose, and hydroxypropyl cellulose at similar test conditions [5,12]. Data on zein films show that WVP increases with plasticizer level and RH surrounding the film [50–52] (Table 3).

The OPs of zein films have been reported, but only at 0% RH [50–53]. Although relatively low, the OP of zein film appears to be at least an order of magnitude greater than for collagen at 0% RH. The lower OP of collagen film likely reflects the highly associated fibrillar nature of collagen. However, the OP of zein film at 0% RH still ranks quite low when compared to synthetic films: one to two orders of magnitude less than LDPE and similar to polyester film [54] (Table 1). In addition, the OP of zein film is one to two orders of magnitude less than methyl cellulose and hydroxypropyl cellulose films at the same test temperature and RH [5]. Zein film OP increases with increased plasticizer level [52]. Given the relatively hydrophilic character of zein, the effect of RH on the OP of zein films should be investigated. At similar plasticizer levels, zein film appears to have TS and E similar to collagen film (Table 2).

Films made from aqueous lattices of ultrapure zein have been reported to have 30–50% greater barrier properties and mechanical strength than films made from ethanolic solutions of commercial zein [45,46].

4. Applications

Besides collagen and gelatin, corn zein is the only other protein that has been promoted commercially as an edible film or coating. The water vapor, oxygen, and lipid barrier properties of zein films have been put to good use in a variety of food products [3,15,29]. Zein is also used in the pharmaceutical industry to coat capsules for protection, controlled release, and masking of flavors and aromas [15].

Zein-based coatings have been used successfully to extend the shelf-lives of almonds, peanuts, pecans, and walnuts [55–58]. Similarly, zein-based coatings have protected confectioneries, dried fruits, and dried vegetables from moisture absorption, oxidation and/or lipid migration [11,37,59,60]. Zein coatings have also been used to coat vitamin-enriched rice, thus protecting vitamins from loss [61]. A zein-based coating was also found effective in maintaining a high concentration of sorbic acid at the surface of

Table 3 Water Vapor Permeabilities of Protein-Based Edible Films Compared to Packaging Films

Film	Film solution/treatment	Film test conditions[a]	Permeability ($g \cdot mm/m^2 \cdot d \cdot kPa$)	Ref.
Zein:Gly (4.9:1)	70°C, 95% EtOH	21°C, 85/0%RH	9.6	50
Cozeen (Zein:Gly=3.9:1)	EtOH	26°C, 50/100%RH	35.2	51
Zein:PEG+Gly (5.9:1)	85°C—15 min, 95% EtOH	25°C, 50/100%RH	46.2–53.1	52
Zein:PEG+Gly (3.6:1)	85°C—15 min, 95% EtOH	25°C, 50/100%RH	48.0–73.7	52
Zein:PEG+Gly (2.6:1)	85°C—15 min, 95% EtOH	25°C, 50/100%RH	70.3–107.3	52
WG:Gly (2.5:1)	75°C—10 min, 52% alk-EtOH	23°C, 0/11%RH	4.8	78, 79
WG:Gly (3.1:1)	70°C, 52% alk-EtOH	21°C, 85/0%RH	53	50
WG:Gly (2.5:1)	100°C, 52% alk-EtOH	26°C, 50/100%RH	108	51
WG:Gly (5.0:1)	40°C, 20–70% EtOH pH=2–6[b]	30°C, 100/0%RH	5.1–8.4	75, 76
WG:BW:Gly (5:1.5:1)	40°C, 45% EtOH pH=4[b]	30°C, 100/0%RH	3.0	82
SPI:Gly (1.7:1)	60°C—10 min, pH=6[b]	25°C, 50/100%RH	262	88
SPI:Gly (1.7:1)	60°C—10 min, pH=12[b]	25°C, 50/100%RH	154	88
SPI:Gly (2.4:1)	100°C—10 min, DW	25°C, 60/100%RH	72	90
SPI:Gly (4.0:1)	85°C DW	28°C, 0/78%RH[c]	39	89
SC	23°C DW	25°C, 0/81%RH[c]	36.7	98
$CaCl_2$-treated SC film	23°C DW/$CaCl_2$	25°C, 0/83%RH[c]	21.2	98
Buffer-treated SC film	23°C DW/pH=4.6	25°C, 0/86%RH[c]	20.9	98
CC:BW (1.7:1)	68°C DW	25°C, 0/97%RH[c]	3.6	98
WPI:Gly (1.6:1)	90°C—30 min DW	25°C, 0/11%RH[c]	6.6	104
WPI:Gly (1.6:1)	90°C—30 min DW	25°C, 0/65%RH[c]	119.8	104
WPI:Gly (4:1)	90°C—30 min DW	25°C, 0/77%RH[c]	70.2	104
WPI:Sor (1.6:1)	90°C—30 min DW	25°C, 0/79%RH[c]	62.0	104
WPI:BW:Sor (3.5:1.8:1)	90°C—30 min DW	25°C, 0/98%RH[c]	5.3	107
LDPE		38°C, 90/0%RH	0.08	48
EVOH (68% VOH)		38°C, 90/0%RH	0.25	49
Cellophane		38°C, 90/0%RH	7.3	26

[a]RHs are those on top and bottom sides of film (top/bottom).
[b]Solution adjusted to pH indicated.
[c]Corrected RH shown for bottom side of film.
WG = wheat gluten, SPI = soy protein isolate, SC = sodium caseinate, CC = calcium caseinate, WPI = whey protein isolate, BW = beeswax, PEG = polyethylene glycol, Gly = glycerin, Sor = sorbitol, LDPE = low-density polyethylene, EVOH = ethylene–vinyl alcohol copolymer, VOH = vinyl alcohol, DW = distilled water.

an intermediate-moisture food [62]. Zein and zein-based coating formulations are marketed commercially for these food uses and related pharmaceutical applications [45,63–65]. Zein coatings have also shown an ability to reduce moisture and firmness loss and delay color change (reduce oxygen and carbon dioxide transmission) in fresh tomatoes [66,67].

Zein-coated paper was judged equal to polyethylene-laminated paper for quick-service restaurant packaging of fatty foods [68]. Zein has also been explored as a replacement for collagen in the manufacture of sausage casings [69] and for the production of water-soluble pouches for dried foods [70].

D. Wheat Gluten Films and Coatings

1. Composition and Structure

Wheat gluten (WG) is defined as the water-insoluble protein of wheat flour which remains after flour dough is kneaded and the starch washed away with dilute salt solution [71]. Thus, WG contains the prolamin and glutelin fractions of wheat flour protein, typically referred to as gliadin and glutenin, respectively. While gliadin is soluble in 70% ethanol, glutenin is not. Gliadin and glutenin are considered globular and exist in approximately equal amounts [3]. Although insoluble in neutral water, WG dissolves in aqueous solutions of high or low pH at low ionic strength [72]. These properties of WG have been used successfully in commercial processes to isolate WG from wheat flour [73].

The amino acid composition of WG accounts for some of its properties [72]. The large amount of glutamine is thought to contribute to extensive hydrogen bonding between gluten chains. The relatively high amounts of nonpolar amino acids, such as proline and leucine, along with the lack of readily ionizable amino acids, such as lysine or glutamic acid, likely cause WG's insolubility in neutral water [72]. Both the gliadin and glutenin fractions of WG contain intramolecular disulfide bonds. Intermolecular disulfide bonds, which link individual glutenin protein chains and result in larger polymers with high molecular weight, have been proposed by some researchers [74]. The fact that glutenin forms stronger films than gliadin or whole WG has been attributed to the larger, more extended nature of glutenin [72]. The gliadin-glutenin complex that makes up the WG protein network is believed to involve both covalent and noncovalent bonding. Smaller globular gliadin protein is packed into a framework provided by larger extended random-coil glutenin to form the WG network [15].

2. Film Formation

Edible films can be formed by drying aqueous ethanol solutions of WG [3,50,51,53, 75–81]. Cleavage of native disulfide bonds during heating of film-forming solutions and then formation of new disulfide bonds during film drying are believed to be important to the formation of WG film structure, along with hydrogen and hydrophobic bonds [3,15].

3. Film Properties

The extensive intermolecular interactions in WG result in quite brittle films, which require a plasticizer such as glycerin (Gly). However, increasing film flexibility by increasing glycerin content reduces film strength, elasticity, and water-vapor barrier properties [76]. Water is also a plasticizer for hydrophilic materials like WG. For similar WG:Gly ratios, increasing the RH surrounding WG films also increases the WVP (Table 3). Increasing the water activity of a WG film increases water-vapor transmission rate and decreases TS [76,80]. Because water sorption decreases as temperatures increases, the WVP and TS of WG films decreased and increased with increasing temperature, respectively [80,81]. At comparable plasticizer content and test conditions, wheat gluten films prepared from alkaline solution appear to be somewhat poorer moisture barriers than corn zein films (Table 3).

The OPs of WG films have been measured, but only at 0% RH [50,51,53,78,79]. Unlike WVP, at comparable plasticizer content and test conditions, the OP of gluten films prepared from alkaline solution is an order of magnitude lower than OP of zein films (Table 1). This suggests that there is an inverse relationship between WVP and OP for edible hydrophilic films. Nonetheless, because of the influence of water activity on WG, the effect of RH on the OP of WG films should be investigated.

At comparable plasticizer content and test conditions, wheat gluten films prepared from alkaline solution appear to have mechanical properties similar to corn zein films (Table 2).

4. Film Composition Effect on Properties

The pH of the film-forming solution affects the WVP and mechanical properties of WG films [75,77]. Gennadios et al. [77] found that WG films produced at alkaline conditions had greater TS than films produced at acid conditions (Table 2), but that E and WVP were not as affected. However, WG films produced by Gontard et al. [75] at acid conditions had significantly lower WVP than WG films produced at alkaline conditions, even lower than the WVP of corn zein films (Table 3). WG and ethanol concentrations of the film-forming solution also affected film properties [75].

The effects of subjecting WG films to several treatments, including lactic acid (tanning agent), calcium chloride (ionic cross-linker), and pH 7.5 buffer (pI of wheat gluten), have been investigated [78]. Lactic acid appeared to act as a plasticizer on mechanical properties, producing significant reduction in TS and increase in E. Calcium chloride increased TS and decreased E, suggesting a cross-linking effect (Table 2). Small changes in WVP and OP for all treatments were observed.

Replacing a portion of the WG with keratin, soy protein, or corn zein generally reduced OP and WVP, but had inconsistent effects on TS and E [78,79]. The biggest effect was from replacing 13% of the WG with keratin, reducing OP and WVP by 85% and 23%, respectively, while reducing TS 35% and increasing E by 32%. Replacing 30% of the WG with soy protein isolate (SPI) reduced OP and WVP by 40 and 9%, respectively, while increasing TS by 69% and reducing E by 16%. Replacing 20% of the WG with corn zein had no effect on OP and decreased WVP by 23%, while increasing TS by 58% and reducing E by 37%.

Adding mineral oil to WG films decreased WVP by 27% and had smaller effects on OP, TS, and E [78]. Addition of acetylated monoglyceride had little or no effect on WVP and OP, but increased TS [79].

Beeswax, diacetyl tartaric ester of monoglycerides, and stearic alcohol reduced the WVP of WG:lipid:Gly = 5:1.5:1 composite films by greater than 50% (Table 3) [82]. Addition of lipid beyond this level (20%) had little effect on WVP. While addition of beeswax produced highly opaque and brittle films, diacetyl tartaric ester of monoglycerides maintained film transparency and increased film strength. Above 37% lipid content, films were quite brittle and impossible to handle.

5. Applications

WG has been explored as a replacement for collagen in the manufacturer of sausage casings [69,83,84]. WG coatings have been developed to enhance adherence of salt and flavorings to nuts, and batters to meats and other foods [15,29].

E. Soy Protein Films and Coatings

1. Composition and Structure

Most of the protein in soybeans is insoluble in water but soluble in dilute neutral salt solutions. Thus, soy protein belongs to the globulin classification [85]. Soy protein is globular in nature and is further classified into 2S, 7S, 11S, and 15S fractions according to relative sedimentation rates [15]. The principal components are the 7S (conglycinin) and 11s (glycinin) fractions, both of which have a quaternary (subunit) structure [86]. Although estimates of the relative amounts of these protein fractions vary, glycinin may account for 60–70% of the total soybean globulins [85].

Soy protein is high in asparagine and glutamine residues. Both conglycinin and glycinin are tightly folded proteins. While the extent of disulfide cross-linking of conglycinin is limited due to only two to three cystine groups per molecule, glycinin contains 20 intramolecular disulfide bonds [85]. Alkali and heating both cause dissociation and subsequent unfolding of glycinin due to disulfide bond cleavage [85].

2. Film Formation

Edible films based on soy protein can be produced in either of two ways: surface film formation on heated soymilk or film formation from solutions of soy protein isolate (SPI) [4]. Soymilk is produced by grinding soybeans with water followed by separation of milk from extracted soybeans. SPI is produced by dilute alkali extraction from defatted soy protein meal, followed by precipitation by pH adjustment to 4.5 [15]. To form films from both soymilk and SPI, (a) heating of film-forming solutions to disrupt the protein structure, cleave native disulfide bonds and expose sulfhydryl groups and hydrophobic groups, and then (b) formation of new disulfide, hydrophobic, and hydrogen bonds during film drying are believed to be important to the formation of soy protein film structure [15]. This process is also favored by alkaline conditions [87].

3. Film Properties

Films based on SPI are transparent and flexible when plasticizer is added, but have poor water-barrier properties [2]. Pretreatment of SPI with alkali to improve solubility and extend protein structure had no effect on film WVP, OP, and TS, and produced only small improvement in E [88]. The pH of the SPI film-forming solution has a significant effect on film properties. Glycerin-plasticized SPI films had lower WVP and OP and higher TS and E at pH 8–12 than at pH 6 (Tables 1–3) [88]. SPI films prepared from solutions at pH 6–11 had lower WVP and higher TS and E than films prepared from pH 1–3 solutions [77].

Water is also a plasticizer for hydrophilic materials like SPI. For identical SPI:Gly ratios, increasing the RH surrounding SPI films also increased the WVP [81]. Because water sorption decreases as temperature increases, the WVP of SPI films decreased with increasing temperature [81]. A combination of decreased plasticizer level and reduced RH surrounding SPI films decreased WVP, increased TS, and reduced E (Tables 2, 3) [89].

Heat treatment of SPI film-forming solution at 85°C promoted intermolecular cross-linking and produced SPI films that were smoother and more transparent, and possessed lower WVP and increased E, compared to those produced from unheated solution [89]. Enzymatic treatment of SPI film-forming solution with horseradish peroxidase both cross-linked and degraded protein chains, with resulting anomalous effect on film properties [89]. Chemical modification of soy protein had little effect on WVP and OP [90]. Heat curing of SPI films at 80 and 95°C reduced WVP, E, moisture content, and water solubility and increased TS (Table 2) [91]. Changes in film properties were attributed to heat-induced cross-linking and lower moisture content within the film, with greatest effect at 95°C.

At comparable plasticizer content and test conditions, SPI films appear to be similar moisture barriers to wheat gluten films prepared from alkaline solution and somewhat poorer moisture barriers than corn zein films (Table 3).

The OPs of SPI films have been measured, but only at 0% RH [88,90]. Like wheat gluten films, the OP of SPI films appears to be an order of magnitude lower than OP of zein films at similar plasticizer levels (Table 1). Because of the influence of water activity on SPI, the effect of RH on the OP of SPI films should be investigated.

SPI films with TS similar to zein and alkaline WG films have lower E, indicating a somewhat less tough film (Table 2).

4. Applications

The use of soy protein in the formation of film coatings on food products has been investigated [15,29]. Soy protein concentrate and SPI have been used successfully to aid batter adhesion and encase meat fibers to aid flavor retention. Soy protein–based coatings showed limited ability to reduce moisture migration in raisins and dried peas [37,92]. Soy protein has been explored as a replacement for collagen in the manufacture of sausage casings [69] and in the production of water-soluble pouches [70].

F. Casein Films and Coatings

1. Composition and Structure

Edible films and coatings can be formed from total milk protein or components of milk protein [7]. Milk proteins are classified into two types: casein and whey protein. Casein consists of three principal components, α, β, and κ, which together form colloidal micelles in milk containing large numbers of casein molecules and are stabilized by calcium-phosphate bridging [93,94]. The casein molecules possess little defined secondary structure, exhibiting instead an open random-coil structure. The negatively charged κ-casein is cleaved by the enzyme rennin used to coagulate milk in cheese production. Once κ-casein is cleaved, the micellar structure is destabilized and the casein precipitates.

Casein, which comprises 80% of milk protein, precipitates when skim milk is acidified to the casein isoelectric pH of approximately 4.6 [95]. Acidification solubilizes the calcium phosphate, thus releasing individual casein molecules, which associate to form insoluble acid casein. The acid casein can be converted to functional soluble caseinates by neutralization through addition of alkali. Sodium and calcium caseinates are most common, but magnesium and potassium caseinates are also available commercially

[93]. The caseinates do not reassociate in any manner, but remain as monomers in an open extended form.

The amino acid composition of casein explains some of its properties. Low levels of cysteine result in few disulfide cross-linkages and, thus, an open random structure. The α-casein possesses relatively even distribution of hydrophilic residues along the protein chain and more charged residues than β-casein [95,96]. β-Casein is hydrophobic in nature, but the last 50 residues of the N-terminal end are charged [95]. The κ-casein is somewhat intermediate to α- and β-casein in nature. The result is caseinates that possess good properties as emulsifiers and foaming agents.

2. Film Formation

Edible films based on various caseinates can be obtained by solubilization in water followed by casting and drying [97,98]. Caseinates form films from aqueous solution without heat treatment due to their random-coil nature. Interactions in the film matrix likely include hydrophobic, ionic, and hydrogen bonding.

3. Film Properties

Glycerin-plasticized caseinate films are transparent and flexible, but have poor water-barrier properties. Treatment with lactic acid or tannic acid improved water-barrier properties [2]. Films can also be prepared from sodium or calcium caseinate without addition of plasticizer. Calcium caseinate film has lower WVP than sodium caseinate films, but treatment of sodium caseinate film with calcium chloride produced greater reduction in WVP, likely due to more effective in situ calcium cross-linking [98]. Adjustment of sodium caseinate film to the isoelectric point of casein insolubilizes the film and also produces significant reduction in WVP (Table 3).

At comparable test conditions, caseinate films appear to be similar moisture barriers to wheat gluten films and soy protein films and somewhat poorer moisture barriers than corn zein films (Table 3). However, taking advantage of the emulsifying ability of caseinate gave a calcium caseinate–beeswax composite film with WVP 90% lower than for pure sodium caseinate film [98] and lower than for zein film (Table 3).

The OP and mechanical properties of caseinate films have not been investigated. Because of the influence of water activity on casein, the effect of RH on the OP and mechanical properties of caseinate films should be investigated.

4. Applications

Casein has been investigated in the formation of film coatings on food products [15]. Laminate coatings that included casein protected dried fruits and vegetables from moisture absorption and oxidation [37]. Caseinate-lipid emulsion coatings were successful in reducing moisture loss from peeled carrots and zucchini [99–101]. Caseinate-acetylated monoglyceride emulsion coatings reduced moisture loss from frozen salmon, but had little effect on lipid oxidation [102]. Casein films were effective in retaining sorbic acid on the surface of a model food system and on the surfaces of intermediate-moisture fruits [38]. Casein has also been investigated for use in the production of water-soluble pouches [70].

G. Whey Proteins Films and Coatings

1. Composition and Structure

Whey protein, which comprises 20% of milk protein, is the protein that remains soluble after casein is precipitated at pH 4.6. Whey protein consists of several component proteins, which are globular and heat labile in nature, including β-lactoglobulin, α-lactalbumin, bovine serum albumin, proteose-peptones, and immunoglobulins [103].

β-Lactoglobulin, which generally makes up 50–75% of whey protein, contains two disulfide groups, one free sulfhydryl group, and hydrophobic groups located in the interior of the globular structure [96]. Heat denaturation above approximately 65°C opens the β-lactoglobulin globular structure, exposes sulfhydryl and hydrophobic groups, and promotes oxidation of free sulfhydryls, disulfide bond interchange, and hydrophobic bonding [96]. The α-lactalbumin, which makes up approximately 25% of whey protein, contains four disulfide bonds. Heat denaturation of whey protein also promotes opening of the α-lactalbumin structure to allow additional disulfide bond interchange.

Whey protein concentrate (WPC) possessing 25–80% protein and whey protein isolate (WPI) containing >80% protein are produced commercially. Generally, ultrafiltration techniques are used to produce undenatured WPC. Ion exchange is used to produce high-purity undenatured WPI.

2. Film Formation

Because of the globular nature of whey proteins, production of films requires heat denaturation to open the globular structure, break existing disulfide bonds, and form new intermolecular disulfide and hydrophobic bonds [104]. The resulting films are transparent and bland, but require plasticizer to overcome brittleness. The intermolecular disulfide bonds also impart insolubility to the resulting films. Whey protein films can also be produced by enzymatic cross-linking using transglutaminase [105].

3. Film Properties

Relative humidity and plasticizer type and amount have significant effect on WVP of WPI films (Table 3) [104]. Reducing the hydrophilic nature of WPI films by introducing lipid decreases WVP substantially (Table 3) [106,107]. Beeswax and fatty acids were more effective at reducing WVP than fatty alcohols. Increasing lipid content of films decreased WVP significantly, as did increasing fatty acid and fatty alcohol chain length. Decreasing mean particle diameter of film-forming emulsions correlated well with a linear decrease in film WVP [107].

At comparable plasticizer content and test conditions, WPI films appear to be poorer moisture barriers than wheat gluten, soy protein, caseinate, and zein films. However, taking advantage of the emulsifying ability of whey protein allows formation of WPI-lipid films with WVP lower than for zein films (Table 3).

Relative humidity also has an exponential effect on the OP of WPI film [108]. Sorbitol-plasticized films with either tensile strength or elongation similar to glycerin-plasticized films had lower OP than glycerin-plasticized films (Tables 1, 2). The OP of WPI films is not available at 0% RH to compare with other protein films. However, the low OP at 50% RH indicates that WPI film is an excellent oxygen barrier at low-to-moderate RH. This again suggests an inverse relationship between WVP and OP.

Interestingly, films with WPI:Gly = 2.3 have TS and E similar to heat-cured films with SPI:Gly = 1.7 (Table 2).

4. Applications

Until recently, no applications of whey protein films had been explored. A spray of WPI solution followed by an antioxidant spray reduced peroxide values in stored frozen King salmon [109]. No effect on peroxide values or moisture loss was found for the WPI solution spray alone or with WPI-acetylated monoglyceride emulsion spray. Whey protein film coating had no effect on the respiration and moisture loss of green bell peppers in high-RH (85%) storage [110]. This result is consistent with the fact that RH has an exponential effect on the oxygen and moisture permeabilities of whey protein films [106,108]. However, whey protein coatings produced significant reductions in oxygen uptake and rancidity of roasted peanuts at low (21%) and moderate (53%) RH [111,112].

IV. CONCLUSIONS

A. Moisture-Barrier Properties

In general, the hydrophilic nature of proteins limits their ability to provide a significant moisture barrier without addition of hydrophobic substances. This is the case even for the less hydrophilic proteins wheat gluten and corn zein. Use of protein-based films and coatings as moisture barriers depends on the formation of protein-lipid composite films. Improved protein-based moisture barriers for food systems depend on successful, practical development of bilayer films where the protein acts as supporting matrix and a continuous layer of supported lipid provides the moisture barrier.

B. Oxygen-Barrier Properties

Consistent with films made from synthetic polymers that can hydrogen bond, protein films are good oxygen barriers at low-to-intermediate RH. Additional data on the effect of RH on oxygen permeability for protein films are needed to identify range of practical use. Nonetheless, oxygen-barrier properties deteriorate with increased RH. Therefore, potential applications include (a) protective coatings for low-moisture food products that are vulnerable to oxidation in conjunction with a simple, moisture-barrier packaging film (e.g., LDPE) and (b) respiration-reducing coatings for fresh fruits and vegetables that are exposed to low RH during storage and transportation. Additionally, development of bilayer films as described above will allow protection of protein film oxygen-barrier properties with a lipid moisture-barrier protein.

C. Aroma/Flavor- and Lipid-Barrier Properties

No research has yet explored the aroma/flavor and lipid permeabilities of protein films. Given the hydrophobic nature of lipids and most aromas/flavors, research and development should be pursued to explore the effectiveness of protein films, which are hydrophilic by nature.

D. Mechanical Properties

Generally, the mechanical properties of protein films are poorer than those of synthetic films. However, they are adequate to provide durability as coatings on food products.

They are also usually adequate to allow the use of protein films as small food pouches or bags.

E. Food Applications

Edible coating formulations must be specific to the intended food use. Formulations must first be wet and spread on the food surface and then form upon drying a film coating that has adequate adhesion, cohesion, and durability to function properly. In addition, edible coatings must provide satisfactory appearance, aroma, flavor, and mouthfeel. Selective application to appropriate foods and good control of environmental conditions will be necessary to ensure microbial stability. The addition of antimicrobials to edible films may widen application possibilities. Finally, any advantage of edible film coatings would have to be provided at an affordable cost. Unfortunately, little published research on food-coating issues, sensory properties, microbial stability, or economics of edible films is available.

REFERENCES

1. J. J. Kester and O. R. Fennema, Edible films and coatings: a review, *J. Food Sci. 40*(12): 47 (1986).
2. S. Guilbert, Technology and application of edible protective films, *Food Packaging and Preservation—Theory and Practice* (M. Mathlouthi, ed.), Elsevier Applied Science Publishers, New York, 1986, pp. 371–394.
3. A. Gennadios and C. L. Weller, Edible films and coatings from wheat and corn proteins, *Food Technol. 44*(10):63 (1990).
4. A. Gennadios and C. L. Weller, Edible films and coatings from soymilk and soy protein, *Cereal Foods World 36*(12):1004 (1991).
5. J. M. Krochta, Control of mass transfer in foods with edible coatings and films, *Advances in Food Engineering* (P. R. Singh and M. A. Wirakartakusumah, eds.), CRC Press, Inc., Boca Raton, FL, 1992, pp. 517–538.
6. K. R. Conca and T. C. S. Yang, Edible food barrier coatings, Activities Report and Minutes of Work Groups & Sub-Work Groups of the R&D Associates, Research and Development Associates for Military Food and Packaging Systems, Inc., San Antonio, TX, 1993, pp. 41–53.
7. T. H. McHugh and J.M. Krochta, Milk-protein-based edible films and coatings, *Food Technol. 48*(1):97 (1994).
8. C. Koelsch, Edible water vapor barriers: properties and promise, *Trends Food Sci. Technol.* 5:76 (1994).
9. J. A. Torres, Edible films and coatings from proteins, *Protein Functionality in Food Systems* (N. S. Hettiarachchy and G. R. Ziegler, eds.), Marcel Dekker, New York, 1994, pp. 467–507.
10. J. M. Krochta, E. A. Baldwin, and M. Nisperos-Carriedo, ed. *Edible Coatings and Films to Improve Food Quality*, Technomic Publishing Company, Inc., Lancaster, PA, 1994.
11. H. B. Cosler, Method of producing zein-coated confectionery, U.S. Patent 2,791,509 (1957).
12. J. L. Kanig and H. Goodman, Evaluative procedures for film-forming materials used in pharmaceutical applications, *J. Pharm. Sci. 51*(1):77 (1962).
13. C. A. Anker, G. A. Foster, and M. A. Leader, Method of preparing gluten-containing films and coatings, U.S. Patent 3,653,925 (1972).
14. W. J. Wolf and J. C. Cowan, *Soybeans as a Food Source*, CRC Press, Inc., Boca Raton, FL, 1975.

15. A. Gennadios, T. H. McHugh, C. L. Weller, and J. M. Krochta, Edible coatings and film based on proteins, *Edible Coatings and Films to Improve Food Quality* (J. M. Krochta, E. A. Baldwin, and M. Nisperos-Carriedo, eds.), Technomic Publishing Co., Inc., Lancaster, PA, 1994, pp. 201–277.
16. R. T. Morrison and R. N. Boyd, *Organic Chemistry*, Allyn and Bacon, Inc., Boston, 1959, pp. 866–883.
17. A. Streitwieser Jr. and C. H. Heathcock, *Introduction to Organic Chemistry*, Macmillan Publishing Co., Inc., New York, 1981, pp. 969–978.
18. W. Bushuk and C. W. Wrigley, Proteins: composition, structure and function, *Wheat: Production and Utilizatin* (G. E. Inglett, ed.), Avi Publishing Co., Inc., Westport, CT, 1974, pp. 119–145.
19. G. S. Banker, Film coating theory and practice, *J. Pharm. Sci. 55*(1):81 (1966).
20. M. Salame, Barrier polymers, *The Wiley Encyclopedia of Packaging Technology* (M. Bakker, ed.), John Wiley & Sons, New York, 1986, pp. 48–54.
21. C. A. Kumins, Transport through polymer films, *J. Polymer Sci: Part C. 10*:1 (1965).
22. B. Bascat, Study of some factors affecting permeability, *Food Packaging and Preservation—Theory and Practice* (M. Mathlouthi, ed.), Elsevier Applied Science Publishers, New York, 1986, pp. 7–24.
23. G. Balian and J. H. Bowes, The structure and properties of collagen, *The Science and Technology of Gelatin* (A. G. Ward and A. Courts, eds.), Academic Press, New York, 1977, pp. 1–30.
24. L. L. Hood, Collagen in sausage casings, *Advances in Meat Research* (A. M. Pearson, T. R. Dutson, and A. J. Bailey, eds.), Van Nostrand Reinhold Co., New York, 1987, pp. 109–129.
25. E. R. Lieberman and S. G. Gilbert, Gas permeation of collagen films as affected by cross-linkage, moisture, and plasticizer content. *J. Polymer Sci.: Symp. No. 41*:33 (1973).
26. C. C. Taylor, Cellophane, *The Wiley Encyclopedia of Packaging Technology* (M. Bakker, ed.), John Wiley & Sons, New York, 1986, pp. 159–163.
27. J. H. Briston, Films, plastic, *The Wiley Encyclopedia of Packaging Technology* (M. Bakker, ed.), John Wiley & Sons, New York, 1986, pp. 329–335.
28. *COOFI*, Brechteen, Mt. Clemens, MI, 1992.
29. R. A. Baker, E. A. Baldwin, and M. O. Nisperos-Carriedo, Edible coatings and films for processed foods, *Edible Coatings and Films to Improve Food Quality* (J. M. Krochta, E. A. Baldin, and M. O. Nisperos-Carriedo, eds.), Technomic Publishing Co., Lancaster, PA, 1994, pp. 89–104.
30. M. M. Farouk, J. F. Price, and A. M. Salih, Effect of an edible collagen film overwrap on exudation and lipid oxidation in beef round steak, *J. Food Sci. 55*(6):1510, 1563 (1990).
31. K. R. Conca, Evaluation of collagen based film as an edible packaging for frozen meat, *Biodegradable Polymers '94*, Ann Arbor, MI, ECM, Inc., 1994.
32. J. Rice, What's new in edible films?, *Food Proc. 55*(7):61 (1994).
33. J. E. Eastoe and A. A. Leach, Chemical constitution of gelatin, *The Science and Technology of Gelatin* (A. G. Ward and A. Courts, eds.), Academic Press, New York, 1977, pp. 73–107.
34. M. Glicksman, Functional properties, *Food Hydrocolloids* (M. Glicksman, ed.), CRC Press, Inc., Boca Raton, FL, 1982, p. 47.
35. *Soft Gelatin Questions and Answers*, Banner Gelatin Products Corp., Chatsworth, CA.
36. G. A. Reineccius, Flavor encapsulation, *Edible Coatings and Films to Improve Food Quality* (J. M. Krochta, E. A. Baldwin, and M. Nisperos-Carriedo, eds.), Technomic Publishing Co., Inc., Lancaster, PA, 1994, pp. 105–120.
37. M. S. Cole, Method for coating dehydrated food, U.S. Patent 3,479,191 (1969).
38. S. Guilbert, Use of superficial edible layer to protect intermediate moisture foods: application to the protection of tropical fruits dehydrated by osmosis, *Food Preservation by Moisture Control* (C. C. Seow, ed.), Elsevier Applied Science Publishers Ltd., Essex, England, 1988, pp. 199–219.

39. A. A. Klose, E. P. Mecchi, and H. L. Hanson, Use of antioxidants in the frozen storage of turkeys, *Food Technol. 6*:308 (1952).
40. L. G. Unger, Zein, *Encyclopedia of Chemical Technology* (R. E. Kirk and D. F. Othmer, eds.), The Interscience Encyclopedia, New York, 1956, pp. 220–223.
41. S. A. Watson, Corn and sorghum starches: production, *Starch Chemistry and Technology* (R. L. Whistler, J. N. BeMiller, and E. F. Paschall, ed.), Academic Press, 1984, pp. 417–468.
42. R. A. Reiners, J. S. Wall, and G. E. Inglett, Corn proteins: potential for their industrial use, *Industrial Uses of Cereals* (Y. Pomeranz, ed.), American Association of Cereal Chemists, Inc., St. Paul, MN, 1973, pp. 285–302.
43. J. E. Turner, J. A. Boundy, and R. J. Dimler, Zein: a heterogeneous protein containing disulfide-linked aggregates, *Cereal Chem. 42*:452 (1965).
44. J. A. Boundy, J. E. Turner, J. S. Wall, and R. J. Dimler, *Influence of commercial processing on composition and properties of corn zein, Cereal Chem. 44*:281 (1967).
45. R. B. Cook and M. L. Shulman, Emerging applications for ultrapure zein, Advances in Ingredient Technology Symposium, Rutgers University, 1992.
46. R. Cook and M. Shulman, Aqueous ultrapure zein latices as functional ingredients and coatings, Corn Utilization Conference V, Bedford, MA, 1994, pp. 1–4.
47. M. Mendoza, Preparation and physical properties of zein based films, University of Massachusetts, Amherst, 1975.
48. S. A. Smith, Polyethylene, low density, *The Wiley Encyclopedia of Packaging Technology* (M. Bakker, ed.), John Wiley & Sons, New York, 1986, pp. 514–523.
49. R. Foster, Ethylene-vinyl alcohol copolymers (EVOH), *The Wiley Encyclopedia of Packaging Technology* (M. Bakker, ed.), John Wiley & Sons, New York, 1986, pp. 270–275.
50. H. J. Park and M. S. Chinnan, Properties of edible coatings for fruits and vegetables, ASAE Paper No. 90-6510, American Society of Agriculture Engineers, St., Joseph, MI, 1990.
51. T. P. Aydt, C. L. Weller, and R. F. Testin, Mechanical and barrier properties of edible corn and wheat protein films, *Trans. ASAE 34*(1):207 (1991).
52. B. L. Butler and P. J. Vergano, Degradation of edible film in storage, ASAE Paper No. 946551, American Society of Agriculture Engineers, St. Joseph, MI, 1994.
53. A. Gennadios, C. L. Weller, and R. F. Testin, Temperature effect on oxygen permeability of edible protein-based films, *J. Food. Sci. 58*(1):212, 219 (1993).
54. J. F. Hanlon. Films and foils, *Handbook of Package Engineering*, Technomic Publishing Co., Inc., Lancaster, PA, 1992, pp. 3–9.
55. H. B. Cosler, Prevention of staleness and rancidity in nut meats and peanuts, *Manuf. Confect. 37*(8):15, 39 (1958).
56. H. B. Cosler, Prevention of staleness and rancidity in nut meats and peanuts, *Candy Ind. Confect. J. 3*:17, 22, 35 (1958).
57. H. B. Cosler, Prevention of staleness, rancidity in nut meats and peanuts. *Peanut J. Nut World 37*(11):10, 15 (1958).
58. J. J. Alikonis and H. B. Cosler, Extension of shelf life of roasted and salted nuts and peanuts, *Peanut J. Nut World 40*(5):16 (1961).
59. H. B. Cosler, A new edible, nutritive, protective glaze for confections and nuts, *Manuf. Confect. 39*(5):21 (1959).
60. J. J. Alikonis, *Candy Technology*, AVI Publishing Company, Inc., Westport, CT, 1979.
61. R. R. Mickus, Seals enriching additives on white rice, *Food Eng. 27*(11):91, 160 (1955).
62. J. A. Torres and M. Karel, Microbial stabilization of intermediate-moisture food surfaces. III. Effects of surface preservative concentration and surface pH control on microbial stability of an intermediate moisture cheese analog. *J. Food Proc. Pres. 9*:107 (1985).
63. C. Andres, Natural edible coating has excellent moisture and grease barrier properties, *Food Proc. 45*(1):48 (1984).
64. Extending shelf life with edible films, *Prepared Foods 156*(3): (1987).
65. *All About Zein*, Freeman Industries, Inc., Tuckahoe, NY, 1993.

66. H. J. Park, M. S. Chinnan, and R. L. Shewfelt, Edible coating effects on storage life and quality of tomatoes, *J. Food Sci. 59*(3):568 (1994).
67. H. J. Park, M. S. Chinnan, and R. L. Shewfelt, Edible corn-zein film coatings to extend storage life of tomatoes, *J. Food Proc. Pres. 18*(4):317 (1994).
68. T. A. Trezza and P. J. Vergano, Grease resistance of corn zein coated paper, *J. Food Sci. 59*(4):912 (1994).
69. A. F. Turbak, Edible vegetable protein casing, U.S. Patent 3,682,661 (1972).
70. L. E. Georgevits, Method of making a water soluble protein container, U.S. Patent 3,310,446 (1967).
71. R. W. Jones, N. W. Taylor, and F. R. Senti, Electrophoresis and fractionation of wheat gluten, *Arch. Biochem. Biophys. 84*:363 (1959).
72. L. H. Krull and G. E. Inglett, Industrial uses of gluten, *Cereal Sci. Today 16*(8):232, 261 (1971).
73. D. A. Fellers, Fractionation of wheat into major components, *Industrial Uses of Cereals* (Y. Pomeranz, ed.), American Association of Cereal Chemists, Inc., St. Paul, MN, 1973, pp. 207–228.
74. R. Lasztity, Recent results in the investigation of the structure of the gluten complex, *Die Nahrung 30*:235 (1986).
75. N. Gontard, S. Guilbert, and J. L. Cuq, Edible wheat gluten films: influence of the main process variables on film properties using response surface methodology, *J. Food Sci. 57*: 190, 199 (1992).
76. N. Gontard, S. Guilbert, and J. L. Cuq, Water and glycerol as plasticizers affect mechanical and water vapor barrier properties of an edible wheat gluten film, *J. Food Sci. 58*(1):206 (1993).
77. A. Gennadios, A. H. Brandenburg, C. L. Weller, and R. L. Testin, Effect of pH on properties of wheat gluten and soy protein isolate films, *J. Agric. Food Chem. 41*:1835 (1993).
78. A. Gennadios, C. L. Weller, and R. F. Testin, Modification of physical and barrier properties of edible wheat gluten-based films, *Cereal Chem. 70*(4):426 (1993).
79. A. Gennadios, C. L. Weller, and R. F. Testin, Property modification of edible wheat gluten-based films, *Trans. ASAE 36*(2):465 (1993).
80. A. Gennadios, H. J. Park, and C. L. Weller, Relative humidity and temperature effects on tensile strength of edible protein and cellulose ether films, *Trans. ASAE 36*(6):1867 (1993).
81. A. Gennadios, A. H. Brandenburg, J. W. Park, C. L. Weller, and R. F. Testin, Water vapor permeability of wheat gluten and soy protein isolate films, *Ind. Crops Prod. 2*:189 (1994).
82. N. Gontard, C. Duchez, J. L. Cuq, and S. Guilbert, Edible composite films of wheat gluten and lipids: water vapour permeability and other physical properties, *Intl. J. Food Sci. Technol. 29*:39 (1994).
83. J. D. Mullen, Film formation from non-heat coagulable simple proteins with filler and resulting product, U.S. Patent 3,615,715 (1971).
84. E. D. Schilling and P. I. Burchill, Forming a filled edible casing, U.S. Patent 3,674,506 (1972).
85. J. E. Kinsella, Functional properties of soy protein, *JAOCS 56*:242 (1979).
86. J. E. Kinsella, S. Damodaran, and B. German, Physiochemical and functional properties of oilseed proteins with emphasis on soy proteins, *New Protein Foods* (A. M. Altschul and H. L. Wilcke, eds.), Academic Press Inc., Orlando, FL, 1985, pp. 107–179.
87. J. J. Kelley and R. Pressey, Studies with soybean protein and fiber formation, *Cereal Chem. 43*:195 (1966).
88. A. H. Brandenburg, C. L. Weller, and R. F. Testin, Edible films and coatings from soy protein, *J. Food Sci. 58*:1086 (1993).
89. Y. M. Stuchell and J. M. Krochta, Enzymatic treatments and thermal effects on edible soy protein films, *J. Food Sci. 59*(6):1332 (1994).

90. H. Li, M. A. Hanna, and V. Ghorpade, Effects of chemical modifications on soy and wheat protein films, Paper No. 93-6529, ASAE, St. Joseph, MI, 1993.
91. A. Gennadios, V. M. Ghorpade, C. L. Weller, and M. A. Hanna, Heat curing of protein films, Paper No. 94-6552, ASAE, St. Joseph, MI, 1994.
92. H. R. Bolin, Texture and crystallization control in raisins, *J. Food Sci. 41*:1316 (1976).
93. J. E. Kinsella, Milk proteins: physicochemical and functional properties, *CRC Crit. Rev. Food Si. Nutr. 21*:197 (1984).
94. J. M. Regenstein and C. E. Regenstein, *Food Protein Chemistry*, Academic Press, Inc., New York, 1984, pp. 48–54.
95. D. G. Dalgleish, Milk proteins—chemistry and physics, *Food Proteins* (J. E. Kinsella and W. G. Soucie, eds.), American Oil Chemists Society, Champaigne, IL, 1989, pp. 155–178.
96. J. R. Brummer, Milk proteins, *Food Proteins* (J. R. Whitaker and S.R. Tannenbaum, eds.), Avi Publishers, Inc., Westport, CT, 1977, pp. 175–208.
97. B. Ho, Water vapor permeabilities and structural characteristics of casein films and casein-lipid emulsion films, University of California–Davis, 1992.
98. R. J. Avena-Bustillos and J. M. Krochta, Water vapor permeability and caseinate-based edible films as affected by pH, calcium crosslinking and lipid content, *J. Food Sci. 58*(4): 904 (1993).
99. R. J. Avena-Bustillos, L. A. Cisneros-Zevallos, J. M. Krochta, and M. E. Saltveit, Optimization of edible coatings on minimally processed carrots using response surface methodology, *Trans. ASAE 36*(3):801 (1993).
100. R. J. Avena-Bustillos, L. A. Cisneros-Zevallos, J. M. Krochta, and M. E. Saltveit, Application of casein-lipid edible film emulsions to reduce white blush on minimally processed carrots, *Postharvest Biol. Technol. 4*:319 (1994).
101. R. J. Avena-Bustillos, J. M. Krochta, M. E. Saltveit, R. J. Rojas-Villegas, and J. A. Sauceda-Perez, Optimization of edible coating formulations on zucchini to reduce water loss, *J. Food Eng. 21*:197 (1994).
102. K. Hirasa, Moisture loss and lipid oxidation in frozen fish-effect of a casein-acetylated monoglyceride edible coating, University of California–Davis, 1991.
103. J. E. Kinsella and D. M. Whitehead, Proteins in whey: chemical, physical and functional properties, *Adv. Food Nutr. Res. 33*:343 (1989).
104. T. H. McHugh, J. F. Aujard, and J. M. Korchta, Plasticized whey protein edible films: water vapor permeability properties, *J. Food Sci. 59*(2):416, 423 (1994).
105. R. Mahmoud and P. A. Savello, Mechanical properties of water vapor transferability through whey protein films, *J. Dairy Sci. 75*:942 (1992).
106. T. H. McHugh and J. M. Krochta, Water vapor permeability properties of edible whey protein-lipid emulsion films, *JAOCS 71*(3):307 (1994).
107. T. H. McHugh and J. M. Krochta, Dispersed phase particle size effects on water vapor permeability of whey protein-beeswax edible emulsion films, *J. Food Proc. Pres. 18*:173 (1994).
108. T. H. McHugh and J. M. Krochta, Sorbitol vs. glycerol-plasticized whey protein edible films: integrated oxygen permeability and tensile property evaluation, *J. Agric. Food Chem. 42*(4): 841 (1994).
109. Y. M. Stuchell and J. M. Krochta, Edible coatings on frozen King salmon: effect of whey protein isolate and acetylated monoglycerides on moisture loss and lipid oxidation, *J. Food Sci. 60*(1):28 (1995).
110. S. Lerdthanangkul and J. M. Krochta, Edible coating effects on postharvest quality of green bell peppers, *J. Food Sci. 61*(1):176 (1996).
111. J. I. Mate and J. M. Krochta, Whey protein coating effect on the oxygen uptake of dry roasted peanuts, *J. Food Sci.* (in press).
112. J. I. Mate, E. N. Frankel, and J. M. Krochta, Whey protein isolate edible coatings: effect on the rancidity process of dry roasted peanuts, *J. Agric. Food Chem. 44*(7):1736–1740.

19

Effects of Processing and Storage on the Nutritional Value of Food Proteins

P. A. Finot
Nestec Ltd. Research Centre, Lausanne, Switzerland

I. INTRODUCTION

The nutritional value of a food protein depends on the distribution of the amino acids that can be absorbed in a bioavailable form. This bioavailability may be modified during processing and storage. Although processing provides an additional value to foods in improving their safety, shelf life, palatability, and nutritive value, the conditions applied during treatment and storage may be detrimental to protein.

Most phenomena involved in the improvement in or the loss of both nutritional and physiological properties of food proteins result from the protein denaturation and chemical modification of amino acids. The latter depends much on the individual amino acid structure and the type of treatment and conditions applied. In the interests of better nutrition, these phenomena must be well understood and controlled by the food industry so as to preserve or improve the intrinsic nutritional value of food proteins and to limit, as far as possible, the decrease in bioavailability of the chemically modified amino acids.

II. AMINO ACID AVAILABILITY AND AMINO ACID REQUIREMENT

Each food protein source has an amino acid composition which is genetically determined and which, until recently, could not be modified. We must meet our amino acid requirements from the food proteins available by a judicious mixture of plant and animal foods, which are more or less rich in proteins that are more or less rich in nutritionally important amino acids. The amino acid requirement depends on the capacity of an organism to synthesize them at the level necessary to maintain optimal health.

The 20 amino acids fall into four categories according to their nutritional importance: (a) 8 "essential" amino acids, which cannot be synthesized by our tissues, (b) 2 "semi-essential" amino acids, which are synthesized by our tissues from an essential amino acid, (c) several "conditionally essential" amino acids, which are normally synthesized by our tissues but, for various physiological reasons, are produced at levels below the requirement, and (d) the "nonessential" amino acids, which are never indispensable.

The recommended daily intake of protein is now well established. Patterns of essential amino acids have been proposed for infants and for preschool children and adults [1]. However, the requirement of clinically defined populations for conditionally essential amino acids is now the subject of many investigations, and the recommended values have not yet been completely established (Table 1). Unfortunately, the amino acids that are the most necessary in our foods are also the most sensitive to destruction during processing and storage. As such, all risks of reduction of essential, semi-essential, or conditionally essential amino acids necessitates evaluation. This is especially true for population groups that have a low protein intake, an elevated requirement for high-quality protein (e.g., infants), or a high requirement for specific amino acids (e.g., patients in clinics).

III. PROTEIN AND AMINO ACID MODIFICATIONS—EFFECTS ON BIOAVAILABILITY

Proteins and individual amino acids react differently to processing and storage conditions. Conventional technologies are used mainly to preserve foods, to isolate food fractions, or to improve functional, organoleptic, or nutritional properties. The treatments applied include heat, pH modification, water removal, enzymes, and chemical modifications. The nutritional consequences of certain processes are shown in Table 2.

The nutritional impact of these modifications must be evaluated in the context of the utilization of the food. Criteria of importance are the "essentiality" of the amino acid(s) involved, the quantitative amino acid loss, expressed as a percentage of the original value, the remaining value expressed in terms of bioavailability, the possible reduction in bioavailability of other amino acids, and the significance of all losses and remaining levels compared to requirements. The last two points merit further comment.

Amino acids located adjacent to a chemically modified amino acid cannot be completely liberated from the protein by the gut enzymes and are therefore not absorbed. In addition, some reactions lead to cross-linkages between protein chains, which contributes to a reduction of total nitrogen digestibility, and hence to decreased availability of all other amino acids. In either case, the bioavailability of the amino acids involved can be estimated by multiplying the amino acid content of the food proteins obtained by chemical analysis by a digestibility factor obtained by a nitrogen balance test in rats or by an in vitro method. The proposed "protein digestibility-corrected amino acid scores" indirectly take these values into account [1].

There are also amino acids which, after modification during processing, are partially or still completely biologically available. Analysis of these amino acids by chemical methods does not reflect their real content, and the protein digestibility-corrected amino acid score is no longer valid. Biological methods are necessary to evaluate the bioavailability of such amino acids.

TABLE 1 Suggested Patterns of Amino Acid Requirement

Amino acids	Suggested pattern of requirement (mg/g crude protein)	
	Infant (0–2 y)[a]	Preschool children and adults
Essential and (semi-essential) amino acids[c]		
Isoleucine	46 (41–53)	28
Leucine	93 (83–107)	6
Lysine	66 (53–76)	58
Methionine + (Cystine)	42 (29–60)	25
Phenylalanine + (Tyrosine)	72 (68–118)	63
Threonine	43 (40–45)	34
Tryptophan	17 (16–17)	11
Valine	55 (44–77)	35
Conditionally essential amino acids		
Histidine	Prematures + infants (mg/g protein)	26 (18–36)
Cystine	Prematures (mg/g protein)	2.1
Glycine	Prematures	>mother's milk
Arginine	Premature/infant	>mother's milk
Glutamine	Immunodepressed patients (g/day)	5–50[b]
Arginine	" (g/day)	up to 10[b]
Cysteine	Source of glutathione (g/day)	>>requirement[b]

[a] Mean values followed by range in parentheses.
[b] Levels actually studied by clinicians.
[c] Safe level of reference protein: infants (0–24 months) 2.2–1.1 g/kg; preschool children, (2–5 y) 1.10 g/kg; schoolchildren, 0.99 g/kg; adults, 0.75 g/kg [1].

TABLE 2 Positive (a) and Negative (b) Nutritional Effects of Process Treatment and Storage on Proteins and Amino Acids

Conditions	Phenomena	Nutritional effects
Processing		
Heat treatment	Protein denaturation	Improvement of intrinsic digestibility (a) Reduction of trypsin inhibitor activity (a)
	Heat-sensitive amino acids	Destruction (b)
	Intramolecular reactions	Cross-linkages (b)
	Reaction with sugars	Maillard reaction (b)
pH modification		
	Solubilization	Protein fractionation/isolation (a) (b) Risk of oxidation (b)
	Acid/alkaline hydrolysis	Improvement of solubility (a) Unspecific peptide bond breakage (b)
	pH-sensitive amino acids	Destruction, cross-linkages (b) Isomerization (racemerization) (b)
Protein fractionation		
	Membrane technology	Protein/peptide enrichment (a)
	pH modification	Change in amino acid composition (a) (b)
Enzymatic reactions		
	Proteases	Peptides (a) (b)
	Oxygenases	Oxidation of amino acids through lipid or polyphenol oxidation (b)
Storage		
	Reaction with sugars	Maillard reaction (b)
	Presence of oxygen	Oxidation (b)
Technologies to improve functional, organoleptic, and nutritional properties		
Functionality	Structural modifications	Reductive alkylation, acylation (b)
	Enzymatic modifications	Proteolysis (a), cross-linkages (b)
Organoleptic	Maillard reaction	Loss of amino acids (b)
Nutrition	Proteolytic enzymes	Better absorption, less allergenic (a)
	Covalent fixation of amino acids	Amino acid fortification (a)

The nutritional significance of amino acid loss depends upon the contribution of the food in question to the whole diet, and amino acid availability must be compared to the daily requirement for the amino acids involved. The following sections will present the potential risks for each of the amino acids with regard to modification of their bioavailability.

Table 3 Biological Availability of Lysine Derivatives

Lysine derivatives	(%) Biological availability
Alkyl derivatives	
Schiff's bases of aliphatic aldehydes	100 (2) (22)
Schiff's bases of reducing sugars	100 (2) (22)
Schiff's bases of aromatic aldehydes (gossypol)	0 (11)
Amadori compounds	0 (22)
"Advanced" Maillard compounds	0 (23)
Lysinoalanine	0 (24,25)
Glycitol-lysine (reductive alkylation)	0 (44)
Oxidized polyphenols (quinones)	0 (42)
Acylated derivatives	
ε-(γ-Glutamyl)-lysine (free)	80–100 (2)
ε-(γ-Glutamyl)-lysine (protein bound)	Same order[a]
ε-(β-Aspartyl)-lysine (free and bound)	0 (2)
ε-(Amino acyl)-lysine (free)	70–100 (2)
ε-(Amino acyl)-lysine (protein bound)	Same order[a] (41)
ε-Formyl-lysine and ε-acetyl-lysine	60–80 (2)
ε-Propionyl-lysine (and long-chain fatty acyl-)	0 (2,26)

[a] Bioavailability probably of the same order as the corresponding free isopeptides.
Source: Refs. 2,11,22,23,24,25,26,41,42,44.

IV. CHANGES AFFECTING LYSINE AVAILABILITY

Lysine is the most readily chemically modified essential amino acid. Its ε-amino group can react with many food components and can be modified by chemical as well as enzymatic systems, including ones that improve the functional properties of food proteins. Such modifications have different effects on the biological availability of lysine, depending on the chemical structure of the derivatives formed (Table 3).

A. Reaction with Reducing Sugars: "Early" Maillard Reaction

The "early" step of the Maillard reaction leads to the formation of an Amadori compound (reaction with an aldose sugar) or to a Heyns compound (reaction with a ketose) through the formation of Schiff's bases. The Schiff's bases of lysine formed with sugars are biologically available, whereas the Amadori and the Heyns compounds are not [2]. Quantitative evaluation of the Amadori compound of lysine can be performed using the furosine method [3]. Furosine, which is not present in foods as such, is produced during acid hydrolysis of Amadori compounds. The yield of its formation has been determined in well-controlled hydrolysis conditions from synthetic-free fructose lysine and from protein-bound fructose lysine as lactulose lysine. Its amount in an acid hydrolysate allows the quantification of the original level of the Amadori compound [3]. The nutritional consequences of the lysine loss must be evaluated in the context of the use of the particular products. Table 4 gives lysine loss values for milk, infant formulas, and pasta. In milk and infant formulas, the lysine loss has a limited nutritional impact, as they contain much more lysine (8.3 and 10–11 g/16 g N, respectively) than the recommended

Table 4 Maillard Reaction on Lysine and Nutritional Consequences

Product	Lysine blockage (%)	Nutritional consequences
Heat-treated milk		
Pasteurization	0	Negligible
UHT	0–2	Negligible
"In can" sterilization	10–15	Negligible
Storage of milk powders		
Low A_w	Slow evolution, up	Negligible reduction if
High A_w	to 50%	lysine <recommended
	depending on T	level
Spray-dried infant formulas		
normal	15–20	negligible
hypoallergenic	15–20	risk for N-terminal AA
Pasta		
Low T	8–12	Negligible
High T	20–35	Negligible

T = temperature.
Source: Regs, 4,6,8–10,30.

level for adults and infants (5.5 and 6.7 g/16 g N, respectively). This means that in infant formulas made of cow's milk proteins, up to 15–20% of lysine can be blocked [4] without any relevant nutritional consequences. In order to avoid nutritional problems, the *Codex Alimentarius* [5] recommends that the protein level "shall not be less than 1.8 g per 100 available calories of nutritional quality equivalent to that of casein or a greater quantity of another protein in proportion to its biological value. The quality of the protein shall not be less than 85% of that of casein."

The evolution of lysine blockage in milk powders has been studied in many model systems. The most important parameters responsible for the evolution of the "early" Maillard reaction are water activity and temperature. In such models, it has been shown that up to 50–60% of lysine can be blocked with no browning development [6]. In practice, milk powder manufacturers especially control the drying conditions of milk powders destined for countries with hot climates and which do not necessarily have facilities to keep the products at a reasonably low temperature. If the residual water is below 2.0%, the Maillard reaction during storage at room temperature is negligible.

The case of hypoallergenic formulas produced from enzymatically hydrolyzed milk proteins (see Chapter 15), must be considered with caution as the level of lysine blockage is not representative of the nutritional loss due to the Maillard reaction. In such products, the amino group of every peptide may react with lactose, which reduces its bioavailability and that of the adjacent amino acids. No routine analytical method is able to quantify this type of reaction, which contributes to the reduction in the protein value. In this case, only animal assays can give a valid answer.

The lysine level in weaning foods made of cereals and milk is of prime importance as such products make a major contribution to protein supply. Depending on the respective proportions of cereals and milk, the lysine content is difficult to maintain at the recommended level, especially in instant products where the Maillard reaction has been exploited to improve their taste. For such products, the process has to be optimized to limit the Maillard reaction in order to maintain the nutritional quality. *Codex Alimentarius* recommends a minimum level of 15% protein, no fortification with lysine when destroyed by the process, and a nutritive value of at least 70% casein [7].

The lysine loss in pasta dried at high temperature (20–35%) compared to the loss at low temperature (8–10%) [8,9] has a very limited nutritional impact, as pasta is not a major source of lysine. The ingestion of 100 g of pasta, which is the mean quantity eaten by the Italian population, provides 25% of required protein and only 8.5% of required lysine. A lysine blockage of up to 35% corresponds to only 3% of the daily lysine requirement, which is negligible. In addition, in Italy pasta is always associated with other sources of lysine, such as cheese, eggs, and/or meat. Animals fed a diet containing pasta dried at low or high temperature, alone or in association with a source of legume protein, did not show any decreased growth [10].

B. Reaction with Carbonyls: "Advanced" Maillard Reaction

Many aldehydes can react with the ϵ-amino group of lysine. Some are present in foods as natural components, e.g., gossypol, a polyphenol present in cottonseed proteins. Others are produced during food processing or storage as by-products of the Maillard reaction or by lipid peroxidation (Table 5).

The stability of the Schiff's bases formed with the ϵ-amino group of lysine depends on the chemical structure of the aldehydes involved. Aliphatic aldehydes and reducing sugars produce reversible Schiff's bases, which are rapidly hydrolyzed at the acidic pH of the stomach, liberating lysine, which thus remains biologically available. In contrast, aromatic aldehydes produce very stable Schiff's bases, which are not hydrolyzed in the stomach; this is the case with gossypol, a mixture of different polyphenols with aldehyde groups [11]. The "advanced" Maillard reaction leads to the formation of carbonyl compounds resulting from the 1-2 and 2-3 enolization of the lysine-sugar adduct (3-deoxyglucosone and 1-deoxyglucosone, respectively) [12] and of their decomposition

TABLE 5 Origin, Structure, and Effects of Aldehydes in Foods

Name	Structure	Effect on	
		Browning	Cross-links
Natural products			
Gossypol	Polyphenolaldehydes	–	+++
Advanced Maillard reaction			
Dihydroxyacetone	$CH_2OH—CO—CH_2OH$	+++	
Monohydroxyacetone	$CH_3—CO—CH_2OH$	–	
5—OH—methylfurfural	CH_2OH—(furfuryl)—COH	–	
Methylglyoxal	$CH_3—CO—COH$	+++	+
Glyceraldehyde	$CH_2OH—CHOH—COH$	±	+
Lipid oxidation			
Hexenal	$COH—CH{=}CH—(CH_2)_2—CH_3$	–	–
Malondialdehyde	$COH—CH_2—COH$	–	+++

–, No effect; +, slight effect; +++, strong effect.
Source: Refs. 11,20,27.

products as presented in Table 5. These highly reactive components are able to react again with the ϵ-amino group of lysine and also with other amino acids, such as methionine, cystine, tryptophan, and arginine, transforming these into unavailable derivatives, the structures of which are not completely known. In parallel, polymerization reactions take place in developing browning, and cross-linkages are created. Both types of reaction contribute to a reduction in nitrogen digestibility [13].

During the "advanced" Maillard reaction, lysine is still the most sensitive amino acid. In sterilized liquid milk, some "advanced" Maillard reaction may occur. This is deduced directly from the production of some browning and indirectly from the difference between initial lysine and lysine not detected as its Amadori compound. In milk powder with high water activity stored at high temperature, an advanced Maillard reaction can take place and intense browning develops. This evolution is autocatalytic as the molecules of water produced by the Maillard reaction help accelerate the reaction. Changes in the transition temperature of lactose appear to be important in this reaction. In such milk powders, lysine loss is much more important than the loss of the other amino acids susceptible to reaction with carbonyls, like tryptophan, methionine, cystine, and others [6].

Two lysine derivatives—ϵ-pyrrol-lysine and ϵ-carboxymethyl-lysine—have been identified as advanced Maillard products (Fig. 1). ϵ-Pyrrol-lysine [14], also called pyrraline, results from the dehydration of the derivative produced by the reaction between lysine and 3-deoxyglucosone: it has been found present in the free state in dried foods [15] and bound to proteins in processed milk [16] and in pasta [17]. Unstable in acidic conditions, it can only be detected after enzymatic hydrolysis. ϵ-Carboxymethyl-lysine, which results from the oxidation of the Amadori compound of lysine [18] (Fig. 1), is acid stable and can be measured in the acid hydrolysates of proteins [19]. Both derivatives are considered to be markers of the "advanced" Maillard reaction but cannot yet be used to quantify its extent.

The free aldehydes produced during lipid oxidation can react with lysine and participate to some extent in a reaction similar to the Maillard reaction. In this context,

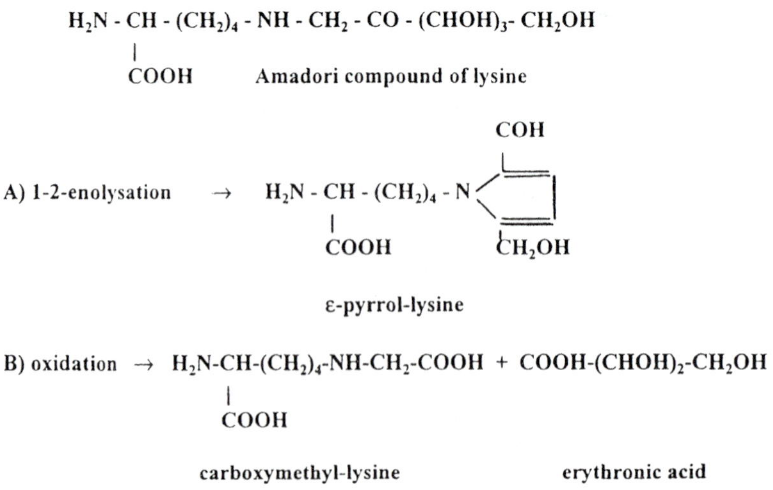

$H_2N-CH(COOH)-(CH_2)_4-NH-CH_2-CO-(CHOH)_3-CH_2OH$

Amadori compound of lysine

A) 1-2-enolysation → $H_2N-CH(COOH)-(CH_2)_4-N$< (pyrrole ring bearing COH and CH_2OH)

ε-pyrrol-lysine

B) oxidation → $H_2N-CH(COOH)-(CH_2)_4-NH-CH_2-COOH$ + $COOH-(CHOH)_2-CH_2OH$

carboxymethyl-lysine   erythronic acid

FIGURE 1 Identified advanced maillard products of lysine.

Table 6 Loss of Lysine and Other Amino Acids and Nitrogen Digestibility in Model Systems Composed of Whey Protein and Oxidizing Linolenate (in % of untreated whey)

	Reactive lysine	Tryptophan by HPLC	Methionine sulfoxide	Nitrogen digestibility
Basal conditions	40	86	93	71
Low water activity	85	100	75	98

Source: Ref: 13.

structural information on the possible cross-linking between proteins and malondialdehyde has been obtained from spectral analysis [20]. In addition, the identification of *N*-ϵ-(2-propenal)-lysine as a major urinary metabolite of malondialdehyde [21] is indirect proof of its formation in food systems. Loss of lysine, methionine, and tryptophan and the nutritional consequences have been studied in protein-oxidizing lipid model systems [13] (Table 6).

C. Reactions with Dehydroalanine: Formation of Lysinoalanine

Dehydroalanine results from the heat of alkaline degradation of cysteine and serine phosphate, through a beta-elimination reaction (Fig. 2). Dehydroalanine reacts with the ϵ-amino group of lysine to produce lysinoalanine and with other amino acids, creating

A) Formation of dehydroalanine

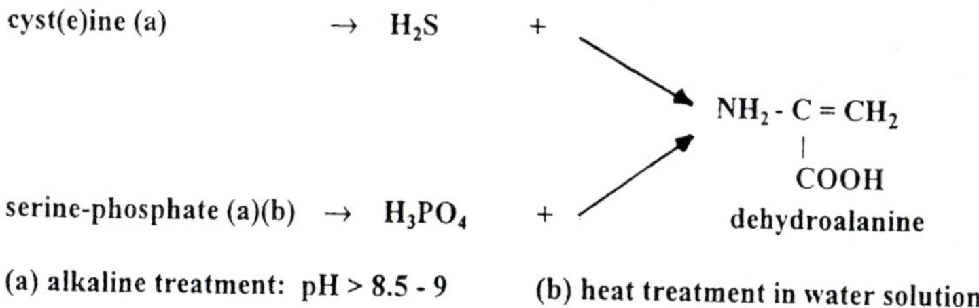

B) Reactions involving dehydroalanine

dehydroalanine + H_2O → D,L-serine

+ NH_3 → β-aminoalanine

+ L-lysine → L-lysino-D,L-alanine

+ cysteine → lanthionine

+ histidine → histidinoalanine (Figure 11)

Figure 2 Formation of dehydroalanine during processing and subsequent amino acid modifications.

cross-linkages between protein chains and contributing to the reduction in nitrogen digestibility [28,29].

The loss of lysine through this reaction is generally negligible as the level of lysinoalanine produced, expressed in ppm, corresponds only to a small percentage of lysine present in the protein. For example, in casein-based milk products rich in serine phosphate, the maximum level of lysinoalanine is around 1200 ppm [29,30], which corresponds to less than 0.1% of total lysine.

If lysinoalanine is produced from serine phosphate, as in alkali-treated casein and caseinates (which are naturally poor in cystine), the nutritional loss results mainly from the reduction of nitrogen digestibility due to cross-linkages. If lysinoalanine is produced from cystine, as in plant protein isolates, the nutritional loss is due to the reduction of nitrogen digestibility but mainly from the destruction of cystine (Table 7). In alkali-treated proteins, the isomerization of some amino acids into their D-isomers may be responsible for the reduction of both digestibility and biological value.

Though the nutritional loss due to reduction in lysine availability is negligible, the presence of a large amount of lysinoalanine as the free amino acid raised some toxicological issues in the 1970s and 1980s. Rats fed for several weeks a diet containing 100 ppm or more of free lysinoalanine develop nephrocytomegaly, a pathological defect involving cytological modification of certain kidney cells [31]. Much higher levels of dietary protein–bound lysinoalanine were necessary to produce a similar effect.

This pathological defect is species specific. It was found to be reversible and not to modify normal kidney functions in rats [29]. In the same species, the L-D isomer of lysinoalanine was found to be much more active in inducing nephrocytomegaly than the L-L isomer [32]. Subsequent work has demonstrated that protein-bound lysinoalanine was very poorly absorbed and that the most active L-D isomer was less well absorbed (~fivefold) than the L-L isomer [33]. This last study demonstrated that the phenomenon observed in rats fed synthetic lysinoalanine was not relevant when feeding the same molecule bound to proteins. Confirmation of this can be found in a study made of premature babies fed formulas processed conventionally and containing up to 1000 ppm of lysinoalanine. Even at these levels, no modification of kidney function was observed, with an average 6–9% excretion of total dietary lysinoalanine in urine [34].

D. Formation of Isopeptides

High-temperature treatment of food proteins produces two isopeptides, $N\epsilon$-(γ-glutamyl)-lysine and $N\epsilon$-(β-aspartyl)-lysine, which result from the transamidation reaction between glutamine or asparagine residues and the ϵ-amino group of lysine [35,36] (Fig. 3). Rat experiments have shown that free synthetic ϵ-(γ-glutamyl)-lysine was 80–100% biologically available [2] and that, incorporated in a food, it appears to be more stable than free lysine to Maillard reaction [37]. In contrast, the free synthetic ϵ-(β-aspartyl)-lysine was found not to be bioavailable [2]. There is much evidence that the ϵ-(γ-glutamyl)-lysine isopeptide has the same bioavailability when bound to proteins as when free, because proteolytic enzymes liberate it in the intestine (38).

Transglutaminase, an enzyme of animal or bacterial origin, has the ability to bind an amino acid on the carboxamide group of glutamine [39]. This enzyme has been proposed as a means of increasing the gelling properties of proteins. Added to a protein solution, this enzyme creates ϵ-(γ-glutamyl)-lysine cross-linkages between protein chains, increasing its gelling properties.

TABLE 7 Effects on Digestibility (D) and Biological Value (BV) of Proteins Containing Lysinoalanine (ppm) Produced from Serine Phosphate or Cystine

Product	Lysinoalanine	D	BV
Soybean oil meal			
Untreated	0	89	71
Alkali-treated[a]	5600	91	45
Casein			
Untreated	0	101	63
Alkali-treated[b]	11500	90	59

[a] Lysinoalanine produced from cystine.
[b] Lysinoalanine mainly produced from serine phosphate.
Source: Ref. 28.

Other ϵ-isopeptides of lysine have been found to be biologically available [2] (Table 2). Covalent attachment of amino acids by enzymatic [40] or chemical means [41] (Fig. 3) on the ϵ-amino group of lysine has been proposed to improve the nutritive value of imbalanced proteins. Animal experiments effectively showed that protein-bound isopeptides of methionine actually improve the nutritional performance of casein [41].

The biological mechanism involved in this process has been questioned, and the enzymatic system involved in the hydrolysis of such isopeptides has been studied [2]. It appears that ϵ-acyl-lysinase or ϵ-aminoacyl-lysinase exist only in the kidney. Therefore, these isopeptides must be liberated by intestinal enzymes, absorbed intact by the gut, and transported to the kidney where they are hydrolyzed. Once liberated, both amino acids return to the bloodstream and are available for protein synthesis.

E. Reaction with Oxidized Polyphenols

Many plant protein sources of good nutritional value, such as sunflower, rapeseed, and potato, contain polyphenolic compounds like caffeic acid or chlorogenic acid, which,

A) Heat treatments

lysyl residues + glutaminyl residues → ε-(γ-glutamyl)-lysine (35)

+ asparaginyl residues → ε-(β-aspartyl)-Lysine (35)

B) Transglutaminase enzyme

lysyl residues + glutaminyl residues → ε-(γ-glutamyl)-lysine (39)

lysyl residues + methionyl-ethylesters → ε-(methionyl)-lysine (40)

C) Chemical reaction: (carbodiimide method)

lysyl residues + BOC-methionine → ε-(methionyl)-lysine (41)

FIGURE 3 Formation of isopeptides of lysine in proteins. (From Ref. 35, 39–41.)

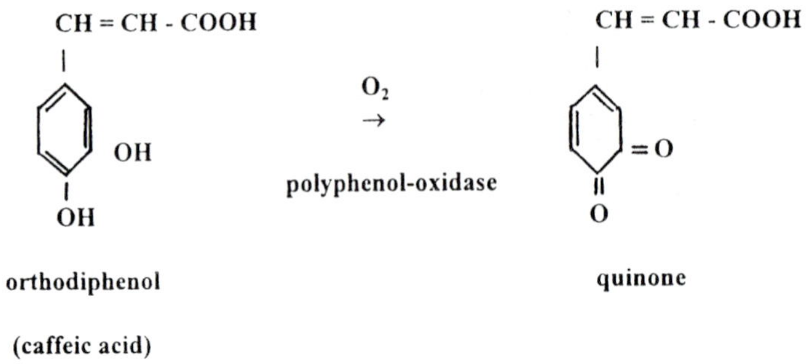

CH = CH - COOH → CH = CH - COOH

OH

O_2 → polyphenol-oxidase

= O

OH O

orthodiphenol quinone

(caffeic acid)

quinone	+ lysine	→	addition compound
	+ cysteine	→	addition compound
			oxidation compounds ?
	+ methionine	→	oxidation compound
	+ tryptophan	→	oxidation + addition compounds
	+ quinones	→	polymerisation

FIGURE 4 Chemical reactions between polyphenols and lysine in food proteins.

under oxidative conditions catalyzed by alkaline treatments or oxidative enzymes (polyphenol oxidase), are oxidized into quinones (Fig. 4). Quinones can bind the ϵ-amino group of lysine and polymerize into brown pigments, either free or bound to proteins, leading to a reduction in the bioavailability of lysine and in nitrogen digestibility [42,13]. The oxidation of polyphenols catalyzes the oxidation of sulfur amino acids, but has practically no effect on tryptophan (Table 8).

F. Acylation

Formylation, acetylation, propionylation, and succinylation of food proteins have been proposed to modify their functional properties [43]. Lysine is the most affected amino acid. Studies carried out with free synthetic acylated lysines have shown that ϵ-formyl-

TABLE 8 Effect of Oxidized Polyphenols on Nutritional Parameters of Casein Treated with Caffeic Acid in Presence of Polyphenoloxidase at pH 7.0 (in % of untreated casein)

		Methionine		Tryptophan		Nitrogen digestibility
	Lysine	intact	as sulfoxide	by HPLC	availability	
Caffeic acid–treated casein	77	79	18	101	85	92

Source: Ref.13.

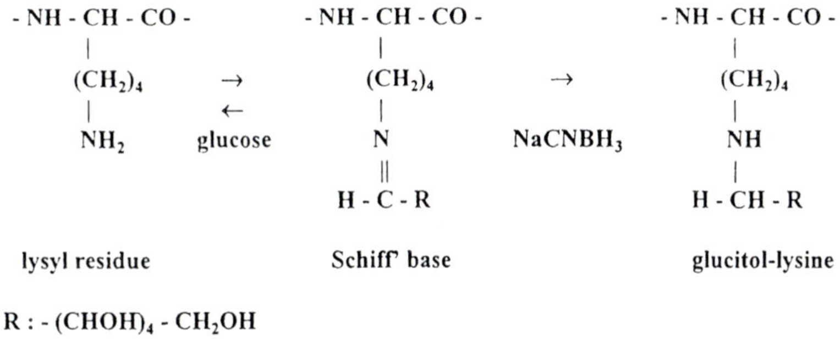

- NH - CH - CO -		- NH - CH - CO -		- NH - CH - CO -
|		|		|
$(CH_2)_4$	→	$(CH_2)_4$	→	$(CH_2)_4$
|	←	|		|
NH_2	glucose	N	$NaCNBH_3$	NH
		‖		|
		H - C - R		H - CH - R
lysyl residue		Schiff' base		glucitol-lysine

R : - $(CHOH)_4$ - CH_2OH

FIGURE 5 Reductive alkylation of lysine with glucose.

lysine and ϵ-acetyl lysine were 60–80% biologically available, while ϵ-propionyl-lysine and long-chain ϵ-fatty acyl-lysine were unavailable. From the nutritional point of view, the attachment of an acyl derivative to the ϵ-amino group of lysine is very different from the attachment of an amino acid to the same position of lysine. In the former case, only formyl and acetyl derivatives were found to be partially biologically available, while the ϵ-amino acyl-lysines were found to have good bioavailability (Table 3).

G. Reductive Alkylation

Reductive acylation of proteins (reaction with formaldehyde or reducing sugars in the presence of a strong reducing agent like sodium borohydride or cyanoborohydride) transforms the ϵ-residues of lysine into alkyl lysines. All alkyl-lysines tested in rats have been found to be biological unavailable (Table 3), because mammals do not possess the enzyme ϵ-alkyl-lysinase. The low nutritional performance of casein covalently bound to glucose through a reductive alkylation reaction proves that, when bound to proteins, alkyl-lysines are not biologically available (Fig. 5) [44].

V. CHANGES AFFECTING METHIONINE AND CYST(E)INE AVAILABILITY

A. Oxidation Reactions

Both sulfur-containing amino acids, methionine and cyst(e)ine, are sensitive to oxidation reactions. This can occur during processing and storage of proteins in the presence of oxygen through reactions catalyzed by metal ions, by pigments sensitive to light (photo-oxidation), by enzymes like lipoxigenases and polyphenol oxidases, or in the presence of oxidizing lipids. The utilization of hydrogen peroxide for packaging sterilization, if not well controlled, is also a risk for oxidation. Both methionine and cyst(e)ine are oxidized into several chemical forms. Old studies made with synthetic oxidized molecules [45–47] (Table 9) showed that at low degrees of oxidation, these oxidized amino acids are biologically available. As demonstrated with tissue extracts or by feeding tests in animals, enzymatic systems are able to reduce these oxidized forms into the corresponding original amino acids. More recent studies demonstrated that protein-bound methionine sulfoxide was completely bioavailable [48] (Table 9).

TABLE 9 Biological Availability in Rats of Oxidized Methionine and Cyst(e)ine

	Formula	% Biological availability
Methionine	R—S—CH_3	100
Methionine sulfoxide	R—SO—CH_3	
free		70–96 (45,46,47)
protein-bound		82–113 (48)
Methionine sulfone	R—SO_2—CH_3	0 (47)
Cystine	R′—S—S—R′	100
Cystine disulfoxide	R′—SO—SO—R′	+ (46)
Cystine disulfone	R′—SO_2—SO_2—R′	0 (46)
Cysteine	R′—SH	100
Cysteine sulfenic	R′—SOH	+ (46)
Cysteine sulfinic	R′—SO_2H	0 (46)
Cysteic acid	R′—SO_3H	0 (47)

+, Promotes growth in rats fed a methionine-deficient diet.
Source: Refs. 45–48.

B. Degradation of Cysteine into Dehydroalanine

In alkaline medium at pH higher than 8.5, cysteine is degraded chemically through a β-elimination reaction into dehydroalanine and H_2S (Fig. 2). This reaction has been found to occur during the solubilization step at pH higher than 8.5–9 during the production of protein isolates. The spinning process proposed to texture plant proteins also required a solubilization step at elevated pH. Such alkaline pH conditions responsible for the destruction of cysteine, mainly used to process plant proteins already deficient in sulfur-containing amino acids, contribute to a reduction in the nutritional value of these proteins, and hence have been abandoned for this reason. The conditions of sodium or potassium caseinate production may also contribute to destruction of the small amount of cystine present if homogeneity in the pH is not ensured. In these caseinates, however, lysinoalanine is mainly formed from serine phosphate. Free dehydroalanine has a high reactivity towards many food constituents (Fig. 2) contributing to a reduction in protein quality by formation of cross-linkages (lysinoalanine and lanthionine). Alkaline treatments can also result in racemization of the amino acids. This aspect is described in a later section.

C. Reaction of Cysteine with Chlorinated Solvents

Chlorinated solvents have been used in the past to remove oil from raw fish or from soybean meal and also to extract caffeine from coffee beans. Proteins defatted by such chlorinated solvents were found to depress growth due to the reduction of cysteine availability [49]. Dichlorovinylcysteine, which has been identified as resulting from the reaction of trichloroethylene with cysteine [50] (Fig. 6) has been found to induce toxicity in calves.

VI. CHANGES AFFECTING TRYPTOPHAN AVAILABILITY

The indol group of tryptophan is susceptible to several chemical modifications.

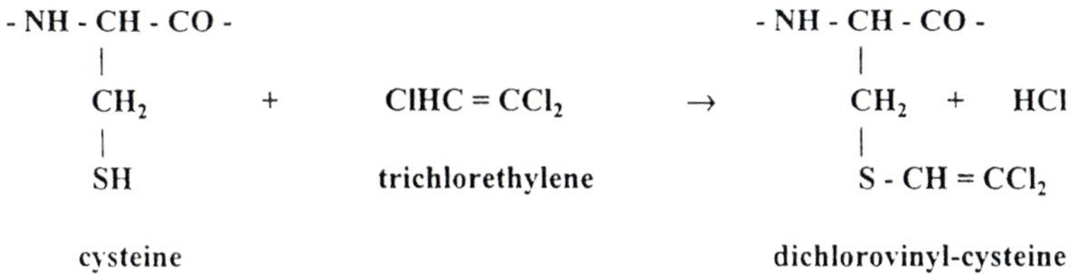

$$
\begin{array}{ccccc}
\text{-NH-CH-CO-} & & & & \text{-NH-CH-CO-} \\
| & & & & | \\
CH_2 & + & ClHC=CCl_2 & \rightarrow & CH_2 \quad + \quad HCl \\
| & & & & | \\
SH & & \text{trichlorethylene} & & S-CH=CCl_2 \\
\text{cysteine} & & & & \text{dichlorovinyl-cysteine}
\end{array}
$$

FIGURE 6 Reaction of cysteine with chlorinated solvents.

A. Oxidation

In the presence of air or oxygen, free tryptophan is thermally degraded to form *N*-formylkynurenine and then kynurenine. In diluted water solution exposed to light, tryptophan is also slowly oxidized. At acidic pH, this oxidation is much more rapid [51]. Bound to proteins, tryptophan is much less sensitive to oxidation. In model systems of oxidation in the presence of hydrogen peroxide (0.2 M at 50°C) or in the presence of oxidizing lipids with an excess of oxygen, the maximum loss has been found to be less than 25% [13,48,52]. During normal conditions of treatment or storage of foods, the loss of tryptophan by oxidation, however, is negligible.

B. Heat Degradation

Heat degradation of tryptophan is associated with oxidation reactions or with reactions in the presence of aldehydes. At temperatures above 200°C, often reached during grilling of meat and fish, tryptophan degrades into α-, β-, and γ-carbolines. Two of the γ-carbolines, as Trp-P-1 and Trp-P-2, are strongly mutagenic in the Ames test. Their formation requires the presence of an aldehyde, which reacts with the α-amino group of tryptophan to produce a new nitrogen-containing molecule [51].

C. Reaction with Carbonyls

The nitrogen of the indole ring of tryptophan can react with free aldehydes to form a Schiff's base [53]. The existence of such a Schiff's base is suggested by the presence of new peaks appearing by HPLC separation of a mixture of free tryptophan and propionaldehyde [54] (Fig. 7). The chemical evolution of this tryptophan derivative is unknown, but if present in a food protein, it is expected to be biologically available. In the early Maillard reaction, the bioavailability of tryptophan is not affected, demonstrating that aldoses do not react irreversibly with the indole ring of tryptophan. The relatively small loss of tryptophan in the advanced Maillard rection confirms that aldehydes do not strongly affect the indole ring of tryptophan.

However, it is well accepted that the instability of tryptophan during acid hydrolysis is due not only to the presence of oxygen but also in large part to the presence of sugars. The formation of Schiff's bases between the two nitrogen atoms of tryptophan and aldehyde groups may be responsible for this degradation.

VII. AMINO ACID ISOMERIZATION: RACEMIZATION

The L-amino acids can be isomerized into D-amino acids (racemization) by heat or by alkaline treatments. Some D-amino acids are also produced by fermentation. This can

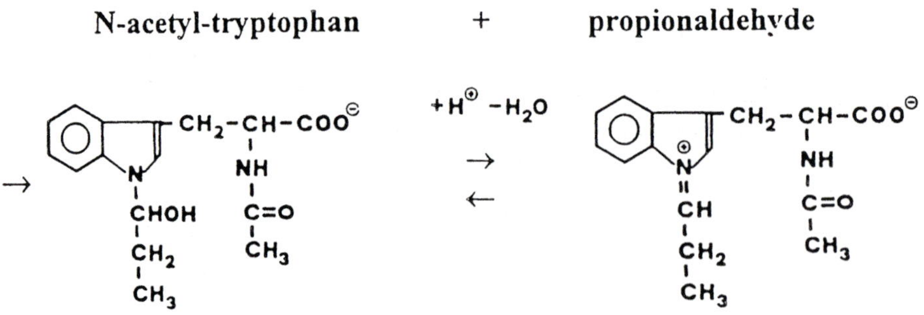

N-acetyl-tryptophan + propionaldehyde

→ CH_2-CH-$COO^{\ominus}$, NH, C=O, CH_3; N–CHOH–CH_2–CH_3 $\underset{\leftarrow}{\overset{+H^{\oplus}\ -H_2O}{\rightarrow}}$ CH_2-CH-$COO^{\ominus}$, NH, C=O, CH_3; $N^{\oplus}$=CH–CH_2–CH_3

FIGURE 7 Reaction of the indol ring of tryptophan with free aldehydes. (From Ref. 54.)

affect the nutritional performance of proteins as most of the essential D-amino acids are poorly or not biologically available, except for D-methionine and D-tryptophan in rats and D-methionine and D-phenylalanine in human beings [55] (Table 10). In addition, proteolytic enzymes of the pancreatic juice and peptidases of the intestinal brush border cannot liberate the racemized or the adjacent amino acids [56]. Both nitrogen digestibility and nutritive value are therefore affected.

A. Effect of Heat Treatments

Severe heat treatment of dried proteins in conditions that are not applied to food systems, except in grain roasting, can racemize amino acids. In plasma albumin (15% water) heated at 121°C for 8 hours, an appreciable isomerization expressed as D/(D+L)(%) was found only with aspartic acid (28.6%) and cystine (29.1%); in chicken muscle heated under the same conditions, only aspartic acid was racemized (22.4%) to a significant extent. Minimal racemization occurred with serine (2.8%) and cysteine (2.6%) and was below 2% with other amino acids [56,57].

TABLE 10 Bioavailability of D-Amino Acids in Rats and Adult Human Beings

	Rats	Humans
D-Methionine	100%	100%
D-Cystine	Not available	
D-Lysine	Not	Not
D-Leucine	Poorly	Not
D-Isoleucine		Not
D-Threonine	Not	Not
D-Valine	Poorly	Not
D-Tryptophan	100%	Not
D-Phenylalanine	Partially	Partially
D-Tyrosine	Partially	
D-Histidine	Partially	

Source: Ref. 55.

In water solution at neutral pH, the racemization rate is very low and can be considered as negligible. Some time ago, an alarmist publication claimed that the reheating of infant formulas with microwaves led to racemization of proline and hydroxyproline, and that this was detrimental for the infant [58]. This led to exhaustive investigations, including the analysis of up to 15 amino acids by several authors, and the demonstration that it was not the case [59–61,63].

B. Effect of Alkaline Treatments

Even at moderate pH (pH 9 at 83°C), all amino acids racemize but at very different rates depending on their structure, in the order cysteine > aspartic acid > serine > all others [55]. The racemization rate of these amino acids differs from one protein to another and is seemingly dependent on the adjacent amino acids in the peptide chain. Under identical conditions, free amino acids are less rapidly isomerized than protein-bound amino acids.

C. Effect of Fermentation

All fermentation methods produce some free D-amino acids that are synthesized by the microorganisms involved [61,62], but only at low levels, which cannot influence the protein quality.

VIII. HEAT-UNSTABLE AMINO ACIDS

Several amino acids, in the dry form and in water solution, are modified by heat treatment. In the dry form, high temperatues such as are obtained in roasting conditions affect many amino acids by deamidation, transamidification, dehydration, β-elimination reaction, and isomerization, as shown in Figure 8. In water solution, however, only a few amino acids appear to be affected.

IX. pH-SENSITIVE AMINO ACIDS

Both γ- and β-amides (glutamine and asparagine, respectively,) are very sensitive to deamidation even at weak acidic pH. They produce glutamic and aspartic acids, respectively, and liberate free NH_3. In practice, this type of reaction occurs in foods at a very

glutamine	→	glutamic acid + NH_3
glutamine + lysine	→	ε-(γ-glutamyl)-lysine
glutamine (amino terminal)	→	pyrolidone-carboxylic acid
asparagine	→	cyclic amides + NH_3
L-amino acids	→	D-amino acids (aspartic, cystine, serine)
serine-phosphate	→	dehydroalanine → lysinoalanine

FIGURE 8 Amino acids unstable at high temperature.

A) Acidic pH

glutamine / asparagine	→	glutamic / aspartic acids + NH_3
tryptophan	→	destruction (presence of oxigen or sugars)

B) Alkaline pH

cyst(e)ine	→	dehydroalanine (see figure 2)
serine-phosphate	→	dehydroalanine (see figure 2)
arginine	→	ornithine + urea
ornithine + dehydroalanine	→	ornithinoalanine
L-amino acids	→	D-amino acids (cysteine, serine, aspartic)

FIGURE 9 Amino acid stability at acidic and alkaline pH.

low level. It is mainly with respect to the analytical procedure that these reaction are important as, due to their acid-sensitive nature, it is impossible to measure them by the standard method of amino acid analysis. This is also the case for tryptophan, which is completely degraded under the conditions used to hydrolyze food proteins. Residual oxygen or carbohydrates can act as catalyzers of this degradation. For the analysis of tryptophan in food proteins, alkaline pH and absence of oxygen are therefore required. Under the hydrolytic conditions for amino acid analysis, methionine sulfoxide as well as cysteine sulfenic and cysteine sulfinic acids are also destroyed.

As shown in Figure 9, several modifications occur at alkaline rather than at acidic pH.

X. ENZYMATICALLY MODIFIED PROTEINS

Many proteases are now available at a reasonable price that can be used industrially: pancreatin, trypsine, and chymotrypsine from hog pancreas and alkalase, neutrase, and others from bacteria.

A. Improvement of Functional Properties

Proteases are used to solubilize soy protein isolates or to avoid the heat coagulation of whey proteins utilized in soy-based or whey-based infant formulas or in clinical products. In contrast, transglutaminase has been propose to increase the gelification properties of soluble proteins [39].

B. Reduction of Allergenicity and Improvement of Amino Acid Absorption

Hog pancreatic proteases are used to produce small peptides in order to reduce the allergenicity of milk proteins in infant formulas [64,65]. The same enzymes or others of bacterial origin are also used for clinical products with the aim to improve the intestinal absorption [66] of patients with malabsorption problems or with small bowel syndrome [67,68].

C. Amino Acid Fortification

Proteolytic enzymes have also been proposed to enrich imbalanced proteins by fixing the limiting amino acid(s) into them via the plastein reaction [69,70]. Finally, transglutaminase has been proposed to improve the functional properties of proteins by the creation of the isopeptide ϵ-(γ-glutamyl)-lysine [39]. This same enzyme has also been proposed to enrich an imbalanced protein with its limiting amino acid in creating γ-glutamyl peptides [71].

XI. EFFECT ON VARIOUS AMINO ACIDS

During processing and storage, various other amino acids are also modified. Although these changes in structure are not quantitatively as important as that observed with lysine, tryptophan, and sulfur amino acids, they can be seen as markers of the processes used and of their severity.

A. Changes Affecting Arginine Availability

The main modification of arginine occurs during alkaline treatment by transformation into ornithine and urea (Fig. 9). During the Maillard reaction side chain of arginine can react with advanced Maillard products such as 3-deoxyglucosone [72] or methylglyoxal [73] to produce novel imidazolone compounds (Fig. 10).

B. Changes Affecting Histidine Availability

The oxidation of histidine by photo-oxidation in the presence of chlorophyll was described in 1968 [74]. Recently, histidine oxidation catalyzed by Amadori compounds and mediated by copper was also observed [75]. The nutritional significance of this reaction, studied in model systems, has not yet been evaluated in food systems.

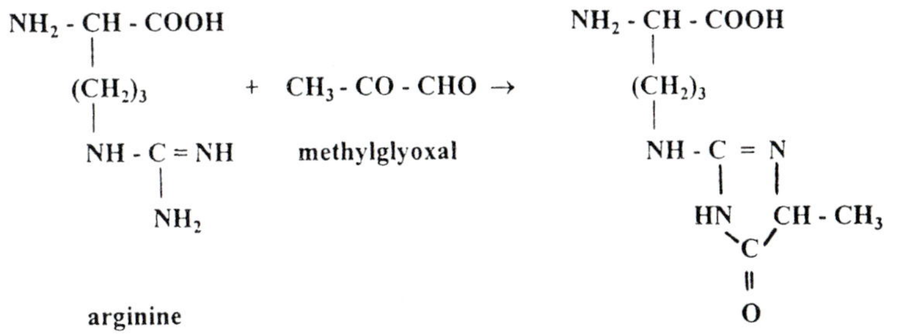

NH_2 - CH - COOH | $(CH_2)_3$ | NH - C = NH | NH_2 (arginine) + CH_3 - CO - CHO (methylglyoxal) → NH_2 - CH - COOH | $(CH_2)_3$ | NH - C = N | HN CH - CH_3 C = O

arginine

methylglyoxal

4-methyl-5-oxo, 4-hydroimidazol-

ornithine

FIGURE 10 Identified advanced Maillard producer of arginine. (From Ref. 65.)

Recently, a new derivative of histidine, histidinoalanine has been identified in heated milk products at levels between 50 and 1800 mg/kg protein, comparable to the level of lysinoalanine [76] (Fig. 11). Its presence, which results from the reaction between histidine and dehydroalanine (produced from serine phosphate) contributes to the formation of cross-linkages in proteins and to the reduction of both histidine availability and nitrogen digestibility.

C. Changes Affecting Phenylalanine and Tyrosine

Both aromatic amino acids, phenylalanine and tyrosine, can be oxidized into their hydroxy derivatives, o-hydroxytyrosine and dihydroxytyrosine, respectively. This oxidation results from a free radical mechanism and has been found to occur during the irradiation of foods [77]. These molecules are then considered as markers of this process.

XII. EFFECTS OF TANNINS

Tannins, which result from the polymerization of different groups of polyphenols, are present in many foods or drinks of plant origin, including tea, cocoa, sorghum, and millet. Since tannins have the property to bind proteins, they affect protein quality. The effect of tea, cocoa, and carob tannins on nitrogen digestibility has been studied in rats [78]. All tannins contribute to decrease nitrogen digestibility. Results indicate that both tea and carob increase fecal nitrogen excretion in rats fed casein-based diets. In contrast, although cocoa powder also increased the fecal nitrogen excretion, the nitrogen digestibility of the casein in the diet was not reduced. This is explained by the fact that tea and carob tannins are free and can bind dietary proteins, while cocoa tannins are already bound to cocoa proteins and do not affect digestibility of dietary proteins. Free tannins of tea have also been found to increase endogenous nitrogen loss by binding with secretory proteins of the intestinal tract [78].

XIII. PROTEIN DENATURATION

Many food-processing methods such as heat, acidic and alkaline, and solvent treatments can denature proteins and therefore change their nutritional and biological properties:

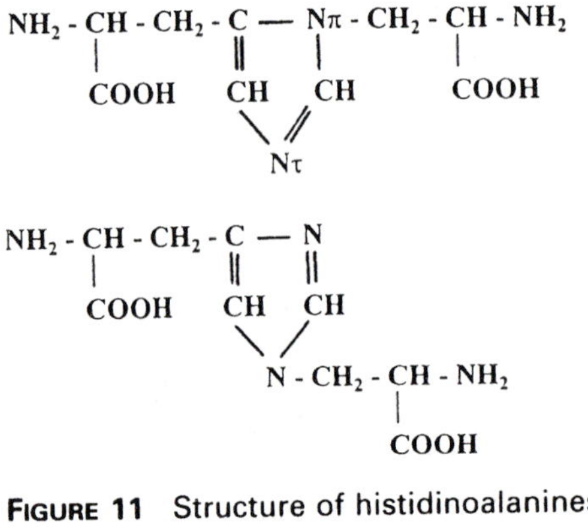

```
NH2 - CH - CH2 - C — Nπ - CH2 - CH - NH2
      |          ||  |           |
    COOH        CH   CH        COOH
                  \ //
                   Nτ

NH2 - CH - CH2 - C — N
      |          ||  ||
    COOH        CH   CH
                  \ /
                   N - CH2 - CH - NH2
                             |
                           COOH
```

FIGURE 11 Structure of histidinoalanines. (From Ref. 68.)

rate of digestion, digestibility, enzymatic, immunological and physiological activities. In some cases, protein denaturation is necessary to inhibit negative effects; in other cases, protein denaturation must be avoided to maintain positive physiological effects.

A. Rate of Digestion/Digestibility

One effect of cooking is to increase the digestibility of food proteins, mainly those from plant sources. Heat treatments destroy the cell matrix and make proteins available for enzymatic attack. In principle, protein denaturation that accompanies heat treatments also contributes to increased protein digestibility. In animal proteins, however, it has been found that raw meat was more rapidly hydrolyzed in vitro than cooked meat [79].

B. Trypsin Inhibitors

Heat treatments also deactivate trypsin inhibitors in plant proteins. This group of proteins can inhibit the activity of trypsin and of chymotrypsin in the intestinal tract. The primary effect of these trypsin inhibitors is to reduce nitrogen digestibility. The secondary effect is to stimulate pancreatic secretion by a negative feedback mechanism controlled by the polypeptide gut hormone CCK, which results in some experimental animals in pancreatic hypertrophy after several weeks of feeding.

Soy proteins contain two main categories of trypsin inhibitors commonly referred to as Bowman-Birk types (molecular weight 6,000–10,000) and Kunitz types (molecular weight 20,000). It is generally accepted, while not definitely demonstrated, that trypsin inhibitors reduce the digestibility of exogenous food proteins and increase the loss of endogenous nitrogen, probably the oversecreted pancreatic enzymes.

Heat treatments carried out in the presence of water appear to inactivate trypsin inhibitors more efficiently than toasting in the absence of water. The trypsin inhibitor activity has been tested in many soy-based commercial products and found to be 73–94% of the initial activity. It is still not known what residual level of trypsin inhibitor activity is necessary to provoke a physiological response in humans and what residual level may be accepted as safe for the human populations [80].

C. Lectins

Lectins are proteins that are widely distributed in the plant kingdom and have the unique property of binding to carbohydrate-containing molecules. One of their properties is their ability to agglutinate the red blood cells from various species of animals. Their interaction with multiple binding sites of intestinal cells leads to hyperplasia of the crypt cells of the small intestine and also to hypertrophy of pancreas by a mechanism similar to trypsin inhibitors. Soybean agglutinin has a molecular weight of 120,000 and is easily denatured by heat treatments, which inactivate it [80].

D. Proteins and Peptides with Physiological Activities

As described in Table 1, many proteins of animal origin have interesting physiological properties that it is worthwhile to preserve. Milk is a good example of a food possessing such active proteins. Most heat treatments applied to milk to kill microorganisms destroy these active components. Minimal processing-preservation technologies, like high-pressure technology, irradiation, high electric field pulses, and microfiltration, have been developed to try to resolve this problem [81].

XIV. BIOLOGICAL TESTS FOR PROTEIN QUALITY

Protein quality is evaluated by using standardized tests (PER and nitrogen balance test) [1] and also by alternative biological tests for the bioavailability of specific amino acids.

A. Protein Efficiency Ratio

This growth test is carried out on weanling rats fed a diet containing the test protein at 10%. After 3–4 weeks of feeding, the protein efficiency ratio (PER) is determined:

$$\text{PER} = \frac{\text{weight gain}}{\text{protein intake}}$$

The results are generally expressed as percentage of the PER of casein, taken as a reference.

B. Nitrogen Balance

In this test, young rats are fed the same 10% test protein diet, and nitrogen excreted in the urine and in the feces is collected and measured. Nitrogen digestibility and the biological value are evaluated as follows:

$$\text{Digestibility (D)} = \frac{I - (F + Fe)}{I}$$

$$\text{Biological value (BV)} = \frac{I - (F + Fe) - (U + Ue)}{I - (F + Fe)}$$

$$\text{Net protein utilization (NPU)} = D \times BV = \frac{I - (U + Ue) - (F + Fe)}{I}$$

where

I = nitrogen intake
F = fecal nitrogen
Fe = endogenous fecal nitrogen
U = urinary nitrogen
Ue = endogenous urinary nitrogen

C. Tests for Bioavailability

Rat and chick tests have been developed to evaluate the bioavailability of lysine, methionine, cystine, and tryptophan. In these tests, the animals are fed a diet limiting in the amino acid to be evaluated for its bioavailability. The growth rate of the animals is directly correlated to the level of the bioavailable amino acid [48].

Microorganisms that need essential amino acids for growth are used to evaluate their bioavailability in foods. *Tetrahymena pyriformis* is used for available lysine and methionine and *Streptococcus zymogenes* for methionine and tryptophan.

D. Reactive Lysine

Many methods have been developed to evaluate "reactive lysine," i.e., lysine residues that still have their ϵ-amino group free and that are expected to be biologically available. The following reagents have been used: fluorodinitrobenzene (Carpenter method), trinitrobenzene-sulfonic acid, isomethyl-urea for guanidination, and acid orange-12 for the dye-binding procedure. The furosine method seems to be most reliable when the Maillard reaction has taken place [82].

XV. CONCLUSION

During processing and storage, the nutritional quality of proteins can be improved or reduced by modifications of nitrogen digestibility and/or amino acid bioavailability. These changes result from the denaturation of food protein and from the chemical modifications of essential, semi-essential, or conditionally essential amino acids. The chemical losses must be evaluated in the context of the food utilization and must take into account the amino acid needs and the dietary contribution of the food concerned. Practically, few amino acids are at risk. Their order of importance are, for the essential and semi-essential amino acids, lysine, tryptophan, cystine, and methionine, and for the conditionally essential amino acids, glutamine and arginine. Analytical methods exist to evaluate the amino acid losses, and biological tests have been developed to assess the nutritional value. These tools are available for the food industry to evaluate the impact of food processing on the nutritional value of their products and to optimize the processing conditions for a better nutrition.

REFERENCES

1. Protein Quality Evaluation, Report of the Joint FAO/WHO Expert Consultation, FAO Food and Nutrition Paper 51, Food and Agriculture Organization of the United Nation, Rome, 1991.
2. P. A. Finot, F. Mottu, E. Bujard, and J. Mauron, N-substituted lysines as sources of lysine in nutrition, *Nutrition Improvement of Food and Feed Proteins* (M. Friedman, ed.), Plenum Publishing Corporation, New York, 1978, p. 549.
3. P. A. Finot, R. Deutsch, and E. Bujard, The extent of the Maillard reaction during processing of milk, *Maillard Reaction in Food. Chemical, Physiological and Technological Aspects* (C. Eriksson, ed.), Pergamon Press, Oxford, 1981, p. 345.
4. T. Henle, W. Heike, and H. Klostermeyer, Evaluation of the extent of the early Maillard reaction in milk products by direct measurement of the Amadori-product lactulose lysine, *Lebensm. Unters. Forsch. 193*:119 (1991).
5. Codex standart for infant formula, Codex Stan 72-1981 (amended 1983, 1985, 1987), *Codex Alimentarius*, Roma, Vol. 4, 1994, p. 17.
6. R. F. Hurrell, P. A. Finot, and J. E. Ford, Storage of milk powders under adverse conditions. 1. Losses of lysine and of other essential amino acids as determined by chemical and microbiological methods, *Br. J. Nutr. 49*:343 (1983).
7. Codex Standard for Processed Cereal-based Foods for Infants and Children, Codex Stan 74-1981 (amended 1885, 1987, 1989, 1991), *Codex Alimentarius*, Vol. 4, 1994, p. 35.

8. P. Resmini, M. A. Pagani, and L. Pellegrino, Valutatione del danno termico nelle pasta alimentare mediante determinazione per HPLC della ϵ-furoylmetil-lisina (furosina), *Tec. Molitoria* (Oct.):821 (1990).
9. M. Miraglia, C. Brera, R. Onori, S. Corneli, E. Quattrucci, R. Acquistucci, and L. Bruschi, Studio di parmetri chimico-nitritionali su paste alimentari essicate a bassa ed alta temperatura, *Riv. Soc. Ital. Sci. Aliment. 23*:11 (1994).
10. L. Arrigo, R. Rondinone, and G. Tomassi, Valutazione della qualita nutritionale di paste alimentari essicate a bassa e alta temperatura, *Riv. Soc. Ital. Sci. Aliment. 23*:27 (1994).
11. L. A. Jones, Gossypol and some other terpenoids, flavonoids and phenols that affect quality of cottonseed protein, *J. Am. Oil Chem. Soc. 56*:727 (1979).
12. J. Mauron, Maillard reaction in food; a critical review from the nutritonal standpoint, *Prog. Food Nutr. Sci.* (C. Eriksson, ed.), Pergamon Press, Oxford, 1981, p. 5.
13. H. K. Nielsen, D. De Weck, P. A. Finot, R. Liardon, and R. Hurrell, Stability of tryptophan during food processing and storage. 1. Comparative losses of tryptophan, lysine and methionine in different model systems, *Br. J. Nutr. 53*:281 (1985).
14. T. Nakayama, F. Hayase, and H. Kato, Formation of ϵ-(2-formyl-5-hydroxymethyl-pyrrol-1-yl)-L-norleucine in the Maillard reaction between D-glucose and lysine, *Agric. Biol. Chem. 44*:1201 (1980).
15. G. H. Chiang, High performance liquid chromatographic determination of ϵ-pyrrol-lysine in processed foods, *J. Agric. Food Chem. 36*:506 (1988).
16. T. Henle, W. Walter, and H. Klostermeyer, Simultaneous determination of protein-bound Maillard products by ion exchange chromatography and photodiode array detection, *Maillard Reactions in Chemistry, Food and Health* (T. P. Labuza, G. A. Reineccius, V. M. Monnier, J. O'Brien, and J. W. Baynes, eds.), The Royal Chemical Society, Cambridge, 1994, p. 195.
17. P. Resmini and L. Pellegrino, Evaluation of the advanced Maillard reaction in dried pasta, *Maillard Reactions in Chemistry, Food and Health* (T. P. Labuza, G. A. Reineccius, V. M. Monnier, J. O'Brien, and J. W. Baynes, eds.), The Royal Chemical Society, Cambridge, 1994, p. 418.
18. M. U. Ahmed, S. R. Thorpe, and J. W. Baynes, Identification of Nϵ-carboxymethyllysine a degradation product of fructose lysine in glycated proteins, *J. Biol. Chem. 261*:4889 (1986).
19. R. Liardon, D. Gaudard-de Weck, G. Philippossian, and P. A. Finot, Identification of Nϵ-carboxymethyllysine, a new Maillard reaction product, in rat urine, *J. Agric. Food Chem. 35*:427 (1987).
20. V. Nair, C. S. Cooper, D. E. Vietti, and G. A. Turner, The chemistry of lipid peroxidation metabolites: crosslinking reactions of malondialdehyde, *Lipids 21*:6 (1986).
21. H. H. Drapper, M. Hadley, L. Lissemore, N. M. Laing, and P. D. Cole, Identification of N-ϵ-(2-propenal)lysine as a major urinary metabolite of malondialdehyde, *Lipids 23*:626 (1988).
22. P. A. Finot, E. Bujard, F. Mottu, and J. Mauron, Availability of the true Schiff's bases of lysine. Chemical evaluation of the Schiff's base between lysine and lactose in milk, *Protein Crosslinking-B. Nutritional and Medical Consequences* (M. Friedman, ed.), Plenum Publishing Corporation, New York, 1977, p. 343.
23. P. A. Finot and E. Magnenat, Maillard Reaction in Food; Metabolic transit of early and advanced Maillard products, in: *Prog. Food Nutr. Sci.* (C. Eriksson, ed.), Pergamon Press, Oxford, 1991, p. 193.
24. M. Sternberg and C. Y. Kim, Growth response of mice and tetrahymena pyriformis to lysinoalanine supplemented wheat gluten, *J. Agr. Food Chem. 27*:1130 (1979).
25. K. R. Robbins, D. H. Baker, and J. W. Finley, Studies on the utilization of lysinoalanine and lanthionine, *J. Nutr. 110*:907 (1980).
26. J. Bjarnason and K. J. Carpenter, Mechanisms of heat damage in proteins. I. Models with acylated lysine units. *Br. J. Nutr. 23*:859 (1969).

27. J. A. Johnson and R. M. Fusao, Alteration of skin surface protein with dihydroxyacetone: a useful application of the Maillard browning reaction, in: *Maillard Reactions in Chemistry, Food and Health* (T. P. Labuza, G. A. Reineccius, U. M. Monnier, J. O'Brien, and J. W. Baynes, eds.), The Royal Society of Chemistry, 1994, p. 114.
28. A. P. De Groot and P. Slump, Effects of severe alkali treatment of proteins on amino acid composition and nutritive value, *J. Nutr. 98*:45 (1969).
29. P. A. Finot, Lysinoalanine in food proteins, *Nutr. Abst. Rev. 53*:67 (1983).
30. R. F. Hurrell and P. A. Finot, Food processing and storage as a determinant of protein and amino acid availability, *Nutritional Adequacy, Nutrient Availability and Needs*, (J. Mauron, ed.), Birkhäuser Verlag, Basel, 1983, p. 135
31. J. C. Woodhard and M. R. Alvarez, Renal lesions in rat fed diets containing alpha protein, *Arch Path. 84*:153 (1967).
32. V. J. Feron, L. Van Beek, L. Slump, and R. B. Beems, Toxicological aspects of alkali treatment of food proteins, *Biochemical Aspects of Protein Food* (Adler-Nissen, ed.), Federation of European Biochemical Society (FEBS). Copenhagen, 1977, p. 139.
33. D. De Weck-Gaudard, R. Liardon, and P. A. Finot, Stereomeric composition of urinary lysinoalanine after ingestion of free or protein-bound lysinoalanine in rats, *J. Agric. Food Chem. 36*:717 (1988).
34. J. P. Langhendries, R. F. Hurrell, D. E. Furniss, C. Hischenhuber, P. A. Finot, A. Bernard, O. Battisti, J. M. Bertrand, and J. Senterre, Maillard reaction products and lysinoalanine: urinary excretion and the effects on kidney function of preterm infants fed heat-processed milk formula. *J. Pediatr. Gastroenterol. Nutr. 14*:62 (1992).
35. J. Bjarnason and K. J. Carpenter, Mechanism of heat damage in protein. 2. Chemical changes in pure proteins. *Br. J. Nutr. 24*:313 (1970).
36. K. P. Weder and U. Scharf, Model studies on the heating of food proteins—heat-induced oligomerization of ribonuclease, II. Isolation of oligomers and comparative studies, *Z. Lebensm. Unters. Forsch. 172*:104 (1981).
37. M. Friedman and P. A. Finot, Nutritional improvement of bread with lysine and γ-glutamyllysine, *J. Agric. Food Chem. 38*:2011 (1990).
38. K. P. Weder and U. Scharf, Analysis of heated food proteins—a new method for rapid estimation of the isopeptides Nϵ-(β-L-aspartyl)-L-lysine and Nϵ-(γ-L-glutamyl)-L-lysine, *Z. Lebensm. Unters. Forsch. 172*:9 (1981).
39. K. Ikura, T. Kometani, M. Yoshikawa, R. Sasaki, and H. Chiba, Crosslinking of casein components by transglutaminase, *Agric. Biol. Chem. 44*:1567 (1980).
40. K. Ikura, M. Yoschikawa, R. Sasaki, and H. Chiba, Incorporation of amino acids into food proteins by transglutaminase, *J. Agric. Biol. Chem. 45*:2587 (1981).
41. A. J. Puigserver, L. C. Sen, A. J. Clifford, R. E. Finney, and J. Whitaker, A method for improving the nutritional value of food proteins: covalent attachment of amino acids, *Nutritional Improvement of Food and Feed Proteins, Advances in Experimental Medicine and Biology*, Vol. 105 (M. Friedman, ed.), Plenum Press, New York, 1978, p. 587.
42. R. F. Hurrell, P. A. Finot, and J. L. Cuq, Protein-polyphenol reactions 1. Nutritional and metabolic consequences of the reaction between oxidized caffeic acid and lysine residues of casein, *Br. J. Nutr. 47*:191 (1982).
43. K. Narayana and N. S. Narasinga Rao, Effect of acetylation and succinylation on the physicochemical properties of Winged bean (Psophocarpus tetragonolobus) proteins, *J. Agric. Food Chem. 39*:259 (1991).
44. Y. Furuichi, T. Oogida, C. Mitsui, M. Matsuno, and T. Takahashi, Nutritional characteristics of a neoglycoprotein, casein modified covalently by glucose, *J. Nutr. Biochem. 1*:196 (1990).
45. M. A. Bennett, Metabolism of sulfur. The replaceability of dl-methionine in the diet of albinos rats with partially oxidised dl-methionine sulfoxide, *Biochem. J. 33*:1794 (1933).
46. M. A. Bennett, Metabolism of sulphur V. The replaceability of L-cystine in the diets of rats with some partially oxidized derivatives, *Biochem. J. 31*:962 (1937).

47. G. H. Anderson, D. V. M. Ashley, and J. D. Jones, Utilization of L-methionine sulfoxide, L-methionine sulfone and cysteic acid by the weanling rat, *J. Nutr. 106*:1108 (1976).
48. H. K. Nielsen, P. A. Finot, and R. F. Hurrell, Reaction of proteins with oxidizing lipids. 2. Influence on protein quality and on the bioavailability of lysine, methionine, cyst(e)ine and tryptophan as measured in rat assays, *Br. J. Nutr. 53*:75 (1985).
49. D. Dhirajlal, W. G. Bergen, O. Mickelsen, and J. T. Huber, Factors influencing the nutritive value of 1,2-dichloethane-extracted fish protein concentrate in rat diets, *Am. J. Clin. Nutr. 24*: 1384 (1971).
50. L. L. McKinney, J. C. Picken, F. B. Weakley, A. C. Eldridge, R. E. Campbell, J. C. Cowan, and H. E. Biester, Possible toxic factor of trichloroethylene-extracted soybean oil meal, *J. Am. Chem. Soc. 81*:909 (1959).
51. M. Friedman and J. L. Cuq, Chemistry, analysis, nutritional value, and toxicology of tryptophan in food. A review, *J. Agric. Food Chem. 36*:1079 (1988).
52. D. de Weck, H. K. Nielsen, and P. A. Finot, Oxidation rate of free and protein-bound tryptophan by hydrogen peroxide and bioavailability of the oxidation products, *J. Sci. Food Agric. 41*:179 (1987).
53. T. Nyhammart and P. A. Pernemalm, Reaction of N-α-acetyl-DL-tryptophan amide with D-xylose or D-xyloses in acidic solution, *Food Chem. 17*:289 (1985).
54. H. K. Nielsen, Nutritional aspects of reactions between oxidizing lipids and protein with emphasis on tryptophan, Ph.D. thesis, Faculté des Sciences de l'Université de Fribourg, Switzerland, 1984.
55. R. Liardon and R. F. Hurrell, Amino acid racemization in heated and alkali-treated proteins, *J. Agric. Food Chem. 31*:432 (1983).
56. M. Friedman, J. C. Zahnley, and P. M. Masters, Relationship between in vitro digestibility of casein and its content of lysinoalanine and D-amino acids, *J. Food Sci. 46*:127 (1981).
57. P. M. Masters and M. Friedman, Racemization of amino acids in alkali-treated proteins, *J. Agric. Food Chem. 27*:507 (1979).
58. G. Lubec, C. Wolf, and S. Bartosch, Amino acid isomerization and microwave exposure, *Lancet 9*:1392 (1989).
59. L. Fay, U. Richli, and R. Liardon, Evidence for the absence of amino acid isomerisation in microwave-heated milk and infant formulas, *J. Agric. Food Chem. 39*:1857 (1991).
60. R. Marchelli, A. Dossena, G. Palla, M. Audhuy-Peaudecerf, S. Lefeuvre, P. Carnevali, and M. Freddi, D-amino acid in reconstituted infant formulas: a comparison between conventional and microwave heating, *J. Sci. Food Agric. 59*:217 (1992).
61. P. Fritz, L. I. Dehne, J. Zagon, and K. W. Bögl, Zur Frage der Aminosäure isomerisierung im Mikrowellenfeld. Ergebnisse eines Modellversuches mit Standardlösungen, *Z. Ernahrungs wiss. 31*:219 (1992).
62. G. Palla, R. Marchelli, A. Dossena, and G. Casnati, Occurrence of D-amino acids in foods. Detection by capillary gas chromatography and by reversed-phase high-performance liquid chromatography with L-phenylalaninamides as chiral selectors, *J. Chromatogr. 475*:45 (1989).
63. D. Jonker and A. H. Penninks, Comparative study of the nutritive value of casein heated by microwave and conventionally, *J. Sci. Food Agric. 59*:123 (1992).
64. R. Jost, J. C. Monti, and J. J. Pahud, Whey protein allergenicity and its reduction by technological means, *Food Technol.* (Oct.):118 (1987).
64. J. J. Pahud, J. C. Monti, and R. Jost, Allergenicity of whey protein: its modification by tryptic in vitro hydrolysis of protein, *J. Pediatr. Gastroent. Nutr. 4*:408 (1985).
66. G. K. Grimble and D. B. Silk, Peptides in human nutrition, *Nutr. Res. Rev. 2*:87 (1989).
67. J. W. Meredith, J. A. Ditesheim, and G. P. Zaloga, Visceral protein levels in trauma patients are greater with peptide diet than with intact protein diet, *J. Trauma, 30*:825 (1990).
68. F. Ziegler, J. M. Ollivier, J. P. Masini, C. Coudray-Lucas, E. Levy, and J. Giboudeau, Efficiency of enteral nitrogen support in surgical patients: small peptides vs non-degraded proteins, *Gut 31*:1277 (1990).

69. M. Yamashita, S. Arai, and M. Fujimaki, Plastein reaction for food protein improvement, *J. Agric. Food Chem. 24*:1100 (1976).
70. G. Y. Hajos, I. Elias and A. Halasz, Methionine enrichment of milk protein by enzymic peptide modification, *J. Food Sci. 53*:739 (1988).
71. K. Ikura, M. Yoshikawa, R. Sasaki, and H. Chiba, Incorporation of amino acids into food proteins by transglutaminase, *J. Agric. Biol. Chem. 45*:2587 (1981).
72. Y. Konishi, F. Hayase, and H. Kato, Novel imidazolone compound formed by the advanced Maillard reaction of 3-deoxyglucosone and arginine residues in proteins, *Biosci. Biotech. Biochem. 58*:1953 (1994).
73. T. Henle, A. W. Walter, R. Haessner, and H. Klostermeyer, Detection and identification of a protein-bound imidazolone resulting from the reaction of arginine residues and methylgloxal, *Z. Lebensm.Untrs. Forsch. 199*:55 (1994).
74. G. Jori, G. Galiazzo, and E. Schoffone, Photodynamic action of porphirins on amino acids and proteins. I. Selective photoxoidation of methionine in aqueous solution, *Biochemistry 8*: 2868 (1969).
75. R. Cheng and S. Kawakishi, Selective degradation of histidine residue mediated by copper (II)-catalysed autoxidation of glycated peptides (Amadori compound), *J. Agric. Food Chem. 41*:361 (1993).
76. T. Henle, A. W. Walter, and H. Klostermeyer, Detection and identification of the cross-linking amino acids $N\tau$- and $N\pi$-(2′-amino-2′-carboxy-ethyl)-L-histidine (histidinoalanine, HAL) in heated milk products. *Z. Lebensm. Unters. Forsch. 197*:114 (1993).
77. N. Chuaqui-Offermanns and T. McDougall, An HPLC method to determine o-tyrosine in chicken meat, *J. Agric. Food Chem. 39*:300 (1991).
78. Y. Shahkhalili, P. A. Finot, R. F. Hurrell, and E. Fern, Effects of foods rich in polyphenols on nitrogen excretion in rats, *J. Nutr. 120*:346 (1990).
79. P. Restani, A. R. Restelli, A. Capuano, and C. L. Galli, Digestibility of technologically treated lamb meat samples evaluated by an in vitro multienzyme method, *J. Agric. Food Chem. 40*: 989 (1992).
80. I. E. Liener, Implications of antinutritional components in soybean foods, *Crit. Rev. Food Sci. Nutr. 34*:31 (1994).
81. B. Güntensperger, International symposium on minimal processing of foods, *Trends Food Sci. Technol. 5*:266 (1994).
82. R. F. Hurrell and K. J. Carpenter, The estimation of available lysine in foodstuffs after Maillard reactions, *Maillard Reaction in Food. Chemical, Physiological and Technological Aspects* (C. Eriksson, ed.), Pergamon Press, Oxford, 1981, p. 159.

20

Extraction of Milk Proteins

J. L. MAUBOIS AND G. OLLIVIER
Institut National de la Recherche Agronomique
Rennes, France

I. INTRODUCTION

During the two past decades, considerable interest has been given to the development of sophisticated extraction procedures for food proteins in order to satisfy the increasing needs of different industrial sectors for tailor-made products. Proteins, in addition to contributing to the basic nutritional status of foods (i.e., the supply to human beings and animals of organic nitrogen and essential amino acids), are exploited as functional ingredients for modifying or enhancing the textural and rheological characteristics of foodstuffs. They can emulsify fat, bind and entrap water, increase the viscosity of liquids, form gels with different opacities, colors, and firmnesses, and entrap and disperse air in foam-type products. Because of the numerous potential applications of proteins in fabricated food products, increased attention is being directed towards large-scale extraction and purification of proteins and peptides from various sources.

Among the food industries, the dairy industry has developed the most advanced industrial-scale extraction procedures for the separation of proteins from fluid milk. This is understandable, in that the industrialization of milk processing started as early as 1880 [1]. Industrial production of casein was instituted at the beginning of the twentieth century [2] as a result of the studies of Hammarsten [3], but until the 1960s, casein was used primarily for industrial applications, such as glues, paper glazing, plastics (galalithe used in button manufacture), and artificial wool (lanital) [2]. Improvement of the bacteriological quality of collected milk and strict observation of the rules of hygiene led, at the end of the 1950s, to the production by several pioneering dairy companies in France, Australia, and New Zealand of food-grade casein and caseinates [1] for use as valuable protein ingredients in many food products, such as baked goods, pastry, sausage, processed cheeses, etc.

A critical development in the extraction procedures of milk proteins occurred in the late 1960s with the advent of membrane-separation technologies. Commercialization of membrane ultrafiltration has spawned a new industry for whey treatment, which met both the need to reduce effluent disposal and the desire to improve the utilization of whey proteins. Until then, the interesting physicochemical, functional, and nutritional properties of whey proteins could not be fully exploited because of the damaging effect of the heat/acid treatments used for their separation from whey. The recent emergence of new membrane-separation technologies, such as microfiltration and nanofiltration, as well as improvements in chromatographic resins offer now to the dairy technologist several kinds of techniques for the extraction and purification of almost all of the main proteins of milk.

Because of the advances in the dairy industry, this chapter is almost entirely dedicated to the separation processes used for the isolation of milk proteins.

II. MILK PROTEINS

It would be impossible to develop a separation process for any group of proteins or any individual protein without prior thorough knowledge of their biochemical and physicochemical properties. Thus, before discussing the recent developments in extraction procedures, a brief description of the milk protein system is important for understanding the principles used in such processes.

Bovine milk contains numerous proteins which are classically divided into two major groups: the caseins, insoluble at pH 4.6 and 20°C, and the whey proteins. Normal milk has a protein content, expressed as N × 6.38, of 30–35 g protein/liter. About 78% of these proteins are caseins, which consist of four principal components, α_{s1}, α_{s2}, β, and κ, in an approximate ratio of 40:10:35:12. In milk, these caseins are organized in the form of micelles, which are large spherical complexes (diameter varying between 50 and 600 nm, average 120 nm) containing 92% proteins and 8% inorganic salts, principally calcium phosphate [4,5]. The structure of the micelles is not yet fully established, and it is still a matter of controversy between the supporters of the submicellar model and those of the coat-core model. The casein micelles dissociate on removing colloidal calcium phosphate either by Ca chelators (phosphate, citrate, EDTA) or by acidification. They are partly responsible for the white color of milk. Their stability results from their zeta potential (approximately −20 mV) and from a steric hindrance caused by the protruding ("hairy") C-terminal segments of glycosylated κ-casein, which prevent close approach of micelles. Removal of these protruding segments by chymosin, the principal enzyme present in the neonate calf stomach, results in coagulation of the altered casein micelles. The integrity of casein micelles is also affected by cooling: At temperatures lower than 4°C, β-casein and Ca phosphate are released into the serum phase of milk. This also occurs when high hydrostatic pressure is applied [6] or at high NaCl concentration.

The whey protein fraction contains numerous proteins. The main components in bovine milk are β-lactoglobulin (β-Lg), α-lactalbumin (α-La), bovine serum albumin (BSA), and immunoglobulins (Ig), which represent approximately 2.7, 1.2, 0.25 and 0.65 g/liter, respectively [7]. There are many other minor proteins, including lactoferrin (Lf), several enzymes (lipoprotein lipase, acid and alkaline phosphatases, lysozyme, xanthine oxidase, lactoperoxidase, catalase, superoxide dismutase, α-amylase, etc.), growth factors, and hormones [7]. The whey protein fraction of human milk is very different from that of bovine milk in that it contains no β-Lg and is very rich in α-La, in Lf, lysozyme, and stimulatory factors (bifidus growth factor, epidermal growth factor, bombesin, insulin-like growth factors, etc.) [8].

III. SEPARATION OF MILK PROTEINS

Figure 1 summarizes the procedures available for the separation of milk proteins. Before discussing the details of the processes shown in this diagram, it is necessary to emphasize some general guidelines the food technologist must keep in mind while developing a strategy for extraction of a protein or a group of proteins from an agricultural product:

1. At each step of the extraction process, the preservation of desirable qualities (i.e., functional, biological, nutritional) must be taken into account in the choice of the separation methods and the physicochemical parameters. Particular attention must be paid to the successive heat treatments, the effect of which is cumulative.
2. Like most liquid foods, milk and its derivatives are very favorable media for spoilage microorganisms. Consequently, pretreatments as well as temperature-time parameters must be chosen in order to control microbial growth.
3. Each separation and fractionation step generates at least one co-product; for environmental reasons, and because of their potential value, the co-products must be considered not only as by-products but also as value-added components of milk. Thus, all extraction procedures must be envisaged in an integrated technological concept that utilizes all of the generated products and effluents.

IV. PRETREATMENT OF MILK

Milk collected by the dairy industry contains lipids (~40 g/liter) organized in fat globules. Most extraction procedures for milk proteins use skim milk as the starting material. Consequently, whole milk is centrifuged in a separator comprising conical discs and running at 5,000 rpm at around 50°C. The separated cream generally represents 10% of the volume of the entering whole milk. Fat separation is never perfect, and the skim milk thus obtained still contains around 0.5 g/liter fat, which can strongly influence the efficiency of downstream processes and the resulting protein products.

Milk in developed countries is normally contaminated by common mesophilic and psychrotrophic microflora and rarely by pathogenic microorganisms. To minimize possible health hazards and to control bacterial growth during milk processing, moderate heat treatments at 63°C for 5–10 minutes or pasteurization at 72°C for 15 seconds is applied to skim milk before further processing. The consequences of this heat treatment are numerous and significant: it decreases the pH, shifts the delicate protein–calcium phosphate equilibrium, causes changes in the micellar structure, which affects its hydration and zeta potential [9], and initiates the Maillard reaction, which permanently modifies the functional and nutritional properties of whey proteins [10]. An interesting recent development in the removal of microorganisms from skim milk is the use of membrane microfiltration technology (Fig. 2). According to this patented process [11], developed by the Alfa Laval Company under the trademark Bactocatch [12], skim milk is circulated under pressure along a ceramic microfiltration membrane (having an average pore size of 1.4 μm) at a temperature between 20 and 55°C. The average transmembrane pressure is 0.55×10^5 Pa. The membrane fouling is controlled by co-current recirculation of the microfiltrate, which exerts a low and uniform transmembrane pressure. Transmission rates of total solids, protein, and fat of skim milk are, respectively, 99.5, 99, and 63% [13]. The average decimal reduction observed between the inlet milk and microfiltered

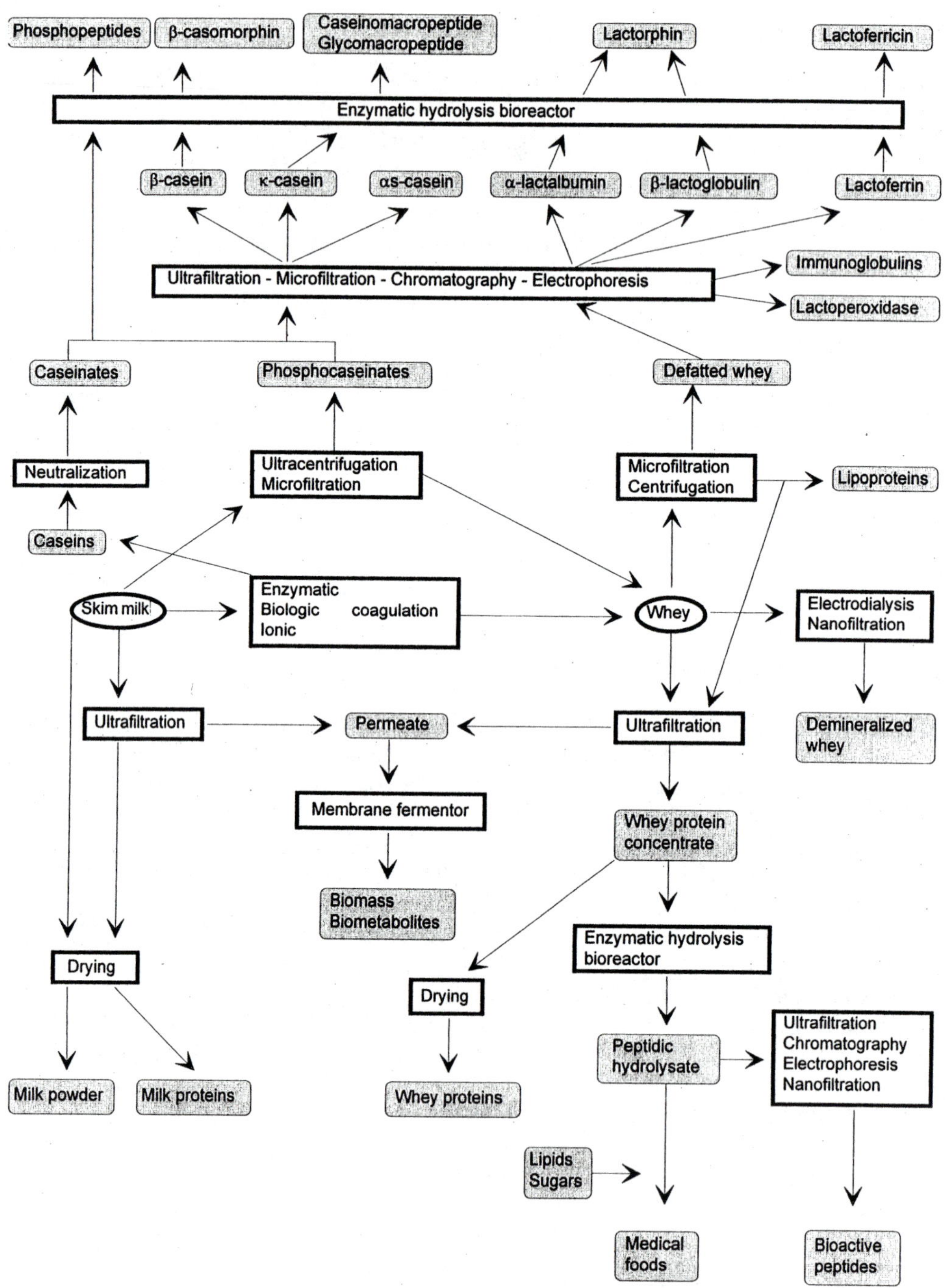

FIGURE 1 Milk protein separation.

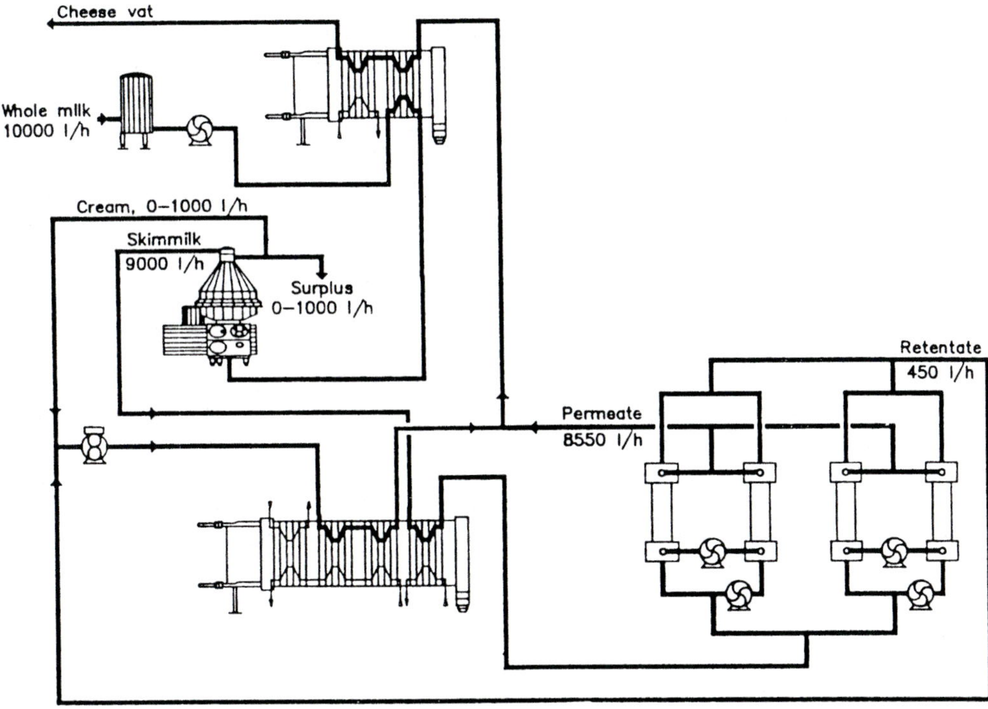

FIGURE 2 Bactocatch treatment of cheese milk. (From Ref. 12.)

milk is 2.6, which corresponds to a concentration of 99.5% of the contaminating microorganisms in 5% of the treated milk volume [14]. A recent development in ceramic membrane manufacture, leading to a narrow pore size distribution, has allowed an increase in the decimal reduction to 3.6 or more with no reduction in its performance efficiency at 500 liters h^{-1} m^{-2} (J. Fauquant and J. L. Maubois, unpublished).

V. ISOLATION OF WHOLE CASEIN

There are two principal established methods for the production of whole casein on an industrial scale: isoelectric precipitation and rennet coagulation [1]. Apart from improvements in equipment (in-line coagulation) and protein enrichment by membrane ultrafiltration of processed skim milk to increase casein yield, there has been no major recent development in the area of rennet casein. In contrast, technology for the production of acid casein has seen several breakthroughs exploiting the advances realized in the knowledge of the casein biochemistry. Destabilization of casein micelles is generally accomplished by adding a dilute mineral acid (HCl or H_2SO_4) to milk heated to 50°C. The casein curd is then separated by straining and washed several times. The resulting co-product, which is a mixture of acid whey and part of the washing water, has a high ash and chloride content, which poses some difficulties during spray-drying. To improve the quality of the co-product as well as the yield of the casein manufacture, use of ion exchangers has been proposed. In this case, a portion of the milk is acidified to pH 2 at

10°C by treatment with a strong hydrogen ion exchanger, which is then mixed with unacidified milk in a proportion to yield a blend with a pH of 4.7 and a temperature of ~30°C [15,16]. The resulting acidified milk is then processed by conventional technologies. An increase in yield of about 3.5% is claimed; this is attributed to precipitation of all of the component 5 and a part of the component 3 of the proteose peptone fractions of the decationized milk portion [16]. The resulting whey has a lower salt content than normal and is partly clarified. It is more suitable for further processing. In spite of its advantages, this process is used, to our knowledge, in only one or two plants in France, probably because of the difficulty of maintaining the ion exchanger under satisfactory bacterial conditions and because of the large volume of effluents generated. Electrodialysis has also been used in an attempt to improve the quality of acid casein whey. In a specific process [17], first the skim milk is partially acidified to pH 5.2–5.4 by substituting the milk cations with H^+ ions, followed by the addition of a required amount of a mineral acid to decrease the pH to the isoelectric pH 4.6. In another, more sophisticated process [18], skim milk is electrodialyzed at 10°C against acidified whey to reduce the pH to about 5; the acidified milk is centrifuged and the sedimented casein is dispersed in water, concentrated, and dried. The product is readily dispersible in water and is claimed to have properties approaching those of native casein micelles [1]; however, to our knowledge, this was not shown in terms of micellar size distribution, structure, or protein-mineral composition of the micelles.

Two other processes have been proposed for the destabilization of casein micelles, but neither is being used commercially. The first one, called cryoprecipitation [19], is based on the effects of a negative temperature treatment on the aqueous environment of the casein micelles. Freezing and storing of milk at about −10°C results in an increase in ionic strength and a simultaneous decrease of pH to approximately 5.8 as a result of precipitation of calcium phosphates, which releases protons. The second process is based on the poor stability of casein when milk is acidified to pH 6.0 and mixed with 10–15% ethanol [20].

Casein micelles may be sedimented by centrifugation at 100,000 *g* for 1 hour, which is widely used on a laboratory scale. A combination of membrane ultrafiltration (UF) of skim milk and ultracentrifugation has been proposed for the industrial production of native phosphocaseinate [21,22]. On a laboratory scale, almost all of the casein micelles in UF retentates containing 17% protein can be sedimented by centrifugation at 75,000 *g* for 1 hour at 50°C. The resulting supernatant has a composition of WPC 35 (whey protein concentrate with a protein/total solid ratio of 0.35). This process has not been scaled up because of the difficulties encountered in continuous desludging of the sediment. Recent developments in composite fiber technology might facilitate development of a continuous ultracentrifuge in the near future.

The most promising technology for the selective separation of casein micelles is undoubtedly membrane microfiltration (MF). When whole or skim milk is circulated along a MF membrane with a pore size diameter of 0.1–0.2 μm, a microfiltrate with a composition close to that of a sweet whey is obtained. Moreover, it is crystal clear and can be sterile if the downstream equipment prevents recontamination. The retentate is an enriched solution of native and micellar calcium phosphocaseinate (PPCN) [23]. Diafiltration against water allows its purification, and it is easily concentrated by vacuum evaporation and spray-dried [24,25]. Native casein has excellent rennet-coagulating abilities. Coagulation time of a 3% PPCN solution is reduced by 53% compared to that of raw milk, and gel firmness at 30 minutes is increased by more than 50% [24]. Although the preparation of micellar casein according to this process is still at the exploratory stage in spite of recent improvements in increasing MF flux (utilization of the uniform transmembrane pressure concept and a new ceramic membrane material), the casein enrichment of cheese milk by MF has recently reached the industrial level. Casein enrichment of cheese milk significantly improves its rennet coagulability and the productivity of cheese plants, especially those making hard cheeses. In addition, partial removal by MF of whey proteins significantly reduces the detrimental effects of heat treatment on rennet coagulability of milk. The extent of formation of κ-casein–β-lactoglobulin complex decreases, and a decrease in the electronegativity of the casein micelles (caused by the cleavage of the caseinomacropeptide) allows the micelles to aggregate even in UHT-treated milk. This result was used by Quiblier et al. [26] to develop a new high- or medium-heat milk powder, with a cheese-making ability similar to that of raw milk [27]. Native casein and its co-product, the whey protein isolate obtained by submitting the microfiltrate to a subsequent ultrafiltration step, are excellent starting substrates for further fractionation and isolation of milk proteins.

VI. FRACTIONATION OF WHOLE CASEIN

There is a considerable interest in developing technologies for the fractionation of whole casein into individual components—α_{s1}, α_{s2}, β, and κ—on an industrial scale. Potential uses include bovine milk–based infant formulas and preparation of biologically active peptides and specific additives. Most published studies are related to isolation of β-casein, the main component of human casein, which contains numerous peptide sequences with physiological activities, such as the well-known β-casomorphin. The proposed methods exploit the temperature-dependent association characteristics of β-casein. When milk is cooled to 4°C, β-casein, which is a very hydrophobic protein, reversibly dissociates from the casein micelles. This property was used by Terré et al. [28] for isolating β-casein at 2°C from milk and Na-caseinate by using 0.14-μm-pore diameter MF membrane and later by Murphy and Fox [29], who used a 300,000 dalton cut-off UF membrane. The yield of β-casein is enhanced by lowering the pH to 4.2–4.6 [30]. However, the purity of the β-casein–enriched product, obtained according to both processes, was not higher than 70%. The main contaminants were α_s- and κ-caseins. In addition, the flux rate at such low operating temperatures was very low (15–20 liters h^{-1} m^{-2}). These disadvantages can be overcome if the starting material is rennet casein as proposed by Le Magnen and Maugas [31]. This approach, which is used commercially, includes a step for solubilization of β-casein from a cold solution of rennet casein (pH 4.6) treated with a calcium sequestrating salt, followed by centrifugation at 3000 g. The β-casein content of the resulting supernatant supposedly has a minimum purity of 90%.

The growing commercial interest in the production of functional peptide fragments of α_s- and κ-caseins will likely lead to future developments in the fractionation of these proteins either from the co-product resulting from fractionation of β-casein or from the native casein micelles dissociated by the combined action of pH, NaCl, and sodium citrate as proposed by Pouliot et al. [32].

VII. WHEY PROTEIN SEPARATION

The whey protein fraction of milk has unquestionably the highest nutritional value for human beings, far ahead of any other source of protein [33]. It contains large amounts of all the essential amino acids, especially isoleucine, leucine, threonine, and tryptophan. Unfortunately, this very heterogeneous group of proteins is present in whey at relatively low concentrations (0.6%) and hence is expensive to separate and to purify. Because of its high water content and relatively high sugar (5%) and in minerals (0.5–0.8%) contents, more than two thirds of whey production was disposed off 20 years ago as waste, with a detrimental effect on the environment: about 1000 liters of whey has the same polluting power as 400 people [34]. Currently, although a significant amount of the whey produced is being processed and utilized, its full utilization has not been achieved [35], and the disposal problem created by the lactose permeate (the end product of whey protein separation) has been reduced by only 15%. As pointed out by Zall [35], further emphasis needs to be put on utilization of whey for the complete use of this product before the next century.

The first whey protein product was derived from the technology used for making whey cheeses, called serac or bruscio in France, ricotta in Italy, and mizithra, anthotiros, or manouri in Greece. This product, improperly called "lactalbumin," is still a source of confusion in the nutritional or pediatric literature, since it is not the same as the well-defined α-lactalbumin, one of the protein components of whey. This product is prepared by heat denaturation of the proteins in acid or rennet whey at ≥90°C and at H ≤ pH 6. Approximately 80% of whey nitrogen coagulates under these conditions and is recovered by centrifugation and spray or roller dried [1]. Since the proteins in "lactalbumin" are extensively denatured and insoluble, they are used mainly for nutritional fortification of foods. To our knowledge, "lactalbumin" production is limited.

With the advent of ultrafiltration technology, the dairy industry took a giant leap forward in the commercial production of whey proteins. This technology, which appeared at the end of the 1960s, enables whey proteins to be separated from lactose, minerals, and other water-soluble, low molecular weight species under mild conditions (10–50°C, pH 4.6–6.5). The membranes used in UF are asymmetrical microporous structures manufactured from synthetic polymers (mainly polysulfone) or ceramics (Al_2O_3 or amorphous carbon with ZrO_2 or TiO_2 surface layers). Their effective skin appears to contain pores with diameters ranging from 1 to 20 nm. They are assembled in four main configurations of modules: tubular, spiral wound, plate and frame, and hollow fiber, with each configuration offering advantages and disadvantages in terms of performances, investment and running costs, installed energy (0.2–0.7 kWh/m^2), membrane lifetime (0.5 to more than 4 years), cleaning cost (tubular and plate and frame configurations are the easiest to clean), and space requirements (tubular configuration requires 3 times more space than plate and frame and 50 times more than hollow fiber).

Performances of any UF process are characterized by two main criteria: selectivity and flux. Selectivity is the degree of retention of a range of molecules of known molecular weight. As the masses of whey molecules to be totally transmitted range from 18 (water) to 342 daltons (lactose), in practice, membranes having a nominal molecular weight cut-off between 5000 and 25,000 are generally used for concentration of whey proteins. The flux is first determined by the intrinsic membrane water flux, which typically ranges from 100 to 500 liters h^{-1} m^{-2} at 20°C and a transmembrane pressure of 400 kPa. However, retention characteristics and flux are affected by the conditions used during ultrafiltration and the interactions of whey components among themselves and with the components of the UF membrane. Numerous data have been accumulated in the last 25 years of the factors governing industrial performances of UF plants treating different varieties of whey. This acquired knowledge about whey and more generally about dairy fluids can be used for developing membrane technology–based processing strategies for any liquid by- or co-product.

Retention by the UF membrane of the macromolecular components modifies the calcium salt–protein equilibria existing in sweet whey. The solubilizing effect of proteins on calcium salts at temperatures higher than 20°C [36] disappears, and consequently the mineral salts precipitate at the surface and (mostly) inside the pores of the UF membrane, causing an intramembrane fouling. However, the most important factor causing fouling is the formation of a secondary membrane on the surface of the active membrane. Convection of proteins and suspended particles immediately builds up this concentration polarization layer, wich increases the hydraulic resistance of the UF membrane and consequently reduces the flux. The treatments that limit the thickness and the compactness of the boundary layer will have a positive influence on selectivity and flux. For example, an increase in velocity of the whey circulating tangentially to the UF membrane will cause an increase in the shear stress, which will result in a reduction of boundary layer thickness. Typically employed flow velocities in industrial equipments are in the range of 1–5 m/sec for turbulent flow systems and less than 1 m/sec for laminar flow systems. Increasing the temperature of whey to a maximum of 50–55°C (beyond which protein denaturation may occur) also increases the flux (3%/°C) depending upon several factors, such as increase of back-diffusion of protein molecules away from the boundary layer, decrease of the viscosity of the concentration polarization layer, and decrease of the viscosity of the UF permeate. Fouling created by the boundary layer is considered to be reversible; it can be removed by a simple rinse with water. On the other hand, the mineral salt deposit as well as adsorption of the lipid materials cause irreversible fouling, which can be minimized by whey pretreatments that remove residual fat; such treatments may increase the average flux from 20 to 151 liters h^{-1} m^{-2}.

Other factors may cause a permanent reduction in flux, such as membrane compaction, presence of antifoaming agents in whey, or use of water or cleaning products containing iron or silica [37,38].

Whey proteins are amphoteric molecules. At pH lower than their isoelectric point ($pH < 5.2$) they can be adsorbed on cation exchangers; at pH above their electric point they can be adsorbed on anion exchangers. To our knowledge, two major ion fractionation processes are commercially used for the separation of whey proteins. The Vistec process uses a carboxymethyl cellulose exchanger in a stirred tank reactor [39,40]. Protein adsorption is realized at pH 3.2, and protein elution by increasing the pH to 7–7.5 [41]. The dilute eluate is then separated from the resin by filtration, concentration by membrane ultrafiltration and evaporation, and spray-dried. The resulting product contains, on a dry basis, 97% protein, 3% ash, 0.2% lactose, and 0.2% fat [41]. The Spherosil process [42] uses porous silica beads coated with a polymer material having either cationic exchange potential (-SO_3H groups) (Spherosil-S) or anionic exchange reactivity [-$N(CH_3)_3$ groups] (Spherosil QMA). Consequently, acid whey (pH about 4.5) is passed through a column of Spherosil-S, and proteins are eluted with ammonium hydroxide (0.1 M), whereas sweet whey (pH about 6.3) is passed through a column of Spherosil-QMA, and the adsorbed proteins are eluted with hydrochloric acid (0.1 M). The advantages of the ion exchange material used in the Spherosil process are that it does not swell with changes in pH and ionic strength, it is noncompressible under moderate pressure, and it is physically and chemically stabile. However, it is expensive and exhibits a low protein-binding capacity of about 79 mg/g [43] compared to 200–500 mg/g for CMC [41]. These adsorption processes recover about 85% of the protein under the best operating conditions. The recovery yield is lower with anionic exchanger because the basic whey proteins such as lactoferrin, lactoperoxidase, and most of the immunoglobulins are not adsorbed by Spherosil-QMA [44]. The recovered concentrates are characterized by high protein and low lactose and lipid concentrations, and they are sometimes named whey protein isolate (WPI) to distinguish them from WPC obtained by UF. Although the functional properties of WPI can be superior to those of WPC on an equiprotein basis, particularly in thermal gelation [41], their production is limited due to higher production cost ensuing from high investment cost, large volumes of required regenerant and rinse solutions, and continuous decrease of the exchange capacity during yearly processing [41,45].

The presence of even a low level of lipids impairs the functionality of WPC, especially the foaming properties, reduces UF flux rates and irreversibly decreases the ion exchange capacity and the performance of enzymatic membrane reactors. All of these problems are minimized by clarification of the whey prior to or during UF. Because the residual lipid-containing materials are large in size relative to the globular whey proteins, Lee and Merson [46], Merin et al. [47], and Piot et al. [48] have proposed that these lipid particles could be removed by microfiltration prior to ultrafiltration. The permeate so produced was almost free of fat and particles, but a portion of the protein was also retained (5–20%, according to the authors). As an alternative method to microfiltration for removing whey lipoproteins, Fauquant et al. [49] proposed aggregation of these components using calcium ions. Conditions identified as yielding the greatest possible clarity with minimal protein loss were as follows: the pH of whey treated with calcium chloride must be adjusted to a value of 7.2 at 2°C in order to prevent premature precipitation of calcium phosphate. The whey is then heated to 55°C for 8 minutes. This treatment led to aggregation of lipid-protein complexes, which were readily separated either by centrifugation or by microfiltration with a 0.1-μm pore size membrane [50]. These conditions were then optimized successively by Pierre et al. [51] and by Daufin et al. [52] in order to avoid $CaCl_2$ addition and to maximize MF separation step. The new proposed procedure includes (a) preconcentration by UF of the whey to 25% of its initial volume at 50°C, (b) removal of microorganisms by MF with a 0.8-μm pore size membrane at 50°C, (c) adjustment of temperature to 55°C and pH to 7.5, this pH value being constantly maintained, and (d) separation of the aggregated phospholipoprotein-calcium salt complex by MF with a 0.1-μm pore size membrane. Such a defatted whey concentrated by UF up to a protein content of 10% yielded meringues with the same appearance and taste as those made from egg white. Another approach for fat removal from whey, which involves selective precipitation of lipid materials by electrostatic complexation with chitosan, has been reported [53].

Whey protein concentrates can perform a number of physical functions in food products. Table 1 summarizes the functionalities of WPC used beneficially in food systems in addition to their unique nutritional value for human beings. Unlike many food proteins, which have minimal solubility near pH 4.5, whey proteins are soluble at these pH values and consequently can be used for protein fortification of acidic beverages (fruit juices and soft drinks). According to the ionic environment and the pH at which heat treatment is applied, WPC forms gels with different appearance, elasticity, and strength [56]. For example, between pH 5 and 9, a 10% protein WPC heated at 70°C forms a white and opaque gel. At pH 11, the gel becomes transparent and elastic. WPC also exhibits good water-holding capacity, which may be useful in products such as meat pastes, sausages, breads and cakes. Depending on the lipid content, WPC exhibits a wide

TABLE 1 Protein Functionality in Foods

Functional property	Mode of action	Food system
Solubility	Protein solvation	Beverages
Water absorption and binding	Hydrogen bonding of water, entrapment of water	Meat sausages, cakes, bread
Viscosity	Thickening, water binding	Soups, gravies, salad dressing
Gelation	Protein matrix formation and setting	Meats, curds, baked goods, cheese
Cohesion–Adhesion	Protein acts as adhesive material	Meat sausages, baked goods, pasta products
Elasticity	Hydrophobic bonding in gluten, disulfide links in gels	Meats, bakery
Emulsification	Formation and stabilization of fat emulsions	Sausages, salad dressing, coffee whitener, soup, cakes, infant formula
Fat absorption	Binding of free fat	Sausages, doughnuts
Foaming	Forms stable film to entrap gas	Chiffon desserts, cakes, whipped toppings

Source: Ref. 54, 55.

range of emulsifying capacities (from very low values to 53 ml of oil per gram of protein) [57] and foaming abilities.

VIII. FRACTIONATION OF WHEY PROTEINS

The techniques for the isolation of individual whey proteins on a laboratory scale by salting-out using sodium sulfate [58] or by ion exchange chromatography [59] have been available for about 40 years. Other published methods concern use of high concentrations (7–30%) of NaCl [60] and ferric chloride [61,62]. However, none of these techniques is commercially desirable because of either a high processing cost or an excessive production of effluent or both. The need for whey protein fractionation existed nevertheless owing to the unique functional, physiological, or other biological properties of many whey proteins [63]. Isolation of the main whey proteins was first attempted in the mid-1980s when Pearce [64,65] proposed exploiting the low heat stability of calcium-free α-lactalbumin.

α-Lactalbumin is a calcium metalloprotein containing one mole of ionic calcium per mole of protein. Removal of calcium from α-lactalbumin by adjusting the pH to about 3.8 or by the addition of a sequestering agent such as EDTA [41], citric acid, or sodium citrate [66] results in much reduced thermal stability compared to the native protein. Subsequent heating around 55°C for a limited period of time leads to a reversible and partially denatured form, which undergoes aggregation. This property of α-

lactalbumin is made use of in two processes developed by Pearce [64] and Maubois et al. [50]. The first one uses whey concentrated by ultrafiltration to about 12% total solids. The pH is adjusted to 4.2 and heated at 65°C for 5 minutes to cause aggregation of α-lactalbumin; during this treatment, both immunoglobulins and the serum albumin also coprecipitate with α-lactalbumin. Separation of the precipitate is realized in a continuous desludging clarifier [65]. The sediment, the α fraction, was shown to be an effective emulsifier and offered the potential for replacing up to 20% of the protein in a processed cheese formulation [65]. The supernatant, the β fraction, was diafiltered through 50,000 dalton cut-off ultrafiltration membrane to yield purified β-lactogobulin. In the second process developed by Maubois et al. [50] whey is clarified by the thermocalcic aggregation process of Fauquant et al. [49] prior to ultrafiltration up to 30-fold concentration. A pH value of 3.8 and heating at 55°C for 30 minutes were used to allow coprecipitation of α-lactalbumin, immunoglobulins, and bovine serum albumin. Separation of a highly purified (95%) soluble β-lactoglobulin was carried out by centrifugation or by membrane microfiltration (pore size 0.1 μm) with a diafiltration step [67]. The sediment or MF retentate yielded a 70% α-lactalbumin fraction.

α-Lactalbumin has a great potential market in nutraceutical foods because of its high content in tryptophan (4 residues per mole, i.e., about 6%) and in infant milk. As already proposed by Maubois [67], a highly simulated human milk formula could be prepared from a mixture of goat native phosphocaseinate, milk permeate, and purified α-lactalbumin and bovine serum albumin. Because α-lactalbumin shows strong affinity for glycosylated receptors existing on the surface of ovocytes [68] and spermatozoides (F. Batelier, unpublished), it may also find use in therapeutic medical foods.

Since no biological function for β-lactoglobulin has been proposed yet, the only main uses of this protein appears to be in gel and foam-type products and in the manufacture of protein hydrolysates.

Whey contains a wide range of (a) minor proteins, such as lactoferrin, immunoglobulins, (b) indigenous enzymes (perhaps 60), such as xanthine oxidase, superoxide dismutase, plasmin, alkaline phosphatase, lipase, lactoperoxidase, etc., and (c) various growth factors, hormones, and peptides. At present, four are of commercial interest, namely lactoperoxidase, lactoferrin, immunoglobulins, and caseinomacropeptide.

Numerous patents have been applied in the last ten years for the isolation by ion exchange chromatography of lactoferrin (also called lactotransferrin) and lactoperoxidase from milk or whey. All these methods use cation exchangers [69]. At the pH of milk or whey, lactoferrin and lactoperoxidase are specifically adsorbed because of their high pI (9.0 and 9.5, respectively). Elution of these two proteins is generally realized through the use of an increasing ionic strength gradient. The growing interest in the isolation of lactoferrin from whey is related to its antibacterial property (attributed to its ability to sequester iron), its activity against free radicals, and the potent biological activities (antiopioid and antithrombotic actions) of some of its peptide fragments. Commercial interest in lactoperoxidase is focused on activation in situ of the antibacterial system present in dairy liquids, which consists of lactoperoxidase, thiocyanate ($^{-}$SCN), and hydrogen peroxide. This antibacterial system can be incorporated in mouthwash solutions or in chewing gums.

Immunoglobulins can be isolated from whey by salting out with ammonium sulfate or sodium sulfate. This method is effective but expensive (including costs of the utilized salt and of the technologies to be used for removing or reusing this salt). Use of membrane ultrafiltration with a molecular weight cut-off of 100,000 or more is preferable, but whey is a poor source of immunoglobulin (Ig) compared to colostrum or milk produced by hyperimmunized cows. Commercial Ig products are mostly used in veterinary medicine or neonatal ruminants and pigs. In fact, because ruminants are born without blood antibodies, they are very susceptible to infection, and it is highly desirable that they receive protection either by suckling colostrum for at least one week or by ingesting an Ig concentrate.

The glycomacropeptide (GMP) is the C-terminal part of κ-casein released in whey by the action of chymosin (EC.3.4.23.4). This 7000 dalton peptide is in fact a family of numerous molecules with different degrees of glycosylation, which can be easily separated by membrane ultrafiltration or by centrifugation from a solution of sodium caseinate added with chymosin [70]. Another process for its separation from whey has been proposed by Tanimoto et al. [71]. This process is based on the tendency of GMP molecules to associate and form a gel when concentrated at pH values higher than 4 (it exists in the monomeric form at pH 3–4). GMP has numerous uses—it induces secretion of cholecystokinin and thus should play a major role in lipid digestion [72], possibly acts on satiety by suppressing appetite, inhibits adhesion of *Escherichia coli* cells to intestinal walls, protects against influenza, and prevents tartar adhesion to teeth [71]. In addition, GMP contains no Phe, Tyr, or Trp, which makes it suitable for use as a nutritional protein supplement for patients suffering from phenylketonuria.

IX. CONCLUSIONS

Extraction procedures developed for the separation and purification of milk components may serve as a model to be applied relatively easily to other food liquids containing molecules with interesting functional or biological properties. Among the possible technologies, membrane technologies appear to be the most suitable, because they are modular and can be performed at moderate temperatures.

Milk is unquestionably a unique source of high-quality proteins, displaying both nutritional and functionality characteristics in foods. Because of the amount of research that has been focused on the improvement of separation processes, it is likely that the general quality and production efficiency of the various milk protein ingredients, i.e., total milk proteins, acid and rennet caseins, and whey protein concentrates, will increase in the near future. The development of microfiltration technology will also facilitate commercialization of native micellar casein and its derived co-product, a unique whey protein isolate. Both products in turn can be used as starting materials for the separation and purification of individual caseins and whey proteins.

Another exciting area for future exploration is the isolation of biologically active peptides [73]. All of the milk proteins appear to contain peptide sequences possessing biological activity. Peptides with opioid activity, called exorphins, have been identified in most of the milk proteins, the best characterized of which are the β-casomorphins. This family of peptides, containing four to seven amino acids with a common N-terminal sequence Tyr-Pro-Phe-Pro, are very resistant to enzymes of the gastrointestinal tract and appear in the contents of the small intestine following ingestion of milk and also in the blood plasma of pregnant and lactating women and newborn infants. All these peptides have been shown to have opioid activities, which include regulation of electrolyte transfer, pain suppression, and sleep induction. However, these peptides are 100–1000 times less effective than morphine [73].

The α_{s1}-, α_{s2}-, and β-caseins contain clusters of phosphoseryl residues, which bind metal ions strongly. It is claimed that these sequences, called casein phosphopeptides, play a major role in the bioavailability of calcium and iron; however, this claim remains controversial [74]. Peptides resulting from tryptic hydrolysis of β- and α_s-caseins have been shown to inhibit in vitro and in vivo the angiotensin-converting enzyme (EC.3.4.15.1) and consequently intervene in blood pressure regulation [75]. Similarly, a peptide released from the N-terminal part of κ-casein caseinomacropeptide has been shown to inhibit blood coagulation by blocking specific receptors on platelets [76]. Finally, several peptides released from enzymatic hydrolysis and from bacterial fermentation [77] of caseins have been shown to possess immunomodulating activities, stimulate the phagocytic activity of human macrophages in vitro, exert a protective effect in vivo in mice against *Klebsiella pneumoniae* infection [78], and stimulate production of cytokines by human lymphocytes [77]. Most of the procedures proposed for the isolation and production of these bioactive peptides utilize enzymatic membrane reactors [70,79].

Further work is obviously needed on the extraction and purification of the many other components present in milk, particularly the three main caseins (α_{s1}, α_{s2}, and κ) and other minor whey proteins. This requires a thorough understanding of the physicochemical properties of these components and development of new sophisticated separation techniques.

REFERENCES

1. P. F. Fox and D. M. Mulvihill, Developments in milk protein processing, *Food Sci. Technol. Today 7*:152 (1993).
2. M. Beau, *La Caséine: Composition, Propriétés, Technique, Commerce*, Dunod, Paris, 1952.
3. O. Hammarsten, *Zur Kenntniss des Kaseins und der Wirkung des Labfermentes*, Uppsala, 1877. Cited by Beau, Ref. 2.
4. H. E. Swaisgood, Chemistry of caseins, *Advanced Dairy Chemistry, Vol. 1: Proteins* (P. F. Fox, ed.), Elsevier, London, 1992, p. 63.
5. H. S. Rollema, Casein association and micelle formation, *Advanced Dairy Chemistry, Vol. 1: Proteins* (P. F. Fox, ed.), Elsevier, London, 1992, p. 111.
6. D. G. Schmidt and W. Buchheim, Elektronenmikroskopische Untersuchung der Feinstruktur von Caseinmicellen in Kuhmilch, *Milchwissenschaft 25*:596 (1970).
7. C. Alais, *Sciences du Lait. Principes des Techniques Laitières*, 4th ed., Sepaic, Paris, 1984.
8. P. F. Fox and A. Flynn, Biological properties of milk proteins, *Advanced Dairy Chemistry, Vol. 1: Proteins* (P. F. Fox, ed.), Elsevier, London, 1992, p. 255.
9. P. F. Fox, Heat-induced coagulation of milk, *Developments in Dairy Chemistry, Vol. 1: Proteins* (P. F. Fox, ed.), Applied Science Publishers, London, 1982, p. 189.
10. J. L. Maubois, J. Léonil, S. Bouhallab, D. Mollé, and J. R. Pearce, Characterization by ionization mass spectrometry of a lactosyl-β-lactoglobulin conjugate occuring during milk heating of whey, *J. Dairy Sci. 78(suppl. 1)*:133 (1995).
11. R. M. Sandblom (Alfa-Laval), Filtering process, Patent SW 74 16 257 (1974).
12. M. Meersohn, Nitrate free cheese making with bactocatch, *North Eur. Food Dairy J. 55*:108 (1989).
13. J. L. Maubois, Nouvelles applications des technologies à membranes dans l'industrie laitière, Proceedings of the XXIII International Dairy Congress, 8–12 Oct., Montréal, Canada, 1990, 2, pp. 1775–1790.
14. E. Trouvé, J. L. Maubois, M. Piot, M. N. Madec, J. Fauquant, A. Rouault, J. Tabard, and G. Brinckman, Rétention de différentes espèces microbiennes lors de l'épuration du lait par microfiltration en flux tangentiel, *Lait 71*:1 (1991).

15. J. P. Rialland and J. P. Barbier, Procédé de traitement du lait par une résine échangeuse de cations en vue de la fabrication de la caséine et du lactosérum, Patent FR 2 480 568 (1980).
16. A. Pierre and M. Douin, Eléments d'étude du procédé Bridel de fabrication de caséine à partir de lait décationisé par échanges d'ions (EI), *Lait 64*:521 (1984).
17. Laiteries Hubert Triballat, Procédé et installation pour la préparation de la caséine à partir du lait et produits ainsi obtenus, Patent FR 2 418 626 (1979).
18. R. Noël (Société Vidaubanaise d'Ingénierie), Procédé de séparation du phosphocaséinate de calcium et du lactosérum d'un lait écrémé et plus généralement d'un composé protéique d'un liquide biologique, Patent PCT WO 92/12642 (1992).
19. D. A. Lonergan, Isolation of casein by ultrafiltration and cryodestabilization, *J. Food Sci. 48*:1817 (1983).
20. M. M. Hewedi, D. M. Mulvihill, and P. F. Fox, Recovery of milk protein by ethanol precipitation, *Ir. J. Food Sci. Technol. 9*:11 (1985).
21. J. L. Maubois, J. Fauquant, and G. Brulé (INRA). Procédé de traitement de matières contenant des protéines telles que le lait, Patent FR 2 292 435 (1976).
22. G. Brulé, J. Fauquant, and J. L. Maubois, Preparation of native phosphocaseinate by combining membrane ultrafiltration and ultracentrifugation, *J. Dairy Sci. 62*:869 (1979).
23. J. Fauquant, J. L. Maubois, and A. Pierre, Microfiltration du lait sur membrane minérale, *Tech. Lait. 1028*:21 (1988).
24. A. Pierre, J. Fauquant, Y. Le Graet, M. Piot, and J. L. Maubois, Préparation de phosphocaséinate natif par microfiltration sur membrane, *Lait 72*:461 (1992).
25. P. Schuck, M. Piot, S. Méjean, Y. LeGraet, J. Fauquant, G. Brulé, and J. L. Maubois, Déshydration par atomisation de phosphocaséinate natif obtenu par microfiltration sur membrane, *Lait 74*:375 (1994).
26. J. P. Quiblier, C. Ferron-Baumy, G. Garric, and J. L. Maubois (INRA), Procédé de traitement des laits permettant au moins de conserver leur aptitude fromagère, Patent FR 2 681 218 A1 (1991).
27. M. El Shiekh, P. Ducruet, and J. L. Maubois, Manufacture of Ras cheese from fresh and recombined milks, *Lait 74*:297 (1994).
28. E. Terré, J. L. Maubois, G. Brulé, and A. Pierre (INRA), Procédé d'obtention d'une matière enrichie en caséine beta, appareillage pour la mise en oeuvre de ce procédé et application des produits obtenus par ce procédé comme aliments, compléments alimentaires ou additifs en industrie alimentaire et pharmaceutique ou dans la préparation de peptides à activité physiologique, Patent Fr 2 592 769 A1 (1987).
29. J. M. Murphy and P. F. Fox, Fractionation of sodium caseinate by ultrafiltration, *Food Chem. 39*:27 (1991).
30. M. H. Famelart, C. Hardy, and G. Brulé, Etude des facteurs d'extraction de la caséine beta, *Lait 69*:47 (1989).
31. C. Le Magnen and J. J. Maugas (EURIAL), Method and device for obtaining beta casein, Patent PCT/FR 91/00506 (1991).
32. M. Pouliot, Y. Pouliot, M. Britten, J. L. Maubois, and J. Fauquant, Study of the dissociation of beta-casein from native phosphocaseinate, *Lait 74*:325 (1994).
33. L. Hambraeus, Nutritional aspects of milk proteins, *Advanced Dairy Chemistry, Vol. 1: Proteins* (P. F. Fox, ed.), Elsevier, London, 1992, p. 457.
34. P. G. Marshall, W. L. Dunkley, and E. Lowe, Fractionation and concentration of whey by reverse osmosis, *J. Food Technol. 22*:969 (1968).
35. R. R. Zall, Sources and composition of whey and permeate, *Whey and Lactose Processing* (J. G. Zadow, ed.), Elsevier, London, 1992, p. 1.
36. G. Brulé, E. Real del Sol, J. Fauquant, and C. Fiaud, Mineral salts stability in aqueous phase of milk: influence of heat treatment, *J. Dairy Sci. 61*:1225 (1978).
37. M. E. Matthews, Advances in whey processing—ultrafiltration and reverse osmosis, *N. Z. J. Dairy Sci. Technol. 14*:86 (1979).

38. R. F. Armishaw, Inorganic fouling of membranes during ultrafiltraton of casein whey, *N. Z. J. Dairy Sci. Technol. 17*:213 (1982).
39. K. J. Burgess and J. Kelly, Technical note: selected functional properties of a whey protein isolate, *J. Food. Technol. 14*:325 (1979).
40. D. E. Palmer, Recovery of proteins from food factory waste by ion exchange, *Food Proteins* (P. F. Fox and J. J. Condon, eds.), Applied Science Publishers, London, 1982, p. 341.
41. R. J. Pearce, Whey processing, *Whey and Lactose Processing* (J. G. Zadow, ed.), Elsevier, London, 1992, p. 73.
42. B. Mirabel, Nouveau procédé d'extraction des protéines du lactosérum, *Ann. Nutr. Aliment. 32*:243 (1978).
43. P. J. Skudder, Evaluation of a porous silica-based ion-exchange medium for the production of protein fractions from rennet- and acid-whey, *J. Dairy Res. 52*:167 (1985).
44. C. M. Barker and C. V. Morr, Composition and properties of Spherosil-QMA whey protein concentrate, *J. Food. Sci. 51*:919 (1986).
45. D. M. Mulvihill, Production, functional properties and utilization of milk protein products, *Advanced Dairy Chemistry, Vol. 1: Proteins* (P. F. Fox, ed.), Elsevier, London, 1992, p. 349.
46. D. N. Lee and R. L. Merson, Prefiltration of cottage cheese whey to reduce fouling of ultrafiltration membranes, *J. Food Sci. 41*:403 (1976).
47. U. Merin, S. Gordin, and G. B. Tanny, Microfiltration of sweet cheese whey, *N.Z. J. Dairy Sci. Technol. 18*:153 (1983).
48. M. Piot, J. L. Maubois, P. Schaegis, R. Veyre, and A. Luccioni, Microfiltration en flux tangentiel des lactosérums de fromagerie, *Lait 64*:102 (1984).
49. J. Fauquant, E. Vieco, G. Brulé, and J. L. Maubois, Clarification des lactosérums doux par agrégation thermocalcique de la matière grasse résiduelle, *Lait 65*:1 (1985).
50. J. L. Maubois, A. Pierre, J. Fauquant, and M. Piot, Industrial fractionation of main whey proteins, *Bull. Fed. Int. Lait. 212*:154 (1987).
51. A. Pierre, Y. Le Graet, G. Daufin, F. Michel, and G. Gésan, Whey microfiltration performance: influence of protein concentration by ultrafiltration and of physicochemical pretreatment, *Lait 74*:65 (1994).
52. G. Daufin, J. P. Labbé, A. Quemerais, F. Michel, and U. Merin, Optimizing clarified whey ultrafiltration: influence of pH, *J. Dairy Res. 61*:355 (1994).
53. D.-C. Hwang and S. Damodaran, Selective precipitation and removal of lipids from cheese whey using chitosan, *J. Agric. Food Chem. 43*:33 (1995).
54. M. E. Mangino, Properties of whey protein concentrates, *Whey and Lactose Processing* (J. G. Zadow, ed.), Elsevier, London, 1992, p. 231.
55. J. E. Kinsella, Relationship between structure and functional properties of food proteins, *Food Proteins* (P. F. Fox and J. J. Condon, eds.), Applied Science Publishers, London, 1982, p. 51.
56. P. Gault, M. Mahaut, and J. Korolczuk, Caractéristiques rhéologiques et gélification themque du concentré de protéines de lactosérum, *Lait 70*:217 (1990).
57. S. Y. Liao and M. E. Mangino, Characterization of the composition, physicochemical and functional properties of acid whey protein concentrates, *J. Food. Sci. 52*:1033 (1987).
58. R. Aschaffenburg and J. Drewry, Improved method for the preparation of crystalline β-lactoglobulin and α-lactalbumin from cow's milk, *Biochemistry 65*:273 (1957).
59. J. L. Maubois, Chromatographie des protéines du lactosérum de brebis, *Ann. Biol. Anim. Biochim. Biophys. 4*:295 (1964).
60. P. Mailliart and B. Ribadeau Dumas, Preparation of beta-lactoglobulin-free proteins from whey retentate by NaCl salting out at low pH, *J. Food Sci. 53*:743 (1988).
61. T. Kaneko, B. T. Wu, and S. Nakai, Selective concentration of bovine immunoglobulins and α-lactalbumin from acid whey using $FeCl_3$, *J. Food Sci. 50*:1531 (1985).
62. T. Kuwata, H. Ohtomo, E. Hori, and Y. Yamamoto, Effects of defatting and desalting on heat stability of whey protein concentrate, *Nippon Shokuhin Kogyo Gakkaishi 32*:493 (1985).

63. J. L. Maubois, Separation, extraction and fractionation of milk protein components, *Lait 64*:485 (1984).
64. R. J. Pearce, Thermal separation of β-lactoglobulin and α-lactalbumin in bovine Cheddar cheese whey, *Aust. J. Dairy Technol. 38*:144 (1983).
65. R. J. Pearce, Fractionation of whey proteins, *Bull. Fed. Int. Lait. 212*:150 (1987).
66. C. Bramaud, P. Aimar, and G. Daufin, Thermal isoelectric precipitation of α-lactalbumin from a whey protein concentrate: influence of protein-calcium complexation, *Biotechnol. Bioeng. 47*:121 (1995).
67. J. L. Maubois, Whey, its biotechnological signification, Proceedings of the 8th International Biotechnology Symposium, July 17–22, Paris, France, 1988, pp. 817–824.
68. B. D. Shur, Alpha-lactalbumin contraceptive, Patent PCT WO 84/04457 (1984).
69. J. P. Perraudin, Protéines à activités biologiques: lactoferrine et lactoperoxydase. Connaissances récemment acquises et technologies d'obtention, *Lait 71*:191 (1991).
70. G. Brulé, L. Roger, J. Fauquant, and M. Piot (INRA), Procédé de traitement d'une matière première à base de caséine, contenant des phosphocaséinates de cations bivalents. Produits obtenus et applications, Patent FR 8 002 280 (1980).
71. M. Tanimoto, Y. Y. N. Kawasaki, H. Shinmoto, S. Dosako, and A. Tomizawa, (Snow Brand), Process for producing kappa-casein glycomacropeptides, Patent EP 0 393 850 A2 (1990).
72. S. Beucher, F. Levenez, M. Yvon, and T. Coring, Effects of gastric digestive products from casein on CCK release by intestinal cells in rat, *J. Nutr. Biochem. 5*:578 (1994).
73. J. L. Maubois and J. Léonil, Peptides du lait à activité biologique, *Lait 69*:245 (1989).
74. D. D. Kitts and Y. Yuan, Caseinophosphopeptides and calcium bioavailability, *Trends Food Sci. Technol. 3*:31 (1992).
75. S. Maruyama, H. Mitachi, H. Tanaka, N. Tomizuka, and H. Susuki, Studies on the active site and antihypertensive activity of angiotensin I-converting enzyme inhibitors derived from casein, *Agric. Biol. Chem. 51*:1581 (1987).
76. P. Jollès, S. Levy-Toledano, A. M. Fiat, C. Soria, D. Gillessen, A. Thamaidis, F. W. Dunn, and J. P. Caen, Analogy between fibrinogen and casein. Effect of an undecapeptide isolated from κ-casein on platelet function, *Eur. J. Biochem. 158*:379 (1986).
77. E. Laffineur, Activité immunomodulatrice *in vitro* de composés produits par la fermentation lactique d'un substrat dérivé du lait, Ph.D. thesis, ENSA, Rennes, France, 1994.
78. F. Parker, D. Migliore-Samour, F. Floc'h, A. Zerial, G. H. Werner, J. Jollès, M. Casaretto, H. Zahn, and P. Jollès, Immunostimulating hexapeptide from human casein: amino acid sequence, synthesis and biological properties, *Eur. J. Biochem. 145*:677 (1984).
79. J. Léonil, D. Mollé, and J. L. Maubois, Study of the early stages of tryptic hydrolysis of β-casein, *Lait 68*:281 (1988).

21

Chemical and Physical Methods for the Characterization of Proteins

P. K. Nandi
Institut National de la Recherche Agronomique, Centre de Recherches de Tours
Nouzilly, France

I. INTRODUCTION

In protein chemistry, the physical characterization of the isolated protein from plant or animal sources mostly employs procedures to determine the apparent molecular weight and charge, size, and shape of the molecule. Food proteins, like their physiological cousins, can be made from subunits, and it is also customary to identify and characterize them. The chemical characterization of proteins generally involves determining amino acid composition and sequence analysis [1].

The first consideration from the viewpoint of characterization of a protein molecule is the amount of pure protein available. As no single method is capable of yielding the information necessary to characterize a protein, enough pure protein should be available to do so. This may not be a major problem for food proteins from conventional sources. At present, most isolation procedures are capable of removing nonprotein contaminants, namely, nucleic acids, carbohydrates, and lipids. Identification of other proteins contaminating the protein of interest is an important step before characterization. For this, one or more fractionation procedures are carried out to demonstrate the presence of a single component. The majority of the methodologies used to characterize a protein are also used to identify other contaminating protein(s) [1,2].

The determination of size, or the physical dimension of the protein molecule, and molecular weight, which is related to the mass of the protein, is based on the following broad areas of study: (a) chemical analysis, the composition of the protein (amino acids and carbohydrates, lipids, or associated metal ligands for biological function); (b) the effect of protein molecules on the colligative properties of the solvent, namely, vapor and osmotic pressures; (c) the transport of protein molecules from one part of the solution to another under mechanical, electrical, or centrifugal force; and (d) scattering of incident radiation, namely, light, x-rays, and neutrons [2]. The size and shape of a protein molecule can be directly visualized by electron microscopy; however, this method and x-ray and neutron diffraction analyses are not routinely practiced in a protein laboratory.

In this chapter, the background of various methodologies used for characterization of an isolated protein are highlighted. Limited examples from various protein fields are cited. The various approaches, either alone or in combination, can yield a wealth of information about different molecular properties as well.

II. CHEMICAL METHODS

A. Composition

Determination of composition provides an estimation of minimum molecular weight of a protein using protein components, namely, amino acid analysis, end-group analyses, and qualitative analysis for prosthetic groups.

The molecular weight of the protein M_{min} (g/mol) can be obtained from the equation:

$$M_{min} = \frac{m}{n} \tag{1}$$

where m is the mass of the protein used and n is the number of moles of the above protein-related components. Although the estimation of mass of a protein can be made very accurately from the measurement of the dry weight of the protein by evaporating the solvent when it is present in a volatile buffer (e.g., ammonium bicarbonate), the protein concentration is normally determined by refractometry, spectrophotometry, and chemical analysis. Of these, refractometry is the most accurate method for determination of concentration, although amino acid analysis can also be used for quantitation of the proteins [2].

Due to the problem of standardization, the accuracy of such measurements is often limited. This is true for glycoproteins, lipoproteins, or proteins with unusual composition. The quantity of material necessary for composition-based molecular weight determination is dependant on the sensitivity of the analytical methods: only microgram or smaller quantities of materials is necessary. This method of molecular mass determination suffers specifically from the assumption that there is only one or an integral number of moles of the analyzed component per mole of protein. This leads to determination of the minimum molecular weight in the case of oligomeric proteins. Often the composition analysis method is coupled with other techniques to obtain the molecular mass of the protein: the combination of amino acid analysis data with sodium dodecyl sulfate gel electrophoresis is useful for determining the mass of a protein [1].

B. Amino Acid Analysis

Amino acid composition of a protein usually provides a method for its quantitative characterization. Hydrolysis of a protein is carried out in 6 N HCl for varying periods of time (20 min to 24 h) at a temperature of 100°C under vacuum. Care is taken in releasing amino acid quantitatively without degradation. However, several amino acids undergo chemical modifications under hydrolyzing conditions, for example, asparagine and glutamine are hydrolyzed to aspartic and glutamic acid, respectively. Tryptophan cannot be recovered in the acid-hydrolysis process. Addition of 2-mercaptoethanol to the hydrolyzing mixtures prevents oxidation of methionine to methionine sulfoxide. Determination of cystine can be accomplished only after its modification by, e.g., oxidation with performic acid, carboxyl methylated, or derivatized with 4-vinylpyridine before hydrolysis. The ala-ala, Ile-Ile, val-Ile, Ile-val, and ala-val bonds require longer hydrolysis, e.g., more than 90 hours, for quantitative hydrolysis of these bonds. Threonine and serine are destroyed over time, and their quantities are estimated from linear extrapolation to zero time of hydrolysis [3]. Due to the degradation of tryptophan in acid, its hydrolysis is performed in alkaline solution. Tryptophan can also be determined spectrophotometrically [3,4].

In addition to the determination of the amino acid composition of a particular protein, this method of chemical analysis can compare the nature of different proteins from the same source. For example, proteins from *Amaranthus hypochondricus* show that albumins and globulins are rich in lysine and valine; glutelins contain relatively large amounts of leucine, threonine, and histidine, and prolamines are rich in sulfur amino acids and phenylalanine [5].

C. Limited N-Terminal Sequence Analysis

Identification of the N-terminus amino acids of proteins is often useful for characterizing the structural and functional domains and is important for the isolation of recombinant DNA clones. Purified proteins obtained either from SDS-polyacrylamide gel electrophoresis (see below) or from reversed-phase HPLC, are subjected to repeated cycles of the Edman degradation. Briefly stated, each degradation cycle consists of three stages: coupling, cleavage, and conversion [6]. In the first stage, the N-terminus of the protein (or polypeptide) is modified with phenyl isothiocyanate in alkaline solution to produce a phenyl-thiocarbamoyl derivative (PTC). The PTC–N-terminal residue is cleaved from the rest of the polypeptide by using either liquid or gaseous trifluoroacetic acid, which forms an anilinothiazoline–amino acid derivative of the original N-terminal residue and is stabilized by conversion to a phenylthiohydantoin amino acid. This derivative is analyzed in a high-performance liquid chromatography (HPLC) column and is identified by its elution time. The N-terminal residue of $(n - 1)$ polypeptide is subjected to another cycle of coupling, cleavage, and conversion steps, and the process is repeated until the desired information about the N-terminal sequence is obtained. In the case of proteins where the N-terminus is blocked, the sequence is determined after generating internal peptides with unblocked terminus by cleaving the protein either chemically or enzymatically. The sequence analysis of individual polypeptides of methionine-rich protein (MRP) of peanut is being utilized to synthesize oligonucleotide probes for screening the peanut cDNA library [7]. From the determination of the N-terminal sequence of D-glutenin of wheat, it has been suggested that these proteins are formed as a consequence of mutation of an ω-gliadin gene [8].

D. Peptide Mapping

This method provides information regarding sequence, that is, the primary structure of a protein. Two proteins with identical primary structures, when cleaved with a site-specific endopeptidase, will yield identical peptide fragments. Peptide mapping consists of four processes to yield the primary structure of a protein: (a) purification of protein, most often by SDS-PAGE; (b) labeling of the protein and hence the peptide fragments, usually by radioiodination, when the quantity of protein is small; (c) specific endopeptidic cleavage of the protein; and (d) the separation and visualization of the resultant peptidic fragments by SDS-gel electrophoresis, two-dimensional thin-layer electrophoresis/thin-layer chromatography or capillary zone electrophoresis or by HPLC [9].

Peptide mapping has recently been used to identify peptides as potential indicators of peanut maturity [10]. As indicated above, the degree of similarity or dissimilarity between the peptide fragments reflects the degree of primary structural similarity between the proteins. Proteins are the major source of peanut flavor precursor, such as peptides and amino acids. Changes in the protein structure, that is, amino acid sequence, would be expected to result in a change in peanut flavor. As peanut maturity can affect peanut flavor, it was assumed that the proteins of mature and immature peanuts would be structurally different. Peptide mapping using arginyl endopeptidase showed that mature and immature peanuts are indeed structurally different, which has opened up the possibility of determining the potential role of different peptides in peanut maturity [10].

III. PHYSICAL METHODS

A. Osmotic Pressure

Boiling point elevation, freezing point depression, and the osmotic pressure of a solution depend on the amount of the solute present, not on its nature. The colligative properties are a result of the modification of the chemical potential of a solvent by the presence of the solute. This relationship is expressed by:

$$\mu_A = \mu_A^\circ + RT \ln (\gamma_A X_A) \tag{2}$$

where μ_A and μ_A° are the chemical potential of a solute in kcal/mol in a solution and at the standard state, respectively, R is the gas constant in absolute scale (8.3144×10^7 kcal/mol), T is the temperature in absolute scale, γ is the activity coefficient of the solute, and X_A is its mole fraction in the solvent. The molecular weight of any solute, for example, protein, can be obtained from the amount of solute that would change the solution activity [1,11]. Thus, the colligative properties, e.g., freezing point depression, boiling point elevation, and osmotic pressure or vapor pressure of the solvent by a protein, could be used to calculate its molecular weight. Due to lack of sensitivity and the extreme temperatures necessary for freezing point depression or boiling point elevation, only vapor pressure and osmotic pressure have been used for protein molecular weight determination.

In an ideal solution, the osmotic pressure Π is related to the number of solvent molecules in a solution by:

$$\Pi = \frac{RTn}{v} \tag{3}$$

where n is the number of moles of solute in a volume v. Π can be related to the molecular weight, M, by $\Pi = cRT/M$, where c equals concentration per unit volume [12].

The osmotic pressure varies with concentration. In general, a plot of Π/c versus c to obtain M does not follow a linear relationship due to solute–solute interaction. The linear relationship could be obtained by an empirical equation $\Pi/c = A + Bc + Cc^2 + \ldots$, where A, B, and C are virial coefficients. The second virial coefficient B is indicative of the quantitative nature of the interaction and dominates over other coefficients.

Complications arising from impurities present in protein solution, dimerization or polymerization, and steric repulsion limit the success of determination of molecular weight by this method. The osmotic pressure method gives an average value of molecular weight, which necessitates equal weighing of all molecules and therefore yields number average molecular weight. Protein concentrations in the range of 0.1–1 mg/ml and volumes of ~10–200 μl are generally required [12,13].

B. Transport Properties

1. Sedimentation Equilibrium

Sedimentation methods can be used to determine both weight-average molecular weights of protein and information regarding its shape. The dimension of the molecule and the nature of its contour are generally used to determine the shape of the protein molecule. Sedimentation equilibrium can be used to determine the molecular weight of the protein. The technique consists of spinning a rotor to produce a protein concentration gradient along the radial axis of the centrifuge cells in an analytical ultracentrifuge. The concentration gradient is optically determined during the spinning of the solution, and the contents of the cell stay at equilibrium. The length of time to reach equilibrium in sedimentation depends on the sedimentation speed, the length of the solution column (usually short, in the millimeter or its fractional range) in the centrifuge cell, the diffusion coefficient, and other properties of the protein molecules and the solvent.

The quantity determined in a sedimentation equilibrium is the reduced molecular weight σ given by:

$$\sigma = M(I - V\rho)\frac{\omega^2}{RT} \tag{4}$$

where V is the partial specific volume of protein (ml/g), ρ is the solution density (g/ml), and ω (radian/sec) is the angular velocity of the rotor related to rotation per minute (rpm) by:

$$\omega = \text{rpm} \times \frac{\pi}{30} \tag{5}$$

In principle, without considering association or nonideality of a protein in solution, the natural logarithm of the plot of concentration as a function of $r^2/2$, where r is the rotor's radial position, yields the value for σ. The reference markers provided in the analytical ultracentrifuge together with optical magnification factor are used to obtain the radial position of the image. This plot also helps in identifying the purity of the protein sample if the foregoing plot yields the same slope independent of the rotor speed. In the analytical centrifuge the concentration gradient value is determined by optical measurements by using absorbance scanner and Rayleigh interferometer [1,14–16].

The concentration distribution at equilibrium can also be determined in a preparative centrifuge or an airfuge, but these techniques suffer from precision due to the disturbance of the concentration gradient as the rotor decelerates and also during fractionation of the solute after centrifugation. Further, unlike in an analytical ultracentrifuge, where attainment of equilibrium can be verified in the same sample during rotation at different time intervals, in the preparative procedures several experiments with varying time intervals are necessary to ensure that the equilibrium has been attained [1]. The equilibrium sedimentation method is also used to determine the oligomerization of the isolated protein [1,14,15].

2. Sedimentation Velocity

This method is used for characterization of the hydrodynamic properties of the protein. The technique depends upon centrifugation of a rotor at a speed greater than 10^4 revolution per minute, which applies a large centrifugal force on the protein molecule. The measured property is the rate at which the boundary between the solution and solute-free solvents move away from the center of rotation. The sedimentation coefficient, s, is defined by:

$$s = \frac{dr}{dt}\frac{1}{r}\,\omega^2 \tag{6}$$

where dr/dt is the velocity of the boundary at a distance r from the center of rotation, t is the time in seconds from the starting of the experiments, and ω has the same significance as before. The value of s is characteristic only of the solution, independent of the speed and dimension of the rotor, and expressed in Svedberg units (10^{-13} s). If the protein sample being examined is a thin zone of molecules, the point of maximum concentration is taken as r. In broad zone or boundary experiments, r is the midpoint value between zero and plateau concentrations.

The value of s depends upon solvent and temperature and is corrected to standard conditions of water at 20°C and zero protein concentration ($s^\circ_{20,w}$) by:

$$s^\circ_{20,w} = s\,\frac{(1 - V_{20,w}\,\rho_{20,w})}{(1 - V_{20,w}\,\rho_{T,b})}\,\frac{(\eta_{T,w})}{(\eta_{20,w})}\,\frac{(\eta_{T,b})}{(\eta_{T,w})} \tag{7}$$

where v, ρ, and η are the partial specific volume, density, and viscosity at experimental temperatures (T) and solution (w, water; b, buffer) conditions, respectively [13,17].

The value of s is related to the shape of the molecule through frictional coefficient and molecular weight by:

$$s = \frac{M(I - V\rho)}{Nf} \tag{8}$$

where N is the Avogadro's number and f is the frictional coefficient of the protein. The shape of the protein can be determined from the ratio f/f_o, where f_o is the frictional coefficient expected for an anhydrous sphere of molecular weight and density obtained from the Stokes equation:

$$f_o = 6\pi\eta r_{sphere} = 6\,\pi\eta[3MV/4\,\pi N]^{1/3} \tag{9}$$

where r_{sphere} is the radius of the equivalent sphere. The ratio f/f_o describes an estimate of the overall shape and asymmetry of the protein. The value of f increases with the departure of the protein from a spherical shape and its degree of hydration. The frictional coefficient depends on the surface area of the protein; a spherical protein would expose minimum area. Therefore, any molecular asymmetry would increase f [1,17]. Due to hydration of the protein, the water associated with the protein molecule moves with the protein in the centrifugal field, thereby increasing the effective radius of the protein. This hydration influence often poses a problem since f does not change markedly with shape. Change from a spherical to a cigar-shaped molecule with a length 10 times the original diameter increases the value of f by only 50% [12].

The availability of an analytical ultracentrifuge is important for the determination of the molecular weight and sedimentation velocity of protein. However, methods of reasonable accuracy have been developed for preparative centrifuge. Ultracentrifugation of a freeze-dried solution of amaranth globulin in an isokinetic sucrose gradient (5–20%) in a preparative Beckman 15-65B ultracentrifuge has been performed [18]. The protein separation was detected by a UV-detector. The sedimentation coefficient of the protein was calculated using proteins of known s values under the same experimental conditions.

For sedimentation experiments, protein in a concentration range of 0.1–1 mg/ml is necessary [1,17]. The principal technical difficulty in sedimentation experiments is convection. The interpretation of the sedimentation coefficient for detergent-solubilized proteins is often complicated due to the concentration of the bound detergent [1]. In addition to the determination of the s value of a particular protein [16,19–22], sedimentation analysis has been utilized to study the effect of physical processes on the protein molecule. For example, dialysis against pure water and subsequent freeze-drying of the major soybean protein 11S glycinin showed both dissociation (a 2S component) and aggregation (a 15S component) of the protein [22].

3. Viscosity

The viscosity of a solution depends on the amount and nature of the macromolecules in it. Proteins adsorb water molecules on their surface and impede the flow of free water molecules, which retards the flow of solution through a capillary tube. The ratio of the time of flow of a protein solution to the time of flow of the buffer in the same viscometer under the same experimental conditions is proportional to the viscosity of the solution and yields the value of the reduced viscosity. The solvent viscosity by definition is taken as 1.0, and subtracting this from the protein solution results in the specific viscosity of the protein. Extrapolation of the reduced specific viscosity (specific viscosity divided by the protein concentration) to zero protein concentration gives the intrinsic viscosity (ml/g). This parameter is independent of size or molecular weight, interaction between protein molecules, and contribution from the free solvent. The differences in intrinsic viscosity arise from the difference in the shape of the protein molecules [12].

A typical globular protein has an intrinsic viscosity of 3 ml/g. Einstein showed that the intrinsic viscosity of any particle is independent of its molecular weight, with a value of 2.5 ml/g for perfectly spherical particles. Collagen, a fibrous protein, has a very high intrinsic viscosity of a 1150 ml/g. Unlike globular proteins, a definite correlation exists between molecular weight and intrinsic viscosity for fibrous proteins. As already mentioned, the viscosity of protein solution is necessary to calculate molecular weight in sedimentation experiments. The apparatus for this classical method is inexpensive, and the experiment is easy to perform and has recently been used for the physiocochemical characterization of rapeseed albumin in different pH and electrolyte solutions [23]. Studies on viscosity have helped in differentiating the muscle protein actomyosin from other species [24].

4. Gel Filtration Chromatography

Gel filtration chromatography is the most widely used, simple, and efficient method of determining the molecular weight of a protein molecule [19]. A thin band of protein solution is applied to the top of a gel bed equilibrated in a buffer in a column, and the protein is eluted through the column with the buffer equilibrating it. Gel is chosen in which the protein of interest is present and does not bind to the gel matrix. The protein molecule portions between the external and internal solvent regions of the gel particle. The elution volume, Ve, is measured to determine the molecular weight (the volume of solvent that must flow through the column before the protein appears). The molecular weight of the unknown protein is estimated from the logarithm of molecular weights of protein standards and their elution volumes in the same column. In reality, Ve is dependent on the logarithm of Stokes radius (hydrated radius) of the protein. The protein standards normally cover a range from Vo (the void volume, determined by using size-graded blue dextran of 2,000,000 molecular weight, whose particle diameter is much larger than the gel pore) to Vi (the internal volume of the pores of the gel particles accessible to the solvent buffer and determined by using a buffer with a different pH or more commonly by a low molecular weight dye, e.g., bromophenol blue). Most often, the term Kav, defined as the fraction of the stationary gel volume accessible to the protein, is used rather than Ve and is defined by:

$$\mathrm{Kav} = \frac{\mathrm{Ve} - \mathrm{Vo}}{\mathrm{Vi} - \mathrm{Vo}} \tag{10}$$

Kav is normally preferred to Ve due to a nearly constant value for Kav for a given gel type from column to column [1,7,17,25–31].

The protein standards used for determination of molecular weight of a protein are compact globular proteins. The principle of gel chromatography requires that the unknown protein be similar in shape and density to the protein standards used [1,31]. The fibrous proteins are expected to behave anomalously in the gel. This can be verified by studying the increased value of Kav for the sample protein at decreased flow rate, which would not affect the Kav values for globular proteins [1].

The estimation of the molecular weights of extensively glycosylated proteins and detergent-solubilized proteins would not be reliable using gel chromatography, as the method is based on the size of the protein. Care should also be taken to avoid possible interaction of proteins with the gel matrix [1,31].

5. Electrophoresis

Nondenaturing Gel Electrophoresis

Electrophoresis is a powerful tool to evaluate the size of the protein molecule. It does not provide quantitative structural data. The mobility of a macromolecule in an applied electric field is proportional to the charge of the particle and the strength of the field. The resulting frictional force is proportional to the velocity, the radius of the particle, and the viscosity of the medium:

$$V = \frac{qE}{f} \tag{11}$$

where q is the net charge of the macromolecule, E is the applied electric field, f is the frictional coefficient, and for a spherical macromolecule with radius r and charge z:

$$V = \frac{zeE}{6\pi\eta r} \tag{12}$$

where e is the charge on an electron and η is the viscosity of the medium.

Electrophoresis under nondenaturing conditions would provide information regarding relative charge for molecules having the same size and shape or about the relative size of molecules with the same charge. Most often electrophoresis in nondenaturing gel is carried out to determine the isoelectric point of the protein in a pH gradient. This method of isoelectric focusing uses a mixture of ampholytes (having different isoelectric points), which under an influence of electric field produces a pH gradient. The added protein migrates until it reaches the pH of its isoelectric point (pI) and can be analyzed by protein absorbance or protein staining and can be isolated by slicing the polyampholyte gels [32]. This procedure has been used to characterize, for example, friabilins and egg yolk proteins [33,34]. Effects of storage, temperature, and humidity on bean protein has also been characterized by electrophoresis [35].

SDS-Gel Electrophoresis

The most popular technique used by protein chemists to determine the molecular weight of a protein is zonal electrophoresis in SDS in a polyacrylamide gel. This anionic detergent binds to all proteins at approximately 1.4 g/g of amino acids. All proteins are converted into similar structures (rods) in the denaturing concentration of SDS and differ only in molecular weight. The negative charge of SDS bound to protein masks the differences of total overall charges between proteins. Since it is assumed that the SDS-protein complex contains a constant fraction of the detergent, the net charge of the protein-detergent complex would be proportional to the molecular weight of the protein. The separation of proteins originates from specifics of interaction of different-sized protein-SDS complexes with the supporting gel in which electrophoresis is carried out. Under the force of an electric field the larger molecules have a lower net mobility. The net mobility is related to the molecular weight, M, by

$$u = b - a \log M \tag{13}$$

where u is the mobility of the protein, b is a constant that includes a term gel free mobility under the applied electric field and is independent of molecular weight, and the parameter a is a function of gel concentration [32]. The reducing agent β-mercaptoethanol is used for incubation of the protein and is also added to the gel to reduce intra- or interchain disulfide linkages. The molecular weight of the protein is determined by electrophoresis of different proteins of known molecular weight in the same gel simultaneously and using Eq. (13) [1,5,32,36].

C. Mass Spectrometry

During the last decade, the method of mass spectrometry generally used to determine the mass of low molecular weight organic compounds has been successfully used to determine the molecular weight of proteins weighing up to more than 100 kDa. The method is highly sensitive and can determine the molecular weight of a protein with a precision of better than 0.01% [37].

Electron ionization and chemical ionization were two procedures utilized for the determination of the molecular weight of organic molecules from 1950 to 1970 [38]. In the first method, gaseous polyatomic molecules are ionized by interaction with a beam of electrons. In the second method, a reagent gas is ionized by electron impact, which, when colliding with molecules of interest for mass determination, produces molecular ions: when these ions are mass analyzed, the chemical ionization mass spectrum of the compound is obtained. For both of these procedures, the sample of interest needs to be present in the gas phase at a pressure of 10^{-5}–10^{-4} torr. During the 1970s and 1980s methods were developed for mass spectra determination of large, nonvolatile, and fragile biomolecules, which involved direct ionization from the solid or solution state [37].

Of the various types of mass spectrometry, electron spray ionization mass spectrometry has often been used for the determination of protein molecular weight [37,39]. Multiple-charged molecular ions are generated from very small droplets of a solution of protein emerging from a tip of a needle by a large electrostatic field gradient, 3 Kv/cm. The positively charged protein molecular ions under the electrostatic field are formed by the attachment of proton, H^+, alkali cations, or ammonium ions. The basic amino acid residues in a protein are distributed so that a multiplicity of peaks, one for each $(P + H_n)^{n+}$ ion, give a value of m/z of around 1000 (P = protein; m = mass, z = charge). The molecular weight of the protein can be determined from each pair of adjacent peaks based on the assumption that the adjacent peaks of a series of plots of relative intensity against m/z values differs by only one charge.

The m/z values are obtained from the center of the peaks and the simplified equation relating the molecular weight M_r of protein and its charge z with the multiple-charged ions at a position, for example, p_1

$$p_1 z_1 = M_r + M_x z_1 = M_r + 1.0079 z_1 \tag{14}$$

where the charge carrying species is a proton (M_x = 1.0079).

The molecular weight of a second multiple protonated ion m/Z p_2 ($p_2 > p_1$) that is j peaks further from p_1 is given by:

$$p_2(z_1 - j) = M_r + 1.0079(z_1 - j) \tag{15}$$

A combination of the above equations results in:

$$z_1 = \frac{j(p_2 - 1.0079)}{(p_2 - p_1)} \tag{16}$$

The value of M_r is obtained by knowing the nearest integer value for z_1, which is the charge of each peak (j = 1 for two adjacent peaks) in the multiple-charge distribution pattern [39].

Recently the method of mass spectrometry has been utilized to determine molecular weights of food proteins from different sources [40,41]. Identification of phaseolin polypeptide subunits of a protein isolate from large lima beans (*Phaseolus lunatus*) is a case in point [41]. The mass spectra results show that a glycosylated subunit of the above protein was similar to a C-terminal segment of the phaseolin polypeptide of *P. vulgaris*, while another glycosylated subunit and its nonglycosylated variant were similar to an N-terminal segment of phaseolin polypeptide of *P. vulgaris*. In the biological field, two members of the s-100 family of Ca^{2+} binding proteins recombinant CAPL and CACY, which are believed to be involved in cell-cycle regulation, differentiation, growth, and metabolic control, have recently been characterized by ESI mass spectrometry [42].

Molecular weight measured by ESI mass spectrometry, when compared with its theoretical molecular weight from the known amino acid composition, generally shows a variation of not more than 0.005% when a quadrapole mass spectrometer is used. The ESI method of molecular weight determination is generally 2–3 orders of magnitude more accurate than molecular weight obtained by electrophoretic method. The molecular weight determined with such high precision has been used in solving the protein structure that requires the determination of molecular weights better than one mass unit, for example, to determine an error in the DNA sequence or posttranslational modification of the gene product [37,43]. This method of molecular weight determination has helped to determine that differences in the conformation and not in posttranslational chemical modification are responsible for the infectious property of prion [43]. This powerful method of MS-ESI has also been used to characterize metal ion binding (Cu^{2+}, Zn^{2+}) with hen egg white lysozyme for their eventual role in the biological function of the enzyme [44].

D. Light Scattering

In the elastic light scattering procedure, a sample of protein solution is radiated with a collimated beam of light of wavelength λ, and the scattered radiation at the same wavelength is measured as a function of the angle between the incident beam and the detector of the scattered light [45]:

$$I = Io(8\pi^2\alpha^2/\lambda^4 r^2)(1 + \cos^2 2\theta) \tag{17}$$

where Io and I are the intensities of the incident and scattered radiation, respectively, α is the polarizability of the molecule, 2θ is the angle between incident and scattered light, and r is the distance between the sample and the detector. The equation can be rewritten to obtain the Rayleigh ratio R_θ, which is the relative intensity of scattered light as:

$$R_\theta = \frac{r^2}{V(1 + \cos^2 2\theta)}(I/Io) \tag{18}$$

where V is the volume of the solution under observation [45].

At a given distance and intensity of the incident light, the intensity of the scattered light is inversely proportional to the wavelength of the incident light and directly proportional to the radius of gyration of the scattering particle:

$$R_\theta = KMc(1 - 16\pi^2 R_G^2 \sin^2\theta/3\lambda^2) \tag{19}$$

where R_θ is the Rayleigh ratio, M is the molecular weight of the scattering particle, c is the concentration, and R_G is the radius of gyration [45]. The term K contains the measurable quantities:

$$K = \frac{2\pi^2 n_o^2}{N\lambda^4}\left(\frac{\delta n}{\delta c}\right)^2 \tag{20}$$

where n_o and n are the refractive indices of the pure solvent and solution, respectively, N is the Avogadro's number and $\delta n/\delta c$ is the refractive index increment with concentration (45).

For extrapolation to infinite dilution of a solution of a macromolecule, the above equation can be rewritten as:

$$\lim_{c\to 0} \frac{Kc}{R_\theta} = \frac{1}{M} + \frac{1}{M}\left(\frac{16\pi^2 R_G^2 \sin^2\theta}{3\lambda^2}\right) \tag{21}$$

Kc/R_θ can be measured at various concentrations as a function of scattering angle 2θ. The data are extrapolated separately to $\theta = 0$ and $c = 0$, and the intersection of the two plots are used to determine the value of molecular weight M [45].

The above light scattering experiments can also be utilized to determine the shape of the molecule under investigation [45]. The initial slope of the curve extrapolated to $c = 0$ yields the value of radius of gyration of the macromolecule R_G. From theoretical consideration, R_G equals $3r^2/5$ for a sphere of radius, r, $r_A^2/5 + r_B^2/5$, for an ellipsoid of revolution of half-axis r_A and perpendicular half-axis r_B, $l^2/12$ for a thin rod of length l, and approximately $d^2/6$ for a flexible polymer where d^2 is the mean square distance between the ends of the flexible polymer [12]. If the general information regarding the shape can be obtained by a direct method, e.g., electron microscopy, the parameters for a quantitative description of that shape can be obtained accurately [45]. The major difficulty with the light scattering method to determine M and R_G is its inherent great sensitivity to any impurities of high molecular weight, which requires careful preparation of samples without aggregates or dust particles.

The other variation of the above is inelastic light scattering, which includes Raman and dynamic light scattering. In the former, radiation is observed at energies equal to that of the incident light plus or minus vibrational quanta [45]. Information regarding the vibrational energy levels of the dissolved proteins (macromolecules) can be obtained from the wavelength distribution of the scattered radiation.

In dynamic light scattering, scattered radiation at wavelengths slightly different from that of the incident beam is measured [45]. Molecules in motion in a solution will alter the wavelength of the radiation due to Doppler effects. Even in the absence of a net motion, the fluctuation in the number of scattering molecules will result in a distribution of wavelengths of the scattered light. A laser light is necessary for the dynamic light scattering study [45]. The spectrum of scattered radiation is:

$$I(S, \nu) = \frac{\langle N \rangle}{\pi} \frac{(S^2D)}{4\pi^2\nu^2 + (S^2D)^2} \tag{22}$$

where ν is the frequency of the light, $\langle N \rangle$ is the average number of solute molecules in the volume being sampled, S is $2\sin\theta/\lambda$, and D is the diffusion coefficient. The half-width of the frequency distribution of the scattered radiation is S^2D. The measurement of half-width as a function of $\sin^2\theta$ yields the diffusion coefficient.

A combination of classical and dynamic light scattering has been successfully used to characterize the human blood platelet $\alpha_{IIb}\beta_3$ complex [46]. This protein is the prototypical component of mammalian cell adhesion receptor, termed integrin, which plays a central role in the hemostatic process. The interpretation of the results based on electron microscopy, chemical methods, and hydrodynamic measurements was contradictory, and the interpretation of these results in terms of macromolecular dimension was hampered by the experimental limitation of size exclusion chromatography for Stokes radius determination and by the large size of the micelles in which this membrane protein receptor was solubilized [46]. The scattering method not only gave accurate values for both the diffusion coefficient and Stokes radius for the $\alpha_{IIb}\beta_3$ complex, the molecular weight determined by the scattering procedures yielded a value of $(2.26 \pm 0.22) \times 10^5$ for the polypeptide moiety of the complex, in excellent agreement with the 2.28×10^5 value calculated from primary structure data [46].

IV. SUBUNIT IDENTIFICATION

Isolated proteins may consist of a single polypeptide chain (monomeric) or be made up of identical or nonidentical polypeptide chains (multimeric), which may remain associated with each other either by intermolecular noncovalent interaction or by covalent disulfide linkage. The simplest approach to identifying the presence of subunits is to run a nondenaturing gel electrophoresis under the buffer conditions at which the protein was isolated and to carry out a second electrophoresis that would dissociate the subunits. For this purpose, a change in the buffer pH or the addition of reagents (e.g., SDS or urea) that denature the protein is used. The basic aim is to determine the apparent size of the protein molecule under associating or dissociating conditions. To investigate the presence of intersubunit disulfide linkage(s), the protein is first run in the gel in the presence of SDS followed by second-dimension gel electrophoresis in the presence of SDS and β-mercaptoethanol or dithiothreitol to reduce the S-S linkage. The presence of intermolecular disulfide-linked subunits is indicated either from the difference in the mobility of the protein bands in the above two electrophoretic experiments or the appearance of multiple components in the second dimension [1,47,48]. The wheat protein glutenin has been characterized by these procedures [49,50]. Column chromatography methods can also be used to identify the presence of subunits in a protein, and a combination of this method, SDS-electrophoresis, and electron microscopy has revealed that the dimer of chloroplast cytochrome b_6-f complex is the structural unit for its biological function [26].

In the noncovalently associated multimeric protein systems, the determination of sedimentation velocity coefficients under dissociating or denaturing conditions would indicate the presence of the subunits when compared with the same experiments with native protein in the isolation buffer. Dissociation would result in the lowering of the value of the sedimentation coefficient [21]. The presence of a single component under dissociating conditions would apparently indicate the presence of identical subunits. The stoichiometry of the identical subunits making up the native protein could be determined from the values of the protein molecular weight obtained from sedimentation equilibrium and the information from gel electrophoresis results [36]. If the protein is made up of different types of subunits, the sum of the molecular weights of the monomers determined from gel electrophoretic experiments compared to the native molecular weight would indicate the stoichiometry of the subunits [1]. The results of electron microscopy and sedimentation velocity have established the nature of oligomeric formation of kinensin, a microtubule-dependent ATPase, involved in the movement of axoplasmic organelles towards synapses [15,51].

ACKNOWLEDGEMENTS

I am grateful to Dr. A. Paraf for discussion and helpful criticism of the manuscript. I am also thankful to Mrs. T. Campone for her excellent secretarial assistance.

REFERENCES

1. T. M. Laue and D. G. Rhodes, Determination of size, molecular weight, and presence of subunits, *Methods in Enzymology, Guide to Protein Purification* (M. P. Deutscher, ed.), Academic Press, New York, 1990, p. 566.
2. D. G. Rhodes and T. M. Laue, Determination of purity, *Methods in Enzymology, Guide to Protein Purification* (M. P. Deutscher, ed.) Academic Press, New York, 1990, p. 555.
3. J. Ozols, Amino acid analysis, *Methods in Enzymology, Guide to Protein Purification* (M. P. Deutscher, ed.) Academic Press, New York, 1990, p. 587.
4. H. Edelhoch, Spectroscopic determination of tryptophan and tyrosine in proteins, *Biochemistry* 6:1948 (1967).

5. A. P. Barba de la Rosa, J. Gueguen, O. Paredes-Lopez, and G. Viroben, Fractionation procedures, electrophoretic characterization, and amino-acid composition of amaranth seed proteins, *J. Agric. Food Chem. 40*:931 (1992).
6. P. Matsudaira, Limited N-terminal sequence analysis, *Methods in Enzymology, Guide to Protein Purification* (M. P. Deutscher, ed.), Academic Press, New York, 1990, p. 602.
7. A. Bolques and S. M. Basha, Isolation and purification of methionine-rich protein from peanut, *J. Agric. Food Chem. 42*:1901 (1994).
8. S. Masci, D. Lafiandra, E. Porceddu, E.J.-L. Lew, H. Tao, D. D. Kasarda, D-glutenin subunits: N-terminal sequence and evidence for the presence of cysteine. *Cereal Chem. 70*:581 (1993).
9. R. C. Judd, Peptide mapping, *Methods in Enzymology, Guide to Protein Purification* (M. P. Deutscher, ed.), Academic Press, New York, 1990, p. 613.
10. S.-Y. Chung, A. H. J. Ullah, and T. H. Sanders, Peptide mapping of peanut proteins: identification of peptides as potential indicators of peanut maturity, *J. Agric. Food Chem. 42*:623 (1994).
11. P. W. Atkins, *Physical Chemistry*, 2nd ed. W. H. Freeman and Company, San Francisco, 1982, p. 228.
12. B. Robson and J. Garnier, *Introduction to Proteins and Protein Engineering*, Elsevier Science Publishers B.V., Amsterdam, 1986, p. 47.
13. I. Tinoco, K. Sauer, and J. Wang. *Physical Chemistry*. Prentice Hall, Englewood Cliffs, NJ, 1985, p. 117.
14. J. C. Hansen, J. Lebowitz, and B. Demeler, Analytical ultracentrifugation of complex macromolecular systems, *Biochemistry 33*:13155 (1994).
15. J. L. Correia, S. P. Gilbert, M. L. Moyer, and K. A. Johnson, Sedimentation studies on the kinensin motor domain constructs K401, K366 and K341, *Biochemistry 34*:4898 (1995).
16. H. T. Pretorius, P. K. Nandi, R. E. Lippoldt, M. L. Johnson, J. H. Keen, I. Pastan, and H. Edelhoch, Molecular characterization of human clathrin, *Biochemistry 20*:2777 (1981).
17. R. Cantor and P. R. Schimmel, *Biophysical Chemistry* Part II: *Techniques for the Study of Biological Structure and Function*, W. H. Freeman and Company, San Francisco, 1980, p. 591.
18. A. P. Barba de la Rosa, O. P. Lopez, and J. Gueguen, Characterization of amaranth globulins by ultracentrifugation and chromatographic techniques, *J. Agric. Food Chem. 40*:937 (1992).
19. W. F. Stafford III, K. Mabuchi, K. Takahashi, and T. Tao, Physical characterization of calponin, *J. Biol. Chem. 270*:10576 (1995).
20. A. Venktesh and V. Prakash, Low molecular weight proteins from sunflower (Helianthus annus L.) seed: effect of acidic butanol treatment on the physicochemical properties, *J. Agric. Food Chem. 41*:193 (1993).
21. V. Prakash and P. K. Nandi, Association-dissociation behaviour of sesame α-globulin in electrolyte solution, *J. Biol. Chem. 252*:240 (1977).
22. W. J. Wolf, Sulfhydril content of glycinin: effect of reducing agents, *J. Agric. Food Chem. 41*:700 (1993).
23. A. Mahajan and S. Dua, Physicochemical properties of rapeseed (Brassica campestris Var. Toria) seed proteins: viscosity and ionizable groups, *J. Agric. Food Chem. 42*:1411 (1994).
24. S. Cofrades, M. Careche, J. Carballo, and F. Jiménez Colmenero, Protein concentration, pH and ionic strength affect apparent viscosity of actomyosin, *J. Food Sci. 58*:1269 (1993).
25. M. Pacaud and J. Derancourt, Purification and further characterization of macrophage 70 kDa protein, a calcium-regulated, actin-binding protein identical to L-plastin, *Biochemistry 32*: 3448 (1993).
26. D. Huang, R. M. Everly, R. H. Cheng, J. B. Heymann, H. Schägger, V. Sled, T. Ohnishi, T. S. Baker, and W. A. Cramer, Characterization of the chloroplast cytochrome b_6f complex as a structural and functional dimer, *Biochemistry 33*:4401 (1994).

27. J. M. Denu, G. Zhou, L. Wu, R. Zhao, J. Yuvaniyama, M. A. Saper, and J. E. Dixon. The purification and characterization of a human dual-specific protein tyrosine phosphatase, *J. Biol. Chem. 270*:3796 (1995).
28. Y. Yada, K. Higuchi, and G. Imokawa, Purification and biochemical characterization of membrane-bound epidermal ceramidases from guinea pig skin, *J. Biol. Chem 270*:12677 (1995).
29. S. Sharma and A. Salahuddin, Purification and some properties of phaseolus mungo lectin, *J. Agric. Food Chem. 41*:700 (1993).
30. S.-T. Jiang and J.-J. Lee, Purification, characterization, and utilization of pig plasma factor XIIIa, *J. Agric. Food Chem. 40*:1101 (1992).
31. R. Cantor and P. R. Schimmel, *Biophysical Chemistry* Part II: *Techniques for the Study of Biological Structure and Function*, W. H. Freeman and Company, San Francisco, 1980, p. 670.
32. R. Cantor and P. R. Schimmel, *Biophysical Chemistry* Part II: *Techniques for the Study of Biological Structure and Function*, W. H. Freeman and Company, San Francisco, 1980, p. 676.
33. S. Oda, Two-dimensional electrophoretic analysis of friabilin, *Cereal Chem. 71*:394 (1994).
34. W. Ternes, Characterization of water soluble egg yolk proteins with isoelectric focussing, *J. Food Sci. 54*:764 (1989).
35. A. Hussain, B. M. Watts, and W. Bushuk, Hard to cook phenomenon in beans: changes in protein electrophoretic pattens during storage, *J. Food Sci. 54*:1367 (1989).
36. J. Bernhagen, R. A. Mitchell, T. Calandra, W. Voelter, A. Cerami, and R. Bucala, Purification, bioactivity and secondary structure analysis of mouse and human macrophage migration inhibitory factor, *Biochemistry 33*:14144 (1994).
37. K. Biemann, Peptides and proteins: overview and strategy, *Methods in Enzymology, Mass Spectrometry* (J. A. McCloskey, ed.), Academic Press, New York, 1990, p. 351.
38. A. G. Harrison and R. J. Cotter, Methods of ionization, *Methods in Enzymology, Mass Spectrometry* (J. A. McCloskey, ed.), Academic Press, New York, 1990, p. 3.
39. C. G. Edmonds and R. D. Smith, Electrospray ionization mass spectrometry, *Methods in Enzymology, Mass Spectrometry* (J. A. McCloskey, ed.), Academic Press, New York, 1990, p. 412.
40. R. Beavis, New methods in the mass spectrometry of proteins. *Trends Food Sci. Technol. 2*: 251 (1991).
41. I. Alli, B. F. Gibbs, M. K. Okoniewska, Y. Konishi, and F. Dumas, Identification of phaseolin polypeptide subunits in a crystalline food protein isolate from large lima beans (phaseolus lunatus), *J. Agric. Food Chem. 42*:2679 (1994).
42. M. Pedrocchi, B. W. Schäfer, I. Duruseel, J. A. Cox, and C. W. Heizmann, Purification and characterization of the recombinant human calcium-binding S100 proteins CAPL and CACY, *Biochemistry 33*:6732 (1994).
43. N. Stahl, M. A. Buldain, D. B. Teplow, L. Hood, B. G. Gibson, A. L. Bulingame, and S. B. Prusiner, Structural studies of scrapie prion protein using mass spectrometry and amino acid sequencing, *Biochemistry 32*:1991 (1993).
44. S. Moreau, A. C. Awadé, D. Mollé, Y. Le Graet, and G. Brulé, Hen egg-white lysozyme-metal in interactions: investigation by electrospray ionization mass spectrometry, *J. Agric. Food Chem. 43*:883 (1995).
45. R. Cantor and P. R. Schimmel, *Biophysical Chemistry* Part II: *Techniques for the Study of Biological Structure and Function*, W. H. Freeman and Company, San Francisco, 1980, p. 838.
46. R. R. Hantgan, J. V. Braaten, and M. Rocco, Dynamic light scattering studies of $\alpha_{IIb}\beta_3$ solution conformation, *Biochemistry 32*:3935 (1993).
47. N. Gavini, L. Ma, G. Watt, and B. K. Burgess, Purification and characterization of a FeMo cofactor-deficient MoFe protein, *Biochemistry 33*:11842 (1994).

48. N. R. Smalheiser and E. Kim, Purification of cranin, a laminin binding membrane protein, *J. Biol. Chem. 270*:15425 (1995).
49. L. Gao, P. K. W. No, and W. Bushuk, Structure of glutenin based on farinograph and electrophoretic results, *Cereal Chem. 69*:452 (1992).
50. W. E. Werner, A. E. Adalsteins, and D. D. Kasarda, Composition of high molecular weight glutenin subunit dimers formed by partial reduction of residue glutenin, *Cereal Chem. 69*: 535 (1992).
51. E. Berliner, H. K. Mahtani, S. Karki, L. F. Chu, J. E. Cronan Jr., and J. Gelles, Microtube movement by a biotinated kinesin bound to a streptavidin-coated surface, *J. Biol. Chem. 269*: 8610 (1994).

22

Applications of Immunochemistry for Protein Structure Control

Alain Paraf
Institut National de la Recherche Agronomique, Centre de Recherches de Tours
Nouzilly, France

C. Y. Boquien
Centre International de Recherches Daniel Carasso
Le Plessis-Robinson, France

I. INTRODUCTION

In 1987, Allen and Smith [1] observed that "although immunoassays have been in clinical diagnosis for some twenty years, they have so far had little impact on food analysis or in Public Analysts' laboratories." At that time, most assays were used either to identify food components or additives or to detect bacterial contaminants (e.g., *Salmonella* or *Listeria*) or toxins. Only a few methods/kits were available. In less than 10 years the situation has changed considerably for many reasons:

1. The number of processing technologies applied to food products has grown, making current analytical methods inadequate.
2. More additives are being used in food products, which require time-consuming methods of analysis that are often not quantitative.
3. While in developed countries the incidence of infectious diseases has lessened due to development of specific vaccines and better hygiene, the occurrence of foodborne diseases has increased considerably.
4. As a result of modern lifestyles, people spend little time preparing meals, often employing inadequate heat treatment.
5. Consumer demand for high-quality food with specific properties (e.g., gluten-free flour or semolina, defatted milk, or fruits with long shelf lives) have increased.

As a general rule any method of analysis developed should be adaptable to new problems arising from new methods of food preparation. Several physical methods, such as x-ray crystallography of antigen-antibody complexes [2], nuclear magnetic resonance (NMR) [2–5], and near infrared reflectance (NIR) [6], can be used in food analysis. For example, NMR can give information on the solids contents of a fat/oil blend [3] or on changes in water distribution during freezing, thawing, cooking, and extrusion processing [4,5]. Recently, differential scanning calorimetric (DSC) analysis of meat cooked at different heating rates [6] showed a correlation between protein denaturation and the textural properties of the meat. But none of these physical methods could be used to study protein structure in complex media with enough accuracy to identify small changes linked to genetic or processing-induced changes.

In contrast to these physical approaches, the structure of proteins in complex media such as food products can be studied by antibodies, mainly monoclonal antibodies. In this chapter we will describe the use of immunological approaches to many aspects of food control, such as quantification of food components and fraudulent additives, detection of allergenic components, quality control, and control of technologies applied to food. The chapter will focus on (a) food proteins as targets for quality control, (b) principles of immunochemistry applied to food products, (c) food analysis by immunochemical methods, and (d) modifications of epitopes to improve food quality, and (e) immunosensors.

II. FOOD PROTEINS AS TARGETS FOR QUALITY CONTROL

A. Detection of Pathogens and Toxins

Salmonella and *Listeria* are among the most frequently cited pathogens responsible for foodborne diseases in developed countries. These bacteria can be found in meat, milk, eggs, and vegetables. Moreover, *Listeria* is also found in soil and in the environment at low temperatures (4–8°C). Thus, detection kits that can be adapted to analysis of food products and soil samples are needed. The main requirement of such kits is that they should never report false negatives or false positives [7]. Unfortunately, various kits of specific antibodies for a specific bacterium often give different results (Table 1) [8].

TABLE 1 Number of Samples Found Positive for *Listeria* Species of 309 Samples Tested

Procedure	No. of positive samples	No. of false-positive[a] samples	Actual No. of positive samples (%)
FDA-MMA	92	0	92 (30)
FDA-LPM	96	0	96 (31)
GENE-TRAK	78	0	78 (25)
Organon-Teknika	111	22[b]	89 (29)

[a] False positive = positive absorbance reading, but negative by all other methods. No *Listeria* species isolated.
[b] Five false-positive milk samples; 17 false-positive vegetable samples.
Source: Ref. 8.

The specificity of diagnosis can be improved by using monoclonal antibodies (mAbs). For instance, specific monoclonal antibodies have been raised either against the surface antigen Act4 of *Listeria* [9] or against the listeriolysin [10] for the identification of *Listeria monocytogenes.*

In the case of salmonellosis, the two species most often encountered in foodborne diseases in humans are *S. enteritidis* and *S. typhimurium.* Specific polyclonal and monoclonal antibodies do exist against these bacteria. However, these antibodies cannot detect *Salmonella* if very few bacteria are present in the food or compartmentalized in a food (for instance, in a muscle or in a lipid globule of milk) or if the bacteria have been injured during food processing.

Similarly many viruses, parasites, and fungi can either by themselves or by action of their toxins or toxic products make food products harmful, and their detection by immunoreagents is therefore extremely valuable. This field is so vast and important that it represents a subject by itself, and we shall not deal with it in this chapter. We will limit our discussion here to some food proteins.

B. Food Components in Raw Material

"Natural" food products, such as milk, meat, potatoes, etc., contain mixtures of proteins, and it is important to carefully select the protein or proteins to be targets for the design of an immunotest. Many factors need to be considered in the selection of a protein antigen.

1. Solubility

The protein should be hydrosoluble so that it can be easily extracted and injected into animals in order to obtain an immune response. However, sometimes the production of antibodies against a protein may be necessary if the protein (the allergen) is poorly hydrosoluble. This is the case for wheat gluten, which requires special techniques to solubilize. For instance, 2 M urea with or without a reducing agent solubilizes wheat gluten [11]. The gluten includes several proteins, which are all hydrophobic and cannot be solubilized directly in water. Common solvents for cereal storage proteins include 2–6 M urea, aqueous alcohols, acids, bases (especially KOH), and ionic detergents such as sodium dodecyl sulfate [12]. However, the solvent used for incubation of water-insoluble flour antigens with plastic or nitrocellulose solid phases affects subsequent detection of antigen in at least two ways. As stated by Skerritt and Martinuzzi, "both the retention of antigen upon the solid phase and the ability of such antigen to bind antibody are altered" [12].

Sometimes the molecules that should be the target for antibodies are hydrosoluble but are surrounded by an enormous amount of lipids. For example, if egg white has been added to *foie gras*, detection of the egg white lysozyme is possible only if the lipids are removed from the product by heating at 60°C followed by centrifugation for 15 minutes at 20,000 rpm [13].

2. Size and Charge of the Molecule

As a general rule, all proteins with a molecular weight of >5000 daltons are antigenic; below this molecular weight, we are dealing with a small peptide, which can be regarded as a hapten (see below).

The charge of the molecule can play an important role. Even complete sequence homology within the antigenic-determinant region does not guarantee cross-reactivity between two native globular proteins because residues elsewhere in the proteins may contribute to their antigenicity [14].

3. Specificity of the Protein

Animals have many proteins with different functions, such as enzymes, transporters of oxygen or metals, globular or fibrous proteins, etc. A protein of animal origin tends to exhibit similar properties and similar structure even if extracted from different animals. It is often important to differentiate between the same protein from two different species to detect adulteration. For instance, cheese made of pure goat milk can be adulterated for economic reasons with cows' milk, in which case, polyclonal antibodies specific to caseins could be used to identify the fraudulent additive qualitatively or quantitatively by radial double diffusion [15].

4. Location of the Protein

When performing immunoassays, two factors should be taken into considerations: the concentration of the target protein and the location of the antigen in a tissue. The concentration should be of the order of several percent of the total proteins. For instance, caseins are good targets for milk, and serum or muscle proteins are suitable for meat. However, we shall see later that different muscle enzymes have been considered as targets for measuring heating temperature applied to meat in relation to the loss of activity. In eggs, ovalbumin has often been considered a good target because it represents 20% of egg white protein and its structure is modified by heat, pH, ionic strength, and protein concentration [16]. In wheat, gliadins are good targets because different wheat varieties contain different types and amounts of gliadins [17].

5. Immunogenicity

Immunogenicity is the ability of a molecule to induce antibody synthesis when injected to an animal. The antigenicity of a molecule is its ability to bind antibodies. Thus, both immunogenicity and antigenicity are important in the choice of molecule to become the target of immunotests for food proteins.

Most animal proteins, including those from meat, milk, and egg proteins, are excellent immunogens. Collagens are poor immunogens and should be used in the soluble form since insoluble cross-linked collagen lacks immunogenicity [18]. Cross-reactivity occurs between collagens types I, II, and III and between collagens from different species. In contrast, lysozymes are excellent immunogens and exhibit two important properties depending on the tissue from which they are isolated. They exhibit different epitopes; a peculiar lysozyme with specific epitopes is present in blood and could be used as a specific marker to identify various species [13,19].

Among vegetable proteins, hydrophilic molecules are often better immunogens than hydrophobic molecules. However, using different adjuvants it is possible to obtain antibodies specific for any kind of food protein.

C. Food Components in Processed Foods

Most food proteins are denatured when exposed to heat. This often affects their reactivity with antibodies. For instance, a 1% solution of ovalbumin heated at 135°C for 15 minutes

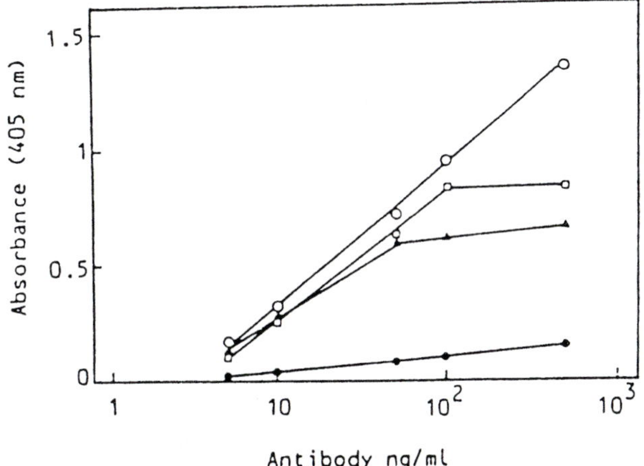

FIGURE 1 Antibody binding of native or heat-treated ovalbumin as measured by direct ELISA. Native ovalbumin, OA (●); noncoagulated, heat-treated ovalbumin, HDOA NC (▲); coagulated ovalbumin solubilized in NaOH 0.1 N, HDOA-NaOH (○) or in PBS + SDS 1%, HDOA-SDS (■). Antigen concentration: 0.1 μg/ml. (From Ref. 20.)

results in coagulation. The coagulated protein can be solubilized in 0.1 N NaOH or in phosphate-buffered saline (PBS) containing 1% SDS. Figure 1 shows that polyclonal antibodies raised against native ovalbumin react more strongly with heat-denatured ovalbumin than with native ovalbumin [20]. This suggests that several antigenic sites in polymerized heat-denatured ovalbumin are topographically more related to each other than those in monomeric native ovalbumin, and therefore more antibody molecules bind to the heat-denatured than to the native protein.

This approach of using enzyme-linked immunoassay (ELISA) has been shown to be useful in the quantitative detection of ovalbumin sometimes added to canned mushrooms as an adulterant to enhance the water-holding capacity of mushrooms [21]. This observation suggests that by raising a collection of monoclonal antibodies it is possible to define thermosensitive and thermoresistant epitopes with the possibility of identifying epitopes destroyed at 50, 65, 85, or 100°C (Fig. 2) [22]. More recently, it was possible by competitive ELISA to recognize different heat-resistant epitopes [23].

III. PRINCIPLES OF IMMUNOCHEMISTRY APPLIED TO FOOD PRODUCTS

A. Antigens

Most food components that can be studied by immunochemistry are proteins, although it is also possible to study polysaccharides and lipids using the same approach. Antigenic structures harbor epitopes that are mostly located on the hydrophilic part of the molecule as in meat, milk, or egg proteins, but epitopes can also be found on the hydrophobic part of the molecule, as in plant proteins like gluten.

Epitopes are recognized by specific antibodies, each epitope being identified by only one monoclonal antibody (see below). Because a majority of protein molecules

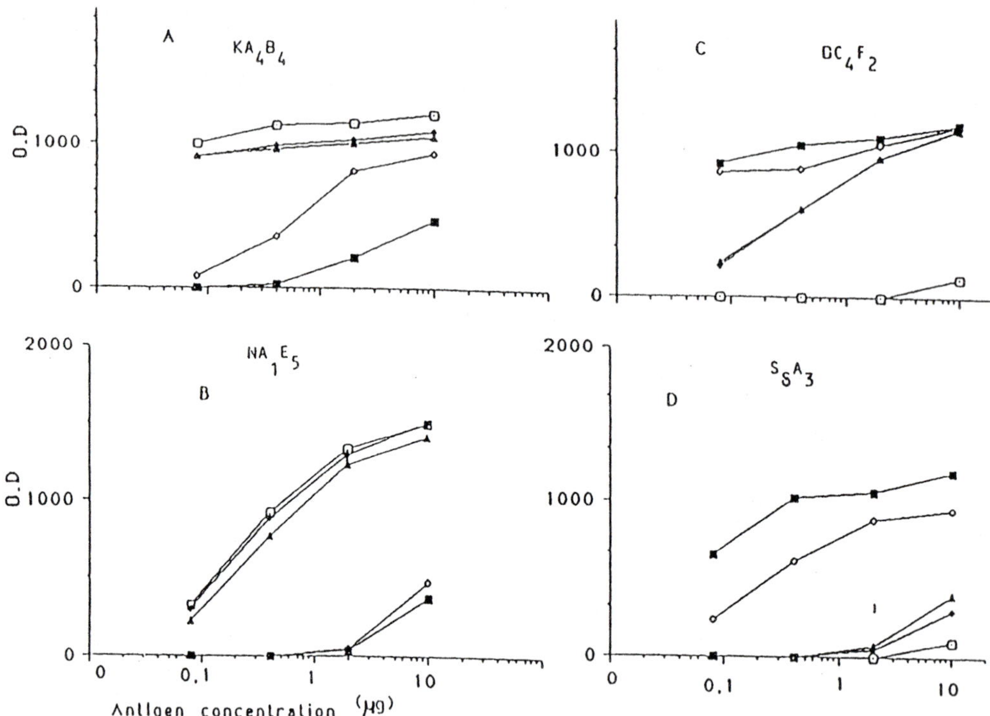

FIGURE 2 Effect of different temperature treatments on ovalbumin binding to four monoclonal antibodies. Binding was evaluated by the sandwich EIA using plates coated with mAb (□) N; (■) 65°C; (○) 70°C; (●) 75°C; (△) 80°C; (▲) 85°C. (From Ref. 22.)

exist in folded globular form as a result of various intramolecular covalent and noncovalent interactions, the structure of an epitope can be either a sequence of amino acids (continuous epitope) or a group of nonsequential amino acids (discontinuous epitope) (Fig. 3) [24]. Laver et al. [25] showed that native proteins generally contain epitopes made up of nonsequential amino acids (foldons), whereas denatured proteins generally contain cryptotopes made up of a sequence of amino acids (unfoldons).

The entire surface of a native protein can be antigenic, and therefore theoretically, a native protein can have several epitopes [26]. In prepared foods, interaction of other food components with proteins may alter their native structure. This may cause the generation of some cryptotopes in addition to the native epitopes. In other words, during commercial processing such as heating, thermoextrusion drying, etc., of foods, denaturation of proteins can lead to partial disappearance of some native epitopes and appearance of cryptotopes. We shall see below that the techniques used for extraction of proteins are critical for retaining the antigenic sites on a protein.

B. Antibodies

Antibodies are immunoglobulins composed of two heavy (H) and two light (L) chains linked by S-S bonds (with the exception of IgM, which has 10 H and 10 L chains). The

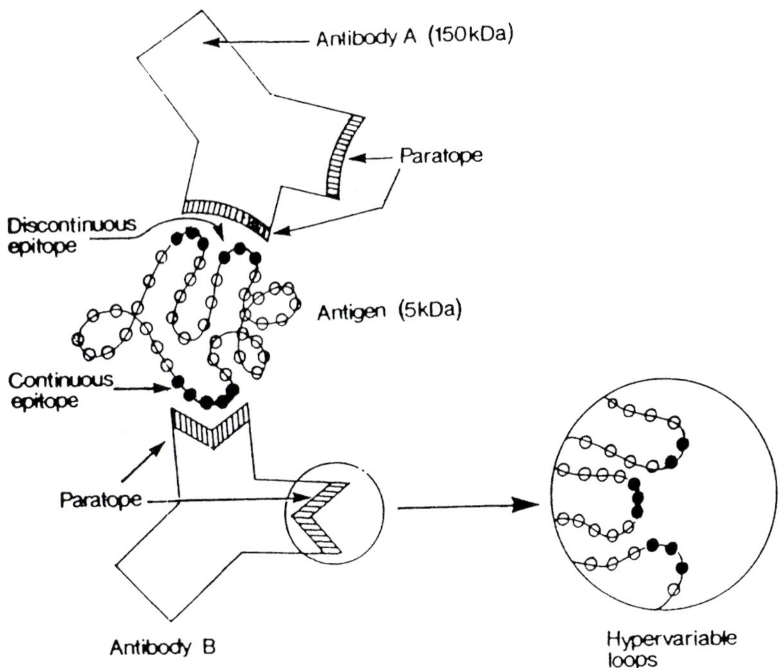

FIGURE 3 Interaction of two antibodies with a continuous and a discontinuous epitope of an antigen. Interacting amino acid residues are indicated in black. (From Ref. 24.)

synthesis of antibodies is induced by antigens, and they possess different conformations with different properties (Fig. 4) [27]. These differences are the bases for different classes of immunoglobulins.

When a protein in a complete Freund's adjuvant is injected intradermally or subcutaneously, mostly IgM will be synthesized during the first 2 weeks, followed by the synthesis of IgG. If the same antigen is injected via the intraperitoneal route or is given by the oral route, mostly IgA will be synthesized, followed by IgM. Some antigens, such as β-lactoglobulin or ovomucin, taken by some human individuals by the oral route cause synthesis of IgE antibodies. Such antigens are called allergens because they can induce different allergic symptoms.

Immunoglobulins are found in the serum of immunized animals at concentrations between 1 and 10 mg/ml and can be used in immunochemical tests, either as such or as partially purified or purified immunoglobulins. It is often advantageous to partially purify immunoglobulins to increase the sensitivity of the test.

1. Polyclonal Antibodies

When an antigen is injected, many B-cell clones are stimulated, leading to the synthesis and secretion of immunoglobulins with different specificities (binding to different parts of the antigen) and/or different affinities. The composition of the mixture can be different not only in different animals, but also from one bleeding to another bleeding in the same

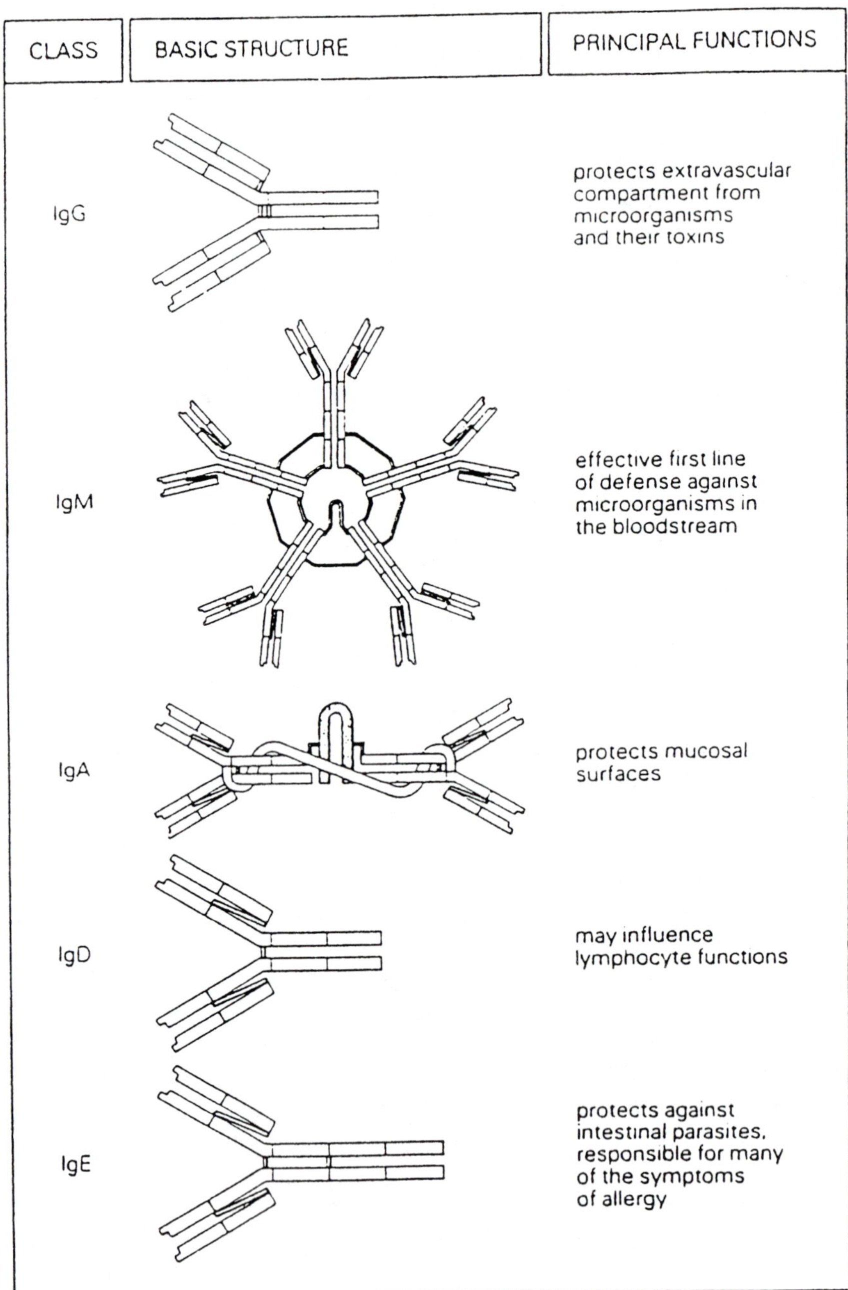

FIGURE 4 Immunoglobulin structure. (From Ref. 27.)

animal. Thus, in order to have a large batch of a uniform mixture, it is important, after analyzing the specificity and affinity of antibodies from different bleedings, to mix sera exhibiting a desired specificity and high affinity.

Immunization can be performed by injecting various kinds of animals (rabbits, mice, goats, horses) by an intradermal or subcutaneous route with a pure antigen or a mixture of antigens using an adjuvant. Many different adjuvants are in use. The complete Freund's adjuvant is a mixture of Arlacel A and paraffin oil containing dead *M. tuberculosis* bacilli cells. The adjuvant is vigorously mixed with an aqueous solution of the antigen in order to obtain a water-in-oil emulsion. The incomplete Freund's adjuvant is the same without the bacilli. For many antigens the first injection is made with complete Freund's adjuvant, while the following booster injections are made with incomplete Freund's adjuvant injected 2–4 weeks apart. IgA immunoglobulins can be stimulated preferentially by injecting antigens by the peritoneal route, while IgE immunoglobulins can be stimulated preferentially by using aluminium hydroxide as adjuvant. The immunization procedure can be different, depending upon the size of the antigen and the hydrophilic or hydrophobic character of the molecule. Some protein antigens are relatively small. For instance, chicken lysozyme with a 14 kDa molecular weight is an excellent immunogen when mixed with an adjuvant. However, as the immune response is dependent on the I-A histocompatibility, the region corresponding to residues [46–61] of lysozyme is a T-cell epitope in mice of the H-2k but not of the H-2d haplotype. This interaction (antigen presentation) is needed for T-cell recognition of the antigen and its subsequent activation and stimulation of antibody-producing B cells [28]. However, it has been shown that immunization with relatively large peptides successfully induces antibody populations having high neutralizing activities towards proteins [29]. Peptides coupled to bovine γ-globulin were more immunogenic than peptides alone and in the case of hen lysozyme peptides, one of them (sequence 57–107) stimulated production of antibodies with higher affinity than the other peptides linked by S-S bonds (1–27, 123–129, 22–33, and 115–125), indicating that the antipeptide antibodies essentially recognize the peptide conformation [30].

Similar proteins from different animal species have many epitopes in common, which when injected into goats, rabbits, or mice induce a relatively restricted immune response to specific epitopes. For instance, mice injected with myosin or myoglobin from bovine will only produce antibodies specific for bovine proteins, which will not bind myosin or myoglobulin from mice. Animals, as a general rule, are unable to raise antibodies against self proteins. However, cross-reactivity has been observed in a number of cases. For instance, rabbits immunized against bovine whey proteins had antisera exhibiting cross-reactivity [31] and cross-reaction occurred even with self proteins [32].

As a result, in recent years, monoclonal antibodies have replaced polyclonal antibodies for many immunoassays. For instance, when a bovine β-lactoglobulin was used as antigen, 30 monoclonal antibodies were obtained, 17 of which were species specific but only one of which was highly specific to the injected antigen (Fig. 5) [33].

Two properties of seed proteins are very important from an immunological point of view: several storage proteins exhibit sequence homologies, and most of these proteins are hydrophobic and alcohol soluble. For instance, according to their electrophoretic mobilities in a stack gel at acidic pH, four groups of wheat gliadins, i.e., α-, β-, γ-, and ω-gliadins (from highest mobility to lowest) have been recognized. α-Gliadin has been reported to be the component of wheat flour that exacerbates coeliac disease in susceptible individuals [34]. Thus, it is important to identify this gliadin. Immunizing rabbits with

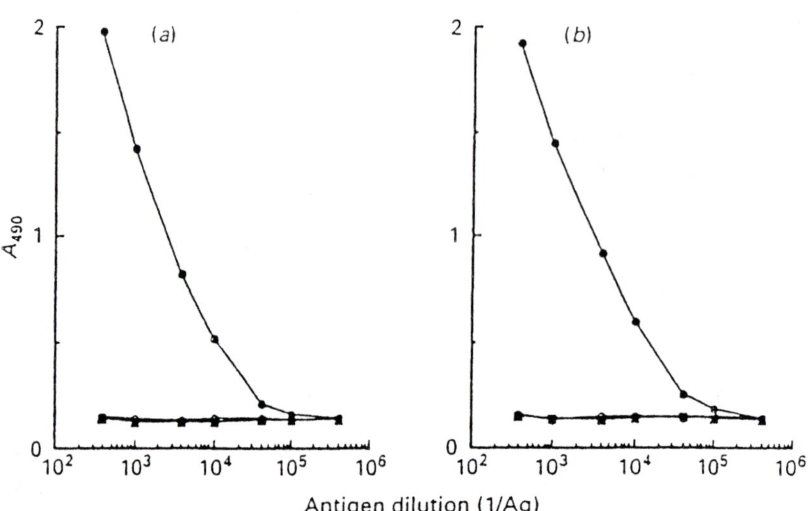

FIGURE 5 Two-site enzyme-linked immunosorbent assay results for the identification of cows', goats', and ewes' (a) milk or (b) whey using capture mAb 17 at 10 μg/ml and conjugate mAb 102 diluted 1:1500—(●), cows'; (○), goats'; (■), ewes'. Each datum point represents the mean of duplicate determinations. (From Ref. 33.)

a wheat α-gliadin fraction led to antisera that (a) were unable to differentiate α- and β-gliadin but distinguished α + β from γ + ω, (b) did not react with gliadins from rye, barley, or oats, and (c) detected α- and β-gliadin in complex foods, such as bread, biscuits, and even "gluten-free" products [35]. However, it is worthwhile mentioning that natural antibodies, similar to those found in the sera of human celiac patients, can be found in rabbits, rats, and guinea pigs fed a normal diet [36]. Thus, antibodies raised against a given storage protein often gives cross-reactions not only with several proteins of a given plant species but also with proteins of related species. This complicates the interpretation of immunological tests made with polyclonal antibodies and explains why monoclonal antibodies are preferred.

The specificity of polyclonal antibodies (see below) is due to the fact that they bind only the antigen used as immunogen. However, in many instances the same molecule (e.g., serum albumin or α-lactalbumin) from different animal species can exhibit many similar epitopes and even different molecules from plants can induce cross-reacting antibodies [37]. Cross-reacting antibodies raised against proteins A and B, for instance, can be separated and purified by immunoadsorption. In this case the purified antigen B is covalently linked to a column either as a glutaraldehyde–cross-linked BSA + B protein mixture or protein B bound to sepharose or latex beads. Then, when the serum is passed through the column, only anti-A antibodies are eluted, while anti-B antibodies are retained on the column, pure anti-B antibodies can be released from the column by eluting with a buffer at pH 2.8. This technique, with variations, is known as immunoaffinity chromatography and can be used for the purification of closely related molecules. For instance, many phytohormones (e.g., >60 gibberellins), exhibit related structures, and the production of antibodies specific for each one would be an impossible task. A solution to this problem is to produce antibodies of defined cross-reactivity that will react

Table 2 The Percentage Cross-Reactions of Monoclonal Antibodies with Storage Proteins from Different Species

	Percentage cross-reaction		
Storage proteins	024	025	026
Degraded soya 11S	105	111[a]	114
Soya 11S	100	100[a]	100
Soya 7S	70	4	3
Pea 11S	130	4	4
Pea 7S	117	3	3
Sesame 11S	344	3	3
Brazil nut 11S	119	2	3
Pumpkin 11S	189	2	3
Oat 12S	513	4	4
BSA	42	2	3

[a] mAb concentration 1 ng ml^{-1}; otherwise 100 ng ml^{-1}.
Source: Ref. 38.

with the majority of gibberellins present in plants. Legume storage proteins, particularly those from soy proteins, are largely used for food, and most stimulate excellent polyclonal antibodies, which often cross-react with proteins of other plants; antibodies raised in rabbit against degraded soya 11S protein cross-reacted with many other proteins (Table 2) [38]. In contrast, two monoclonal antibodies obtained from mice immunized against soya 11S protein were shown to be soya specific. Since each monoclonal antibody reacts only with a simple epitope, they are much preferable to polyclonal antibodies for identifying specific molecules as well as for recognizing how various processing methods, such as heating, thermoextrusion, and fermentation, can modify protein structure.

2. Monoclonal Antibodies

A monoclonal antibody is a unique immunoglobulin produced by a single cell clone [39]. A cell clone is obtained by fusing two types of cells, e.g., a splenic antibody-secreting cell from the mouse immune system and a long-lived cancerous immune cell, using polyethylene glycol into a hybrid cell or hybridoma. Such a hybrid can secrete the antibody from the immune donor and divide indefinitely, synthesizing the same immunoglobulin with the same specificity, i.e., binding to the same epitope. Approximately one epitope is expressed per 10 kDa of a protein's molecular mass (about 100 amino acids), although as few as 10–20 amino acids are thought to be capable of defining an epitope. Epitopes may be formed by sequential, linked amino acid residues or may be defined by amino acids that become closely associated after folding of the protein (conformational epitopes) (Fig. 3) [24]. Monoclonal antibodies also exhibit as many classes as have been detected in polyclonal antibodies. The IgM class is often polyspecific and has a tendency to aggregate and hence should be discarded. The IgG class is the best for use in immunochemical tests. Monoclonal antibodies of either the IgA or the IgE class can be employed for specific uses. Compared to polyclonal antibodies, monoclonal antibodies exhibit special qualities that are extremely useful in food analysis. Because different monoclonal antibodies raised against an antigen bind to different parts of the antigen, they can be used to identify related proteins and processing-induced changes in the antigen.

Detection of Adulteration Using Monoclonal Antibodies

Immunoassays have been used in foods to detect, for example, the fraudulent addition of vegetable proteins to meat, the addition of inexpensive cows' milk to goats' or sheeps' milk in the production of specific cheeses, or the presence of fungal or bacterial contamination or toxicants such as pesticides, hormones or antinutritional factors. Such tests have been reported to have sensitivities in the range of 0.4–40 ppb. Immunoassays are also used to characterize thermostable antigens, whose structure is not modified during the heating procedure.

Food Intolerance and Food Allergies

Detailed biochemical and immunochemical studies can yield insights into the possible biochemical basis for food intolerance reactions, such as coeliac disease. Such studies are needed in order to understand which peptides or amino acids are responsible for food intolerance reactions and the ways to suppress their activity.

Characterization of Fermentation

Starter cultures used in the fermentation of milk, meat or grain were for many years ill defined. Immunochemical tests have been used to identify the relative growth of different bacteria.

C. Antigen-Antibody Interactions

Antigens bind to antibodies mainly via hydrogen bonding, electrostatic, van der Waals, and hydrophobic interactions. As these are noncovalent bonds, the antigen-antibody interactions are thermodynamically reversible. The antigen-antibody affinity constant K_A results from the application of the mass acting law to the following equation at equilibrium:

$$\mathrm{Ag + Ab \leftrightarrow Ag - Ab\ complex}$$

$$K_A = \frac{\mathrm{[Ag - Ab]\ bound}}{\mathrm{[Ag][Ab]\ free}}$$

High-affinity antibodies have a K_A between 10^8 and 10^{11} M^{-1}. There are numerous immunological tests (Tables 3 and 4) [27]. We will mention only those that are used in food quality control.

1. Immunochemical Tests in Liquid Phase

When an antibody is mixed with an antigen in liquid phase, a cloudy precipitate appears, which can be centrifuged and quantitated. The antigen can be any kind of food extract, including the antigen corresponding to the known antibody used. Such a technique can be used to detect a fraudulent additive, such as milk in sausage. It is easy to perform and does not require special apparatus. However, this technique is not very sensitive and is only semi-quantitative. Three different techniques can be used in food analysis:

1. The ring test: this test is performed in a glass capillary tube (2 mm internal diameter), where the immune serum is first pipetted at the bottom of the tube,

TABLE 3 Immunoassays Requiring Precipitating Ag-Ab Systems

Immunoprecipitation (Ag-Ab) obtained by		Techniques
Diffusion		
The antigen is not separated by electrophoresis	In liquid medium	Ramon, Heidelberger:
	In liquid medium	Quantitative precipitation curve
	In liquid medium	Ring test (quantitative)
	In gel medium	Nephelometric assays
	In gel medium	Oudin: tubes
	In gel medium	Ouchterlony: plates
		Mancini: radial immunodiffusion
The antigen is first separated by electrophoresis in gel	In gel medium	Grabar and Williams: immunoelectrophoresis
	In gel medium	Overlay or immunofixation detection
Electrophoresis		
The antigen is not separated by electrophoresis	In gel medium	Laurell rockets: Ag migrates in an Ab containing gel
	In gel medium	Bussard: Electrophoresis or counterimmunoelectrophoresis
		Ag and Ab migrate toward each other
The antigen is first separated by electrophoresis in gel	In gel medium	Ressler, Laurell, Clarke, and Freeman: two-dimensional immunoelectrophoresis

Source: Ref. 27.

while the lighter food extract solution is added on the top. If the antigen is present in the food, a ring-shaped precipitate appears after a few minutes at the interface of the serum and food solution. If nothing is seen after 20 minutes, the reaction should be done with diluted antigen solutions.

2. Agglutination: this test can be applied to bacteria with antisera against surface antigens, especially to identify pathogenic species. When the antiserum is mixed with a suspension of bacteria, precipitation of antibody-bacteria complex occurs. The agglutination is faster at 37°C than at 4°C, but its highest sensitivity is obtained at 4°C after 24 hours. The technique is easy to perform but is being increasingly replaced by ELISA or DNA probes.

The ring and agglutination tests yield excellent results with polyclonal antibodies, but cannot be performed with monoclonal antibodies.

3. Nephelometry: antigen-coated microspheres agglutinate in the presence of antibodies. The resulting turbidity is measured with a nephelometer. The competition for the antibody sites between the antigen bound to the microspheres and the free antigen inhibits agglutination of the microspheres and thus allows the nephelometric measurement of the free antigen [40]. This technique could

Table 4 Immunoassays That Can Be Used with Nonprecipitating Ag-Ab Systems: One Ligand (Ag or Ab) Is Immobilized on a Solid Phase

Solid phase nature	Technique
1. Solid sample	Immunohistology, immunofluorescence
2. Cells	Agglutination with Ab:
	Active (e.g., bacteria, animal, or plants cells)
	Passive hemagglutination
3. Beads (cellulose, latex, etc.)	Agglutination of ligand-coated beads (visual observation for quantitative or semiqualitative estimation, particle counting or nephelometric)
magnetic beads (magniogel, etc.)	Magnetic beads do not require sedimentation or centrifugation for washing step but use a magnet to separate the bound from the soluble phases in a very rapid manner
4. Salt precipitation of Ag-Ab complex	Farr technique: the free labeled Ag is not salt precipitable (hapten)
	The labeled Ag-Ab complex is precipitated and quantified
5. Plastic tubes	Radioimmunoassays (RIA)
Plates	Enzyme immunoassays (EIA, ELISA, etc.)
6. Paper discs in tubes	RIA, e.g., PRIST, RAST, etc.
7. Filtration membranes: nitrocellulose, nylon, PVDF, etc.	Blotting, immunoprints, etc.

Source: Ref.27.

be applied to milk samples without any pretreatment after more than 30,000-fold dilution and on cheese after a 750-fold dilution in a buffer.

2. Immunochemical Tests in Solid Phase

ELISA

The principle is to bind either the antigen or the antibody to a solid phase (Fig. 6) [27] and then to detect the occurrence of an immunological reaction using enzyme-labeling reagents. In a simple assay, the antigen (food extract) is bound to a 96-well plastic plate at different dilutions and an antibody (labeled with an enzyme) is added. In order to increase the sensitivity, the antibody is not labeled and an antiantibody labeled with an enzyme is added. The substrate, which becomes colored by the action of the enzyme, is added. If the antigen is present, the specific antibody binds and a color appears; if the antigen is absent, no antibody binds and no color appears (Fig. 7) [27].

In sandwich ELISA, first an antibody is bound to the plastic, which captures the antigen, and the antigen is detected as above. This technique is more sensitive than the one described above and can identify 1–10 ng/ml of antigen (Fig. 6) [27].

In competitive ELISA, first an antibody is bound to the plastic, and competitive binding between a labeled antigen at a known concentration and the antigen present in the food is studied. The sensitivity is of the order of 10 ng/ml.

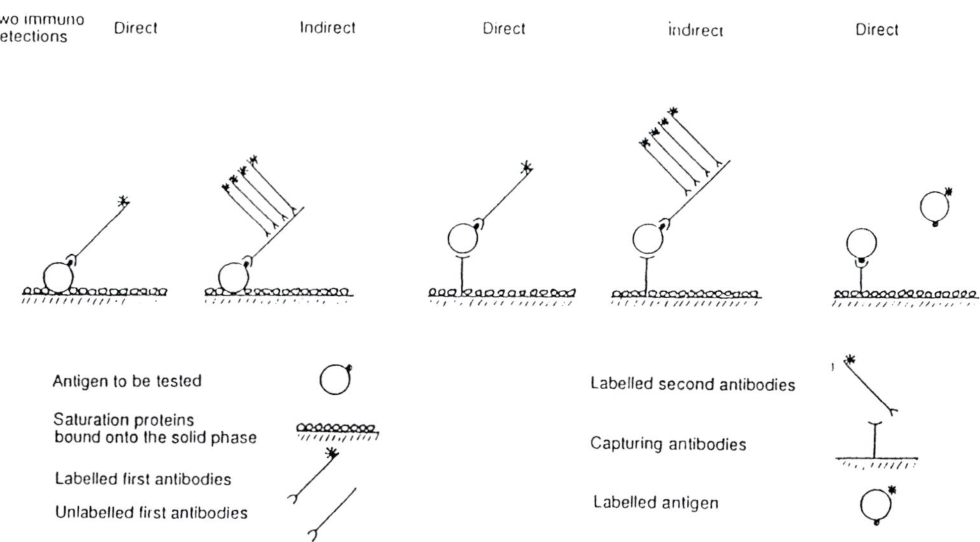

FIGURE 6 ELISA principle. (From Ref. 27.)

The sandwich ELISA is probably the best technique for quantitative detection of antigen in food products. Its sensitivity can be increased by allowing more time for the test.

Immunoblot

Antigen is electrophoresed in agar gel followed by transfer onto a cellulose sheet by electrophoresis. Then the sheet is soaked in a polyclonal or monoclonal antibody solution. After treatment with a labeled anti-immunoglobulin, a thin colored line appears where the antigen is located. Many other different ELISA techniques have been described, which can exhibit different sensitivities depending on the nature of the antigen present in a food product.

Immunosensors

A "biosensor" is an analytical device consisting of an immobilized biological component in intimate contact with a transduction device that converts the signal from a biological element into a quantifiable electrical signal. The potential wide acceptance of immunosensors is due to their general applicability—any compound can be analyzed as long as specific antibodies are available—and to the specificity and selectivity of the antigen-antibody reaction and the high sensitivity of the method, depending on the detection method used [41].

Several biosensors can be used to detect, in a short period of time, the presence of pathogenic microorganisms, pesticides, environmental contaminants, natural toxicants, or food additives. Biacore is a device often used in food analysis. It measures changes in surface plasmon resonance, which is proportional to changes in mass. When an antibody binds to an antigen, the increase in mass is measured in resonance units [42]. Biacore presents many advantages compared to other methods: (a) it does not require labeling of either the antigen or the antibody, (b) using monoclonal antibodies it is possible to

I - Immunochemistry and Immunoassays

1. Plate coating with Ag (2 h at 37°C)

2. Saturation or blocking with antigen unrelated proteins (15 min)

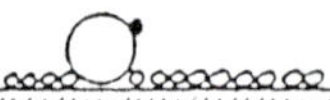

3. Incubation with test antibody (2 h at 37°C)

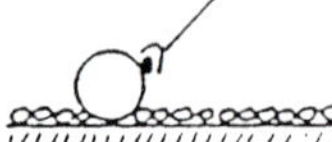

4. 5 successive washings (5 min)

5. Addition of the labeled second antibody (2 h at 37°C)

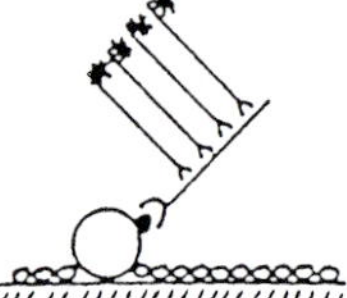

6. 5 successive washings (5 min)

7. Substrate S added, enzymatic reaction (5–15 min)

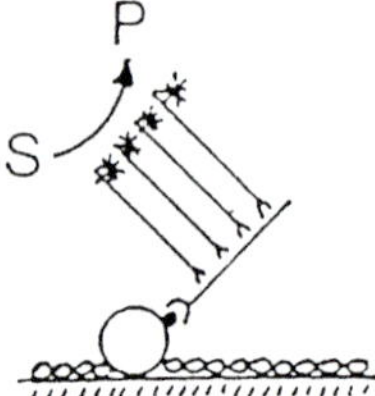

8. Densitometry of the accumulated enzymatic product P

FIGURE 7 ELISA. Simple method with indirect immunodetection. (From Ref. 27.)

map epitopes of a given protein, and (c) binding constants of antibodies can be quickly evaluated [43]. We should mention, however, that the apparatus is expensive and requires an experienced technician.

D. The Choice of Epitopes to Be Recognized

1. Principles of Epitope Recognition

An epitope is defined as that part of a molecule specifically recognized by the binding sites or paratopes of antibody molecules [44]. It is customary to distinguish "continuous epitopes," where a short linear stretch of residues is recognized by the antibody, from "discontinuous epitopes," where a set of noncontiguous residues in the sequence are recognized by the antibody. The different approaches that have been used for mapping protein epitopes are summarized in Table 5 [44], where Van Regenmortel discusses the limits of interpretation of each technique used to define an epitope. While for a long time immunologists thought that a protein exhibited a small number of epitopes, it is now thought that even if some epitopes are immunodominant, a protein can induce a

Table 5 Methods Used to Localize Epitopes in Proteins

Method	Type of epitope recognized	Criterion for allocating residues to epitope	Average number of residues identified in epitope
1. X-ray crystallography of antigen-Fab complexes	Discontinuous epitope reacting with homologous antibody	Van der Waals contact in epitope paratope interface	15
2. Study of cross-reactive binding of peptide fragments with antiprotein antibodies	Continuous epitope cross-reacting with heterologous antibody	Residual binding of linear fragment above threshold of assay	3–8
3. Study of cross-reactive binding of protein with antipeptide antibodies	Continuous epitope cross-reacting with heterologous antibody	Induction of cross-reactive antibodies	3–8
4. Determination of critical residues in peptide by systematic replacement with other amino acids	Continuous, cross-reacting epitope containing critical residues interspersed with irrelevant residues	Abrogation of cross-reactivity by substitution of critical and functional relevant residues	3–5
5. Study of cross-reactivity between homologous proteins or point mutants	Discontinuous epitope	Abrogation of cross-reactivity substitution of critical residue	1–3

Source: Ref. 44.

large number of antibodies specific for many overlapping epitopes. For instance, in the immune response against hen egg white lysozyme, antibodies covered at least 80% of the lysozyme's surface, and their pattern of overlap suggested a continuum of potential antibody epitopes [45]. Since food proteins are always mixed with sugars, lipids, and other proteins, all of these molecules interact with each other and modify the structure of the protein and induce new conformations with new epitopes. Moreover, when different processing methods are applied to raw food, they induce new conformation with new epitopes or cryptotopes. We shall describe below how a protein in a pure solution can be treated to mimic such changes and used to induce specific antibodies that will make it possible to identify which processing methods/conditions have been applied to food products.

2. Molecular Recognition by Polyclonal Antibodies

The dairy industry is faced with problems related to the composition of proteins in milk. Chemical and physical methods, such as Kjeldahl or infrared spectroscopy, are available for measuring the total protein content of milk. Various immunological methods have been used, among which ELISA and nephelometry are probably the most used by the dairy industry. By binding antibodies against either α-casein or κ-casein [40], α-lactalbumin, or β-lactoglobulin [46] to acrylamide microsphere it was possible to design an enhanced nephelometric immunoassay for detecting these proteins in milk. An ELISA for bovine α-lactalbumin and β-lactoglobulin in serum and tissue culture media [47] gave a slight cross-reactivity with caseins or bovine serum albumin (less than 0.001%). This test has been used to control the physiological capacity of heifers.

Identification of meat species has been tried using polyclonal antibodies to differentiate, for instance, meats from cattle, buffalo, kongoni, topi, and wildebeest [48]. However, antisera against bovine muscle antigen cross-reacted with fresh buffalo meat antigens. Such cross-reactions are often found when the protein used as target antigen is also found in different species, with some differences in the amino acid sequence. Many studies have focused on skeletal muscle fiber types for basic research. Myoglobin as the target antigen showed species-dependent antibody response [49]. Moreover, it was found later that antibodies to beef myoglobin raised in sheep were able to distinguish between beef and sheep myoglobins, although these two proteins differ in only 6 of 153 amino acid residues. By contrast, antibodies to beef myoglobin raised in rabbits, dogs, and chicken bind almost equally well to beef and sheep myoglobins [50]. Thus, in order to elicit specific polyclonal antibodies against a target food protein, it is important to select the appropriate animal.

Moreover, the nature of the surface determinants on the beef myoglobin molecule that direct the distinctive antibody response of sheep have been further defined. The synthetic C-terminal peptide 140–153 of beef myoglobin contains four of the six amino acid substitutions between sheep and beef myoglobins. When the beef myoglobin molecule is inoculated to rabbits or mice, few antibodies are directed against the C-terminal peptide, whereas when injected in sheep, most antibodies are directed against two topographic domains, which include the 140–153 sequence [51]. Thus, the C-terminal 140–153 sequence seems to be the epitope of choice to obtain specific antibodies against the beef myoglobin.

Lysozyme is an enzyme, found in various tissues and secretions, which hydrolyzes 1,4-β linkages between *N*-acetylmuramic acid and *N*-acetyl-D-glucosamine in the bacterial cell wall. It is synthesized in a number of tissues in the same animal and has been detected not only in mammals, birds, fish, and invertebrates, but also in bacteriophages, bacteria, fungi, and plants. Due to its wide distribution, this molecule should be a good target for food identification. The c-type of lysozymes found in hen egg white (molecular mass 14.5 kDa), whose sequence and properties are well known, has been found to be present in the egg white of only two orders of birds, the Galliforms and the Anseriforms [52]. In contrast, the lysozyme originally found in goose egg white (g-type, molecular mass 21 kDa) has been detected in a large number of avian species [53]. There is only slight amino acid sequence homology between g-type and c-type lysozymes, located around the active site of the enzyme. However, a partial common domain structure for both types of lysozyme has been determined by crystallography [54]. In earlier studies, although immunological cross-reactivity has been found between different c-type lysozymes for many years, no cross-reactivity was found between c-type and g-type lysozymes [55]. However, in recent studies using different immunological techniques, injection of either g-type or c-type lysozyme into rabbits and mice was found to induce both specific and cross-reacting antibodies (Fig. 8) [56]. This observation shows that while the c-type lysozyme from duck (anseriform) exhibits common epitopes with the g-type lysozyme from goose, the c-type lysozyme from chicken (galliform) does not cross-react with the goose lysozyme.

The existence of conformational similarities is supported by crystallographic observations, showing some common domain structures between g-type and c-type lysozymes, despite the fact that the amino acid sequences are almost entirely different. Knowing that g-type and c-type lysozymes have different locations, it is understandable that depending upon tissues present in a food, different epitopes should be taken as targets

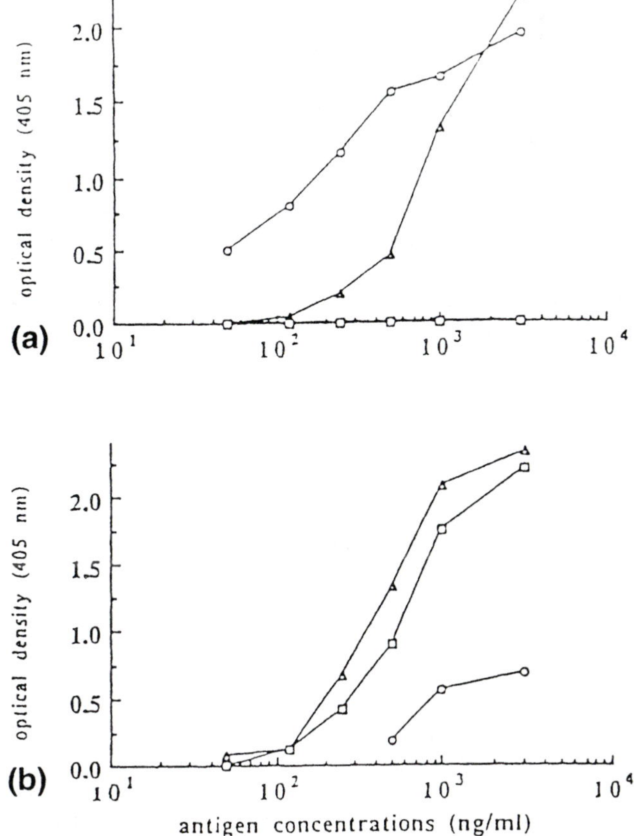

FIGURE 8 Antigen concentration curves. Checkerboard titrations of hyperimmune rabbit sera (a) anti-GEWL 1/10 000 and (b) anti-DEWL 1/3000) against different concentrations of lysozyme (goose; ○; duck; △; and hen, □) as treated by indirect ELISA. (From Ref. 56.)

for immunological tests. Thus, epitopes present on c-type duck egg white are not sufficiently different from those of g-type goose lysozyme to differentiate egg white from duck from that of goose, although c-type lysozyme from the chicken egg did not exhibit any cross-reactivity with g-type lysozyme found in the egg white of the goose. These results demonstrate that polyclonal antibodies against one protein often exhibit cross-reactivity against other proteins that have some sequence homologies or similar tertiary structure. We shall see later that it is possible to select, among a population of antibodies, those that are specific for a given epitope mimicking the properties of monoclonal antibodies. However, this approach is time consuming, and the direct approach of preparing monoclonal antibodies is more convenient for selecting specific reagents for a given epitope.

3. Molecular Recognition by Monoclonal Antibodies

The average affinity of monoclonal antibodies is often lower than the affinity of polyclonal antibodies. Compared to polyclonal antibodies, monoclonal antibodies exhibit a very narrow specificity. They bind a unique epitope, while polyclonal antibodies bind many epitopes and often the entire surface of the molecule. However, there are some exceptions, especially among closely related molecules. Different monoclonal antibodies were obtained against the triple-helical structure of collagens, showing that it is possible to identify conformational epitopes of the molecule. Most were directed against human collagen and then the specificities compared to epitopes found in sheep, calf, and rat. Table 6 lists some epitopes recognized by monoclonal 4F3-C7/col3 and 4F7-E11/col3 that are common to the four species and others that are more specific for human collagen (3D2-C7/col3) or human and sheep collagens (IE7-D7/col13 or 5F6-D8/col 13) [57].

An antibody raised against the αs2-11P casein also binds α_{s2}-9P-casein, α_{s1}-A-casein, α_{s1}-β-casein, α_{s1}-C-casein, and α_{s}-casein, but would not bind β-casein or κ-casein or α-lactalbumin [58]. Among 14 monoclonal antibodies raised against bovine κ-casein, only two were specific for the immunizing antigen. The remainder exhibited cross-reactivity with bovine α- and β-caseins or with nonmilk proteins [59].

TABLE 6 Species Specificities of Anti-(Type III Collagen) mAb

Antibody		Species reactivity			
		Man	Sheep	Calf	Rat
1. 4D3-C4/Col3	IgG2b'*k*	++	++	++	−
2. 4F3-C7/Col3	IgM'*k*	++	++	++	++
3. IE7-D7/Col3	IgG1'*k*	++	−	−	++
4. 5F6-D8/Col3	IgG2b'*k*	++	−	−	++
5. 3D2-C7/Col3	IgG1'*k*	++	+	−	−−
6. 3B9-E9/Col3	IgG3'*k*	+	+	−	−
7. 2G8-B1/Col3	IgM'*k*	++	++	++	−
8. 4F7-E11/Col3	IgG1'*k*	++	++	++	+

Reactivity was assessed by ELISA and by immunoblotting after nondenaturing PAGE: ++, strong reactivity; +, moderate reactivity; −, no reactivity.
Source: Ref. 57.

One out of six monoclonal antibodies was specific for bovine β-lactoglobulin [60]. Similar observations were made of meat, egg, and plant proteins, indicating that only some epitopes were specific for the immunizing protein. In the case of plant proteins, it is not surprising that cross-reactivity is often observed between several proteins, as many proteins can exhibit large sequence homologies. In the case of the glutenin complex, many monoclonal antibodies raised against gliadins and high molecular weight glutenin subunits cross-reacted with low molecular weight glutenin subunits [61].

For many years it has been claimed that peptides are able to mimic epitopes on globular proteins, although claims to the contrary are also available. Several hundred monoclonal antibodies to rat cytochrome *c* were examined by ELISA for binding the intact protein and cyanogen-bromide–cleaved peptides. The vast majority of monoclonal antibodies that bound the native protein did not bind peptides, whereas most of the antibodies specific for denatured forms did bind peptides. Therefore, the earlier claims that peptides can mimic epitopes of globular proteins must have been a result of the partially denatured form of the antigen. In other words, the assumption that a protein is in its native state when assayed for antibody binding in solid-phase assays is simply not valid [62]. Thus, depending on the ELISA test used, the protein can be in its native state (in sandwich ELISA) or in a denatured form (in direct ELISA). Similar observations have been made for enzymes [63], viral proteins [64], and food proteins [22].

Several conclusions can be made from the above observations. The choice of target proteins is of prime importance and should be based on many factors, such as frequency and specificity, solubility, stability of epitopes, etc. One of the main difficulties in immunological tests applied to food is the denaturing effect of the extraction procedure employed for solubilizing the protein. Identification of a food protein can be easily done by either polyclonal or monoclonal antibodies. Monoclonal antibodies are much more discriminating than polyclonal antibodies. Immunological tests, such as direct or indirect ELISA tests, can induce denaturation. However, such denaturation can be overcome.

IV. FOOD ANALYSIS BY IMMUNOCHEMICAL METHODS

A. Extraction Procedures

Foods are complex mixtures containing proteins, polysaccharides, lipids, DNA, and RNA with complex molecules, such as glycoproteins, lipoproteins, and nucleoproteins, in a complex "milieu." Immunotests may have different goals, such as the identification of (a) components in raw foods, (b) components generated in foods as a result of heating (thermoextrusion), fermentation, partial proteolysis, pH modifications, etc., and (c) additives, toxins, etc.

Most of the proteins in meat or milk are hydrosoluble, whereas the components of seed proteins such as wheat or rice are hydrophobic. In the case of insoluble proteins, appropriate solubilizing methods that allow partial purification and partial renaturation of the protein in order to reconstitute native epitopes should be developed. Moreover, new technologies, such as low-fat foods, and processing techniques (e.g., microparticulation), may modify the structure of food proteins considerably, which may also require new extraction techniques. In the following sections, the extraction procedures applied to eggs, meat products, grains such as wheat, and soybeans are described. Since milk proteins are hydrosoluble, extraction procedures applied to them will not be discussed.

1. Extraction Procedures for Meat and Egg Proteins

Heating is commonly used in the manufacture of meat products. Heating causes the denaturation of both sarcoplasmic and myofibrillar proteins, resulting in a loss of extractability. Proteins of beef or pork muscle are often extracted with saline solutions such as 3% NaCl + 0.44% sodium tripolyphosphate [65] or 16.4% salt, 5.7% sugar, 0.2% $NaNo_2$, 3.3% Na tripolyphosphate.

To solubilize the meat heated at relatively low temperatures (50–70°C), other solvents have been used. For example, 0.01 M phosphate buffer was used for the extraction of myoglobin and α-lactic-dehydrogenase in order to measure the extent of heat treatment given to the proteins between 54 and 70°C.

In meat products, thermostable antigens have been shown to be of use in species identification of both fresh and heated meat. Different techniques used to isolate and partially purify thermostable muscle antigens have been described [66].

Various methodologies, such as electrophoresis, isoelectric focusing (IEF), SDS-PAGE, and immunological tests, have been used to identify fish products. For immunological tests, depending on the target protein chosen, different extraction procedures have been used. For the water-soluble sarcoplasmic proteins, salt solutions are preferred, whereas for myofibrillar proteins, extraction with SDS or urea solutions followed by dialysis is usually used [67]. Several extraction methods have been compared for the electrophoretic identification of raw and cooked shrimp. Extracts prepared using saline, 1% SDS, or 8 M urea showed highly species-specific banding patterns for raw shrimp, while patterns for SDS extract provided information on species variation in cooked shrimp. The highly specific monoclonal antibody 4H2 10D3 recognized the M protein (17.7 kDa) in native and heat-denatured rock shrimp extracts [68].

Egg proteins are often added to food products for different purposes. The extraction procedure for egg white proteins depends on the food in which the additive has been introduced. When egg white is added to mushroom preserves, heat-induced coagulation of ovalbumin blocks drainage of water from the mushrooms (the price of canned mushrooms is based on drained weight). The entrapment of water in mushrooms as a result of the addition of coagulated egg white can increase the weight 10–20%. Extraction of the mushrooms with 0.1 M NaOH makes it possible to determine the exact amount of egg white added. Egg white is also added to *foie gras* for economic and rheological reasons. Immunological testing of this product is difficult because of its high fat content and low levels of added egg white. The extraction procedure involves two steps: First, the *foie gras* is heated for 1 hour at 60°C and the oil supernatant is removed. Second, the proteins, including lysozyme (often chosen as the target protein), are extracted with phosphate-buffered saline solution [13]. Extraction procedures for other fraudulent additives will be discussed in Section IV.C.

2. Extraction Procedures for Vegetables

While most proteins are hydrophilic and thus could be extracted using saline solutions, proteins from vegetables, especially grain proteins like glutenin and prolamine, are hydrophobic. However, some grain proteins, especially the albumin-type proteins, are hydrophilic and can be solubilized in water or saline solutions. Certain cereal proteins cause allergic reactions in certain individuals, and variations in the amount and composition of gluten storage proteins affect wheat flour quality. It is thus understandable that quantitative detection of proteins such as glutenins or gliadins in wheat flour or in meat additives or such as hordeins in barley is of prime importance. As most storage proteins are hydrophobic, they cannot be solubilized directly in water or saline. Table 7 shows typical solvents used for solubilizing wheat proteins and their effects on immunoassays. For instance, $Na^{125}I$-labeled gliadin losses due to KOH can be avoided by the addition of carbonate. Quantitative immunoassays are of value only if quantitative extraction is performed, leading to a complete extraction of these hydrophobic proteins. It should be noted that among the extractants used, the highest sensitivity of detection was achieved with 2 M urea in carbonate buffer, which also was the best solvent for gliadin.

The glutenin complex, defined as the disulfide-linked gluten proteins, is considered to be the major fraction responsible for dough strength and loaf volume of leavened (pan) bread. Extraction of gluten proteins from flour or whole meal for the first step of two-dimensional SDS-PAGE requires the absence of a reducing agent; the best results are obtained using 4% SDS in Tris-HCl buffer. Most of the highly mobile gliadin-binding antibodies were bound to small clusters of α- and β-gliadins; only a few bound to certain γ-gliadins. These antibodies did not bind HMW-GS in blotting assays [65].

Gluten properties are important determinants of bread quality. Immunochemical studies could help to define the viscoelastic properties of different varieties of wheat and to establish the relationship between epitopes defined by a panel of monoclonal antibodies and bread-making qualities. But the conclusion might vary depending on several properties of the test used. Skerritt [69] showed that among the extracts obtained with urea, HCl, SDS, or SDS-DTT propanol–acetic acid mixture and KOH as extractants, the extract obtained by vortex mixing with 0.5% SDS followed by extraction with 2.5% SDS–50 mM DTT showed the best correlation between antigenicity and bread-making qualities. Sandwich ELISA with "capture" monoclonal antibody exhibited the best performance with a specific combination of two monoclonal antibodies for each protein under study.

Various hordein proteins in barley have been found to undergo structural changes at different rates during malting. By extraction of barley and malt samples with 1 M

TABLE 7 Solvent Effects on Binding and Polyclonal Antibody: Detection of Gliadin on Plastic Microwell Plates

Solvent	%125 I-labeled gliadin bound[a]	Limit of detection[b] (ng)	A410 at 1 μg gliadin[c]
Urea (2M)	1.9 ± 0.2	10	1.28
Ethanol (70%)	14.0 ± 2.3	3	0.43[d]
Isopropanol (55%)	6.9 ± 1.1	30	0.11[e]
KOH (1%)	0.9 ± 0.1	300	1.18
Carbonate buffer (CB) pH 9.6	1.5 ± 0.3	3	1.70
Urea/CB	1.4 ± 0.3	1	1.63
KOH/CB	1.6 ± 0.2	3	1.74

[a] Total amount of gliadin added was 10 μg gliadin/200 μl/well.
[b] Lowest antigen concentration yielding an absorbance greater than 0.10 above blanks.
[c] Corrected for background (zero gliadin) in the same solvent.
[d] Maximum A410 (0.47) at 0.5 μg gliadin.
[e] Maximum A410 (0.36) at 0.5 μg gliadin.
Source: Ref.12.

urea and 1% mercaptoethanol, and using specific monoclonal antibodies, it was possible to define different stages of structural modifications of proteins in nine barley cultivars with different malting qualities [70].

Soybean is a traditional crop in East and Southeast Asian countries. Many soybean-based foods have been developed and soy protein products such as isolates, concentrates, flours, etc., have been introduced in different processed foods (especially meat) in these countries as well as in North America and in Europe. Different methods of extraction have been studied for quantitative preparation of a protein extract for immunological tests. There is some concern that residual soybean protein inhibitors (SBPI) could be present in commercially available soybean products, even though these products are often subjected to heat. These inhibitors are extracted with solvents such as 0.01 N NaOH or $CaCl_2$ in Tris solution.

Meat products are generally extracted with solvents such as chloroform-methanol (2:1 v/v), acidified ethanol (ethanol:water:HCl, 80:20:0.01 by volume) or acetone [71]. Extraction with carbonate buffer solution under sonication has been reported to be effective for extraction of SBPI from frankfurters [72].

In conclusion, the extraction procedure should extract 100% of the target protein in order to obtain a quantitative immunoassay. When the extractant is a salt solution, denaturation of the protein is minimal and native epitopes can be easily recovered. This is often the case for meat and egg products. However, in the case of thermally processed products, in which hydrophobic interactions between different proteins might occur, extractants such as SDS or urea may be necessary to solubilize the proteins and induce important structural modifications in the target protein. In addition, a dialysis step is necessary to remove the denaturing agent and to recover the native epitopes. Thermostable epitopes exist on some specific muscle antigens, which can be used as the target for the preparation of specific polyclonal or monoclonal antibodies [73]. Food products containing proteins such as wheat, barley, rice, and soy proteins, which are hydrophobic (glutenin and prolamin), may require alcohol, acetone, SDS, or urea to solubilize. Unfortunately, these solvents cause extensive denaturation of the proteins. Renaturation, which often requires high dilution and a long time (several days to several weeks), can be achieved in saline solution.

B. Epitopes Before and After Application of Processing

Polyclonal antibodies are not suitable for studying protein structure, as they represent a mixture of antibodies with different specificities covering the entire surface of the molecule. However, some early studies made with polyclonal antibodies have shown that protein unfolding could be detected [74]. On the other hand, monoclonal antibodies can be used to detect the sequence of structural changes in different parts of a protein molecule during and after processing.

1. Native Epitopes

To obtain monoclonal antibodies specific for the native state, a native antigen should be injected. For instance, the tryptophan synthase of *Escherichia coli* is constituted of several subunits, among which the epitopes of β_2-subunit have been thoroughly studied. Friguet et al. [75] showed that the monoclonal antibodies which rapidly inactivated the β-subunit recognized the epitopes present on the native protein, whereas those that reacted very slowly in solution recognized only the antigenic determinants normally

TABLE 8 Relationships Between mAb Specificity, Thermostability, and Polymerization of Ovalbumin

mAb	Ig	Native ova.	mH DOA	pHDOA	Plakalbumin
Na1E5	IgG1	1	0.3	0.0	0.0
IA2A2	IgG1	0.6	0.8	0.0	0.0
MB4D8	IgG1	1.5	1.7	0.2	0.0
QA6B5	IgG1	1.5	1.7	0.1	0.0
KD1D3	IgG1	1.4	1.8	0.0	0.0
KA4B4	IgG1	1.2	1.7	0.1	0.1
S8A3	IgG1	0.0	0.0	1.5	0.9
S3C5	IgGM	0.0	0.0	1.3	0.9
S10A1	IgGM	0.0	0.0	1.4	0.8
S10C3	IgGM	0.0	0.0	1.2	0.5
S8A1	IgGM	0.0	0.0	1.5	1
S5A1	IgGM	0.0	0.0	1.5	0.9
S7A3	IgGM	0.0	0.1	1.5	0.8
S7B4	IgGM	0.0	0.1	1.5	0.5
G1C6	IgG1	0.0	0.0	1.5	1
				1.5	1.2
1BAE5	IgG1	0.0	0.5		
BC4F2	IgM	0.0	0.4	1.5	0.9
AD6C4	IgG1	1.4	1.8	0.5	0.7

Binding was evaluated by a sandwich ELISA using plates coated with mAb at 5 $\mu g\ ml^{-1}$ which was allowed to react with native monomer, heat-denatured monomer (mHDOA), heat-denatured ovalbumin polymer (pHDOA), or plakalbumin at 0.4 $\mu g\ ml^{-1}$.
Source: Refs. 22,23.

hidden in the native protein, but became exposed upon coating the protein in the solid-phase procedure. Thus, when mice were immunized with a native protein, several monoclonal antibodies exhibited specificities directed against the denatured form of the β-subunit. Similarly, polyclonal [20] and monoclonal [22] antibodies obtained against the native form of ovalbumin exhibited specificities against the heat-denatured form of ovalbumin. Moreover, polyclonal antibodies had a greater affinity for the denatured ovalbumin than for the native molecule. Among the 18 monoclonal antibodies obtained, 6 of 6 antinative molecules were of the IgG isotype, while 7 of 9 antidenatured molecules were of the IgGM isotype; among the 3 monoclonal antibodies that bound both the native and the denatured forms of ovalbumin, one exhibited the IgM isotype (Table 8) [22,23]. These observations seem to indicate that the heat-aggregated ovalbumin essentially induced IgM isotype antibodies, while the native ovalbumin stimulated only antibodies of the IgG isotype.

The structure of chicken egg white lysozyme has been thoroughly studied, and several important conclusions have been drawn from the patterns of 49 Balb/c monoclonal antibodies during the immune response [45]. Newman et al. [45] showed that:

1. In the late immune response, the monoclonal antibodies recognized nonoverlapping antigenic regions that could include overlapping antibody epitopes, but when monoclonal antibodies of all periods were considered together, their specificities interconnected, suggesting a continuum of potential sites.

2. The apparent topographical organization of specificity patterns changed during the transition from the early to secondary immune response, and the distinction between individual antigenic regions became more pronounced later in the immune response.
3. Average avidity of antibodies did not increase during the course of the immune response to hen egg white lysozyme.

Thus, in order to get specific monoclonal antibodies against the native hen egg white lysozyme, mice should be hyperimmunized to cover as many epitopes as possible and monoclonal antibodies with maximum avidity should be chosen.

The U.S. Department of Agriculture uses several assays to determine if ready-to-eat meat products have been cooked to the proper end-point temperature: coagulation test, bovine catalase test, and acid phosphatase activity. It has been shown that during heat treatment lactate dehydrogenase (LDH) activity decreased and finally disappeared. The activity of polyclonal antibodies raised in rabbits against native LDH from either turkey or chicken progressively decreased with increase in temperature during the heat treatment [76]. There was no discrimination when meat was cooked at either 70.9 or 72.1°C. In contrast, four monoclonal antibodies raised against native chicken muscle LDH were able to discriminate between uncooked muscle and muscle heated to 70.9°C [77].

Both monoclonal and polyclonal antibodies can easily discriminate between animal species but not between plant proteins. In wheat, glutenins are largely responsible for gluten elasticity and gliadins are responsible for the extensibility and viscosity of wheat dough. Therefore, it would be desirable to have an immunological test that could discriminate between glutenins and gliadins. The α-, β-, γ-, and ω-gliadins of wheat exhibit sequence homologies with those of barley and rye. Thus, immune cross-reactivity has often been observed between all of these cereals using polyclonal antibodies. Only preparations of monoclonal antibodies have been able to solve this problem. For instance, a panel of monoclonal antibodies raised against a gliadin fraction prepared from the hard wheat cultivar Avalon exhibited good specificity [78].

2. Neo-epitopes and Cryptotopes

When a native molecule is subjected to either physico-chemical or enzymatic treatment, its structure creates neo-epitopes which do not cross react immunologically with epitopes of the native molecule. Moreover, cryptotopes do exist on the native molecule; but these are buried in the interior of the native molecule and become available to antibodies only when the structure of the native molecule is modified as a result of, for example, non-specific binding to plastic, heating, or cleavage by a protease. For instance, some monoclonal antibodies produced against the native ovalbumin do not bind ovalbumin, but will bind plakalbumin (molecule obtained from ovalbumin treated with the protease subtilisin). Specific monoclonal antibodies for plakalbumin have been isolated from mice immunized with native ovalbumin [23].

Neo-epitopes and cryptotopes have been found in a number of food proteins extracted from food products subjected to different technologies, such as heat treatment, thermo-extrusion or enzymatic treatment.

Heat Processing

Meat: Heat is applied to meat products (e.g., beef sausages, ground beef, canned hams, luncheon meats) in order to destroy pathogens, such as microbes, viruses, or parasites. The U.S. Department of Agriculture Food Safety and Inspection Service has defined end-point temperatures and different targets to design various control tests. Reverse-phase high-performance liquid chromatography (RP-HPLC) has been proposed to measure differences in water-soluble extracts from porcine muscle extracts heated at different temperatures for different times [79], and several other physico-chemical techniques have been used to define end-point temperatures.

Enzymes have also been the target of tests to define the end-point temperature. The APIZYM system and, recently, transaminase [80] and glucosaminidase activities [81] have shown some interesting properties. However, several formulations and processing variables may alter the relationship between enzyme activity and end-point temperature. For instance, when LDH was used as the target for monoclonal antibodies, it was able to accurately identify the end-point temperature of turkey thigh rolls to within $\pm 1.1°C$ between 68.3 and 72.1°C [82]. However, it was recently found that LDH activity in cooked turkey thigh rolls was higher than that in breast rolls due to the presence of heat-stable isoenzymes in thigh muscle, making the test unsuitable for temperatures between 68.9 and 71.1°C. Instead of LDH, serum proteins, particularly turkey serum albumin, were found to be better suited for use at those treatment temperatures [83].

Milk: Milk products are often heat treated to destroy pathogens, to diminish allergenic properties, and to modify functional properties. To study the consequences of heating procedures applied to milk products, β-lactoglobulin has often been taken as a model target and immunochemical tests have been used to understand heat-induced intermolecular reactions and structural modifications in this protein. For instance, when β-lactoglobulin is heated in the presence of lactose, it loses allergenic properties, presumably because of the Maillard reaction [84]. Severe heating has been shown to destroy the anaphylactic sensitizing capacity of β-lactoglobulin and α-lactalbumin and to diminish their antigenicity [85]. Structural changes in β-lactoglobulin due to retinol binding have been probed by using five different monoclonal antibodies; two major changes have been observed at 67 and at 80°C, corresponding respectively to a change in the random-coil region Lys^{8}-Trp^{19} followed by a structural change at the helical region Thr^{125}-Lys^{135} [86].

Eggs: Chicken egg white is often used as an additive in food products, and ovomucoid, ovalbumin, and lysozyme have been used as targets to evaluate the foaming and gelation properties of egg white and to detect its presence as a fraudulent additive in other food products.

Ovomucoid has three domains (DI, DII, and DIII) and possesses allergenic and trypsin-inhibitory activities: mAb 23E5 binds more efficiently to DIII-1 (DIII free from carbohydrate) and mAb 32A8 binds more efficiently to DIII-2 and inhibits the trypsin-inhibitory activity [87]. The antigenic reactivity and trypsin-inhibitory activity diminish progressively with increasing heating time, as measured by radioimmunoassay with rabbit antisera for ovomucoid heated in boiling water [88].

Ovalbumin, the major secretory product of oviduct cells, is a 43,000 dalton glycoprotein. Heating markedly decreases its allergenicity. This has been attributed to structural changes involving an increase in β-sheet structure and the disappearance of α-helices [89].

The polyclonal antibodies obtained from animals injected with native ovalbumin can be used to identify the presence of ovalbumin in heated food products. Unfortunately, this approach does not provide enough information on structural changes in ovalbumin. Eighteen monoclonal antibodies raised from mice immunized with either native ovalbumin or ovalbumin that had been heat-denatured at 100°C (HDOA) were used to study changes in antigenic sites induced either by heat or by subtilisin treatment. Using ELISA, three major groups of antigenic sites were found: group I, thermolabile native epitopes; group II, epitopes specific to heat-denatured epitopes; and group III, relatively thermostable native epitopes. Plakalbumin (PK) behaved similarly to group II epitopes of HDOA formed as a result of exposure of hydrophobic residues at 75°C (Table 8) [22,23]. Moreover, these monoclonal antibodies were able to distinguish heating procedures between 50 and 85°C [90].

Reducing sugars undergo Maillard reaction with protein amino groups and form relatively stable intermediates known as Amadori products. By immunizing mice with lactose-ovalbumin Maillard adduct and screening with lactose-bovine serum albumin, monoclonal antibodies specific for lactose-protein amino carbonyl products were obtained. These monoclonal antibodies can be used as tools for the detection of lactose-protein Maillard adducts in milk proteins [91]. Thus, not only the protein but also the polysaccharide present in food products can be a target for immunological analysis.

As ovalbumin characters: Lysozymes are extremely interesting molecules that are used in food research for several reasons:

1. They have different sequences in different mammals and birds [92].
2. Different lysozymes are expressed in different organs and tissues of the same animal [13].
3. Lysozyme structure is modified by heat [93].
4. Lysozyme is used as an additive in foods for its bactericidal activity [94].
5. Lysozyme is probably one of the best studied proteins with known crystallographic structure.

Plant proteins: Vegetables are the main source of cheap proteins and can be used either alone or as additives in meat products. They exhibit specific gelling and foaming properties. However, one factor that limits their use in foods is the presence of trypsin inhibitors. Since commercial heat treatment only partially inactivates trypsin inhibitors, a need exists to characterize the residual inhibitory activity remaining after processing. Purified Kunitz trypsin inhibitors (KTI) show an excellent correlation between antigenic and inhibitory activities. However, in flour samples the correlation was not good, due to the presence of Bowman-Birk inhibitors which are not recognized by anti-KTI monoclonal antibodies. Using two monoclonal antibodies to identify epitopes on the Kunitz soybean inhibitor heated at 121°C for 50 minutes in the presence of either glucose, maltose, lactose, or starch, it was found that the sugars decreased the inhibitory activity by 60–80%, whereas starch was less effective. The decrease was rapid, occurring within 10 minutes in the presence of glucose. As suggested by Oste et al. [95], such a treatment might also suppress allergenic properties of that molecule.

Immunological tests are also useful in the detection of extender properties and dough quality, prediction of grain, discrimination of wheat flours or whole meals on the basis of differences in dough strength, or identification of specific high molecular weight glutenins with bread-making quality. The choice of antibody combination and experimental conditions have been described for the analysis of a wide variety of bread wheats

[96]. More recently, a simplified and sensitive method for detecting dough strength has been described using a competition-sandwich ELISA format [97].

Among the various components of wheat flours, gluten storage proteins have an important role in bread-making qualities. While the group of Skerritt [97] studied high molecular weight glutenin subunits, the group of Morgan [98] has focused on the studies of low molecular weight glutenin. After having developed a library of monoclonal antibodies specific for low molecular weight glutenin subunits (which represent 40% of the total gluten proteins), one monoclonal antibody, mAb IFRN 0067, was shown to be specific for two glutenin polypeptides belonging to the B group of low molecular weight subunits [98]. Comparison of baking performances measured in alveograph parameters by test bake loaf and by immunoblotting showed a high correlation between the presence of the specific epitope and baking quality. Notably, the binding of IFRN 0067 showed a significantly high correlation with loaf volume.

In conclusion, heat processing is applied to complex food products, including meat, milk, eggs, and vegetables. It is extremely important not only to identify proteins present in these complex media using simple tests, but also to be able to correlate structural changes with the heating conditions and eventually to predict the quality of the raw material for specific uses. Monoclonal antibodies, with or without polyclonal antibodies, may respond to these different requirements. In most cases it is necessary to make a library of monoclonal antibodies by immunizing mice either with the native molecule or with the heat-denatured molecule. The choice of the molecule that will be the best target for immunological tests requires preliminary studies to identify the properties of the molecule correlated with structural modifications taking place in the complex food during heating. For instance, while thermostable antigens should be chosen if the goal is to detect fraudulent additives in meat products, molecules with thermosensitive epitopes should be used for the detection of time/temperature processing. Similar approaches should be used to correlate the presence of specific epitopes on target molecules with specific qualities required for milk proteins or plant proteins.

Thermoextrusion

These techniques are mainly applied to legume proteins. Extrusion usually involves conditioning the material (e.g., defatted soy flour) to a moisture content of 15–40%; this is then fed through a feeder/hopper into the hollow barrel of the extruder, which is usually heated, where a rotating set of screws forces the material towards the die. This pressure increase is accompanied by an increase in temperature. The usual residence time in the extruder is 30–60 seconds, after which the "molten" material passes into the die section before it is squirted out into the atmosphere, where the spontaneous evaporation of water takes place [99]. Chemical reaction, e.g., Maillard reaction, also occurs during extrusion processing. Thus, depending on the conditions used, denaturation of proteins usually occurs in extruded food products, and therefore special extraction procedures need to be identified. Because extraction with a urea + DTT solution causes complete loss of antigenicity, it is not preferred. Instead, although the yield of proteins, e.g., legumin, is often less than 20%, extraction with 0.16 M citrate-phosphate buffer is often preferred to other procedures [100].

An extrusion treatment is characterized by specific mechanical energy (SME), moisture content, shear rate at the die, and product temperature. Structural modifications in the legume fraction are often used as indicators of protein denaturation during processing. As shown in Figure 9, a link between the inhibition of an ELISA reaction (IER)

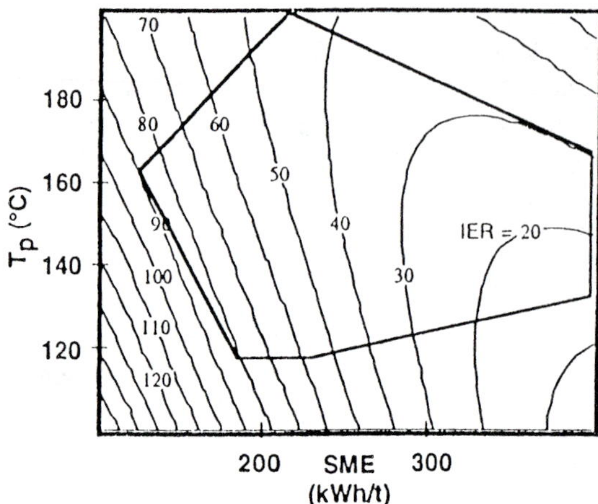

FIGURE 9 Pea flour extrusion. (From Ref. 101.)

and SME and product temperature at the die (Tp) suggests that IER is a good indicator of the severity of treatment [101].

In pasta products a major problem is discrimination between common wheat proteins and those of durum wheat. This is possible in pasta heated at low temperature, even though many of these proteins exhibit sequence homologies; however, at high temperatures none of the described methods is suitable. Monoclonal antibodies raised against different wheat proteins can identify the proteins in pasta [102]. Friabilin was shown to be a reliable target for the detection of common wheat in pasta [103]. Various applications of immunochemical techniques on barley, malt, and beer have been described [104].

C. Fraudulent Additives

The sensory properties of processed food are influenced by interactions among its various components during processing. For instance, milk has been used as such or transformed into different products, such as fermented milk, yogurt, cottage cheese and different kinds of cheeses, cream, and butter. In the last 20 years new technologies have been introduced in milk processing to produce several daily products and dairy ingredients. The use of ultrafiltration, combined with ultracentrifugation, led to the separation of many milk proteins, each expressing specific properties, which could be added to different food products for "modifying" their structures and tastes. Although such uses of dairy proteins in nondairy products are allowed in some cases, quantitative assays are required to determine the exact amount of dairy proteins in order to prevent undue adulteration and to comply with labeling laws. Immunoassays, using either polyclonal or monoclonal antibodies, can be used successfully in this regard.

Adulteration of ewes' milk with cows' milk is a well-known fraudulent practice because of the price difference. Analytical methods, such as gas chromatography or electrophoretic techniques, have been used to detect this adulteration. However, these methods are expensive, not sensitive, and sometimes not applicable to skim milk. Specific polyclonal antibodies have been successfully used in the detection of cow casesins. The immunotests are simple to perform, sensitive enough to detect as little as 0.1% cows' milk in ewe milk, and applicable to pasteurized or sterilized milk and skim milk as well as to ewes' cheese. As caseins are highly thermostable, the method is suitable for analyzing heat-treated milk products [105]. Many other immunological techniques have been described for caseins, two of which are worth mentioning: The cows' milk identification test (COMIT), which can be applied in the field and uses polyclonal antibodies in agar plates with antisera discs, can detect 3% cows' milk in ewes' milk but is not quantitative [31]. The other method uses two monoclonal antibodies specific for identifying two different epitopes on β-lactoglobulin, and the test could detect 1 part cows' milk per 100,000 parts goats' or ewes' milk [33].

Polyclonal antibodies and, more recently, monoclonal antibodies have been used to distinguish meat species in different foods. Antisera to heat-stable adrenal antigens identified species used in thoroughly cooked beef sausages by agar gel diffusion [106].

An antiserum against porcine muscle extract could detect pork meat by an ELISA in processed mutton or beef meat heated at 70, 100, and 120°C for 30 minutes. The detection limit was 1% (Table 9) [107], and false positives or negatives did not exceed 5%. The use of sheep instead of rabbit to produce the antiserum avoided cross-reactivity between beef or mutton with pork. Among many antisera, cross-reactivity was found between beef and sheep or between chicken and turkey, requiring an immunoadsorption step. Myoglobulin was used as a target protein to avoid such cross-reaction. Separating muscle proteins by electrofocusing, followed by electroblotting, detection of less than 10% pork, horse, or sheep meat was possible in a beef-based meat product that had been heated to an internal temperature of 120°C for 5 minutes. However, the method was not suitable for detection of chicken or turkey probably due to the low levels of myoglobin in chicken and turkey [108].

Identification of seafood species is often difficult, and morphological or electrophoretic methods have been rejected. Hybridoma technology has been employed in attempts to identify fish species and fish stocks; the method often failed because of cross-reactivity when applied to heated fish meat. Rock shrimp–specific protein M, determined from a comparison of the SDS-PAGE protein banding pattern of rock shrimp with those of 23 other seafoods, was isolated to develop monoclonal antibodies. A monoclonal antibody was obtained that identified specifically the protein M extracts of native or heated rock shrimp by immunodot blotting or by ELISA. The presence of rock shrimp could be detected at concentration levels as low as 4.3 ng in sample mixtures containing various seafoods or meat samples [68].

Processed foods based on traditional meat products may also contain foreign (nonmeat) proteins added as binders to improve texture and the retention of water or fat or

Table 9 Cross-Reactivity as Percent Binding of Various Muscle Species with Respect to Pork

Species	70°C for 30 min	100°C for 30 min	120°C for 30 min
Beef	17.3	12.0	3.8
Sheep	24.9	15.3	2.1
Pork	100.0	100.0	100.0
Chicken	20.0	13.1	5.3
Turkey	16.9	11.9	4.5

Source: Ref. 107.

as meat extenders or analogs. In analyzing such a product, the foreign proteins must be distinguished from the meat proteins [109]. ELISA techniques have been successfully used in quantitative identification of soy protein isolate added to beef burgers [109]. It is important to identify by a quick and easy test any kind of plant protein used as a binder in meat products. In this case, the use of polyclonal antibodies has several drawbacks:

1. Plant proteins often exhibit common antigenic properties, and therefore polyclonal antibodies cross-react with proteins from several different plants.
2. Polyclonal antibodies were difficult to use in sandwich ELISA tests, which are more sensitive and specific than the direct ELISA tests.
3. It is difficult to produce a polyclonal serum with consistent qualities every time.

V. MODIFICATION OF EPITOPES TO CONTROL ALLERGENIC PROPERTIES OF FOODS

The number of people, especially neonates and infants, who are plagued with food allergies is increasing. Like Matsuda and Nakamura [110], we shall confine the term allergy to the type I allergic reaction due to an IgE immune response. The normal seric level of IgE immunoglobulin in a normal human is of the order of ng/ml, while in an allergic individual it can reach 1 μg/ml. Thus, it is very important to quantify IgE molecules in human sera and identify their specificity. Among the most frequently encountered allergenic molecules are those listed in Table 10. We will use several examples to demonstrate how immunohistochemistry could help to identify allergens and allergenic individuals and identify specific epitopes involved in allergenicity.

A. Allergenic Epitopes

IgE is central to the induction of allergic diseases through its binding to the high-affinity receptor on mast cells and basophils; cross-linking by allergens of the bound IgE leads to the release of various inflammatory mediators [111]. For a long time, its structure was ill defined due to the very low amount found in serum. However, this structure was identified when a rare myeloma synthesizing IgE was discovered, exhibiting a heavy chain with four constant domains as IgM, while the two C-terminal domains are highly homologous to those of IgG. Due to its extremely high affinity for its receptor, many therapeutic strategies under study have been unsuccessful so far.

One of the most common food-allergy diseases is coeliac disease, caused by gluten absorption. The low cost of wheat gluten and the fact that it has specific physicochemical properties makes it desirable as an extender. However, in Australia, for instance, only <0.3% protein, which corresponds to a gliadin content of 1.2 mg/g, from wheat, rye, barley, and oats is acceptable as an extender. Thus, a quantitative test sensitive enough to detect this low level is needed [112]. In many cases it has proven to be difficult to quantify gluten for regulatory or quality control purposes, especially after foods have been cooked or processed. Skerritt and Hill [113], by choosing a thermostable epitope on ω-gliadin and related prolamines from wheat, rye, and barley, designed a sandwich immunoassay with a detection limit of 10–15 μg/g. It was shown that different extractants could give different results. Those obtained with 40% ethanol were more accurate [113].

TABLE 10 Structural and imunological properties of major food allergens

Source food	Potent allergen	Structural and immunological properties
Cows' milk	αS1-Casein	23 kDa phosphoprotein with sequential epitopes
	β-Lactoglobulin	18kDa protein belonging to the lipocalin family; not present in human milk
	Maillard adducts	Amino-carbonyl reaction products between lactose and protein amino groups
Egg white	Ovalbumin	43 kDa phosphoglycoprotein belonging to the serpin family; amino acid region 323–339 contains allergenic and antigenic epitope(s)
	Ovomucoid	28 kDa glycoprotein; trypsin inhibitor; heat-stable allergen
Soybean	Glycinin	320–360 kDa legumin-like protein composed of 6 acidic and 6 basic subunits
	2S-globulins	Mixture of low molecular weight proteins including trypsin inhibitors
	32 kDa allergen	Oil-body–associated allergen with sequence similarity to a house dust mite allergen
Peanut	65 kDa allergen	ConA-reactive glycoprotein; heat-stable allergen
	63.5 kDa allergen (Ara hl)	Very similar to 65 kDa allergen but ConA-negative
	17 kDa allergen (Ara hll)	Immunologically cross-reactive with Ara hl; glycoprotein rich in Glu/Gln
Castor bean	2S storage proteins	11–12 kDa glutamine-rich albumins belonging to the amylase/trypsin inhibitor family
Rice	16 kDa allergen	14–16 kDa albumins belonging to the amylase/trypsin inhibitor family
Codfish	Allergen M	12 kDa protein belonging to the calcium-binding parvalbumin family; amino acid region 41–64 contains allergenic epitope(s)
Shrimp	Tropomyosin	34–38 kDa water-soluble proteins with acidic isoelectric point (pl 4.5–5.8)

Source: Ref. 110.

Allergenicity has also been found in formulas containing soybean, which induced an increase in the levels of IgE and IgG in children prone to protein-induced enterocolitis [114] and in adults [115]. In products that have been processed extensively either by fermentation (such as miso, mold-hydrolyzed vegetable proteins, tempeh, and tofu) or sprouting, ELISA tests required higher protein concentrations to be positive, suggesting that fermentation and sprouting might alter specific epitopes linked to allergenicity of the molecules.

Egg proteins are among the most allergenic food proteins, especially ovalbumin. Carboxymethylation followed by treatment with trypsin did not eliminate allergenic and antigenic reactivities [116]. The N-terminal decapeptide of ovalbumin was synthesized and shown to react with functional structures on the IgE molecule from the sera on individuals allergic to eggs [117]. Also, the ovalbumin peptide 323–339 was able to bind to human specific IgE from serum pools of patients allergic to eggs [118]. It was assumed from these results that an α-helix conformation played an important role. This is consistent with the demonstration that certain model peptides must be at least nine residues in length in order to form an α-helix configuration [119]. More recently, five peptides located in the region 11–70 were studied and allergenicity shown to be distributed over the whole region 11–70 [120]. Another approach using a panel of monoclonal antibodies led to similar conclusions [121].

Cows' milk allergy (CMA) can be established by the demonstration of elevated levels of cows' milk–specific antibodies, especially IgE antibodies in the serum or tissues. The prevalence of CMA is a subject of continuing debate, but approximately 1% of infants have this allergy. β-Lactoglobulin and casein are the most common allergens, rather than other cows' milk proteins. The intestinal absorption of antigens is greatest during early infancy, which would explain the higher frequency of CMA in neonates. Elevated IgE antibodies can be found in normal individuals, and Taylor [122] showed that with intracutaneous injections, the skin tests were more frequently positive in allergic children without CMA than children with CMA. Therefore, the identification of cows' milk–specific IgE is not sufficient for a diagnosis of CMA, unless oral challenge testing is positive and an impressive history of adverse reaction is provided [122].

B. Removal of Allergenic Epitopes

A monoclonal IgE antibody directed against bovine milk β-lactoglobulin was produced which, when injected intravenously followed by feeding with aggregated β-lactoglobulin, induced immediate hypersensitivity. Accumulation of liquid within the small intestine and diarrhea were evident 30–90 minutes later, with increased permeability of the venulae from the submucosa and serosa and edema within the villae [123]. This monoclonal antibody also bound native β-lactoglobulin but formed a stable immune complex only with the aggregated form. Thus, a unique epitope on aggregated β-lactoglobulin is enough to support the binding of the IgE molecule and induce allergic reaction. Two other reports have described similar allergenic reactions with ovalbumin [124] and mellitin [125]. In all cases such allergic diseases have been shown to be due to binding of IgE to basophils and mast cells followed by a sequence of events involving not only different parts of the IgE molecule [126] but also different cytokines, which in turn can regulate IgE synthesis [111].

Heating procedures have been shown to destroy many epitopes and, as a consequence, the ability to destroy the allergenic property of a molecule. However, some epitopes are thermoresistant, which might be the origin of allergenicity. Although heating often destroys allergenicity, it also impairs the nutritional and functional properties of proteins. For instance, β-lactoglobulin is denatured between 70 and 75°C [86] as measured by monoclonal antibodies. However, heated (90°C) bovine whey protein isolate lost its immunogenicity when elicited by the oral route (oral tolerance). This has been attributed to the degradation of heat-denatured whey protein isolate into nonimmunogenic forms by enzymatic digestion in the gastrointestinal tract [127]. Similar observations have been made with other foods, such as eggs, where ovomucoid is a major allergen.

Thermolysin under high pressure (1000–2000 kg/cm^2 and 30°C for 3 hr) completely destroyed β-lactoglobulin, which is considered to be the major allergen in milk [128]. The extent of digestion was expressed as the percentage ratio of the absorbance of the enzymatic digest to that of the original solution of β-lactoglobulin. The binding activity of five monoclonal antibodies also disappeared under the same conditions. However, an antibody specific for α-lactalbumin was not affected by proteolytic digestion under pressure.

Proteolysis is among the most frequently studied methods of destroying the allergenicity of food products. As early as 1979, Takase et al. [129] made enzymatic hydrolysate of bovine casein in order to obtain polypeptide fractions with a molecular weight of less than 1000. This fraction completely lost its antigenicity as measured by passive cutaneous anaphylaxis. Extensive hydrolysis was shown to be essential for rendering the milk proteins immunologically unreactive for feeding allergy-prone infants. Hydrolysis of a commercial acid casein showed extensive loss of antigenicity during the first 10% hydrolysis time and only small changes after 90% hydrolysis time [130]. More recently, whey protein concentrate was hydrolyzed using the technical food-grade enzyme corolase 7092 in order to abolish the allergenicity of whey protein. The minimal molecular mass to elicit immunogenicity and allergenicity of whey protein hydrolysates appeared to be between 3000 and 5000. In these hydrolysates specific IgE binding has almost disappeared [131].

Another approach to suppress allergenicity is to chemically modify the allergens. Glycosylation of ovalbumin through the Maillard reaction by heating for 2–7 days at 50°C and 65% relative humidity decreased its allergenicity [132]. Many other studies have shown the possibility of suppressing allergenic properties of β-lactoglobulin while improving solubility, heat stability, foaming, and emulsifying properties. To achieve low allergenicity, conjugation of a protein with polysaccharides is thought to be more effective than conjugation of a protein with low molecular weight molecules. This is due to the fact that large molecules cover allergenic epitopes more effectively than low molecular weight molecules. When 13 stearic acid molecules were covalently attached to β-lactoglobulin, it became resistant to hydrolysis by α-chymotrypsin and pepsin and somewhat resistant to hydrolysis by pancreatin; also, it did not induce passive cutaneous anaphylaxis and lost most of its allergenicity [133]. Both direct and competitive ELISA showed that low and medium levels of fatty acid incorporation changed the protein structure, which exposed more antigenic or allergenic sites, while the high–fatty acid incorporated protein (13.1 β-LG) exhibited a low antibody-binding ability. Similar results were obtained with S-carboxymethylation of β-lactoglobulin [134], where an anti-S-carboxymethyl β-lactoglobulin did not precipitate the native β-lactoglobulin.

Immunochemical tests can be used for the identification of enzyme inhibitors common in plants. The presence of inhibitors of digestive enzymes in soy proteins, for instance, impairs the nutritional quality and possibly the safety of this legume as well as that of lima beans, kidney beans, and (at lower levels) potatoes and cereals. Thus, is it important not only to detect and quantitate such inhibitors, but also to measure their activity either in crude material or in food after heat treatment. Various tests, such as trypsin inhibitor assay, proteolytic digestion assay, half-cystine and methionine content, SH content, and immunoassays for inhibitors of trypsin, have been designed [135] to

Table 11 Suppression of Autoimmunity by Oral Tolerance

Condition	Protein fed
Animal models	
EAE	MBP, PLP
Arthritis	Type II collagen
Uveitis	S-antigen, IRBP
Diabetes (NOD mouse)	Insulin, glutamate decarboxylase
Myasthenia gravis	Acetylcholine receptor
Thyroiditis	Thyroglobulin
Transplantation	Alloantigen, MHC peptide
Human disease trials	
Multiple sclerosis	Bovine myelin
Rheumatoid arthritis	Chicken type II collagen
Uveoretinitis	Bovine S-antigen
Type I diabetes (planned)	Human insulin

Source: Ref. 140.

measure their activities. Monoclonal and polyclonal antibodies against the Kunitz trypsin inhibitors were used in sandwich ELISA with a detection limit of 30 ng/ml. Moreover, it was possible to detect the residual activity of KTI in heat-treated samples by ELISA. The activity determined by ELISA correlated strongly with the activity determined enzymatically.

Autoimmune diseases where the immune system inappropriately reacts against the body's own components, may be affected by foods [136], and, conversely, introduction by the oral route of specific antigens can protect against different allergic diseases. For instance, for centuries American Indians avoided contact sensitivity to poisons in plants by chewing the leaves [137]. This suppression of allergenicity by the oral route may involve different immune mechanisms, depending upon the dose of antigen absorbed. A high-dose feeding regimen was described [138], which is close to the high-dose tolerance state described by Mitchison [139]. A low-dose feeding regimen induces an antigen-driven active suppression, with increased secretion of transforming growth factors-β (TGF-β) and IL-4 with minimal energy. Such mechanisms are also involved in autoimmune diseases. In fact, it has been possible to suppress many autoimmune diseases induced by different antigens; the mechanism of suppression depended on the dosage used (Table 11) [140].

VI. Immunosensors

A. Definition

An immunosensor is an affinity-based biosensor. Biosensors are analyzing devices utilizing selective biological agents closely bound to a transducing device. The biological agent reacts with the analyte and the resulting change is converted by the transducer into a quantifiable signal. This signal can be correlated with the analyte concentration.

In an immunosensor, the biological agent is either an antibody or an antigen, and the reaction with the analyte gives rise to an antigen-antibody complex.

B. Different Types of Immunosensors

The immunosensors can be classified according to the detection system used, which may be direct or indirect. With the former, the immunocomplex is directly determined by measuring physical changes induced by its formation. With the latter, a label is incorporated in the immunocomplex. The determination of the label is used to monitor the formation of the immunocomplex.

1. Direct Detection

There are three types of transducers that allow direct detection of an antigen-antibody complex: potentiometric, piezoelectric, and optical transducers.

Potentiometric Transducers

In this case, a selective biological agent is immobilized on an electrode. The binding of the analyte to this agent causes a change in the electrode potential, which is determined by comparing with a reference electrode. These immunosensors are not very sensitive [141].

Piezoelectric Transducers

Piezoelectric devices are based on the ability of some crystals to vibrate when set in an electric field. The oscillating frequency changes with the crystal mass according to the Sauerbrey equation:

$$\Delta F = \frac{-2{,}3\ 10^{-6}\ F^2\ \Delta m}{A}$$

where ΔF is the frequency change of the crystal from the reference frequency F, Δm is the mass difference from the reference mass, and A is the crystal area.

In a piezoelectric immunosensor, crystals are coated with a highly specific and biologically active compound that is involved in the antigen-antibody reaction. The frequency measurement is generally done in a gas phase. The resonance frequency is determined before and after exposure to the test sample. This is a microgravimetric assay since the frequency change is directly correlated to the mass deposited on the crystal. No label is used in this method. It is also possible to oscillate the crystal in the liquid phase. However, the frequency shifts observed depend not only on the mass deposited but also on some interfacial interactions [142].

While the basic method involves binding of the test sample to the coated crystal, variations of this technique can be also used in competition immunoassay or in sandwich immunoassay. The latter allows an amplification of the mass difference and overcomes nonspecific binding.

Piezoelectric immunosensors can be used to detect low molecular weight compounds as well as large macromolecules and microorganisms (*Candida albicans*, *Saccharomyces cerevisiae*). It is also possible to detect traces of compounds in the gas phase, such as pesticides.

Optical Transducers

The two optical transducers described below allow direct detection of immunocomplexes, which is a major advantage. Other advantages common to all optical transducers are the possibility of use in remote-sensing applications and their potential use in the detection of more than one analyte.

Surface Plasmon Resonance Immunosensor: This system is based on the reflection of laser light at a metal/liquid interface. With an incident laser light, electronic oscillations, called plasmons, are generated in the metal. These plasmons penetrate into the liquid at the surface of the metal. Change of the refractive index in the vicinity of the metal surface changes the angle of incidence of plasmons, which can be detected by a shift in the intensity of the reflected beam [143].

Specific ligands (antigen, antibody, etc.) are immobilized on the metal surface of a surface plasmon resonance immunosensor. The binding of an analyte on the ligand produces a local change in the refractive index, which is directly proportional to the amount of analyte bound to the immobilized ligand.

Spectral Interferometry: By using a diode array spectrometer, the spectral distribution of light reflected from the surface of a thin interference layer can be analyzed to determine the optical path length in the layer.

Spectral interferometry has been applied to determine low refractive index. During a solid-phase immunoassay, the formation of a complex antigen-antibody leads to development of a protein layer at the solid phase with a refractive index different from that of the solution. Spectral interferometry can be used in on-line monitoring of changes in thickness of the protein layer and consequently of the immunological reaction.

Brecht et al. [144] described an interferometric immunoassay carried out on thin film of synthetic silica. A bifurcated multifiber light guide was used to illuminate the interference layer and collect the reflected light. Interferograms were recorded with a diode array refractometer. The increase in protein layer thickness vs. time (Fig. 10) was monitored, and the concentration was determined from initial reaction rates.

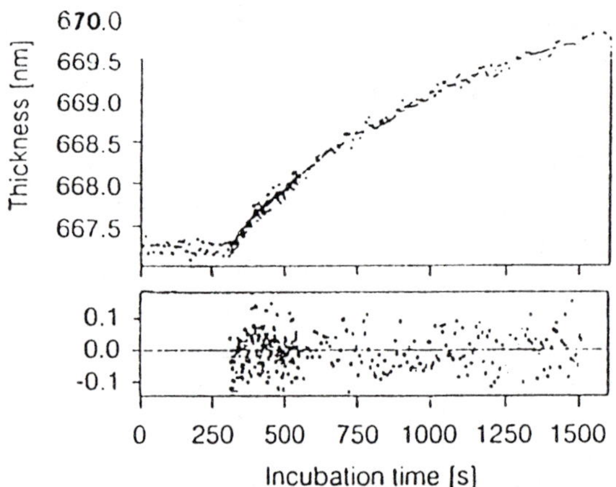

FIGURE 10 Protein layer thickness vs. time plot for an interferometric immunoassay. Top: Raw data (●) and curve fit (——). Bottom: Residuals from fit. (From Ref. 144.)

2. Indirect Detection

In immunosensors with indirect detection, a sensitive detectable label is incorporated and the immunocomplex is thus determined through measurement of the label. These immunosensors have generally been adapted from heterogeneous immunoassays. Otherwise, a step would be required to separate bound species from free species. Labels can be either fluorochromes or, very often, enzymes. With enzymes there is an amplification of the signal, since one molecule of the enzyme is able to convert many molecules of substrate into a detectable product. Immunosensors using enzymes as labels are similar to ELISA assays integrated to a detector. Only a few immunosensors have been adapted from homogeneous assays. The transducers used in indirect detection systems are electrochemical or optical.

Electrochemical Transducers

Two categories of electrochemical transducers exist: potentiometric and amperometric transducers. In potentiometric transducers, electrodes are sensitive to ions and the potential difference determined is proportional to the logarithm of the concentration of ions. This potential difference is measured against a reference electrode maintained at zero current flow.

Immunosensors with potentiometric transducers use enzymes as labels, which produce or consume these ions. For example, Brown and Meyerhoff [145] developed an immunosensor with adenosine deaminase as a label. This enzyme produces ammonium ions, which are then detected by potentiometry.

An amperometric transducer determines an oxido-reduction current flow. This current flow is proportional to the concentration of the analyte that is oxidized or reduced.

Many immunosensors have been set up by modifying a Clark-type oxygen electrode. Enzymes producing oxygen (e.g., catalase) or consuming oxygen (e.g., glucose oxidase) can be used as labels in these immunosensors. Boitieux et al. [146] proposed an immunosensor to measure α_1-fetoprotein using antibodies coupled to a catalase. Oxygen production by the catalase was measured on-line by an oxygen electrode. The repeatability of measurements was good (coefficient of variation $< 2.2\%$).

Electron acceptors other than oxygen have also been used in amperometric immunosensors, for example, ferrocene with glucose oxidase as the label.

Optical Transducers

Detection systems based on light determination have been integrated into immunosensors. If a colorimetric reaction allows detection of an immunocomplex, then the concentration can be determined by measuring the absorbance. Light also can be absorbed by a fluorochrome, and the fluorescence emission is then measured. Furthermore, light can be produced by reactions of chemi- and bioluminescence. We will focus on the detection of fluorescence and of chemi- and bioluminescence.

Detection of fluorescence: Bright et al. [147] described an immunosensor based on fluorescently labeled F(ab′) antihuman serum albumin antibody fragments covalently immobilized to the distal end of a fiber-optic probe. When the analyte (human serum albumin) was present, it bound to the antibody fragments and shielded the fluorescent label from the solvent water; this promoted a three- to fivefold increase in the fluorescence from the label.

Detection of chemi- and bioluminescence: Luminescence reactions occur when molecules in an excited state emit light while returning to the ground state. Chemical reactions of luminescence belong to chemiluminescence. When these reactions are catalyzed by enzymes and occur in living organisms, it is called bioluminescence. Chemiluminescence is based on luminol oxidation by horseradish peroxidase in the presence of hydrogen peroxide. Bioluminescence reactions involve the use of luciferase (e.g., from firefly) in the presence of luciferine and a cofactor such as ATP. Phenol derivatives, such as *p*-iodophenol, enhance the intensity of the chemiluminescent reaction and its life span.

Liu et al. [148] described a solid-phase chemiluminescent sandwich immunoassay using a membrane-based reactor. The antibody used for the detection of antigen was labeled with horseradish peroxidase. The chemiluminescent reaction was carried out in the presence of luminol, hydrogen peroxide, and *p*-iodophenol. This reaction took place in a thin-layer flow cell as immunoreactor, placed directly in front of a photomultiplier tube.

C. Characteristics of Immunosensors

Immunosensors should find numerous applications in the food industry if they have certain characteristics.

1. Specificity

The molecular recognition of antibodies for antigens is highly specific and generally confers a high specificity to the immunosensors. The quality of antibodies when setting up an immunosensor is obviously of major importance.

Bright et al. [147] immobilized labeled F(ab′) anti–human serum albumin fragments onto a fiber-optic probe. Some cross-reactivity existed with canine, bovine, porcine, mouse, and chicken serum albumins. The same results were obtained when the experiment was carried out in solution. However, interference was only 60% of that obtained with the immunosensor.

2. Sensitivity

Immunosensors exhibit variable sensitivity. Table 12 gives the detection limits for antigens, which vary from 1 ng/ml to 1.2 μg/ml and depend on the immunosensor and the antigen determined.

3. Rapidity

One major advantage of immunoassays is rapidity, due to the speed at which the antigen binds to the antibody. The formation of this antigen-antibody complex is an equilibrium reaction, and equilibrium can be reached in less than one hour [145,147]. It is not always necessary to wait until equilibrium is reached in order to accurately determine the amount of analyte present in the sample. The initial reaction rate, which is correlated to the analyte concentration, can be used to determine the concentration. Brecht et al. [144] evaluated the initial reaction rate with a 20-minute incubation time. The coefficient of variation was 15%, which increased to 28% when the evaluation time was reduced to 2 minutes.

A complete assay cycle comprised of a measurement followed by a regeneration of the immunosensor (see below) can take as little time as 16 minutes (Table 12).

TABLE 12 Characteristics of Different Immunosensors

Immunosensor	Antigen determined	Detection limit (ng/ml)	Time (min)[a] for an assay cycle	Ref.
Direct detection				
Interferometric	Goat IgG	100	—	144
Indirect detection				
Amperometric	α_1-fetoprotein	1	24	146
Fluorescence detection	Human serum albumin	1200	16	147
Chemiluminescence detection	Mouse IgG	6	16	148

[a] An assay cycle comprised of a measurement followed by a regeneration.

4. Reusability

An immunosensor has to be reusable. Since formation of a antigen-antibody complex is a reversible reaction, it is possible to find physicochemical conditions that allow dissociation of the antigen from the antibody. However, it is important that the functionality of the antibodies bound to the solid phase is intact after dissociation.

To regenerate a fiber-optic–based immunosensor, Bright et al. [147] used 0.1 M phosphoric acid for 10 minutes followed by rinsing with phosphate-buffered saline. The regenerated sensor was then ready for a new assay. Thus, each cycle consisted of an assay followed by a regeneration. The sensor could be used for up to 50 cycles since the response was still 50% of the initial value. However, the immunosensor had to be assayed regularly to test the stability of the signal. The activity of antibodies bound to the support was destroyed after 75 chaotropic cycles.

An immunosensor used for determining α_1-fetoprotein has been reused by regenerating the solid phase with glycine-HCl buffer (10 mM, pH 2.3) containing 0.5 M NaCl followed by rinsing with phosphate buffer (50 mM, pH 7.2) containing 0.15 M NaCl [146]. After incubation with the analyte, 100% of the initial signal was recovered. No significant loss of signal was observed after 20 successive cycles. About 98% of the initial signal was still retained after 30 cycles (Fig. 11). The surface plasmon resonance immunosensor can be regenerated, and 50–100 cycles can be performed on a single surface depending on regeneration conditions [149].

Although these examples indicate that reuse of immunosensors is simple, some studies show contradictory results. For example, Brown and Meyerhoff [145] regenerated a potentiometric immunosensor with 0.1 M guanidine-HCl buffer (pH 3.0) containing 10% glycerol. After 8–10 cycles, a considerable decrease in the binding ability of the immobilized antibodies occurred, probably because of partial denaturation.

In a flow injection analysis (FIA) sandwich immunoassay, Liu et al. [148] regenerated the membrane with a 0.1 M phosphate buffer, pH 2.2, for 10 minutes. A loss of membrane capacity occurred after regeneration, due probably to denaturation or loss of immobilized antibody. After three regenerations, the output signal decreased by about 20%.

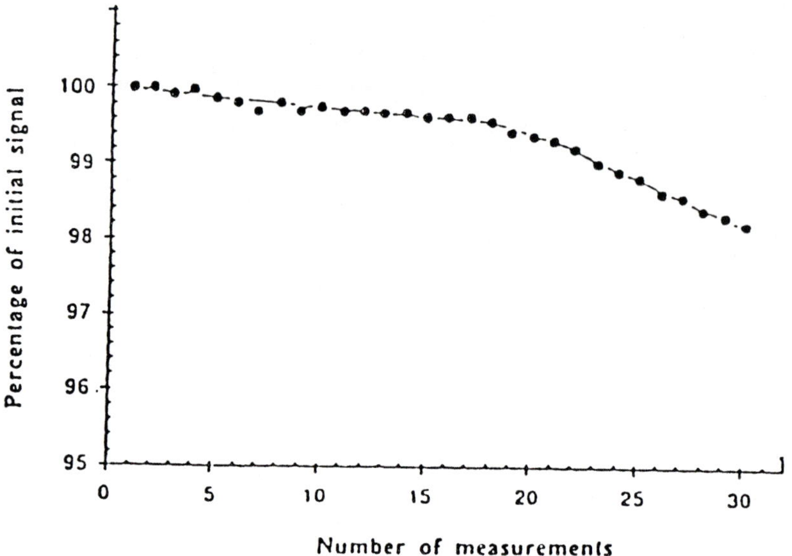

FIGURE 11 Variations of the immunosensor signal as a function of the measurement cycles. (From Ref. 146.)

5. Long-Term Stability

Bright et al. [147] developed a fiber-optic sensor based on F(ab′) antibody fragments immobilized onto quartz plates. Studies on long-term stability showed that the immunosensor response was stable to up to 4 months of storage in PBS at 4°C, after which the response decreased significantly and was completely lost after 6 months. This was attributed to denaturation of F(ab′) antibody fragments.

6. Ability to Be Calibrated

Calibration of affinity-based biosensors is relatively easy since the biochemical binding event is reversible. The reuse of immunosensors is feasible in certain cases, and the production of a standard curve using solutions of known concentrations is possible. After calibration, unknown samples can be assayed and the concentration subsequently read off this standard curve.

Specific devices have been developed to make possible the calibration of nonreusable immunosensors. Robinson et al. [150] proposed an immunosensor with one assay and two reference regions. The sensor did not need any standard before use since the calibration was carried out in the sample. The effects of sample matrix on the biochemical components of the immunosensor are thus decreased. The signal produced by the reference regions was completely independent of the concentration of the analyte present in the sample. It was possible to scale the assay signal with the signal given by one of the reference regions. Also, a simple comparison of these two signals showed if the sample was positive or negative. Semi-quantitative data can be obtained without any need to run a standard curve before measurement.

7. Combination with Flow Injection Analysis System

All the steps involved in solid-phase immunoassays (immune reactions, washings, detection of the antibody-antigen complex) can be carried out with an FIA system. FIA has two major advantages: it allows handling of small sample volumes and assays are very fast (about 10 min).

In an optical immunosensor, Brecht et al. [144] used an FIA system consisting of a multichannel pump, injection valve, air trap, and controller unit with a photometer (Fig. 12). They compared two flow regimens: continuous and alternating (flow and reverse flow). Sample consumption reached 10 ml with the first regime and less than 2 ml with the second. Sensitivity of the immunoassay was similar in both flow regimens. Samples could be reduced to 200 μl using a 15-μl flow cell.

In an FIA sandwich immunoassay, Liu et al. [148] emphasized the influence of the flow rate of the mobile phase containing the sample to be analyzed. The flow rate is directly correlated to the residence time; consequently, the higher the flow rate, the lower the output signal.

D. Future Developments of Immunosensors in the Food Industry

Immunosensors should find applications in the control of protein structure in the food industry. The main applications have been reviewed in this chapter. They include the quality control of food components in raw material and in food products modified by technology. Other applications not described extensively in this chapter are also given in the literature, one of which is the control of pathogens.

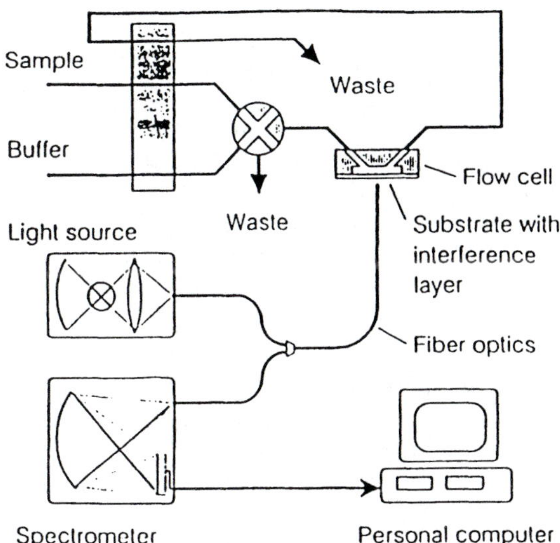

FIGURE 12 Optical set-up and fluid handling system for interferometric measurements in an FIA system. (From Ref. 144.)

Since immunosensors are based on the reaction of a selective biological agent (antibody or antigen) with an analyte, they exhibit the characteristics common to all immunoassays, particularly specificity and sensitivity, that make them very attractive compared to chemical and physical methods.

A major characteristic of the immunosensor is its ability to be reused (unlike an ELISA assay). This allows it to be cost effective, since antibodies immobilized on the solid phase are able to stay active during numerous cycles of use and there is no need to use new antibodies for each assay. There is also the possibility of automation as well as calibration of the immunosensor and quantitative determination of antigens.

Among all of the immunosensors presented here, those based on homogeneous immunoassays are of considerable importance. They do not need any separation step to eliminate free species from bound species, and their use is thus simplified. Moreover, since the measurements can be carried out in real time, a kinetic analysis of the interaction between antigen and antibody is possible. The future development of these immunosensors is linked to that of the physical methods (interferometry, surface plasmon resonance, fluorescence polarization [151], etc.) which they are based on. More generally, the use of these physical methods has also been extended to the analysis of protein-protein, protein-DNA, and ligand-receptor interactions.

In this chapter, we have emphasized the necessity of setting up efficient sample preparation procedures. This is of major importance before applying these immunosensors in food analysis, particularly to reduce nonspecific interactions.

Finally, since any analyte for which antibodies are available can be analyzed by an immunosensor, it should find wide application in the food industry in the future. Moreover, its cost and the simplicity of operation by a technician also make it attractive as a useful technique for food analysis.

VI. CONCLUSION

We have seen in this chapter that immunochemical methodologies involve polyclonal antibodies, which are able to recognize the whole molecules present in a food, and monoclonal antibodies, which bind a specific small surface of the molecule. As many protein molecules, such as gliadins or glutenins, can express the same epitope(s), several of these molecules may exhibit cross-reactivity in immunological tests such as ELISA. It is obvious that in order to detect a specific epitope, a given monoclonal should be used as an immunological captor, while polyclonal antibodies will bind any protein expressing one or several epitopes present on different molecules. In sandwich ELISA, different information linked to the structure of the protein can be obtained using different monoclonal antibodies as protein captors, and the proof of protein binding is reached using either other monoclonal or polyclonal antibodies.

One main difficulty in the application of immunochemistry to food products is the choice of extraction procedure to solubilize the target protein; the procedure should keep the epitopes intact. Immunochemical tests also can be applied to identify fraudulent additives and allergens. Immunosensors can be conveniently used in the on-line control of food quality during processing.

An immunochemical approach can be used as a starting method in combination with other physical methods, such as nuclear magnetic resonance, near infrared reflectance, differential scanning calorimeter, etc., for determining protein structures in complex media. To be able to identify relationships between protein structure and function, not only in the food industry but also in others (e.g., clothing, pharmaceutical), it is necessary to combine the work of physicists and chemists with that of immunochemists and nutritionists.

ACKNOWLEDGMENTS

Helpful discussion and suggestions from Drs. M. Van Regenmortel, P. Nandi, and T. Haertlé and typing and chapter organization from T. Campone are deeply acknowledged.

REFERENCES

1. J. C. Allen and C. J. Smith, Enzyme linked immunoassay kits for routine food analysis, *Trends Biotechnol. Tibtech 5*:193 (1987).
2. T. O. Fischman, G. A. Bentley, T. N. Bhat, G. Boulot, R. A. Marriuzza, S. E. V. Phillips, D. Tello, and R. J. Poljak, Crystallographic refinement of the three dimensional structure of the Fab 1.3 lysozyme complex at 2.5-resolution, *J. Biol. Chem. 266*:12915 (1991).
3. H. Watanabe and M. Fukuoka, Measurement of moisture diffusion in foods using pulsed field gradient NMR, *Trends Food Sci. Technol. 3*:211 (1992).
4. K. L. McCarthy, R. Kauten, and C. K. Agemura, Application of NMR imaging to the study of velocity profile during extrusion processing, *Trends Food Sci. Technol. 3*:215 (1992).
5. S. Ablett, Overview of NMR applications in food science, *Trends Food Sci. Technol. 3*:246 (1992).
6. M. Riva and A. Schiraldi, A DSC investigation of the effects of heating rate on cooking indexes of ground meat, *J. Food Sci. 1*:43 (1994).
7. P. K. Wolberg and R. L. Green, Detection of bacteria by transduction of ice nucleation genes, *Tibtech 8*:276 (1990).
8. J. E. Heisick, F. M. Harrell, E. H. Peterson, S. McLaughlin, D. E. Wagner, I. V. Wesley, and J. Bryner, Comparison of four procedures to detect *Listeria* spp in foods, *J. Food Prot. 52*:154 (1989).
9. C. Kocks, E. Gouin, M. Tabouret, P. Berche, H. Ohayon, and P. Cossart, *L. monocytogenes* induced actin assembly requires the act A gene product, a surface protein, *Cell 68*:521 (1992).
10. P. Cossart, M. F. Vicente, J. Mengaud, F. Baquero, J. Perez-Diaz, and P. Berche, Listeriolysin O is essential for virulence of *Listeria monocytogenes*: direct evidence obtained by gene complementation, *Infect. Immun. 57*:3629 (1989).
11. J. H. Skerritt and R. A. Smith, A sensitive monoclonal-antibody-based test for gluten detection: studies with cooked or processed foods, *J. Sci. Food Agric. 36*:980 (1985).
12. J. H. Skerritt and O. Martinuzzi, Effects of solid phase and antigen solvent on the binding and immunoassay of water soluble flour proteins, *J. Immunol. Meth. 88*:217 (1986).
13. F. Hemmen, A. Paraf and S. Smith-Gill, Lysozymes in eggs and plasma from chicken, duck and goose: choice and use of mAbs to detect adulterants in "Foie gras," *J. Food Sci. 58*: 1291 (1993).
14. J. G. R. Hurrel, J. A. Smith, P. E. Todd, and S. J. Leach, Cross reactivity between mammalian myoglobins: linear VS spatial antigenic determinants, *Immunochemistry 14*:283 (1977).
15. E. Gombocz, E. Hellwig, and F. Petuely, Immunologischer Nachweis von Kuhmilchasein in Shafkäsen, *Z. Lebensm. Unters. Forsch. 172*:178 (1981).
16. G. Varshney and A. Paraf, Use of specific polyclonal antibodies to detect heat treatment of ovalbumin in mushrooms, *J. Sci. Food Agric. 52*:261 (1990).

17. K. Kobrehel, D. Agaga, and J. C. Autran, Possibilité de détection de la présence de blés tendres dans les pâtes alimentaires ayant subi des traitements thermiques à haute température, *Ann. Falsif. Exp. Chim 78*:109 (1985).
18. R. Timpl, Antibodies to collagens and procollagens: *Methods enzymol. 82*:472 (1982).
19. H. Saunal, F. Hemmen, A. Paraf, and M. H. V. Van Regenmortel, Cross-reactivity and heat lability of antigenic determinants of c-type duck lysozyme (DEWL) and g-type goose lysozyme (GEWL), *J. Food Sci. 60*:1019 (1995).
20. C. Breton, L. Phan Thanh, and A. Paraf, Immunochemical properties of native and heat denatured ovalbumin, *J. Food Sci. 53*:222 (1988).
21. C. Breton, L. Phan Thanh, and A. Paraf, Immunochemical identification and quantification of ovalbumin additive in canned mushrooms, *J. Food Sci. 53*:226 (1988).
22. G. C. Varshney, W. Mahana, A. M. Filloux, A. Venien, and A. Paraf, Structure of native and heat denatured ovalbumin as revealed by monoclonal antibodies: epitopic changes during heat treatment, *J. Food Sci. 56*:224 (1991).
23. W. Mahana, P. K. Nandi, and A. Paraf, Antigenic properties of ovalbumin following heat denaturation at different temperatures: comparison with enzymatic denaturation, *Food Agric. Immunol. 3*:73 (1991).
24. M. H. V. Van Regenmortel, Which structural features determine protein antigenicity? *Trends Biochem. Sci. 11*:36 (1986).
25. W. G. Laver, G. M. Air, R. G. Webster and S. J. Smith-Gill, Epitopes on protein antigens: misconceptions and realities, *Cell 61*:553 (1990).
26. D. C. Benjamin, J. A. Berzofsky, I. J. East, F. R. N. Gurd, C. Hannum, S. J. Leach, E. Margoliash, J. G. Michael, A. Miller, E. M. Prager, M. Reichlen, E. E. Sercarz, S. J. Smith-Gill, P. E. Todd, and A. C. Wilson, The antigenic structure of proteins, *Annu. Rev. Immunol. 2*:67 (1984).
27. A. Paraf and G. Peltre, *Immunoassays in Food and Agriculture,* Kluwer Academic Publishers, 1991, p. 373.
28. B. P. Babitt, P. M. Allen, G. Matsueda, E. Haber, and E. R. Unanue, Binding of immunogenic peptides to Ia histocompatibility molecules, *Nature 317*:359 (1985).
29. M. Shapira, M. Jibson, G. Muller, and R. Arnon, Immunity and protection against influenza virus by synthetic peptide corresponding to antigenic sites of hemaggutinin, *Proc. Natl. Acad. Sci. 81*:2461 (1984).
30. J. Seki, A. Ota, Y. Suzuki, N. Sakato, and H. Fujio, Inducibility of protein reactive antibodies by peptide immunization: comparison of three epitope peptides of hen eggwhite lysozyme, *J. Biochem. 111*:259 (1992).
31. T. Garcia, R. Martin, E. Rodriguez, P. E. Hernandez, and B. Sanz, Development of a cow's milk identification test (COMIT) for field use, *J. Dairy Res. 56*:691 (1989).
32. D. Levieux, Heat denaturation of whey proteins: comparative studies with physical and immunological methods, *Ann. Rech. Vet. 11*:89 (1980).
33. D. Levieux and A. Venien, Rapid, sensitive two site ELISA for detection of cows' milk in goats' or ewes' milk using monoclonal antibodies, *J. Dairy Res. 61*:91 (1994).
34. M. J. Kendall, P. S. Cox, R. Schneider, and C. F. Hawkins, Gluten subfractions in coeliac disease, *Lancet 2*:1065 (1982).
35. P. J. Ciclitira and E. S. Lennox, A radioimmunoassay for α and β gliadins, *Clin. Sci. 64*: 655 (1983).
36. R. R. A. Coombs, M. Kieffer, D. R. Fraser, and P. J. Frazier, Naturally developing antibodies to wheat gliadin fractions and to other cereal antigens in rabbits, rats, and guinea pigs on normal laboratory diets, *Int. Arch. Allergy Appl. Immun. 70*:200 (1983).
37. T. W. Okita, H. B. Krishnan, and W. T. Kim, Immunological relationships among the major seed proteins of cereals, *Plant Sci. 57*:103 (1988).

38. J. M. Carter, H. A. Lee, E. N. C. Mills, N. Lambert, H. N. S. Chan, and M. R. A. Morgan, Characterization of polyclonal antibodies against glycinin (11 S storage protein) from soya (glycin max), *J. Sci. Food Agric. 58*:75 (1992).
39. G. Köhler and C. Milstein, Continuous cultures of fused cells secreting antibody of predefined specificity, *Nature 256*:495 (1975).
40. C. Collard-Bovy, E. Marchal, G. Humbert, and G. Linden, Microparticle-enhanced nephelometric immunoassay—1) Measurement of αs-casein and κ-casein, *J. Dairy Sci. 74*:3695 (1991).
41. S. Oh, Immunosensors for food safety, *Trends Food Sci. Technol. 4*:98 (1993).
42. R. J. Fisher and M. Fivash, Surface plasmon resonance based methods for measuring the kinetics and binding affinities of biomelocular interactions, *Curr. Opin. Biotechnol. 5*:389 (1994).
43. M. H. V. Van Regenmortel, M. Altschuh, J. L. Pellequer, P. Richalet-Secordel, H. Saunal, J. A. Wiley, and G. Zeder-Lutz, Analysis of viral antigens using biosensor technology, *Methods companion Meth. Enzymol. 6*:177 (1994).
44. M. H. V. Van Regenmortel, Structural and functional approaches to the study of protein antigenicity, *Immunol. Today 10*:266 (1989).
45. M. A. Newman, C. R. Mainhart, C. P. Mallett, T. B. Lavoie, and S. J. Smith-Gill, Patterns of antibody specificity during the balb/c immune response to hen eggwhite lysozyme, *J. Immunol. 149*:3260 (1992).
46. E. Marchal, C. Collard-Bovy, G. Humbert, G. Linden, P. Montagne, J. Duheille, and P. Varcin, Microparticle-enhanced nephelometric immunoassay—2) Measurement of α-lactalbumin and β-lactoglobulin, *J. Dairy Sci. 74*:3702 (1991).
47. F. C. Mao and R. D. Bremel, Enzyme linked immunosorbent assays for bovine α-lactalbumin and β-lactoglobulin in serum and tissue culture media, *J. Dairy Sci. 74*:2946 (1991).
48. E. K. Kang'ethe, J. M. Gathuma, and K. S. Lindqvist, Identification of the species of origin of fresh, cooked and canned meat and meat products using antisera to thermostable muscle antigens by Ouchterlony's double diffusion test, *J. Sci. Food Agric. 37*:157 (1986).
49. S. S. Twinning, H. Lehmann, and M. Z. Atassi, The antibody response to myoglobin is independant of the immunized species, *Biochem. J. 191*:681 (1980).
50. H. M. Cooper, J. J. East, P. E. E. Todd, and S. J. Leach, Antibody response to myoglobins: effect of host species, *Mol. Immunol. 21*:479 (1984).
51. H. M. Cooper, P. E. E. Todd, and S. J. Leach, Antibody response to the c-terminal peptide sequence in beef myoglobin, *Mol. Immunol. 23*:1289 (1986).
52. P. Jolles and P. Jolles, What's new in lysozyme research? *Mol. Cell. Biochem. 63*:165 (1984).
53. F. Schoentgen, P. Jolles, and P. Jolles, Complete amino-acid sequence of ostrich (struthio camelus) eggwhite lysozyme, a goose-type lysozyme, *Eur. J. Biochem. 123*:489 (1982).
54. M. G. Grütter, L. H. Weaver, and B. W. Matthews, Goose lysozyme structure: an evolutionary link between hen and bacteriophage lysozymes? *Nature 303*:828 (1983).
55. A. Hindenburg, J. Spitznagel and N. Arnheim, Isozymes of lysozymes in leucocytes and eggwhite: evidence for the species-specific control of eggwhite lysozyme synthesis, *Proc. Natl. Acad. Sci. USA 71*:1653 (1974).
56. F. Hemmen, W. Mahana, P. Jolles and A. Paraf, Common antigenic properties of a g-type (goose) and a c-type (duck) eggwhite lysozyme: antibody responses in rabbits and mice, *Experientia 48*:79 (1992).
57. J. A. Werkmeister and J. A. M. Ramshaw, Multiple antigenic determinants on type III collagen, *Biochem. J. 274*:295 (1991).
58. C. T. Leung, K. M. Kuzmanoff, and C. W. Beattie, Isolation and characterization of monoclonal antibody directed against bovine αS2 casein, *J. Dairy Sci. 74*:2872 (1991).
59. K. M. Kuzmanoff, J. W. Andrese, and C. W. Beattie, Isolation of monoclonal antibodies monospecific for bovine α-casein, *J. Dairy Sci. 73*:2741 (1990).

60. K. M. Kuzmanoff and C. W. Beattie, Isolation of monoclonal antibodies monospecific for bovine β-lactoglobulin, *J. Dairy Res. 74*:3731 (1991).
61. J. H. Skerritt and L. G. Robson, Wheat low molecular weight glutenin subunits. Structural relationship to other gluten proteins analysed using specific antibodies, *Cereal Chem. 67*: 250 (1990).
62. R. Jemmerson, Antigenicity and native structure of globular proteins: low frequency of peptide reactive antibodies, *Proc. Natl. Acad. Sci. 84*:9180 (1987).
63. L. Djavadi-Ohaniance, B. Friguet, and M. E. Golberg, Structural and functional influence of enzyme-antibody interactions: effects of eight different monoclonal antibodies on the enzymatic activity of *Escherichia coli* tryptophan synthase, *Biochemistry 23*:97 (1984).
64. E. L. Dekker, C. Porta, and M. H. V. Van Regenmortel, Limitations of different ELISA procedures for localizing epitopes in viral coat protein subunits, *Arch. Virol 105*:269 (1989).
65. Y. Goto, L. J. Calciano, and A. L. Fink, Acid-induced folding of proteins, *Proc. Natl. Acad. Sci. 87*:573 (1990).
66. D. Levieux, A. Levieux and A. Venien, Immunochemical quantification of heat denaturation of bovine meat soluble proteins, *J. Food Sci. 60*:678 (1995).
67. E. K. Kang'ethe and K. S. Lindqist. Thermostable muscle antigens suitable for use in enzyme immunoassays for identification of meat from various species, *J. Sci. Food Agric. 39*: 179 (1987).
68. H. An, P. A. Klein, K. J. Kao, M. R. Marshall, W. S. Otwell, and C. Wei, Development of monoclonal antibody for rock shrimp identification using enzyme-linked immunosorbent assay, *J. Agric. Food Chem. 38*:2094 (1990).
69. J. H. Skerritt, A simple antibody-based test for dough strength I. Development of method and choice of antibodies, *Cereal Chem. 68*:467 (1991a).
70. J. H. Skerritt and R. J. Henry, Hydrolysis of Barley endosperm storage proteins during malting II quantification by enzyme and radio-immunoassay, *J. Cereal Sci. 7*:265 (1988).
71. N. M. Griffiths, M. J. Billington, A. A. Crimes, and C. H. S. Hitchcock, An assessment of commercially available reagents for an enzyme-linked immunosorbent assay (ELISA) of soya protein in meat products, *J. Sci. Food Agric. 35*:1255 (1984).
72. K. Yasumoto, M. Sudo, and T. Suzuki, Quantitation of soya protein by enzyme-linked immunosorbent assay of its characteristic peptide, *J. Sci. Food Agric. 50*:377 (1990).
73. M. A. McNiven, B. Grimmelt, J. A. MacLeod, and H. Voldeng, Biochemical characterization of low trypsin inhibitor soybean, *J. Food Sci. 57*:1375 (1992).
74. M. L. Marin, C. Casas, M. I. Cambero, and B. Sang, Study of the effect of heat (treatments) on meat protein denaturation as determined by ELISA. *Food Chem. 43*:147 (1992).
75. B. Friguet, L. Djavadi-Ohaniance, and M. E. Goldberg, Some monoclonal antibodies raised with a native protein bind preferentially to the denatured antigen, *Mol. Immunol. 21*:673 (1984).
76. C. H. Wang, M. M. Abouzied, J. J. Pestka, and D. M. Smith, Antibody development and enzyme-linked immunosorbent assay for the protein marker lactate dehydrogenase to determine safe cooking end-point temperatures of turkey rolls, *J. Agric. Food Chem. 40*:1671 (1992).
77. M. M. Abouzied, C. H. Wang, J. J. Pestka, and D. M. Smith, Lactate dehydrogenase as safe end-point cooking indicator in poultry breast rolls: development of monoclonal antibodies and application to sandwich enzyme-linked immunosorbent assay (ELISA), *J. Food Protect. 120*:124 (1993).
78. E. N. C. Mills, S. R. Burgess, A. S. Tatham, P. R. Shewry, H. W. S. Chan, and M. R. A. Morgan, Characerization of a panel of monoclonal anti-gliadin antibodies, *J. Cereal Sci. 11*: 89 (1990).
79. R. J. McCormick, D. H. Kropf, G. R. Reeck, M. C. Hunt, and C. L. Kastner, Effect of heating temperature and muscle type on porcine muscle extracts as determined by reverse phase high performance liquid chromatography, *J. Food Sci. 52*:1481 (1987).

80. W. E. Townsend and C. E. Davis, Transaminase (AST/GOT and ALT/GPT) activity in ground beef as a means of determining end-point temperature, *J. Food Sci. 57*:555 (1992).
81. W. E. Townsend, G. K. Searcy, C. E. Davis, and R. L. Wilson, End-point temperature (EPT) affects N-acetyl β D glucosaminidase activity in beef, pork and turkey, *J. Food Sci. 58*:710 (1993).
82. C. H. Wang, A. M. Booren, M. M. Abouzied, J. J. Pestka, and D. M. Smith, ELISA determination of turkey roll end-point temperature: effects of formulation, storage and processing, *J. Food Sci. 58*:1258 (1993).
83. C. H. Wang, J. J. Pestka, A. M. Booren, and D. M. Smith, Lactate dehydrogenase, serum protein and immunoglobulin G content of uncured turkey high rolls as influenced by end point cooking temperature, *J. Agric. Food Chem. 42*:829 (1994).
84. H. Otani and F. Tokita, Contribution of the sugar moiety in the browning product between β-lactoglobulin and lactose as an antigenic determinant, *Jpn. J. Zootech. Sci. 53*:344 (1981).
85. P. J. Kilshaw, L. M. Heppell, and J. E. Ford, Effects of heat treatment of cows' milk and whey on the nutritional quality and antigenic properties, *Arch. Dis. Child. 57*:842 (1982).
86. S. Kaminogawa, M. Shimizu, A. Ametani, M. Hattori, O. Ando, S. Hachimura, Y. Nakamura, M. Totsuka, and K. Yamauchi, Monoclonal antibodies as probes for monitoring the denaturation process of bovine β-lactoglobulin, *Biochim. Biophys. Acta 998*:50 (1989).
87. S. Kaminogawa, A. Enomoto, J. I. Kurisaki, and K. Yamauchi, Monoclonal antibodies against hen's egg ovomucoïd, *J. Biochem. 98*:1027 (1985).
88. Y. Konishi, J. I. Karisaki, S. Kaminogawa, and K. Kamauchi, Determination of antigenicity by radioimmunoassay and of trypsin inhibitory activities in heat or enzyme denatured ovomucoïd, *J. Food Sci. 50*:1422 (1985).
89. A. Kato and T. Takagi, Formation of intermolecular β-sheet structure during heat denaturation of ovalbumin, *J. Agric. Food Chem. 36*:1156 (1988).
90. A. Paraf, A role for monoclonal antibodies in the analysis of food proteins, *Trends Food Sci. Technol. 3*:263 (1992).
91. T. Matsuda, H. Ishiguro, I. Ohkubo, M. Sasaki, and R. Nakamura, Carbohydrate binding specificity of monoclonal antibodies raised against lactose-protein Maillard adducts, *J. Biochem. 111*:383 (1992).
92. T. B. Lavoie, W. N. Drohan, and S. J. Smith-Gill, Experimental analysis by site-directed mutagenesis of somatic mutation effects on affinity and fine specificity in antibodies specific for lysozyme, *J. Immunol. 148*:503 (1992).
93. D. Kenett, E. Katchalski-Katzir, and G. Fleminger, Use of monoclonal antibodies in the detection of structural alterations occurring in lysozyme on heating, *Mol. Immunol. 27*:1 (1990).
94. A. Pellegrini, U. Thomas, R. von Fellenberg, and P. Wild, Bactericidal activities of lysozyme and aprotinin against gram-negative and gram-positive bacteria related to their basic character, *J. Appl. Bacteriol. 72*:180 (1992).
95. R. E. Oste, D. L. Brandon, A. H. Bates, and M. Friedman, Effect of Maillard browning reactions of the Kunitz soybean trypsin-inhibitor on its interaction with monoclonal antibodies, *J. Agric. Food Chem. 38*:258 (1990).
96. J. H. Skerritt, A simple antibody-based test for dough strength II. Genotype and environmental effects, *Cereal Chem. 68*:475 (1991b).
97. J. L. Andrews, M. J. Blundell, and J. H. Skerritt, A simple antibody-based test for dough strength III. Further simplification and collaborative evaluation for wheat quality screening, *Cereal Chem. 70*:241 (1993).
98. G. M. Brett, E. N. C. Mills, A. S. Tatham, R. J. Fido, P. R. Shewry, and M. R. A. Morgan, Immunochemical identification of LMW subunits of glutenin associated with bread-making quality of wheat flours, *Theor. Appl. Genet. 86*:442 (1993).
99. D. A. Ledward and R. F. Tester, Molecular transformations of proteinaceous food during extrusion processing, *Trends Food Sci. Technol. 5*:117 (1994).

100. L. Quillien, T. Gaborit, J. Guéguen, J. P. Melcion, and A. Kozlowski, Evaluation par la technique ELISA de l'effet dénaturant de la cuisson-extrusion sur la légumine dupois, (*Pisum sativum*), *Sci. Aliment. 10*:429 (1990).
101. G. della Valle, L. Quillien, and J. Gueguen, Relationships between processing conditions and starch and protein modifications during extrusion-cooking of pea flour, *J. Sci. Food Agric. 64*:509 (1994).
102. J. C. Autran, Protein analysis of wheat by monoclonal antibodies and nuclear magnetic resonance, *Modern Methods of Plants Analysis—Seed Analysis* (H. F. Linskens and J. F. Jackson), Springer Verlag, Berlin, 1992, p. 109.
103. M. Bony and W. Stimson, Immunochemical detection of the common wheat specific albumin friabilin, Final report of the contract No. 5266/1/5/333/89/10 BCRF (10) 1990–1993, p. 51 Bruxelles.
104. P. Vaag and L. Munck, Immunochemical methods in cereal research and technology, *Cereal Chem. 64*:59 (1987).
105. P. Aranda, R. Oria, and M. Calvo, Detection of cows' milk in ewes' milk and cheese by and immunodotting method, *J. Dairy Res. 55*:121 (1988).
106. A. R. Hayden, Use of antisera to heat stable antigens of adrenals for species identification in thoroughly cooked beef sausages, *J. Food Sci. 46*:1810 (1981).
107. W. N. Sawaya, M. S. Mameesh, E. El Rayes, A. Nusain, and B. Dashti, Detection of pork in processed meat by an enzyme linked immunosorbent assay using antiswine antisera, *J. Food Sci. 55*:193 (1990).
108. F. W. Janssen, G. H. Hägele, A. M. B. Voorpostel, and J. A. de Baaij, Myoglobin analysis for determination of beef, pork, horse, sheep and kangaroo meat in blended cooked products, *J. Food. Sci. 55*:1528 (1990).
109. C. H. S. Hitchcock, F. J. Bailey, A. A. Crimes, D. A. E. Dean, and P. J. Davis, Determination of soya proteins in food using an enzyme-linked immunosorbent assay procedure, *J. Sci. Food Agric. 32*:157 (1981).
110. T. Matsuda and R. Nakamura, Molecular structure and immunological properties of food allergens, *Trends Food Sci. Technol. 4*:289 (1993).
111. J. F. Gauchat, S. Henchoz, G. Mazzel, J. P. Aubry, T. Brunner, H. Blasey, P. Life, T. Talabot, L. Flores-Romo, J. Thompson, K. Kishi, J. Butterfield, C. Dohinden, and J. Y. Bonnefoy, Induction of human IgE synthesis in B cells by mast cells and basophils, *Nature 365*:340 (1993).
112. J. H. Skerritt, A sensitive monoclonal antibody-based test for gluten detection: quantitative immunoassay, *J. Sci. Food Agric. 36*:987 (1985).
113. J. H. Skerritt and H. S. Hill, Monoclonal antibody sandwich enzyme immunoassays for determination of gluten in foods, *J. Agric. Food Chem. 38*:1771 (1990).
114. A. W. Burks Jr., H. L. Butler, J. R. Brooks, J. Hardin, and C. Connaughton, Identification and comparison of differences in antigens in two commercially available soybean protein isolates, *J. Food Sci. 53*:1456 (1988).
115. A. M. Herian, S. L. Taylor, and R. K. Bush, Allergenic reactivity of various soybean products as determined by RAST inhibition, *J. Food Sci. 58*:385 (1993).
116. S. Elsayed, A. S. E. Hammer, M. B. Kalvenes, E. Florvaag, J. Apold, and H. Vik, Antigenic and allergenic determinants of ovalbumin. Peptide mapping, cleavage at the methionyl peptide bonds and enzymic hydrolysis of native and carboxymethyl OA, *Int. Arch. Allergy Appl. Immun. 79*:101 (1986).
117. S. Elsayed, E. Holen, and M. B. Haugstad, Antigen and allergenic determinants of ovalbumin, II-The reactivity of the NH_2 terminal decapeptide, *Scand. Immunol. 27*:587 (1988).
118. G. Johnsen and S. Elsayed, Antigenic and allergenic determinants of ovalbumin. III MHC Ia-binding peptide (OA 323-339) interacts with human and rabbits specific antibodies, *Mol. Immunol. 27*:821 (1990).

119. W. Mayer, R. Okonomopulos, and G. Jung, Synthesis and conformation of a polyoxyethylene bound unidecapeptide of the alametixein helix and (2-methyl alanyl-L alanine), *Biopolymers 8*:425 (1979).
120. S. Elsayed and L. Stavseng, Epitopes mapping of region 11-70 of ovalbumin (Gal d I) using five synthetic peptides, *Int. Arch. Allergy Immunol. 104*:65 (1994).
121. H. Kahlert, A. Petersen, W. M. Becker, and M. Schlaak, Epitopes analysis of the allergen ovalbumin (Gal d I) with monoclonal antibodies and patients IgE, *Mol. Immunol. 29*:1191 (1992).
122. S. L. Taylor, Immunologic and allergic properties of cows' milk proteins in humans, *J. Food Prot. 49*:239 (1986).
123. D. A. Granato and P. F. Piguet, A mouse monoclonal IgE antibody anti-bovine milk β-lactoglobulin allows studies of allergy in the gastrointestinal track, *Clin. Exp. Immunol. 63*: 703 (1986).
124. I. Böttcher, G. Hämmerling, and J. F. Kapp, Continuous protection of monoclonal mouse IgE antibodies with known allergic specificity by a hybrid cell line, *Nature 275*:761 (1978).
125. T. P. King, L. Kochoumian, and J. Alison, Mellitin specific monoclonal and polyclonal IgE and IgG antibodies from mice, *J. Immunol. 133*:2668 (1984).
126. B. J. Sutton and H. J. Gould, The human IgE network, *Nature 366*:421 (1993).
127. A. Enomoto, M. Konishi, S. Hachimura, and S. Kaminogawa, Milk whey protein fed as a constituent of the diet induced both oral tolerance and a systemic humoral response, while heat denatured whey protein induced only oral tolerance, *Clin. Immunol. Immunopathol. 66*: 136 (1993).
128. M. Okamoto, R. Hayashi, A. Enomotto, S. Kaminogawa, and K. Yamauchi, High pressure proteolytic digestion of food proteins: selective elimination of β-lactoglobulin in bovine milk whey concentrate, *Agric. Biol. Chem. 55*:1253 (1991).
129. M. Takase, Y. Fukuwatari, K. Kawase, I. Kiyosawa, K. Ogasa, S. Susuki, and T. Kuroume, Antigenicity of casein enzymatic hydrolysate, *J. Dairy Sci. 62*:1570 (1979).
130. M. I. Mahmoud, W. T. Malone, and C. T. Cordle, Enzymatic hydrolysis of casein: effect of degree of hydrolysis on antigenicity and physical properties, *J. Food Sci. 57*:1223 (1992).
131. E. C. H. Van Beresteijn, R. A. Peeters, J. Kaper, R. J. G. Meijer, A. J. P. Robben, and D. G. Schmidt, Molecular mass distribution, immunological properties and nutritive value of whey protein hydrolysates, *J. Food Prot. 57*:619 (1994).
132. Y. Kato, T. Matsuda, K. Watanabe, and R. Nakamura, Alteration of ovalbumin immunogenic activity by glycosylation through Maillard reaction, *Agric. Biol. Chem. 49*:421 (1985).
133. E. M. Akita and S. Nakai, Lipophilization of β-lactoglobulin: effects on allergenicity and digestibility, *J. Food Sci. 55*:718 (1990).
134. H. Otani and A. Hosono, Antigenic reactive regions of S. carboxymethylated β-lactoglobulin, *Agric. Biol. Chem. 51*:531 (1987).
135. M. Friedman, M. R. Gumbmann, D. L. Brandon, and A. H. Bates, Inactivation and analysis of soybean inhibitors of digestive enzymes, *Food Proteins* (J. E. Kinsella and W. G. Soucie, eds.), American Oil Chemists Society, Urbana, IL, 1989 p. 296–328.
136. F. W. Scott, J. Cui, and P. Rowsell, Food and the development of autoimmune disease, *Trends Food Sci. Technol. 5*:111 (1994).
137. A. M. Mowat, The regulation of immune responses to dietary protein antigens, *Immunol. Today 8*:93 (1987).
138. A. Friedman and H. L. Weiner, Induction of energy or active suppression following oral tolerance is determined by antigen dosage, *Proc. Nat. Acad. Sci. 91*:6688 (1990).
139. N. A. Mitchison, The dosage requirements for immunological paralysis by soluble proteins, *Immunology 15*:509 (1968).
140. H. L. Weiner, Oral tolerance, *Proc. Natl. Acad. Sci. 91*:10762 (1994).
141. M. Aizawa, Immunosensors, *Biosensor Principles and Applications* (L. J. Blum and P. R. Coulet, eds.). Marcel Dekker, New York, 1991, p. 249.

142. A. A. Suleiman and G. G. Guilbault, Piezoelectric (PZ) immunosensors and their applications, *Anal. Lett. 24*:1283 (1991).
143. A. H. Severs, R. B. M., Schasfoort, and M. H. L. Salden, An immunosensor for syphilis screening based on surface plasmon resonance, *Biosensors Bioelectron. 8*:185 (1993).
144. A. Brecht, G. Gauglitz, and J. Polster, Interferometric immunoassay in a FIA-system: a sensitive and rapid approach in label-free immunosensing, *Biosensors Bioelectron. 8*:387 (1993).
145. D. V. Brown and M. E. Meyerhoff, Potentiometric enzyme channeling immunosensor for proteins, *Biosensors Bioelectron. 6*:615 (1991).
146. J.-L. Boitieux, M.-P. Biron, G. Desmet, and D. Thomas, Dissociation of immunocomplexes by ionic shock for the development of immunosensors: application to measurement of α_1-fetoprotein, *Clin. Chem. 35*:1026 (1989).
147. F. V. Bright, T. A. Betts, and K. S. Litwiler, Regenerable fiber-optic-based immunosensor, *Anal. Chem. 62*:1065 (1990).
148. H. Liu, J. C. Yu, D. S. Bindra, R. S. Givens, and G. S. Wilson, Flow injection solid-phase chemiluminescent immunoassay using a membrane-based reactor, *Anal. Chem. 63*:666 (1991).
149. M. Malmqvist, Biospecific interaction analysis using biosensor technology, *Nature 361*:186 (1993).
150. G. A. Robinson, J. W. Attridge, J. K. Deacon, A. M. Thomson, C. A. Love, S. Whiteley, M. Pugh, and P. B. Daniels, The calibration of an optical immunosensor—the FCFD, *Biosensors Bioelectron. 8*:371 (1993).
151. W. J. Checovich, R. E. Bolger, and T. Burke, Fluorescence polarization—a new tool for cell and molecular biology, *Nature 375*:254 (1995).

Index

ISBN 0-8247-9820-1
EAN
9 780824 798208

90000>